The Origin of the Solar System

Soviet Research 1925–1991

The Origin of the Solar System

Soviet Research 1925–1991

Aleksey E. Levin ▪ Stephen G. Brush
University of Maryland
Institute for Physical Science & Technology
College Park, Maryland

AIP PRESS
American Institute of Physics New York

AIP Press
American Institute of Physics
500 Sunnyside Boulevard
Woodbury, NY 11797-2999

Library of Congress Cataloging-in-Publication Data
Levin, Aleksey E.
The origin of the solar system : Soviet research, 1925–1991 / Aleksey E. Levin, Stephen G. Brush.
p. cm.
Includes bibliographical references and index.
ISBN 1-56396-281-0 (alk. paper)
1. Solar system -- Origin. 2. Solar system -- Research -- Soviet Union -- History. I. Brush, Stephen G. II. Title.
QB501.L49 1994 94-5329
523'.2--dc20 CIP

10 9 8 7 6 5 4 3 2 1

Contents

IV. Rotation of Planets

V. Formation of the Earth and Other Terrestrial Planets

VI. Origin of the Moon and Other Planetary Satellites

VII. Formation of the Giant Planets

VIII. Asteroids, Comets, and Meteorites

IX. Other Planetary Systems

Preface

The inspiration for this book came from the realization that Soviet scientists had strongly influenced Western research on the origin of the solar system in the 1970s and 1980s, yet the original Russian publications were not easily accessible. Even when English translations exist, they are not available in many libraries. We decided that a comprehensive collection of Soviet research on planetary cosmogony, going back to the seminal work of Otto Schmidt, would be useful to scientists as well as to historians of science. The end of the Soviet period makes 1991 a natural stopping place.

The two introductory essays by the editors describe, respectively, the historical context of Soviet science in which this research developed, and the historical context of Western planetary cosmogony into which it was introduced. We then present a selection of Schmidt's writings, including some articles not previously published in English. The rest of the book is a compendium of articles and abstracts from the last four decades, arranged first by topic and then chronologically within each topic. At the end is a bibliography of Soviet publications—not complete by any means, but a fairly good indication of what is available in major American libraries—and a comprehensive index.

Since 1957 the American Institute of Physics has played a major role in making available, to the international scientific community, Soviet research in astronomy and astrophysics, through the publication of a cover-to-cover translation of *Astronomicheskii Zhurnal* (along with *Pisma*, "Letters to the Editor"). We have drawn heavily on these translations, which include most of the important articles published by Victor Safronov and his group in the 1960s and 1970s. Limitations of space and copyright fees have forced us to omit many other relevant translations, but the abstracts and bibliography should make it easier for the interested reader to locate them.

We thank E. Ruskol, V. S. Safronov, and A. V. Vityazev for valuable written and oral comments on our introductory essays. Additional information about Soviet publications has been provided by V. Frenkel, V. Kirsanov, and the staff of the Schmidt Institute of Earth Physics. We received financial support from the National Science Foundation, and from the Department of History and the Institute for Physical Science and Technology of the University of Maryland, College Park.

A. E. Levin

S. G. Brush

College Park, Maryland, November 1994

I. Introduction

Andrei Borisovich Makalkin. *Senior Research Scientist, Schmidt Institute of Physics of the Earth, Moscow.*

Evgenia L. Ruskol. *Senior Research Scientist, Schmidt Institute of Physics of the Earth, Moscow.*

Victor Sergeyevich Safronov. *Chief Scientist, Schmidt Institute of Physics of the Earth, Moscow.*

Tamara Vladimirovna Ruzmaikina. *Senior Research Scientist, Schmidt Institute of Physics of the Earth, Moscow.*

Andrei Vasilyevich Vityazev. *Head of the Laboratory "Origin of the Earth and Planets" at the Schmidt Institute of Physics of the Earth, Moscow.*

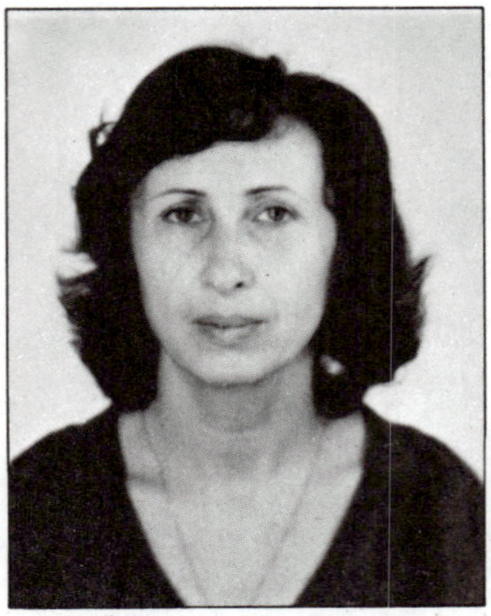

Galina Victorovna Pechernikova. *Senior Research Scientist at the Schmidt Institute of Physics of the Earth, Moscow.*

Otto Iulevich Schmidt *(1912 photograph).*

The Otto Schmidt School and The Development of Planetary Cosmogony in the USSR

Aleksey E. Levin

The problem of the origin of the planets of the solar system would appear to belong to the domain of astronomy—at least, if it is regarded from a purely intellectual standpoint. Therefore, it seems remarkable that for two centuries, that is, from the earlier cosmogonic schemes by Georges Louis Leclerc de Buffon (1745) and Immanuel Kant (1755) up to the 1930s and 1940s, this problem was most typically formulated and developed either by scientists who were not professional astronomers at all, or by those who had received their basic training in another field. To say nothing of such founding fathers of planetary cosmogony as Kant (trained as a philosopher) and Pierre-Simon de Laplace (trained as a mathematician), this rule holds reasonably well for a great many other major contributors to the area, such as Jules-Henri Poincaré, Thomas Chamberlin, Harold Jeffreys, James Jeans, Hannes Alfvén, Carl Friedrich von Weizsäcker, Harold Clayton Urey, and—what is of special importance in the context of this study—Otto Yulievich Schmidt, the founder of the Soviet school of planetary cosmogony whose centennial anniversary was celebrated in the fall of 1991.

One could offer more than one hypothetical explanation for this puzzling fact. For instance, the science of astronomy as it had originated from the classical Greek theories of celestial movements was generally noncommittal concerning the nature of the moving objects, and that cherished indifference could not help hindering astronomers from asking questions about the origin of celestial bodies. The very excellence of the computational techniques and the mathematical explanatory models of observational astronomy and celestial mechanics worked precisely in the same direction. The excellent and very powerful arsenal of analytical methods developed by Euler, Clairaut, D'Alembert, Lagrange, and Laplace was directed at the solution of purely mechanical problems and did not provide technical resources and conceptual means which could stimulate the modeling of essentially nonmechanical (electromagnetic, thermal) processes of planetary formation. It hardly may be attributed to pure chance that Laplace's famous hypothesis was introduced only as a qualitative picture and made public as a section of a semipopular treatise.

It certainly can be plausibly suggested that the absolute majority of professional astronomers of the 18th and 19th centuries tended to refrain from inventing cosmogonic models thanks to the lack of relevant factual data and/or technical methods. This explanation, however, looks much less convincing in relation to the first half of our century, since that period witnessed an astounding growth of cosmogonic studies initiated by nonastronomers or by late entrants in the field of astronomy who utilized more or less the same intellectual resources which could have been used, and sometimes were actually used by astronomers themselves. Whatever may be the causes of this trend, the formation of the Schmidt school of planetary cosmogony appears to correspond to it perfectly.

The formation of this school looks especially interesting and, in a sense, unique due to the fact it was triggered and very strongly influenced by the peculiarities of the intellectual, emotional, and, strange as it may sound, political aspects of the life story of only one principal actor, that is, O. Yu. Schmidt himself. It would be an exaggeration to maintain that he had no precursors or colleagues in the history of Russian and Soviet science, but his personal role is nevertheless far more important than that of any other contributor to the development of planetary cosmogony in Russia during the first half of this century. Schmidt managed to establish himself as the recognized intellectual and institutional leader of his school, and also as a popular hero and the powerful symbol of Soviet cosmogony in general. And, what is of utmost importance from the point of view of our own times, his role—unlike the role of a great many other "heroes" of the Stalin period—still looks genuine and generally progressive. The importance of this circumstance during the current years of the great iconoclastic upheaval in the Soviet Union can hardly be understated. Thus Schmidt's activities should and will be given special consideration in this introduction.

During the pre-Soviet period, Russian science did not produce any original cosmogonic research. Actually, prior to the great reforms of Alexander II (1818–81) no cosmogonic problems could even be openly formulated thanks to the conservative attitudes of the state authorities exercising rigid control over publications and universities as well as the uncompromisingly negative position of the Russian Orthodox Church. To give a good illustration, it suffices to say that during the reign of Nicholas I, one Moscow resident of common origin was expelled to Kazakhstan for his comparatively innocent ideas concerning the possibility of space travel, and because of a really amazing coincidence, he was sent to a small village whose name has become internationally famous more than a century later since it was given (as a sort of cover) to the Soviet cosmodrome from which Yuri Gagarin's mission in Vostok I was launched! Although this ideological taboo was gradually weakened and finally abolished in the second half of the 19th century,[1] the Russian astronomical community continued to remain disinterested in cosmogonic problems. Before the Revolution of 1917, Russian astronomical studies were developing predominantly at only three university centers possessing fine observatories and professionally trained cadres. First was St. Petersburg with its university and the glorious Pulkovo Observatory nearby; then Moscow, the motherland of higher education in Russia, whose university observatory was founded in the early 1830s; and finally Kazan, whose university has had its own observatory since 1814. The Pulkovo Observatory founded by Friedrich von Struve in 1839 quickly became a world center for research in astrometry and stellar astronomy, the orientation of which was supplemented by the end of the century by research in stellar spectroscopy, celestial mechanics, and cometary astronomy. In Moscow, cometary studies, photometry, and gravimetric research dominated, and the Kazan astronomers were mainly interested in lunar astronomy, stellar catalogs, gravimetry, and celestial mechanics. The Kazan astronomy professor Marian Kovalsky (1821–84) seems to have approached the area of cosmogony closer than any other Russian astronomer. Kovalsky, the founder of the important Kazan school of celestial mechanics, developed the first theory of the motion of Neptune and was also the first figure in the history of astronomy to formulate the problem of galactic rotation. The distance between the works of other Russian astronomers and the problems of the origin of planets was even greater.

Turning to the areas of research outside the area of astronomy, it is necessary to stress the potential importance of P. N. Lebedev's very elegant and precise measurements of light pressure, and particularly his measurements of the radiation pressure of light on gaseous substances completed by 1910 (the cosmogonic role of light pressure is known to have been first suggested by Svante Arrhenius). The Russian mathematician A. M. Liapunov in his series of theoretical studies undertaken in 1905–1912 gave a general theory of the figures of equilibrium of rotating liquids, and, in particular, he identified George Darwin's mistake concerning the problem of the stability of pear-shaped figures which invalidated Darwin's hypothesis related to the formation of binary stars (Liapunov's results were justified by Jeans, who discovered Darwin's calculational mistake in 1917). From a more general point of view, the high and broad research culture of the Russian schools of mathematicians and specialists in theoretical mechanics also played a very significant role, particularly in relation to the intellectual development of Schmidt himself.

If we are to discuss the concrete studies pertaining to the problem of the formation of the Earth, a few examples can also be presented. The Russian physicist Alexander Sokolov (1854–1928), a pupil of Stoletov, Kirchhoff, and Helmholtz, a professor of Moscow University and a scientist with very diversified research interests, happened to study the thermal balance of the Earth and performed calculations of the Earth's age as deduced from the model of a contracting gaseous nebula, which, according to his figures, lay within the range of 23–155 million years. The same problem was also occasionally discussed by some Russian geologists. For example, one of the most visible Russian geologists and paleontologists, professor Nikolai Andrusov of the University of Kiev, published a detailed review "On the Earth's Age" in 1912,[2] wherein he analyzed and compared the hypotheses by Kant, Laplace, Chamberlin and Moulton, and See; in the next year, the same magazine published a separate article on the See hypothesis.[3] It is worth noting also that the cosmogonic section of F. R. Moulton's "Introduction to Astronomy" (1906) was quickly translated into Russian and published as a separate booklet, as well as Arrhenius's "Werden der Welten" (1907) and "Die Vorstellung vom Weltgebäude im Wandel der Zeiten" (1908).

During the 1920s, original Soviet research in planetary and solar cosmogony was dominated by the works of Vassily G. Fesenkov, whose approach, methodology, and results will be given special consideration a bit later. In the same decade, however, cosmogonic problems chanced to attract the attention, albeit only briefly, of another Russian scientist whose activities in the field may probably be characterized as an "unwritten chapter" in the history of cosmogony. His life story looks so unusual and his research capabilities so high, that one cannot help feeling pity that his interest in cosmogony turned out to be sporadic rather than lasting. To provide the proper historical context this scientist's role in the development of Soviet cosmogony should be examined carefully if only for the reason that his activities intersected somehow with the activities of the principal contributors to Soviet cosmogonic research.

Vladimir Aleksandrovich Kostitzin (1883–1963) was born in Efremov, a small town in the Tula province. In 1904 or 1905 he graduated from the Mathematical Section of the Department of Physics and Mathematics of the University of Moscow, took part in the Revolution of 1905, and after its defeat emigrated from Russia. He spent several years in Switzerland where he began doing research in the area of the mathematical problems of geophysics. He came back to Russia in 1914 and joined the Russian Army in order to participate in World War I; after the February Revolution, he was even appointed as political commissar in one of the military regiments. After the Civil War had ended, Kostitzin joined the faculty of the First Moscow State University as professor of mathematics, and he also worked at the Soviet agency *Glavnauka* (the main scientific administration of the Commissariat for Public Education). His visibility in the scientific establishment of the Soviet capital seems to have increased very rapidly. In 1924, he founded the State Research Institute of Geophysics and became its director, and he also held the positions of permanent fellow of the State Astrophysics Institute and permanent fellow of the Research Institute of Mathematics and Mechanics affiliated with the Department of Mathematics and Physics of the First Moscow State University. However, in 1927 he moved to Paris, probably because of family reasons (his wife and occasional coauthor, a specialist in parasitology, had moved to Paris first, and Kostitzin eventually decided to follow her, although he never renounced his Soviet citizenship). While in France, Kostitzin's research flourished. He published some theoretical studies reflecting his former geophysical interests, particularly an important albeit brief monograph on the evolution of the atmosphere and the theory of glaciations,[4] whose general approach reflects the influence of the ideas of the great Russian geochemist and philosophic thinker Vladimir Vernadsky, whose close friend and associate Kostitzin had been in Moscow. Kostitzin also developed a close cooperation with Vito Volterra; this triggered his own research in theoretical ecology and mathematical biology whose results are represented in two books.[5] The importance of his contribution to mathematical biology and theoretical ecology is fully recognized nowadays.[6] This contribution determines Kostitzin's place in the history of science probably more than his other works.

In the context of this study, however, it is of immediate interest that during the first half of the 1920s, Kostitzin was an associate of the (then only) Soviet cosmogonist Vassily Fesenkov,[7] who held the office of Director of the State Astrophysics Institute. Kostitzin himself published six original studies in theoretical astrophysics in 1922–28, and more than that, he seems to have contemplated research in cosmogony, possibly thanks to Fesenkov's influence.[8] Although he never produced any original results on the subject, he was highly instrumental in the dissemination of the findings of Western cosmogonists. In 1923, Svante Arrhenius's "Der Lebenslauf der Planeten" was published in Russia in Kostitzin's translation. In the same year, a more relevant book, a collection of chapters taken from several classic foreign cosmogonic studies, was published,[9] Kostitzin being the compiler and chief editor. The collection included, in addition to excerpts from Kant and Laplace, works by Hervé Faye, George Darwin, and Henri Poincaré. It is also of interest that both books were released by the State Publishing House, whose director was then O. Yu. Schmidt. Although Schmidt's personal role in promoting these books cannot be established with any certainty, it is probably safe to say that he at least noticed and possibly even happened to read them.

Kostitzin's introduction to the collection provides a detailed critical review of the "classic" state of cosmogonic studies as well as his own estimate of the level of cosmogonic research achieved at that time. By and large, he regarded this level only as the first step towards the solution of the problems of planetary cosmogony. In his own words,

> "We consequently see that we still lack the solution of the problem of the origin of our world, and probably we will lack it for a rather considerable time; however, we have basic materials for our future studies as well as the first indications concerning their direction, and, what is most important, we firmly believe that the problem may be finally solved."[10]

Taking into account that this opinion was expressed in the early 1920s, it looks reasonably realistic.

Kostitzin had accepted the model of the formation of the Earth from a gas-dust nebula, and he even used it qualitatively in his study of the evolution of the atmosphere.[11] Although he never developed his own original approach to cosmogonic issues and left the field rather soon, his personality is interesting from at least two points of view. First, a number of rather intriguing analogies between Kostitzin's and Schmidt's biographies will be immediately evident after the latter's life story has been discussed. Second, Kostitzin, like Schmidt, was a brilliant mathematician, and also, unlike Schmidt, a professional geophysicist and astrophysicist with very good connections in astronomical circles. Therefore, one can guess that if he had deepened his cosmogonic research, he could have achieved important results some 20 years before Schmidt. Although questions of the type "what could have happened if..." are not very fashionable in historical studies, still Kostitzin's unrealized career in planetary cosmogony seems to remind us once again that the actual developments of historical events tend to get entangled with potential ones, and sometimes in a rather bizarre way.

Kostitzin happened to publish one more monograph devoted to cosmogonic problems, a popular book on the origin of the universe which was issued in 1926.[12] However, the peak point of his interest in cosmogony was obviously passed through in 1923, and after he had moved to Paris he never tried his hand at this sort of research again, although he kept doing intensive research in his new field of theoretical biology.[13] Anyway, Kostitzin's thorough familiarity "with all the resources of analysis and its applications"[14] as well as his capability to combine successfully "mathematical and experimental studies"[15] in the treatment of the problems of his research were not developed during his Parisian period. In the same way, his profound and stable interest in the general problems of evolution and the consistently evolutionary style of his thinking also emerged prior to his move to France. It is true that in the 1930s, "he paid particular attention to the interactions between geophysical and evolutionary phenomena,"[16] but it does not seem unreasonable to surmise that the same evolutionary thinking and the same excellent command of the conceptual and technical resources of mathematics, geophysics, and astrophysics could have become extremely fruitful if Kostitzin had devoted his effort to the modeling of the evolution of the primordial gas-dust cloud whose existence he believed in. Thus we are once again returning to the hypothesis that the period of Kostitzin's active interest in planetary cosmogony may be reasonably regarded as a potentially promising but hidden "shoot" of Soviet cosmogonic research.

Besides the two above-mentioned books of translations published by Kostitzin, four cosmogonic works by Jeans and one by Jeffreys were published in the USSR in 1924–32. However, the only Soviet author who happened to publish original research papers in the cosmogony of the solar system was Vassily Fesenkov, whose biography can be instructively compared with the biographies of Kostitzin and Schmidt.

Fesenkov was born in 1889 in the city of Novocherkassk.[17] Astronomy had been his hobby since he was nine years old, a circumstance which determined his future profession. In 1911, he graduated from the University of Kharkov where his tutor in astronomy was Ludwig Struve.[18] Even prior to graduation, Fesenkov performed very difficult and time-consuming calculations of a cometary orbit which were rewarded by a gold medal. After graduation, Fesenkov spent two years in France where he received excellent training in mathematics, mechanics, and computational astronomy. Nevertheless, it was during the period of his post-graduation research when Fesenkov developed a very strong interest in astrophysics which turned out to become his principal research field. Despite his initial mathematical training, Fesenkov tended to utilize sophisticated computational methods only rather modestly in his later work, preferring observations and qualitative reasoning. He happened to be a very prolific author, and the total number of his publications, including popular works, exceeds 650 titles, but only a small part of them contains serious mathematical calculations. These features of his individual research style influenced the development of his cosmogonic notions too.

After defending his doctoral dissertation at the University of Paris, Fesenkov came back to Russia where he remained in the center of the activities of the national astronomical and astrophysical community until the end of his long life (he died in 1972). He taught at several universities; in 1923–31 he was director of the State Astrophysics Institute;[19] for almost 40 years he was the chief editor of *Astronomicheskii Zhurnal*; he organized the Astronomical Council of the USSR Academy of Sciences and was its first chairman; from 1945 and up to his demise he headed the Meteoritic Committee of the Academy of Sciences, which in 1935 had elected him a full member.

Fesenkov's earliest work devoted to planetary cosmogony, "On the angular momentum of the solar system from the point of view of the Laplace cosmogonic hypothesis," was published in 1917,[20] but the first version of what would be called later his own cosmogonic model was not issued until five years later.[21] During his life, Fesenkov published more than 40 professional studies and popular essays in the field of cosmogony, corresponding to 6% of the total bulk of his publications.

It is worth stressing that Fesenkov's cosmogonic ideas hardly deserve the title of developed theory, at least if evaluated in terms of comparatively modern standards. Altogether they rather comprise a more or less eclectic complex of only qualitative hypotheses and explanations, quite insufficiently grounded on precise factual data, and especially lacking in detailed numerical calculations. Also, Fesenkov's understanding of the origin of planets was not stable. Initially, his views were expressed in mechanical terms. By and large, he adhered to the Laplacian rotational hypothesis, except for the fact that he regarded the primordial nebula as consisting not of hot gases alone but of gas and dust which he believed to have been comparatively cool. In his opinion, the initially homogeneous cloud without significant gradients of temperature undergoes an irreversible evolution thanks to gravitational effects and the effects of collisions between dust particles. These collisions result in the heating of dust particles and their partial evaporation, which in turn lead to the radial stratification of the nebula. The central area of the nebula then becomes the site of its hot gaseous core, whereas the meteoritic fractions of the cloud tend to move outward and their concentration within the peripheral zone of the cloud increases. The same physical processes trigger the uniform rotation of the central core, which thus transforms into a hot revolving gaseous cloud of the Laplacian type. Fesenkov did not accept the Laplacian mechanism of the formation of rings within the revolving cloud; this mechanism was replaced in his scheme by another process. Fesenkov, by his own admission, invoked the mechanism first developed by Emden as a means for the explanation of sun spots. Accordingly, the radial structurization of the nebula was interpreted by Fesenkov in terms of the formation of radial convective currents. The interaction between these currents brings about the vortices whose axes tend to align with the axis of the rotating nebula. In turn, these vortices under some specific conditions are able to produce the rotating clots of nebular matter which give birth to the planets of the solar system.

This explanatory scheme, in accordance with the above-made remark, was presented in purely mechanical terms and, additionally, it did not attribute any special role to the Sun. Moreover, Fesenkov's model implies that the planets had been formed *before* the Sun itself came into existence. As such a picture was in plain contradiction with the accepted estimates of the comparative ages of the Sun and the Earth, Fesenkov had to reconsider his views. First he appeared to be more or less inclined to accept the Jeans hypothesis,[22] but after its demise he began to look favorably at the concept of a simultaneous formation of the Sun and the planets from the primordial gas-dust cloud. According to his new scheme, there first emerges a central rotating clot of matter, whose mass exceeds the mass of modern solar system by roughly a factor of 10, and whose angular momentum equals the momentum of the present solar system. The central zone of the clot becomes the site of some specific sequence of nuclear reactions. As their nuclear fuel burns down, the clot experiences contraction, its stability decreases, and therefore it partially loses its mass and momentum, which are transferred to the peripheral area. Gradually the central core gives birth to a normal star, that is, to the Sun, whose mass accounts for only 10% of the mass of the initial clot. All but a tiny part of the mass of the clot disperses in the surrounding space, but this fraction, which tends to concentrate within the equatorial plane of the nebula, gives birth to the planets. The distances of these growing protoplanets from the central zone are determined by the criteria of their stability in relation to the disturbing action of tidal forces, which means that the outer planets emerge first with the planets of the terrestrial group following suit. Initially, all of the growing planets had the same chemical composition, but later on the inner planets had lost their helium and hydrogen.

Such are the principal features of the cosmogonic model suggested by Fesenkov. As a whole, this model was shaped during the 1930s and early 1940s, although Fesenkov kept introducing some modifications even much later (for instance, being initially a steady partisan of the "hot" formation of the planets and thus an opponent of the cosmogonic model developed by Schmidt, he displayed a more conciliatory approach towards Schmidt's views in the 1960s). One can easily observe that some components of Fesenkov's prewar model look rather much like the nowadays paradigmatic ideas, whereas some differ from them significantly. However, any detailed comparisons between Fesenkov's qualitative insights and modern highly mathematicized cosmogonic theories would look rather artificial.

Thus, the first attempt of a Soviet scientist to develop a comprehensive theory of the origin of the solar system turned out to become rather a failure than a success. Unsurprisingly, Fesenkov's views have never attracted serious attention from Western cosmogonists. This failure looks rather enigmatic, at least superficially—after all, Fesenkov was a brilliant astronomer and astrophysicist, his professional training was excellent, and additionally, during some stages of his professional career he possessed a considerable degree of institutional and administrative leverage over the development of astronomical and astrophysical research in the Soviet Union. A tentative explanation of this puzzle would be in order.

One impeding factor looks evident, namely, Fesenkov's already mentioned reluctance to use mathematical research tools. A specific defect of his research methodology may be understood as another hindering factor. Fesenkov was very persistent in promoting his belief that the formation of the Sun and the planets had been a unified physical process and thus had to be regarded in this capacity by cosmogonists. Generally accepted as it has turned out to be, this idea was rather counterproductive in the 1920s and 1930s when nuclear physics was comparatively underdeveloped and a variety of the subtle aspects of stellar evolution could not be understood properly. In particular, Fesenkov's approach did not allow him to solve the problem of planetogenesis independently of the much less clear puzzle of the formation of the primordial nebula itself. Because of this reason, Fesenkov continually maintained that the very consideration of the problem of the origin of the planetary system was rather premature. Consequently, Fesenkov was never inclined to place planetary cosmogony in the focus of his research interests. Fesenkov's research activities related to this area never demonstrated something like the brainstorm which in several years brought about Schmidt's principal results. Finally, Fesenkov did not make an effort to provide a firm institutional basis for his cosmogonic research, for example, in the form of a targeted research seminar or a special research team. Therefore, his cosmogonic publications have remained rather an episodic moment of his professional career.

During the 1930s, some research in the area of planetary cosmogony was undertaken at the P. K. Sternberg State Astronomical Institute whose sector of cosmogony was headed by N. D. Moiseev.[23] The results by N. Rejn and subsequent findings by N. Parijsky contributed to the critical consideration of the Jeans encounter hypothesis, but by and large they were of only secondary significance, being preceded by H. N. Russell's refutation of that hypothesis published in 1935.

It would probably be useful to sum up the development of Soviet planetary cosmogony during the pre-Schmidt era. First, this development had never reached a definite state either concerning the theoretical results or with regard to the establishment of ongoing research programs by individual scientists. Some participants were coming into the field and leaving it, and some research findings were being sporadically published, but no general research program was formed and backed by a growing number of participants. There were no recognized institutional and intellectual leaders of cosmogonic research, and their organizational and institutional support was also quite insufficient and unstable. Using Thomas Kuhn's vocabulary, one can say that at the time Soviet cosmogony had reached only its preparadigmatic stage.[24]

The period of the formation of the original cosmogonic paradigm in the Soviet Union may be identified in a more or less straightforward way. It began in the spring of 1944 and was completed seven years later. Its initial point was marked by Academician Schmidt's first reports on his cosmogonic ideas delivered to the professional audience at the Sternberg Institute in Moscow.[25] The end of this seven-year period may be reasonably related to the First Conference on Cosmogony, which gathered in Moscow in the spring of 1951. This meeting, whose participants by and large endorsed Schmidt's approach and gave it not only scientific, but also ideological blessing (which was vitally important then!) signified the official acceptance of Schmidt's views as a sound and safe basis for the future development of Soviet cosmogony. During that crucial period the circle of Schmidt's associates and pupils had been formed whose members have managed to continue its founding father's endeavor long after his demise. Similarly, the still existing (albeit under another name) departmental unit whereat Schmidt and his principal co-workers joined their research efforts was also organized during that period.

Academician Schmidt's personal biography looks so unusual and so intrinsically connected with the development of his cosmogonic theory that it should be outlined first. He was born in Moscow on 30 September 1891.[26] His father was then a petty trade clerk of peasant origin whose ancestors had moved to the Baltic provinces from Germany (this factor of Schmidt's German origin would play a very important part in his life one day in the future!). In 1909, Schmidt graduated with a gold medal from a gymnasium in Kiev, and four years later from the mathematical division of the Department of Physics and Mathematics of the Saint Vladimir Imperial University in Kiev, where he majored in theoretical algebra under the tutorship of Professor Dmitry Grave. Even prior to his graduation, Schmidt had already published three research papers, one of them abroad.[27] After graduation, Schmidt was offered a position of the so-called professorial stipendiat (the Russian equivalent of a graduate student), although, unlike Fesenkov, he was not sent abroad because of the war. In 1916, he passed the set of final examinations which enabled him to defend his master's dissertation[28] and began to teach mathematics at the University. In the same year he published an important monograph on the theory of groups,[29] which was awarded by the University the grand gold medal and the Rakhmaninov Prize.[30]

Thus Schmidt, like Kostitzin, was primarily trained in pure mathematics, but his initial career in the field, unlike Kostitzin's, was developing in a perfectly normal way. However, in 1917 Schmidt's life trajectory changed rather drastically. Early in that year, Schmidt took a second job as the deputy director of the ration cards department of the Kiev City Council where he reportedly demonstrated excellent managerial skills. After the February Revolution, Schmidt became strongly engaged in different public activities at the University too, although he had never revealed any interest in politics earlier. In July of 1917, he was sent to Petrograd on municipal business and also as a representative of the University at a conference on the problems of higher education. Captured by the turbulent development of the political events in the capital, Schmidt decided to stay in Petrograd as a staffer of the Food Ministry. After the Bolsheviks had seized power, the Ministry was abolished and replaced by the People's Commissariat of Food (Narkomprod), one of the most powerful agencies of the new government. Schmidt rapidly established a career there and became a very influential and politically trusted official. In particular, in December of 1918 he joined the Bolshevik Party.

It would be rather difficult to list all of Schmidt's different activities and positions during the 1920s and 1930s. During the first postrevolutionary years, he held important offices at the People's Commissariats of Food, Finance, and Public Education; in 1921–24 he was director of the State Publishing House, and the editor-in-chief of the Grand Soviet Encyclopedia in 1924–48. During the period of the existence of the Communist Academy, which was merged with the USSR Academy of Sciences in 1936, Schmidt was one of its leaders and the chief editor of its official magazine and of some other periodicals as well; he was a member of the State Academic Council and director of the Institute for Economic Studies. From 1929 and up to his death, he was professor of mathematics of the Moscow State University where he founded the chair of higher algebra. In 1930s, he became internationally famous as an Arctic researcher, especially after the first Soviet air expedition to the North Pole in 1937, which he headed. He was appointed director of the Institute for Arctic Research in 1930, and in 1932–39 he was the head of the Main Directorate of the Northern Naval Route. He was elected full member of the Soviet Academy of Sciences in 1935, and in 1939 he was made the Academy's first vice-president with broad executive power. Thanks to his initiative, the Academy's Institute of Theoretical Geophysics was established in 1937. This institute existed under Schmidt's directorship until 1948 when it was merged with the Institute of Seismology, and thus the new Institute of Geophysics of the USSR Academy of Sciences was formed.[31]

This long list of occupations and achievements looks really impressive; moreover, Schmidt, unlike a number of other politically promoted academicians of 1930s, was by and large a genuine scientist. However, it would be erroneous to believe that his life was nothing but milk and honey. After his return from the North Pole Schmidt was awarded the Gold Medal of the Hero of Soviet Union and made an icon of official propaganda. Nevertheless, early in 1939 he was abruptly fired from his office at the Main Directorate of the Northern Naval Route. This dismissal had been preceded by some nasty hints in the media, which was a bad sign during those years of the Grand Purge. Schmidt is said to have been accused by his second-in-command at the polar station, I. D. Papanin (who headed the expedition after Schmidt had returned to Moscow in June of 1937 and was evacuated together with his men only in February of 1938), of poor organization of the rescue expedition and inefficient management of the Directorate.[32] These charges were likely to find an attentive ear, since Stalin and his cronies were strongly dissatisfied with the slow pace of the rescue operation (caused, not by Schmidt's lack of competence or ill will, but by very complicated ice and weather conditions) and regarded this slowness as diminishing the propaganda effect of the whole expedition.[33] Besides, Schmidt reportedly tried to help one of his arrested co-workers and thus excited animosity from the side of the powerful prosecutor in the show trials of 1936–38, Andrei Vyshinski.[34] Schmidt was possibly at risk of being purged himself, but probably he was saved by his domestic and international fame. Anyway, his promotion to the position of the Academy's vice-president (he was formally elected on 28 February 1939) meant that Schmidt was safe, at least temporarily (since nobody could feel permanently safe under Stalin's rule).

As a leader of Soviet polar research and director of the Institute of Theoretical Geophysics, Schmidt was not an absolutely unlikely candidate for participation in cosmogonic activities, but actually he joined the field under the influence of rather special factors. After the German invasion in 1941, Schmidt—together with many of the Academy's scientists and members—was evacuated to Kazan, from where he virtually directed the Academy, since the Academy's president V. L. Komarov was seriously ill and, anyway, lacked the managerial skills which were necessary for running the Academy during the war. Schmidt was broadly regarded as the heir apparent, but his status was drastically changed by Stalin's personal order issued in April of 1942 to relieve Schmidt of his office at the Academy. The dictator's reasons can be only guessed, but Schmidt's German name is likely to have played a crucial role. Stalin's personal idiosyncrasy also cannot be ruled out.

Being suddenly extracted from his customary circle of numerous administrative obligations, Schmidt was able to allocate much more time for pursuing his personal research interests and for running the Institute of Theoretical Geophysics. First he hesitated about the choice of his future research, contemplating such distant areas as atmospheric physics and linguistics.[35] However, finally his old interest in cosmogony happened to prevail.

Schmidt's initial concern for the problem of the origin of the planets reportedly arose during his student years.[36] In March of 1925, Schmidt made some interesting notes in his personal diary which are included in this collection; these notes confirm his continuing interest in the problems of cosmogony, particularly his "dream" to explain the Bode law in terms of the stability of orbital motions. The same notes clearly reveal Schmidt's desire to concentrate his research efforts on cosmogonic problems. Evidently, he had too many different time-consuming obligations at the time (amongst them, professorship at the Second Moscow State University and permanent fellowship at the Research Institute of Mathematics and Mechanics of the First Moscow State University, where, we remind the reader, Kostitzin also worked in the mid-1920s), so the realization of this intention had to be postponed. Schmidt also happened to pay attention to questions of cosmogony in some of his general works published during the 1930s. Thanks to the just-outlined circumstances, he became able to devote his complete efforts to them only early in the 1940s.

Intellectually, Schmidt's interest in cosmogony could stem from different sources. He had some aptitude for epistemological thinking, and the problem of the origin of the planets appears to have attracted him from this direction too—his article "Hypothesis" published in the first edition of the Grand Soviet Encyclopedia provides a good example. On the other hand, Schmidt's geophysical studies are likely to have been instrumental as well. Early in the 1920s, while being a member of the so-called Special Commission of the Kursk Magnetic Anomaly, he obtained some important theoretical results pertaining to the mathematical treatment of gravimetric observations.[37] Schmidt's regular participation in the Soviet polar expeditions tended to enhance his interest in geophysics, which, moreover, began to acquire, so to say, historical, or evolutionary character: Schmidt came to the conclusion that the currently developing geological, geochemical, and geophysical processes could be partially interpreted in terms of the processes which had given birth to our planet.[38] "Aspiring after the understanding of the most crucial problems of geophysics, O. I. Schmidt realized that all of them were related to the 'cause of causes,' that is, to the problem of the origin of the Earth."[39] The organization of the Institute of Theoretical Geophysics in 1937 can be seen, with the advantage of hindsight, as an excellent institutional basis for Schmidt's future cosmogonic research.[40]

It is difficult to say when Schmidt resumed his cosmogonic studies after his demotion in April of 1942, but they were developing at a rapid pace. Already in November of 1943, still in Kasan, Schmidt presented his first report about his initial results to the academic councils of the Institute of Theoretical Geophysics and the Astronomical Institute.[41] In the spring of 1944, after coming back to Moscow, Schmidt began to discuss his findings at some research institutions in the capital.[42] Simultaneously he began to publish his research articles on the subject, eight technical papers for specialists and one popular essay in 1944–46.[43] This productivity looks very impressive, especially if we take into consideration that from 1944 Schmidt was permanently suffering from deteriorating tuberculosis which would lead to his death in September of 1956. At the beginning of 1945,[44] or possibly in December of 1944,[45] Schmidt formed a special department at the Institute of Theoretical Geophysics with the intention to make it an organization basis for his cosmogonic research, and he also became its first head. At the time it was named the Department of the Evolution of the Earth; its current title is the Laboratory of the Origin of the Earth and the Planets of the O. Iu. Schmidt Institute of the Physics of the Earth of the USSR Academy of Sciences. Schmidt's first co-workers at the department were the astronomer B. Iu. Levin, the specialist in celestial mechanics G. F. Khilmi, and the planetologist S. V. Kozlovskaia. By the end of the 1940s, the department became larger. The future principal contributor to the modernization of Schmidt's approach, V. S. Safronov, joined its staff, as well as E. L. Ruskol, E. A. Liubimova, and S. V. Maeva,[46] who would also contribute considerably to the progress of the rapidly developing Soviet cosmogonic research. In the summer of 1946, the department established its own research seminar whose sessions became regular. The number of these sessions amounted to 140 by the end of 1953 (when Schmidt was forced to stay permanently in bed because of his illness), and Schmidt himself delivered 18 reports there.[47]

During his life Schmidt published quite a few technical, popular, and even philosophic papers presenting and defending his and his associates' findings.[48] The absolute majority of his calculational results are currently regarded as outdated or simply erroneous. However, his general methodological scheme, the "hard core" of his research program if we are to use Imre Lakatos's terminology, has demonstrated its validity in many ways.[49] These aspects of Schmidt's contribution to the modern cosmogony deserve some special elaboration.

If Fesenkov's research methodology can be called holistic, then Schmidt's was apparently reductionist and separative. This methodology accented the fruitfulness of the division of the whole problem of the origin of the solar system into three relatively autonomous components. The first part comprises the problems related to the origin and the composition of the primordial physical medium which gave life to the Earth and the other bodies of the system. The second part would be then the modeling of planetogenesis from the initial differentiation of the primordial medium up to the formation of the bodies of the system. The third stage of cosmogonic research, in Schmidt's approach, should include the examination of the subsequent evolution of the planets, particularly of the Earth itself. Schmidt especially insisted on the importance of the elaboration of the geological, geophysical, and geochemical implications of the theories of the formation of our planet which could link the cosmogony of the Earth and the domain of geosciences proper. Schmidt also believed that the second component could be dealt with irrespective of any final solution of the problems pertaining to the first one.[50] Thus, the methodological considerations which had hindered the development of Fesenkov's approach were not present in the framework of Schmidt's research methodology.

It is worth stressing that Schmidt himself never produced any comprehensive (at least, in the modern meaning of the term) solution of the principal problem of planetary cosmogony. Facing the problem of the selection of the primordial medium from which the processes of planetogenesis had to originate, he resorted to the mechanically elementary but physically not too convincing model of a swarm consisting of hard particles of different sizes revolving around the already formed Sun in accordance with Kepler's laws. The set of deductions and calculations which were produced by Schmidt on the basis of this surmise comprises what should be justifiably called Schmidt's *theory*.[51] Subsequently, this theory has been substantially reconstructed and modified by Schmidt's associates and their junior colleagues, as well as by many other members of the international community of planetary cosmogonists. As a result, plausible scenarios of planetogenesis in the system of a star of solar mass surrounded by a revolving gas-dust disk are now available.[52] The initial attitude of Soviet astronomers and astrophysicists towards Schmidt's ideas would probably have been more favorable if he had not amended his "meteoritic" (in his own terms) model by an additional and logically unnecessary component designed specially in order to explain the origin of the primordial medium. The explanation was given in terms of the gravitational capture of the meteoritic swarm by the Sun during its passage or passages through some interstellar dust cloud or clouds.[53] As Schmidt himself stressed, this hypothesis allowed him to get around the painful problem of the spatial distribution of angular momentum in the solar system.[54] Probably, Schmidt also believed that this hypothesis was expedient not only physically, but also ideologically. Emphasizing the regular, nonsporadic character of stellar passages through the nebulas of interstellar matter and the considerable probability of the gravitational capture of this matter by the passing stars, and particularly by the Sun, Schmidt was able to maintain that the very formation of our planetary system had been a regular rather than a unique process. The idea of the regularity of planetary formation was much more acceptable from the point of view of the dogmas of the Soviet official Marxism of the period, which Schmidt, whose epistemological thinking was rather unsophisticated, seems to have believed in wholeheartedly. Probably he was sincere when he kept accusing Jeans of the compatibility of the latter's theory with the religious and idealistic worldview which, according to the standards of Soviet propaganda, ruled supreme in the West. In this way, Schmidt could also hope to avert open attacks from the side of the party ideologists and their unscrupulous disciples within scientific circles, a strategy which was vitally important at the time of the flourishing aggressive obscurantism of the last period of the Stalin era.[55] Schmidt succeeded in performing the calculations which proved a nonzero probability of gravitational capture in a three-body system, results which were later given a broader justification by G. F. Khilmi and some other Soviet specialists in planetary mechanics. Nevertheless, the hypothesis of gravitational capture is regarded as outdated and redundant nowadays. It is interesting that Schmidt was inclined to present it as the most important aspect of his approach,[56] and this idiosyncrasy was likely to stimulate his conflicts with a number of Soviet astronomers.

Both Schmidt's *theory*, that is, his meteoritic model of planetary formation, and Schmidt's *hypothesis*, that is, his interpretation of the origin of the primordial Keplerian cloud in terms of its gravitational capture,[57] had been generally formed by the mid-1940s. On 31 January 1947, Schmidt reported his findings at the session of the Second All-Union Geographic Congress in Leningrad.[58] The same year in March, a special session was devoted to Schmidt's speeches to the conference, which was attended by astronomers from the Pulkovo Observatory as well as by the staffers of the Institute of Theoretical Astronomy of the Academy of Sciences and of the Astronomical Institute of the Leningrad State University. By and large, the conferees approved Schmidt's approach and results, and the consensus was reached that the general features of the new cosmogonic scheme had been already determined and that the Moscow guest's studies promised rich results in the future. It is interesting that the published account of the conference[60] contains some hints that its participants were already aware of the plans to convene an all-union conference on cosmogony in Moscow in the near future, which contradicts later statements that the first preparatory steps towards the organization of such a meeting were taken only in the second half of 1950,[61] and that Schmidt himself was informed of this decision only at the beginning of 1951.[62]

Anyway, the First All-Union Conference on Cosmogony was convened only four years later, after Schmidt had already published the first and the second editions of his principal cosmogonic treatise *Four Lectures on the Theory of the Origin of the Earth*.[63] But it is also true that the published materials of the next professional discussion of Schmidt's theory, which took place at the Sternberg State Astronomical Institute in February of 1948, contain no references to any planned major discussion of Schmidt's findings.[64] In the spirit of the time, that meeting was sponsored not only by the administration of the Sternberg Institute, but also by its Party cell (!); it was attended, in addition to the staffers of the institute, by Schmidt, Levin, and Khilmi, and also by Schmidt's most radical adversary, Academician Fesenkov. Fesenkov's attitude looked extremely critical. He described Schmidt's research approach as purely mechanical and not taking into account the effects of electromagnetic radiation as well as other relevant factors; he also charged his opponent with basing his theory on a number of unjustified presuppositions and hypotheses. N. D. Moiseev also opposed Schmidt's results, particularly his model of gravitational capture. He regarded this model as baseless and Schmidt's whole cosmogony as nothing more than a development of Kant's ideas, which in his view did not even deserve the title of a "new cosmogonic theory." It is worth stressing that by and large the astronomers of the Sternberg Institute have always tended to disapprove Schmidt's works.

The prehistory of the First All-Union Conference on Cosmogony, which took place in Moscow on 16–19 April, 1951, is still unclear and hardly can be traced in detail without extensive work in Soviet archives. Technically, the conference was convened by the Academy's Division of Physical and Mathematical Sciences, but is evident that it was authorized, and perhaps even initiated, by much higher authorities. This meeting was not the first and not the last amongst other "show" discussions of scientific matters in the late 1940s and early 1950s. The notorious "special session" of the Academy of Agricultural Sciences in August of 1948, at which T. D. Lysenko, who is known to have secured Stalin's personal support, presented a completely obscurantist report on the "Situation in Biological Science," is widely known, but it was not the only event of the kind. For example, in June of 1951 (that is, *after* the Conference on Cosmogony) the All-Union Conference on the Theory of Chemical Structures in Organic Chemistry was convened,[65] and one more conference on the importance of Academician I. P. Pavlov's teaching for psychiatry and neurology took place in October.[66]

These and other similar meetings could be regarded as scientific discussions only if judged by very superficial appearances. In fact, all of them were nothing but highly politicized public shows, chauvinistic and anti-Western. Although the main attacks are known to have been directed against the followers of "harmful views" in the areas of biosciences and medical sciences, a similar "discussion" of the philosophic aspects of modern physics was also prepared in 1949, being canceled at the last moment thanks to interference from some leading Soviet nuclear physicists.[67] However, in the midsummer of 1950, that is, less than a year before the conference on planetary cosmogony, two Soviet academies, the Academy of Sciences and the Academy of Medical Sciences, organized the so-called "joint session" which resulted in a devastating blow against Soviet physiology.[68] According to some newly published data,[69] this session was launched due to Stalin's direct order issued in 1949. It is also worth recalling that in 1950 Stalin initiated another notorious "scholarly" campaign related to some aspects of linguistics.

Against this background, the published materials of the conference on planetary cosmogony of 1951 look rather unexciting. I confess that this assessment of mine differs from the recollections of the late Soviet astrophysicist I. S. Shklovsky, posthumously published in 1988. Paying full tribute to Schmidt's talents and personal decency, Shklovsky nevertheless reproaches Schmidt for his alleged "Lysenkoism" in promoting his ideas within the Soviet astronomical community.[70] I am afraid that I cannot concur with these and other similar (and occasionally even more harsh[71]) judgments. My own research, including my contacts with Academician Schmidt's former associates, have led me to deem such opinions to be exaggerations, memory aberrations,[72] or possibly the reflections of some ancient personal or institutional conflicts, not uncommonly mixed with the attempts to present the author's former activities in a favorable way.[73] Although it is certainly true that Schmidt occasionally (and in his articles published by "ideological" periodicals regularly) resorted to the standard, virtually ritual invective against the allegedly intrinsic and incurable flaws and defects of "bourgeois cosmogony," neither he nor his immediate associates used this kind of polemics against their opponents in the Soviet Union. This fact should be regarded as crucial. As regards Schmidt's polemical style and methods, his conduct differed quite considerably from the conduct of a great many other leaders of different clans in Soviet science, to say nothing of such an odious figure as Trofim Lysenko. Consequently, Shklovsky's characterization of Schmidt's activities as "Lysenkoism in astronomy"[74] appears completely unfair and unjustified.[75] It is also worth stressing that Fesenkov's own report, delivered to the conference in Fesenkov's absence by P. G. Kulikosvsky, was quite aggressive in relation to Schmidt's views, whereas Schmidt's references to Fesenkov's works and judgments of them were always correct and nonadversary. In particular, Fesenkov rather groundlessly maintained that Schmidt's theory was completely incompatible with the established data on the physical properties of the solar system[76] and surprisingly challenged the very possibility of applying deductive reasoning to cosmogony.[77]

It is true that in the course of the conference of 1951, even Schmidt's strongest opponents refrained from direct ideological attacks against his views. However, after the first edition of the *Four Lectures* had been released in 1949, such accusations became dangerously intensive in the astronomical milieu. Schmidt was blamed not only for his alleged mechanism, but also for more serious (according to the contemporary Soviet standards) flaws such as positivism and agnosticism.[78] It was probably Schmidt's desire to neutralize these invectives and to secure a broader constituency for his approach that induced him to use his influence for the promotion of a multidisciplinary conference on planetary cosmogony. Directing his efforts towards this aim, Schmidt could reasonably hope that those geoscientists less committed to the defense of astronomical dogmas and united by the astronomers' corporative solidarity would be more receptive to his ideas and more inclined to discuss them in a businesslike manner.

Technically, the conference was convened by the Academy of Sciences, but most certainly, it was authorized by the Party authorities. Reportedly, Schmidt used a double leverage working through the Academy's Managing Vice-President A. V. Topchiev, a very mediocre chemist who supervised the Academy as a creature of the Party bureaucracy, and a certain Griaznov, an officer of the Department of Science of the Party Central Committee.[79] In total, there was an attendance of about 300 people at the conference, which lasted for four days. In addition to Schmidt himself and Academician Petrovsky who presided over the conference, 40 participants delivered their speeches. In accordance with the accepted procedures, the conference passed a resolution on the future development of cosmogonic research in the Soviet Union. Yuri Zhdanov, Stalin's son-in-law and the chief of the Department of Science of the Central Committee, blessed the conference with his presence too,[80] although he delivered no address there. It seems evident that the final resolution was also endorsed by the Central Committee.

Schmidt's principal report need not be discussed here, since all of its main points are reproduced in his article, published after the conference in a Society philosophic magazine, which has been included in this collection. The report contained plenty of the ritualistic "refutations" of Western cosmogonic theories, as well as equally ritualistic (and, for that matter, modeled upon Lenin's sacral statements) recitations about the alleged complete incapability of "bourgeois science" to create bold and broad theoretical concepts, particularly in cosmogony. Such phraseological camouflage hardly could be avoided in any case, and by and large, Schmidt resorted to this sort of rhetoric without abusing his opponents (I am stressing once more, only foreign opponents) too much. For that matter, quite similar curses may be easily found in Academician Petrovsky's introductory speech as well as in Fesenkov's lecture. For example, Fesenkov presented as a "model example of the correct development of scientific theory" something that had never existed at all, the so-called "teaching" by Ivan Michurin, a self-trained breeder of fruit trees, whose effort did not actually result in any stable new varieties of fruits but who nevertheless was hailed as a great hero of Soviet science.[81]

It is rather difficult to say whether the ideological framing of Schmidt's report genuinely reflected his own views, or was motivated, at least partially, by other plausible factors, such as his preparedness to observe the rules of the politically scientific spectacles of his times and the natural desire to secure a safe niche for himself and his followers. Admittedly, different motivations were mixed there, and their proportion is hard to discover. Most probably, Schmidt was not a cynical person. Having been once captured by the Bolsheviks' ideas, he remained truly devoted to them during his whole life. It is certainly true that Marx's and Lenin's original ideas were adjusted recurrently by Stalin and his cronies to suit their immediate political needs, and the officially permitted and disseminated "Marxism–Leninism" was nothing but a parody of any theoretical thinking. But is equally true that there always were some gaps between Stalin's general formulas and slogans and their concrete interpretations and applications. In the 1920s and during the first half of the 1930s, Schmidt belonged to the core of the official interpreters of the Party's "General Line" in relation to the area of natural sciences. As a member of the Presidium and Chairman of the Section of Natural Sciences of the Communist Academy, which existed in Moscow until 1936 and even aspired to compete with the Academy of Sciences, Schmidt never questioned the official dogmas of the ideologically aggressive institute, which maintained the "degradation" and "futility" of current western science.[82] Schmidt was neither heretic nor saint, he was perfectly aware of the subtleties of the different power games in the Soviet Union, including those in Soviet science, and he certainly shared all of the principal biases of his environment and its political quasiculture. But, as I keep stressing, Schmidt was unlikely to engage in purposeful demagoguery and he never tried to harass his opponents, not a small thing in his times.

The textual analysis of the published materials of the conference reveals that in addition to the speeches of Academicians Petrovsky, Schmidt, and Fesenkov, 11 other talks contained some standard ideologically propagandistic themes. Moreover, such themes are identifiable only in the talks presented by astronomers and astrophysicists, but not by geoscientists, whose reports look more technical and businesslike. Some users of this phraseology resorted to it only superficially, but sometimes such a usage apparently surpassed the framework outlined by Schmidt's report. For example, the Leningrad astronomer M. S. Eigenson accused "the bourgeois West" of "cosmogonic idiocy,"[83] which was quite consistent with his notoriously servile behavior.[84] But there were examples of the opposite kind too, such as the talk of Pulkovo director A. A. Mikhailov, who offered a correct and even friendly review of Western cosmogonic studies, including the cursed works by Jeans.[85]

It is worth examining the distribution of opinions on the Schmidt theory within different groups of participants in discussion. Although the published materials do not identify the specialties of discussants, they still can be found out on the basis of other sources. The group of specialists in astronomy, astrophysics, and celestial mechanics contained 28 people, and the group of geoscientists 10 people. Also, there was one theoretical physicist and one more discussant whose specialty I have failed to discover. Putting these two persons aside, we have two main groups of 28 and 10 people. Eleven members of the first group voiced their support of the Schmidt theory, whereas only five, including Fesenkov and his pupil Alla G. Masevich, stated their strong objections against this theory as a whole. Seven persons expressed their disagreement with Schmidt's hypothesis of the capture of galactic matter only, and five discussants talked about subjects unrelated to the Schmidt theory. Putting them aside too, we come to the conclusion that among those astronomers and astrophysicists who cared to discuss Schmidt's theory, almost half (11 out of 23) supported Schmidt's views without much reservation, whereas the proportion of the adversaries of his approach was more than twice as small.[86]

It is only natural that the astronomers were less enthusiastic in relation to the hypothesis of capture since it really lacked convincing factual and theoretical evidence. However, some astronomers pretended to refute Schmidt's whole theory while arguing solely against the Schmidt capture hypothesis. But some objections against Schmidt's views were quite sound, for example, those expressed by V. A. Ambartsumian,[87] a leading Soviet astrophysicist and a future full member of the Academy of Sciences. Disregarding his objections against the hypothesis of capture, Ambartsumian recognized the considerable potential of Schmidt's model of planetary formation, paying particular tribute to Schmidt's explanation of planetary rotation.[88] By and large, Ambartsumian delivered a well-balanced evaluation of Schmidt's works, which was rather surprising, since his attitude to Schmidt's results was known to be rather jealous, and there were good reasons to believe that he would want to ally himself with Fesenkov.[89]

Another spectrum of opinions is revealed by the talks of geoscientists. Nobody in this group cared to refute the hypothesis of capture, which is quite understandable, since the problem lies completely within the realm of astronomy and celestial mechanics. Four geoscientists evaluated Schmidt's theory with approval, only one voiced his negative attitude, and five more talks turned out to be neutral or irrelevant to this theory. While reading these texts, one cannot help feeling that Schmidt's ideas were peripheral to the concerns of Soviet geologists.

The concluding statement adopted by the conference looks very interesting.[90] Preceded by a standard set of phrases about the inferiority of Western science and the great achievements of Soviet cosmogony, the endorsed assessment of Schmidt's views and results appears, by and large, objective and well-balanced. It is true that some aspects of Schmidt's approach were criticized, but on purely scientific grounds and without any ideological allegations—for example, such aspects as the lack of proper explanation of the origin of comets and asteroids or the difficulties related to the understanding of the chemical composition of giant planets. The conference recommended a reasonable program of future interdisciplinary cosmogonic research and approved the idea of publishing collections of cosmogonic articles regularly. By and large, the final document looks quite businesslike and ideologically neutral, in sharp contrast with similar documents issued by other show conferences of the period. As a result, the ideological attacks against Schmidt were effectively neutralized. In all probability, the draft of the final document had been stamped by the proper authorities in charge of science at the Central Committee of the Communist Party.

The final document also deals with Fesenkov's works, in only one paragraph. If Schmidt's alleged "Lysenkoism" had really taken place, he could have been expected to criticize his rivals harshly and to demand their total surrender, as the real Lysenko always did. Nothing of that kind actually happened. The document paid due homage to Fesenkov's pioneering role in Soviet cosmogony and urged him to develop his idea quantitatively,[91] advice that looks quite reasonable. The importance of V. A. Ambartsumian's works on stellar cosmogony was also recognized, albeit rather briefly.[92]

The materials of the conference were published at the end of 1951. This publication being not only a scientific event but also a symbolic confirmation of Schmidt's political credibility, the Soviet media published in the winter of 1952 a number of popular expositions of Schmidt's views authored by his sympathizers.[93] It is of special interest that occasionally they called Schmidt's theory not only good, but also a "teaching," which was a much higher form of recognition—this title was normally applied to the views of Marx, Engels, Lenin, and, first and foremost, Stalin (and concerning natural sciences, to the works by I. P. Pavlov, Michurin, and Lysenko). In 1952–53, Schmidt delivered more than 15 popular talks about his theory,[94] and these activities were terminated only by the deterioration of his health in the fall of 1953, which made him a prisoner of his bedroom. In the same years, he published several popular articles with the same content. After Moscow State

University had moved to its new campus on Leninskie Gory in 1953, Schmidt was appointed head of the newly established Geophysical Division of the Department of Physics of the university. He used this opportunity to organize the Chair of the Evolution of the Earth, which he wanted to use as an institutional basis for training specialists in planetary cosmogony.[95] In the autumn semester of 1953, he even began to offer a special course on cosmogony, but was able to deliver only eight lectures.[96]

Although the first conference on cosmogony strengthened Schmidt's institutional positions, he never tried to monopolize the area of cosmogony research. In particular, when the Presidium of the Academy of Sciences instituted a special intra-Academy Commission on Cosmogony in September of 1952, not Schmidt but Ambartsumian was appointed its chairman,[97] and the second most important role in the commission was played not by any of Schmidt's people but by A. G. Masevich. Had Schmidt been a genuine follower of Lysenko's tactics, he would certainly have acted in a different way.

After Schmidt's death, his pupils (especially V. S. Safronov) and the pupils of his pupils produced a great many very important results in planetary cosmogony. These studies have been centered in the Soviet Union at Schmidt's former department at the O. Iu. Schmidt Institute of the Physics of the Earth of the USSR Academy of Sciences. By and large, Schmidt's methodological scheme has been followed by his co-workers and pupils.

In judging Schmidt's overall achievements, it will be only fair to say that he managed to function rather efficiently in the capacity of an intellectual, institutional, and even ideological (what was especially important then and there) leader of Soviet planetary cosmogony. The hard core of his research program has become the basis of the fruitful studies of his pupils and followers in the Soviet Union. The permanent activities of Schmidt's former department at the Institute of the Physics of the Earth, where Schmidt is still warmly remembered,[98] have contributed greatly to Soviet cosmogonic research and made its development uninterrupted and cumulative.

Not only has the hard core of the Schmidt research program proved sound, but at least some of his theoretical approaches have also. For example, he was the first to suggest and develop a very general approach to the problem of planetary rotation based on the analysis of the equations of the conservation of energy and angular momentum in the process of planetary condensation. Although Schmidt's conclusion about some direct links between the scale of thermal dissipation and the direction of the rotation of a planet or its satellite has turned out erroneous, his general presentation of the problem has proved fruitful. The same assessment may be given to another of his beloved ideas: an emerging planet is scooping the matter out of some neighboring "feeding zone," and the consideration of this process permits one to evaluate the pace of planetary growth. Schmidt's own figures turned out to be unrealistically high, but, again, his general approach did survive. On the other hand, Schmidt's evaluation of the initial temperature of the young Earth was too low, mainly because he failed to take into proper account the thermal consequences of the collisions between growing planets and rather large (more than 1000 m across) celestial bodies. Models of moderately hot young planets experiencing partial melting, degassing, and differentiation already in the early stage of their growth are being explored nowadays, but these studies are still incomplete.

The currently existing consensus concerning the main features of the formation of the Sun and the planets is based on a scenario which is much more sophisticated than Schmidt's initial model. This scenario completely lacks the hypothesis of capture, whereas the notion that the planets were formed from a gas-dust disk plays a central role there. According to the scenario, the disk emerged together with the Sun in the process of the gravitational collapse of a fragment of a molecular cloud. This cloud existed in one of the galactic zones rich with young forming stars, and the collapse began about 4.57 billion years ago. The collapse of the slowly revolving primordial nebula brought about the formation of a compact core in its center surrounded by a thick gas-dust disk. The central core was gravitationally attracting and accumulating the matter from the nebula and thus gradually transforming into a young Sun. The dust component of the initial disk tended to move towards its central plane and accumulate there. In this way, a thin dust subdisk was being formed. After the density of the subdisk had reached the critical limit, it became gravitationally unstable, which resulted in its disintegration into spheroidal dust clots. These clots tended to collide and to merge together, thus increasing their density and scooping the matter out of the neighboring space. The largest of them tended to experience a faster growth and to give birth to planets or planetary cores. In this way, the planets of the terrestrial group were formed, as well as the cores of Jupiter and Saturn and probably the outer planets too. The cores of the future giant planets subsequently were accreting gaseous matter from space and thus have become what they are today.

Even this very brief outline of the basic scenario demonstrates that it combines some ideas developed by Schmidt with some views of his principal rival Fesenkov. The latter's belief that the Sun prior to its formation was something like a rotating clot surrounded by a material disk turned out to be accepted, although the later studies did not support his views concerning the speed of this rotation and the sources of the matter from which the disk had been formed. On the other hand, the viability of Schmidt's general methodology and at least some of his model approaches is also evident nowadays. The problem of the present distribution of the angular momentum of the solar system was finally explained without resorting to the hypothesis of capture (in terms of the outward transfer of angular momentum due to magnetic tensions or turbulence within the initial disk), but this explanation has used to a considerable degree the technical and conceptual resources developed by Schmidt himself and especially his co-workers and pupils. So both Fesenkov's and Schmidt's approaches deserve a place in the history of cosmogony.

[1] Characteristically, the first Russian (and the only prerevolutionary) edition of the "cosmogonical" section of Laplace's *Exposition du systeme du monde* was published in Russia in 1861, that is, the same year when the Russian serfs were emancipated.
[2] Priroda (Moscow) **1912** (March), 394.
[3] Priroda (Moscow) **1913** (April), 399.
[4] V. A. Kostitzin, *Évolution de l'atmosphère, circulation organique, époques glaciaires* (Hermann, Paris, 1935).
[5] V. A. Kostitzin, *Simbiose, parasitisme et évolution*, (Hermann, Paris, 1934); *Biologie mathématique*, (Colin, Paris, 1937) [this book was translated into English: *Mathematical Biology*, translated by Th. H. Savory (Harrap, London, 1939)].
[6] See F. M. Skudo and J. R. Ziegler, "Vladimir Aleksandrovich Kostitzin and Theoretical Ecology," Theor. Popul. Bio. **10** (3) December 395 (1976).
[7] *Istoriko-astronomicheskie issledovaniya* (Nauka, Moscow, 1989), Vol. 21, p. 307.
[8] The institutional framing of Soviet cosmogonic research was initiated just then too. The State Astrophysics Institute was structured into five research departments, one of theoretical astronomy and cosmogony.
[9] *Klassicheskie kosmogonicheskie gipotezy* (Classical Cosmogonic Hypotheses), edited by V. A. Kostitzin (Gosizdat, Moscow, 1923).
[10] Reference 9, p. 32.
[11] See V. A. Kostitzin, Ref. 4, p. 8.
[12] V. A. Kostitzin, *Proiskhozhdenie vselennoi* (Origin of the Universe) (Gosizdat, Moscow, 1926).
[13] See the list of his publications of the 1930s in F. M. Skudo and J. R. Ziegler, Ref. 6, pp. 411–12.
[14] Vito Volterra, Preface, in V. A. Kostitzin, *Mathematical Biology*, Ref. 5, p. 9.
[15] Reference 14, p. 11.
[16] See F. M. Skudo and J. R. Ziegler, Ref. 6, p. 410.
[17] On Fesenkov's life and professional career, see G. F. Sitnik, "The life and activities of Vassily Grigorievich Fesenkov" (in Russian), in V. G. Fesenkov, *Izbrannye trudy. Solntse i Solnechnaya systema* (Selected Works: The Sun and Solar System) (Nauka, Moscow, 1976), pp. 5–16; N. B. Divari, "Vassily Grigorievich Fesenkov as an Outstanding Astrophysicist" (in Russian), Vest. Akad. Nauk SSSR **1989** (3), 101; see an enlarged version of this article in *Istoriko-astronomicheskie issledovaniya*, (Nauka, Moscow, 1989), Vol. 21, pp. 302–326.
[18] It is possibly worth noting that amongst L. Struve's pupils also was one more leading Soviet astrophysicist Boris Gerasimov, a tragic victim of Stalin's purges (see Otto Struve, "About a Russian Astronomer", Sky Telesc. **16** (8), 379 (1957); Dict. Sci. Biogr. **5**, 363 (1972).
[19] In 1931, the State Astrophysics Institute was merged with the Institute of Astronomy and Geodesy of the Moscow University, and thus the still existing P. K. Sternberg State Astronomical Institute was instituted. V. G. Fesenkov headed the Sternberg Institute in 1936–39.
[20] V. G. Fesenkov, Soobshch. Kharkovsk. Matem. Obsh. Ser. 2, **15**, (5-6), 278–287 (1917).
[21] V. G. Fesenkov, Tr. Grafo, **1**, 49–185 (1922); see also "Evolution du systéme solaire" Astronom. Nach. **216** (5179) 361–67 (1922).
[22] V. G. Fesenkov, "On the Origin of the Solar System" (in Russian), Astronom. Zh. **7** (2), 130–151 (1930).
[23] Nikolaj D. Moiseev (1902–55), the founding father of the so-called Moscow school of celestial mechanics, the author of important numerical methods.
[24] For the sake of brevity, I cannot explore here an intriguing problem of the ideological debates related to the domain of natural sciences in the Soviet Union during the 1930s. Suffice it to say that the most active participants in these debates did not pay much attention to cosmogonic research.
[25] See B. Iu. Levin, "Creator of the Theory of the Origin of the Earth" (in Russian), in *Otto Iulievich Schmidt: Zhizn'i Deiatel'nost'* (The Academy of Sciences Publishing House, Moscow, 1959), pp. 64–94, on p. 66.
[26] For biographical materials about Schmidt see *Otto Yulievich Schmidt. Zhizn' i Deiatel'nost'* (The Academy of Sciences Publishing House, Moscow, 1959); Igor Duel', *Kazhdoi Graniu* (Znanie, Moscow, 1981); *O. Iu. Schmidt i Sovetskaya Geofizika 80-kh Godov*, (Nauka, Moscow, 1983).
[27] Otto Schmidt, "Sur les produits directs", Bull. Soc. Math. France, **41**, 1913.
[28] Schmidt left the university system in the summer of 1917 before he had had time to finish his master's thesis. Soon, during the Civil War, all the scientific degrees were abolished in Russia and restored only in 1934.
[29] O. Yu. Schmidt, *Abstraknaia Teoriia Grupp*, Kiev, 1916.
[30] Ivan Rakhmaninov (1826–97) was one of the leading professors of the Kiev University, the dean of the Department of Physics and Mathematics in 1868–75 and the rector of the University in 1881–83; he worked in the areas of applied mechanics, general dynamics, and differential geometry. At the end of the 1850s, he spent six months in the United States which was rather unusual for Russian scientists at the time.
[31] In 1956, the Institute, in turn, was divided into three new ones, one of them being the O. Iu. Schmidt Institute of the Physics of the Earth.
[32] V. S. Safronov, private information.
[33] Igor Duel', Ref. 26, p. 178.
[34] S. V. Kozlovskaia and S. O. Schmidt, private information.
[35] S. V. Kozlovskaia, private information.
[36] B. Iu. Levin, Ref. 25, p. 64.

[37]As Schmidt himself happened to note (*O. Iu. Schmidt i Sovetskaya Geofizika*, Ref. 26, p. 69), his own research on the subject was clearly related to some of Kostitzin's earlier results—one more proof of the fact that both scientists' research interests in the 1920s had much in common.

[38]This methodological principle was quite transparently stressed in Schmidt's report at the First Conference on the Problems of Cosmogony delivered on 16 April 1951 [O. Yu. Schmidt, "The Problem of the Origin of the Earth and Planets" (in Russian), in *Trudy Pervogo Sovetschaniia po Voprosam Kosmogonii*, Izdatelstvo Akademii Nauk SSSR, Moscow, 1951, pp. 27–28]. In his own emphatic words, planetary cosmogony "is obliged to serve geosciences by means of research clarifying the process of the evolution of the Earth beginning with its origin. Only this kind of historical consideration is able to identify the forces acting in the depth of the Earth and to provide geologists with a basis for their theoretical constructions, for example, those related to the causes of mountain formation." It would be of interest to find out to what degree this principle was related to the Marxist dogma, especially emphasized by Engels, that all the principal regularities and laws of the physical as well as the social world should be understood and interpreted historically.

[39]V. S. Safronov, "O. Yu. Schmidt and Cosmogony" (in Russian), Ukrain. Matemat. Zh., **23** (5), 707–716 (1971), on p. 707.

[40]It is worth noting that according to some very recent data, Schmidt began to work on his meteoritic theory as early as in 1938, that is, only one year after the Institute of Theoretical Geophysics had been established [see *Nauka i Tekhnika SSSR, 1917–1987* (Nauka, Moscow, 1987), p. 125].

[41]Igor Duel', Ref. 26, p. 52.

[42]B. Iu. Levin, Ref. 25, p. 66.

[43]B. Iu. Levin, Ref. 42.

[44]B. Iu. Levin, Ref. 42.

[45]Igor Duel', Ref. 26, p. 62.

[46]V. S. Safronov, Ref. 40, p. 708.

[47]B. Iu. Levin, Ref. 25, p. 67.

[48]The very first of Schmidt's cosmogonic studies "Meteoric Theory of the Origin of the Earth and Planets" has been included in this collection. His first popular presentation of his cosmogonic ideas, "A New Theory of the Origin of the Earth," was published in *Priroda* (Moscow) **1946** (7).

[49]It is my great pleasure to express my gratitude to the Director of the Laboratory of the Origin of the Earth and Planets of the O. Yu. Schmidt Institute of the Physics of the Earth, Dr. A. V. Vitiazev, for his very friendly hospitality and cooperation as well as for a number of discussions of Schmidt's role, conduct, and achievements.

[50]O. Yu. Schmidt, *Chetyre Lektsii o Teorii Proiskhozhdeniia Zemli* (Four Lectures on the Theory of the Origin of the Earth), 3rd ed. (Izdatelstvo Akademii Nauk SSSR, Moscow, 1957), p. 28.

[51]See, for example, B. Iu. Levin, Ref. 25, p. 68; V. S. Safronov, Ref. 40, p. 709.

[52]V. S. Safronov and A. V. Vitiazev, "Origin of the Solar System" (in Russian), *Itogi Nauki i Tekhniki*, Ser. Astron. (VINITI, Moscow, 1983), Vol. 24, pp. 5–93; V. S. Safronov, *Proiskhozhdenie Zemli* (Origin of the Earth) (Znanie, Moscow, 1987); A. V. Vitiazev, "New Results of Planetary Cosmogony and the Reconsideration of the Ideas on the Early Earth" (in Russian), in *Planetnaia Cosmogoniia i Nauki o Zemle* (Nauka, Moscow, 1989), pp. 6–12. It may be also noted that Schmidt himself came to the acceptance of the existence of the gas-dust component of the protoplanetary cloud by the beginning of the 1950s.

[53]Initially, Schmidt preferred to consider the model of the capture of a meteoric swarm, the approach of which may be found in the first edition of his "Four Lectures" released in 1949. However, after the important study by L. Gurevich and A. Lebedinsky, included in this collection, had been published, he accepted the idea of the capture of a gas-dust cloud, which was reflected in the second edition of the "Lectures" (1950).

[54]See, for example, Schmidt's report at the First Conference on the Problems of Cosmogony, in *Trudy Pervogo Soveshchaniia po Voprosam Cosmogonii* (Izdatelstvo Akademii Nauk SSSR, Moscow, 1951), p. 22.

[55]On the 'Zhdanovshchina', see, for example, Loren R. Graham, *Science, Philosophy, and Human Behavior in the Soviet Union* (Columbia University Press, New York, 1987).

[56]V. S. Safronov, Ref. 40, p. 710.

[57]Anyway, Schmidt's views on this problem did not remain unchanged. According to B. Iu. Levin's recollections (B. Iu. Levin, Ref. 25, p. 29), Schmidt, during his last years, considered as the most probable not purely gravitational capture, but capture as a result of nonelastic collisions.

[58]O. Yu. Schmidt, Izvest. Geograf. Obshch. **1947**, No. 3.

[59]Priroda (Moscow) **1947** (10), 90.

[60]Reference 59.

[61]B. Iu. Levin, Ref. 25, p. 70.

[62]Igor Duel', Ref. 26, p. 156.

[63]The title is related to the fact that the book was based on the four lectures on cosmogony which Schmidt had read at the Institute of geophysics of the USSR Academy of Sciences in November of 1948.

[64]O. Yu. Schmidt, Astronom. Zh. **25** (4), 280–84 (1948).

[65]A. Sonin, "Uneasy Years of Soviet Chemistry" (in Russian), Znanie-Sila **1988** (10), 64.

[66]See the verbatim records of this seccion in the book *Fiziologicheskoe Uchenie Akademika I. P. Pavlova v Psikhiatrii i Nevrologii* (Gosmedizdat, Moscow, 1952).

[67]The general analysis of this period and the relevant developments within Soviet science and philosophy may be found in Loren R. Graham, *Science and Philosophy in the Soviet Union*, (Allen Lane, London, 1971). On the planned but canceled conference on the philosophic aspects of physics see Priroda (Moscow) **1990**, No. 3-5.

[68]Nauka i Znanie **1988** (3), 129; **1988** (4), 147; **1989** (1), 94.

[69]Nauka i Znanie **1988** (3), 131.

[70]Energiia, **1988** (6), 52.

[71]For example, the renowned Soviet author Andrei Bitov maintained that Stalin, like Hitler, had preferred to have his own cosmogony whose principal priests were Schmidt and Fesenkov (Novy Mir **1989** (4), 140). This rather strong statement is unlikely to be historically correct for several reasons; for instance, Fesenkov's cosmogonic views had been formed by the mid-1920s, when Stalin was too preoccupied with building up his political constituency to pay any attention to the problem of the origin of the Earth.

[72] Professor Shklovsky characterized Schmidt's views as "most controversial, and in their reasonable part borrowed" [I. S. Shklovsky, "Cosmogonic Poem" (in Russian), Energiia, 1988, no. 7, pp. 51–55, on p. 53]. However, Shklovsky, while delivering his speech to the First Conference on the Problems of Cosmogony (*Trudy*, pp. 168–176), voiced his objections only against Schmidt's hypothesis of capture, but recognized the "outstanding achievements" of Schmidt's cosmogonic model.

[73] I cannot help giving an amazing example of this trend, albeit unrelated to the events which Schmidt participated in. The Soviet historian of psychology Michael G. Jaroshevsky, who today heads a special research group at the Institute for the History of Science and Technology of the USSR Academy of Sciences, instituted with the aim to research the problems of "persecuted science" in the Soviet Union, was one of the first ideological fighters against cybernetics at the beginning of the 1950s; see, for example, his essay "Cybernetics as a 'Science' of Obscurants," Literatur. Gaz., 5 April 1952, p. 4.

[74] Energiia **1988** (7), 52.

[75] Schmidt forcefully insisted in his report at the First Conference on the Problems of Cosmogony that all of the controversies among Soviet cosmogonists are nevertheless the "disputes within one materialistic camp" (*Trudy*, p. 17). Schmidt's unwillingness to capitalize ideologically on his debates with his opponents deserves respect, especially in comparison with the much less noble conduct of a great many other Soviet scientists during the last years of the Stalin era and even much later.

[76] O. Yu. Schmidt, "The Problem of the Origin of the Earth and Planets" (in Russian), in *Trudy Pervogo Sovetschaniia po Voprosam Kosmogonii*, Izdatelstvo Akademii Nauk SSSR, Moscow, 1951, p. 41.

[77] Reference 76.

[78] Igor Duel', Ref. 26, pp. 154–55.

[79] V. S. Safronov, private information.

[80] V. S. Safronov, private information.

[81] O. Yu. Schmidt, "The Problem of the Origin of the Earth and Planets" (in Russian), in *Trudy Pervogo Sovetschaniia po Voprosam Kosmogonii*, Izdatelstvo Akademii Nauk SSSR, Moscow, 1951, p. 38.

[82] See *Za povorot na fronte estestvoznaniia* (Moscow-Leningrad, Moscow, 1931).

[83] O. Yu. Schmidt, "The Problem of the Origin of the Earth and Planets" (in Russian), in *Trudy Pervogo Sovetschaniia po Voprosam Kosmogonii*, Izdatalstvo Akademii Nauk SSSR, Moscow, 1951, p. 222.

[84] Robert A. McCutcheon, "The 1936–37 Purge of Soviet Astronomers," Slav. Rev. **50**, 100–117 (1991).

[85] O. Yu. Schmidt, "The Problem of the Origin of the Earth and Planets" (in Russian), in *Trudy Pervogo Sovetschaniia po Voprosam Kosmogonii*, Izdatalstvo Akademii Nauk SSSR, Moscow, 1951, pp. 345–47.

[86] For example, A. A. Mikhailov stated his belief that Schmidt's theory was bound to share eventually the fate of the cosmogonic model developed by Jeans, that is, it would be rejected by the professional community as a whole, but some of its components would be included in future theories.

[87] O. Yu. Schmidt, "The Problem of the Origin of the Earth and Planets" (in Russian), in *Trudy Pervogo Sovetschaniia po Voprosam Kosmogonii*, Izdatalstvo Akademii Nauk SSSR, Moscow, 1951, p. 327.

[88] O. Yu. Schmidt, "The Problem of the Origin of the Earth and Planets" (in Russian), in *Trudy Pervogo Sovetschaniia po Voprosam Kosmogonii*, Izdatalstvo Akademii Nauk SSSR, Moscow, 1951, p. 328.

[89] E. Ruskol, private information.

[90] O. Yu. Schmidt, "The Problem of the Origin of the Earth and Planets" (in Russian), in *Trudy Pervogo Sovetschaniia po Voprosam Kosmogonii*, Izdatalstvo Akademii Nauk SSSR, Moscow, 1951, pp. 364–69.

[91] O. Yu. Schmidt, "The Problem of the Origin of the Earth and Planets" (in Russian), in *Trudy Pervogo Sovetschaniia po Voprosam Kosmogonii*, Izdatalstvo Akademii Nauk SSSR, Moscow, 1951, p. 365.

[92] O. Yu. Schmidt, "The Problem of the Origin of the Earth and Planets" (in Russian), in *Trudy Pervogo Sovetschaniia po Voprosam Kosmogonii*, Izdatalstvo Akademii Nauk SSSR, Moscow, 1951, p. 365.

[93] See, for example, A. Lebedinsky, "Novel Results in the Teaching on the Origin of Planets" (in Russian), Izvestia, 17 February 1952, p. 2; B. Levin, "Novel Results in the Science of the Origin of Celestial Bodies" (in Russian), Komsomolsk. Pravda, 1 March 1952, p. 3; V. V. Radzievsky, "Novel Results in the Teaching on the Origin and Development of the Solar System" (in Russian), *Bloknot Agitatora*, No. 9 (63) (Yaroslavl', 1952). The recurrent usage of the word "teaching" in relation to a supposedly strictly scientific theory is noteworthy and quite typical for the Soviet official phraseology of the period.

[94] B. Iu. Levin, Ref. 25, p. 71.

[95] B. Iu. Levin, Ref. 94.

[96] B. Iu. Levin, Ref. 94; see also Igor duel', Ref. 26, pp. 183–4.

[97] *Vopr. Kosmog.* **2**, 337 (Izdatelstvo Akademii Nauk SSSR, Moscow, 1954).

[98] In particular, a conference dedicated to Schmidt's centennial anniversary was convened in Moscow in the fall of 1991.

[99] *Planetnaia Kosmogoniia i Nauki o Zemle* (Nauka, Moscow, 1989), p. 5.

Planetary Cosmogony in the West and Safronov's Theory

Stephen G. Brush

Before the 1970s, most planetary scientists in the United States and Western Europe paid little attention to Soviet research on the origin of the solar system, even though (as can be seen from this book) many of the important papers had appeared in English translation in *Soviet Astronomy AJ* and other journals. I will review here some of the main themes in Western planetary cosmogony, with special emphasis on those that provided a favorable environment for the reception of the ideas of Otto Schmidt's followers.[1]

MONISTIC VERSUS DUALISTIC THEORIES; HOT VERSUS COLD

From a modern viewpoint the most fundamental feature of a theory on the origin of the solar system is the relation it postulates between the formation of the planets and the birth of the Sun. Cosmogonies that assume both arose from a single process may be called "monistic"; those that take for granted the prior existence of the Sun and attribute the rest of the solar system to the action of an alien entity such as another star are "dualistic."

According to current ideas about the structure and evolution of the galaxy, the probability of stellar encounter is so small that relatively few planetary systems could have been formed by such a dualistic process. But if our system was formed by a monistic process it is reasonable to suppose that the formation of stars in general (or at least the formation of single stars) is accompanied by the formation of planets. Thus to estimate the abundance of life in the galaxy, and the chance of our communicating with other civilizations, one needs to know whether the formation of our own system was monistic or dualistic.

Monistic versus dualistic is an astronomical classification of cosmogonies; of more interest to Earth scientists is the initial temperature: hot or cold? Planets formed by condensation from a hot gas would presumably have different physicochemical properties and geological histories than those assembled from cold solid particles.

Two major theories proposed in the 18th century employed opposite astronomical assumptions but had the same geological consequences. The French naturalist George-Louis Leclerc, Comte du Buffon, invoked a huge comet to eject planet-forming material from the sun; the comet became a star in later dualistic theories. The French mathematical astronomer Pierre Simon de Laplace, in 1796, imagined the planets to have formed from rings spun off by the hot extended, rotating contracting atmosphere of the sun; this, combined with William Herschel's 1811 scenario for star formation from nebulas, became the *nebular hypothesis*, the paradigm of monistic cosmogonies. Both Buffon and Laplace assumed that Earth and the other planets started as hot gaseous spheres that gradually cooled and solidified. This assumption became the basis for most 19th-century geological theorizing.

In the 1840s a major advance in physics showed that one could have a hot primeval Earth without assuming a hot primeval nebula. The law of energy conservation, and in particular the thermodynamic transformation of mechanical to thermal energy, indicated that an extended cold cloud, collapsing by its own gravitational contraction, would become hot enough to power the Sun's present output for 20 million years or so (the Helmholtz–Kelvin model), while allowing the planets to start their existence as gaseous or molten balls.

The nebular hypothesis was closely related to the evolutionary world view popular among 19th-century natural and social scientists. It gave a plausible explanation of how the complex present could have developed from a simple past through the gradual operation of identifiable causes. (It was only in the 20th century that "evolution" came to mean Charles Darwin's random, harshly competitive, undirected process.)

The German social philosopher Friedrich Engels, later to become a respected authority on science for Soviet dialectical materialists, praised Kant for rejecting the "petrified outlook on nature" of the 18th-century scientists who thought the world had been created for a purpose. Kant's proposal that "the earth and the whole solar system appeared as something that had *come into being* in the course of time," as developed by Laplace and Herschel, was for Engels the proper basis for a scientific conception of planetary evolution.[2]

Engels saw no conflict between planetary evolution and Darwin's theory of biological evolution, which he also supported. But the British physicist Lord Kelvin, calling himself an evolutionist, used the hypothetical hot origin of the Earth and Joseph Fourier's heat conduction theory to estimate the Earth's age as less than 100 million years, perhaps only 20 million years. This caused serious difficulties for Darwin, who had thought much longer periods were available for biological evolution by natural selection. The resulting controversy in the late 19th century helped to establish the superiority of physics over the other sciences, despite the fact that Kelvin's estimates were refuted after the discovery of radioactivity.

The American geologist Thomas Chrowder Chamberlin challenged Kelvin's result in 1899, pointing out that there was no convincing evidence for a hot origin; the Earth might have been formed by accretion of cold particles, in which case Kelvin's calculational method could not put any limit on its age. Chamberlin was able to use another argument from physics to undermine the hot-origin hypothesis: according to the kinetic theory of gases, if the Earth had ever been hot enough to vaporize iron, molecules of lighter elements would have acquired velocities great enough to escape from the Earth's gravitational field entirely. Thus the Earth would have lost its atmospheres and oceans.

Chamberlin called his primeval particles "planetesimals" and, with F. R. Moulton, worked out a detailed cosmogony. He recognized that planets would not necessarily have retrograde rotation (as had earlier been inferred from a simplistic application of Kepler's Third Law), but could have direct rotation (if one takes account of Kepler's First and Second Laws). This problem was ignored by other Western planetary scientists until the 1960s, but was discussed by Gurevich and Lebedinskii in 1950, and Chamberlin's conclusion was confirmed by Artem'ev and Radzievskii in their 1965 paper presented in this book.

Having shown that the nebular hypothesis failed to explain the distribution of angular momentum within the solar system, along with other deficiencies, Chamberlin and Moulton introduced a dualistic hypothesis. They postulated that the tidal force exerted by a passing star released hot gases from the Sun. These gases, which originally formed the arms of a spiral nebula, eventually cooled and condensed to planetesimals. The resulting planets are thus "children" of the Sun, deserted by their wandering father whose present location is unknown.

A similar dualistic theory was proposed independently by the British physicist-astronomer James Jeans and the British geophysicist Harold Jeffreys in the 1910s. Jeans also provided another argument against the nebular hypothesis: if the matter in the present solar system were spread uniformly throughout its volume, gravitational forces would not be strong enough to start the condensation process (Jeans's stability criterion). As a result, the alternative "tidal" or "encounter" theory was generally accepted by astronomers until about 1935. But Jeans and Jeffreys did not adopt Chamberlin's cold-origin hypothesis; Jeffreys argued that even if solid planetesimals could be formed, their mutual collisions would quickly vaporize them.

Schmidt, in his 1951 essay translated in this book, denounced the Jeans encounter theory as an unscientific attempt to pander to "the religiously-idealistic world views reigning in bourgeois society." Jeans himself, in his popular writings, inferred from his theory that our own existence is a result of pure chance, not design. The formation of the solar system is an unlikely event that has happened only because there are so many stars "wandering blindly through space for millions of millions of years" that two of them must eventually come together to produce a planetary system; just as a monkey, banging the keys of a typewriter for millions of years, will eventually type a Shakespeare sonnet.[3]

The tidal-encounter theory was abandoned because of technical criticism by the American astronomer Henry Norris Russell and others in the 1930s. One objection was dynamical: the tidal interaction could not put material into orbit with the required angular momentum. Another was thermal: according to Eddington's theory of stellar structure, based on radiation rather than convection, the gases extracted from the Sun would have a temperature of about a million degrees. Atomic speeds would be so great that the gases would escape into interstellar space before they could condense. (This objection was worked out in detail by Lyman Spitzer.)

REVIVAL OF THE NEBULAR HYPOTHESIS

Although it appeared in the early 1940s that both possible theories, monistic and dualistic, had been refuted, new ideas provided a plausible basis for reviving a monistic theory. Russell's own earlier work inspired one of those ideas. Confirming the discovery of the British-born astronomer Cecilia Payne in her 1925 Harvard dissertation research, he found in 1929 that the Sun is mostly hydrogen and helium. If one assumed that the primeval nebula had the same composition as the present-day Sun, it must have been much more massive than the present-day planets and must have subsequently lost most of its hydrogen and helium. The enhanced density would make it easier (by the Jeans criterion) for processes such as gravitational instability, viscosity, and turbulence to start the condensation of gases and dust.

Another new idea was "magnetic braking." If the early Sun had a strong magnetic field and the early nebula was ionized, lines of magnetic force would be trapped in the plasma and transfer momentum to it. In this way one might explain why most of the angular momentum of the present solar system is in the giant planets, rather than in the Sun as one would have expected from the nebular hypothesis.

These two ideas were exploited in the 1940s by C. F. von Weizsäcker in Germany and Hannes Alfvén in Sweden, respectively. The net result was a movement away from the dualistic cosmogonies dominant in the first third of the 20th century toward monistic cosmogonies. While it was generally agreed that the planets could not have been formed from material pulled out of the Sun, it was not universally accepted that the Sun and planets came from the same nebula. Thus Alfvén, and Schmidt in the USSR, postulated that a previously formed Sun captured material from interstellar space—either from a single "protoplanetary cloud," as Schmidt's followers called it, or several different clouds (Alfvén). Such theories could be called dualistic, although they ascribed a different role to the two actors in the creation drama; the planets are adopted rather than natural children of the Sun.

Most theorists accepted T. C. Chamberlin's hypothesis that the Earth and perhaps all the planets formed at fairly low temperatures by the accumulation of small solid planetesimals rather than by the cooling and contraction of a hot gaseous ball. There was considerable disagreement as to whether there had been a high-temperature stage before the formation of planetesimals.

Among those who developed nebular theories in the United States, Gerard Kuiper (born in the Netherlands) and Harold Urey were the most influential in the 1950s. Kuiper had initially judged the origin of the solar system a problem not yet soluble by direct attack, so he turned instead to what he considered an easier problem: the origin of double stars. He then developed a picture of the solar system as an "unsuccessful" double star. Kuiper postulated a massive solar nebula, about one tenth the mass of the Sun, i.e., about 100 times the present mass of the planets, and assumed that it would form large protoplanets by gravitational collapse (Jeans instability). After the planets formed, the excess material would be blown away by the Sun's radiation pressure. Kuiper's theory was discussed by Soviet astronomers at the International Astronomical Union (IAU) meeting in Moscow in 1958; see the articles by E. Ruskol (1958) and B. Yu. Levin (1959), in Part III of this book, for an indication of how this and other Western theories were received in the USSR.[4]

Urey started from Kuiper's theory but soon rejected the protoplanet hypothesis, assuming instead that numerous smaller objects of asteroidal and lunar size were first formed and later accumulated into planets. He was primarily interested in explaining the chemical properties of solar system constituents and in elaborating the consequences of his assumption that the Moon was formed before the Earth and later captured by it. By the time of the Moscow meeting Urey and Kuiper were bitter enemies; Ruskol recalls observing a sharp oral argument between them (1990 interview with S.G.B.).

The British theoretical astronomer Fred Hoyle developed a theory of the early history of the Sun, based on Alfvén's magnetic braking hypothesis. He postulated that planetesimals would be formed in a disk surrounding the Sun and then pushed outward by the gas flowing from the Sun as magnetic forces transferred angular momentum to the disk. He presented his theory at the 1958 IAU meeting in Moscow; it was strongly criticized by Safronov; see his 1966 paper (Part III).

METEORITES AND STARS

After 1960, "isotopic anomalies" (deviations from the "normal" relative abundances) found by precise measurements on meteorite material played a major role in Western theories of the initial stage of the evolution of the solar system. Although these anomalies have little bearing on most of the traditional problems of planet and satellite formation, they were believed to offer important clues to the initial stages of formation and contraction of the solar nebula as related to nuclear processes in the Sun and other stars. It is remarkable that Safronov's group paid very little attention to this area of research, preferring to concentrate on the later stages of planetary development—and in fact to move ahead of Western theorists in that area.[5]

The best known inference from isotopic anomalies is the "supernova trigger" hypothesis, based in part on the excess ^{26}Mg found in the Allende meteorite. The hypothesis goes back to 1960, when J. A. Reynolds found an excess of ^{129}Xe in the Richardton meteorite. Theorists reasoned that a short-lived isotope (in this case ^{129}I) must have been synthesized in a supernova, ejected into the interstellar medium, and incorporated into a meteorite parent body that cooled down enough to retain xenon gas, all within a period of only about 100 million years. Since a supernova explosion also produces a shock wave that might compress rarefied clouds to densities high enough for them to become unstable against gravitational collapse, the isotopic anomalies might indicate that a supernova *caused* the solar system to form.

If a supernova is *necessary* to produce a planetary system, then one loses an attractive feature of monistic cosmogonies, namely, the inference that the same process that forms a star generally forms a planetary system as well; hence planets and life are widespread in the universe.

The supernova trigger hypothesis was not taken seriously until the establishment of the Allende ^{26}Mg anomaly by Lee, Papanastassiou, and Wasserburg at Caltech in 1976. This anomaly was attributed to the isotope ^{26}Al, which has a half-life of only 700,000 years and thus was synthesized less than a few million years before the formation of the solar system.

The major advocate of the supernova trigger hypothesis was A. G. W. Cameron, a Canadian-born astrophysicist who has worked in the United States during the past three decades, and has been the most influential North American theorist since 1960. An expert on nucleosynthesis in stars, he could speak authoritatively on the significance of isotopic anomalies. Taking full advantage of the fast but cheap computers available in the 1960s and 1970s, he developed a series of numerical models for the condensation of solar nebula, experimenting with a range of different physical assumptions. He was one of the first to discover that the mathematical collapse of a cloud does not ordinarily lead to a large central body surrounded by smaller bodies unless special processes are invoked. In contrast to Alfvén, Hoyle, Mestel, and others, Cameron concluded that turbulent viscosity rather than magnetic braking is primarily responsible for the transfer of angular momentum from Sun to planets.[6] In 1976 he revised his models to incorporate the theory of accretion disks developed by Lynden-Bell and Pringle; at the same time he concluded, contrary to his early views, that the planets were probably formed from giant gaseous protoplanets. Cameron was also one of the major proponents of the theory that the Moon was formed by impact of a Mars-size planet on the Earth (see below).

In the 1960s Cameron also supported the ideas of the Japanese physicist C. Hayashi about the early evolution of the Sun; Hayashi argued in 1961 that before a star reaches the main sequence it must go through a convective stage in which it is highly luminous. This stage was thought by some theorists to be associated with the strong mass outflow (greatly enhanced solar wind) observed for T Tauri stars. The young star's emission of radiation and matter would sweep away the excess nebular material after the planets had formed, in particular, the extra hydrogen postulated to be present in the original nebula in accordance with the present abundance of hydrogen in the Sun. But in 1969 Richard Larson concluded, on the basis of his Caltech Ph.D. thesis calculations of the dynamics of a collapsing protostar, that a star of solar mass would have a much smaller luminosity than that predicted by Hayashi's theory. Theorists in the 1970s assumed that the Sun may have been formed without reaching very high temperatures until after the planets had been formed, and the role of a hypothetically hyperactive early Sun in planetary cosmogony was considerably diminished. In 1983 Cameron and his colleagues revived the idea of a highly luminous early Sun and again proposed nebular models in which the temperature would be high enough to evaporate small bodies out to a distance corresponding to the orbit of Mars.

Cameron abandoned the supernova trigger hypothesis in 1984 because of new evidence suggesting that ^{26}Al is copiously produced by other mechanisms, for example, in novas or in the outer envelopes of red giant stars. The γ-ray telescope on the High Energy Astronomical Observatory satellite (HEAO-3) found large amounts of ^{26}Al throughout the galaxy—much more than could have been produced by supernovas, according to current theories. Donald Clayton (Rice University) and others argued that since the supernova hypothesis was unnecessary to account for the abundance of ^{26}Al, there was no longer adequate justification to invoke it in explaining the origin of the solar system.

SAFRONOV'S PROGRAM: ACCRETION OF PARTICLES IN THE PROTOPLANETARY CLOUD

By 1960 the hypothesis that planets formed by gravitational collapse of massive gaseous protoplanets, advocated primarily by Kuiper, had been abandoned by most theorists (though it was later to be revived by Cameron). The most popular alternative was accretion of solid particles, with or without the presence of gas during the later stages of planetary formation.

Although several scientists such as Urey in the United States and A. E. Ringwood in Australia discussed the chemical aspects of the formation of terrestrial planets, there were few attempts before 1970, outside the USSR, to develop quantitative physical models of the accretion process itself. This seems odd in view of the fact that powerful theoretical methods for treating very similar processes were widely known in the physical sciences. The kinetic theory of gases, formulated by James Clerk Maxwell and Ludwig Boltzmann, had been actively developed for 100 years; it provided systematic techniques for computing the properties of systems of colliding particles and could be modified to take account of inelasticity of collisions, combination and fragmentation of particles, their nonspherical shape, spatial inhomogeneities, external fields, etc. Physical chemists had worked out approximate theories to describe coagulation and chemical reactions in fluid media. Astrophysicists were familiar with the application of stochastic models to systems of interacting stars. And, when analytic techniques could not adequately handle more complicated "realistic" models, computers were available to grind out numerical solutions. It appears to me that most of the theoretical research on planetary accretion done in the 1970s and 1980s—with the possible exception of some projects requiring very fast, large-memory computers—could have been done at least 10 or 15 years earlier if anyone had been interested.

In fact, the only person[7] who seems to have been seriously interested in pursuing this research during the 1960s was Viktor Sergeyevich Safronov at the O. Yu. Schmidt Institute of Earth Physics in Moscow. Following the ideas of Schmidt and other Soviet cosmogonists (see the preceding essay by A. Levin), Safronov worked out in considerable detail the dynamical and thermal aspects of a model of colliding, accreting, and fragmenting solid particles.[8] Although a few of his papers appeared in English translation shortly after their original publication,[9] Safronov's achievements were not generally recognized in the West until 1972, when an English-language version of his 1969 book *Evolution of the Protoplanetary Cloud* became available and when his detailed model for the accumulation of the planets was discussed at the Nice (France) symposium on the origin of the solar system.[10] Since then the Safronov model or one of its variants has been the most popular explanation for the formation of the terrestrial planets. It has also played a major role in the leading theories of the origin of the giant planets and their satellites, asteroids, comets, and meteorites.

Safronov urged a division of labor in cosmogony: the history of planets after their formation could be treated separately from the problem of the initial formation of planets in the protoplanetary cloud (or "solar nebula" in Western terminology), and that problem in turn was distinct from the problem of the origin of the protoplanetary cloud itself. He preferred the monistic hypothesis of common formation of the Sun and protoplanetary cloud (PPC) over Schmidt's assumption that the PPC was formed elsewhere and later captured by the Sun, but considered himself a proponent of Schmidt's ideas since his model pertained only to the second stage. Thus Safronov's theory did not compete with those of Hoyle and Cameron in trying to explain the formation of the Sun. He did dispute Cameron's assumption that the PPC was very massive (more than one solar mass in addition to the Sun's own mass), preferring a low-mass PPC (about 0.05 solar masses). He also criticized the models of von Weizsäcker, Cameron, Hoyle, and others based on turbulence since they did not provide a mechanism for stirring the material.[11] He did, however, acknowledge the role of turbulence in later papers (e.g., Safronov and Ruzmaikina, 1985).

Starting with a relatively low-mass gas-dust cloud in which any primeval disordered motions had been damped out, Safronov assumed that dust particles would settle to the central plane and grow to centimeter size. As suggested by Edgeworth in 1949, and by Gurevich and Lebedinskii in 1950 (see below), the dust layer would break up into many condensations by local gravitational instability. These condensations would then combine and contract.[12]

Coagulation theory goes back to the work of the Polish physicist Marion von Smoluchowski on Brownian movement at the beginning of the 20th century; Safronov may have read an influential 1943 review article on Brownian movement by the Indian-American astrophysicist S. Chandrasekhar.[13] In Safronov's first model, fragmentation by collisions was neglected; the coagulation coefficient was assumed to be proportional to the sum of the masses of two colliding bodies. The number of particles with mass m was found to vary approximately as $m^{-3/2}$ over a broad range of values of m; for large m, an exponential damping factor became important. Fragmentation did play a role, especially when the relatively velocity of two colliding particles was high. But if the relative velocity was very small, the particles would tend to move in similar orbits and collide so rarely that growth could not occur. Safronov argued that as the particles grew, encounters that did not lead to collisions would increase the relative velocities. The relative velocities most favorable for growth are those somewhat less than the escape velocity, which of course depends on the mass of the particles. The average relative velocity would tend to increase as the particles grew, so that it would remain in the range favorable for further growth.[14,15]

Safronov also concluded that when one body in a region happened to become significantly larger than the others, it would start to grow even faster because its effective cross section for accretion of other bodies would be enhanced by gravitational forces. In this way a single planet could emerge in each "feeding zone" within the PPC and then sweep up the rest of the material in that zone.

In a 1959 paper Safronov noted the importance of high-speed impacts of a few large bodies in the formation of the Earth, a feature he attributed to B. Yu. Levin,[16] and he emphasized this point again in 1964.[9] He estimated that the formation of the Earth was essentially completed in 10^8 years, and that in spite of the large impacts the surface temperature of the growing Earth was only a few hundred degrees. Using an equation derived by Ye. A. Lyubimova in 1955, he found that heating by contraction would raise the central temperature to about 1000 K at the end of the formation process if the bodies were smaller than hundreds of meters across; radioactive heating would later raise this to several thousand degrees. Thus the 19th-century scenario—cooling from an initial temperature of several thousand degrees—was completely reversed. Here Safronov's model was in agreement with Western studies of the thermal history of the Earth (e.g., Urey's).[17]

Using a theoretical relation between the impacts of small bodies on the accreting planets and the resulting inclination of their axes of rotation, Safronov estimated from the observed obliquities that the largest bodies striking the Earth during its formation had masses about 1/1000 that of the present Earth. Thus the large tilt of the Uranian axis was ascribed to impact of a body having 1/20 the mass of that planet.

Some scientists supposed that the initial temperature of the Earth was only a few hundred degrees, and that planets further from the Sun started out much colder—perhaps cold enough to freeze hydrogen and helium from the PPC. But Safronov argued in 1962 that the gas-dust layer was so thin that the Sun's radiation would go not only through it but along its surface, so that it could be scattered into it through a boundary layer. This effect would keep the temperature from falling below 30 K at the distance of Jupiter and 15 K at the distance of Saturn. Thus these planets could not condense hydrogen directly but could only accrete it gravitationally after reaching a sufficiently large mass at a later stage of their growth.

A major limitation of Safronov's theory, according to other theorists, was that the estimated time for formation of the outer planets, using the equations derived for the terrestrial planets, was about 10^{11} years. In addition to the obvious disadvantage of requiring a time longer than the present age of the solar system (4.5×10^9 years) to form these planets, it is inconvenient not to have a fairly massive proto-Jupiter present while Mars is being formed, if one wants to attribute the small size of Mars (relative to Earth) to interference from its giant neighbor.

To alleviate this difficulty Safronov assumed that the outer regions of the PPC originally contained a much larger amount of material, much of which was ejected by gravitational encounters with the growing embryos of massive planets. This hypothesis would accelerate the early stages of the accretion process, while gravitational trapping of gas would accelerate the later stages. But the *ad hoc* or qualitative nature of these hypotheses damaged the credibility of his theory.[18]

The extremely low initial temperature of the Earth also created a problem if one wanted to explain the segregation of iron into the core. Safronov was temporarily attracted by the idea that the Earth's core is not iron but silicate, chemically similar to the mantle but converted to a metallic fluid by high pressure. This was the hypothesis of the Soviet geochemist V. N. Lodochnikov, published in 1939, independently proposed in 1948 by the British astronomer W. H. Ramsey, and widely discussed in the 1950s. As pointed out by B. Yu. Levin in 1962, the Lodochnikov–Ramsey hypothesis has cosmogonic advantages, which Safronov recognized. But the hypothesis was refuted by high-pressure shock wave compression experiments by L. V. Al'tshuler and his colleagues in the USSR and by R. G. McQueen and his colleagues at Los Alamos Laboratory in the United States; it was also undermined by theoretical calculations of W. Elsasser. Silicate compounds simply do not have high enough density at core pressures to account for the observed density of the Earth. So Safronov was forced to accept either the traditional iron core or a compromise iron-oxide core with a correspondingly higher internal temperature.[19]

Safronov's program lacked the glamour of more ambitious schemes that promised to explain the formation of the Sun as well as the planets from a simple initial state, and it encountered difficulties in explaining the properties of the present solar system. Yet he was successful in building up a body of basic theory that turned out to be useful as a starting point for other cosmogonists.

THE AMERICANIZATION OF SAFRONOV'S PROGRAM

The *Science Citation Index* gives a rough measure of the visibility of selected publications in the Western scientific community. Of course one cannot get any information about the nature of the reception or influence of these publications from citation counts alone, and citations not listed in the *Index* may turn out to be more important than those that are.[20] Bearing in mind these caveats, I still think it is significant that the total number of citations (excluding those by Safronov and other Soviet scientists) of all of Safronov's publications from

1961 through 1971 was only 25. For comparison, one paper by Cameron published in 1962 was cited 101 times in this period (excluding self-citations). Starting in 1982, the year when Safronov's *Evolution of the Protoplanetary Cloud* was first available in English translation, going through 1972, that book was cited 107 times by non-Soviet scientists, and Safronov's earlier papers were cited 31 times. So his visibility in the West was more than five times as great in the second 11-year period, primarily because of the English translation of his book.

Looking at the papers that cited Safronov in the 1970s one finds that almost all of them contain favorable remarks, even when disagreeing on specific technical points. Here are some examples of the Western response to Safronov's work, in chronological order.

(1) Peter Goldreich and William R. Ward at Caltech, in a note added in proof at the end of an influential paper on the formation of planetesimals by gravitational instability in a dust disk without the need to invoke "stickiness," acknowledged that Safronov had given a similar discussion, which they had apparently read only after finishing their own work.[21] Safronov subsequently received partial credit in the Western literature for what nevertheless was most often called "Goldreich–Ward instability."

(2) R. J. Dodd and W. Napier at the Royal Observatory in Edinburgh reported that numerical simulations based on Safronov's model confirmed his conclusion that a dominant nucleus arose which quickly incorporated lesser objects; the simulations gave correct values for the rotation rates of terrestrial planets but not for Jupiter and Saturn.[22]

(3) Stuart J. Weidenschilling at the Massachusetts Institute of Technology supported the suggestion by Schmidt and Safronov that matter ejected from Jupiter's zone could deplete the zones of Mars and the asteroids.[23]

(4) Joseph Burns at Cornell University suggested that the angular momentum of Mars could be ascribed to the impacts of the last few bodies falling on it, as in Safronov's theory.[24]

(5) William K. Hartmann and D. R. Davis at the Planetary Science Institute (Tucson, Arizona) acknowledged that they had been "influenced by some of the early Soviet accretion theories, published in the 1950s and 60s" in developing their ideas about lunar origin, although they had not studied Safronov's 1972 book in detail.[25]

(6) S. Fred Singer at the University of Virginia agreed with Safronov that the observed obliquities of the planets could be explained by late impacts.[26]

(7) P. Farinella and P. Paolicchi at the Osservatorio Astronomico di Brera (Merate, Italy) found from their theory results on the mass distribution consistent with those of Safronov.[27]

(8) In an elaborate calculation of the thermal evolution of the Earth and Moon based on Safronov's model, William M. Kaula at the University of California at Los Angeles (UCLA) found that accreting planetesimals would add enough heat to the Earth to bring about core segregation if not vaporization; he also inferred from his results that an impact origin of the Moon was more likely than binary accretion (see below).[28]

(9) J. N. Goswami and D. Lal at the Physical Research Laboratory in Ahmedabad (India) stated that their observations of particle tracks in chondrites provided evidence for Safronov's accretion model and against Cameron's gas-collapse model.[29]

Many other scientists simply quoted and used Safronov's results without bothering to discuss their validity.

In 1976, George Wetherill at the Carnegie Institution of Washington announced the first results of his calculations on a further development of Safronov's theory. Wetherill's work was motivated in part by photographs of Mercury's surface taken by the *Mariner 10* spacecraft on 29 March and 21 September 1974, analyzed by Bruce Murray's group at the Caltech Jet Propulsion Laboratory. It appeared that Mercury, like the Moon, had suffered a "late heavy bombardment" after its formation. Hence it was likely that there was a high flux of asteroid- or Moon-sized bodies throughout the inner solar system, 4 to 4.5 billion years ago.[30]

Wetherill's research, unlike Safronov's, made extensive use of computer simulation. Although he confirmed many of Safronov's results, he found one important difference. When the Earth was half formed, its "feeding zone" (region from which it accreted planetesimals) merged with that of Venus. The resulting perturbations produced higher relative velocities and thus reduced the cross section for capture of planetesimals by massive bodies. This would prevent runaway growth of the largest embryo in each zone. The second-largest body in the Earth's zone could then have a mass as large as 1/20 of the Earth's, rather than only 1/1000.[31]

Such large bodies, though having only a transient existence in the last stage of their accretion, would produce substantial heating by their impacts on the terrestrial planets.[32] Since Safronov accepted the conclusion that the Earth had been heated by large impacts during its formation, Wetherill could say that every current theory predicted high initial temperatures for the formation of planets.[30]

In 1978, Richard Greenberg and colleagues (listed in notes 33, 35) at the Planetary Science Institute in Tucson, Arizona announced another numerical simulation project based on a modification of Safronov's theory. They supported the idea that large bodies were prevalent in the early solar system by showing that planetesimals as large as those generated in Wetherill's scheme could have been produced without invoking perturbations by proto-Venus. Further numerical results generally supported Safronov's analytic work but contradicted his conclusion that relative velocities of planetesimals would tend to be comparable with the escape velocity of the dominant body. More of the total mass of the system was found to be in smaller planetesimals, which would collide mostly with each other and therefore tend to have smaller velocities; hence when they did collide with a larger body they would be more likely to accrete and promote its runaway growth.[33] Somewhat similar conclusions were reached in the USSR by B. J. Levin.[34] One consequence of this result, according to the Tucson group, was that Uranus and Neptune could grow "in a reasonably short time, well below the actual age of the system, without the need for *ad hoc* assumptions about excess mass or artificially-low relative velocities among the icy planetesimals."[35]

Although theorists are still not sure how centimeter-sized particles grew to kilometer-sized bodies,[36] given the existence of such bodies Wetherill has shown that a modified Safronov model may be able to explain the existence of four terrestrial planets starting from 500 bodies each of mass 2.5×10^{25} kg (one-third the lunar mass). This result is clearly stochastic and depends on the existence of large impacts. Several runs gave three or four planets, but none reproduced precisely the observed distribution of masses and distances. So the best theory of the formation of terrestrial planets is not quite capable of explaining the simplest properties of those planets as known 200 years ago; research continues.[37]

How did Safronov's model come to be the basis for many contemporary calculations on planet formation? According to planetary cosmogonist Stuart J. Weidenschilling, scientists rarely say explicitly in print that they "accept" a theory, whereas they may state that they reject it (or at least one aspect of it) in order to justify a different course of investigation. But they do "vote with their feet" by addressing questions relevant to a particular model. Thus one "accepts" a theory if (a) it poses interesting questions for further work; (b) the questions are relevant to one's own expertise; (c) "some combination of data, analytical techniques, and/or computational ability... allow progress toward answering those questions"; and (d) funding is available. From this point of view Safronov's theory is attractive because it has a content "sufficiently stable for meaningful work, but with many areas for progress. The dynamical questions are accessible to the rapidly growing power of computers. This in itself has kept the Safronov model dynamic; without computers it would have reached a dead end at the limits of analytic modeling in the early 1970s.... I regard the existence of computers to be the greatest single factor in the 'acceptance' of Safronov's general model in the sense of inspiring further work."[38] (The last remark refers primarily to acceptance in the United States.)

ORIGIN OF THE MOON: EARLY SELENOGONIES

Through its influence on research by Hartmann, Wetherill, and other planetary scientists, Safronov's theory indirectly supports the "giant impact" theory of the origin of the Moon, currently accepted as the best working hypothesis in the United States. Yet that theory does not yet seem to be regarded favorably in the USSR;[39] Soviet cosmogonists have supported an alternative theory developed by Evgenia Ruskol (Safronov's wife). To understand this situation we need to go back a few decades to review the history of *selenogony*, the theory of the Moon's origin.[40]

Modern theories of the origin of the Moon go back to 1878. In that year George Howard Darwin, son of the evolutionist Charles Darwin, proposed that the Moon had once been part of the Earth and was ejected from it by an instability triggered by the action of the Sun's tidal force. The additional hypothesis that the scar left by the Moon's departure became the Pacific Ocean was proposed by Osmond Fisher in 1882.

Darwin's theory was not based on any direct evidence that the Moon was once part of the Earth, but rather on an interpretation of the observed "secular acceleration" of the Moon. In the 18th century astronomers thought that the Moon was gradually moving faster in its orbit around the Earth. That would imply (by Kepler's Third Law) that it was approaching the Earth. But in the 19th century quantitative analysis of the gravitational actions of the other planets on the Earth and Moon indicated that the Moon was actually moving more slowly than in the past, and that the apparent acceleration was due to a slowing down of the Earth's rotation. The physical cause was identified as dissipation by lunar tides in the Earth's oceans. Darwin pointed out that since the angular momentum lost by the Earth–Moon system is conserved, the angular momentum lost by the Earth must be transferred to the Moon. As a result the Moon's orbit is gradually receding from the Earth; conversely it must have been closer in the past. Making specific assumptions about the mechanical properties of the Earth, he traced the lunar orbit back to a state in which the Moon moved around the Earth as if rigidly fixed to it, in a period of 5 h 36 min, with its center about 600 miles from the Earth's surface. Before that state there was no unique solution.

The major objection to the hypothesis that the Moon was ejected from the Earth was dynamical: a body with the combined mass and angular momentum of Earth and Moon, rotating in about five hours, would not be unstable against spontaneous fission. Darwin was aware of this objection but proposed to circumvent it by invoking a resonance of the Sun's tidal action with the free oscillations of the proto-Earth.

Another objection was that the Moon could not go into orbit as a single body because it would initially be inside the critical distance known as the "Roche limit" (based on the formula derived by the French planetary scientist Éduoard Roche in 1849). Like Saturn's rings, it would therefore be broken up into many smaller bodies by the tidal forces. Darwin argued that this flock of small bodies would still produce tidal dissipation, which would expand its orbit out beyond the Roche limit, so it could eventually recombine into a single satellite.

Roche himself was the major 19th-century proponent of the hypothesis that the Moon was condensed from a ring spun off by the rotating gaseous proto-Earth, just as the Earth was condensed from a ring spun off by the solar nebula in the nebular hypothesis. This became known as the co-accretion or "sister" hypothesis, the Moon being regarded as Earth's sister who grew up alongside the Earth.

A variant of the sister selenogony was the hypothesis proposed in 1893 by the American geologist Grove Karl Gilbert, that the Moon formed from a ring of small solid particles; the final stage of the process would produce the craters on the Moon's face.

A third hypothesis, advocated early in the 20th century by the American astronomer T. J. J. See and others, created the Moon in some other part of the solar system and later brought it to be captured by the Earth; the Moon was thus Earth's "wife."

Darwin's hypothesis, which described the Moon as Earth's "daughter," remained the most popular in the early 20th century, although some astronomers rejected it because the Russian mathematician A. M. Lyapunov disproved a related theorem about the stability of rotating fluids. In 1930, Harold Jeffreys criticized the Darwin selenogony, arguing that viscosity in the Earth's mantle would damp out the motions required to build up the postulated resonant vibrations. Advocates of other hypotheses subsequently cited Jeffreys and/or Lyapunov as having "refuted" the fission hypothesis.

In the 1950s the capture hypothesis was revived by Harold Urey in the United States and Horst Gerstenkorn in Germany. Urey's theory was developed as part of a general theory of the origin of the planets, mentioned above. Gerstenkorn worked out a quantitative dynamical theory of the capture process, following the approach of G. H. Darwin but with different assumptions about the initial state and the mechanical properties of Earth and Moon.

Urey's theory was largely qualitative; he was less interested in the dynamics of the capture process than in the nature of the Moon itself as a key to the early history of the solar system. He argued that the Moon was a frozen relic, a surviving example of bodies that used to populate the solar system. It had always been cold since its capture and thus preserved on its face a record of events that left no trace on the surface of the geologically active Earth. This was a powerful argument for exploring the Moon: analysis of the lunar surface not only should be able to tell us the conditions prevailing at the time and place of the Moon's formation, but might reveal facts about the Earth's history that could not be learned by studying the Earth itself. If, on the other hand, Darwin were right and the Moon is just a piece of the Earth, it would not be worth the trouble to go there.

Gerstenkorn's theory raised different kinds of questions. What is the range of initial conditions for which capture is dynamically possible? Could the Moon have been captured from a retrograde orbit? How could the lunar orbit have acquired its present eccentricity and inclination? How much energy had to be dissipated during the capture process, and would this energy have been enough to melt the Earth or at least produce some effects that could be detected today? Extensive calculations by MacDonald, Goldreich, Singer, and others in the 1960s indicated that while capture of the Moon was not dynamically impossible, it would be extremely difficult to satisfy all the conditions necessary to produce the present lunar orbit.

It might appear that the only way to test the capture theory would be by mathematics: to see if the known astronomical facts about the Moon could be deduced from a plausible initial state. The larger the set of possible initial states that could be shown to lead to the given final (i.e., present) state, the more likely that the hypothesis is correct.

From Urey's point of view this kind of test is irrelevant. Even if the probability that any given Moon-sized body would be captured by the Earth were very small, there were so many bodies in the early solar system that there was a reasonable chance of capturing one of them. (Urey did not pursue that argument to the conclusion that an actual collision could also have occurred.) In any case such calculations cannot be used to compare capture with other hypotheses since different kinds of adjustable parameters are involved in the hypotheses (viscosity of proto-Earth, conditions in the primeval nebula, etc.). Instead, the real test must be chemical: if the Moon is unlike the Earth is must have been formed elsewhere; if it is like the Earth it was at least formed in the same part of the nebula (co-accretion) if not actually inside the Earth (fission).

Ruskol's Selenogony

During the 1960s and 1970s, one of the most active selenogonists was the Russian scientist Evgenia Ruskol.[41] Although her theory has been classified under the heading co-accretion, she rejected Kuiper's protoplanet theory and concluded as early as 1960 that "the Earth–Moon system did not originate as a double planet from some double embryo."[42] Her hypothesis, based on the cosmogonic ideas of O. Yu. Schmidt and V. Safronov, was that a swarm of particles formed around the growing Earth, with a radius of 100 to 200 times the radius of the Earth. Other particles would be captured from heliocentric orbits (in a "supply zone" ranging from 0.8 to 1.3 A.U. from the Sun into the circumterrestrial swarm by inelastic collisions. It is this specific mechanism—inelastic collisions of particles, most of which are less than 100 km in radius—which distinguishes Ruskol's hypothesis from others that postulate a circumterrestrial swarm destined to evolve into the Moon. The swarm grew most rapidly when the Earth's mass was 0.3–0.5 of its present mass. Since, according to Safronov's calculations, the rates of growth of the largest body in a system and the second-largest body start to diverge when their sizes reach a critical value, the Moon could have started growing very early without having grown as rapidly as the Earth and, hence, could still be much less massive while being almost as old. The difference between the ages of the Moon and Earth should be less than 100 million years.

Ruskol suggested that the Moon was initially formed at a distance of 5 to 10 Earth radii; its subsequent recession was due to tidal evolution. In order to extend the time scale to more than 4 billion years, it was necessary to assume that the effective lag angle of the Earth–Moon tidal interaction (tidal effective Q) was smaller in the past than at present. This assumption was justified, according to Ruskol, by evidence that the Earth's interior "was generally heated from a lower temperature, while releasing the water of the oceans to the surface."[43]

An obvious objection to all co-accretion theories is the apparent difference in composition between Earth and Moon: since the Earth is denser it presumably has a substantial iron core, whereas the Moon has little or no core. Ruskol suggested that this difficulty could be overcome "if we admit that the Earth's core is composed of metallized silicates," so that Earth and Moon have essentially the same chemical composition; the greater density of the Earth is due to a phase transition occurring at high pressures. This is the "Lodochnikov–Ramsey hypothesis" mentioned above, which was eventually abandoned by both Soviet and Western scientists.[44] Thus another explanation for the Earth–Moon density difference had to be found (see below).

Ruskol's theory was not widely known in English-speaking countries in the 1960s, although it was vigorously supported by B. Yu. Levin at a Caltech–Jet Propulsion Laboratory conference in 1965,[45] and several of her papers appeared in cover-to-cover translation journals (including some reprinted in this book). Following the publication of an English translation of Safronov's 1969 book in 1972, Western scientists took up Safronov's theory (see above) and hence also its application to selenogony by Ruskol.

On the eve of *Apollo 11*: Three selenogonies

According to B. Yu. Levin, at the Pulkovo symposium "all participants agreed that revival of the hypothesis of the separation of the Moon from the Earth was impossible." Levin seemed annoyed that certain scientists such as A. E. Ringwood in Australia and Donald Wise and A. G. W. Cameron in the United States had advocated the fission hypothesis without mentioning that "important errors are found in Darwin's and

Jeans' calculations which seemed to prove the possibility of the smooth separation of a rotating fluid mass."[46] (His compatriot Lyapunov was one of those who had denied such a possibility.)

Some scientists did revive the fission hypothesis in the 1960s, because it was congruent with certain theories about the early development of the Earth. Ringwood and Cameron used ejection of the Moon to get rid of Earth's hypothetical primeval atmosphere. Wise argued that the traditional objections to fission had been weakened by recent developments, while Soviet photographs suggesting that the far side of the Moon differed from the near side provided new evidence in favor of fission. The American planetary scientist John O'Keefe proposed a high-temperature fission process that would deplete the Moon in volatile substances.

It was difficult to find any conclusive test of selenogonic hypotheses before the return of the first lunar samples, and indeed one of the announced reasons for going to the Moon was to find out where it came from. But few theorists actually published specific predictions about what would be found in those samples. I have found only four: Urey's discussion based on his capture theory, and Ringwood's, O'Keefe's, and Wise's based on fission hypotheses. Urey expected to find evidence of water on or near the surface. O'Keefe argued that the Moon would be poorer than Earth in water and other volatile substances and would also be deficient in siderophile elements such as nickel. (Roughly speaking, these predictions reflected the consequences of a cold or hot origin, respectively.) Wise predicted that the near side of the Moon should have the same composition as Earth's mantle, while the far side should have a less dense proto-Earth's crust. Ringwood predicted a thermal history in which a temperature maximum started near the surface and gradually moved toward the center, possibly exceeding the melting point for a brief period about 10^9 years after formation. This implies loss of volatiles from the crust but not from the deep interior, and in fact the density should decrease with depth.

Apollo's Impact on Selenogony

In July 1969 the *Apollo 11* mission brought back the first lunar samples from Mare Tranquillitatis. Preliminary analysis of these samples indicated a high concentration of refractory elements (Ti, Zr, etc.); low concentrations of volatiles (Pb, Bi, Tl); strong depletion of siderophile elements, especially Ni and Co; and an absence of hydrated minerals, showing a scarcity of surface water.

The group led by Edward Anders at the University of Chicago quickly concluded that these results, together with a strong depletion of Au and Ag, provided good evidence against the fission theory. Ringwood argued that the scarcity of volatile metals, siderophiles, and water did rule out the original fission hypothesis but not his own high-temperature version.

Urey's capture theory seemed to be refuted by the *Apollo* data, in particular by the scarcity of water and the evidence for an early high-temperature stage. After extensive discussions with O'Keefe, Urey decided to abandon his capture theory and eventually leaned toward fission, though he was not very enthusiastic about that or any other theory.

Although it was frequently stated during the 1970s that the fission theory had been refuted, cosmochemists were finding increasing similarities between lunar and terrestrial composition. The early conclusions about excess refractory abundance and depletion of siderophiles were later judged to have been somewhat exaggerated. Moreover, oxygen isotope abundances were found to be the same in lunar and terrestrial material.

The major opposition to lunar–terrestrial similarity came from the Anders group, which favored co-accretion after fractionation in the solar nebula; they argued that the Moon was formed in a circumterrestrial orbit from material that had condensed at higher temperatures than the Earth.

The American sociologist Ian Mitroff interviewed many of the *Apollo* scientists in the early 1970s. He asked them to evaluate five hypotheses of lunar origin: *fission*, *capture*, and three versions of simultaneous formation: *double planet*, *condensation* from a hot silicate atmosphere of the primordial Earth, and *accretion* from planetesimals of type I carbonaceous chondrites. On a scale from 1 ("agree strongly") to 7 ("disagree strongly"), no theory had an average score better than about 4 ("neither agree nor disagree") before *Apollo 11*; accretion was slightly more favored than the others. No theory enjoyed a statistically significant improvement in its average rating as a result of analysis of *Apollo* samples; fission and accretion suffered significant losses. The double-planet theory received the best rating, 3.73, but even that was weaker than "agree moderately." Fission dropped to 5.6 ("disagree moderately"), leading Mitroff to conclude that this theory "was the first to fall by the wayside."[47]

Mitroff's finding that co-accretion was more popular than fission or capture (though not enthusiastically supported) is confirmed by occasional remarks in popular writings by scientists in the 1970s and early 1980s.

But the consensus of the lunar science community in this period was that *none* of the pre-*Apollo* theories offered a convincing explanation of the origin of the Moon.

Fission. In addition to the original angular momentum difficulties of this theory, new calculations on viscous rotating fluids indicated that they could not be spun fast enough to cause fission; instead they simply lost matter from equatorial regions. Rotational instability could be produced by planetesimal accretion only if one planetesimal were about one-tenth of the mass of the proto-Earth, in which case the fission model would become an impact model. Fission models were deemed incapable of explaining why the Moon is substantially richer in both iron and refractory elements than the Earth's mantle.

Capture. This hypothesis lost its original advantage of being able to explain Earth–Moon compositional differences when it was shown that capture was dynamically impossible unless the Moon were formed at about the same heliocentric distance as the Earth, and even then it would be rather unlikely. Disintegrative capture was also unlikely. On the other hand, even if the Earth could have captured a Moon formed far away from the Earth (in order to account for chemical differences), one would then have difficulty accounting for the *similarity* of oxygen isotopic composition.

Co-accretion had difficulty in explaining the compositional differences between Earth and Moon, even with a postulated "composition filter" to separate iron from silicates; moreover, it could not account for the angular momentum of the Earth–Moon system.

Selenogony seemed to have reached an impasse. Other areas of planetary science were also slowing down. Cutbacks in funding for space science, especially in the United States, made it difficult to acquire new data except from the Voyager missions to the giant planets.

THE GIANT-IMPACT HYPOTHESIS

What happened next—the emergence of the giant-impact hypothesis—bears a superficial resemblance to a Kuhnian revolution. Selenogony before 1969 had been dominated not by a single theory but by a paradigm: the evolutionary cosmogony exemplified by the 19th-century nebular hypothesis, supplemented by relevant results of physics, chemistry, astronomy, and geology. Within this paradigm, cosmogonic processes had to be deterministic and uniformitarian, even if their net result was the formation of a qualitatively new system. Thus fission, a catastrophic event, could occur only when certain physical conditions were present, and its result was predetermined. Two-body interactions, as in the capture theory, or the earlier tidal theory of the origin of the solar system, should be treated as deterministically as possible; actual collisions or extremely improbable initial states should be avoided. Mainstream cosmogonists were unwilling to postulate random catastrophic events, for reasons that may be called philosophical.

Thus, when the accepted paradigm was afflicted with insuperable difficulties, so that the very existence of the Moon became an "anomaly" in the Kuhnian sense, the constraints of the old paradigm were discarded and the first steps were taken toward a new one. The new hypothesis, which is not yet a fully developed theory, suddenly attracted the enthusiasm of many scientists in what even its proponents describe as a "bandwagon" effect. Since the hypothesis explicitly invokes a random catastrophe, it is difficult to show that it is objectively superior to theories that exclude such catastrophes on philosophical grounds; if the criteria for testing hypotheses change, the paradigms are at least partially incommensurable in Kuhn's sense.[48] This is not to say that the new criteria are less strict; on the contrary, because of the availability of better computers, proponents of any hypothesis are now expected to demonstrate quantitatively that their mechanism will actually work with reasonable physical assumptions, where previously one could get away with qualitative arguments.

I used the phrase "superficial resemblance" to warn the reader that the Kuhnian revolution is only an abstract historiographic model. One cannot expect to find a real historical event that is accurately described by the model any more than one can expect to find a perfectly rigid sphere in nature. Moreover, most historians and philosophers of science insist that Kuhnian revolutions have (or should have, respectively) nothing to do with how science works. Nevertheless, many Earth scientists affirm that the establishment of plate tectonics in the 1960s was a Kuhnian revolution, and the issue will inevitably arise whenever any radical change in accepted theories occurs. It is therefore worthwhile to point out some Kuhnian and non-Kuhnian aspects of the rise of the giant-impact hypothesis. In view of these circumstances we can then raise the question, have Soviet cosmogonists (especially those in Safronov's group) influenced, followed, or been left behind by a paradigm switch that has occurred in the United States?

The most obvious non-Kuhnian feature is that all discussions and calculations on the giant-impact hypothesis employ the same established principles of physics that were used to develop the previous theories, and the major dynamical problem that the new hypothesis was designed to solve is precisely the one that was considered of paramount importance in traditional cosmogony. As the American theorist Howard Baker pointed out more than 30 years ago, "that the Moon was forcibly separated from the Earth by some extraneous force is indicated in its excess angular momentum about the Earth," and this force must have been exerted by a close gravitational encounter, if not an actual collision, with some large heavenly body.[49]

But Baker's hypothesis, published as a pamphlet by the Detroit Academy of Natural Sciences, was completely ignored by the scientific community; as far as I can determine it was unknown to mainstream selenogonists in the 1970s. Aside from the fact that *most* scientific papers, even those published in respectable journals, are never cited by anyone except their authors, one may attribute the neglect of this work to the general dislike of scientists in the 1950s for catastrophic theories of solar system history, as shown by their reaction to Immanuel Velikovsky's books. (There were many objective reasons for rejecting Velikovsky, but the emotional tone of the criticism indicates that the argument was partly on a metascientific level). The dominant paradigm defined such theories as unscientific. By the 1980s they could command serious consideration, at least if supported by scientists who had already established their reputations—the Alvarez hypothesis linking the cretaceous–tertiary extinctions to an asteroid striking the Earth is an obvious example.

Before the epoch of planetary exploration, a giant impact on the Earth might have seemed very unlikely. But Mercury's cratered surface, revealed by *Mariner 10* in 1974, suggested that the terrestrial planets were bombarded by somewhat smaller bodies for hundreds of millions of years after their formation. This made it much more plausible than before that the Earth could have been struck by an object large enough to eject a substantial amount of material from its mantle.

In August 1974 William K. Hartmann presented to an IAU Colloquium at Cornell University the hypothesis that the Moon was formed from material ejected into a circumterrestrial disk by a large (>1000-km radius) body that struck the Earth. A. G. W. Cameron, in the audience, remarked that he had been working on a similar hypothesis with an even larger impacting body, comparable in size to Mars. Hartmann worked with D. R. Davis to develop a theory published in 1975, while Cameron collaborated with W. R. Ward to obtain results that they summarized in a three-page abstract published in 1976. Both theories were directly related to mainstream planetary cosmogonies and were sponsored by scientists with established reputations.

Hartmann had long been interested in lunar craters and the time variation of their size distribution. During the 1960s he had also been impressed by Safronov's papers on planetary formation by accretion of solid planetesimals, including the hypothesis that the tilt of the Earth's rotation axis ("obliquity of the ecliptic" in astronomical terminology) was primarily due to the impact of the last large body that was added during its accretion. Moreover, measurement of the flux of impacting bodies as a function of time showed that the pre-mare cratering rate was enormous and included basin-forming bodies much bigger than the crater formers in the last few billion years. Those were debris left after planetary formation. Extrapolation back to an earlier epoch suggested that bodies as large as the Moon itself could plausibly have been moving in the vicinity of the Earth.

Hartmann and Davis attempted a numerical reconstruction of the size distribution of bodies that could have grown during planetary accretion and that were left behind near the end of planet formation, assuming a process starting with accretion of small particles. They found that "the probability of the planet interacting with a large body is much larger than has been considered in some past descriptions of planet growth."[50] For certain assumptions, they found that among Earth-sized planets the second-largest bodies could be of radius 500–3000 km, and there could be tens of bodies larger than 100-km radius. Half of the kinetic energy of a planetesimal about 1200 km in radius, arriving at the Earth's surface with a velocity of 13 km/s, could eject two lunar masses to near-escape speeds. Assuming that the collision occurred after the Earth's core had started to form, one would expect the ejected material to be depleted in iron, as in the fission theory.

The advantages of impact over fission, according to Hartmann and Davis, are (1) that an energy source to raise the material off the Earth is provided; and (2) that "the theory is not purely evolutionary" (i.e., the outcome for a given planet is randomized, not purely deterministic), "depending on a chance encounter so that it does not require prediction of similar satellites for Mars or other planets." After the material is ejected it forms a cloud of hot dust, enriched in refractory elements and rapidly depleted in volatiles. The subsequent evolution follows the "widely admired" theory of Ringwood.

Although Cameron was present at the original presentation of the Hartmann–Davis hypothesis he had already begun to develop a similar hypothesis independently and was not directly influenced by the Safronov planetesimal accretion model, since at that time he preferred a gaseous protoplanet model instead. According to his later account, he and Ward "were led to the suggestion of a collision origin of the Moon through the following consideration. The angular momentum of the Earth–Moon system is less than sufficient to spin the Earth to rotational instability; we were nevertheless interested in determining the mass of the body which, striking a tangential blow to the proto-Earth, could impart the angular momentum of the Earth–Moon system to the proto-Earth.... The required projectile turned out to be about the mass of Mars.... That defined the basic scenario of our lunar formation process."[51]

William Kaula was one of the earliest supporters of the giant-impact theory. In 1977 he mentioned it as a promising explanation for the early differentiation needed to account for the Moon's bulk composition, although the probability of such an impact still seemed quite low. Subsequent calculations of thermal evolution based on Safronov's model led him to favor the impact hypothesis.

Additional support for giant impact was provided by George Wetherill's calculations on the accretion of planetesimals, mentioned above. Wetherill found from his version of Safronov's model that a substantial fraction of the total mass in each region would reside in bodies only one order of magnitude smaller than the dominant planetary embryo at a fairly late stage of the process. It was therefore an essential feature of this process that a terrestrial planet would be hit by an object as large as Mars during the final stage of its growth, although Wetherill emphasized that impact was only one of several processes that could be expected to provide material for the formation of the Moon in his model.

A consensus in favor of the giant-impact theory emerged at a conference on the origin of the Moon held in Hawaii in October 1984. According to David Stevenson this was "not because of any dramatic new development or infusion of data, but because the hypothesis was given serious and sustained attention for the first time. The resulting bandwagon has picked up speed (and some have hastened to jump aboard)."[52] Hartmann, one of the organizers of the conference, said his idea "had languished" since its publication; "when I went to a planning session for the conference to look over the abstracts for the proposed papers, I found, to my amazement and joy, that eight or ten of the abstracts—independently of each other—were about the impact idea."[53]

But when theorists started to work out the details of the impact hypothesis they found that it might not perform one of the functions that made it seem attractive: getting the Moon out of the Earth. Contrary to what had been generally assumed, Cameron stated at the 1984 Hawaii conference that, in the collisions he had simulated, most of the material in the disk came from the impactor rather than the Earth. This might not make much difference if the impactor was chemically similar to the Earth, but that seemed unlikely unless it was formed at the same distance from the Sun and had a mass comparable to the Earth's. Otherwise the hypothesis would be vulnerable to the same objection as the capture theory: the chemical composition of the Moon must simply be postulated rather than predicted or derived from specified processes acting on known terrestrial material. If one thinks of the impactor as a planet chemically like Mars, then it would have a composition significantly different from Earth and Moon. This point is sometimes overlooked by those who support the giant-impact theory because it makes the Moon chemically like the Earth.

Detailed calculations of the physical processes postulated by the impact theory were started only recently. The results so far indicate that the impact mechanism is capable of placing material into orbit around the Earth, at the cost of melting the early Earth and thus coming into conflict with geochemical evidence. But other scientists say this objection can be overcome and that the geochemical evidence on balance favors impact.

MODIFYING THE CO-ACCRETION THEORY

While the impact theory seemed to be more attractive than any other single hypothesis in the last 10 years, many lunar scientists favored some kind of modified co-accretion theory. Following the ideas of Ruskol (see above), one assumes that a swarm of planetesimals begins to collect around the Earth during its formation, and later accumulates into a single satellite. Unlike the classical co-accretion theories, the newer theories assume that a large part of the selenogenic material is captured by the swarm after it has formed, and processed by physical or chemical mechanisms in order to give the Moon a somewhat different composition from the Earth. No single incoming body contributes more than half of the Moon's substance—unlike the classical capture theory and Cameron's latest impact model. The same flux of planetesimals feeds the growing Earth with mass and angular momentum. The theories differ among each other in quantitative details such as the time of arrival of the planetesimals during the Earth's formation, their sizes and extent of prior differentiation into iron cores and silicate mantles, the nature of their processing by the circumterrestrial swarm, and the explanation given for the failure of the same mechanism to produce similar satellites for other planets.

In the early 1970s Ruskol argued that her model would provide more effective depletion of volatile elements at the edge of the swarm where particles would be exposed to the solar wind than at the center where the Earth itself formed; and that because of the long time scale for formation of the Earth and Moon ($\approx 10^8$ years) such processes would produce the necessary effect better than in other models which are based on rapid or catastrophic processes.

Having abandoned the Lodochnikov–Ramsey hypothesis, Ruskol had to account for the much greater iron content of the Earth. She ascribed the segregation of iron into the Earth rather than the Moon to the fact that silicate particles tend to break up into fine debris in collisions while iron-rich particles are more likely to survive and grow. While the Earth accretes all the particles that strike it, the circumterrestrial swarm can more easily capture the finer silicate particles.

In 1975 Alan W. Harris, working with Kaula at UCLA, showed how Ruskol's theory could be modified to explain how a large satellite such as Earth's Moon could be formed by the same mechanism that produces small satellites around other planets. According to Harris and Kaula, the Moon must have started growing in the Earth orbit when the Earth was between 0.1 and 0.2 of its present mass. (This fraction was revised to about 0.4 by Harris in 1978.) Moreover, the planet's Q (specific dissipation function) must be less than about 1000. "The high Q of the outer planets does not allow a satellite embryo to survive a significant portion of the accretion process, thus only small bodies formed very late in the accumulation of the planet remain as satellites. The low Q of the terrestrial planets allows satellite embryos of these planets to survive during accretion, thus massive satellites such as Earth's Moon are expected."[54]

The problem is then to explain why other terrestrial planets do not have massive satellites. Harris and Kaula concluded that they did originally have them but lost them because of tidal friction, as suggested earlier by Joseph Burns and by W. R. Ward and M. J. Reid.

Another possibility is that a satellite of Venus was lost because of the rapid growth of its orbital eccentricity, an effect predicted by the theory if tidal dissipation in the satellite is not very large; the escape would have carried away most of the angular momentum of the system. Reimpact could possibly lead to the present retrograde rotation of Venus. Alternatively, if the satellite orbit expanded while tidal dissipation slowed the rotation of the primary to the point where day and month are equal, and solar tidal forces then made the day longer than the month, the effect of tidal dissipation on the orbit must reverse and the satellite will be drawn in again to crash on the primary, as pointed out in 1905 by G. H. Darwin. Ward and Reid suggested in 1973 that solar tidal action could have produced this effect on satellites of Mercury and Venus. Thus, as argued by Harris and Kaula, the present lack of satellites of those planets does not refute a scenario that predicts formation of satellites about other terrestrial planets.

Nevertheless, Kaula now finds it more plausible to explain the absence of a Venerean satellite as an example of the normal development of an Earth-like planet, while the presence of Earth's Moon is an exception due to random impact. He suggests a correlation between possession of a satellite and the evolution of life for these two planets. Venus is inhospitable to life because of its hot dense atmosphere; Earth once had a similar atmosphere. "The creation of the Moon was a necessary condition for us to be here today. This event blasted away the excess atmosphere and allowed the Earth's surface temperature to drop below the boiling point of water."[55]

Results of possible relevance to the compositional sorting of material in the circumterrestrial swarm were reported by the Anders group in 1977. They found that, among highland samples from Apollo 16 and 17, the "most moonlike" (high refractories, low metals) bodies fell first, followed by bodies of progressively less Moon-like composition (as indicated by the Ir/Au ratio). This correlation suggested that the basin-forming objects were genetically related to the Moon in a way that would be expected from Ruskol's theory.

Another kind of evidence for co-accretion was brought forward by S. K. Runcorn in Britain: his theory of ancient lunar magnetism led to the conclusion that the rotational pole of the Moon has been changed by impacts of bodies that were once satellites of the Moon, and these satellites originated from the ring of bodies that formed the Moon.

Wasson and Warren, at UCLA, were skeptical about Ruskol's mechanism for separating iron from silicates in the circumterrestrial swarm, and proposed instead that the differentiation had already taken place in larger ($\gtrsim$50-km radius) bodies. These would not be captured whole by the smaller particles in the swarm, but silicate fragments broken out of them could be captured; the iron cores would pass through the swarm and could be stopped only by a direct hit on the Earth. A group calling itself the "Tucson Lunar Origin Consortium" proposed an integrated dynamical and geochemical approach, based on the assumption that "As the

Earth grew by planetesimal bombardment, a circumterrestrial cloud of particles was created from a combination of impact-ejected mantle material and planetesimals captured directly into orbit around the Earth. Such a swarm continued to capture planetesimals and to receive ejecta until the bombarding population thinned, the Earth stopped growing, and the Moon accreted in orbit. If Earth-mantle material dominates the swarm, the model resembles the fission hypothesis; if small planetesimals dominate the model represents the 'growth in Earth orbit' end-member; if the swarm were dominated by a single large planetesimal, we would essentially have a capture model. A model intermediate between these extremes appears most promising."[56]

While recognizing the possible stochastic nature of a process dominated by large impacts, the Consortium treated the circumterrestrial swarm as a composition filter, "preferentially capturing small weak silicate bodies while passing large iron planetesimals (cores of broken parents)."

What has been the response of the Soviet cosmogonists to the giant-impact model? Have they modified their co-accretion theory in the way suggested by Western theorists?

Soviet planetary cosmogonists do not seem very enthusiastic about the new catastrophic paradigm, even though it is in some respects an outgrowth of their own work. In a recent survey of protosatellite swarms, the Ruskol–Safronov group acknowledged that the Moon *might* be an irregular satellite formed by a catastrophic event, but seems reluctant to abandon or revise its co-accretion theory in favor of giant impacts.[57] In the next few years they may have to decide whether to put aside philosophical reservations and follow Western scientists—as Soviet physicists did in adopting the quantum worldview of Bohr and Heisenberg—or to hang onto the old paradigm, as did many Soviet geologists who followed Beloussov in rejecting plate tectonics. Alternatively they could follow the example of the Tucson Consortium and seek a compromise that gains some of the advantages of the impact hypothesis without risking the consequences (scientific and otherwise) of a single catastrophic event.

[1] This essay is based primarily on my article "Theories of the Origin of the Solar System 1956–1985," Rev. Mod. Phys. **62** (1) 43–112 (1990), where references to primary and secondary sources may be found. I thank V. S. Safronov, E. Ruskol, and A. V. Vityazev for several additions and corrections.

[2] F. Engels, *Dialectics of Nature*, translated and edited by Clemens Dutt with a preface and notes by J. B. S. Haldane (International, New York, 1940, 1963), pp. 8–15, 40, 186. In his note on p. 15 Haldane, basing himself on theories current in the early 20th century, says "Laplace's theory is fairly certainly incorrect." For a more recent assessment of Engels's views on science see Robert S. Cohen, "Engels, Friedrich," Dict. Sci. Biog. **15**, 131–47 (1978).

[3] J. H. Jeans, *Eos: or the Wider Aspects of Cosmogony* (Kegan Paul, London, 1928), pp. 82–83; *The Mysterious Universe*, new revised edition (Macmillan, New York, 1932), pp. 1–5.

[4] Kuiper surveyed the Soviet theories in "The Formation of the Planets, Part II," J. R. Astronom. Soc. Can. **50**, 105 (1956); while agreeing with them on some points he complained that their assumptions were not clearly explained (at least in the translations available to him).

[5] For reviews of theories of star formation from the Soviet viewpoint see the articles by Safronov and Vityazev (1983,1985) and Safronov and Ruzmaikina (1985), cited in the bibliography at the end of this book. My statement about the neglect of isotopic research refers to the 1960s and 1970s; the earliest Soviet publications mentioned in the Safronov–Vityazev article are Vityazev's (1982,1983).

[6] At the 1972 conference on the origin of the solar system held at Nice (France), Safronov criticized Cameron's turbulent viscosity model and Cameron replied; see *L'Origine du Système Solaire*, edited by H. Reeves (Editions du Centre National de la Recherche Scientifique, Paris, 1972, reprinted 1974), pp. 361–66.

[7] The earliest comparable work in the West is that of Stephen H. Dole, "Computer Simulation of the Formation of Planetary Systems," Icarus, **13**, 494 (1970), but his calculation was much less ambitious than those of Safronov.

[8] Schmidt's theory was generally ignored in Western countries, with the exception of Kuiper's discussion cited in Note 4. In one of the very few American comments on Schmidt's theory, J. M. Witting wrote that it "appears to be on solid ground as far as the boundary conditions [facts to be explained] are concerned; none are violated, and the theory is able to explain many of the dynamical boundary conditions well and completely." It was summarized in a British popular astronomy journal by L. Randic, "Schmidt's Theory of the Origin of Visual Binary Stars and of the Solar System," Observatory, **70**, 217 (1950). The British astronomer R. A. Lyttleton presented it in his *Mysteries of the Solar System* (Clarendon, Oxford, 1968), pp. 28–32.

[9] In addition to the translations published in *Soviet Astronomy–AJ*, reprinted in this book, Safronov published three papers in English before he wrote his monograph: "On the turbulence in the Protoplanetary Cloud," Rev. Mod. Phys. **30**, 1023 (1958), (reprinted here); "On the Gravitational instability in Flattened Systems with Axial Symmetry and Non-Uniform Rotation," Ann. Astrophysique **23**, 979 (1960), "The Primary Inhomogeneities of the Earth's Mantle," Tectonophys. **1** 217 (1964).

[10] V. S. Safronov, *Evolution of the Protoplanetary Cloud* (Israel Program for Scientific Translations, Jerusalem, 1972); "Accumulation of the Planets," in *L'Origine du Système Solaire*, edited by H. Reeves (Editions du Centre National de la Recherche Scientifique, Paris, 1972), pp. 89–113. Dale P. Cruikshank, who had known Safronov for several years, received a copy of the Russian edition of *Evolution* and recommended to NASA's technical translation office that it be translated into English (letter to S.G.B., 27 June 1990).

[11] The turbulence was supposed to be supported by thermal convection in the radial direction, but Safronov ("On the Turbulence...", *op. cit.*, Note 9) "showed that for such convection, an inadmissibly high gradient of temperature was needed (two orders of magnitude higher than what might be reasonable)." V. S. Safronov, interview with A. E. Levin, 3 November 1989.

[12] "Safronov found conditions under which the instability sets in. He generalized to the flattened rotating disk the dispersion equations found by Chandrasekhar and by Bel and Schatzman for an infinite uniformly rotating medium, and by Bel and Schatzman for an infinite nonuniformly rotating medium. For the infinitely thin disk such an equation was found later by Goldreich and Ward (1973)." (See below, Note 20.) *Ibid.*

[13] "Smoluchowski found the 'classic' exact solution of the coagulation equation (without fragmentation) for the case of constant coagulation efficiency. Safronov found the exact solution for the other, more useful, case in which coagulation efficiency is proportional to the sum of the masses of the two colliding bodies. This value is intermediate between that for small bodies (geometrical cross-section) and that for large bodies (gravitational cross-section)." *Ibid.*

According to Vityazev (private communication), "the scientist who gave Safronov the idea to use the coagulation equation was the well-known physicist Davidov. He had a graduate student, Pshenai-Severen who was the husband of Lubimova. This man defended his dissertation on coagulation at our Institute. His article was published in Doklady Akad. Nauk in 1954. What is Safronov's genuine achievement, and a really first-class result, is that he obtained the solution of this equation for the situation when the coagulation coefficient is proportional to the sum of masses. Since then, one more analytic solution has been obtained, viz., for the case when the coefficient is proportional to the product of the masses. It seems to me that it will be possible to obtain even more general results corresponding to the sum of these coefficients, and in this case it will certainly be included in future textbooks on physical-chemical kinetics."

[14] "Asymptotic power law solutions of the equation (in the form m^{-q}) were also found for more realistic forms of the coagulation coefficient as well as for a more general equation including fragmentation of particles with smooth dependence of its intensity on m. It was found that an inverse power law with $q \approx 1.8$ can satisfactorily describe the distribution of masses for all m except the largest ones. Fragmentation is important for small bodies and particles especially when the relative velocities become high. But gravitation of the largest bodies prevents their fragmentation.

"Safronov investigated the dynamical evolution of a swarm of preplanetary bodies. Their relative velocities are increased due to gravitational perturbations at close encounters, but their inelastic collision decrease the velocities. During the accumulation the bodies' growth and their average velocities increase proportionately to their radii. For the power law mass distribution with $q<2$ the total mass is concentrated in the larger bodies and the velocities are a few times smaller than the escape velocity on the surface of the largest body." (Safronov, interview with A. E. Levin, 3 November 1989.)

[15] See also the analysis by G. W. Wetherill, "Numerical Calculations Relevant to the Accumulation of the Terrestrial Planets," in *The Continental Crust and its Mineral Deposits*, edited by D. W. Strangway (Geological Association of Canada, Toronto), Special Paper 20 (1980), p. 5; D. E. Fisher, *The Birth of the Earth* (Columbia University Press, New York, 1987), pp. 224–26 (1980).

[16] V. S. Safronov, "On the Primeval Temperature of the Earth," Bull. (Izvestiya) Acad. Sci. USSR, Geophysics Series No. 1, 85 (1959); see p. 87, paragraph before Eq. (5).

[17] "Later (1969,1978) he has developed a method of determination of the heating of the Earth by impacts of large bodies (up to hundreds of kilometers in size). The energy is liberated at a depth of about one or two diameters of the impactor and the heating was proportional to the radius of the largest impactor. The upper mantle of the Earth could be heated by large impacts to the temperature of the onset of melting." Safronov, interview with A. E. Levin, 3 November 1989.

[18] According to Vityazev (private communication), "The circumstance that the equation describing the growth of mass with a small coefficient theta should not be regarded as a weak point of Safronov's theory. It is true that small theta as well as small surface density do[es] not correspond to the physical situation for the larger part of the disk. But this problem was studied by Safronov himself in his book published in 1972 and in more detail in our new book (Vityazev *et al.*). As our analysis shows, this parameter theta is bound to be large when the eccentricities are large (when the masses of the planetary embryos are large enough). That some part of the matter is ejected when eccentricities are high is also a mathematical fact (disregarding the physical nature of the Oort cloud or Hill's cloud) which has been confirmed by numerical calculations performed, for example, by Sergei Ipatov at the Institute for Applied Mathematics. That means that to maintain that this is only an *ad hoc* hypothesis is true only in the same sense that a hypothesis about the very existence of a protoplanetary cloud is *ad hoc*."

[19] Safronov, Evolution, p. 152; "The Initial State of the Earth and Certain Features of its Evolution," Izvestiya Acad. Sci. USSR, Earth Phys. **1972**, 444–47. On the Lodochnikov–Ramsey hypothesis see S. G. Brush, "Chemical History of the Earth's Core," Eos **47**, 1185 (1982).

Vityazev (private communication) recalls that "I had to fight with Prof. [Boris] Levin for a long time because of his strange preoccupation with Lodochnikov's hypothesis. The beginning was rather funny. It was in 1970 and the preparation for the International Geological Congress was underway. I had just come to the Institute [of Earth Physics] from the University and Levin invited me to prepare a joint report about the formation of the core and its calculated features. I agreed willingly since that report would be my first delivered to an international congress. Since time was short, I first wrote a rather vague draft and only then began to study the problem in detail. Finally I came to the conclusion that although the metallized-silicate hypothesis was internally logically consistent, nevertheless it didn't correspond to reality. I gave my opinion to Prof. Levin and suggested that we should present to the Congress a joint report about the differentiation of the Earth with an iron core. But Levin insisted that he could not change his point of view. Nevertheless (and that was characteristic of him) he permitted me to submit my own report independently. In 1971 this work was published in *Fizika Zemli* and only since then has the model of evolution of the Earth with an iron core been developed at our laboratory. Later Safronov, Ruskol and others adopted that model."

[20] *Science Citation Index* does not cover several major conferences on the origin of the solar system, and does not cover systematically the papers presented at conferences on planetary science. For a comparison of the assessment of scientific publications by citation counts and by other methods see S. G. Brush, "The Most-Cited Physical-Sciences Publications in the 1945–1954 *Science Citation Index*," Current Contents, 14 May 1990, pp. 7–17. Part 3 of this article (*ibid.*, 22 October 1990, pp. 7–16) discusses publications in astronomy and the earth sciences.

[21] P. Goldreich and W. R. Ward, "The Formation of Planetesimals," Astrophys. J. **183**, 1051 (1973) (quotation from p. 1061).

[22] R. J. Dodd and W. McD. Napier, "Direct Simulation of Collision Processes. II. The Growth of Planetesimals," Astrophys. Space Sci. **29**, pp. 51 (1974).

[23] S. J. Weidenschilling, "Mass Loss from the Region of Mars and the Asteroid Belt," Icarus **26**, 361 (1975).

[24] J. A. Burns, "The Angular Momentum of Solar System Bodies: Implications for Asteroid Strengths," Icarus **25**, 545 (1975).

[25] W. K. Hartmann and D. R. Davis, "Satellite-sized Planetesimals and Lunar Origin," Icarus **24**, 504 (1975).

[26] S. F. Singer, "The Early History of the Earth–Moon system," Earth-Sci. Rev. **13**, 171 (1977).

[27] P. Farinella and P. Paolicchi, "Conservation Laws and Mass Distribution in the Planet Formation Process," Moon **17**, 401 (1977, pub. 1978).

[28] W. M. Kaula, "Thermal Evolution of Earth and Moon Growing by Planetesimal Impacts," J. Geophys. Res. **84**, 999 (1979).

[29] J. N. Goswami and D. Lal, "Formation of the Parent Bodies of the Carbonaceous Chrondrites," Icarus **40**, 510 (1979).

[30] B. C. Murray, R. G. Strom, N. J. Trask, and D. E. Gault, "Surface History of Mercury: Implications for Terrestrial Planets," J. Geophys. Res. **80**, 2508 (1975); B. C. Murray, M. J. S. Belton, G. E. Danielson, M. E. Davies, D. E. Gault, B. Hapke, B. O'Leary, R. G. Strom, V. Suomi, and N. Trask, "Television Observations of Mercury by Mariner 10," in *The Soviet-American Conference on Cosmochemistry of the Moon and Planets, Moscow, 1974*, edited by J. H. Pomeroy and N. J. Hubbard, Report NASA SP-370 (National Aeronautics & |Space Administration, Washington, 1977), pp. 865–85; G. W. Wetherill, "Pre-Mare Cratering and Early Solar System History," *ibid.*, pp. 553–67; "Late Heavy Bombardment of the Moon and Terrestrial Planets," in *Proceedings of the 6th Lunar Science Conference, Houston, 1975*, edited by R. N. Merrill (Pergamon, New York, 1975), pp. 1539–61; "The Role of Large Bodies in the Formation of the Earth and Moon," in *Proceedings of the 7th Lunar Science Conference Houston, 1976*, edited by R. B. Merrill (Pergamon, New York, 1976), pp. 3245–57.

[31] Vityazev (private communication) disputes the statement that Safronov did not make extensive use of computers, pointing to the work of Pechernikova, Safronov's graduate student, who "began to produce numerical calculations [from] the coagulation equations already in 1972." Since 1979 Vityazev, Pechernikova and others have studied the emergence of large bodies (comparable to the Moon and Mars) in the accretion of planets, taking account of "mega-impacts" (in the American terminology); see the review by Safronov and Vityazev (1983,1985), Sec. 6.

[32] V. S. Safronov and S. V. Kozlovskaya, "Heating of the Earth by the Impact of Accreted Bodies," Izvestiya Acad. Sci. USSR, Phys. Solid Earth **13**, 677 (1977); V. S. Safronov, "The Heating of the Earth During its Formation," Icarus **33**, 3 (1978); "Initial State of the Earth and its Early Evolution," in *Evolution of the Earth*, edited by R. J. O'Connell and W. S. Fyfe (Geological Society of America, Boulder, 1981), pp. 249–55; G. W. Wetherill, "The Formation of the Earth from Planetesimals," Sci. Am. **244** (6), 163 (1981).

[33] R. Greenberg, W. K. Hartmann, C. R. Chapman, and J. F. Wacker, "The Accretion of Planets from Planetesimals," in *Protostars and Planets*, edited by T. Gehrels (University of Arizona Press, Tucson, 1978), pp. 599–622; R. Greenberg, J. Wacker, C. R. Chapman, and W. K. Hartmann, "Planetesimals to Planets: A Simulation of Collisional Evolution," Icarus, **35**, 1 (1978); R. Greenberg, "Growth of Large, Late-Stage Planetesimals," *ibid.* **39**, 141 (1979); R. Greenberg, "Collisional Growth of Planetesimals," Moon and Planets **22**, 63 (1980).

[34] B. J. Levin, "Some Problems Concerning the Accumulation of Planets," Sov. Astron. Lett. **4**, 54 (1978); "Relative Velocities of Planetesimals and the Early Accumulation of Planets," Moon Plan. **19**, 289 (1978).

[35] R. Greenberg, S. J. Weidenschilling, C. R. Chapman, and D. R. Davis, "From Icy Planetesimals to Outer Planets and Comets," Icarus **59**, 87 (1984). More recent results are reported by R. Greenberg and B. Rizk, "Incipient Runaway Growth of Planetesimals: Why the Biggest Bodies Were Not All the Same Size," Lunar Plan. Sci. **18**, 362 (1987); S. J. Weidenschilling, "Evolution of Grains in a Turbulent Solar Nebula," Icarus **60**, 553 (1984); "Accumulation of Solid Bodies in the Solar Nebula," Gerlands Beitr. Geophys. **96**, Sonderheft 21 (1987); D. R. Davis, C. R. Chapman, S. J. Weidenschilling, and R. Greenberg, "Collisional History of Asteroids: Evidence from Vesta and the Hirayama Families, Icarus **62**, 30 (1985). See also the research of the Tucson group on the origin of the Moon, cited below in Note 56.

[36] A. P. Boss, "Low-Mass Star and Planet Formation," Publ. Astron. Soc. Pac. **101**, 767 (1989).

[37] G. W. Wetherill, "Occurrence of Giant Impacts During the Growth of the Terrestrial Planets," Science **228**, 877 (1985); "The Formation of the Solar System: Consensus, Alternatives, and Missing Factors," in *The Formation and Evolution of Planetary Systems*, edited H. Weaver and L. Danly (Cambridge University Press, New York, 1989), pp. 1–30, and other articles in this volume; G. W. Wetherill and L. P. Cox, "The Range of Validity of the Two-Body Approximation in Models of Terrestrial Planet Accumulation," Icarus, **60**, 40 (1984); **63**, 290 (1985); G. W. Wetherill and G. R. Stewart, "Accumulation of a Swarm of Small Planetesimals, *ibid.* **77**, 330 (1989); G. W. Wetherill, "Comparison of Analytical and Physical Modeling of Planetesimal Accumulation," *ibid.* **88**, 336 (1990).

[38] S. J. Weidenschilling, letter to S. G. Brush, 15 December 1988.

[39] Vityazev (private communication) says that he, in contrast to Safronov and Ruskol, concluded that the accretion model is inadequate to explain the formation of the Moon without invoking giant impacts. However, this conclusion is stated only rather tentatively in the last paragraph of the paper by Pechernikova, Maeva, and Vitayev (1984); see the translation reprinted in Part VI of this book. I have not yet seen any detailed development of the giant-impact model by Vityazev's group comparable to what has been done in the United States.

[40] This section is based on S. G. Brush, "Early History of Selenogony," in *Origin of the Moon*, edited by W. K. Hartmann, R. J. Phillips, and G. J. Taylor (Lunar and Planetary Institute, Houston, 1986), pp. 3–15; "Nickel for Your Thoughts: Urey and the Origin of the Moon," Science **217**, 891 (1982), which provide further details and references.

[41] The following material is based on my article "A History of Modern Selenogony: Theoretical Origins of the Moon, From Capture to Crash 1955–1984," Space Sci. Rev. **47**, 211 (1988), which should be consulted for further details and references.

[42] E. L. Ruskol, "The Origin of the Moon," in *The Moon* (IAU Symposium No. 14, Pulkovo Observatory, December 1960), edited by Z. Kopal and A. K. Mikhailov (Academic, New York, 1962), pp. 149–55; quotation from p. 154.

[43] E. L. Ruskol, *Origin of the Moon*, translated from Russian, Report No. NASA TT F-16,623 (National Aeronautics and Space Administration, Washington, DC, 1975), p. 41.

[44] E. L. Ruskol, "On the Past History of the Earth–Moon System," Icarus **5**, 221 (1966) (quotation from p. 225). Ruskol informs me that her "agreement" with the Lodochnikov–Ramsey hypothesis in 1962–66 "may be better referred to Prof. B. Yu. Levin who was [an] active proponent of it [at] the time.... in explicit form I abandoned this hypothesis only later, in 1971. But I had never written that I hold this hypothesis" (letter to S.G.B., 18 January 1985).

[45] B. J. Levin, "The Structure of the Moon," in *Proceedings of the Caltech–JPL Lunar and Planetary Conference, September, 1965*, edited by H. Brown *et al.*, JPL Tech. Memo. 33-266 (Jet Propulsion Laboratory, Pasadena, 1966), pp. 61–76.

[46] Levin, Ref. 45, p. 61.

[47] I. I. Mitroff, *The Subjective Side of Science* (Elsevier, New York, 1974), p. 156.

[48] T. S. Kuhn, *The Structure of Scientific Revolutions*, 2nd ed. (University of Chicago Press, Chicago, 1970).

[49] H. B. Baker, "The Earth Participates in the Evolution of the Solar System," Occasional Papers Detroit Acad. Nat. Sci. No. 3, pp. 12 and 16 (1954).

[50] W. K. Hartmann and D. R. Davis, "Satellite-Sized Planetesimals and Lunar Origin," Icarus **24**, 504 (1975) (quotation from page 511).

[51] A. G. W. Cameron, "Formation of the Prelunar Accretion Disk," Icarus **62**, 319 (1985) (quotation from p. 319).

[52] D. J. Stevenson, "Origin of the Moon–The Collision Hypothesis," Annu. Rev. Earth Plan. Sci. **15**, 271 (1987).

[53] W. K. Hartmann, quoted by H. S. F. Cooper, Jr., "Letter from the Space Center," New Yorker **63** (16) (June 8) 71 (1987) (quotation from p. 80).

[54] A. W. Harris and W. M. Kaula, "A Co-Accretional Model of Satellite Formation," Icarus **24**, 516 (1975) (quotation from p. 516).

[55] W. M. Kaula, Lecture at International Union of Geodesy and Geophysics, XIX Assembly, Vancouver, 1987.

[56] R. Greenberg, C. R. Chapman, D. R. Davis, M. J. Drake, W. K. Hartmann, F. L. Herbert, J. Jones, and S. J. Weidenschilling, "An Integrated Dynamical and Geochemical Approach to Lunar Origin Modelling," in *Papers presented to the Conference on the Origin of the Moon, Kona, Hawaii, October, 1984* (Lunar and Planetary Institute, Houston, 1984), p. 51.

[57] V. S. Safronov, G. V. Pechernikova, E. L. Ruskol, and A. V. Vitjazev, "Protosatellite Swarms," in *Satellites*, edited by J. A. Burns and M. S. Matthews (University of Arizona Press, Tucson, 1986), pp. 89–116. Safronov (private communication) says he is not against the giant-impact hypothesis, but thinks that the people who proposed that hypothesis are the ones who should develop it. Cf. Note 39.

II. Selected Writings of Otto Schmidt, 1925–54

On the Stability of Planetary Motions

O. Yu. Schmidt

Translated from: unpublished notes, Izbrannye Trudy: Geofizika i Kosmogoniya
(Izdatelstvo Akademie Nauk SSSR, Moscow, 1960), p. 181

I. ON THE STABILITY OF PLANETARY MOTIONS

Concerning the problem of three bodies, the *trends* of motion. If the initial velocities do not lie on the plane of these bodies, will their motions tend to be confined within some plane, or will another dynamic trend possibly arise? Whether the motions in different configurations are equally stable? Is it necessary to take a fourth body into consideration in order make judgements concerning the problem of stability? This is my topic for reading during the near future—to select from an enormous bulk of literature dealing with the three-body problem.

1925, March

* * * * *

It is my old dream—to approach the Bode law from the side of the stability of motion (the same—concerning the quanta). Let us consider the typical, real case of the perturbations of planetary motion. These perturbations will depend on the perturbing force and on the parameters of the initial motion.

The problem is: Which should be the equations describing the links between the parameters in order to make the effect of the perturbing force minimal (the first coefficient in the series=0)? Is it possible that a discrete set of values will arise?

This method has real justification in case the minimum is very sharp. Then, in practice, the motion will rather "soon" reach one of those "stable" states.

They are possibly stable even in a more precise sense, that is, a small perturbation brings them back rather than carries them away.

Of what kind are the perturbations? In reality, an atom always experiences an influence from the side of "remote masses" (the other atoms of the same body), and a planet is influenced by the stars. These perturbations are small, but permanent, and they seem to be of the utmost importance in respect to my problem. It is also worth exploring the effect of a weak *short-lived* perturbation, too.

(Since 1920, I have made a number of more detailed exploratory attempts in these directions. I could not spare much time to them, and my effort has not been systematic. Now, I want to work on the problem more seriously.)

1925, March

* * * * *

Origin of the Planetary System (unpublished notes, 1942)

O. Yu. Schmidt

Translated from: unpublshed notes, Izbrannye Trudy: Geofizika i Kosmogoniya Izdatelstvo Akademie Nauk SSSR, Moscow, 1960, pp. 183–4

The following are the principal defects of the former hypotheses (although their historical significance is indisputable, especially that of Kant's):

(1) the intent to build the whole Universe on the basis of the *model* "the Sun—the planets (the Earth)";

(2) the power of biased concepts (initially highly heated and gradually cooling Earth, etc.);

(3) reliance on some sporadic and rather improbable constellations of events (such as neighboring stars), which makes our planetary system a kind of *exception* to the general pattern.

These weak points are shared, albeit to a *different* degree, by all of the existing hypotheses.

The development of our astronomical knowledge, and especially that related to the structure of the galactic system, allows us to take a new road. This is my starting point: the Galaxy *does exist*, and its basic features are such and such. The Sun exists within this Galaxy, too. What may be expected to happen then?

As it will turn out, a planetary system must come into being *necessarily*, and its properties will mainly become just those, which we observe in reality.

Hence, we have not a *hypothesis*, but a *theory*. The realm of hypotheses will embrace only such matters, as the origin of the Galaxy, the evolution of the whole Universe, etc., although even there our theory allows us to achieve a new understanding of many things.

What are the properties of the Galaxy that I can begin with? The Galaxy is an accumulated mass shaped like a lens. Its different components (the stars, the dark masses of different sizes, meteorites, splinters, gas) revolve *in the same direction* around their common center, in accordance with Newton's laws; as a rule, their axes are only insignificantly inclined from the line perpendicular to the central plane. The concentration of mass is especially high within that plane.

Our Sun is nothing but a star belonging to the Galaxy; it moves with a small inclination relative to the central plane.

What is going on then?

While advancing along its orbit, the Sun will eventually intersect the central plane. As a result, it will capture some portion of the galactic matter which is especially heavily concentrated in this area, and then the captured matter will accompany the Sun as its satellites. (Possibly—repeated captures!)

I shall especially discuss the biases against "capture" which have become established in science.

After capture:

(1) The captured bodies, which formerly revolved more or less within the same plane, will remain revolving within one plane—hence, one basic feature of our solar system may be easily explained.

(2) Those having revolved prior to their encounter with the Sun in one direction, *will keep revolving in the same direction* (that will be proved mathematically, together with the limits of possible—and, consequently, actual—violations of this pattern, such as reverse rotation).

(3) The plane of a new orbital rotation (around the Sun) will not be more parallel to the central plane of the Galaxy. This is a result of a compensation of angular momentum in relation to the plane which is perpendicular to the Galaxy, due to the Sun's inclined orbit capturing a portion of the galactic matter.

This idea also may be and should be mathematically justified and corroborated by means of direct calculations (see further).

(4) The mathematical theory of the capture demonstrates that capturing has its own limits. The bodies passing too close to the Sun cannot be caught easily, their hyperbolic velocity being too high. Equally, the bodies which are too distant from the Sun, are attracted weakly. (All of that should be examined mathematically in full detail.)

Therefore, the captured mass will be comparatively small near the Sun (except the zone of its immediate *absorption*), further it will be *denser*, and then again less concentrated—just as it really is.

(5) The captured masses may be "ready-made" accumulations of the kind of planets, swarms (comets), solitary meteorites, or anything else. However, they are nonuniform [another possible version: they are distributed nonuniformly; editor's note by Aleksey Levin], and they, in turn, will be capturing each other. It is possible to show mathematically that every planet will be surrounded by a certain empty zone, which implies the existence of a *certain* law of planetary distances whose first tentative approximation was the Bode law.

Consequently, all of the principal features of the planetary system can be thus explained *without additional hypotheses* ("ad hoc"), and without arbitrary conjectures or surmised irregular events.

* * * * *

The principal themes were wandering in my head long ago (partially, even during my youth). My temporary studies in the area of celestial mechanics (in the 1920s) were of the same direction, but I have been able to work out this theory systematically only beginning in the spring of 1942, when I received the necessary time (and even then, only partially), and could busy myself with a (gradual) development of its mathematical aspects and geophysical consequences.

Kazan, June 1942

Amendment

I have forgotten that:

(6) The planet's and the Sun's rotation around their axes results from the absorption of neighboring satellites. Therefore, their equators approximate the ecliptic plane. However, there were also inclined satellites amongst those captured. Even a modest degree of inclination can produce a considerable angular momentum. While being transmitted in the course of absorption, it generates a huge momentum oriented not-perpendicularly to the plane of ecliptic.

That implies irregularities of axes. What about the Sun and the big planets, where the acts of absorption were numerous—there the probability that the orientation of the equatorial plane corresponds to the stable plane of the system is comparatively higher. The theory fits the facts once more. The special features of the most distant planets do not contradict this theory, although it has not explained them individually.

June 1942

* * * * *

A METEORIC THEORY OF THE ORIGIN OF THE EARTH AND PLANETS

O.Yu. Schmidt
Comptes Rendus (Doklady) de l'Académie des Sciences de l'URSS, Vol. 45, p. 229-233(1944)

1. For the present there is no generally recognized theory of the origin of the Earth and planets. All former theories have been found to be irreconcilable with facts. And in working out a new theory one has to work it out from the new facts that have become known to astronomers within these two decades and been not yet turned to full account in cosmogony.

Two of these facts, the rotation of the Galaxy and the presence near its central plane of big masses of obscuring matter (dust-like and meteoric); lie at the basis of the theory put forward in the present paper. This theory rests on the following conception: the Sun on its galactic course once crossed a dark cloud (of dust and meteorites) and; having entrained part of its matter; compelled the captured particles to revolve about it as a centre. Eventually these particles united into large formations—the planets.

The purpose of the theory is to put this conception into the language of equations and formulae; and upon deriving some quantitative conclusions check them against. the known data of observation.

In the face of Galactic Rotation, which has since 1926 been a definitely established act, it is no longer possible to support the simplified scheme (models) which heretofore were placed at the basis of the theory of stellar motion. It will not do to-day either to compare the stars to gas molecules or reduce the problem to a case of interaction between two (or three) isolated objects. Two stars while performing their complicated motions in the Galaxy;—which are each a combined result of attraction by the central galactic masses and of the actions of external masses and the neighboring individual stars—can happen to come in a position where the external forces tend to draw them so closely together that capture and formation of a double star takes place.

In an earlier paper [1] the author has advanced a theory in which the formation of visual binary stars is supposed to occur by such a capture under conditions of a rotating Galaxy. In it have been adopted the following simplifying assumption considered as typical, i.e., corresponding to statistical average:

a) The stars move along circular orbits being subject to attraction by the mass of the Galaxy which is conceived to be concentrated in centre.

b) The orbits lie in different planes forming small but significant angles between each other.

c) The exact instant at which capture takes place is when the two stars pass the position of the distance between the orbits.

On these assumptions the formula

$$1/(1\mp e) - 1/2 = a/\rho \qquad (1)$$

has been derived (1); where e and a are the eccentricity and the semi-major axis of the new formed double star's orbit; respectively, and the quantity ρ does not depend upon the elements of the relative orbit, but solely upon the galactic parameters of the stars; namely

$$\rho = R\frac{\underline{\mathit{M}}}{M}\,\frac{1}{2\sin^2\alpha/2} \qquad (2)$$

Here R is the distance to the centre of the galaxy: M is the total mass of the two stars; M is the mass of the inner part of the Galaxy: α is the angle between the orbital planes of the stars previous to capture. The quantity ρ is the limiting value of the capture distance d; above which no capture is possible. If $d>\rho/2$; then e in formula (1) is to be taken with the upper sign; while the lower sign corresponds to $d<\rho/2$. With $d=\rho/2$ the orbit is a circumference.

Formula (1) has provided a simple explanation of all the principal laws established by the statistics of visual binaries; and; moreover: there has proved to be a remarkable numerical agreement between theory and observation.

The above-stated simplifying assumptions (circular orbits, etc.,) are not, of course; as a general case exactly fulfilled in nature. Therefore formula (1) is an approximate relation and is only true in the mean for a large number of objects.

This formula and the same concept of capture will now be used in analyzing the capture of meteoric matter by a solitary star; the Sun.

2. The interstellar and interplanetary matter comes up in different cosmogonic theories of which the most recent are those by B. Lindblad [2] and H. Alfvén [3]. According to the author's theory, planets arise from interstellar matter captured by the Sun in its galactic motion.

Essential to this theory is the fact that the plane of the Sun's galactic orbit is inclined at a certain angle (of about 3°) to the central plane of the Galaxy. If the Sun while at one of the nodes of its orbit passes through a cloud of matter (meteorites; dust); meteorites are captured according to the same laws as underlie the formation of double stars. A meteorite thus captured moves about the Sun along Kepler's orbit. As has been explained in the cited paper on double stars, the plane of the relative motion of the binary system is determined by the direction of the two following straight lines: the shortest distance between the original orbits, and the relative velocity of the two stars previous to capture. Supposing, in accordance with the latest data; that the meteoric cloud is all nearly in one plane (the central plane of the Galaxy); the captured meteorites will move in a swarm also nearly in one plane; thus forming a flat lens. The characteristic of their orbits is given by formula (l) which for simplicity we can use with the upper sign only, giving e values ranging from—1 to +1. Within these limits all values of e are equiprobable, as has been shown in the above-mentioned paper. To be sure; all meteorites are not exactly in the position of the shortest distance between the original orbits when their capture takes place. Some are captured when they have already passed this position; and others again when they have not yet reached it. This goes somewhat to increase the difference between the planes of the new orbits; so that a slightly thicker lens is formed, and we may be certain to meet with statistical deviations from formula (1).

The initial velocity of the relative motion is the relative velocity of the meteorite at the instant of capture.As the position of the meteorite at that moment can as well be to one side of the Sun as to the other; both direct and inverse motions will arise. The odds in favor of either direction would be equal if only the cloud was one of perfectly uniform density. If, however; the density of the cloud in the region crossed by the Sun is not uniform (say; the Sun passes near the edge of the cloud); then more meteorites are captured on one side than on the other. Accordingly, by far the larger number of meteorites forms a swarm revolving about the Sun in one direction only. The rest revolve in the opposite sense and these are bound to collide sooner or later with some portion or the main sw arm. The colliding meteorites lose their rotational momentum and fall on the Sun.

3. Within the swarm; the forces of interaction between the individual meteorites come into play. An exchange of energy and momentum takes place (while the total energy and amount of rotation remain invariable), which again results in some of the meteorites falling on to the Sun. But the main mass segregates by and by into larger bodies and this is the formation of planets. What we observe at this date as separate meteorites is the remainder of this process. The segregation of the meteoric swarm proceeds through smaller particles falling upon the larger ones in the same way as we know meteorites and meteors to fall on the Earth. As the orbits of the meteorites do not differ greatly from one another the relative velocities are rather small, and the impact caused by the fall must not, as a rule, lead to volatilization or scattering in every direction, which contingency has been often used as an argument against various meteoric hypotheses. This point has been cleared up in Lindblad's paper [2].

We speak of meteorites and of dust as two distinct phenomena, but the difference is merely in size rather than principle. Dust clouds are likely to include larger particles upon which dust particles settle. In a meteoric swarm formed about the Sun this process of condensation must go on at a higher rate because of the greater density of the swarm as compared to that of a dust cloud. In the case of dust particles and small meteorites the effect of light pressure needs to be inquired into. It is known [5,6] that a particle of small; but finite, size revolving about the Sun is losing momentum under the influence of light pressure and in the long run, if isolated; will fall on the Sun. But ours are not isolated particles; and; therefore; only a small number of them will settle down on the Sun. The rest, by dint of continuous interaction; will have this loss of momentum distributed among themselves; and most of them, moreover, will rather soon be united into larger bodies, thus giving rise to planets. So the pressure of light makes only for a more thorough intermixing of meteorites; thereby accelerating slightly the formation of planets, but does not in the least affect the ultimate outcome.

4. What can we conceive to be the main features of a planetary system formed according to this scheme?

As the swarm of meteorites has the form of a flat lens, the planets in progress of formation have their orbits lying approximately in one plane which is the central plane of the lens. As the meteorites are revolving mainly in one direction, the planets will be revolving in this one direction also.

The meteorites of the swarm move along ellipses for which the distribution of eccentricities and axes is given by formula (1). At the beginning the major axes have a dominant direction (along the shortest distance between the galactic orbits of the Sun and the cloud), but are soon deflected from it owing to reciprocal perturbation in all directions. For that reason no planet arising from an aggregate of a very large number of meteorites can have a preferential direction in which its orbit would be elongated. Only a circular orbit; or nearly so; is possible in this case.

Thus, what we know to be the main features of our planetary system follows directly from the suggested theory.

The angular momenta of the planets; which were the stumbling block of the former theories, are obtained on this theory directly from the galactic momentum of meteorites.

The further development of the theory will bring within its compass such problems as the rotation of the Sun; the rotation of planets; the formation of satellites, the age of the planetary system.

Certain features of the internal structure of the Earth can also be derived from this theory. Some of these problems are discussed in the papers that are now being prepared for the press. The present communication will only give the first quantitative results.

5. From formula (l) may be derived some consequences concerning the distribution of mass and momentum in the planetary system . Suppose for simplicity sake that the portion of the meteoric mass fallen to the Sun can be neglected.

Theorem I. *The total momentum of the system is equal to that of a single planet supposed to have a mass equal to the total mass of the planets revolving in a circular orbit at a distance* $\rho/2$ *from the Sun.*

As has been mentioned above, all values of e between

—1 and +1 are equiprobable, the total mass of the meteorites m is distributed evenly over these values, and, the range of e equaling 2, we have $dm = (m/2)de$. The momentum of a particle with a mass dm, and eccentricity

e and semi-axis a is $k\sqrt{M\,a(1-e^2)}\,dm$, where k^2 is the gravitation constant, M is the mass of the sun. As it follows from (1) that

$$a(1-e^2) = \frac{\rho}{2}(1+e)^2$$

we obtain for the momentum the expression

$$k\sqrt{M}\sqrt{\rho/2}\,(1+e)\,dm = k\sqrt{M}\sqrt{\rho/2}(m/2)(1+e)\,de \quad (3)$$

Integration over e from —1 to +1 gives the total momentum of the system

$$k\sqrt{M}\sqrt{(\rho/2)}m$$

and this establishes Theorem I.

Let us further consider the problem of the boundary of the solar system Believing the planets to have been formed from meteorites, part of which still continues in a free state, is obviously incompatible with the idea of the planetary system having a strict boundary. Individual meteorites can deviate from formula (l) which is only true in the mean, and there is no limit therefore to the distance at which some particular meteorites may be supposed to perform their revolutions about the Sun. Yet the main mass of meteorites obey formula (1), and their momentum cannot possibly exceed the maximum, which this formula admits. This maximum momentum corresponds to $e = +1$. Substituting this value in formula (3), we obtain

$$k\sqrt{M}\sqrt{(\rho/2)}\,2dm = k\sqrt{M}\sqrt{2\rho}\,dm$$

which gives the momentum on a circular orbit of radius 2ρ. Taking now for the "practical boundary" of the solar system the distance beyond which the existence of but quite an insignificant portion of the total planetary mass is possible, we can establish the following theorems:

Theorem II. *The practical boundary of the planetary system runs at a distance of 2ρ from the Sun.*

Theorem III. *The practical boundary of the planetary system is as far away from the Sun as four times the radius of the circular orbit which a planet, having the total mass and momentum of all planets taken together, may be imagined to describe about the Sun.*

This theorem is an immediate consequence of Theorems I and II and is remarkable for not involving the quantity ρ.

6. Though the motion of the Sun is no circular one, and neither formulae (1) and (2) nor the theorems deduced from them can possibly apply to it accurately, they are yet useful as a first approximation in describing the structure of our planetary system and verifying how far the results obtained agree with the observational values.

If the distance from the Earth to the Sun, the orbital momentum of the Earth and the mass be put at unity, then the total mass of the planets is known to be 446, and the total momentum,1176 (cf. [7]; Table 1). The ratio of total momentum to total mass is 2.64. For the imaginary "total" planet to have this momentum—mass ratio it must be supposed to revolve at a distance of $(2.64)^2$ from the Sun,which is 7 astronomical units. The practical boundary of the solar system lies near the orbit of Neptune, that is, about 30 astronomical units away from the Sun. Dividing 30 by 7 we get 4.3 instead of 4, which is the figure required by Theorem III. The agreement is satisfactory.

A distance of 7 astronomical units corresponds, according to Theorem I, to $\rho/2 = 7$ or $\rho = 14$. Now, what is the value of ρ obtained directly from formula (2)? It depends upon how far from the centre of the Galaxy the capture of meteorites by the Sun took place. The present distance of the Sun from the centre of the Galaxy is estimated by the majority of workers as about 7500 parsecs. An osculating orbit of the Sun [4]gives a maximum distance of 14200 parsecs. Assume; moreover; that the internal mass of the Galaxy is 10^{11} times the mass of the Sun and that the orbit, of the Sun is inclined at an angle $\alpha = 3^\circ$ Then formula (2) gives the following limits of the variation of ρ as a function of R (in astronomical units):

$$11.3 < \rho < 21.4$$

Within these limits lies also the value of $\rho = 14$ which we have obtained before.

Thus, our formulae, though of approximate nature and based on simplifying assumptions, prove to yield such quantitative estimates as come close to the observed data .

Institute of Theoretical Geophysics.
Academy of Sciences of the USSR.

Received
22 Sept. 1944.

REFERENCES

1. O.Yu. Schmidt, C.R. Acad. Sci. URSS, 44, No. 1 (1944).
2. B. Lindblad, Nature, January 26 (1935).
3. H. Alfvén, Stockholms Observatoriums Annaler, 14, Nos. 2, 5 (1942, 1943).
4. P. P. Parenago, Astron. Zhur., 16, No. 4 (1939).
5. J. H. Poynting, Phil. Trans. Royal Society London, Ser. A, 202 (1903).
6. H. P. Robertson, Mont. Not. Royal Astron. Society, 97, No. 6.(1937)
7. H. N. Russell, The Solar System and its Origin (1935). (Russian translation, 1944)

On the Origin of Comets (abstract)

O. Yu. Schmidt

Comptes Rendus Acad. Sci. USSR **49**, 404 (1945)

The theory of the planets' origin put forward by the author naturally leads to the following ideas as to the origin of the comets. The latter, like the meteorites now observed, are *remnants of a meteoric swarm* that was once formed around the Sun, by capture, out of a cloud of dark matter, which the Sun had crossed on its way in the central plane of the Galaxy. The main mass of this swarm has been used up in the formation of the planets, but a part has remained and continued revolving around the Sun.

THE COSMOGONIC SIGNIFICANCE OF THE POSITION OF THE ECLIPTIC PLANE IN THE GALAXY

O. Yu. Schmidt
Comptes Rendus de l'Académie des Sciences de l'URSS
Vol. 52, No. 7 (1946).

The plane of the ecliptic is inclined to the galactic plane at an angle of 60°.5, and its line of nodes lies in galactic longitude 153°.2-333°.2. Are these figures of casual nature? They appear, indeed, to have been looked upon as a matter of chance. However, the theory advanced by the author to account for the origin of the Earth and planets implies that the position of the ecliptic plane in the Galaxy is not due to chance, but has rather a definite cosmogonic significance.

§1. According to this theory, the Sun, while revolving with the Galaxy, had at one time crossed a cloud of dark (meteoric) matter lying in the galactic plane, and captured part of its meteorites which began to revolve around the Sun and eventually became the material for the formation of the planets. In an earlier paper the author has shown [1] that the maximum concentration of orbit planes in the swarm of meteorites thus captured by the Sun was along a line (described as Y-axis) lying in the galactic plane perpendicular to the direction of the Sun's motion at the moment of capture. Therefore the planets, if formed from meteorites, should have their orbit planes either running through the Y-axis or very close to it. As for meteorites with long orbits, which escaped absorption by the planets to make up long-period comets, they too, according to [1], must have the Y-axis as the line of maximum concentration of their orbit planes. And again it is the Y-axis about which were concentrated the orbit planes of those meteorites which in the end, as a result of mutual perturbations, fell to the Sun and imparted to it their angular momenta. So the rotation of the Sun, which may be accounted for by the fall of these meteorites, should likewise take place in a plane running through the Y-axis.

Thus, on this theory, the planes of the planet orbits, the major part of the planes of long-period comets, and the plane of the Sun's equator should all intersect almost along a single straight line lying in the galactic plane. Does this conclusion square with the actual state of things? The following table gives the galactic longitude of the lines of nodes of the planets' orbit planes and of the Sun's equator in the galactic plane:

Node of ecliptic plane in the Galaxy --	153°.2 and 333°.2
Node of Jupiter's orbit -----------------	153°.6 and 333°.6
Node of Sun's equator ------------------	151°.5 and 331°.5
{Line of concentration of orbits -------- of long-period comets [1]}	155°1 and 335°.1

These data bear out the theory. In fact, there is close coincidence between the nodes of all these planes in the Galaxy.

§ 2. The Y-axis, according to the author's theory, lies in the galactic plane at right angles to the velocity with which the Sun was moving when it captured the meteorites near the central plane of the Galaxy. The capture could not have occurred unless the Sun was crossing one of the nodes of its galactic orbit in the galactic plane. If we knew the galactic orbit of the Sun, it would be possible for us to decide whether the Y-axis, as derived from the position of the node, coincided with the direction defined in § 1.

Approximately the galactic orbit of the Sun has been computed by P. P. Parenago[2]. His remarkable work gives an elliptic orbit with eccentricity =0.30. Future modifications of the estimate of the distance of the Sun from the centre of the Galaxy may, of course, affect the scale of the computed orbit, but they will not affect its position in space. Parenago speaks of his orbit as osculating in our epoch. However, Schwarzschild [3], Rosseland [4] and Chandrasekhar [5] have found that the time of relaxation for stars is of the order of 10^{12} - 10^{14} years. Even allowing for the fact that these authors have simplified the problem to the case of two bodies, thereby excluding the possibility of mutual capture, we still obtain a time of relaxation of far greater length than the few milliard years that have elapsed since the inception of the planetary system, *i. e.*, since the meteorites were captured by the Sun. So the odds are greatly in favor of the Sun having preserved its orbit during these several milliard years without any appreciable change.

In Parenago's orbit the galactic longitude of the direction of the Sun's velocity at the ascending node of the orbit is 60°, and the galactic longitude of the Y-axis perpendicular to this direction is, accordingly, 150—330°, which is very close to the node of the ecliptic. The 330° direction is close also to what the direction to the centre of the Galaxy is at present, because it is not long since the Sun last time passed through its ascending node in the galactic plane. It turns out therefore that the direction of the node of the ecliptic in the galactic plane is actually determined by the position of the node of the galactic orbit of the Sun.

§ 3. Neither has the inclination of the ecliptic to the galactic plane assumed its value by chance. Let us consider, from the theoretical point of view, the distribution of orbit poles for the Sun captured swarm of meteorites within each of the quadrants formed by the planes XY and ZY in the system of coordinates adopted in the preceding paper (1). For the angular coefficients of the orbit poles with the coordinate axes we have formerly (1, 8) obtained the expressions:

$$\cos A=\xi \; ; \cos B=\eta\sqrt{1-\xi^2}/\xi \; ;$$

$$\cos\Gamma=[\sqrt{\xi^2-\eta^2}\;\sqrt{1-\xi^2}\,]/\xi \qquad (1)$$

If the pole is in the first quadrant, then $\xi>0$,

$\sqrt{1-\xi^2}\;\sqrt{\xi^2-\eta^2}>0$, where $0\le\xi\le1$ and $-\xi\le\eta\le\xi$. Within these limits all values of ξ and are equally probable.

When meteorites fuse into planets the total angular momentum remains unchanged. Let us compute the theoretical direction of the total angular momentum for the poles lying in the first quadrant. Every meteorite will be supposed to have the same (mean) value of angular momentum, and the distribution of directions is given by formulae (1). Integration over η (from $-\xi$ to $+\xi$) and over ξ (from 0 to 1) will give the components of the total angular momentum parallel to the coordinate axes. On account of symmetry the Y-axis component is obviously equal to zero. For the two other components integration gives

$$\iint \cos A\,d\xi\,d\eta = 2/3;$$

$$\iint \cos\Gamma\,d\xi\,d\eta = \pi/6$$

The total angular momentum, therefore, is a vector lying in the XZ-plane and making with the XY-plane an angle b defined by the established components: $\mathrm{tg}b=\pi/4$; $b=\mathrm{arctg}\,\pi/4=38°9'$. This angle may differ in sign, according to the quadrant, but its absolute magnitude remains constant.

§4. We shall now investigate the conditions in favor of direct or retrograde motion of the captured meteorites. Let the plane of the drawing (Fig. 1) pass through the Sun S and be parallel to the XY-plane. Denote the velocity of the Sun by v_1 (it will be remembered that v_1 by definition is directed along the Y-axis). According to the author's theory, the meteorite is captured by the Sun when both are at the respective ends of the shortest distance between their galactic orbits. As this distance is at right angles to the direction of v_1, the projection of the meteorite on to the plane of the drawing will lie on the Y-axis at a point P. Denote by v the galactic velocity of the meteorite, and by L its velocity with respect to the Sun. Let v' and L' be the projections of these vectors on to the plane of the drawing. It may be seen from the drawing that when point P is to the right of S, the motion is direct if L' points towards positive x, and retrograde if L' points the other way. In other words, the motion of right-hand side meteorites is direct or retrograde, according as the projection of the meteorite velocity on to the Sun velocity is larger or smaller than the Sun velocity. The reverse of this condition holds for meteorites that pass the Sun on the left-hand side.

If in the case of direct motion the galactic velocity v of a right hand side meteorite is directed downward (from the XY-plane), the velocity L is likewise directed downward, and the positive pole will be directed onward (into the first quadrant). If the velocity v is directed upward, the pole is in the second quadrant

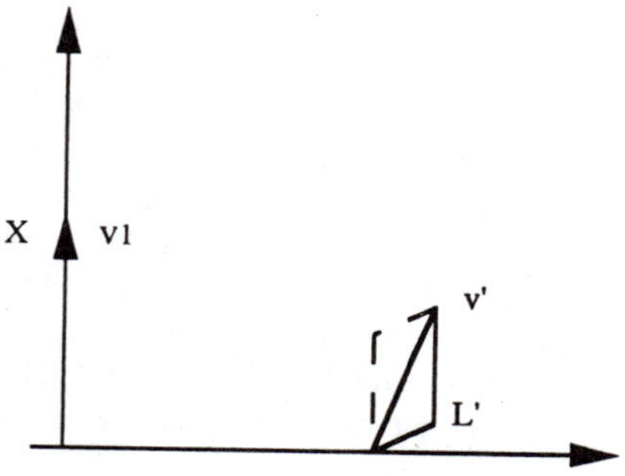

Figure 1

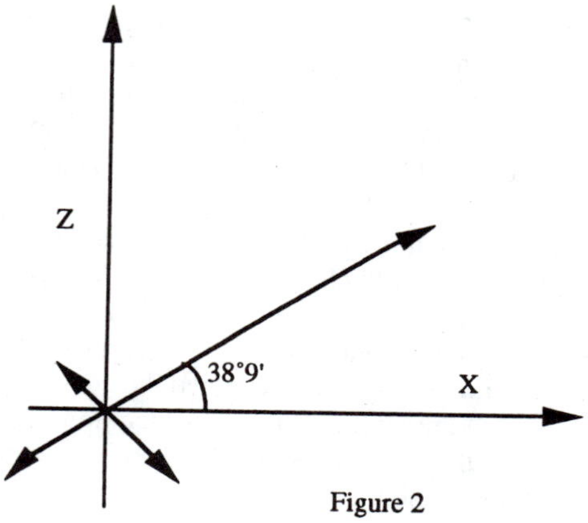

Figure 2

Such are the simple conditions which determine the division of the meteorite orbits into direct and retrograde, and govern the distribution of poles in the quadrants. They may be illustrated by the following scheme.

Motion	Downward	Upward
Direct Motion		
1st Quadrant	Right and Larger	Left and Smaller
2nd Quadrant	Left and Smaller	Right and Larger
Retrograde Motion		
3rd Quadrant	Left and Larger	Right and Smaller
4th Quadrant	Right and Smaller	Left and Larger

§5. There is every reason to suppose that one of the quadrants may predominate over the others. In fact, the present position of the Sun, as has been mentioned in § 2, is close to its ascending node, so that the positive direction of the X-axis is above the galactic plane. And as the meteoric cloud revolves almost exactly in the central plane of the Galaxy, all but an insignificant number of its meteorites must have been moving with velocities directed downward from the XY-plane. So the meteorites that were moving "upward" need not be considered. Further, if we know nothing of the velocity distribution of the meteorites in the galactic cloud before they were captured by the Sun, we nevertheless may assert that there was one principal direction, *viz.* the one along which the cloud made its galactic rounds, and one general velocity on which differences in the velocity of individual meteorites were superimposed. The projection of this general velocity on the velocity of the Sun must be either larger or smaller than the Sun velocity.

Therefore, for a major part of the meteorites only one of the abovementioned conditions will hold: say, "larger". Finally, it may be noted (what has already been emphasized in [7]) that if the stream of meteorites had been equally dense to either side of the Sun, no planetary system would have been brought to life (with direct and retrograde motions divided in equal numbers the meteorites would have gradually fallen to the Sun). However, on account of the flocculent structure of galactic clouds the more probable case is that the density to one side of the Sun (to the right, for instance) was greater than to the other.

Thus, the theory requires that one of the quadrants should predominate. But we are denied the possibility to reveal this quadrant by theoretical considerations. In actual fact, all the orbit poles of the planets are in the first quadrant (see table).

Galactic Coordinates of Planetary Orbit Poles

	l	*b*
Sun (Equator)	61°. 5	22°. 5
Mercury	58°. 4	24°. 0
Venus	62°. 5	26°. 0
Earth	63°. 2	29°. 5
Mars	62°. 0	28°. 0
Jupiter	63°. 6	28°. 2
Saturn	64°. 5	27°. 2
Uranus	63°. 0	28°. 7
Neptune	64°. 6	28°. 2
Pluto	69°. 9	13°. 5

§ 6. Thus, in each quadrant the vectors of angular momenta give a resultant lying in the plane XZ and making an angle of 38°9' with the plane XY. The vectors in the first quadrant, characterized by the conditions "right", "larger" and "downward", strongly predominate. The meteorites passing (in weaker numbers) to the left of the Sun under otherwise equal conditions will have their resultant vector in the third quadrant (Fig. 2, plane XZ). It is directed opposite and is smaller in magnitude. The case of "upward" motion we discard on grounds mentioned above. There remains a certain number of possible meteorites with characteristics "downward" and "smaller". Those of them which are "right" will give poles in the fourth quadrant, while the "left" ones will give poles in the second quadrant.

The distribution of the resultant vectors over the four quadrants gives the picture represented schematically in Fig. 2. From our analysis as well as from the drawing it is evident that the total resultant vector (summed up over all the quadrants) will lie in the first quadrant and make with the plane XY an angle that is slightly less than the limit angle of 38°9'.

§ 7. The planets, according to the proposed theory, have arisen from the fusion of a large number of meteorites. The angular momenta of their orbital motions are geometrical sums of the angular momenta of the respective meteorites. It follows, accordingly, from the above analysis that the poles of the planetary orbits must be inclined to the central plane of the Galaxy (which differs but little from the XY-plane) at angles slightly smaller than 38°9'. This conclusion of the theory is fully corroborated by the data of the above table. The position of the planetary orbits is liable to change in time owing to mutual perturbations, yet Laplace's plane, which practically coincides with the orbital plane of Jupiter, maintains invariable its position in space, and the fluctuations in the position of the other orbit planes are so insignificant as to detract nothing from the validity of our conclusion. In addition, it may be remarked that the limit angle of 38°9', equal to arc tg $\pi/4$, which we have found theoretically, depends upon no parameter and is a constant for any planetary system originated in the same way as the solar system.

Institute of Theoretical Geophysics.

Received
Academy of Sciences of the USSR. 17 April, 1946.

REFERENCES

1. O. Yu. Schmidt, C. R. Acad. Sci.URSS, Vol. 49, No. 6 (1945).
2. P. P. Parenago, Astron. Zhurnal, Vol. 16, No. 4 (1939).
3. K. Schwarzschild, Probleme der Astronomie, Seeliger- Festschrift, Berlin, 1924.

4. S. Rosseland, Astrophysik auf atomtheoretische Grundlage, Berlin, 1931.
5. S. Chandrasekhar, Astrophys. J.,Vol. 93 (1941).
6. O. Yu. Schmidt, C. R. Acad. Sci.URSS, Vol. 49, No. 1 (1945).
7. O.Yu. Schmidt, C. R. Acad. Sci. URSS, Vol. 45, No. 6 (1944).

ON THE LAW OF PLANETARY DISTANCES

O. Yu. Schmidt
Comptes Rendus (Doklady) de l'Academie des Sciences de l'URSS Vol.52, No.8, pp. 666-672 (1946)

1. The distances from planet to planet increase as we move away from the Sun. Is this variation of distance subject to mathematical law, and if it is, just what is the underlying physical reason? The question has long attracted the attention of astronomers. Well-known is "Bode's law," which was made public in 1772. In terms of the distance between the Earth and the Sun (put at unity) the distance from the Sun to Mercury is approximately 0.4, and the distances to the other planets, according to Bode's law, are expressed by the formula $0.4 + 0.3.2^n$, where n is the number of the planet (n =0 for Venus, n=1 for the Earth, and so on). In the table below the figures obtained according to Bode are compared with the actual distances:

	Bode's law	Actual distance
Mercury	0.4	0.39
Venus	0.7	0.72
Earth	1	1
Mars	1.8	1.52
	2.6	......
Jupiter	5.2	5.2
Saturn	10.0	9.54
Uranus	19.6	19.49
Neptune	38.8	30.07
Pluto	77.2	39.5

In many cases the coincidence is striking, indeed. But there are also considerable departures. We find no planet between Mars and Jupiter, though the law requires that one should be present there. The asteroids fill the gap badly, for their total mass is far less than that of any individual planet. Unsatisfactory also is the figure for Neptune, and if we refer it to Pluto in order to obtain a better coincidence, we will find it even more difficult to explain why the little Pluto should be admitted to full membership in the series, when much the more massive Neptune is excluded from it.

For close on two centuries Bode's law has continued to be a subject of discussion. Some scientists considered it a law of nature, unaccounted-for but none the less real. Others (their number appears to be stronger) looked upon it as a chance coincidence of two sequences of numbers. Recently Weizsäcker [1] has made an attempt to deduce Bode's law in a simplified form by approximately doubling the distance on transition from one planet to the next. But the premises on which his conclusion is based seem very artificial.

Nor is it all that can be said against Bode's law. Its most essential drawback is that the planets are arranged in a single row without taking account of the fact that they actually fall into two sharply different groups. It is in fact an important feature of the solar system that Jupiter and the planets farther away from the Sun are of much larger mass, compared to the nearer planets - from Mercury to Mars - not to speak of their differences between these groups. One may hardly expect that an adequate law of planetary distances can be based on the neglect of this distinction.

The author's theory of the origin of planets [2] yields a reasonable interpretation at once of the existence of two groups of planets, and of the distance relationships within each group separately.

First, we shall consider the qualitative aspect of the problem, then proceed to its quantitative treatment, and, lastly, compare the results of the theory with the known facts.

2. On the author's theory the planets have arisen from a swarm of meteorites captured once by the Sun, while it was crossing the central plane of the Galaxy. Afterwards, as a result of collisions, smaller meteorites settled on those of larger mass, thus contributing to the eventual formation of several large bodies, the planets.

Let us examine this process in greater detail. The relatively large nuclei of the future planets, which had segregated at the early stages of the process, must, on account of symmetry, have been revolving in the central plane of the swarm along circular orbits. Collision between such a nucleus and a meteorite occurs when the meteorite which may move in an elliptical as well as circular orbit happens to arrive at its node in the central plane just mentioned at the time that the nucleus is also there [3]. Adding its mass to that of the planetary nucleus, the meteorite also imparts to it its angular momentum of revolution about the Sun. Thus, the angular momentum of a planet is the sum of the angular momenta of the meteorites of which it is composed, and the position of its orbit (its distance from the Sun) is determined by the value of this total angular momentum. All the time mutual perturbations make the meteorites slightly change their orbits, and the neighbour meteorites come to fill the place of those "scooped out" by the planets.

Let us see now what is likely to happen in the natural course of events to two neighbouring planet nuclei that are in progress of growth. If close to each other, they will soon exhaust the store of meteorites that stand a chance to get between them. With no meteorites to be captured from these quarters, the nuclei, provided they do not fuse together, will further increase in mass and momentum at the expense of meteorites from outside of the exhausted interval. This means that one of the planets will now aggregate meteorites revolving nearer to the Sun, and accordingly, having smaller angular momenta in the mean, as compared to the meteorites that will add to the other planet. As a result, the angular momentum per unit mass will gradually decrease in one planet, and increase in the other, and the difference between their orbital radii will grow correspondingly. This will continue until the planet is drawn into the region where it will have to compete with its neighbour from the other side, which will exert upon it an opposite influence. It appears therefore that the planetary distances have been controlled by the mechanism of the planet growth from meteorites. A mathematical treatment of the results of this control will be given in § 4.

3. Before we proceed to it we shall dwell on the fate of the planets initiated in the neighbourhood of the Sun. In capturing meteorites such a planet had to compete not only with its neighbour farther away from the Sun, but with the Sun itself. Certainly, it was no match for the latter. The major part of the meteorites was bound to fall to the Sun and not to the planet on account of two factors. First, owing to perturbations, part of the meteorites may have assumed orbits with perihelion distances shorter than the Sun's radius, and such meteorites were destined to fall to the Sun on the next revolution. Secondly, the pressure of Sun light made the particles of matter gradually lose their orbital momentum [4] with the result that numbers of meteorites approached the Sun in a spiral, and eventually fell upon it. The magnitude of this effect (time of approach) depends on the size of the particle and its initial distance from the Sun.

It will be obvious that the influence of the two factors was particularly strong on meteorites revolving in the vicinity of the Sun. The fall of these meteorites, as will be shown elsewhere, is responsible for the rotation of the Sun on its axis. So in the region about it the Sun itself came in for by far the bigger share of the meteorites present and thus prevented their formation into bodies of respectable size. From the remains of the meteoric mass in the proximity to the Sun only small planets could arise, and the first planet whose mass corresponded to the total mass of the meteorites revolving in its domain could only be formed at a distance from the Sun, where the influence of the above-mentioned factors was so weakened as not to affect the result materially.

This is the reason why we have today two groups of planets: the so-called terrestrial planets, not very different from the Earth in size, and the major (distant) planets of much larger dimensions.

4. We are going now to derive a law of planetary distances from the theory. To begin with, let us consider the major planets, to which we shall assign numbers in order of distance from the Sun, putting $n=0$ for Jupiter.

The total angular momentum of the system rests invariable, though individual meteorites may gain or lose momentum by mutual approach. Of course, small changes are more probable than great. At this juncture the law of distribution of these probabilities is of no importance to us, and it will suffice to assume that increase and decrease in angular momentum by the same amount are equally probable.

We shall speak of domains of meteorites belonging to particular planets and of the boundaries between these domains as of definite notions. For an individual meteorite the chances are in favour of its being brought in the end on to that particular planet whose angular momentum per unit mass differs least from that which the meteorite had upon the formation of the meteorite swarm about the Sun. Together, all the meteorites whose angular momenta in this sense are nearest to the angular momentum of an n-th planet will be described as the "domain of the n-th planet." A meteorite which stands equal chances to fall to the n-th or to the $(n+1)$-th planet is said to be the boundary between the domains of these planets. As a matter of fact, all meteorites of a domain do not necessarily fall to the planet controlling this domain. Some may land on an alien planet. But as their own planet is also likely to capture some meteorites from foreign domains, there will be a tendency to equalize the balance, and we may assume, for the sake of simplification, that every planet will in time receive all the meteorites revolving in its domain, and no others. Neither shall we take account of the small angle that may possibly exist between different momentum vectors, so that they might be added arithmetically.

No further simplification is required for the mathematical deduction of the law.

Let u_n be the angular momentum per unit mass of the n-th planet and m_n the total mass of the meteorites in the respective domain. The angular momentum of the meteorite revolving at the boundary between the domains n and $n+1$ will be denoted by u'_n. The mass of an individual meteorite will be expressed as a differential dm, and its angular momentum per unit mass will be denoted by u. Then, in virtue of the law of conservation of angular momentum, we can write the following expression for the total angular momentum in the domain n

$$u_n m_n = \int_{u=u'_{n-1}}^{u=u'_n} u \; dm \qquad (1)$$

By the definition of the boundary, the angular momentum u'_n differs from u_n just as much as from u_{n+1}, i.e.,

$$u_{n+1} - u'_n = u'_n - u_n$$

$$u'_n = (u_n + u_{n+1})/2 \qquad (2)$$

The paper cited above [2] contains a deduction of the relation

$$a = \frac{\rho}{2}\,\frac{1+e}{1-e} \qquad (3)$$

which connects the semi-major axis of the orbit with its eccentricity e for every meteorite captured by the Sun. For the meaning of the quantity ρ (the limit distance at which capture occurs) the reader is referred to that paper. Here it will suffice to mention that ρ may taken to be constant in the mean throughout the meteoric swarm. In the cited paper it was shown also that e in this formula can be used with + as well as with — sign, and that all its values from —1 to + 1 are equally probable. In virtue of the latter circumstance the mass of meteorites with e ranging from e_1 to e_2 is proportional to the magnitude of this interval. If the total mass of the swarm be denoted by m, we shall have

$$dm = \frac{m}{2}\, de \qquad (4)$$

because the interval of variation of e from -1 to $+1$ equals two. For any member of the system the angular momentum per unit mass is known to be

$$k\sqrt{M}\ \sqrt{a\,(1-e^{2})} \qquad (5)$$

where M is the mass of the Sun, k^2 is the constant of gravitation. Let us so select the units

as to have $k\ \sqrt{M} = 1$. For an n -th planet moving in a circular orbit at a distance R_n from the Sun the

angular momentum per unit mass is $\sqrt{R_n}$

For the angular momentum u of a meteorite we have, by (3) and (5), the expression

$$u = \sqrt{a\,(1-e^2)} = \sqrt{\rho/2}\ (1+e) \qquad (6)$$

Denoting eccentricity of the boundary meteorite orbit by e'_n we have respectively

$$u'_n = \sqrt{\rho/2}\ (1+e'_n) \qquad (7)$$

Making use of the formulae (4) to (7), we can rewrite the equality (1) as follows

$$u_n m_n = \int_{e'_{n-1}}^{e'_n} \rho\ (1+e)\ m\ de$$

or, after integration,

$$u_n m_n = \frac{m}{2}\sqrt{\rho/2}\ \ [(1+e'_n)^2-(1+e'_{n-1})^2]/2 \qquad (8)$$

On the other hand, by virtue of (4),

$$m_n = \frac{m}{2}\ (e'_n - e'_{n-1}) \qquad (9)$$

From (8) and (9) follow

$$u_n = \sqrt{\rho/2}\ [(1+e'_n)+(1+e'_{n-1})]/2 \qquad (10)$$

Hence using [2] in order to express $u'_{n-1}$and u'_n through the angular momenta of the planets, we get

$$u_n = (u_{n-1}+u_{n+1})/2$$

And as $u_n = \sqrt{R_n}$ for the planets, therefore

$$\sqrt{R_n} = (\sqrt{R_{n-1}}+\sqrt{R_{n+1}})\ /2 \qquad (11)$$

This equality can be written also in the form

$$\sqrt{R_{n+1}} - \sqrt{R_n} = \sqrt{R_n} - \sqrt{R_{n-1}} \qquad (12)$$

To put it into words:

The difference between the square roots of their distances from the Sun is a constant for any pair of successive planets .

This theorem involves the law of planetary distances as derived from the author's theory. We can, in fact,

denote $\sqrt{R_0}$ by a, and the constant difference between the successive square roots by b, to obtain

$$\sqrt{R_n} = a + bn \qquad (13)$$

which means that *the square roots of the distances between the successive planets and the Sun form an arithmetical progression.*

This is precisely the author's law of planetary distances. We have derived it for the distant planets, i.e., for the region where the direct absorption of meteorites by the Sun, discussed in § 3, is a factor of minor importance. The following considerations will show, however, that it holds also for the nearer planets. In the course of time the Sun had absorbed the main mass of smaller particles from the regions of the nearer planets, so that only those of relatively larger size remained, and their orbits were less sensitive to the influence o light pressure. Moreover, once the meteorites with the longest orbits had fallen to the Sun, the remaining orbits, being more circular, were also less liable to be affected by perturbations. Therefore the action of the two factors mentioned in § 3 was growing weaker as time went on, and eventually the matter in the regions lying nearer to the Sun. On this consideration we may expect the above theorem and the law of planetary distances, as expressed by formula (13), to hold for the nearer planets as well, though, of course, with modified coefficients a and b.

5. Let us compare our conclusions with the actual data (the values of R are given in astronomical units).

	Jupiter	Saturn	Uranus	Neptune	Pluto
$\sqrt{R}$ theoret.	2. 28	3. 28	4. 28	5. 28	6. 28
$\sqrt{R}$ Actual	2. 28	3. 09	4. 38	5. 48	6. 29
R theoret.	5. 20	10. 76	18.32	27.88	39.44
R Actual	5. 20	9. 54	19. 19	30. 07	39. 52
Departure	0	+13%	-5%	-7%	0

From the figures of the last row the square root differences appear to be strictly constant. Yet they fluctuate within a rather narrow range about a mean value equal to 1.00. We may look upon this coincidence as satisfactory, for the law only expresses the average tendency in the action of millions of meteorite falls, the process which has not even come to an end as yet.

We shall now compare the law $\sqrt{R_n} = a + bn$ with the actual data. For a we take actual value

of $\sqrt{R_0}$ for the first planet of the series (Jupiter), and for

b the mean value of the differences $\sqrt{R_{n+1}} - \sqrt{R_n}$, i.e., 1.00.

	Mercury	Venus	Earth	Mars
$\sqrt{R}$ theoret..	0. 62	0. 82	1. 02	1. 22
$\sqrt{R}$ Actual	0. 62	0. 85	1. 00	1. 23
R theoret..	0. 39	0. 67	1. 04	1. 49
R Actual	0. 39	0. 72	1. 00	1. 52
Departure	0	-7%	+4%	-2%

Here, in contrast to Bode's law, Neptune and Pluto comply with the general rule

Let us turn to the nearer planets. For them the actual differences $\sqrt{R_{n+1}} - \sqrt{R_n}$ are

0.23 0.15 0.23

the mean value being 0.20. The actual and theoretical figures are brought together in Table 2.

	Jupiter	Saturn	Uranus	Neptune	Pluto
R	5. 20	9. 54	19. 19	30. 07	39. 52
$\sqrt{R}$	2. 28	3. 09	4. 38	5. 48	6. 29
$\sqrt{R_{n+1}}-\sqrt{R_{n-1}}$		0. 81	1. 29	1. 10	0. 81

Institute of Theoretical Geophysics.
Academy of Sciences of the USSR.

Received
29 April 1946.

REFERENCES

1. C. F. Weizsäcker, Z. f. Astrophysik, Vol.22 (1944).
2. O.Yu. Schmidt, C. R. Acad. Sci. URSS, Vol. 45, No.6 (1944).
3. O. Yu. Schmidt, ibid., Vol. 46, No.9(1945).
4. H.P. Robertson, Mont. Not. Roy. Astron. Soc., Vol.97, No.6 (1937).

ON THE ORIGIN OF SUN'S ROTATION

O.J.Schmidt
Doklady Akademii Nauk SSSR **54**, pp.15-18 (1946)

1. The Sun moves round on its axis rather slowly. Its period of rotation is 24-25 days in the region of the equator, and increases slightly with the latitude. The cause responsible for solar rotation has not been disclosed so far.

Some stars rotate with notable speed. From the evidence furnished by the broadening of spectral lines, G.A. Shajn, O. Struve and others[1-2] have assigned rotational velocities of 250 km./sec to certain stars in the spectral classes O, A, and B. Most frequent among the stars of these classes are rotational velocities of the order of 60 km./sec. On the other hand no star of the G class, where our Sun belongs too, shows any evidence of rotation. So far as the broadening of the lines in the spectrum is concerned, and the rotational velocity of the Sun is only about km./sec. Therefore, whatever may be the cause that lies behind the rotation of the stars in the O-B classes there is no sufficient ground to suppose that the same cause might also be responsible for the rotation of the Sun, or conversely.

Because the plane of the Sun's equator closely approaches the plane of the ecliptic(making with it an angle of $7^{\circ}10'.5$), many authors of cosmogonic theories were inclined to believe that the rotation of the Sun was somehow connected with the origin of the planets. In the cosmogonies of Kant and Laplace the Sun and the planets are supposed to have been parts of a single rotating nebula. Against this assumption, however, incontrovertibly speaks the fact that whereas the mass of the solar system is largely concentrated in the Sun, the share of the Sun in the total angular momentum of the system is quite insignificant(about 2 percent). On the tidal theories the planets are supposed to have been formed from the matter torn off from the Sun by some star while it was moving past it. Part of this matter might have fallen back to the Sun. The small rotational momentum of the Sun is no obstacle to these theories, but the large rotational momentum of the planets has proved an insuperable barrier.

On the theory suggested by the present author[2], the Sun at one time, while crossing the central plane of the Galaxy, met there with a cloud of obscuring matter--cosmic dust and meteorites. In this encounter a portion of the cloud was captured by the Sun and compelled to revolve about it in the form of a meteoric swarm. Within this swarm a process of aggregation set in, the smaller particles fusing with the larger ones by attraction. In this way in the course of time the planets came into existence. Some meteorites had escaped this process and continued either in the individual state, or as small accumulations held together by mutual gravitation (comets). Others, again, had fallen to the Sun and imparted to it their angular momentum. Computations based on this theory (see below) show that both the direction and the magnitude of the rotational momentum of the Sun can be explained by the fall of these meteorites. So the Sun's rotation appears to be a side-effect of the process which has led to the formation of the planetary system.

2. In deducing the law of planetary distances [4] we discussed the rôle of two mechanisms in bringing part of the meteorites to the Sun. First, as a result of mutual perturbations some meteorites would assume orbits whose perihelion distances were shorter than the radius of the Sun. Secondly, there was the influence of light pressure. Robertson [5] has shown theoretically how the sunlight falling on a body of small but finite size should make it lose a portion of its momentum with the consequent contraction of its orbit until at last it is drawn on to the Sun.

An essential point in our problem is the difference in the operation of these two mechanisms. In either case the Sun acquires the mass of the meteorites, but while it also acquires its momentum in the first case there is a considerable loss of momentum in the second. The effect of light pressure is stronger, the smaller the size of the particle (down to a certain limit) and the closer it is to the Sun. If the ratio of small particles be large enough, the fall by light pressure will so greatly predominate over that by mutual perturbation that the latter effect may be neglected in the computations.

3. Let us now establish the theoretical position of the pole of the Sun's equator. We shall use galactic coordinates. In one of the earlier works [6] we have discussed the theoretical distribution of orbit poles in the meteoric swarm, from which may be derived the position of the poles of planetary orbits. We have found that on addition of meteorites the resulting momentum should have definite galactic longitude--along the direction of the Sun's motion at the moment of capture. As for the galactic latitude of the resulting momentum, it must depend on the proportion of meteorites in direct and retrograde motion. Supposing all the meteorites to have been in direct motion at the time of capture, the galactic latitude of the resulting momentum should be $+38^{\circ}9'$ (or arctan $\pi/4$). On the other hand, it should be $-38^{\circ}9'$ if the motion of the meteorites had been entirely retrograde. As direct motions vastly predominate over retrograde motions, the momenta of the planets should have positive latitude, somewhat below $38^{\circ}9'$. In actual fact the poles of the planetary orbits (with the exception of Pluto's) range in galactic longitude within the narrow limits of $58^{\circ}.4$ to $64^{\circ}.6$ (see the table in the paper cited) [6] and the galactic latitudes of the orbit poles from the Earth to Neptune lie within $27^{\circ}.2$-- $29^{\circ}.5$.

The same must be true of the Sun. In fact, the galactic coordinates of the pole of the Sun's equator are

$$l = 64^{\circ}.5, \quad b = 22^{\circ}.5$$

The longitude is seen to come within the range of planetary longitudes and the latitude satisfies the requirement of the theory not to exceed the limit angle $38^\circ 9'$. Yet it is considerably lower than the latitude of most planetary orbit poles. As the latitude for Venus is $26^\circ.1$ and for Mercury, $24^\circ.0$ one may think of a regular decrease of latitude from the Earth through Venus and Mercury down to the Solar equator. We shall presently see that this phenomenon fits within the frame of our theory.

Within the swarm the meteorites are liable to collide or undergo mutual perturbation of orbits. In direct collisions, if one of the colliding meteorites is in direct motion and the other in retrograde motion, both will lose their momenta. In the case of perturbation there will be a decrease of momentum in either meteorite. "Retrograde" meteorites, as the less numerous, will more often encounter "direct" meteorites than their own kind, and so they will gradually lose momentum and approach the Sun. Together with them their equal number of "direct" meteorites will also be drawn to the Sun,. This means that the relative number of "retrograde" meteorites will steadily decrease in the regions far away from the Sun, and increase correspondingly in the nearer regions. It has been made clear above that increase in the proportion of "retrograde" meteorites is bound to bring on a decrease in the latitude of the resulting momentum, which should be more marked, the shorter the distance between the Sun and the growing planet, and must be most pronounced for the Sun itself. This indeed corresponds to the actual state of things.

4. We are now to make an approximate estimate of the mass and momentum which the Sun has received from meteorites. For the sake of simplicity the meteorites will be supposed to move in one plane. In the paper on the law of planetary distances [4] it was proved that each planet receives in the main such meteorites whose angular momenta per unit mass lie within a definite domain. The boundary between the domains of two neighboring planets is marked by the arithmetical means of the angular momenta (per unit mass) of these planets. On our theory [4] a meteorite's angular momentum per unit mass is expressed, in units suitably chosen, by the formula

$$\sqrt{\rho/2}\,(1+e)$$

where e is eccentricity of the meteorite orbit and may be positive as well as negative. The value of ρ which is taken to be constant for the solar system has been estimated in [3] at $\rho \approx 14$ astronomical units.

To simplify, we suppose that the meteorites that have landed on the Sun or been absorbed in the formation of the nearer planets are all those whose momenta were lower than the boundary momentum of the domain of Jupiter. The value of the latter can be derived from the available observational data. In fact, the orbital radii of Jupiter and Saturn are approximately equal to 5.20 and 9.54 astronomical units, respectively. The half sum of their square roots is 2.69. Therefore for the upper boundary meteorite of the domain of Jupiter we have, according to [1]

$$\sqrt{7}(1+\varepsilon_1) = 2.69$$

Hence $\varepsilon_1=0$. On the same theory Jupiter's angular momentum per unit mass, i.e.,$\sqrt{5.2} = 2.28$, should in turn be the arithmetical mean of the angular momenta of the boundary meteorites. Denoting by ε the eccentricity of the lower boundary we get

$$\sqrt{7}(1+\varepsilon) = 1.87$$

Hence $\varepsilon = -0.29$.

According to our theory, the total mass of the meteorites with orbit eccentricities lying between two values e and e' is proportional to their difference $e'-e$. The mass of Jupiter is therefore proportional to 0.29, and the total mass of the meteorites absorbed by the Sun and the nearer planets (the asteroids inclusive) is proportional to the difference between the numbers -0.29 and -1 (since the lower boundary of all the eccentricities is -1), or to 0.71. So the mass of these meteorites is to the mass of the Jupiter as 7 to 3, or, in other words, is about 740 times as large as the mass of the Earth.

Using the formulae derived on our theory one will easily find that the total angular momentum of the meteoric mass under consideration is approximately 690 times the angular momentum of the Earth. As a matter of fact, the Sun has received but a small portion of it because of the loss occasioned by light pressure.

The angular momentum actually imparted to the Sun can be estimated by the following method, suggested by B.J. Levin. Under the influence of the pressure exerted by the solar radiation a meteorite will be gradually losing momentum and, its orbit growing continuously shorter, it will be moving towards the Sun in a spiral until finally it will touch the Sun and merge in it. At the point of contact with the Sun's surface its angular momentum per unit mass will be equal to the one it would have on a circular orbit of a radius equal to the radius of the Sun (about 1/220 astronomical units). As angular momentum is proportional to the square root of the orbit radius, therefore the angular momentum of our meteorite will be $\sqrt{220} \approx 15$ times as small as the angular momentum of the Earth per unit mass. A mass 740 times the Earth's mass would impart the Sun a momentum equal to 740 : 15 = 50 times the Earth's momentum.

For the actual angular momentum of the Sun no exact figures are available. From observations of the rotational velocity at the equator it might be taken to be 40 times as great as the angular momentum of he Earth, provided we assume the density of the Sun to be constant. If the latter should vary with the depth according to the generally adopted law, the Sun's angular momentum would be only 20 times the Earth's one. On the other hand, its value would be higher if, as some authors believe, the angular velocity of the Sun increased with depth.

The value obtained on our theory is 50. So the coincidence in order of magnitude is satisfactory.

It appears that the theory advanced by the author is capable to account both for the direction and magnitude of the Sun's rotational momentum. Furthermore it gives the reason for the gradual decrease of orbit poles in galactic latitude-- from the Earth to Venus, from Venus to Mercury, and finally, to the Sun.

Institute of Theoretical Geophysics
Academy of Sciences of the USSR.

Received
17th August 1946.

REFERENCES

1. G. Shajn and O. Struve, Monthly Notices of the Royal Astronomical Society 89, No. 3 (1929).
2. O. Struve, Astronomical Journal, 72, No. 1 (1930).
3. O.J. Schmidt, Comptes Rendus Acad. Sci. URSS, 45, No. 6 (1944).
4. O.J. Schmidt, ibid., 52, No. 3 (1946).
5. H.P. Robertson, Monthly Notices of the Royal Astronomical Society, 97, No. 6 (1937).
6. O.J. Schmidt, Comptes Rendus Acad. Sci. URSS, 52, No. 7 (1946).

ON THE POSSIBILITY OF A CAPTURE IN CELESTIAL MECHANICS

O. Yu. Schmidt
Doklady Akademii Nauk SSSR, pp. 213-216 (1947)

1. The possibility of a capture of one heavenly body by another (or mutual capture of two bodies) has been repeatedly used as a mechanism for explaining different phenomena of cosmogony; but it has usually been met with sharp objections. These objections had either shown the impossibility of such capture or its utter improbability. Leaving the question of estimation of the probability of a capture till the next paper, let us now busy ourselves with an elucidation of its very possibility.

2. By a "capture" is usually understood the following phenomenon: two heavenly bodies, which initially had been moving independently of each other, at some anterior time, under the influence of mutual attraction and various other contributing causes, left their initial orbits and, from that time on, began to rotate about their mutual center of gravity following elliptical (or perturbed elliptical) orbits. It is possible to give still another definition of a capture. We say that there is a capture if, up to the moment t_0, the distance between the bodies had been always greater than a certain distance ρ and, after the moment $t_0 + \tau$, the value of the distance will always remain less than ρ. This definition of a capture is, so to speak, eternal. And, of course, it may be weakened for the case of a "temporary" capture.

In the case of the Newtonian attraction, the capture is impossible for two mutually gravitating bodies. For the capture to succeed, it is necessary to have at least three bodies. Thus, the question of a capture involves into the famous problem of three bodies. In this field, the most detailed investigation belongs to Chazy [1] who proposes to find the motion when the time increases to infinity (*allure finale*). Chazy cites three classes of these "final" states of the system: (1) the motions are limited when all three bodies remain in a bounded region near their common center of gravity; and oscillatory when the bodies alternately approach one another and now recede, not having any upper limit for their distances, but, at the same time, not receding to infinity; (2) the motions are hyperbolic-elliptic when, as the time increases to infinity one of the bodies recedes to infinity and the other two remain at a finite distance and, thus their relative motion, in the limiting case, becomes elliptical. (3) The motions are hyperbolic when, in time, all three bodies recede to infinity.

In Chazy's investigations, an important role is played by the sign of the constant h in the energy integral when the motion is referred to the center of gravity of the system. When $h < 0$, the first two of the above mentioned classes of motion are possible; when $h > 0$, the second and third classes are possible. For the case of $h < 0$, Chazy proved an important theorem in 1929 [1]:

If in the distant past ($t \to -\infty$), the motion belonged to one of the above mentioned classes, then in the limiting future time ($t \to +\infty$) it will also belong to the same class. This result received great publicity. Sometimes it is considered to be a proof of the impossibility of a capture. This conclusion is untenable because it forgets two things: first, that Chazy's theorem is valid only for the case of $h < 0$, and second, that the approach of the heavenly bodies from infinity, when they follow independent paths, presupposes $h > 0$. True, in 1932 Chazy [2] investigated this last case and came to the same conclusion, but he did not give a rigorous proof of it, as he did in his first memoir, merely limiting himself to reasoning by analogy. Below will be given an example refuting Chazy's statement for the case of $h > 0$, namely, an example of a motion which was "hyperbolic" in the past and becomes "hyperbolic-elliptic" in the future.

3. The phenomenon of capture lies at the basis of my theory of the origin of the planetary systems and double stars [7, 8]. Therefore, I must prove, before anything else: the possibility of a capture in principle. For this, it will be sufficient to find in the three body problem even one example leading to a capture. There are no such examples in the scientific literature.

Becker [3] computed several orbits of the characteristic "exchange" when a star, passing by a double star, tears away one of the pair and substitutes itself in its stead. But this case does not solve our problem.

When three heavenly bodies come from infinity, it is very difficult to make a correct judicious guess concerning the course of their encounter and the possibility of a capture because this judgment would be based upon insufficient evidence - the bodies' known velocities and their positions at a certain moment of time. It is practically impossible to match such a set of conditions. But the problem will be considerably simplified if we investigate, simultaneously with capture, a case of breaking apart of a pair of bodies by an action of a third body. The rupture of a double star has been attracting the attention of a number of investigators: Öpik [4], Lyttleton [5], Chandrasekhar [6]. But all these authors neglected one important circumstance, viz, that every case of rupture is in one-to-one correspondence with a case of capture. Indeed, as is well known, the equation of celestial mechanics allows a change in sign in the time variable. If at the initial time one system has the coordinates x_i, and the velocities v_i, and after a time lapse τ this system comes to position with the coordinates x_{i2} and velocities v_{i2}, then another system, which begins its motion with the same coordinates x_{i2}, but with the opposite velocities $-v_{i2}$, and arrives, after the time τ, at the same position x_i, in which, in the beginning, was the first system, then its velocities will be exactly equal to $-v_{i1}$. For example, if two stars originally were rotating around each other in an elliptic path,and then a third star passing by, ruptured the system in such a manner that, after a certain time, all three stars will be found going in different directions with relative hyperbolic velocities, then, by fixing the positions of the stars at one of such consequent moments, and by assigning to them velocities of the same magnitude as in the first case, but in a reverse sense (like a kino-film reeled in the opposite direction) and, having started with a mutually - hyperbolic motion of three stars we will finish with an elliptic motion for the two of them, that is, with the formation of a double star via capture.

Selection of the conditions necessary for rupture of a double star is much easier . In fact, the very first numerical example offered by me gave a clear picture of the rupture of a double star. Viewed in reverse, it gave, therefore, an example of formation of a double star via capture. My computations were checked by O. P. Kramer and V. I. Kovrigin, under the general leadership of N. N. Paritsky *. At present, the author is engaged in a new series of computations for the next paper.

4. Here we are going to discuss a motion of three bodies of the same mass, equal to the mass of the Sun. This mass is taken as the unit of mass, the astronomical unit (A.U.) is taken for the unit of distance, and $(1/2)\pi$ of a year is chosen for the unit of time. Under these conditions, the gravitational constant is equal to 1, which simplifies the calculations. Here we are going to deal with a relative motion with respect to one of three bodies to which we assign the coordinates of the origin 0 of our coordinate system. The equation of relative motion of bodies with the coordinates x_1y_1 and x_2y_2 have the form:

$$x_1'' = [-2x_1/(r_{10})^3] - [x_1/(r_{12})^3] + [x_2/(r12)^3] - [x_2/(r_{20})^3]$$

$$y'' = [-2y_1/(r_{10})^3 - [y_1/(r_{12})^3] + [y_2/(r_{12})^3] - [y_2/(r_{20})^3]$$

and an analogous pair of equations for the x_2'' and y_2''. Denotations of the distances between the bodies are self-evident. For the sake of simplicity, the case of a plane motion is taken. The initial data for the time $t = 0$ were such that the unperturbed orbit of the body A, under attraction of the body 0, would be an ellipse with the major semi-axes equal to 200 A.U. (which corresponds to a period of 2000 years) and with an eccentricity equal to 1/2; the unperturbed orbit of the body B - a hyperbola (for its initial data consult Table 1). From this point of departure: the positions and velocities of the relative motion, forward ($t>0$) as well as backward ($t < 0$), were found by the ordinary means. In Fig. 1, the trajectories of motion are given. In Fig.2, on a larger scale, is given that part of the trajectory of the body A where the rupture of the pair happened. The plotted points correspond to those moments of time for which positions and velocities were computed. In the Table 1, the initial data ($t = 0$) are given, and for the sake of illustration, some results of the computation are given also.

We see that with increasing t, there was an actual breaking apart of the double star. Both the figure and the simple reasoning convince us that after the rupture the motion remains hyperbolic with respect to the body 0. From our equations we see that as soon as r_{12} and r_{20} become greater than r_{10}, the acceleration of the body A becomes and remains less than the value of $5/(r_{10})^2$ that is less than the acceleration corresponding to attraction by a single body placed in the origin 0 and having a quadruple mass. From Table 1 and 2, one can see that the velocity of the body A given in the last lines is considerably greater than would be a parabolic velocity taken for the sake of comparison. Thus, the body A remains on a hyperbolic path. Reverse motion gives us a capture with formation of a double star and an elliptic orbit.

* I express deep gratitude to N. N. Paritsky and to all participants in this project.

The example discussed is no exception. It is clear that small changes in the initial data, including a three-dimensional case, will give us only small changes in the region of rupture (capture). Many initial data leading to a capture have positive values, not zero.

5. Thus, a capture in a three body problem is possible as well as a rupture of a double star, and, in fact, this capture (rupture) will not be temporary, but permanent. The motion of a system of three bodies, hyperbolic-elliptic in the past, may become hyperbolic in the future, and vice versa.

The probability of a capture, though small, is, nevertheless, positive, not zero. Therefore considering that our galaxy is made up of approximately 10^{11} stars, and that at least several billion years have passed since its birth, one can be reasonably sure that in the Galaxy there are cases where this possibility had been realized, that is, at least some portion of the visual double stars was formed via capture. This is not a mere hypothesis now; it is a fact. From this, however, we cannot draw an immediate conclusion that all visual double stars were formed this way. But we can build a theory for those double stars which have their origin in capture, that is we can derive those laws and formulas which the eccentricities, the lengths of axes and the inclination of orbits to ecliptic must obey, and then compare these formulas with the observational results. This was done in my works [7]; and it was found that theoretical and empirical laws are in close agreement.

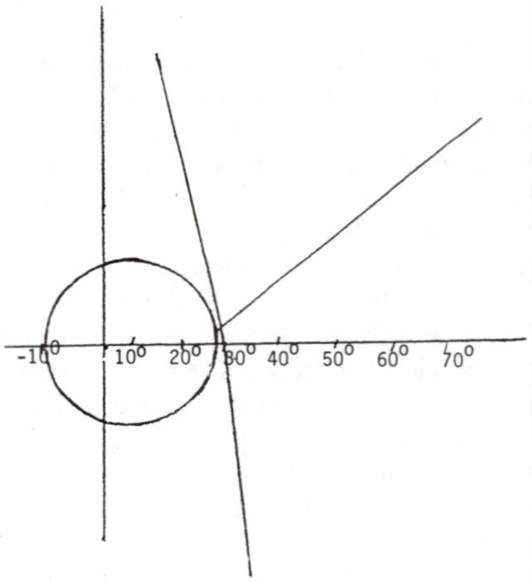

Figure 1

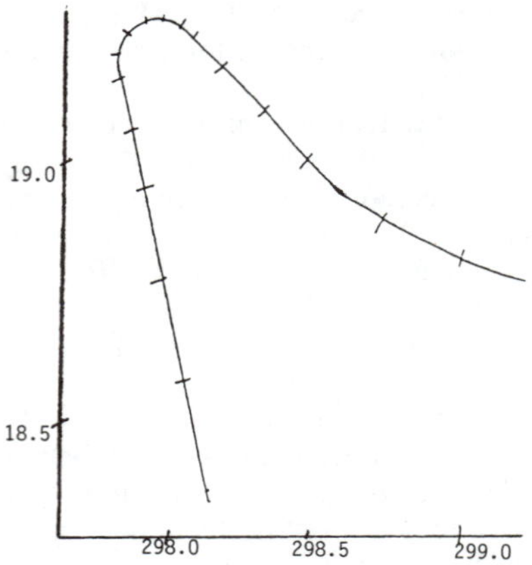

Figure 2

Table 1

t	x_1	x_1'	y_1	y_1'
-5824	-84.69	-0.072	67.38	-0.148
0	291.500	0.0195	-49.958	0.0561
1252	297.963	0.0041	19.315	0.0096
1254	297.974	0.0071	19.331	0.0053
1256	297.992	0.0110	19.337	0.0004
129764	7004.0	0.128	10975	0.084

Table 2

t	x_2	x_2'	y_2	y_2'
-5824	187.77	-0.045	-6529.7	0.851
0	320.00	0.000	-1200.00	0.9549
1252	309.232	-0.0363	2.374	1.0110
1254	309.156	-0.0394	4.400	1.0153
1256	309.073	-0.0434	6.435	1.02022
129764	-28636	-0.226	116430	0.905

Thus, not a small portion, but all or an overwhelming majority of the visual double stars had their origin in capture.

Geophysical Institute of the USSR Academy of Sciences
Received August 8, 1947.

REFERENCES

1. J. Chazy. Jour, de Math., vol. 8 (1929).
2. J. Chazy. Bull. Astronomy, vol. 8 91932).
3. L. Becker. Monthly Notices, Roy. Astr. Soc.,vol. 80, No. 6 (1920).
4. E. Opik. Proc. Am. Acad. of Arts and Sci. vol. 67, No. 6 (1932).
5. R. A. Lyttleton. Monthly Notices. vol. 96, No. 6 (1936).
6. S. Chandrasekhar. Aph. J. vol. 99, No. 1 (1944).
7. O. Yu. Schmidt. Reports of the USSR Academy of Sci. vol. 44, No.1 (1944); vol. 49, No. 16 (1945).
8. O. Yu. Schmidt. Reports USSR Academy of Sciences.vol. 45, No. 245 (1944); vol. 46, No. 9 (1945); vol. 49, No. 413 (1945); vol. 52, No. 581 (1946); vol. 52, No. 673 (1946); vol. 54, No. 15 (1946).

THE PROBLEM OF THE ORIGIN OF THE EARTH AND PLANETS

O. Yu. Schmidt

Voprosy Filosofii, No. 4, pp. 120-33 (1951)

PLANETARY COSMOGONY IN THE USSR AND THE CAPITALIST COUNTRIES

Our conference bringing together astronomers, physicists, geologists, and geophysicists will help us to answer the question that inspires the interest of the broadest masses of our people, that is, how the Earth originated. Let us begin with a brief review of the historical development of this problem and its current state in the USSR and in the science of the capitalist countries. Scientific cosmogony originated from Kant and Laplace. Their great historic achievement, transcending the limits of cosmogony, is that they, while dealing with the problem of the origin of the Solar system, for the first time in the history of natural sciences consciously used the principle of development under the influence of natural forces as the basis for their theories, without appealing to the idea of divine creation. The hypotheses by Kant and Laplace were not purely speculative. They were based upon then-established facts and managed to explain satisfactorily a considerable number of the features of the planetary system.

Besides the very concept of natural development, the theories' valuable contribution to scientific thinking consisted in the initial presupposition shared by both namely in the idea that the primary substance from which the planets have been subsequently formed had existed in a scattered state, either as something like a gas (Laplace), or as some vaguely specified 'particles' which could be interpreted as gas as well as dust or even more sizable solid bodies (Kant). We should not discard this legacy. It is well known that the broadly used term 'meteoric hypothesis' refers to the scattered solid phase of the primary matter, but this does not mean that this substance was formed by the same meteorites which can be observed in our museums.

Works of genius that they were, Kant's and Laplace's theoretical constructions were restricted historically by the level of eighteenth century science. Not only was the variety of established facts negligible in comparison with modern knowledge, but some very important components of theoretical sciences were lacking as well. The energy conservation law, and the regularities of energy transformations without which modern cosmology cannot be imagined, had not become parts of scientific knowledge yet. Thermodynamics and statistical physics did not exist either, thus Kant and Laplace turned out to be more or less helpless when dealing with their 'particles.'

The nineteenth century was not a creative period for cosmogony. The Laplace hypothesis was completely dominant. All scientific works on the subject of cosmogony bore some marks of following its general direction. There were virtually no attempts to reevaluate the Laplace theory, and even those objections which were voiced (for example, the noted discrepancy between the spatial distribution of mass and angular momentum which the traditional cosmogony was not able to explain) were generally ignored. Some attempts to develop new hypotheses based on the approach by Kant and Laplace, such as Ligondes were of no importance for the progress of science.

Thus, whereas the nineteenth century witnessed a powerful development of physics, chemistry, biology, geology, etc., cosmogony suffered from stagnation. One may note only isolated positive achievements of Western European science pertaining to some mechanisms of cosmogonic relevance, such as the well known "Roche's limit" and especially G. Darwin's theory of tidal forces and tidal friction. These results, although sometimes used uncritically, are still valid.

The geologists and geophysicists of the nineteenth century accepted the Kant-Laplace hypothesis without any reservations, including the notion of the initially-fiery liquid and then gradually-cooling Earth suggested by volcanic phenomena. In view of this fact it is especially important to emphasize that in Russia the opinion of a famous astronomer was voiced who tried to explain volcanism not by the cooling of the initially hot Earth but by its heating produced by internal factors. That was F. A. Bredikhin's voice.

The state of cosmogony by the beginning of the twentieth century was summarized in H. Poincaré's well known book. Poincaré paid attention to almost every hypothesis known at the time, although his personal preferences were with the Laplace theory since, in his own words, the old lady still looked more hale than her younger rivals despite all her wrinkles. Poincaré commented upon different cosmogonic theories, but he was interested in them only as starting points to formulate and solve purely mathematical problems. His book displays clearly the author's idealistic world view denounced by Lenin.

The defects of the Kant-Laplace hypothesis had become quite apparent by the beginning of the twentieth century. There were new ideas, such as See's and Moulton-Chamberlin's, and then the twenty year reign of Jeans began. His was a genuine monopoly. No other hypotheses were suggested during the period of domination of the Jeans hypothesis.

The reason for the unique popularity of Jeans' hypothesis was not its scientific merits (it had none) and equally not the obvious talents of its author, but rather the fact that it proved to look the most acceptable to the religiously-idealistic world views reigning in bourgeois society. After the undisguisedly materialistic and "storming-the-skies" creations by Kant and Laplace there were numerous attempts to construct a quasi-scientific picture which would not be too transparently at odds with the biblical legend about the beginning of the world.

In particular, such was the intent behind Faye's hypothesis. However, such efforts could not be successful since science had already gone too far in its development. Thus the church was ready, even if not to support openly, at least to bless silently a theory that, although not identical with the biblical mythology, was still compatible with the keystone of the Christian religion, that is, with the idea of the unique position of the Earth. The same feature of the Jeans theory which discredited it in the eyes of the astronomers, precisely the negligible probability of the process of planetary formation, became its principal merit in the opinion of the laymen unwilling to reject religion. Jeans' hypothesis proved to be the most acceptable compromise. To be sure, the very rarity of the formation of planets in his scheme is not in itself idealism (after all, rare events do exist in nature), but it did open the door to idealism in cosmogony. That was the reason for such unanimous support of Jeans' hypothesis by bourgeois media and schools. In the U.S.S.R. it was subjected to systematic criticism.

After the downfall of the Jeans hypothesis brought about by Russell's objections (1935), and especially after N.N. Pariisky's decisive blow (1943), planetary cosmogony began to undergo considerable revival. During the 1940's the scientists of the capitalist countries suggested more than ten new hypotheses. But this revival does not display the genuine growth of scientific knowledge. The majority of these new hypotheses were sporadic, immature, and short-lived.

It is very characteristic that amongst them one can find such products of the epoch as the Milne-Haldane hypothesis explaining the origin of the planetary system in terms of the blow allegedly produced by only one fantastically energetic quantum, and F. Hoyle's hypothesis according to which the Sun had been the part of a binary star system whose second component turned into supernova and exploded. Hoyle now teaches the continual creation of new matter from nothing.

Some authors of critical reviews of the current state of cosmogony in the West (Jeffreys, D. ter Haar, W. J. Luyten) have arrived at a negative assessments of all existing cosmogonical theories.

But what are the reasons for the recent situation when the theoretical thought of the cosmogonists in the capitalist countries is struggling helplessly in a blind alley despite all their richly equipped observatories and remarkable achievements in collecting novel facts ? The explanation lies in the general crisis of bourgeois science which is still able to produce positive results within some localized areas but already fails to embrace the most significant and fundamental problems which are especially influenced by its philosophical and consequently methodological shortcomings. These negative features are highly visible in the sphere of cosmogony.

And indeed, whereas the materialist who is searching for the objective truth sees his obligation in systematic, consistent and also quantitative development of his hypothesis, the idealistically minded scientist is satisfied if he manages to produce a "generalization" of the available factual data, even if this generalization is nothing but an impressive-looking vague qualitative scheme. Whereas materialistically thinking scientists see their duty in making careful comparisons between their theories and facts and cannot regard the theories as completed until they can deal with even isolated factual counterclaims, we observe the idealists to treat contradictions with startling indifference. One idealist generalizes somehow from a certain collection of those much praised "factual data" whereas another chooses quite a different factual basis for his generalizations, and they are equally unconcerned about any contradictions since they do not believe in the existence of.objective truth anyway. This results in their inclination toward a superficial elegance, in their careless calculations, even in the absence of elementary logic in their reasoning.

The above-mentioned hypotheses display all these features especially clearly. Indeed, Luyten has good reasons to state that such hypotheses should not be taken too seriously. However, the same features are present in many superficially more sound hypotheses, although there they are not immediately evident.

Such is the hypothesis suggested by the German physicist Carl Von Weizsäcker (1941). It is safe to say that this hypothesis is the most fashionable nowadays in the West. Chandrasekhar, ter Haar, and some other scientists have adopted it and tried to develop it further. There is a positive feature in Weizsäcker's and his followers' works since they use thermodynamics and statistical physics on a larger scale than the authors of earlier cosmogonic theories. But nevertheless Weizsäcker's gnoseological presuppositions manifest themselves immediately in the very first basic point of his hypothesis: in order to explain the origin of the planets and the regularities in their distances from the Sun, he assumes in advance that there had existed a cloud with vortical regions whose number he believes to have been precisely equal to the number of the large planets of our planetary system. These vortices were rotating clockwise, but they supposedly gave birth to something like ballbearings which revolved in the opposite direction, and the planets allegedly originated from those ballbearings. The very artificial character of this hypothesis is immediately apparent. Moreover, Weizsäcker does not conceal his idealistic philosophy.

However, it would be wrong to suppose that all cosmogonic theories developed in the bourgeois countries are nothing but pure idealism. Even bourgeois scientists are being permanently pushed by the development of natural sciences towards materialistic conclusions, even if against their own wishes. Therefore one can find some isolated valid ideas even in the most modern cosmogonic hypotheses. Thus, for example, there are positive aspects in the ideas proposed by Alfvén (Sweden), Urey and Kuiper (America), Edgeworth (England), as well as in the afore-mentioned critical reviews. But all these grains of scientifically-progressive truth have to be searched for and elicited from the heap of contradictory and mostly idealistic constructions.

Let us list some positive aspects of the most recent development of cosmogony in the bourgeois countries.

The details of the evolution of gaseous as well as dust-like or mixed initial states have been examined. The specific mechanism of the accretion of particles has become of interest. Berlage, Chandrasekhar, and some other researchers, especially Von Weizsäcker, have been using the apparatus of thermodynamics and statistical physics in trying to solve this problem. Also, those principal features of the planetary system which have to be explained on a cosmogonical basis have been defined much more precisely than in the preceding epoch (ter Haar and some other scientists). Unlike the conceptions of the nineteenth century, nowadays considerable attention has been paid to the relative distances between planetary orbits, and the fundamental significance of the distribution of angular momentum has been stressed. However, the authors of almost every hypothesis openly admit that they cannot explain this distribution, and Edgeworth utilizes the theory of the capture of interstellar dust-like matter several years after this hypothesis had been developed in the USSR and he does not provide any proofs of the possibility of such a capture.

The understanding of the necessity of establishing close connections not only between planetary cosmogony and astrophysics but also between cosmogony and geosciences (geophysics, geology, geochemistry) has also increased recently. This trend became visible at the meeting which took place at Santa Fe in 1950 where the representatives of all these sciences participated. Geologists and geochemists noted there the considerable probability of a cool early Earth on the basis of geological and geochemical data. Here in the USSR this idea was being suggested persistently by Academician Vernadsky long ago. The examination of the accretion (freezing) of gaseous matter on specks of dust, which appears to be of fundamental importance for cosmogony, has been begun as well.

Naturally, all these results of the cosmogonic studies undertaken by Western scientists should be taken into consideration by Soviet specialists in cosmogony, and those which prove to be useful should be utilized. However, Soviet cosmogony has not followed the road of the imitation of the science of the capitalist countries. It has found its own way which has a number of clearly recognizable individual features.

In the USSR, planetary cosmogony has been developing together with the overall growth of our Soviet science and the victorious march of the cultural revolution. The scientific explanation of the origin of the Earth is no longer the concern of only a small group of scientists in our country. This problem, being a keystone of materialistic philosophy, inspires the intensive interest of the broadest masses. The Soviet people, awakened by the October Revolution and deeply imbued with the creative teaching of Marxism-Leninism, do not want to tolerate any ambiguities in this question and demand that their scientists provide a clear and definite solution. This implies the great importance of the popularization of cosmogonic knowledge. Academician Fesenkov should be especially commended for his efforts undertaken since the 1920s to raise and popularize cosmogonic problems.

One may also note another characteristic feature of Soviet cosmogony, that is, its rigorous precision. A good study directed to the detailed criticism of cosmogonic studies in the capitalist countries (predominantly of their logical and mathematical aspects) was made at the Sector of Cosmogony of the P.K. Schternberg State Astronomical Institute under the direction of N.D. Moiseev in the 1930s. One can only express regret that this sector's life was so short. N.F. Rein's scrupulous critical work should be noted. Slightly later, N.N. Pariisky once and for all refuted the Jeans hypothesis by his well known critical analysis. He has also demonstrated that tidal friction was unable to move planets from the vicinity of the Sun to their recent positions during any plausible time period. Thus, one more harmful illusion has been destroyed.

Before the October Revolution scientists in Russia had not produced special works on cosmogony, but the generally high level of Russian mathematics, astronomy, and physics provided the background for such studies in the future and helped to establish the tradition of consistently materialistic science that Russia is famous for.

In addition to purely astronomical works (Bredikhin, Baklund, Belopolsky, etc.), the works by P.N. Lebedev on light pressure, by A.M. Liapunov on the equilibrium figures of rotating fluids, and by I.V. Metschersky on the mechanics of bodies with changing masses are of especial importance to cosmogony.

During the Soviet period, V.G. Fesenkov produced a number of works on cosmogonic problems. He used different physical considerations for solving these problems and demonstrated their importance for the explanation of the origin of the Earth. The general idea behind all Fesenkov's studies is that the creation of planets was tied with the evolution and even with the origin of the Sun. However, Fesenkov more than once stated that his studies were not aimed at the construction of a developed and all-embracing cosmogony of planets and that any attempt at such construction was, from his point of view, premature.

Recently (1943-44) Soviet cosmogony became much more active since the cosmogonic theory whose exposition constitutes the main content of my report began to be developed and made public. Its discussion and criticism not only inspired its further development, but also stimulated counter-hypotheses.

As a result, ours is the only country where a cosmogonic theory providing the explanation of all principal features of the planetary system has been suggested and elaborated. In particular, the explanation of the distribution of angular momentum between the Sun and planets based on the capture hypothesis has been given only in the USSR, and this explanation provides also the demonstration of the possibility and positive probability of gravitational capture. Further, the only idea able to compete, at least to a degree, with the capture hypothesis in respect to the explanation of the angular momentum problem (namely, the idea that the sun is losing its momentum in the process of emission, be it light emission (Tsiolkovsky) or corpuscular emission (V.A. Krat and V.G. Fesenkov)), appeared nowhere but in our country. Our country is the first where the origin, the direction of rotation, and the regularities of distances of the planets have been explained.

Only in the USSR has the genuine solution of the problem of the mechanism responsible for the formation of regions of condensed matter which in turn can develope into planets been given (the works by L.E. Gurevich and A.I. Lebedinsky) whereas in the West this problem was only partially dealt with.

The development of planetary cosmogony in the USSR is directly tied to the general progress of Soviet science, especially the progress of astronomy, physics, geology, geophysics, and geochemistry. Nobody should forget that our country is the motherland of geochemistry and that the USSR is the world leader in some branches of geology and geophysics as well as in meteoritic studies.

Of greatest importance are the works by V.A. Ambartsumyan and his school from which a bold materialistic conception of star formation has originated. The recent works by G.A. Shine on star rotation and on the content of the isotopes of carbon in the stars, studies of interstellar matter (G.A. Shine, V.A. Ambartsumian, P.P. Parenago, B.V. Kukharkin) and its ties with stellar evolution (B.A. Vorontsov-Veliaminov), and also works on the general problems of stellar structure and stellar evolution (N.A. Kozierev and some other scientists) are of utmost significance.

The works by Soviet scientists on the problems of cosmogony, stellar as well as planetary, still have a number of shortcomings, and they in no way may be regarded as perfected. But even despite all their shortcomings these works are revealing more and more clearly and definitely the distinctive features of the science of the Stalin epoch. The first and the most fundamental feature of this science is its conscious and consistent materialism. This materialism is genuinely conscious and deliberate unlike that spontaneous materialism which only occasionally breaks through the idealistic thicket in the capitalistic West. Any differences of opinion amongst Soviet cosmogonists are ones inside the same camp, the camp of materialists, and therefore all our disputes are of a creative character.

The second feature of Soviet science is that its cosmogonic branch also reflects that higher level of thought which is being cultivated by dialectical materialism. Looking at the cosmogony which studies the origin and development of large concentrations of matter we can see especially clearly that such problems cannot be properly formulated and solved outside the materialistic dialectic. Undoubtedly, we are better and better understanding the complicated network of causes and effects; we are identifying the main links with more and more certainty; we are grasping more and more deeply the whole process of development with all its knots and transitions into new quality; we are perceiving the different forms of the realization of the laws of nature in every specific area more and more distinctly.

The third feature of Soviet science constitutes its boldness in the formulation of new problems and new initial notions. This boldness differs fundamentally from that light-mindedness and schemes-making which are so widespread amongst foreign cosmogonic hypotheses, as was mentioned already. When we introduce novel hypotheses we always ground them upon facts. Then we work out our hypotheses and present them in a numerical form, thus transforming them into theories which we meticulously compare with all the huge quantity of facts established by different sciences.

The principal features of the Soviet social system display themselves in the collective character of our work as well as in its broad audience. Our discussions demonstrate the great concern of Soviet science for its basic principles and for its philosophical and methodological basis, although we, specialists in cosmogony, still find ourselves far behind the high level recently achieved in discussions on biological and linguistic matters which have become historic landmarks in scientific developments.

Soviet science is optimistic, which again is an important feature. From our point of view, the origin of the Earth is not only a most urgent puzzle of the natural sciences, but also a problem ripe for being solved, a problem whose solution has been prepared by the whole process of the earlier development of science of astronomy, physics, and geosciences. We believe the problem of the origin of the Earth to be soluble not only in the philosophical sense according to which all the phenomena of objectively existing nature are knowable in principle, but equally in the sense that we are able to solve it ourselves, today. It is necessary only that all our forces be concentrated and unified, and then this great problem will be solved in our country and within our time.

This optimism has brought about the creation of the novel meteoritic theory of the origin of the planetary system in the USSR.

THE MODERN METEORITIC THEORY OF THE ORIGIN OF PLANETS

In the cosmogony of the planetary system the most highly developed theory is the meteoritic theory that has been emerging in the USSR during the last few years. Its current state is presented in my book "Four lectures on the theory of the origin of the Earth" whose second edition was published in 1950. I do not intend to give you now a comprehensive account of the theory, but I mean to outline its principal methods and results paying special attention to its connections with some other scientific problems.

The theory is a conclusion from a huge collection of facts, some of which have been known since long ago, and some have been discovered by modern science.

Kant and Laplace, as long ago as they lived, came to the opinion resulting from their dealing with the facts, that all the planets had come into existence in the course of some common process. Indeed, there are many similarities in the planetary movements: all of them are quasi-circular and unidirectional, all lie more or less in the same plane, etc. Still, there are also significant distinctions. For instance, it is immediately evident that the planets may be naturally grouped into two sets, those nearest to the Sun and those remote, and the former are comparatively small and dense whereas the latter are much heavier but have considerably lower density. The satellites of the planets also demonstrate some regular differences: the near satellites of Jupiter and Saturn revolve in the same direction as their planets, while remote satellites revolve in the opposite direction. These deviations show that the common process of planetary formation proceeded differently in different circumstances. The principal task of cosmogonic theory is just to understand this process in its totality, to discover these specific circumstances, and to explain the origins of all observable regularities and the causes for similarities and dissimilarities.

It is necessary to underline specially that the cosmogony of planets should deal with the origin of the planetary system but not with the origins of any celestial bodies whose masses are smaller than the typical stellar mass, as some astronomers believe. It seems possible that all these comparatively small bodies come into existence in the Universe in different ways, but those constituting the Solar system with all its regularities could have only one mode of formation, precisely that which could generate these and only these regularities. The most essential fact about Jupiter is not that it is smaller than the Sun, but that it constitutes a part of some specific planetary system.

Examination of the principal planetary regularities indicates that the planets have been formed in the process of the accretion of a great many tiny particles formerly revolving independently around the Sun. And indeed, had the planets torn themselves away from the Sun as one compact piece of solar matter, or had they been captured by the Sun already formed, their orbits would have been elliptical but with occasionally quite different amounts of eccentricity, as the observation of binary stars clearly demonstrates. Quite the contrary, when the independently revolving particles join each other, their orbits undergo a process of natural averaging which results in the formation of symmetrical, that is, circular orbits.

It has been already noted in the historical introduction that some hypotheses regarded the primordial elementary particles as gaseous whereas the other hypotheses believed them to have been made of solid matter. There are convincing reasons to believe these particles to have been solid objects, dust-like and even more sizable bodies. On the one side, such particles are able to serve as the centers of condensation for gaseous matter, and on the other hand, their collisions being inelastic, a good part of their kinetic energy transforms into heat which in turn is radiated into space. This emission of surplus kinetic energy is, as we soon shall demonstrate in more detail, the necessary condition for the evolution of a meteoritic swarm towards the formation of a planetary system.

Therefore, the theory of the origin of the planetary system regards the primordial swarm as formed basically of solid particles plus some amount of gas. According to the terminology which is accepted in cosmogony, this theory should be called meteoric, or planetesimal. However, it differs from a great many other meteoric theories principally in the respect that in the USSR the fundamental processes of the evolution of meteoric swarms such as the above-mentioned energy transformation and the leveling of the dynamic and physical qualities of particles in the process of their accretion, have been clearly exemplified and used for an explanation of factual data. These are the leading ideas of our researchers which were either lacking in other hypotheses, or could be found there only in an embryonic stage.

Thus, modern science makes us believe that the Sun was surrounded by a giant swarm of diffused (that is, mostly dust-like) matter whose subsequent evolution has resulted in the formation of the system of planets. The swarm had a huge angular momentum which was later transformed into the orbital and rotatory moment of the planets.

The dynamic and physical factors responsible for the flattening of the swarm have been also identified. This flattening was brought about the increase of the density of the matter which lay in the vicinity of the central plane perpendicular to the vector of the angular momentum of the system. This evolution was driven by the aforementioned motor, that is, by the gradual loss of part of the kinetic energy because of multiple collisions. In 1950, L.E. Gurevich and A.I. Lebedinsky demonstrated in their important work that the swarm would become more and more flat and dense in the course of its evolution, and after some critical density had been achieved, the process of intensive formation of regions of condensation would begin. These regions will be colliding, dispersing, splintering and again merging together into what will finally result in the formation of planets.

According to our theory, the Sun directly participated in this evolution. The regions of the primordial swarm that were closer to the Sun lost matter first, since some particles were falling on the Sun because of light pressure, some were being drawn away by the same factor, and additionally, the volatile components of the solid particles of the swarm were evaporated by the solar heat. Conversely, far from the Sun these processes did not take place and moreover, the flattened swarm being sufficiently opaque and hence very cool, the gases were freezing out on the surface of the solid particles. It becomes thus evident that in the immediate neighborhood of the Sun the material for the formation of major planets was lacking and only relatively small planets could be formed whereas the giant planets could come into existence only far from the Sun. Consequently, the division of the planets into the two groups has received a simple explanation. Additionally, the distinctions between these groups in respect of their chemical composition may be explained as well. Far from the Sun, solid particles had retained all their gaseous components and even absorbed gases from space. This means that the composition of the particles was coming into correspondence with the chemical composition typical of the whole universe, dominated by such light and most abundant elements as H, C, N. O. On the contrary, the Earth-like planets were formed from particles which contained comparatively less of the light elements, of which only oxygen could be confined relatively easily due to its chemical affinity with heavier elements. Thus, it is only natural that the density of the Earth and its neighbors is considerably more than the density of Jupiter, Saturn, etc. There is one more characteristic detail: nitrogen, which can be found in the Universe almost in the same quantities as oxygen, is yet only insignificantly present in the earth minerals; this is precisely because of its chemical inertness which prevented its chemical bonding with the heavy components of the Earth. However, nothing hampered its existence in the form of frozen ammonia (together with ice, frozen methane, and some other light chemical compounds) in the Jupiter region. And one more significant detail: the former cosmogonic hypotheses which surmised that the matter to form the planets had been ejected by the Sun in the state of extremely hot gas, assumed the Earth to have been unable to retain hydrogen and other light gases due to its relatively small mass, contrary to the situation on the major planets. However, this opinion has not been confirmed by the latest studies of the speed of hydrogen dissipation from the atmosphere (I.S. Shklovsky) and by newly identified factual data, such as the existence of the methane atmosphere on one of Saturn's satellites, Titan, which looks like the Moon in respect of its mass but has an atmosphere similar to Saturn's. The explanation provided by our theory is free from all of these contradictions.

We are continuing to develop the explanation of the regularities of the Solar system. It has been demonstrated above that the circularity of planetary orbits may be very naturally explained on the basis of the idea of averaging. The fact that all the planets are rotating in the same direction and within the same plane may be explained equally easily, and on the same basis. However, this physical effect displays itself in a more complicated manner in other phenomena. The theory that I am outlining now clarified for the first time the meaning of the observed regularities of planetary distances. As has been found, the distances between the planets and the Sun depend upon the distribution of the particles in term of their specific angular momentum. This has resulted in the conclusion that the square root of the radius of a planetary orbit has to be approximately equal to some average between the square roots of the radius of the two nearest orbits. The principle of averaging may be different since it depends upon the mass distribution. However, this dependence is only weak, thus the arithmetic mean provides a good approximation. Also, one should not forget about the subdivision of the planets into two groups which has been explained above. \

The joint analysis of the processes of averaging and energy transformation explains the rotation of planets which could not be explained by the earlier cosmogonies. The purely mechanical reasoning used by their authors resulted in a paradoxical conclusion that the planets had to rotate in the direction opposite to their actual rotation. We have managed to solve this puzzle by demonstrating that the partial transformation of kinetic energy into heat should be taken into account. Naturally, this heat is radiated into space. If all the kinds of energy and angular momentum are taken into joint consideration, the result is that the theoretical prediction for the rotation of planets corresponds to their real rotation, providing that the loss of energy is sufficiently high.

The formation of satellites may be explained in the same way. Moreover, it has been shown for the first time that the nearby satellites should revolve around their planet in the same direction as the planet's own revolution around the Sun whereas the remote satellites should most probably revolve in the opposite direction, as do the remote satellites of Jupiter and Saturn.

The explanation of the basic features of the Solar system, outlined above, is well substantiated and, as far as I know, has not encountered fundamental objections. The case is somewhat different with the spatial distribution of angular momentum which was mysterious for a long time. As is well known, the planets are responsible for 98 per cent of the total angular momentum whereas the Sun contains 99 per cent of the total mass. None of the earlier cosmogonies was able to explain this paradox. The hypotheses by Kant, Laplace, and Jeans were killed mostly by this contradiction which they failed to interpret soundly.

The solution of this puzzle should be sought in the origin of the primordial swarm which appeared in the neighborhood of the Sun. If we surmise that the matter forming the swarm had been captured by the Sun from the outside, that is, from the clouds of galactic dust, all the difficulties concerning the angular momentum of the system will be eliminated. There is a great number of dust clouds in the Galaxy, thus this hypothesis had some factual basis. There were problems, however, connected with the mathematical justification of the very possibility of the capture. There was a common belief in science about the impossibility of the capture, which impeded the use of this idea in cosmology. Moreover, one could find a number of publications which pretended to provide the proofs of such impossibility. Our idea of solar capture first was purely hypothetical as well. However, the possibility of capture in respect to the classic three body problem was given an exact proof in 1947. It was also proved that the probability of the capture is positive, that is, greater than zero. The history of these studies is presented in our "Four lectures" and in special publications.

Thus, the hypothesis that the swarm originated from the capture of galactic matter has been given a firm foundation.

Our calculations dealt with the case when all three interacting bodies had masses of the same magnitude as the mass of the Sun, but the theory demonstrates that capture is possible regardless of the mass distribution. In particular, a large body can capture a small one, even something like s speck of galactic dust. Still, some additional calculations are desirable, particularly connected with the capture of small particles, which can give us new examples of such capture thus making our understanding more specific and realistic.

The capture of dust (meteoric) matter may take place at any stage of the evolution of the Sun, but its probability very strongly depends upon the state of the spatial region in the vicinity of the Sun at the moment of capture.

I maintained in my report delivered at the meeting of the Division of Physico-Mathematical Sciences in 1949 that "Under circumstances now observable within the galactic vicinity of the Sun, the probability of captured matter depends upon the Sun during its route through the Galaxy having experienced various conditions including being surrounded by clouds of dense matter, and at an earlier stage of the Sun's evolution its chances to be passing through such clouds were relatively higher. [1]

The effectiveness of capture, that is the amount of captured matter, depends upon the density of the dust-like substance in the region. Obviously, the higher the density of the cloud the more effective is the capture. The actual probability of capture in its classic form depends upon the stellar density, particularly upon the distribution of the relative velocities of stars: the smaller it is, the more often the capture takes place. Relative velocities inside galactic star-clusters average to about one thousand meters per second or even smaller values - in comparison with twenty to thirty thousand inside the neighborhood of the Sun. Therefore the capture takes place inside clusters with much higher chances than outside (the difference constitutes several orders of magnitude).

It follows that we who are working on the problems of planetary cosmogony are greatly interested in following the scientific achievements connected with studies of all the kinds of stellar clusters. Every success in the field is of great importance to us (aside from its direct significance in the framework of stellar cosmogony), since our notions concerning capture thus become more specific and the possibilities of numerical estimations get more promising.

Some other modes of capture, besides the purely gravitational one, have been identified as well (Agekian, Radzievsky). We do not know yet which mode was realized in the case of our Sun and to what degree, but the final result can be only one, namely the formation of a swarm of particles in the vicinity of Sun. Since there are several possible modes of capture, the total probability of this phenomenon becomes higher.

In addition to the variety of physical conditions in space, that is in different regions of the Galaxy, different time periods again may differ from each other. Our theory does not surmise that the physical conditions in the past were different in comparison with those observed now, but if that really took place, it means a change of the numerical probability of capture.

In view of this, it is necessary to pay special attention to the problem of the origins of stars, particularly the Sun. I will touch on this question only in respect to its ties with our principal subject. If the Sun had developed from dispersed matter and in a medium of higher density with rather low relative velocities, then its chances to capture interstellar matter in an amount sufficient to form the planetary swarm were, of course, much higher until the Sun left this maternal medium. And if stars come into existence under such conditions not one by one but in clusters, then the chances for capture within the same cloud again become higher, and to a considerable degree

It is clear therefore that we follow the latest achievements in the studies of the problem of stellar formation with great care, in particular the works by V.A. Ambartsumian.

The idea of the capture of dust matter not only explains the origin of the swarm which gave life to the planets but also provides an explanation for the distribution of the angular momentum of the system. V.A. Krat and V.G. Fesenkov have recently suggested another hypothesis to explain the same problem, namely the idea that the corpuscular radiation of the Sun brought about a loss of not only good part of its mass but also an even greater share of its momentum, but they have failed to provide proof that the phenomenon can explain quantitatively the existing distribution of momentum. Calculations do not support this hypothesis if we assume that the Sun evolved along the main sequence. Therefore, we have to go even further backward, to the epoch when the Sun was not a main sequence star yet. But then we face new difficulties, and, what is most important, we have no factual data to support the idea. Thus, the distribution of angular momentum is not explained but only relegated to the dim past. We should therefore regard corpuscular radiation as an interesting additional factor, but not the principal one. As to the Sun losing its mass, this loss, as is well known, results only in the increase of the orbital radii in the same proportion, thus the regularities of the planetary system do not change in the process of the Sun's evolution.

At the moment we have yet too little reliable knowledge about the evolution of the Sun, although some interesting hypotheses do exist. When such reliable knowledge is accumulated, it, naturally, will affect planetary cosmogony by introducing some additions or corrections. On the other hand, however, every theory of solar evolution must pay attention to the fact that the planetary system has existed for several billion years. Not every theory of the evolution of the Sun is compatible with this fact.

The meteoric theory created in the USSR together with our idea of capture constitute an integrated theory explaining the origin of planets as well as all the principal regularities of the Solar system. Such completeness has been achieved in cosmogony for the first time. In this theory physical, mechanical, and chemical properties are explained together and in their close natural interconnection. I will further mention some problems yet unsolved.

Our theory achieved its recent stage of completeness and consistency not immediately but through a process of continual development in the course of which more and more scientists of different specialties have been taking part in its elaboration. The general problem being so broad and multidimensional, the joint effort of the representatives of many different scientific disciplines is necessary to deal with it. It goes without saying that success may be attained only through collective enterprise and that acute criticism is a necessary condition for the development of scientific thought. The participants in the development of our theory were engaged in its critical discussion to no lesser degree than its opponents, trying to come to a deeper understanding of the fundamental processes of the formation of the Solar system.

My first work published during the initial period of these studies contained a lot of poorly grounded tentative ideas, and even some actual mistakes. Therefore I weeded out a good part of the earlier studies when I was working on the cumulative presentation of the theory given in my "Four Lectures." For example, it was erroneous to assume that the galactic gravitational field had affected the capture, and my arguments concerning the distribution of the orbits of comets were also groundless. These mistakes were eliminated in the course of the subsequent development of the theory under the influence of criticism and its own growth. I have to underline thankfully that no other cosmogonic theory was made the object of such intensive criticism as ours.

The collective of the Department of the Earth's Evolution at the Geophysical Institute of the Academy of Sciences of the USSR constituted the main core of the participants in our cosmogonic studies. It is very characteristic that such a department was first created just in our country and in our Academy. We work there together with B. Yu. Levin and G.F. Khilmi. Their role in the development of the theory is very great and is not limited to their published works. We have been accustomed to discussing every problem in detail, paying little attention to whom should have been given credit for every next step, since only our collective movement forward is important. The role of the younger members of our collective (V.S.Safronov, S.V. Kozlovskaya, E.A. Lyubimova) is also important.

However, no isolated collective, however devoted to the work it might be, is able to comprehend a problem of such a large scale. It is very important that our problem attracted the attention of a broad circle of scientists. I have already mentioned and shall mention further some works of significance by T.A. Agekian, V. V. Radzievsky, E.N. Lyustikh, and some other researchers.

An outstanding work by astrophysicists from Leningrad, L.E. Gurevich and A.I. Lebedinsky should be mentioned specially. This work was undertaken not with the aim of supporting one or another meteoric theory, but it constitutes a part of the same endeavour directed to the development of contemporary Soviet cosmogony whose general results were presented above. The strong and free scientific cooperation that we have established with our comrades from Leningrad permits a continual exchange of scientific ideas and results and provides the best opportunity for achieving a comprehensive solution of our mutual problem.

The theory outlined above is grounded upon established observational facts and theoretical achievements of different sciences. In the course of its development it faces some problems yet unsolved by neighbouring sciences. Such is the problem of the internal structure and composition of planetary cores. Its complete solution depends upon the theory of matter under the conditions existing deep inside the planets, especially giant ones, where the pressure may be as high as hundreds of thousands or even millions of atmospheres. Under such pressures the structure of matter changes, the electronic shells of atoms experience deformations. Besides, theoretical studies of the behavior of matter at such pressure have been initiated only recently and failed to produce reliable results so far, whereas experimental data simply do not exist. We would like to attract our physicists' attention to this problem.

There is another still-unsolved problem pertaining to physical chemistry, that is, how does the process of gas condensation upon small particles actually take place under conditions of very low temperature and density? Some works exploring this subject exist, but nobody paid attention to this problem in the USSR. For a long time we have been calling upon our physicists and astrophysicists to work it out, which would be important not only to planetary cosmogony.

The understanding of the laws of this condensation is also necessary if we want to come to know the particulars of the mineralogical composition of meteorites. In the framework of our theory it is only natural to believe meteoric bodies to have originated from the same captured swarm that gave life to the planets. In my earliest works I made no distinction between the particles of the primordial swarm and the meteorites of our time, which was a mistake. In reality, the history of meteorites is very complicated. They have resulted from the evolution of the swarm of primordial particles that formed planets, which means that they also experienced multiple splintering and sticking together. Their structure, which is usually heterogeneous indicates that. However, the slow changes of crystalline structure taking place under the conditions of very low temperature, sometimes abruptly raised by collisions or by their passing near the Sun, have not been explored yet. The solution of this problem being absent, some researchers are inclined to explain the structure of meteorites by surmising them to be the remnants of the disintegration of some planet, or of a large asteroid whose existence had come to an end long ago. This is a temporarily--useful working hypothesis but nothing more, since it does not explain why different parts of the disintegrated body were combined again without mixing (for example, ferrous fragments turned out to become predominantly separated, etc.). The partisans of this view sometimes regard it as a refutation of our theory of the origin of planets, but that is nothing but an illusion. Even if it were proved that the meteorites had been formed inside planets, our cosmogonic theory remains unchanged; only the possibility of making estimates of the age of the Earth on the basis of the amount of meteoric matter falling from space would have been eliminated.

This numerical estimate was made by myself in 1945; however even then I stated clearly that the result (7 billion years) should be regarded as only approximate. This figure does not contradict the results based on the laws of radioactive disintegration.

Now we will consider the problems of the origin of the Earth and of those forces and processes which are active inside its depths. This field belongs to the geosciences; these problems will be solved by geologists themselves, and it is not cosmogony which should be expected to solve them. However, it is cosmogony's duty to assist the geosciences by explaining the evolution of the Earth beginning with the very moment of its formation. Only such historical exploration of the problem can identify the forces now acting inside the Earth and provide geologists with the basis for their theoretical constructions, for instance, concerning the causes of the formation of mountains.

For the time being, we have concentrated our effort upon only two processes, namely the thermal history of the Earth and the differentiation of its matter according to its density throughout the body of the Earth. The controversy over whether the Earth is cooling down or warming up still persists. Following V.I. Vernadsky, we are convinced that the Earth's initial stage was cool, and that only in the course of time has our planet's temperature been rising because of the disintergration of radioactive elements. Our theory implies this conclusion unambiguously. But it is very difficult to overcome the geologists' habit of thinking of the Earth in terms of its gradual cooling. As a matter of fact, it became clear soon after the discovery of radioactivity that no special cosmogonic considerations were needed to demonstrate that the increase of heat due to radioactive disintegration was much more than its loss because of its emission into space, which means that even had the Earth been initially hot it would not have been able to be cool further after the formation of its solid crust, but, on the contrary, it would have begun to warm up. The process of its cooling will begin when the generation of heat becomes less than its loss, due to the decrease of the amount of radioactive materials. According to our calculations, that stage will come only in many billions of years.

Therefore, the so-called contraction theory connecting the formation of mountains with the cooling and wrinkling of the Earth must be completely rejected. On the contrary, we regard the Earth's heating as a powerful source of energy for this process. We are engaged at the moment in meticulous and laborious theoretical calculations of the expected temperatures at different distances from the Earth surface's down to its centre at the present time and in the past. This heat can support not only the expansion of the Earth but also some other endothermic (heat consuming) transformations, recrystallization, etc. In respect to this problem, we need contacts with geochemists which are still insufficient.

We believe that there is another energy source, namely the gradual differentiation of the Earth's substance. The earlier Earth consisted of particles of different size and composition which were all mixed together. Although the Earth is firm, after the temperature had reached one thousand degrees Centigrade or so, the substance inside the Earth became plastic, so the lighter components tend to move gradually upward, and the mechanical stresses accumulating thereby were eliminated by means of discrete impulses. Naturally, this process may be developing only very slowly due to high density and viscosity, and even nowadays it has not been completed. Just because of its slowness, the process affects the Earth during a very long time; it opens prospects for theoretical investigation of the vertical movements of matter including the causes of deep-focus earthquakes. As is evident from E.N. Lyustikh's calculations, gravitational differentiation is able to produce an amount of energy each year which is of the same magnitude as that provided by radioactive disintegration.

Thus, our cosmogonic theory is ready to suggest some new initial ideas pertaining to the evolution of the Earth which, we hope, may be useful for geologists and geophysicists, especially in the framework of theoretical geotectonics.

Of course, we cannot give a full presentation of the theory we are working on in the framework of a short article. This theory is not a speculative construction, but it is grounded upon the classical heritage and the careful analysis of all the previous development of cosmogony. It was especially instructive to clarify the causes for earlier failures. We used two methods in our effort to avoid mistakes or, when they still inevitably took place, to correct them quickly. First, we always tried to develop our explorations up to the stage where definite numerical results could be achieved which could be compared with observational data. This is probably one of the distinctive features of our work. Its second feature is that we worked not in isolation but in permanent contacts with numerous scientific collectives. We are much obliged to the critical discussions of our theory which has helped us to identify and correct its mistakes and weak points.

The theory has gone through a long development, and it will be developed even further. In the course of this development some parts will be changed or even eliminated, but the basic direction has proved to be sound. The explanatory completeness achieved, completeness considerably exceeding that of former hypotheses, as well as the absence of insuperable contradictions, make us believe that our theory is on the right track and that it captures some essential aspect of objective truth.

However, the theory is far from being completed yet. It still contains a lot of lacunae and insufficiently elaborated components. We have not yet enriched our theoretical arsenal to the necessary degree by the results of the latest physics and astrophysics, and we give too little attention to such neighboring sciences as geology, geophysics, and geochemistry. The further success of planetary cosmogony considerably depends on the solution of the problems faced by other sciences.

The convocation of our conference may become a turning point for planetary cosmogony. We who are working on this problem will be enriched by new ideas and assisted by penetrating criticism. Also, we hope that the conference will stimulate the interest of workers of astronomy, physics, and geosciences concerning cosmogony, and that new forces will join the mutual enterprise. One may hope that the problem of the origin of the Earth will find its place in the university curriculum and that new cadres will be trained.

We live in a remarkable time. Our Party and government have created exceptionally favourable conditions for scientific development. The broadest masses hopefully follow the development of the problem of the Earth's origin. We are absolutely confident that by means of combined effort, overcoming the difficulties of the problem and our own shortcomings, we will be able to achieve its solution in our time and in this country, in the Stalin epoch.

REFERENCES

1. Izvestiya AN SSSR, Seria Fizicheskaya, Vol. 44, 1950, p. 43.

NEW ACHIEVEMENTS IN THE SCIENCE OF THE ORIGIN OF THE EARTH

O. Yu. Schmidt
Translated from Bol'shevik, No. 5, pp. 23-33 (1952)

The content of astronomy is intrinsically connected with such fundamental world-view problems as questions about the eternity and boundlessness of the Universe, the evolution of matter, the origins of the stars, planets, our Sun, our Earth, etc.

Astronomy is divided by a watershed of uncompromising ideological battle between the forces of progress and reaction, between materialism and idealism, between science and religion. Indeed, it is no coincidence that in this struggle astronomy suffered from especially heavy losses in comparison with the other natural sciences: let us only remember the execution of Giordano Bruno, Galileo's trial, and his forced renunciation of the propagation and further development of Copernicus' materialistic teaching made under pressure from the Roman Catholic Church.

Like all other sciences, astronomy grew out of practical needs. All cattle-breeding and agricultural peoples needed some astronomy for proper establishment of the "calendar" for their different seasonal activities. Because of its special economic importance this branch of knowledge was monopolized quickly by the rising caste of priests and further by the church which attempted to use astronomy in order to impose its spiritual influence and dominance on the society. And the most different religions maintained the idea of the creation of the world and the Earth by the divine will as a cornerstone of their dogmas. Centuries pass, the productive forces of society develop, science achieves decisive refutations of these views, but religion and church nevertheless cling to ideas rooted in the primitive ignorance of mankind and in the social oppression of people, since church and religion themselves are nothing but instruments in the hands of the exploiter classes and so they want to obscure the consciousness of the working masses. Nowadays, the forces of imperialist reaction try to interpret idealistically every new scientific discovery and to "reconcile" science and religion thereby defending the pillars of bourgeois society and the bourgeois world.

Materialism's struggle against idealism is developing especially severely and persistently in respect to the problem of the origin of the Earth. This problem is not solely astronomical, it is faced also by geology, geochemistry, and paleogeography which is the science of the changes of the Earth's surface from the most ancient epochs up to our time. It is impossible to understand, without scientific knowledge about the origin of the Earth, the process of its later development, the forces acting inside our planet and their different effects, the causes of geological phenomena (such as mountain formation, for instance the regularities of the distribution of mineral deposits, the causes of volcanic activity, earthquakes, etc).

The origin of the Earth is one of the most important problems of natural sciences, like the problems of the origin of life on the Earth and the origin of man. The total process of the development of human knowledge about the Earth as a part of the Universe cannot be examined in this article. However, in the first place it is necessary to emphasize the enormous revolutionary significance of the discovery made by Copernicus which inflicted a mortal blow on the Ptolemaic system (formerly reigning during many centuries), the same system which was regarded by the Christian theologians as their fulcrum. Secondly, one should also note the importance of the contribution to materialistic ideas about the Earth and the Universe made by the great Russian scientist M.V. Lomonosov. He maintained the conservation of matter and motion long before Lavoisier, and his ideas about the evolution of the Earth and nature constituted the basis of genuinely scientific, materialistic teaching about the Universe. However, no developed scientific theories of the origin of the Earth and solar system existed before the hypotheses of Kant and Laplace.

Kant and Laplace presented the first scientific hypotheses intended to explain materialistically the origin of the Earth and solar system. Both were based on the idea that the solar system as we know it has developed naturally from some other preceding state of matter, that is from some "primordial nebula" consisting of small particles, either gaseous ones (Laplace), or particles of any state, possibly even small solid bodies (Kant). F. Engels paid high tribute to Kant's and Laplace's contribution to cosmogony (the science about the origin and development of heavenly objects); he especially noted that they for the first time had weakened the conception of a nature without any temporal history.

Proceeding from the facts known in their time, Kant and Laplace managed to explain some important features of the solar system. But their resources were limited by the level of scientific knowledge in the eighteenth century. For example, Kant was unable to explain the origin of the undirectional rotation of all the planets. Laplace avoided this difficulty by assuming the initial rotation of the whole nebula. According to his basic idea, the nebula that was to form the solar system had consisted of a hot gas. The nebula was contracting in the process of its gradual cooling, which resulted in the eventual separation of gas rings. These isolated rings cooled further and finally condensed in the form of planets, whereas the central densest region of the nebula transformed into the Sun. But Laplace failed to realize that the Sun should have rotated quickly according to his hypothesis, whereas its real rotation is very slow (one complete turn in 25 days). Laplace did not take into consideration that his gaseous rings would be dispersing, and not condensing. A number of other points of those hypotheses were later demonstrated in the future to contradict the facts. The planets are known to contain 98 percent of the total angular momentum of the system, whereas the Sun contains 99 percent of its total mass. None of the former cosmogonies were able to explain this fact, including those of Kant and Laplace, which turned out to become the main reason why they were eventually abandoned. Viewing these hypotheses concretely-historically, however, we should not forget that they were bold and objectively materialistic attempts to penetrate deep into the past of nature which contained some grains of objective, absolute truth and thus constituted an important stage in the development of our knowledge.

Our approach to this scientific legacy differs from that of the astronomy of the capitalist countries. Instead of developing further the materialistic framework of the Kant-Laplace theory on the basis of newly available scientific knowledge, the astronomers of the nineteenth and early twentieth century tried only to patch up the discrepancies in the details of this theory, and after failing to reach this aim they momentarily turned away from the theory abandoning its positive, materialistic content. A number of new, often very fantastic cosmogonic theories were suggested in the capitalist countries in our century, and none proved to be sound and long-lived.

The hypothesis by the English idealist and astrophysicist Jeans happened to survive longer than its rivals. Jeans maintained that long ago the Sun had been approached by another star and thus a huge tidal wave was generated on the solar surface due to which a cigar-like jet consisting of very hot gases was ejected from the Sun; this "cigar" further broke up into several pieces whose number correspond to the number of presently existing planets. The hypothesis was impressively presented, and soon it dominated the popular literature and found its place in textbooks (unfortunately, including our Soviet textbooks). However, nobody but Soviet scientists put this hypothesis under systematic criticism, and in 1948 N. N. Parisky killed it completely. He demonstrated on the basis of indisputably precise mathematical calculations that even if this "cigar" had come into existence, its fragments never could have revolved around the Sun at a distance equal to that of the Earth, to say nothing of the more remote planets like Jupiter and the other. Nobody supports the Jeans hypothesis nowadays. However, what was the reason for its former popularity and twenty-year dominance? Its scientific level was rather low, its correctness was never proved, but it perfectly coexisted with the ruling ideology of the capitalist world. The hypothesis presented the origin of the Earth as a consequence of an extremely rare event which was more or less acceptable from the viewpoint of the ideology of idealism and religion. Since the literary belief in the biblical myth about the creation of the Earth from "nothing" by God is too outdated nowadays, the idea that our planet may be almost unique and possibly even the sole cradle of life in the Universe sounds very attractive.

After Jeans, bourgeois science fared even worse. Its latest stage is notorious for such products of the degeneration of bourgeois theoretical thinking as the Milne hypothesis attributing the origin of the planetary system to a collision between the Sun and a sole quantum of enormous energy, or the hypothesis developed by Hoyle, according to which the Sun had been part of a binary system whose second component eventually exploded into fragments which gave birth to the planets. This is the same Hoyle who teaches the creation of matter from "nothing" and who wrote in his book published in England in 1950 that Marxism was more dangerous than the atomic bomb. This is the way of ideological reaction which joins its effort with political reaction.

Of course, it would be wrong to believe that the science of cosmogony in the West is nothing but pure idealism. The very facts studied in the course of scientific research inevitably lead scientists to scientific materialistic conclusions, even against their own will. Therefore, even the modern cosmogonic theories in the capitalist countries occasionally contain some valid ideas. But these grains of scientifically-progressive knowledge must be searched for and extracted from a heap of contradictory and basically idealistic constructions.

V.I. Lenin demonstrated in his immortal book "Materialism and empiriocriticism" that the researchers of nature, captured by idealistic philosophy, are bound to fall down into the slough of agnosticism advocating the unknowability of the essence of our world. The cosmogonic theories proposed by reactionary bourgeois scientists exemplify this conclusion, since their authors do not really intend to reach the objective truth and to explain scientifically the origin of the Earth, the only thing that they want is to invent a variety of means to imagine the origin of our planet, disregarding whether their theories correspond to reality or not. In the area of cosmogony they tend to construct speculative, subjectively-idealistic schemes reflecting no reality and contradicting established facts. On the other hand, the view that the cosmogonic problem is unsolvable is also being propagated. One of the latest books by the English astronomer Smart maintains openly that the solution of this problem cannot be achieved by science. But beyond the realm of science lies religion which the reactionaries in science appeal to. The same trend implies another characteristic feature of the constructions by bourgeois scientists, that is, their sceptism and the rejection of comprehensive scientific theory. These scientists, instead of attempting to grasp all established facts in the framework of one general theory, content themselves with the explanation of isolated details, and they are never bothered by the contradictions inevitably produced by such "methodology."

In this way, cosmogony displays the degradation of bourgeois science, which is still able to make new discoveries in its observatories or to analyze some isolated phenomena, but which can no longer integrate the accumulated data into the framework of unified theories. Only the science based on the ideas of dialectical materialism is the genuine heir and successor of the whole former development of science, able to accept its achievements critically, to develop them further, and to synthesize them. First and foremost, such is Soviet science, the most progressive science in the world.

The development of cosmogony, as well as other areas of natural science, has been based in the USSR on the strong materialistic tradition which is the glory of Russian science. Russian experimenters of nature, beginning with Lomonosov, always tended to observe nature carefully and thoughtfully, to search for natural causes of phenomena, and not to restrict their effort to the description of visible events alone. The activities by the revolutionary democrats Gersten, Chernyshevsky,

Dobroliubov, and further Pisarev paved the way for openly materialistic scientific works by I. M. Sechenov in physiology, F. A. Bredikhin in astronomy, D. I. Mendeleev and A. M. Butlerov in chemistry. This direction was followed by the great scientists I. P. Pavlov and I. V. Michurin. But only the victory of the Socialist revolution made possible the absolute dominance of materialistic ideas in science.

Already the works by M. V. Lomonosov, F. A. Bredikhin, D. I. Mendeleev, A. M. Butlerov, P. N. Lebedev, V.I. Vernadsky, etc. contained many profound ideas leading to the creation of correct cosmogony. Interest in cosmogony increased immediately after the victory of the Great October Socialist Revolution. In the 1920s and 30s, Soviet scientists had seriously contributed to the critical examination of the cosmogonic hypothesis suggested in the capitalist countries at different times. Academicián V. G. Fesenkov, who had begun his systematic researches in cosmogony in the 1920s, was the first Soviet scientist to suggest new cosmogonic ideas. He examined a variety of solutions of some problems connected with the origin of the Earth, planets, stars, etc. However, Fesenkov himself and some other Soviet astronomers repeatedly maintained that they regarded as premature the construction of a comprehensive cosmogonic theory. We believe such a viewpoint erroneous.

Soviet science has to be fighting not only against idealism and foreign reactionary science, but also some echoes of their influence in this country as well. It is a matter of common knowledge how intensive and victorious was the struggle by progressive Soviet biologists and physiologists against this anti-scientific influence carried out under the direct leadership of the Communist Party. This erroneous idealistic influence has not yet been completely overcome in cosmogony. Unashamedly idealistic statements face concerted resistance in our country, but sometimes we happen to be unable to discern in time the implicit manifestation of these alien ideological and methodological principles. For example the uncritical emulation of foreign authors had once made fashionable a disrespectful attitude to the first classic cosmogonic hypotheses by Kant and Laplace in our country. These hypotheses had been called "speculative" although such an assessment sharply contradicted the opinions shared by the founders of Marxism. Not uncommonly, some people failed to understand the genuine role played by hypothesis in the process of the development of the natural sciences, although Engels had come to the conclusion based on the history of science that "hypotheses is a mode of the development of natural sciences so far as these sciences are able to think." Many scientific theories and discoveries are known to come into existence on the basis of the verification of scientific hypotheses by practice, by new factual data.

Especially harmful to the progress of science was the propagation of the idea that the time to solve the problem of the origin of the Earth had not come yet, and that it was necessary either to wait until the problem of the origin of stars had been solved (although that is a more difficult question), or to wait until the technical means for discovering planetary systems belonging to some other stars became available.

But the fact is that the science of our time, and especially Soviet science, has accumulated already so much data in astronomy, geology, and other sciences that a solution of the problem of the Earth's origin is now at hand. Soviet scientists are creating a really progressive and materialistic science about the origin and development of the Universe, struggling thereby against fruitless, reactionary, and idealistic theories and hypotheses; theirs is the science which maintains that our world is material in its essence, that it undergoes permanent motions and changes, and that it is infinite in space as well as in time. As our scientists master more completely the great teachings of Marx-Engels-Lenin-Stalin, they are becoming more experienced in its practical usage for the solution of the concrete problems of the natural sciences, their scientific horizon is broadening, and they are becoming able to demonstrate more courageous efforts to solve the most important and complicated scientific problems.

The great Stalin urged our scientists to bring forward new ideas and notions in a bolder way rejecting and discarding outdated views. And now the Soviet people urges their scientists to explain how in reality the planet originated, whereon we live and work, the planet which we are reshaping in the process of our buildup of the Communist society.

The problems of cosmogony are being studied in our country by scientific collectives which include scientists of different specialties, not only astronomers, but also geophysicists, geologists, geochemists, mathematicians, etc. The Soviet scientists working in the area of cosmogony face the following problem: they should contemplate profoundly the scientific data on the origin of the Earth, and they should do it on the basis of the complete information accumulated by science and under the guidance of the methodology of materialistic dialectics.

A new theory of the origin of the Earth has been worked out since 1943 by a collective of Soviet scientists on the initiative of the author of this article. The development of the theory has been considerably influenced by sharp comradely critique which turned out to be very helpful in stimulating the author to reject some erroneous parts of his earlier works. In addition to the author's specialized scientific works, the basic ideas of the theory were presented in his small book "Four lectures on the theory of the origin of the Earth" whose second edition was published in 1950.

In 1951 the Academy of Sciences convened a conference on the problems of planetary cosmogony wherein the representatives of a variety of disciplines (such as astronomy, physics, mathematics, geology, geophysics, and geochemistry) participated. This conference was a new demonstration of the fruitfulness of criticism in science and one more confirmation of Stalin's statement that "no science can develop and progress without the confrontation of different opinions, without the freedom to criticize." More than forty scientists of different specialties participated in the discussion of the problem of the origin of the Earth from all possible sides, putting the new theory under detailed critical scrutiny. The conference came to a decision to approve the work by our collective directed to explaining the formation of the planets and their satellites from the cloud of dispersed matter; it also emphasized that these studies had delivered a serious blow to agnosticism. The conference also noted some lacunae and shortcomings in our studies and made a number of suggestions concerning future research in the area of cosmogony. The conference's concluding statement, endorsed by the Presidium of the USSR Academy of Sciences, says in its closing phrase that: "The conference urges all astronomers, physicists, geophysicists, geologists, geochemists to intensify the studies directed to the solution of the problems of planetary cosmogony and it is confident that this most difficult problem of the natural sciences will be solved in our country where the leadership of the Party and government has created all the prerequisites for the successful development of progressive science."

Now, we outline the basic ideas of the new theory.

The first question which the theory should answer is: what was the initial state of matter which gave birth to the Earth and all the other planets? All of the data accumulated by science about the Universe, especially during the modern epoch, lead to the conclusion that the planetary system developed from a primordial swarm of particles surrounding the Sun and consisting of gases, dust, and even more sizable bodies. The size of the swarm exceeded the size of the presently existing planetary system. The swarm was rotating but, contrary to Laplace's opinion, not as a whole. Every particle rotated around the Sun independently, being affected by its gravitation in accordance with Newton's law, but one direction was predominant. This conception stems from a careful examination of the currently existing solar system. The system contains nine planets, all of which, beginning from Mercury and ending by Pluto, revolve around the Sun in the same direction; the planetary orbits only slightly differ from circular ones, and they are located more or less in the same plane. The system also contains some less huge bodies, such as asteroids, comets, and meteoroids. They also revolve around the Sun, but their orbits are, as a rule, elliptical, their orbital planes are different, and some of these smaller bodies revolve in the direction which is opposite to the direction of the planetary revolution.

Paying attention to these dissimilarities, earlier cosmogonic hypotheses assumed that every group of celestial bodies (planets, comets, etc.) had its own particular origin. However this conclusion is erroneous. Metaphysical and unsound is the methodology explaining the phenomena of the solar system without taking into account their mutual interdependence. Our theory maintains the existence of a uniform process of development which generated all the bodies of the solar system. Uniform as it was, the process was taking place under different conditions, which circumstance has resulted not only in a number of similarities amongst these bodies, but also in some dissimilarities.

Let us note that the largest bodies of the system (i.e., planets) have almost circular orbits; the heavier a planet is, the closer its orbit is to a circular one (the heaviest planet is Jupiter, and its orbit is the best approximation to a perfect circle). The conclusion is that all the bodies of the solar system came into existence though a process of the coalescence of many smaller objects which had previously been revolving around the Sun in a variety of different orbits, including considerably elongated ones. The motions of these particles underwent the process of natural "averaging" in the course of their coalescence, and this "averaging" of the orbits stretched along different axes could only result in the formation of more or less symmetrical, that is almost circular orbits. In this way, the circular orbits of planets may be explained. As a result of the same "averaging" of the individual motions of particles, the planets formed from these particles move in the same direction and almost within the same plane.

The same mechanism explains the fact of a considerable difference between two groups of planets. The first group consists of the four planets closest to the Sun (Mercury, Venus, the Earth, and Mars), all comparatively small, but rather dense. The four more remote planets of the second group (Jupiter, Saturn, Uranus, and Neptune) are much heavier, but their density is smaller (close to the density of water).

To explain this phenomenon, let us consider the chemical composition and the physical state of the particles of the pre-planetary swarm. Astronomers in the USSR and some other countries have obtained a number of new results about stellar nebulas and interstellar matter in general during the last two decades. As is now known, interstellar space contains a lot of dispersed matter, usually in the form of giant clouds. These clouds are made of gases and dust; hydrogen predominates over all the other gases, and it is followed by some other light elements, such as oxygen, carbon, and nitrogen. There are also some compounds of these elements, such as methane (marsh-gas), and some others. The solid particles contain different elements and their compounds. There are no reasons to believe the chemical composition of the protoplanetary swarm to have differed significantly from that of interstellar matter.

The amount of matter contained in that domain of the swarm which neighbored the Sun was gradually diminishing, due to a number of causes. The principal cause was the evaporation of certain particles which, while moving along their elongated orbits, penetrated into the solar atmosphere. Even those particles which survived during their passing through the vicinity of the Sun evaporated water (that is, ice) and other volatile substances, such as methane, ammonia, and carbon dioxide. According to the studies by A.I. Lebedinsky and L.E. Gurevich, inside Jupiter's orbit the gases were so heated by the Sun that they were beginning to move very quickly and eventually scattered in different directions all over the domain of the swarm. On the contrary, the region beyond the orbit of Jupiter was shielded from the Sun's radiation by the inner part of the swarm, there reigned low temperatures, and the gases, far from evaporating, became frozen on the surface of particles. Because of these factors, the amount of matter to survive inside the domain of the inner planets turned out rather small, enough only to form comparatively small planets like the Earth, and the giant planets were formed only at larger distances from the Sun, beginning from Jupiter's orbit. It is understandable now why the Earth and the neighboring planets consist mainly of dense refractory substances, whereas the outer planets contain a great many light substances.

This example helps us to see that the new theory, contrary to its predecessors, does not restrict its analysis to the mechanical aspects of the planetary system (that is, the planetary motions alone), but utilizes physics, chemistry, mechanics, geology, and other sciences in order to explain these phenomena.

The analysis of the evolution of the gas-dust swarm permits us to understand a number of subtler phenomena, inconceivable on the basis of former hypotheses. Consider, for example, the daily rotation of planets. Laplace's and all other hypotheses lead to the apparently paradoxical conclusion that the planets had to rotate in the direction opposite to the real one. Our studies revealed the source of Laplace's error which happened to be the circumstance that he had failed to take into account the process of the partial transformation of kinetic energy into the energy of heat. The upshot is that numerous collisions between the particles of the swarm were generating thermal energy which in turn radiated into surrounding space. and dispersed there. If all of the relevant losses of energy are taken into account mathematically and the full balance of the different forms of energy and angular momentum properly calculated, the result is that the discrepancy is eliminated if the magnitude of the losses is sufficiently high.

The partial transformation of mechanical into non-mechanical energy demonstrates itself not only in respect to the phenomena of planetary rotation; according to our theory, it proves to be the motive form of the whole process of the evolution of the solar system. The principal role of this transformation was brilliantly envisaged by Engels in his Dialectics of Nature.

Our theory explains the formation of planetary satellites, the regularities of distances between different planets and the Sun, and a number of other phenomena.

This means that Soviet science has been able to explain the principal structural features of the solar system, the origin of its planets and other bodies, etc., from one unifying standpoint. The next task is to discover how and from where the pre-planetary swarm had originated and come into existence. The author of this article has hypothetically suggested that it was formed when the Sun captured a part of one of those gas-dust clouds numerous in our stellar system (Galaxy). G.F. Khilmi and I have developed a theory of this capture and demonstrated that it becomes possible under some special conditions. The argument in support of the hypothesis of capture is that it permits explaining the special scale of the solar system and the existence of the prevailing direction of the motions of particles in accordance with the data of modern science. Nevertheless, it is still only a hypothesis. Another viewpoint exists as well, to be more specific, that the matter to build the planets was ejected by the Sun.

All Soviet cosmogonists agree that the conditions for the swarm formation (by means of capture or otherwise) were especially favorable just when the Sun had been formed. Concerning this point, the above mentioned conference declared in its final statement that "the problem of the role played by capture at that stage should be worked out further." The problem of the origin of stars, including the Sun, is also waiting for its solution. Soviet scientists have already approached it rather closely. V. A. Ambartsumian's idea that stars are being formed in groups and the process of star formation is taking place even now may be used as a basis for a materialistic theory of the origin of the stars.

Having outlined the new theory of the origin of the planets, now let us turn to the planet which is of special interest to all of us, i.e., to our Earth. How did it grow gradually from tiny particles? What time did the whole process take? What was the early Earth like and how did it develope further? How has its history affected its current state?

Meteoroids happen to fall onto Earth even now; those meteoric bodies which are insufficiently durable or massive are splinter and evaporate in the atmosphere, failing to reach the Earth's surface (the phenomenon commonly known as "falling stars"). Those are the last stages of the process of the development and growth of the Earth, but they are considerably weakened already, since the majority of the tiny bodies of the swarm had been entrapped by different planets long ago. Earlier this process was much more intensive; the Earth was continually bombarded with big and small meteoroids and thus its surface was becoming more and more pulverized. It is possible to calculate the development of this process and to specify its duration, i.e., to identify the Earth's age. According to my calculations, this age roughly amounts to six or seven billion years. This is the "astronomical" age of the Earth., and it is in good correspondence with the "geological" age of the Earth's crust calculated on the basis of the data about radioactive disintegration. According to such calculations, the age of the crust turns out to be some three to four billion years, which is reasonable, since the crust must be younger than the Earth.

The problem of the initial temperature of the Earth is very important. In our theory, the temperature of the falling particles depends solely on their heating by solar radiation, i.e., it is just the same temperature which the particles acquire while traveling through space. The latter is well known to average about 1^0K. Collisions between the particles and the Earth bring about, naturally, some local warming, but this heat is quickly radiated back into space, and the Earth's temperature increases only slightly. Consequently, the Earth began its life as a cold body.

However, meteoric particles contain, among others, some radioactive elements (uranium, thorium, etc.). These elements generate heat permanently. While being generated inside tiny particles, this heat could not accumulate, it was continuously emanating. However, after the Earth had reached a critical size, the radioactive heat began to accumulate in its depths (the Earth is a poor heat conductor, and so the thermal flow from its inner parts to the surface is very slow). The initially cool Earth was gradually warming up.

It is worth noting that this warming up is almost insensible on the surface where we live, and the temperature there depends solely on the Sun. However, the temperature raises noticeably inside deep mines, and the greater the depth, the higher its temperature. The temperature of the inner parts of the Earth amounts to one thousand degrees. Now we are occupied in calculating the process of the Earth's heating. Radioactive substances are disintegrating slowly, but steadily, so their total amount inside the Earth is permanently decreasing, which means that the total store of energy which can transform into heat has its upper limit. The temperature in the deeper parts of the Earth, not too far from its center, seems to be increasing still, but nearer to the surface it has reached its peak already, and now it is slowly decreasing. Some parts of the crust contain a higher concentration of radioactive elements, and the temperatures there amount to 1000^0- 1300^0 C, so the minerals become melted and magma comes into existence (sometimes it pours out through volcanic craters). Yet, it is rather a local phenomenon.

In respect to this problem the new theory is in sharp disagreement with the opinion reigning in geology that initially the Earth was initially a very hot, even fiery liquid, and it only gradually cooled, being covered by its forming crust (initially this view was based on the Kant-Laplace hypotheses according to which the earliest Earth had passed through a gaseous and liquid state).

The volcanic phenomena were believed in the last century to attest to the liquid state of the Earth's depths. However, further, more precise observations resulted in another outcome: now we know definitely, that the seismic waves from underground (sometimes very deep) earthquakes pass through the Earth as through a firm, but not liquid, body (except possibly only the core where, due to an extremely high pressure amounting top millions of atmospheres, the matter acquires new properties and so it may not be regarded as a solid or liquid substance). There are also some other geophysical data refuting the hypothesis about the initially melted Earth. The real source of the Earth's internal heat was indicated by the discovery of the considerable percentage of radioactive elements contained in all types of rocks (this discovery was made at the beginning of our century).

Nevertheless, the notion of the initially liquid Earth is still widely spread. A great many people learned this idea at school, and now they are reluctant to discount it. Meanwhile, geological data are contradicting this theory. Academician V. I. Vernadsky, a great geochemist, was a life-long partisan of the theory of the "cold" Earth, which he defended on the basis of geological and geochemical data. Some outstanding scientists, for example, Academician L. S. Berg, a geographer, supported Vernadsky on this point, but the majority of geologists recognized his views only very slowly. Our theory provides a cosmogonic justification of Vernadsky's viewpoint.

The process of the internal development of the Earth was already taking place during the period of its growth. As was noted above, the Earth's matter was being formed from the particles of the initial swarm whose chemical composition and density were different (modern meteoroids also differ in terms of their composition). At the first stage, all these blocks and boulders were gathering together chaotically, so the composition of the Earth was on the average the same at different depths. However, when the temperature deep inside the Earth rose to several hundred degrees (such heating makes the matter plastic), there began the exfoliation and differentiation of the Earth's matter due to the action of gravity: the lighter components moved to the surface, whereas the heavier ones experienced a downward movement. In this way, the lighter substances of the Earth's crust accumulated nearer to the surface. Simultaneously, different substances were interacting chemically and thus rocks were forming.

The gravitational stratification of the Earth's matter is a very slow process since it is hindered by high pressure and high viscosity. Even now it has not been completed. Blocks composed of lighter and heavier substances coexist in the Earth's depths, but the lighter ones tend to move upward. Due to very high friction, this motion may be considerably delayed, and the resulting stresses eventually bring about rather violent shifts of the rocks, which we perceive as earthquakes. The motions of some parts of the Earth's matter take place (under the influence of gravity and temperature differences) near the Earth's surface, too. This process has been insufficiently studied.

The earlier geological theories, which were based on outdated cosmogonic views, looked for the causes of mountain formation in the process of the gradual cooling of the planet, which was believed to produce wrinkles on the face of the Earth. This primitive picture is not confirmed by factual data about the geological structure of mountain areas which are now well studied. Many hypotheses were suggested with the aim of overcoming these contradictions, but no one of them has gained general recognition. Progressive scientists take into consideration local distinctions concerning the radiation of radioactive heat in search for an explanation of mountain formation. Our theory adds two more relevant factors: the general warming of the Earth and gravitational stratification.

Naturally, the problem of mountain formation should be solved by geologists themselves. However, modern cosmogony is able to be of assistance to them, since it can provide them with data on the pre-geologic history of our planet, which will join its geological history embracing last two billion years. Cosmogonic theory itself seems to have no practical and technical relevance, but in reality it is able to help geologists to build adequate geological theories whose practical importance is evident (for example, in respect to geological surveys).

The formation of the sea and atmosphere also constitutes a part of the problem of the origin and development of the Earth. Some meteoroids of recent origin contain water. There was especially much water (ice) in those tiny bodies which revolved around the Sun in elongated orbits and thus spent a lot of time far away from the Sun; we mentioned them already while discussing the chemical composition of the planets. Some of them, when approaching the Sun and remaining still cool, fell onto the Earth and released their water and other light substances (methane, ammonia, and carbon dioxide), which process gave origin to the Earth's atmosphere, and also to water on the Earth's surface.

In order to explain the origins of life on earth, it is very important to take into consideration that even at an earlier stage the Earth possessed an atmosphere and water, and this water contained in solution some elementary compounds of carbon, such as methane. The most persuasive and well-developed hypothesis about the origin of life on our planet is A. I. Oparin's. According to it, life came to existence in the course of the process of gradual growth and structurization of chemical compounds consisting of carbon and hydrogen dissolved in the water on the Earth's surface. The new theory may be of some use with regard to this problem too: it maintains that methane and some other elementary hydrocarbonic compounds from the very beginning existed on the Earth's surface in the form of their water solutions, which means that life could appear on the Earth during a very remote epoch and then be developing during several billion years bringing about an astounding variety of more and more complex and perfect life forms.

The fact that our unifying theory explains all the practical features of the solar system makes us believe that this theory is developing in the right direction. The new theory has disproved the opinion, influenced by idealistic philosophy, that this problem cannot be solved at present. But our theory is far from completion. It contains some shortcomings and lacunae; some of its ideas should be checked and clarified; a great effort should be made towards the utilization of cosmogonic results in the areas of geology and geophysics.

Soviet scientists are engaged in fruitful studies of the laws of nature; they are fighting intensively against bourgeois, idealistic theories in all domains of science, including cosmogony. Soviet science has gained considerable success in respect to the problem of the origin of the Earth.

Soviet science, the most progressive science in the world, never stops and never can stop. It is developing on the basis of the genuinely scientific philosophy which is dialectical materialism; it is closely connected with practice. It rests upon the powerful growth of the culture of Soviet people. It is directed and inspired by the great chorus-leader of science, Comrade Stalin.

ON THE ORIGIN OF THE EARTH AND PLANETS (REPLIES TO READERS' LETTERS)

O.Yu.Schmidt

Voprosy Filosofii, no. 5, p. 267-270 (1953)

Letters, sent to me by the readers of my articles published in Bol'shevik and Problems of Philosophy, express their requests to clarify some problems remaining rather vague to them, and also their critical remarks and the exposition of their own viewpoints.

Some manuscripts presenting their authors ideas reveal their basic mistakes, their fundamentally erroneous approaches to the problems of natural sciences. Some authors try to change scientific theories drastically, basing their effort not upon profound understanding of factual data and the objective regularities of natural phenomena, but rather on speculative notions and the dogmatic usage of the wrongly interpreted laws of dialects. The authors of such manuscripts simply discard the laws of nature established by science, neither suggesting anything in return, nor deeming necessary to study the interconnections between different phenomena and processes reflected in these laws. Such an approach to scientific problems leads to breaking with materialism and embracing idealistic philosophy.

The materialistic dialectics of Marxism, disclosing the most general laws of the development of nature and society, should be used to provide a method and philosophic basis for the scientific understanding of our world and the laws governing the development of its objects and phenomena. Dialectics are necessary for the investigation of natural and social phenomena, but they cannot replace this investigation.

Only if he is in command of Marxist dialectic methodology, can a researcher understand the numerous phenomena of nature, realize their real connections with the surrounding conditions, identify the basic factors conditioning the development of the object under study, and eliminate the factors playing no significant role in respect to this object.

But one cannot, interpreting dialectics scholastically, impose any properties upon nature and the objects under examination without taking into consideration the real facts and the results ofscientific observations. It would be completely inappropriate, for example, to maintain, using as the only justification the "conflict" of opposites constituting the basis of all the processes of development, that in the depths of the Earth a struggle between opposing forces should take place under every circumstance which brings about some "pulsation" of the Earth (that is, the Earth's volume periodically increases and decreases, depending on which force gains advantage at the moment). Undoubtedly, the Earth's depths are the arena of a struggle among many opposite forces and trends. Thus, for example, there is a trend towards warming up of the globe due to radioactive heating, which is opposed by the trend towards the Earth's cooling caused by the flow of heat into outer space. The forces tending to increase the Earth's size because of its warming are opposed by the force of gravity, whose action tends to produce the Earth's contraction. However, it is impossible to maintain in advance, without studying the concrete qualities and quantities of the opposing forces and trends, that their struggle has to result in the "pulsation" of the globe, but not in a lasting unidirectional development due to the domination of only one of those trends and forces.

F. Engels characterised the right approach to scientific researches in this way: "All of us agree, that in every scientific area - in the area of nature as well as in the area of history - one should proceed from the factual data at hand; consequently, a natural scientist should proceed from different material forms and different forms of the motion of matter, which means that in the area of theoretical natural sciences one cannot invent some connections and impose on them factual data, it is necessary to extract such connections from available facts, and when the connections have been discovered, their reality should be specially proved as far as possible, on the basis of experience." (Dialectics of Nature)

There is an example of such speculative "invention of connections": some people maintain that the hypothesis of the joint formation of the Sun and the planets is consistent with dialectical materialism, whereas the hypothesis about a later formation of the planets in the neighborhood of the already existing Sun appears to oppose its principles, since it "separates" the processes of the formation of the Sun and its planets. Dialectical materialism insists that all the ties between phenomena should be based on factual knowledge, but not upon speculative assumptions and hypotheses. Regarding this example, a proper solution of the problem of the origin of the Sun and planets will be achievable only on the ground of reliable factual data, whose examination will allow us to understand exactly when and how the gas-dust cloud surrounding the Sun, which eventuallly gave life to the planets, had come into existence. This cloud could be formed, as a result of the same physical process, with the Sun, but it could be captured by the already formed Sun from space.

If the first hypothesis turns out to be true, this will mean that a unified process of the simultaneous formation of the cloud and the Sun happened to merge without any break into the process of planetary formation. In this case, the Sun and its planets had come into existence together, and it would be erroneous to isolate those processes from each other. But, should the second option be adequate, that is, should the examination of phenomena demonstrate that the gas-dust cloud and the Sun came into being due to different processes, possibly even divided by a lengthy period of time, then it will be equally incorrect to consider an unification of these processes, since it is inconsistent with their real development.

At the present time, a conflict of opinions on the problem of the origin of the gas-dust cloud around the Sun is still going on in Soviet science. In my works, I advocate the hypothesis that the cloud was captured from interstellar space by the already formed Sun. Interstellar space, which contains the stellar system (Galaxy) including our Sun, is filled with a dispersed substance whose state is gaseous or dust-like and which is concentrated in some domains in denser gas-dust clouds.

The probability of a solar capture of the gas-dust cloud is negligible under the recent conditions of the Sun's motion and interactions with interstellar nebulas. Therefore, I do not insist on my initial assumption that the capture took place when the Sun was moving around the galactical center under the physical Sun.

According to modern data, a process of the formation of the clusters of stars from gas-dust nebulas is taking place in the surrounding space. After being formed, young stars first move amongst the fragments of their maternal nebula, which circumstance favor's capture.

Some astronomers reject the role of capture and favor the hypothesis of the simultaneous formation of the cloud and the Sun. However, the process of the formation of the stars and the Sun is unstudied as yet, so they cannot formulate their hypothesis concretely.

The problem of the origin of the cloud will hardly be convincingly solved until the origin of the Sun and stars is examined. However, this should not hinder the development of the study of the origin of planets. The results of these studies will be taken into account in the course of work on the Sun's origin. Both cosmogonies, stellar and planetary, should assist each other in the understanding of the origins of all the bodies of the solar system. The intention to postpone working out the questions of planetary cosmogony until the more fundamental problems of stellar cosmogony are solved is, in fact, an echo of agnostism, that is, of scepticism in respect to the power of science. The history of science demonstrates that general and particular problems are usually worked on together, thus mutually fertilising each other and helping the progress of science. Just waiting for a solution of the general problems of stellar cosmogony without working on special problems would mean to stop any progressive movement and to solve no problem at all.

Whereas the hypothesis about the formation of the gas-dust cloud in the neighborhood of the Sun by means of its gravitational capture provokes controversies, the principal theses of my theory of the formation of the planets from this cloud are shared by the majority of Soviet astronomers. It is the first cosmogonic theory which permits a unified explanation of all of the principal features of the solar system.

Stellar masses are hundreds of thousand of times more massive than the Earth, and so in the depths of stars typical temperatures amount to millions and even dozens of million of degrees, and pressures can amount to dozens and hundreds of million atmospheres. Nuclear reactions take place under such conditions, or, to put it another way, atomic nuclei begin to experience different transformations. For a great many stars the reactions in which hydrogen transforms into helium are especially important. These reactions result in gigantic releases of atomic energy that accumulated earlier inside the nuclei. This energy is able to keep the star burning for a duration of billions of years, despite constant emission of large quantities of light from its surface into the surrounding space. Unfortunately, we do not know yet how and where this emitted light transforms into some other forms of matter, which is a considerable flaw in modern scientific knowledge.

The planetary masses are incomparably less in magnitude then the stellar ones, so pressures and temperatures existing inside planets are far from being as high as those in the depths of the stars. This means that nuclear reactions of the stellar type cannot take place inside the Earth. However, there is another source of nuclear energy which does exist there. The Earth's matter contains a small percentage of radioactive elements (uranium, thorium, the radioactive isotope of potassium, etc.). Their nuclei disintegrate, transforming into the nuclei of other elements and releasing energy. This type of energy production plays no significant role in the Sun and other stars, since the released energy is negligibly small in comparison with the nuclear fusion energy.

However, under the natural conditions existing inside the Earth only the process of nuclear fission takes place, which means that the storage of the radioactive elements continually decreases. These elements were formed at the time when the matter of the present Earth existed under completely different conditions. Those conditions were those in which the atomic nuclei experienced very strong interactions from the other nuclei, that is, the conditions similar to ones artificially created by man with the help of some technical devices, which are used nowadays by nuclear physicists.

An idealistic hypothesis, fashionable these days in the West, maintains that our Universe was created two billion years ago as a result of blowing up a tremendously huge "primeval atom." This concept has led to hypotheses about the generation of all the atoms, including radioactive ones, during the very first minutes of the process of expansion, when the matter of the Universe was supposedly extremely hot and dense. Some Western scientists' attempts to discover this state of matter in our Universe turned out to bring no result. Moreover, it was demonstrated that the idea about the elements creation from the super-dense state of matter is inconsistent with their observable properties. Nowadays, Soviet scientists are engaged in successful research on the processes developing inside interstellar gas-dust clouds. Due to the motions of these clouds, consisting to a considerable degree of electrically charged atoms and dust particles, very strong electromagnetic fields, capable to accelerate charged atoms up to very high speeds, are being generated. One may hope that future studies of such fields and the atoms experiencing acceleration inside them will contribute to our understanding of the origin of radioactive elements as well as cosmic rays.

It is possible to calculate, on the basis of our knowledge about the present percentage of radioactive elements and their decay rate, that some five to seven billion years ago the cosmic matter, which our Earth is made from, existed under such unusual conditions that radioactive elements may have been created. These calculations are in good agreement with the estimations of the "age" of the Earth's matter made by other methods.

- - - - - - - - - -

The conception of the initially extremely hot, melted Earth has been reigning over the Earth sciences for the duration of more than two centuries. Long ago people learned that the depths of the Earth were rather hot during our geologic period. This fact is confirmed by a great quantity of data (temperature measurements inside deep mines demonstrate the steady increase in temperature as one moves downward; volcanic eruptions produce liquid lava whose temperature amounts to 1200^{0}C). Moreover, it has been established that many kinds of rocks were formed as a result of cooling and hardening an initially melted substance. After Laplace's hypothesis, which seemed to lead to the idea of the initially hot Earth, had been suggested, the conception of the "fiery-liquid initial stage" became widely adopted. According to the scientists' views, that was the first stage in the development of natural phenomena, which nowadays exist around us upon the Earth's surface.

In 1906, I.V. Stalin, in his work <u>Anarchism or Socialism</u> ,presented the development of nature as an example of development in general. In his own words:

> "Everything in our world changes and develops, but <u>how</u> does this change proceed, and in <u>what form</u> does this development take place? We are aware, for instance, that long ago the Earth existed as a melted fiery mass of matter, further it cooled step by step, then plants and animals came into existence, the development of animals was followed by the development of a certain kind of monkey, and finally this developmental process resulted in the emergence of man. Such was the general way of the development of nature." (Collected works, Vol. 1, p. 310-311)

This mode of the development of nature - which begins from the formation of the Earth itself, further brings about the emergence of life on its surface, and finally results in the emergence of man - has been corroborated by all the data of modern science. However, science itself also develops; as Comrade Stalin teaches us, novel facts and discoveries compel us to clarify, correct, and sometimes even reconsider former scientific views and theories. "Science is called science just because it does not recognize any fetishes, is not afraid to raise its hand against everything that is archaic and outdated, and lends an attentive ear to the voice of experience and practice." (I.V. Stalin, <u>Problems of Leninism</u>, p. 540.)

The radioactive elements were discovered in 1896. Later, in 1903, scientists discovered that radioactive decay was accompanied by the release of energy capable of warming up the surrounding substance. As was established in 1906, all rocks contain some small amounts of radioactive elements. Calculations demonstrated that the radioactive decay of these elements had to bring about the release of heat whose magnitude exceeded the amount of thermal energy radiated by the Earth into space. One could expect this discovery to revolutionize the problem of the Earth's thermal history immediatedly. However, the belief in the Earth's gradual cooling from the initially melted state had become so deeply enrooted in contemporaneous scientific thinking, that scientists agreed to adhere to a very artificial assumption, according to which the radioactive elements exist in the Earth's crust solely, and are completely absent in its depths.

Academician V.I. Vernadsky had been fighting against the theory of the hot origin of the Earth for almost forty years until his death, calling for the recognition of the radioactive origin of the Earth's internal heat. His struggle was unsuccessful, since his ideas could not be reconciled with the dominant cosmogonic theories. Vernadsky's views gained recognition only after the complete failure of Jeans' hypothesis. During those years I had developed a new cosmogonic theory abou the Earth's initially cool state.

During the Earth's formation in the process of the accretion of cool firm particles, the temperature inside its depths did not exceed several hundred degrees. Further, the accumulation of radioactive heat raised this temperature up to several thousand degrees. Billions of years ago, the amount of these elements in the Earth's substance was several times more than now, thus the amount of the emanated heat was equally more. Therefore,the Earth's temperature had been increasing rather quickly during the early period of its existence.

The Earth's internal heat is continuously conducted away from its depths towards its surface and dissipates into the cool cosmic space surrounding the Earth. However, the processes on the surface remain uninfluenced by this heat, since its magnitude is negligable in comparison with the solar heating of the surface, which is five thousand times more. The thermal flow from the Earth's depths is likely to have experienced some changes during the duration of the Earth's history, but it played no role in its climate.

- - - - - - - - - -

The Sun and the Earth, as well as the other planets, are tied together by gravitational forces, and so they constitute a unified system of celestial bodies. Gravity is the principal factor in the Earth's motion in space. This force causes the Earth to revolve around the Sun and, at the same time, to travel along with the Sun amongst the other stars of our Galaxy. The Sun's motion is also basically regulated by gravitational forces from the side of the other stars as well as nebulas.

Magnetic and electrical forces play an insignificant role in the Earth's and other planets' motions, although the Earth has its own magnetic field, and the Sun possesses magnetic as well as electric fields. Therefore, the astronomers, while examining the planetary motions, take into account only gravitational forces and exclude magnetic and electric forces from consideration. This does not mean, however, that the phenomena, influenced considerably or even principally by electrical and magnetic forces, do not exist in space. Thus, one should take into account light pressure in the course of the analysis of the motions of relatively small bodies (whose size does not exceed one kilometer) belonging to our solar system. In every concrete case the problem of which forces should be taken into consideration and which may be excluded from consideration has to be solved on the basis of precise calculations characterising numerically the role of different forces.

Such calculations demonstrate that gravitational forces play the principal role in the interactions amongst sizable celestial bodies, like planets and stars, thus all the other forces may be as a rule (but not under every circumstance), excluded from consideration. Therefore, astronomers rely on the Newtonian law of gravity while dealing with the motions of such bodies. This does not mean, however, that this law is an absolute truth. As every law of science, the law of gravity is a generalization of some regularities which take place when the bodies move and interact under certain special circumstances. Studyng the gravitational interactions between extremely remote bodies, or between the bodies moving with extremely high speeds, we come to the conclusion that this law must be corrected. Such a correction is provided by the general theory of relativity. However, that is just a correction, not a complete change of the law of gravity, since the new modified law brings about virtually the same results while being used for the calculation of the gravitational forces from the same conditions which had been taken into account in the process of the initial establishment of the law.

It is necessary to emphasize once again that the question, whether some forces and processes should or should not be taken into consideration in a given case under study, has to be solved on the basis of the precise scientific analysis of their role in the case.

On the Origin of the Asteroids

O. Yu. Schmidt
Dokl. Akad. Nauk SSSR **96**, 449 (1954)
Translated by Ludmila Sgarlata

Asteroids are minuscule planetlike bodies revolving about the Sun in orbits lying, principally, between those of Mars and Jupiter. Asteroids usually are regarded as bodies filling the "gap" in the well-known empirical rule of Titius and Bode which gives the distances of the planets from the Sun. The hypothesis of Olbers on the origin of asteroids by explosion of a primordial planet, which supposedly existed in the past between Mars and Jupiter, still has the adherents, in spite of its failure to provide the necessary proof.

Recently, Fesenkov[1] made an attempt to revive this hypothesis of Olbers by assuming that the primordial planet had an orbit with a large eccentricity and that, during one of its revolutions about the Sun, it came so close to Jupiter that Jupiter's attraction disturbed the pressure equilibrium in the deep interior of this primordial planet, and, because of this, the temperature of this planet increased by a jump, and the planet exploded.

However, this variant of Olbers's hypothesis also is in contradiction with a number of facts. First, the orbits of the planets do not possess large eccentricities, but are nearly circular; this phenomenon, as has been shown,[2,3] is in agreement with the law of formation of planets through accretion of small particles. Second, the effect of Jupiter's attraction could not lead to an explosion since the adiabatic expansion of the planetary matter is accompanied by its cooling, not heating. Third, the variety of planetary orbits cannot be explained by a single explosion. One can understand, to a certain degree, the tendency to look for the origin of the asteroids in disintegration of a single relatively large planet, if one remembers the researches first of Roche and then of Jeans, who proved the impossibility of the formation of small bodies from the gaseous medium. But now, when it has been proved that the planets originated from a medium which consisted of gas and dust, and when it has been shown that formation from such a medium of gas and dust of bodies of the size of the asteroids is a possibility,[4] there is no reason in insisting on the formation of asteroids via an explosion or disintegration of a primordial planet.

In light of the theory of the origin of the planets which we shall develop, there is no necessity in any special hypotheses for the origin of the asteroids since their peculiarities follow from the general laws set by the theory.

The works of Soviet cosmogonists[3–5] have explained the cause of the separation of the planets into two groups sharply different in their planetary masses and densities. One group, the group of the so-called "terrestrial" planets (Mercury, Venus, Earth, and Mars), consists of comparatively small planets of large densities (4 to 5 g/cm^3). The second group includes the giant planets (Jupiter, Saturn, Uranus, and Neptune), much more massive (14.5 to 318 times as massive as the Earth), but less dense (0.7 to 1.5–2 g/cm^3). In the above-mentioned works, the causes of uneven distribution of the planetary matter were given, that is, of the relative scarcity of matter in the inner region of the primordial planetary swarm compared with the region of the giant planets. The most effective of these causes was the different amount of heat received by the particles of the primordial planetary swarm from the Sun compared to the amount of heat received by the matter in the region of the giant planets. This explains the absence of the frozen volatile particles in the composition of the primordial particles.

It can be shown that, in the case of a transparent space, a black or gray body heated equally all round, will have a mean surface temperature given by the formulas

$$T_1 = \frac{218\ \mathrm{K}}{\sqrt{R}} \quad \text{and} \quad T_2 = \frac{393\ \mathrm{K}}{\sqrt{R}},$$

where K is the temperature in degrees Kelvin reckoned from the absolute zero and R is the distance of the body from the Sun, expressed in astronomical units (A.U.). The first formula relates to small bodies heated from different sides (e.g., dust particles). The second formula refers to larger bodies always having the same face toward the Sun (for instance, a perfectly radiating asteroid or a planet without an atmosphere). In the case of a perfectly radiating surface, the heat received from the Sun does not have time enough to spread all over the body and is immediately reflected.

Let us investigate the temperature of the particles of the primordial swarm in those stages of its evolution when the space could have been considered transparent. By applying the first formula to the solid particles in the region of the Earth's orbit ($R = 1$), we can see that they were heated to a temperature above 0 °C, so that such prevalent volatile substances as methane and ammonia could not freeze to the surface of solid particles, but were in a gaseous state. At Jupiter's distance from the Sun, the temperature was −150 °C, and both methane and ammonia, depending on their partial pressure, either froze to the surface of the particles or slowly evaporated. But in this stage of the evolution of the primordial asteroidal swarm, when it was opaque, and when the temperature of the particles in this zone was still considerably lower, freezing was occurring at practically any partial pressure. This was the cause of the above-mentioned difference in the densities and masses of the two groups of planets.

The orbits of most of the asteroids lie in a belt at a distance of approximately 2.8 A.U. from the Sun. According to the above-given formula, the temperature of the particles is near −100 °C. At still lower temperatures, when the primordial swarm was opaque, both methane and ammonia could have been in a solid state in the form of "ice," so that in its constitution the region of the asteroids was close to that of the giant planets. However, the larger bodies formed through accretion of small particles and we must apply the second formula. For the distance R equal to 2.8 A.U., it gives $T = -38$ °C. This means that, although in the region of the asteroids, at some time in the past, particles of "ice" could have existed, after their aggregation into larger bodies, these "ice formations" must have begun to evaporate, so that into the composition of the planet in this region only the more refractory matter could have entered, as in the case of the Earth, with an addition, perhaps, of a large quantity of water. Thus, from the point of view of the process of formation of the large bodies, the belt of asteroids belongs to the region of the terrestrial group of planets impoverished in the matter which could compose such bodies. If a single large planet was formed in the region of the asteroids, it would be small, like the Earth and Mars, not like Jupiter. But a single planet could not form there. The process of formation of the planets in this region stopped at an intermediate stage of formation of smaller bodies. This was conditioned by the proximity of massive Jupiter and the above-mentioned peculiarities of temperature and composition of the bodies formed in that zone.

Formation of the planets begins with the appearance of numerous bodies of asteroidal dimension ("embryo" planets), growing by way of accretion of the particles and breaking up in mutual collisions. The laws governing the process of growth, reflected in the laws governing the planetary distances, do not allow for the formation of two large bodies describing coplanar orbits close to each other.[3] In the early stages of the evolution of the protoplanetary swarm perturbations of the growing planet Jupiter exercised great influence on the movements of the bodies that originated in the region of the asteroids by increasing the mean eccentricities and inclinations of their orbits, thus preventing their further recombination. When bodies describe nearly circular coplanar orbits, the rate of their growth is greater than the rate of their disintegration. But with the increase of eccentricities and inclinations, that is, with the increase of relative velocities, the process of disintegration begins to take the upper hand.

Limiting the position of the belt of asteroids, which led to the fact that change in the temperature of the particles during the process of their unification into larger bodies was accompanied by substantial changes in their chemical composition, helped Jupiter's perturbations take their effect. Evaporation of the volatile matter from the bodies in the process of formation either led to their disintegration or, weakening their durability, led to their breaking up in collisions. By the same token, evaporation slowed down the process of formation of large bodies in the region of the asteroids and gave time for perturbations caused by Jupiter to change their orbits.

At the present time, the combined mass of all the known asteroids is probably one-thousandth that of the Earth. In consequence of the breaking of the asteroids (in mutual collisions and in collisions with still smaller bodies of the meteors) and due to the falling of their small fragments to the Sun (as a result of radiational friction), the combined mass of the asteroids is constantly diminishing; in the past it was larger than now. However, there are reasons to believe that it was always rather small. In the first place, massive Jupiter has been absorbing and uniting all those particles which moved into its zone from the neighboring zones. In the second place, by perturbing the motions of the bodies and particles moving in the region of the asteroids, Jupiter has prevented their unification, thus giving time for the effect of radiational friction to display itself by shifting small particles from the region of the asteroids in the direction of the Sun.

The very large role played by these factors in the region of the asteroids points to the fact that even in the region of Mars, lying considerably farther from Jupiter, the influence of so massive a neighbor as Jupiter conditioned the relatively small mass of Mars.

Thus the asteroids cannot be considered a result of an explosion or disintegration of a large planet. They are bodies for which the process of unification of the particles of the primordial swarm stopped at an intermediate stage because of the position of their region between the planetary families of different composition and masses. This process then reversed —from the predominance of unification of the particles and bodies to the predominance of breaking into pieces and disintegration.

Geophysical Institute of the USSR Academy of Sciences
Received 4-5-1954

[1] V. G. Fesenkov, Vopr. Kosmog. **1**, 92 (1952).
[2] O. Yu. Schmidt, Izv. Akad. Nauk SSSR Ser. Fiz. **14**, No. 1 (1950).
[3] O. Yu. Schmidt, *Four Lectures on the Theory of the Origin of the Earth* (USSR Academy of Sciences, Moscow, 1950).
[4] L. E. Gurewich and A. J. Lebedinskii, Izv. Akad. Nauk SSSR Ser. Fiz. **14**, No. 6 (1950).
[5] B. Yu. Levin, Priroda (Moscow) **1949**, No. 10.

On the Origin of Comets (abstract) (unpublished notes, 1954)

O. Yu. Schmidt

From: "O proiskhozhdenii komet," unpublished notes, *Izbrannye Trudy: Geofizika i Kosmogohiya* (Izdatelstvo Akademii Nauk SSSR, Moscow, 1960), pp. 204–7

First, Schmidt lists some basic facts about the comets, including those about cometary tails. He maintains that the visibility of the tails is connected with the existence of radiating gases there, and that the gases themselves are mainly produced in the course of the evaporation of some volatile frozen substances when a comet approaches the Sun. Thus, the problem of the origin of these "ices" arises. In accordance with his theory of the origin of the planetary system, Schmidt insists that the gases had formed a component of the protoplanetary swarm which happened to give life to the planets as well as to the comets. The comets, wherein the concentration of the "iced" matter is considerable, could originate only from the outer zone of the swarm, which was more distant from the Sun. "Thus, according to our theory, the comets are not some unique, esoteric bodies, but, quite the contrary, they display that form of matter which was *typical* at a certain stage of the development of the solar system." Cometary orbits cannot remain perfectly stable, since there are a number of physical factors tending to change them. Particularly, one of these factors is intrinsically connected with Schmidt's theory of the origin of planets. The comets, while advancing along their orbits, will tend to absorb and accumulate dust and gas dispersed in space. This process results in changing their masses, energies, and angular momentums, which, in turn, brings about some modifications of their orbits. In the last paragraph Schmidt stresses his belief that his cosmogonic theory "explains all of the principal features of the structure and the composition of comets, as well as the character of cometary orbits."

III. The Protoplanetary Cloud: General Theories of Planet Formation

The Formation of the Planets. I. Gravitational Condensation

L. E. Gurevich and A. I. Lebedinskii
Translated from: Izv. Akad. Nauk USSR Ser. Fiz. **14**, 765 (1950)

In this part it is shown that if the Sun had been surrounded by a cloud possessing the angular momentum and mass of the contemporary planetary system, then in the course of time, part of the matter of this cloud which had converted into solid particles was bound to flatten. On attaining a certain critical thickness, small in comparison with the dimensions of the cloud, it disintegrated into a large number of separate condensations with masses not exceeding the masses of the asteroids. In Sec. 1 the quasistatic theory of gradual flattening of the cloud due to a decrease in energy of its constituent particles as a result of inelastic collisions is given. For every stage in the process of flattening of the system of particles it is possible to assign a certain "temperature" which is, generally speaking, dependent on the distance from the Sun, R, and connected in a well-defined way with the dispersions of velocities of the particles and the thickness of the cloud. In Sec. 2 it is shown that on the attainment of a critical thickness H_{kp} by the cloud it becomes unstable and disintegrates into condensations of radius r on an order exceeding H_{kp}. In Sec. 3 it is shown that at the moment of the beginning of condensation, the particles moved with relative velocities on the order of meters per second. The dimensions and masses of condensations which possessed the greatest magnitude in the zone of Jupiter ($m \approx 3.20^{22}$ g) are calculated.

INTRODUCTION

One of the basic problems of cosmogony is the mechanism of the formation of planets from the cloud of diffuse matter that surrounded the Sun and possessed the mass and the total angular momentum approximately equal to the values of these magnitudes for the contemporary planetary system.

The origin of the primeval cloud has, so far, not been explained. Different hypotheses explain it in different ways.

But the state of a greater part of the matter of future planets, quite independently of the origin of the cloud, was bound to become solid just before their formation, because the gravitational condensation of gases was not possible at the temperature which they had to have at the distances of planets from the Sun.

If the cloud was thrown out from the Sun in a gaseous state, then the matter of future planets must have condensed into solid particles before their formation.

The particles of the cloud collided among themselves in an inelastic way and so a part of their energy was converted irreversibly into heat and emitted with radiation. Consequently, the kinetic energy of the particles was reduced, and therefore dispersion of velocities was reduced also. But as the angular momentum remained unchanged, the system was becoming more and more flat, quite independently of the transformations which accompanied the collisions of the particles. Such processes as the breaking down of particles as well as their partial or complete vaporization with subsequent condensations causing supplementary irreversible loss of energy only accelerated the condensation of the system. Generally speaking, these processes could be accompanied by partial dissipation of the heated gas which had vaporized during the mutual collisions of particles. However, upon sufficient flattening of the cloud the relative velocities of the colliding particles were much less than the parabolic velocity, and therefore the energy generated during the collisions of the particles was not sufficient for their dissipation out of the cloud. Due to this, the gas was again bound to soon condense into solid particles. This is evident in case of the vapors of refractory substances, and for such (substances) as methane, ammonia, and other volatile substances (this will be proved in part III of the work).

Upon sufficient flattening of the cloud, gravitational condensation is bound to take place, which results in the formation of condensations with masses not exceeding the masses of the asteroids. Further, these condensations could be united into larger bodies.

These qualitative concepts have long since been worked out in works on cosmogony. A majority of them we find in Poincaré; recently Edgeworth[1] evaluated the upper limit of the mass of condensation by using the criteria of Maxwell and Jeffreys which determine stability in relation to tidal forces. In this he proceeded from the hypothesis of the capture of a dust cloud by the Sun, which agrees with the hypothesis put forward by Schmidt[2] in 1943, as well as from the qualitative representations pointed out by these authors above.

The evaluation of Edgeworth is extremely inaccurate due to two causes. First, the criterion used by him (which is related to tidal forces) is not applicable to this case; second, the condensed masses were very flattened (oblate) formations, whereas Edgeworth regarded their dimensions in different directions as comparable.

In this work we shall examine the process of the formation of planets from the dust cloud. In the first part, an account of the theory of the primeval process of condensation is given, and in the other parts an explanation of certain peculiarities of the solar system has been given on the basis of this theory.

We arrive at the conclusion that *the only possible equilibrium state of matter, in the dust cloud surrounding the Sun, was the unification of this matter into a few large planets*, if the mass of the cloud was sufficiently great; if, on the other hand, the mass of the cloud had been small, a ring would have formed around the Sun, like that of Saturn which, apparently, originated in exactly the same way.

1. THE PROCESS OF THE FLATTENING OF THE CLOUD

We assume that the flattening of the cloud was taking place quasistatically, so that at every given moment the particles were distributed in equilibrium with respect to their velocities and to the altitude of the cloud.

Due to the rotation of the cloud, the particle velocity distribution function at a definite distance from the sun is known[3] to have the form:

$$f(v)=Ae^{-(\epsilon-\omega\kappa)/\theta}, \tag{I.1}$$

where ϵ is the energy of a particle, κ is the component of its angular momentum along the Z axis, coinciding with the axis of rotation of the cloud, and A, ω, and θ are the parameters of the distribution.

We shall examine the sufficiently flattened systems, wherein the equilibrium distribution along the vertical coordinate Z sets in rapidly but the exchange of angular momentum and energy of particles located at different distances R from the Sun takes place slowly. Therefore the parameters of distribution are functions of R, but not of Z. θ is 2/3 of the average kinetic energy of particles which we shall henceforth call the "temperature" of particles. The parameter ω is the average angular velocity of rotation of the cloud at the distance R from the Sun, so that $R\omega$ is the local average velocity. In stellar statistics it is called the velocity of the centroid.

In cylindrical coordinates

$$\begin{aligned}\epsilon-\omega\kappa&=(m_1/2)(\dot{R}^2+\dot{Z}^2+R^2\dot{\phi}^2)-m_1\omega R^2\dot{\phi}-m_1\Phi(R,Z)\\&=(+m_1/2)[\dot{R}^2+\dot{Z}^2+R^2(\dot{\phi}-\omega)^2]\\&\quad-\tfrac{1}{2}m_1\omega^2R^2-m_1\Phi(R,Z)\\&=\epsilon'-\tfrac{1}{2}m_1\omega^2R^2-m_1\Phi(R,Z),\end{aligned} \tag{I.2}$$

where m_1 is the mass of the particle, and $-\Phi(R,Z)$ is the gravitational potential. Therefore

$$f(R,Z,v)=A(R)\exp\left[-\left(\frac{\epsilon'-\frac{1}{2}m_1\omega^2R^2-m_1\Phi(R,Z)}{\theta(R)}\right)\right]. \tag{I.3}$$

$\Phi(R,Z)$ is the sum of the potential of the Sun and the potential of the cloud:

$$\Phi=\Phi_1+\Phi_2, \tag{a}$$

where

$$\Phi_1=\frac{\gamma M_\odot}{\sqrt{(R^2+Z^2)}}=\frac{\gamma M_\odot}{R}-\frac{1}{2}\frac{\gamma M_\odot Z^2}{R^3}+\cdots, \tag{b}$$

because upon sufficient flattening of the cloud, the potential at small altitudes, $|Z|\ll R$, is dominant.

We shall calculate Φ_2 by regarding the cloud as flat. Upon sufficient flattening, the cloud may be considered as a highly flattened ellipsoid of rotation of which the major semi-axis is equal to R_0, whereas the minor one (whose length is equal to the half-thickness of the cloud) is $H_0\ll R_0$. In conformity with Ref. 4 the potential inside a homogeneous compressed ellipsoid of rotation is described by the following formula:

$$\begin{aligned}\Phi_2=(2\pi\gamma\rho)\,\frac{(1+l^2)}{l^3}\,&[H_0^2l^2\arctan(l)-\tfrac{1}{2}R^2(\arctan\{l\\&-[1/(1-l^2)]\})-Z^2[l-\arctan(l)]],\end{aligned} \tag{I.4}$$

where

$$l=\frac{\sqrt{(R_0^2+H_0^2)}}{H_0}. \tag{c}$$

When $H_0\ll R_0$,

$$\begin{aligned}\Phi_2&=-(2\pi\gamma\rho)[Z^2+(\pi/4)(H_0/R_0)R^2-(\pi/2)H_2R_0]\\&=-2\pi\gamma\rho Z^2+\Phi_2'(R).\end{aligned} \tag{I.5}$$

The potential depends on Z in the same way as an infinite homogeneous plane layer does. In reality, the density of the cloud depends both on R and Z. However, the dependence on R is much weaker than on Z, and, therefore, it is possible to apply Eq. (I.5) to the cloud without much error regarding ρ as a slowly changing function of R. As regards the dependence of ρ on Z, the gravitational condensation, which will be examined below, takes place within the densest parts of the cloud (near the plane of its symmetry), and the error arising out of the application of expression (I.5) to it will relate to those less dense parts of the cloud which play only a minor role in the processes under study.

When $H_0\ll R_0$, it is possible to substitute in Eq. (I.5) $\beta(\sigma/H)$ for the averaged density, where H is the effective half-thickness of the cloud at a distance R from the Sun, σ is the total mass of a column of a unitary cross section, perpendicular to the plane of symmetry of the cloud, and β is the numerical coefficient which we shall define below, together with H. It is clear that σ also depends on R.

Then we shall get

$$\Phi_2=\frac{-2\pi\gamma\sigma(Z^2)}{H}+\Phi_2'(R), \tag{I.6}$$

but the complete potential is

$$\Phi=-[(2\pi\beta\gamma\sigma)/(H)+(\gamma M_\odot)/(2R^3)]Z^2+\Phi''(R). \tag{I.7}$$

Let us note that the simplifications that we have derived and the formulas obtained are completely analogous to those which are used in stellar astronomy for the study of the distribution of stars in the direction perpendicular to the galactic plane.

The average velocity ωR entering into Eq. (I.3) also changes in the course of the flattening of the cloud. To explain this, we write down averaged equations of the motion of particles for the stationary state[5]:

$$\frac{\partial(\rho\overline{V_R^2})}{\partial R}+\rho\,\frac{(\overline{V_R^2}-\overline{V_\phi^2})}{R}=(\rho)\,\frac{\partial\Phi}{\partial R} \tag{I.8}$$

and

$$\frac{\partial(\rho\overline{V_Z^2})}{\partial Z}=(\rho)\,\frac{\partial\Phi}{\partial R}. \tag{I.9}$$

In our case of Maxwell's distribution,

$$\overline{V_R^2}=\overline{V_Z^2}=\overline{(V_\phi-\omega R)^2}=\frac{\theta}{m_1}. \tag{d}$$

But

$$\overline{(V_\phi-\omega R)^2}=\overline{V_\phi^2}-\omega^2R^2 \tag{e}$$

and hence

$$\overline{V_\phi^2}=\frac{\theta}{m_1}+\omega^2R^2. \tag{f}$$

Now Eq. (I.8) gives

$$\frac{\theta}{m_1}\frac{\partial\ \ln\rho}{\partial R}-\omega^2R^2=\frac{\partial\Phi}{\partial R}. \tag{g}$$

But as

$$V_0(R)=\sqrt{(-R\partial\Phi/\partial R)} \tag{h}$$

is the circular velocity, so

$$\omega^2R^2-V_0^2=(\theta/m_1)(\partial\ \ln\rho/\partial\ \ln R). \tag{I.10}$$

Equation (I.9) gives the barometric formula

$$\begin{aligned}\rho(Z)&=\rho_{\max}\exp[(m_1\Phi)/\theta]\\&=\rho_{\max}\exp\{-[m_1/\theta][(2\pi\beta\gamma\sigma)/H\\&\quad+(\gamma M_\odot)/(2R^2)]Z^2\}\\&=\rho_{\max}e^{-aZ^2}\end{aligned} \tag{I.11}$$

thus obtained irrespective of Eq. (I.3), but in accordance with it. We shall now determine the effective half-thickness of the layer

$$\begin{aligned}H=1/\sqrt{a}=\{&(\theta/m_1)[(2\pi\beta\gamma\sigma)/(H)\\&+(\gamma M_\odot)/(2R^3)]^{-1}\}^{1/2}.\end{aligned} \tag{I.12}$$

Henceforth we shall employ the notation

$$\omega_0^2=(4\pi/3)\gamma\rho_0=(\gamma M_\odot)/R^3, \tag{I.13}$$

where ω_0 is the angular velocity of circular motion, and ρ_0 is the density which would have been effective if a sphere with radius R were filled with the mass of the sun. These characteristic dimensions are used below in many equations and therefore are convenient notations.

From Eq. (I.12) it follows that

$$\theta=2\pi\gamma Hm_1(\beta\sigma+\tfrac{1}{3}\rho_0H), \tag{I.14}$$

that is, the thickness of the cloud decreases in proportion to the dissipation of its energy, and when the θ's are small, the thickness is directly proportional to θ. Hence the average value of the square of the components of velocity:

$$\overline{V_Z^2}=\overline{V_R^2}=\overline{V_\phi^2}-\omega^2R^2=\theta/m_1=2\pi\gamma H(\beta\sigma+\tfrac{1}{3}\rho_0H). \tag{I.15}$$

The total mass of the column of a unitary cross section is

$$\sigma=\rho_{\max}\int_{-\infty}^{\infty}e^{-aZ^2}dZ=\rho_{\max}\sqrt{\pi/a}=\sqrt{\pi}\rho_{\max}H. \tag{I.16}$$

On the other hand, we shall define the average density of matter of the cloud within the limits $-H<Z<+H$, which is essential in the processes examined below, with the help of the equation

$$\bar{\rho}=\rho_{\max}\int_0^1 e^{-x^2}dx=[(0.843\sqrt{\pi})/2]\rho_{\max}, \tag{i}$$

whence

$$\sigma=(2/0.843)H\bar{\rho}=2.37H\bar{\rho}=(1/\beta)H\bar{\rho}. \tag{I.17}$$

Therefore

$$\beta=0.42. \tag{I.18}$$

Henceforth instead of $\bar{\rho}_{\max}$ we shall write simply ρ.

2. CONDITIONS OF GRAVITATIONAL CONDENSATION

Upon the flattening of the cloud and the decrease of the velocity of irregular motion of its particles their gravitational condensation becomes possible. This condensation is obstructed by the tidal forces of the Sun and by the relative motion of particles which is formed by their irregular relative motion which depends on temperature θ, and the regular relative motion determined by the dependence of ωR on the distance R. The condensation of a certain mass m becomes possible when its total energy (the sum of potential and kinetic energies) in absolute magnitude exceeds its kinetic energy or when the absolute value of potential energy is 2 times the kinetic energy, which is the same thing.

We may use the factor 2, which comes from the virial theorem, only for gravitational forces, not for tidal forces. However, by using it we shall obtain, in all cases, a sufficient condition for condensation.

We shall show that condensation becomes possible only after the cloud has transformed into a very flattened system, and that the separate clots of matter formed are of such volumes that their horizontal dimensions considerably exceed their vertical dimensions along the Z axis.

The horizontal dimensions of clots along the direction of the radius of the cloud and along the perpendicular to the radius are different from each other because tidal forces and the energies of relative motion in these directions have different values. However, the vertical dimension of condensation due to the strong flattening of the cloud is much less than both horizontal dimensions, whereas there are no reasons to regard the latter ones as different from one other by an order of magnitude. Therefore in this part of the report we limit ourselves to an examination of rotation with semiaxes h and r satisfying the inequality

$$h \ll r \ll R. \tag{j}$$

The potential energy of such a strongly flattened ellipsoid of rotation with semiaxes r and h is

$$U = \tfrac{1}{2} \int \rho \Phi \, dv = -\frac{\pi}{2} \int_{-h}^{+h} \Phi \xi^2 dz. \tag{k}$$

Substituting expression (I.4) for Φ and the equation of the ellipsoid

$$\xi^2 = r^2(1 - z^2/h^2) \tag{l}$$

for ξ^2, under the symbol of the integral, we obtain as the result

$$U = \frac{3}{20} \frac{\gamma m^2}{h} \frac{1+l^2}{l^3} \left[-(1+5l^2) \frac{h^2}{r^2} \arctan l + \frac{h^2}{r^2} + 2\left(\arctan l - \frac{l}{1+l^2} \right) \right], \tag{m}$$

where

$$l = \frac{\sqrt{r^2 - h^2}}{h} \tag{n}$$

and m is the mass enclosed inside the ellipsoid, which is equal to $(4\pi/3)r^2 h$.

For a very flat ellipsoid

$$l = (r/h) \gg 1. \tag{o}$$

That means that the average energy per unit of mass inside the delineated volume is equal to

$$U = \frac{-9\pi}{40} \frac{\gamma m}{r} = \frac{-3\pi^2}{10} \gamma \rho r h = \frac{-3\pi^2}{10} \beta \gamma \sigma r \frac{h}{H}. \tag{I.19}$$

This expression is only slightly different from the corresponding formula for a sphere, $-\tfrac{3}{5}(\gamma m/r)$ (in relation to I.18), and therefore the conclusions based on it are qualitatively true, irrespective of the degree of compression.

The energy of irregular motion per unit of mass

$$\frac{\overline{\epsilon'}}{m_1} = \frac{3}{2} \frac{\theta}{m_1} \tag{I.20}$$

where m_1 is the mass of the particle.

The potential energy of any point of the delineated volume in the field of the tidal forces of the Sun with reference to the center of the volume is

$$\Phi' = \tfrac{1}{2}\omega_0^2(2y^2 - x^2 - z^2),$$

where y is measured from the center of the volume in the direction of the radius vector drawn from the Sun, and x is in a direction perpendicular to that.

Because the process of condensation, as we will see, takes place with a great flattening of the cloud and so, according to (I.14), at a low temperature θ, then, as is indicated by (I.10), ωR is very close to the circular velocity in the unflattened field of the Sun and the cloud, which only slightly differs from the circular velocity in the field of the Sun alone due to the insignificance of the mass of planets in comparison with the mass of the Sun. Therefore let us take

$$\omega R = V_0 = \sqrt{\gamma M_\odot / R} = \omega_0 R. \tag{I.21}$$

The energy of a well-ordered relative motion per unit of mass is $\tfrac{1}{2}(\Delta v)^2$, where Δv is the difference between the circular velocities at the point (x,y) and in the center of the volume, is equal to

$$\epsilon'' = \tfrac{1}{2}(\Delta v)^2 = \frac{1}{2}\left[\left(\frac{y\,dV_0}{dR} \right)^2 + \left(\frac{xV_0}{R} \right)^2 \right], \tag{p}$$

where the first component is determined by the change of magnitude V_0 of circular velocity due to the shift along the radius at a distance y, and the second component is determined by the alteration in its direction due to the displacement perpendicular to the radius at a distance x.

Taking V_0 from Eq. (I.21), we obtain

$$\epsilon'' = (\tfrac{1}{2})(\Delta v)^2 = \tfrac{1}{2}\omega_0^2(\tfrac{1}{4}y^2 + y^2). \tag{q}$$

We shall sum up these magnitudes and integrate over the isolated volume keeping in mind that $Y = \xi \cos \phi$ and $dV = \xi d\xi dz d\phi$. Then

$$(1/V) \int (\Phi' + \epsilon'')dV = (\omega_0^2/2V) \int (\tfrac{9}{4}y^2 - z^2)dV = (\tfrac{9}{10}r^2 - \tfrac{1}{10}h^2)\omega_0^2. \tag{I.22}$$

Hence the criterion for the possibility of condensation has the form

$$(3\pi^2/10)\beta\gamma\sigma r(h/H) > (3\theta/m_1) + (\tfrac{9}{20}r^2 - \tfrac{1}{5}h^2)\omega_0^2. \tag{I.23}$$

Substituting Eq. (I.13) for ω^2, and replacing θ according to formula (I.14) we obtain

$$(3\pi^2/10)\beta\gamma\sigma r(h/H) > 6\pi\gamma H(\beta\sigma + \tfrac{1}{3}\rho_0 H) + (3\pi/5)\gamma\rho_0 r^2 - (4\pi/15)\gamma\rho_0 h^2. \tag{I.24}$$

We will rewrite this criterion by canceling γ, in such a form:

$$A\sigma r(h/H) > B\rho_0 r^2 + C\sigma H + D\rho_0 H^2 - E\rho_0 h^2, \tag{r}$$

where

$$A = (3\pi^2/10)\beta, \quad B = 3\pi/5, \quad C = 6\pi\beta, \quad D = 2\pi, \quad E = 4\pi/15, \tag{s}$$

or introducing $h/H = \eta$:

$$A\sigma r\eta[1 - (C/A)(H/\eta r)] > B\rho_0 r^2[1 + (D/B)(H^2/r^2) - (E/B)\eta^2(H^2/r^2)], \tag{t}$$

whence

$$r < \frac{A}{B}\frac{\sigma\eta}{\rho_0}\frac{1 - (C/A\eta)(H/r)}{1 + [(D - E\eta^2)/B](H^2/r^2)}. \tag{I.25}$$

In order that this inequality could be satisfied when $\eta \leqslant 1$ it is necessary that

$$r > (C/A)(H/\eta) = (C/A)h \tag{I.26a}$$

so that

$$\frac{C}{A}\frac{H}{\eta} < r < \frac{A}{B}\frac{\sigma}{\rho_0}\frac{\eta - (C/A)(H/r)}{1 + [(D - E\eta^2)/B](H^2/r^2)}. \tag{I.26b}$$

The coefficient C/A approximately equals 7. Therefore the first inequality shows that a spherical volume cannot condense. Clots originate only from very flat formations for which horizontal dimensions exceed the vertical by one order of magnitude.

Taking into account inequality (I.26b) brings about the conclusion that the formation of clots becomes possible only when the cloud is dense enough, that is, when its density exceeds ρ_0 by one order of magnitude.

Now we shall calculate the real dimensions of clots formed in the process of gravitational condensation at the moment of a sufficient flattening of the cloud.

Let us take

$$r = \xi H. \tag{u}$$

Then the inequalities take the form

$$H < \frac{A}{B}\frac{\sigma}{\rho_0}\frac{\eta^{\xi} - C/A}{\xi^2 + (D - E\eta^2)/B}, \quad \xi > \frac{C}{A\eta}. \tag{I.27}$$

The right-hand side of the first inequality (I.27) attains the maximum when $\eta = 1$; then the inequalities (I.26b) take the form

$$\xi = \xi_m = C/A + \sqrt{(C/A)^2 + (D - E)/B}. \tag{v}$$

At this value of ξ, the critical half-thickness of the layer and the half-thickness of a clot also reach the maximum value $h_m = H_m$, and the size of the corresponding clot

$$r_m = \xi_m H_m = (A/2B)(\sigma/\rho_0), \tag{I.28}$$

i.e., it is determined by the gravitational energy and the energy of ordered relative motion.

Substituting the values of coefficients, we obtain

$$r_m = (\pi\beta/4)(\sigma/\rho_0). \tag{I.29}$$

Under these conditions and when $\beta = 0.42$,

$$r_m/H_m = \xi_m = 12.9, \tag{I.30}$$

which justifies our approximation of a strongly flattened ellipsoid. Henceforth we shall deal only with the clots of maximum dimensions and therefore we shall denote their dimensions simply as r and h.

Our initial double inequality (I.26a) limits the dimensions of condensation both from above and below. The limitation from below is imposed by the condition of gravitational instability, and it is equivalent to the criterion of Jeans. The limitation from above is imposed by the condition of stability in relation to tidal forces and well-ordered relative motion. This condition is close to the criteria of Roche, Maxwell, Poincaré, and Darwin. Thus the inequality (I.26b) is equivalent to these two criteria which have been repeatedly used separately beforehand.

In the work cited, Edgeworth[1] used one inequality—a criterion of the type of Roche's, due to which he had to make an arbitrary assumption about the form of clots. Edgeworth regards the condensation as spherical, which, as we have seen, is grossly incorrect, because a spherical volume cannot condense due to the condition $r > (20/\pi)h$. Besides, the coefficient in the inequality of Jeffreys that he used is also incorrect because it was calculated by Jeffreys for a totally different case.

3. THE PROPERTIES OF CONDENSATION AND THE VELOCITY OF FLATTENING OF THE CLOUD

Condensation at a given place in the solar system will start at the time when the half-thickness H of the cloud in this place decreases to the critical value H_{kp}. Under this condition the masses of the greatest possible dimensions r_m will condense. Now we can verify our initial assertion that condensation takes place on a considerable flattening of the cloud. For this we evaluate r_m. According to Eq. (I.29)

$$r = (\pi^2\beta/3)(\sigma R^2/M_{\odot}). \tag{w}$$

Assuming that $\sigma(R)$ is distributed within the cloud in the same way as the mass of planets is distributed in the contemporary solar system, and taking into consideration that the distances between every pair of neighboring big planets are of the order of magnitude of their distances from the Sun, we come to the conclusion that $\sigma(R)$ at such a distance from the

Sun, at which a given planet is located, is now on the order of the mass of the planet divided by the doubled area within its orbit:

$$\sigma(R) \approx M_n/2\pi R^2. \quad \text{(x)}$$

Hence

$$r = (\pi\beta/6)(M_n/M_\odot)R \quad \text{(I.31)}$$

and by Eq. (I.30), the critical half-thickness of the cloud

$$H_{\text{cr}} = \frac{\pi\beta}{6\xi}\frac{M_n}{M_\odot}R = 0.017\frac{M_n}{M_\odot}R. \quad \text{(I.32)}$$

Finally, the average velocities of the irregular motion of the particles at the initial moment of condensation are determined according to Eqs. (I.15), (I.29), and (I.30) by the expression

$$\begin{aligned}\sqrt{v^2} &= [6\pi\gamma H_{\text{cr}}(\beta\sigma + \tfrac{1}{3}\rho_0 H)]^{1/2}\\ &= \sqrt{\pi/(2\xi_{\text{cr}})}\,\beta(M_n/M_\odot)V_0(R)\\ &= 0.15(M_n/M_\odot)V_0(R). \quad \text{(I.33)}\end{aligned}$$

The relation of the mass of large planets to the mass of the Sun is of the order 10^{-3}. The circular velocity $V_0(R)$ is of the order 10^6 cm/s. Therefore the velocities of the relative motion of particles at the moment of the formation of clots in the region of large planets are on the order of meters per second. The condensations arising under such conditions would move around the Sun along almost circular orbits.

We now evaluate the mass of an isolated condensation m_n within the zone of the nth planet by taking into consideration Eqs. (I.27) and (I.31). It is equal to

$$\begin{aligned}m_n &= (4\pi/3)r^2 h\rho\\ &= (4\pi\beta/3)r^2\sigma\\ &= (\pi^2\beta^3/54)M_n(M_n/M_\odot)^2\\ &= 0.014M_n(M_n/M_\odot)^2. \quad \text{(I.34)}\end{aligned}$$

The formula (I.34) gives an evaluation only by the order of magnitude: first, because we replaced the area of the ring by the area of the circle, and second, in the case of Mars and the ring of asteroids, because a considerable portion of matter in their zones was probably captured by Jupiter due to which the corresponding planets almost vanished.

According to Eq. (I.34), at the distance of Jupiter from the Sun, the masses of condensations were maximal and equaled, by the order of magnitude, 3×10^{22} g. At the distance from the Sun of the Earth, the masses of condensations were 10^{15} g. After transforming into solid bodies with the density of the Earth they would accordingly have had radii of 100 and 0.4 km.

The initial radii of condensations according to Eq. (I.31) at the distance of Jupiter and at a distance of the Earth were 1.7×10^{15} and 100 km, respectively.

We will now evaluate the frequency of collisions of the particles which determined the irreversible evolution of the cloud. As long as the cloud had a spherical form, particles moved in all possible directions and their relative velocities were comparable with circular velocities, V_0.

The free-path time is

$$\tau_0 = 1/(\eta V_0 \pi r^2) = \tfrac{4}{3}(\rho_1 r_1)/(\rho V_0), \quad \text{(I.35)}$$

where ρ_1 and r_1 are the corresponding density and radius of the particles.

If at every collision of the particles some part, α, of their relative kinetic energy was converted into heat, then the time of sufficient flattening of the cloud when $\rho_1 = 3$ g cm^{-3}, $\rho = 10^{-15}$ g cm^{-3}, $V_0 = 3\times10^5$ cm s^{-1} is

$$\tau = \tau_0/\alpha \approx 10^{10} r_1/\alpha. \quad \text{(I.36)}$$

This time diminished according to the degree of flattening of the cloud. The evaluation (I.36) gives a considerable time of flattening, being strongly overestimated due to two causes: first, the value 10^{10} was obtained because we have taken the velocities of the particles at the boundary of the cloud where they had the lowest value, and second, we have examined the initial stages of flattening when the cloud was almost spherical. Nonetheless, even this overstated evaluation leads to the conclusion that the characteristic time cannot be too long; even in case of dimensions of particles on the order of tens of kilometers it does not surpass one billion years.

The energy dissipation of thin dust in the absence of any forces precluding collisions between the specks of dust is, from the point of view of cosmogony, practically a momentary process.

If the specks bear electric charges of the same sign, this will work as a factor impeding their collisions. In the absence of such charges, the stabilization of a dust cloud of a finite thickness is possible in the presence of irregular disturbing forces, which, as we shall see later, do really exist and play a considerably greater role than the possible electrical repulsive forces.

In the case of the absence of the stabilizing forces, as it happened in the case of the rings of Saturn, for example, the velocity dispersion of specks was bound to disappear very rapidly during a time less than the time of their fall upon the plane of symmetry of a cloud. In the latter case, the specks would have fallen on the plane of symmetry with a "temperature" $q = 0$, and the gravitational condensation of the dust would have been conditioned only by the relation between gravitational and tidal energies and also by the energy of the well-ordered relative motion.

4. CONCLUSION

The theory of condensation examined in this part of our work is based on the hypothesis of quasistatical flattening, and on the following additional premises.

(1) All the dust which would be affected by the process of condensation had existed inside the cloud from the very beginning "in a cooked state," but was not feeding the cloud regularly during the process of condensation. We cannot give up this hypothesis, since the origin of the primordial cloud still looks unclear.

(2) All the particles forming the cloud were identical. This assumption is not too important, and our theory may be easily modified for a more general case of different particles.

(3) If the cloud was not homogeneous in terms of its density and the condensation took place in its different zones not simultaneously, then the processes of condensation were developing independently in different areas. Actually, the clots which had come to existence in some domain of the cloud prior to all the other condensations, could, as we will demonstrate in the second part of our paper, travel later to other zones and to cause perturbations there. These perturbations, in turn, were able to hamper the flattening of the cloud, thus precluding the attainment of the critical density; on the other hand, such migrating clots could play the role of the centers of condensation, that is, they could grow due to the accretion of surrounding smaller particles. In that case, the process of condensation was developing according to Schmidt's model.[2] These phenomena are examined in more detail in the next part of this work.

Disregarding all the differences between the real process of condensation and the quasistatic scheme developed in this part of our work, the scheme is still valid as a general conceptual tool which furnishes a basic understanding of the process of condensation in all of the possible cases, and in every specific case it should be only slightly modified accordingly to its particular features. This is the reason why we have begun our discussion of the evolution of the cloud with the general quasistatic scheme.

[1] Edgeworth, Monthly Notices RAS, no. 6, 385 (1949).

[2] Schmidt, O. Yu., Dokl. Akad. Nauk, **45**, no. 6, 705 (1944); *Four Lectures on the Theory of the Origin of the Earth* (in Russian), Izd. Akad. Nauk. SSSR, Moscow & Leningrad, 1949.

[3] Landau, L. and Lifshits, E., *Statistical Physics* (in Russian). G.I.T.T.L, Moscow and Leningrad, 1940.

[4] Sretenskii, L., *Theory of the Newtonian Potential* (in Russian), p. 116. Gostekhizdat, Moscow and Leningrad, 1946

[5] Parenago, P., *Course of Stellar Astronomy*, p. 329. Gostekhizdat, Moscow and Leningrad, 1946.

The Formation of the Planets. II. The Law of Planetary Distances and the Rotation of Planets

L. E. Gurevich and A. I. Lebedinskii
Translated from: Izv. Akad. Nauk SSSR Ser. Fiz. **14**, 776 (1950)

As is shown in this part of our paper, from the initial clots which, according to the conclusions in the first part of this work, are the product of condensation of the primeval cloud, some secondary clots of greater masses had to be formed. These secondary clots gave birth to a planetary system whose properties were close to the properties of the contemporary solar system. As is demonstrated in Sec. 1, in a system consisting of a finite number of bodies, the orbits of these bodies were bound to become elliptical as a result of their mutual perturbations. Two processes took place in the course of the growth of these bodies: an increase in the eccentricities of their orbits due to the growth of mutual perturbations, and a decrease in these eccentricities in the acts of coalescence. The first process prevailed during the initial stage of the growth of clots, thus leading to their coalescence into planets; in the final stage, the second process predominated which caused the formation of the almost circular orbits of planets. In Sec. 2, the law of planetary distances has been inferred from an examination of eccentricities of orbits, as well as the "law of geometrical means" which connects the distances of two planets from the Sun, located symmetrically in relation to Jupiter. These conclusions agree satisfactorily with observations. In Sec. 3, some data of observations are discussed and a comparison between the law of planetary distance derived from Sec. 2 and the Bode–Titius law has been made. In Sec. 4, it has been shown that the ratio of the rotational and orbital angular momenta of a planet is a function only of the mass of the planet, and the agreement of this conclusion with observations has been stated.

1. SECONDARY CONDATION AND THEIR ORBITS

In the first part of this work[1] it was shown that if the Sun had been surrounded by a cloud possessing the mass and angular momentum of the contemporary planetary system, then the cloud was bound to transform into a flattened system which afterwards disintegrated into a large number of separate clots.

In this part of the report we shall study the process of the formation of planets from those concentrations of matter.

The initial axial symmetry of the cloud was disappearing in the course of the coalescence of condensations and the decrease in the number of bodies in which the cloud disintegrated. Due to the multiple interactions between the newly formed bodies, their circular orbits experienced perturbations and became elliptical. Simultaneously with the radial component of velocity,*

$$V_r \approx \epsilon V_0, \tag{a}$$

it inevitably emerged, due to the tendency to the equipartition of energy amongst the degrees of freedom, that the z component of velocity approximately equals the average of the radial component

$$v_z \approx i V_0, \tag{b}$$

where

$$V_0 \approx \sqrt{(gM_\odot)/R} \tag{c}$$

is the velocity of circular orbital motion around the Sun, ϵ is the eccentricity, and i is the inclination of the orbit.

Hence, on average, ϵ has to be roughly equal to i, $\epsilon \approx i$. The growth of eccentricity of the orbits increased the relative velocities of perturbing encounters between bodies, whereas the alteration of relative velocities affected the angle of deviation from the initial orbit which was bound to change as a result of the encounter, i.e., changed the perturbations of the eccentricity.

During the encounter of bodies not accompanied by their coalescence, the initially small eccentricities of their orbits grew gradually. After the coalescence of bodies moving along elliptical orbits, the resulting body on the average was bound to move along an orbit with a smaller eccentricity.

The greater the masses of encountering bodies and the less their dimensions, the more the maximal angle of scattering and the maximum alteration of eccentricity due to the bodies drawing together without coalescing.

Indeed, suppose an encounter not leading to coalescence between two bodies of the same mass m moving with relative velocity V and let one of them be deflected as a result of the encounter by an angle $\vartheta \ll \pi/2$. Then the change in velocity

$$|\Delta v| \approx 2v \sin \vartheta/2 \approx v\vartheta. \tag{d}$$

The duration of the encounter of two bodies moving along elliptical orbits is less than the time of their revolution around the Sun, and therefore, it is possible to examine their mutual interaction without taking into consideration the gravitational field of the Sun.[2] In such a case, the order of magnitude of the angle of deflection from the original direc-

*In this part of the report the same notations as in the first part have been used.

tion due to an encounter is determined by the ratio of the potential energy of mutual interaction $\gamma m^2/v$ at a distance Γ to the kinetic energy $(m/2)v^2$ of relative motion:

$$\vartheta \approx (\gamma m)/(rv^2), \tag{II.1}$$

and hence

$$\Delta v \approx v\vartheta \approx (\gamma m)/(rv). \tag{e}$$

The alteration in the eccentricity of the orbit is

$$\Delta\epsilon \approx \Delta v/V_0 \approx (\gamma m)/(rvV_0). \tag{II.2}$$

The order of magnitude of the relative velocity when $\epsilon \ll 1$ is determined by the greatest of the two magnitudes

$$v' = \epsilon U_0 \tag{II.3a}$$

and

$$v'' = rdV_0/dR. \tag{II.3b}$$

Substituting v' in Eq. (II.2) we obtain

$$\epsilon\Delta\epsilon \approx (\gamma m)/(rV_0^2) = (m/M_\odot)(R/v). \tag{f}$$

These formulas are applicable only when $\vartheta < 1$, and this means that an increase in eccentricity up to large values would require a series of collisions. Putting aside the coalescence of concentrations which tended to intensify mutual perturbations, we shall assume that the most essential alteration in eccentricity comes as the result of an encounter between the bodies whose minimal distance is on the order of magnitude equal to their diameter $2r_0$. The probability of such a close approach is close to the probability of direct collision leading to coalescence. One may expect that coalescence was preceded by a few such encounters. Therefore the eccentricities ϵ acquired by bodies before their coalescence are comparable with the alteration $\Delta\epsilon$ during a single intimate encounter:

$$\epsilon \approx \sqrt{(m/2M_\odot)(R/r_0)}. \tag{II.4}$$

Along with the process of mutual perturbation of the orbits there was a compression of individual clots. When their densities increased from the initial value,

$$\rho = (\beta\sigma)/H_{\rm cr} = (8\xi/\pi)\rho_0 = 16\rho_0, \tag{II.5}$$

determinable by expressions (I.17), (I.29) and (I.30), to the final value which may be taken equal to 3 for the region of interior planets, and equal to 1 for the region of exterior planets.

Now by taking ϵ from Eq. (II.4) and substituting it in Eq. (II.3a), it is possible to compare Eqs. (II.3a) and (II.3b):

$$\frac{v'}{v''} = \frac{R}{r_0}\epsilon = \left[\frac{m}{2M_\odot}\frac{R^3}{r_0^3}\right]^{1/2}, \tag{g}$$

or according to Eqs. (I.31) and (I.34),

$$\frac{v'}{v''} = \sqrt{\frac{2}{\pi}} = 0.8. \tag{h}$$

Later on, this ratio grew in proportion to the compression of the condensations, and so the magnitude determined by expression (II.3a) always predominated.

In the initial stage of compression and coalescence of condensations, according to Eqs. (II.4), (I.31), and (I.34),

$$\epsilon \approx \frac{\beta\sqrt{\pi}}{3\sqrt{2}}\frac{M_n}{M_\odot} = 0.17\frac{M_n}{M_\odot}, \tag{II.6}$$

which gives in the zone of Jupiter $\epsilon = 2\times10^{-4}$. Thus in the zone of Jupiter primeval condensations were bound to move with respect to one another with velocities on the order of ten-thousandths of their orbital velocities and therefore were inevitably bound to combine.

Hence for the space of time $(1/\epsilon)P$ (where P is the period of revolution of the condensation around the Sun) all condensations located inside the ellipsoidal core with cross section πrH and length $2\pi R$ had combined into a bigger "secondary" condensation. The mass of such a secondary condensation is

$$\mu = 2\pi R\ \pi rH\rho, \tag{i}$$

or according to Eqs. (I.17) and (I.29),

$$\mu = \frac{\pi^2\beta^2}{2}\frac{\sigma^2}{\rho_0}R = \frac{\pi^2\beta^2}{6}\frac{M_n^2}{M_\odot}. \tag{II.7}$$

This conclusion is correct in the case when $\epsilon R \ll r$, i.e., the radial displacement of condensations is less than their magnitudes. In our case $\epsilon R \approx r$; hence Eq. (II.7) gives the correct order of magnitude.

The mass μ of secondary condensations at a distance of Jupiter from the Sun is $0.6\times10^{27}g$, but at the distance of the Earth from the Sun $\mu = 6\times10^{21}g$.

In the course of the coalescence of condensations situated in the same circular orbit, their specific angular momentum was conserved, and thus the centrifugal force on the equator decreased. On the other hand, the force of gravity was increasing due to the increase in mass. Therefore even if original condensations were situated at the limit of centrifugal stability, the secondary condensations formed from them possessed a considerable reserve of centrifugal stability.

In fact, let the specific moment

$$k = \omega^2 r = \text{const}, \tag{j}$$

where ω is the angular velocity of rotation. Hence the centrifugal force is

$$\omega^2 r = k^2/r^3, \tag{k}$$

the gravity per unit of mass is equal to $\gamma\mu/r^2 = \gamma\rho r$, and therefore the ratio of these forces $k^2/\gamma\rho r^4$ decreases with an increase in dimensions r and mass ρr^3.

In the course of the coalescence of condensations and the increase of their centrifugal stability, they tended to contract and to acquire a spherical form. On attaining a certain density ρ', the dimensions of condensations are

$$r=\left[\frac{3\mu}{4\pi\rho'}\right]^{1/3}, \qquad (1)$$

and according to Eq. (II.4),

$$\epsilon=\sqrt{\left[\frac{\pi}{6}\right]^{1/3}\frac{\mu^{2/3}\rho'^{1/3}}{M_{\odot}}R}. \qquad \text{(II.8)}$$

After substituting μ for Eq. (II.7), we obtain

$$\epsilon=\left[\frac{\pi^4\beta^4}{288}\frac{\rho'}{\rho_0}\right]^{1/4}\left[\frac{M_n}{M_{\odot}}\right]^{2/3}=0.46\left[\frac{\rho'}{\rho_0}\right]^{1/3}\left[\frac{M_n}{M_{\odot}}\right]^{2/3}. \qquad \text{(II.9)}$$

The condensations underwent double alterations, both of which led to an increase in their eccentricity. First, the coalescence of condensations was accompanied by the growth of their masses, and second their contraction led to the increase of their densities. Therefore, the formula (II.9) which does not take into account the first process gives the lower limit of eccentricity.

In the zone of Jupiter where $\rho'=1$, we have $\epsilon=0.2$. In reality the eccentricities could become even higher than is given by Eq. (II.9), because at a strong compression (contraction) of the bodies, their long-distance approaches, which we have not taken into account, did play an important role. These eccentricities are not small, and our approximate formulas, strictly speaking, are not applicable. But, anyway, it may be maintained that the eccentricities of condensations in the course of time were becoming comparable with unity.

Proceeding from almost circular orbits of condensations, we have come to the conclusion that mutual perturbations increase their eccentricity. On the other hand, two condensations whose initial eccentricities had been roughly equal to unity, after approaching each other, evidently continued their motion along orbits with lesser eccentricities. In the stationary state, the average eccentricity has to be near 0.5.

For condensation in the zone of the Earth the formula (II.9) gives an eccentricity $\epsilon\approx10^{-3}$. However, this result is obtained on the assumption that only the condensations of equal masses participate in collisions. In reality, when the eccentricity distribution was in equilibrium, there existed a nonzero probability that more massive condensations would move from the zone of Jupiter into the zone of the Earth. These condensations, because of their considerable mass, created the same eccentricities in the condensations of the terrestrial zone as in the condensations of the zone of Jupiter. In general, eccentricities were on the whole bound to be conditioned by the perturbations produced by the most massive bodies, if only the latter's share was not too small. Therefore, the condensations of the Jupiterial zone created the same eccentricities as in the zone of Jupiter, i.e., with $\epsilon\approx0.5$, within a certain range from Jupiter where their amount was sufficiently high, but at greater distances in both directions from Jupiter eccentricities tended to decrease. The condensations of the Jupiterial zone could perturb not only the condensations within other zones but also uncondensed dust there, thus securing a nonzero thickness of the dust cloud or hindering the process of its falling.

Condensations of the zone of Jupiter which had moved into the zones of the interior planets did not succeed during the short time of their passing near the perihelion (which was on the order of 1 year) in losing a considerable quantity of volatile matter. In the course of one passage, even under conditions of conversion into latent heat of vaporization of the whole solar radiation falling on the surface of the body, a layer of thickness of about 10 M could be vaporized. In fact, an atmosphere of great optical thickness was formed at once around the body with a mass of 10^{27} g as a result of the process of vaporization. The temperature of this atmosphere quickly decreased with depth because a cold vaporizing surface was under it. The atmosphere reflected and reemitted back into space a large portion of the stream of solar radiation, and therefore from a solid surface of a body in practice could vaporize only a layer of thickness essentially less than 10 meters. The dissipation of the atmosphere went on slowly because the parabolic velocity on the surface of the body with mass 10^{27} g and density $\rho'\approx1$ was of the order of 4.5×10^5 cm/s; this already precludes the possibility of the rapid dissipation of CH_4 and NH_3, whereas only H_2 may vaporize and dissipate from the surface itself.

The passage of condensations of the zone of Jupiter across any zone of interior planets might lead to the capture of condensations. In this case almost all of the volatile substances which were part of the condensations within a certain time were bound to vaporize because of the action of solar heat, due to which the mass of the captured condensations was reduced at least a few tenfold. Such captured condensations from the zone of Jupiter could serve as the nuclei of the unification of the local condensations, the process of which led to the formation of planets. In the process of this unification the mass of the condensations increased manyfold, and due to the averaging of the orbits which had accompanied the unification,[3] the eccentricity was decreasing considerably, and thus the orbit of the formed planets happened to be almost circular. In the case of the Earth and Venus the mass of condensations (which had already lost their volatile substances) initially captured from the zone of Jupiter constituted no more than 1% of the total mass; therefore its presence could not exert an influence on the elements of the orbit of the planet. In the case of Mercury, whose mass exceeds the unvaporized mass of captured condensations by only about tenfold, the elements of the orbit of the modern planet were bound to depend essentially on the elements of the orbit of the nucleus of condensation which had become a part of the planet.

2. THE THEORETICAL LAW OF PLANETARY DISTANCES

Secondary condensations moved along intersecting orbits and so tended to combine. The region where this coalescence took place may be taken as a ring of radius R and width $2\epsilon R$. Planets were formed as a result of the coalescence of secondary condensations. The boundaries of the region inside which condensations coalesced into a planet (we shall call it the zone of the planet) cannot be determined accurately on the basis of our theory, and we can only estimate their position. For the purpose of clarity, we shall consider that if the internal boundary of the zone was at a distance R from the Sun, then its external boundary was at a distance $R(1+2\epsilon)$.

The planet nearest to the Sun was formed from the substance of the ring stretching from the internal edge of the cloud (situated at a certain distance from the Sun, R_1) to the distance $R_1(1+2\epsilon_1)$ from the Sun (where ϵ_1 is the average eccentricity of the orbits of the condensations which existed within this ring). The distance of the first planet to the Sun is equal to the average of these two distances, i.e., to $R_1(1+\epsilon_1)$. The next planet was formed in the ring at a distance from $R(1+2\epsilon_1)$ to $R(1+2\epsilon_2)(1+2\epsilon_2)$ and so on: the nth planet was formed in the ring with internal radius $R_1(1+2\epsilon_1)\cdots(1+2\epsilon_{n-1})$ and with the outer radius $R(1+2\epsilon_1)\cdots(1+2\epsilon_N)$, and the average of these distances is equal to the distance of the nth planet from the Sun

$$R_n=R_1(1+2\epsilon_1)\cdots(1+2\epsilon_{n-1})(1+2\epsilon_n). \qquad \text{(m)}$$

If all the average eccentricities are equal to one another, then

$$R_n=\frac{1+\epsilon}{1+2\epsilon}R_1(1+2\epsilon)^n, \qquad \text{(II.10)}$$

whence the total number of planets is

$$N=\frac{\ln\left[\frac{1+2\epsilon}{1+\epsilon}\frac{R_0}{R_1}\right]}{\ln(1+2\epsilon)}, \qquad \text{(II.11)}$$

where R_0 is the radius of the cloud.

The actual formation of these zones is related to the fact that planets were not formed simultaneously. Jupiter was formed first, having exhausted the zone from R_6 (see below) to $R_6(1+2\epsilon_6)$, and only afterwards did the process of the formation of planets extend in succession to zones more and more distant from Jupiter.

When $\epsilon=0.5$, we obtain $N=7$, that is, the correct order of the number of planets. For this Eq. (II.10) gives

$$\ln R_n=a+n\ \ln(1+2\epsilon)=a+0.30ln. \qquad \text{(II.12)}$$

This formula should be valid for the region on both sides of Jupiter wherein condensations from the zone of Jupiter could most probably travel according to the distribution of their eccentricities and major semiaxes. Beyond the boundaries of this region a considerable decrease of eccentricities had to begin, and it may be expected that it will be in a certain sense symmetrical on both sides with respect to the zone of Jupiter.

The decrease in eccentricities led to the decrease in the width of the ring corresponding to every planet, and hence to the decrease in the distances between planets as compared to the figures predicted by formula (II.12). This decrease was also bound to be symmetrical on both sides of Jupiter. A precise character of this symmetry cannot be established on the basis of our reasoning. But it turns out that a satisfactory approximation may be obtained in the following way.

Let us number the planets according to their distances from the Sun. Mercury—1, Venus—2, Earth—3, Mars—4, ring of asteroids—5, Jupiter—6, Saturn—7, Uranus—8, Neptune—9, Pluto—10.

Let us assume

$$\ln R_n=\ln R_6+0.301(n-6)+\beta(n-6)^3+\gamma(n-6)^5+\cdots \qquad \text{(II.13)}$$

Where 6 is the number of Jupiter. In this case the following relation is bound to hold, which we shall call "the rule of geometrical means":

$$R_{6-k}R_{6+k}=R_6, \qquad \text{(II.14)}$$

i.e., the distance of Jupiter from the Sun must be the geometrical mean of the distances of any pair of planets situated symmetrically in respect to Jupiter. We give the corresponding observational data (Table 1), assuming that the ring of asteriods corresponds to Saturn.

Here

$$\Delta=\ln R_{6+k}+\ln R_{6-k}-2\ \ln R_6. \qquad \text{(n)}$$

From Table 1 it is evident that deviations of distances from this rule do not exceed 6%. Further, all differences happen to be of the same sign. This shows that Jupiter itself is displaced by 0.2 A.U. in the direction of the Sun from the "center of symmetry" of the solar system, i.e., from the point satisfying Eq. (II.14) and situated at a distance of $R'=5.41$ A.U. from the Sun.

The deviations

$$\Delta'=\ln R_{6+k}+\ln R_{6-k}-2\ \ln R_6' \qquad \text{(o)}$$

indicated in the last column of Table 1 correspond to deviations of distances from Eq. (II.14) not exceeding 1.5%.

The law of geometrical means, confirmed by observations, has a simple energetic sense: specific gravitational energies in the zones, symmetrical relative to Jupiter, differ from the specific energy in the zone of Jupiter equalfold.

TABLE 1.

Number of planet, counting from Jupiter	Names of planets of symmetrical pairs	Δ	Δ'
4	Venus–Pluto	+0.012	−0.006
3	Earth–Neptune	+0.023	+0.005
2	Mars–Uranus	+0.018	+0.000

Formula (II.13) is also well corroborated by the observational data for all planets except Mercury if the following values of the coefficients are taken:

$$\ln R_6=0.724, \quad \beta=5.25\times10^{-3}, \quad \gamma=\delta=\cdots=0. \tag{II.15}$$

The corresponding data are given in Table 2.

Mercury does not satisfy formula (II.13). The point corresponding to Mercury does not lie on the even curve but proves to be displaced away from it by several tenths on the logarithmic scale. This discrepancy cannot be explained away by the position of Mercury at the margin of the primeval cloud because the marginal position could only increase rather than decrease its distance from the Sun in comparison with Eq. (II.13).

It is possible, as we have already indicated above, that the elements of the orbit of Mercury (the major semiaxes, eccentricity, and inclination) have anomalous values due to the fact that the essential part of the mass, energy, and angular momentum of that planet consists of the mass, energy and momentum of the condensation captured initially by Venus from the zone of Jupiter, then was partly vaporized, and afterwards partly served as a nucleus for the combination of condensations of the zone of Mercury into a single planet.

The absence of planet 5 in the solar system and the small mass of Mars are probably explained by the fact that a considerable portion of substances from their zone was swallowed up by Jupiter. The remnants of secondary condensations in the zone of the nonexisting fifth planet are most probably represented by the contemporary asteroids. If that is so, then it is possible to calculate from the mass of the present asteroids, by means of Eq. (II.7), the mass of the fifth planet. Taking their mass $\mu=10^{21}$ g, we obtain the mass of the planet on the order of the mass of interior planets.

The law of planetary distances was first deduced theoretically by Schmidt,[4] who proceeded from the basic rule conditioning the process of planetary formation, that is, from the law of conservation of angular momentum, without analyzing, in this case, the actual mechanism of the process of coalescence of particles into planets.

Unlike Schmidt, we do not analyze the problem of the distribution of angular momentum in the process of the formation of planets, but we investigate the mechanism of the unification of condensations. We also stress the importance of taking into account the eccentricities of orbits. Essentially, we examine an alternative limit case of the general theory which takes into consideration both the mechanism of coalescence bound up with the eccentricities of orbits of condensations as well as the redistribution of momenta. This redistribution creates rotation of the formed planets, and so only a theory which simultaneously determines both the rotation of planets and their distances from the Sun can serve in the capacity of such a general theory.

The second difference of our derivation from that by Schmidt consists in our rejection of his division of planets into interior and exterior ones, implying different quotients in the law of planetary distances. The rule of geometrical means, established by us, makes it reasonable to consider that the law of planetary distance must be the same for all planets.

3. COMPARISONS WITH THE BODE–TITIUS LAW OF PLANTARY DISTANCES

If R_n is determined with the help of the following formula with four indeterminate coefficients,

$$\ln R_n=\alpha(n-6)+\beta(n-6)^3+\gamma(n-6)^5+\delta, \tag{p}$$

then the method of least squares applied to all planets except Mercury leads to the following algebraic expression for $\ln R_n$:

$$\ln R_n=0.726+0.293(n-6)-5.6\times10^{-3}(n-6)^3 +5.3\times10^{-5}(n-6)^5. \tag{II.16}$$

The mean-square error is equal to 0.014. This formula corresponds to our theory in the case of the average value of the orbital eccentricities of secondary condensations in the zone of Jupiter $\epsilon=0.48$.

The relation between the logarithms of distances of planets from the Sun and their number is plotted in Fig. 1. The observational data are represented by circles. The continuous thick line is drawn according to formula (II.16).

In Fig. 1 the dotted curve represents the law of Bode–Titius

$$\ln R_n=\ln[0.4+0.3(2^{n-1})]=\ln[0.4+0.3(10^{0.301(n-1)}]. \tag{II.17}$$

For values of $n\geqslant1$ the curve was plotted according to the points calculated for sufficiently small fractional values Δn. Such interpolation is impossible between Venus and Mercury, because Eq. (II.17) gives the distance of Mercury if we ascribe the value $n=-\infty$ to that planet. Therefore between Venus and Mercury the curve is constructed approximately.

The law of Bode and Titius correctly gives the distance from the Sun to those planets which were known at the time

TABLE 2.

Number of planet	2	3	4	6	7	8	9	10
$\ln R_{calc}$	−0.144	−0.037	+0.164	0.724	1.020	1.284	1.485	1.592
$\ln R_{obs}$	−0.143	0.000	+0.183	0.716	0.974	1.283	1.478	1.597

of its formulation. It correctly characterizes the gradual growth of the coefficient of the exponential law which is multiplied by the index (which means from the point of view of our theory that the eccentricities of orbits of secondary condensations tend to grow in proportion to their approach to Jupiter) as well as its apparent approaching a certain constant value. However, a further decrease of this coefficient (i.e., the decrease of eccentricity in the case of moving away from Jupiter to the periphery of the solar system) could not be expressed by the Bode law because it becomes noticeable only at the distance of Neptune from the Sun. Therefore Uranus proved to be the last planet to which we can still apply Bode's law. The circumstance that Bode's law gives the correct distance of Pluto if the number for Neptune is attributed to it is only an accidental coincidence from this point of view. The conformity of the distance of Mercury with Bode's law (II.17) is attained by using an artificial method whose very necessity testifies that there is no actual concurrence.

It is to be noted that in our approximation as well as the approximation of Bode's formula, Saturn, to some extent, deviates from the general law. It is possible that this is explained by the fact that Saturn is the only planet whose number is adjacent to that of Jupiter.

We shall now compare formally the mean-square errors of laws (II.16) and (II.17). Abstracting from their physical meaning, it may be said that both laws describe eight planets out of nine existing naturally, only if in Bode's Law the number of Neptune is attributed to Pluto. For this case Eq. (II.17) gives the mean-square error 0.013, which is almost the same value as 0.014 indicated above in the case of Eq. (II.16). These figures probably display generally the maximal accuracy of any law of planetary distances because in reality there takes place a scattering of points and some anomaly of the distance of Saturn.

Let us now clarify the extent to which formula (II.16), with coefficients determined by the method of least squares, satisfies the law of geometrical means. For this in Fig. 2 along the axis of ordinates we put down values $Z_n = \ln R_n - 0.293(n-6) - k0.726$, i.e., the deviations of the points of Fig. 1 from a straight line which are compared with the curve

$$Z' = -5.6\times10^{-3}(n-6)^3 + 5.3\times10^{-5}(n-6)^5 \quad \text{(q)}$$

obtainable from the analytic law (II.16).

In the same figure the crosses having the same meaning as in Table 1 are drawn. The deviations for interior planets are taken with the opposite sign and shown in points with abscissas corresponding to exterior planets symmetrical to them. A comparison of positions of these crosses and circles makes it possible to visualize fully the measure of this symmetry of deviations, because the perfect concurrence of crosses with circles would have corresponded to complete symmetry. It appears that the half-sums of deviations for

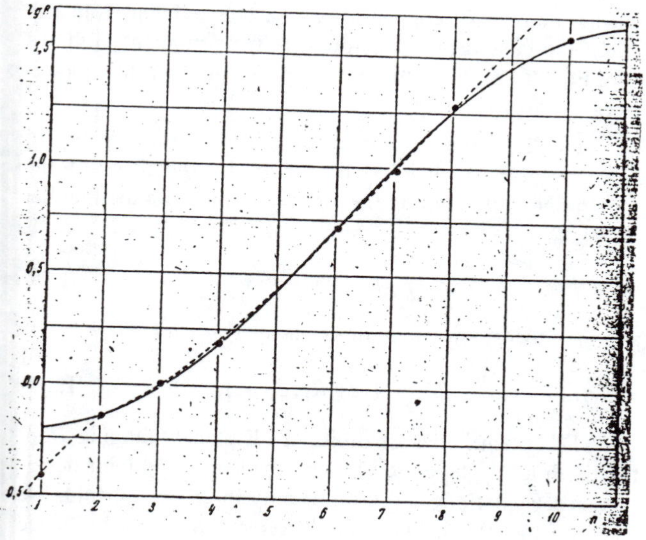

FIG. 1. Relation between logarithms of distances of planets from the Sun and their numbers. Data observed are represented by points. The continuous curve is constructed according to the formula

$$\ln R_n = 0.726 + 0.293(n-6) - 5.6\times10^{-3}(n-6)^3 + 5.3\times10^{-5}(n-6)^5.$$

The dotted curve represents the law of Bode and Titius:

$$\ln R_n = \ln[0.4 + 0.3\times10^{0.301(n-1)}].$$

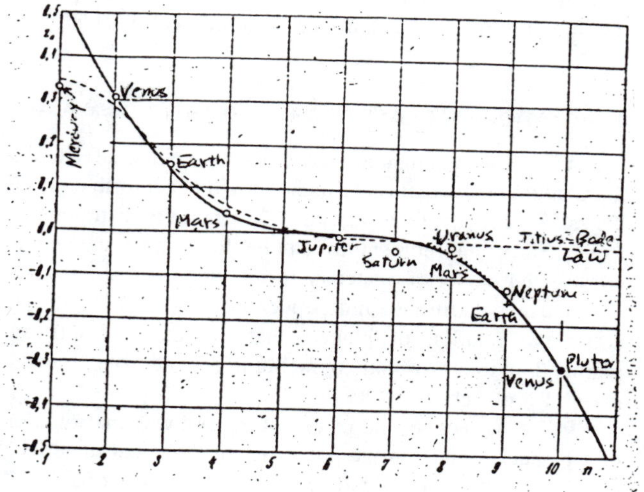

FIG. 2. Law of geometrical means. The observed data are represented by circles for exterior and crosses for interior planets for which deviations from the law of geometrical means are given in points with abscissas correspondingly symmetrical to exterior planets. The curve is constructed according to the formula

$$z_n = -5.6\times10^{-3}(n-6)^3 + 5.3\times10^{-5}(n-6)^5.$$

TABLE 3.

Names of planets	Δ''
Venus, Pluto	+0.002
Earth, Neptune	+0.014
Mars, Uranus	+0.008

planets of different symmetrical pairs, which in the case of symmetry would have been equal to zero, have values represented in Table 3.

All these sums appeared to be of the same sign. This shows that the exponential law (II.16) is not the best basis for approximation. Better agreement with observations would have given the modification of the basic curve in Fig. 2 represented by a dotted line. In this case the mean-square error would have been equal to 0.010.

Comparing Tables 1 and 3 we see that $\Delta''\approx\frac{1}{2}(\Delta+\Delta')$, where $|\Delta'|\ll|\Delta|$.

All these facts have the following meaning. Generally speaking, the law (II.13) can contain these errors: (1) exponential approximation for an unknown dependence of distances of planets upon their numbers may be inaccurate, (2) the distance of Jupiter may not agree with the geometric mean from the distances of symmetrical planets, and (3) it may be that the law of geometric means in itself is inaccurate. The obtained result means that the first two errors are the basic ones, and the method of least squares distributes the total deviation roughly in equal parts between them. The law of geometrical means proves to be comparatively the most accurate if it is formulated in terms of mutual equality of all the three products of distances of symmetrical planets.

4. ROTATION OF PLANETS AROUND THEIR OWN AXES

A certain conclusion related to the rotational angular momentum of the planets (which we will henceforth call, for the sake of brevity, their intrinsic momentum to distinguish it from their orbital momentum) can be drawn from the above-constructed model of planetary formation.

It can be affirmed independently of a detailed theory about the rotation of planets that the ratio of intrinsic momentum to its orbital momentum depends only on the mass of a planet.

In fact it may be said in a most general way that rotation of a planet depends on the properties of "original material," which had entered into its formation, and of the properties of the planet which had formed. The "original material" means separate condensations. We know their masses μ; eccentricities of their orbits ϵ, characterizing their velocities and the dimension of the region from which they agglomerated into a planet; the number N of the condensations which formed the planet; and the density n of their distribution determining together with their velocities the kinetics of their collisions and therefore the kinetics of the formation of the planet. We shall now evaluate the intrinsic momenta K of individual condensations.

Condensations were formed from particles moving along almost circular orbits, so that the order of the magnitude of their initial linear velocity of rotation was determined by the difference

$$\Delta v_0=\frac{dV_0}{dR}\,r \tag{r}$$

of circular velocities of their constituent particles, whereas their intrinsic momenta according to Eqs. (I.31) and (I.34) are

$$K\propto mr^2\,\frac{dV_0}{dR}\propto m\,\frac{\sigma^2}{\rho^2}\,\frac{dV_0}{dR}\propto M_n\left[\frac{M_n}{M_\odot}\right]^4 R^2\,\frac{dV_0}{dR}$$
$$\propto\sqrt{\gamma M_\odot R}M_n\left[\frac{M_n}{M_\odot}\right]^4\propto k_0 M_n\left[\frac{M_n}{M_\odot}\right]^4, \tag{II.18}$$

where k_0 is the specific orbital momentum. As regards the essential properties of the planet pertaining to the problem under examination, they include only its mass M_n, its density ρ_n, and the radius R of its almost circular orbit.

Our theory of formation of condensations leads to the conclusion that all the indicated magnitudes which characterize the condensations, except density, are functions of three magnitudes M_n, R_n, and ρ_n. Concerning the mass m and the number of condensations N, it is evident from Eq. (I.34); the conclusions regarding intrinsic momenta follows from Eq. (II.18). Finally, n also depends on the same magnitudes.

Actually, the volume of zone is equal to $2\epsilon R\ 2\pi R\ 2H\approx 8\pi\epsilon^2R^3$, where half the thickness of the zone in the z direction is $H\approx iR\approx\epsilon R$. Therefore the number is

$$n\approx N/(8\pi\epsilon^2R^3) \tag{II.19}$$

and the intrinsic momentum of planet is

$$K_n=f(M_n,R_n,\epsilon_n,\rho_n). \tag{II.20}$$

The eccentricities of orbits of all planets ranging from the Earth to Neptune are virtually the same, and their densities, except that of Saturn, differ only three- to fourfold. We shall see that this insignificant difference does not influence the momenta of planets, i.e., ρ_n virtually does not enter into the expression (II.20).

This circumstance can be qualitatively explained in the following way. The momentum of a planet can depend upon its density only due to tidal forces. But these forces were essential only at the initial stage of the formation of planets when their dimensions were great and the densities were small and determined by Eqs. (I.17), (I.29), and (I.30),

$$\rho_n\propto\rho_0\propto M_\odot/R_n^3, \tag{s}$$

i.e., depended only upon R_n. On subsequent compression, tidal forces were decreasing so rapidly that the deceleration of rotation of planets practically stopped. Only the two planets nearest to the Sun formed and were the exception, Mercury and Venus, whose rotation is very slow. For the remaining planets it is possible to neglect the action of tidal forces at the latest stages of their compression when Eq. (s) is already incorrect, as the retarding action of tidal forces very swiftly decreases when the distance R_n grows. Therefore in Eq. (II.20) only considerably changing dimensions M_n and R_n are effective, whereas the mass of the Sun $M_\odot$ and gravitation constant γ may enter as parameters.

But on the right-hand side of Eq. (II.20) the dimension of specific orbital moment can be possessed only by the magnitudes

$$K_n=\sqrt{\gamma M_\odot R_n}, \quad \sqrt{\gamma M_n R_n}=\sqrt{\gamma M_\odot R_n}\,(M_n/M_\odot)^{1/2}, \tag{t}$$

i.e., the indicated functional dependence can be written down as

$$K_n=f(M_n,k_n), \tag{u}$$

but in such a case the only possibility consistent with the dimensional considerations seems to be the ratio

$$K_n/K_{0n}=f(M_n/M_\odot), \tag{II.21}$$

where K_{0n} is the orbital momentum of the planet.

In Fig. 3, $y=\ln(K_n/K_{0n})$ is placed along the Y axis and $x=\ln(M_n/M_\odot)$ along the X axis. Further, Mercury and Venus, whose rotation was probably slowed by tidal friction due to their closeness to the Sun, as well as Pluto whose

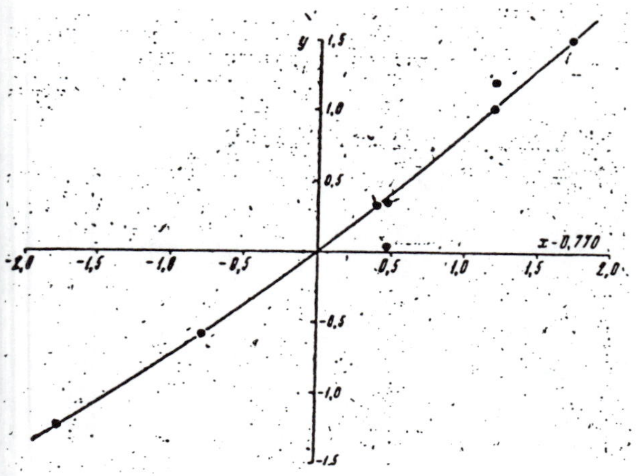

FIG. 3. Dependence of the relation of the intrinsic moment of planet to its orbital moment upon the mass. Neptune and Saturn have each been represented by double points: their meaning is explained in the text. The curve is constructed according to formula

$$\ln\frac{K_n}{K_{0n}}=0.796\ln\frac{M}{M_\odot}-4.64\left(\ln\frac{M}{M_\odot}-0.770\right)^2+\text{const.}$$

period of rotation is unknown, have been excluded from consideration. The intrinsic momenta were calculated, without taking into account the internal structure of planets, according to the formula

$$K_n=\tfrac{2}{5}M_n\omega_n r^2 \tag{v}$$

(r_n is the radius of a planet), which may be the ca'ise of some dispersion of points. Neptune and Saturn are each represented by a couple of points: for Neptune label II corresponds to the period 7^h8^m (Ref. 5) determined photometrically and label I the period 15^h8^m (Ref. 6) determined spectroscopically; for Saturn, label I corresponds to the value K_n calculated from the observable value of the radius and II to the value which would have been obtained if Saturn, with its mass unchanged, had the density equal to the density of Jupiter.

The curve in Fig. 3 is determined by the equation

$$y=0.796x-4.64(x-0.770)^2+\text{const}, \tag{II.22}$$

where the three constants are determined in such a way that the curve could pass through the points of Jupiter, Uranus, and Mars. It should be noted that the value for Earth differs from the value calculated according to Eq. (II.22) by 0.007, which may indicate the absence of the transition of a substantial part of the intrinsic momentum of the Earth to the orbital momentum of the Moon.

If the values labeled by II are regarded as the real values for Saturn and Neptune, then all of the six planets lie well on the even curve. Then the mean square deviation amounts to 0.016, i.e., the dependence of the intrinsic momenta of planets upon their mass and distance from the Sun is as accurate as is the law of planetary distances.

In fact this regularity is an empirical fact verified with such accuracy which cannot be completely explained by our theory.

A complete theoretical deduction of this ratio can only result from the construction of the theory of rotation of planets. Such a theory has been outlined by Schmidt in his work[7] where a different result was obtained, namely, that K/K_0 is proportional to $\sqrt{M_n/R_n}$.

Alfvén[8] obtained a dependence of the same type, which is a particular case of our dependence, Eq. (II.21), by examining a concrete process, namely, the process of gravitational capture of particles by a growing planet without direct collisions. The cross section of capture depends upon the same variables M and R which figured in our deduction, and so Alfven obtained the dependence $K/K_0=aM^{3/2}$, which is a particular case of our general dependence.

REFERENCES

[1] L. E. Gurevich and A. I. Lebedenskii, Izv. Akad. Nauk SSR Ser. Fiz. **14**, 765 (1950). (see preceding paper)
[2] L. E. Gurevich and B. Y. Levin, Astron. Zh. **27**, 273 (1950).
[3] O. Yu. Schmidt, Dokl. Akad. Nauk SSSR **45**, *6*, 245 (1944).
[4] O. Yu. Schmidt, Dokl. Akad. Nauk SSSR **52**, *8*, 673 (1946).
[5] E. Schönberg, Ann. Acad. Sci. Fenn. Ser. A **XVI** (5), 21 (1921).
[6] J. H. Moore and D. H. Menzel, Publ. Astron. Soc. Pac. **15**, 236 (1928); **15**, 234 (1928).
[7] O. Yu. Schmidt, Izv. Akad. Nauk SSSR Ser. Fiz. **14**, *1*, 29 (1950).
[8] H. Alfvén, Stockholm Observ. Ann. **14**, 148 (1943).

The Formation of the Planets. III. Structure of the Primordial Nebula and Separation of Planets into Outer and Inner Ones

L. E. Gurevich and A. I. Lebedinskii
Translated from: Izv. Akad. Nauk SSSR, Ser. Fiz. **14**, 790 (1950)

In this part of the paper the structure of the primordial cloud is examined and the causes of the differences between the properties of the inner and outer planets are elucidated. As is demonstrated in Sec. 1, neither gas nor dust had to be dissipating within the limits of the currently existing solar system, whereas hydrogen was bound to dissipate beyond its border. Consequently, the size of solar system is determined by the condition of the nondissipation of the hydrogen cloud. It is shown in the second section that Schmidt's and Fesenkov's hypotheses both lead to the conclusion that the nearest vicinity of the Sun had to contain virtually no substance of the cloud. The dust cloud was shaped like a strongly flattened ellipsoidal toroid at the moment of its gravitational condensation, which explains the absence of any planets between Mercury and the Sun. In Sec. 3 we explain that the dustlike particles, confined within the inner zone of the toroid, heated by solar radiation, could be composed only of nonvolatile substances. The internal planets which consist of just such substances were formed precisely inside that zone. In contrast, the greater part of the cloud had a temperature close to that of interstellar space, and thus its particles contained, in addition to other components, some volatile compounds rich in hydrogen. Possibly, the particles included even molecular hydrogen. Due to the low temperature of the particles, virtually all gases which had existed in the spherical cloud gradually condensed on the particles of the flattened dust system. This circumstance explains the large masses of the outer planets, and especially the particular features of their chemical composition. In Sec. 4, the kinetics of the formation of the planetary system is considered. The actual process of that formation differed from the scheme given in parts I and II of the paper [L. E. Gurevich and A. I. Lebedinskii, Izv. Akad. Nauk SSSR Ser. Fiz. **14**, 765 (1950)] in the respect that the cloud was not homogeneous, but consisted of bodies ranging from dustlike particles to condensations with masses amounting to 10^{27} g.

INTRODUCTION

The matter of the protoplanetary cloud experienced some radial redistribution while the planets were forming. However, this redistribution was not strong enough to eradicate completely the vestiges of the initial structure of the cloud. That means that the mass distribution of the currently existing planets allows making judgements about the density distribution corresponding to the primordial cloud which had given life to the planets.

The surface density σ in the primordial cloud reached its peak at some finite distance from the Sun which approximately equaled the distance between the Sun and Jupiter or Saturn. This density was decreasing in the direction of the Sun, that is, the cloud, after it had become flattened to a considerable degree, was shaped like an ellipsoidal toroid with the sun located in its center. The outer radius of the cloud was of the order of the magnitude of Pluto's orbital radius.

There are two differences between the inner and outer planets: the latter have much greater masses and lesser densities. That allows us to surmise that they are primarily composed of hydrogen and also some compounds rich in hydrogen.

In this paper, we shall examine the problem of the inner and outer borders of the cloud. We shall also consider the problem of the bordering zone dividing the domains of the inner and outer planets.

1. ON THE OUTER BORDER OF THE PRIMORDIAL CLOUD

If the primordial cloud's dimensions were the same as those of the modern solar system, and if the cloud was in the state of radiational equilibrium with the Sun, then it could not dissipate. Indeed, the temperature of the cloud's surface is

$$T=\left(\frac{L_{\odot}}{8\pi\kappa R_0^2}\right)^{1/4}, \qquad \text{(III.1)}$$

and the average velocity of thermal motion corresponding to this temperature is

$$v=\sqrt{\frac{8kT}{\pi m_1}},$$

where $L_{\odot}$ is the luminosity of the Sun, R_0 is the radius of the cloud, m_1 is the mass of a particle, and κ is the radiational constant. On the other hand, the velocity necessary for a particle to escape from the cloud is

$$V=\sqrt{\frac{2\gamma M_{\odot}}{R_0}},$$

where γ is the gravitational constant, and $M_\odot$ is the mass of the Sun. The process of dissipation develops quickly when $V<\beta v$, that is, when

$$R_0>\left(\frac{\pi\gamma M_\odot m_1}{4k\beta^2}\right)^2\sqrt{\frac{8\pi\kappa}{L_\odot}}\,. \tag{III.2}$$

If $\beta=5$,[1] then (for atoms of hydrogen) the radius R_0 turns out to be of the order of the current size of the solar system, which means that this size is determined by the spatial border of the dissipation of the cloud.

According to Eq. (III.2), the dimensions of the primordial cloud are proportional to $(M_\odot^4/L_\odot)^{1/2}$, that is, to a quantity which is almost constant, because of the relations between the mass and luminosity of the Sun. Consequently, if, in accordance with Fesenkov's hypothesis,[2] the mass and luminosity of the Sun during the epoch of planetary formation differed significantly from their modern magnitudes, the dimensions of the solar system would not be affected. The ejection of dust particles by light pressure also could not effectively take place, if the proportion of the substances able to condense at the temperature of the cloud (50–100 K or so) was of the same order of magnitude as inside the stars, or even more. In such a case, the majority of dust particles increased their sizes during their flight through the cloud due to the absorption of colliding atoms to such a degree that the effect of light pressure became insignificant.

Indeed, the motion of a dust particle under the action of light pressure and the "friction" from the medium is described by the equation

$$\dot{u}=\frac{1}{mc}Fs_1-\frac{1}{\tau}u,$$

where u and $\dot{u}$ are, correspondingly, the speed and acceleration of the "directed" radial motion of the particle, F is the density of the flow of radiation, S_1 is the particle's cross section, and c is the speed of light. Finally,

$$\tau=\frac{m_1}{\rho_{\text{gas}}vs_1}$$

is the time interval during which the particle passes its energy to the atoms of gas whose density is ρ_{gas}; v is the average velocity of the chaotic motion of particles.

The stationary speed of the particle described by this equation is

$$u=\frac{\tau Fs_1}{m_1c}=\frac{F}{vc\rho_{\text{gas}}},$$

and it travels through the cloud during the period of time T,

$$T=\frac{R_0}{u}=\frac{R_0vc\rho_{\text{gas}}}{F}=\frac{3vcM_{\text{gas}}}{L_\odot},$$

where M_{gas} is the total mass of gas in the cloud, $M_{\text{gas}}=(4\pi/3)R_0^3\rho_{\text{gas}}$. During that time the particle increases its mass according to the following law:

$$\frac{dm_1}{dt}=s_1vx\rho_{\text{gas}},$$

here x is the weight proportion of the substances which can be absorbed by the dust particle. Since

$$s_1=\pi r_1^2$$

and

$$m_1=\tfrac{4}{3}r_1^3\rho_1$$

(where ρ is the mass density of particles), then

$$\frac{dr_1}{dt}=\frac{vx\rho_{\text{gas}}}{4\rho_1}.$$

Consequently, the increase of the radius during the time of the particle's travel through the cloud is

$$\Delta r_1=\frac{vx\rho_{\text{gas}}}{4\rho_1}T=\frac{3}{4}\frac{v^2cM_{\text{gas}}\rho_{\text{gas}}}{\rho_1M_\odot}x.$$

If the temperature of gas is the same as of that of the dust, then

$$v^2=\frac{3k}{m},$$

where m is the mass of an atom, and

$$\rho_{\text{gas}}=\frac{3M_{\text{gas}}}{4\pi R_0^3},$$

and so

$$\Delta r_1=\frac{27}{32\pi^{5/4}}\frac{ckM_{\text{gas}}^2}{m\rho_1\kappa^{1/4}L_\odot^{3/4}R_0^{7/2}}.$$

If we take $R_0=6\times10^{14}$ cm, $M_{\text{gas}}=2\times10^{30}$ g, $\rho_1=3$ g cm^{-3}, $m=10^{-22}$ g, then we will come to the formula

$$\Delta r_1=1.5x, \tag{III.3}$$

which corroborates our statement if $x>10^{-4}$, which is more than sufficient, since x could not be less than 10^{-2} even if hard particles had formed exclusively from metals and refractory substances whose proportion in the cloud was of the same order as in the atmospheres of stars. In reality x was considerably more than 10^{-2}, since the particles contained occluded gases, and, moreover, on the fringe of the flattened cloud, where, as we will see, the temperature of dust was very low, the particles were composed not only of metals, but also of such volatile substances as methane, carbonic acid, and ammonia.

The colliding particles experienced strong heating, and thus their substance could undergo a partial transformation into the gaseous state. However, if the velocity dispersion of the particles was not too high, then the energy of the evaporating gases could not be sufficient for their dissipation, and thus those gases eventually condensed again on other particles.

To sum up, the primordial cloud surrounding the Sun had to consist of gas as well as dust, and two components could not dissipate from the cloud to a considerable degree. This conclusion justifies the hypothesis suggested in the formerly published parts of this work, that the mass of the cloud roughly equaled the combined mass of the currently existing planets.

2. THE DISTRIBUTION OF NONVOLATILE SUBSTANCES WITHIN THE CLOUD

The mass distribution of the modern planets roughly characterizes the mass distribution inside the primordial cloud: the surface mass density σ did not decrease monotonically with distance from the Sun, but reached its peak at the distance of Jupiter.

A mass distribution of that kind may be quite naturally interpreted in the framework of O. Yu. Schmidt's hypothesis.[3] Indeed, the mass density distribution corresponding to a cloud, captured and revolving around the Sun, whose different parts only rather slowly exchange their angular momentum, should have a maximum at a certain distance from the center star.

However, a similar density distribution could also result from another mode of the formation of the cloud, namely, if the cloud had emerged from the outer shell of the nebula which had given life to the Sun, and later, in the course of contraction, separated from that nebula. According to Fesenkov's hypothesis,[2] the Sun was emitting corpuscular radiation and thus losing its angular momentum. Let us examine how this radiation affected the planetary shell. We will first assume that the shell itself was spherical. Then, since the effective thickness of the shell δ considerably exceeded the free path length of the particles l, all the emitted particles had to be absorbed by the shell. In fact,

$$l=\frac{1}{ns_1}=\frac{16\pi}{3}\frac{R^2\rho_1 r_1\delta}{M}=\frac{16\pi}{3}\frac{10^{28}\times 3}{3\times 10^{30}}r\delta=0.2r_1\delta,$$

where n is the number of particles per unit volume, ρ_1 and r_1 are, correspondingly, their density and radius, M is the mass of the cloud, and R is the distance from the Sun. When $r_1<5$ cm,

$$l<\delta.$$

Consequently, the planetary shell was affected by two forces which compressed it along the radial dimension. The particles emitted by the Sun bore a radial momentum directed outward which they transmitted to the shell. However, as it has been known after the work by Krat[4] and Martynov's calculations,[5] the Sun could contract to its present size and slow down its rotation correspondingly only due to the loss, because of its corpuscular emission, of a considerable proportion of its mass, exceeding manyfold the total mass of the planets. Since the Sun's corpuscular emission was being completely absorbed by the spherical planetary shell, the latter, in turn, was bound to be losing a huge mass by emitting particles outward. Impacted by the force of reaction, the shell experienced contraction.

The two mechanisms of the shell's contraction described above affected it in another way, too: they greatly impeded its rotation. The particles coming from within possessed the specific angular momentum which corresponded to a distance equal to the Sun's radius, and thus was much smaller than the specific angular momentum of the shell. In contrast, the specific momentum of the particles escaping from the shell outward exceeded its own specific momentum. As a result, the shell had to be contracting towards the Sun. Its final distance from the Sun would have been much less than the distance between the Sun and the currently existing planets.

In the course of our analysis, however, we have not taken into account the possibility that the shell could flatten rather quickly, which changed the physical situation considerably. Then, indeed, a major portion of the matter emitted by the Sun could escape without any interaction with the shell, and only some minor share of the radiated particles could not avoid passing through the shell. Consequently, the shell's rotation was impeded to a much lesser degree in comparison with the spherical structure, precisely, as the ratio of H to R, where H is the semithickness of the shell at the distance R from the Sun. When the thickness is close to its critical value, that ratio, as was demonstrated in part I of the study [see Eq. (I.32)] is less than the ratio of the total mass of the planets to the mass of the Sun. That means that the hindering turned out rather small far from the Sun. At smaller distances, where the value of H/R was higher, the shell, in practice, had to fall onto the Sun, thus producing an empty area around the Sun. We will not discuss the dimensions of the solar system in this case, since special calculations would be necessary. What is important from our standpoint is the conclusion that an empty space containing no matter at all had to exist in a vicinity of the Sun, regardless of the origin of the primordial planetary cloud.

3. CHEMICAL COMPOSITION AND TOTAL MASS OF OUTER AND INNER PLANETS

There is one more cause bringing about the observed mass distribution of the planets and also perfectly independent of the origin of the planetary cloud. Even if the number density of matter had decreased with the increase of the distance from the Sun, the dust component of the cloud which we are interested in could still have a maximum, since in the neighborhood of the Sun, the chemical composition of the particles, and consequently the composition of the planets formatted from these particles, had to differ from the composition of the particles far from the Sun. Fesenkov[6] pointed out the following, cosmogonically important fact: the heavy chemical elements with high melting temperatures may be found on the Earth roughly in the same proportion as on the Sun and in the interstellar space. On the other hand, the light and volatile elements, such as hydrogen, helium, or neon, are virtually absent on the Earth, but they and their compounds are abundant in the atmospheres of the giant planets. Levin[7] suggested an idea which may be used to explain that fact, whatever the solution of the problem of the origin of the cloud might be. He paid attention to the following circumstance: on the periphery of the solar system, the particles, captured from interstellar space, could retain such volatile substances as water, carbon dioxide, ammonia, and methane. He explained on this basis the differences between the chemical composition of the outer and inner planets. ter Haar[8] believes, on the basis of the assumption that the temperature inside the flattened dust cloud decreased in reverse proportion to $\sqrt{R}$, that the volatile substances could condense on the surface of dust particles. However, the latter idea is incompatible with the former: at the temperatures of the order of 100 K, ammonia, methane, and other similar substances condense only at a high density of gas, which definitely could not take place in the solar system.

Levin's hypothesis is correct, but, as we will now demonstrate, redundant, since even within the solar system the transition of any volatile substances into the solid phase is possible, by means of occlusion as well as in the process of freezing. As long as the cloud was spherical, the temperature of its particles could not be less than its surface temperature $T \approx 100$ K. However, after the cloud had flattened, but the gaseous system still remaining spherical, the situation changed. The gaseous cloud was perfectly transparent to solar radiation. Even if its mass had exceeded the total mass of the planets a hundredfold, its transparency would not have been less than that of the terrestrial atmosphere, because its mass per unit surface would have been of the same order of magnitude as for the atmosphere of the Earth (if we assume the distance between the main part of the mass of the cloud and the Sun to be equal to the distance between the Sun and Jupiter).

Because of the transparency of the gas, the flow of solar radiation had to pass outward almost without being absorbed. In contrast, the state of radiational equilibrium had to set in within the dust cloud. This cloud was perfectiy opaque, since when its flattening reached the stage corresponding to the beginning of gravitational condensation, the ratio H/R, in accordance with Eq. (I.32), became roughly equal to 10^{-2} $M/M_\odot$. Consequently, the mass, contained within a radial column with the cross section's surface equal to one unit surface, may be evaluated in the following way:

$$\frac{M}{2\pi RH} \cong 10^2 \frac{M_\odot}{2\pi R^2} \approx \frac{2\times10^{35}}{6\times10^{28}} \approx 3\times10^6 \text{ g cm}^{-2}.$$

This amount of dust is perfectly opaque, if the size of dust particles

$$r_1 \ll \frac{3\times10^6}{\rho_1} \approx 10^6 \text{ cm}.$$

So, the optical thickness of the cloud along the radial dimension was $\tau_R^0 \gg 1$, and thus the temperature of its fringe, remote from the Sun, was extremely low, virtually the same as the temperature of interstellar space. Further, the cloud was also opaque along its vertical dimension, thus the corresponding optical thickness τ_2^0 was again $\tau_2^0 \gg 1$. If r_1 was less than 1 cm, that was sufficient.

We will introduce the notion

$$B(R,Z) = \kappa T^4,$$

where $\kappa = 5.77\times10^{-5}$ erg cm^{-2} s^{-1} g^{-4} is the radiation constant, and T is the temperature. Then the equation for the temperature stationary distribution is

$$\nabla_\tau^2 B = \frac{\partial^2 B}{\partial \tau_R^2} + \frac{\partial^2 B}{\partial \tau_z^2} = 0.$$

Since τ_z^0 is very high, $B(R,H) < B(R,0)$, and one can substitute approximately $\Delta^2 B/\Delta \tau_z^2$ for $-B/(\rho_2^0)^2$.

As a result, we will have

$$\frac{\Delta^2 B}{\Delta \rho_R^2} = \frac{B}{(\tau_z^0)^2}.$$

Assuming that τ_z^0 does not depend on R, we arrive at the formulas

$$B = B_1 e^{-\tau_R/\tau_z^0},$$
$$(TR) = T_1 e^{-\tau_R/4\tau_z^0}, \tag{III.4}$$

where T_1 is the temperature at some distance R_1 from the Sun. Thus, we see that temperature was falling very rapidly when the distance from the Sun was increasing. Because of stellar radiation, it could not approach absolute zero, but the temperature of interstellar dust.

As, according to Eq. (I.32), the maximal flattening of the dust cloud is of the order of 1.6×10^{-2} $M_n/M_\odot$, so the ratio

$$\frac{\tau_2^0}{\tau_R^0} \approx 2\times10^{-2} \frac{M_n}{M_\odot} \approx 10^{-5}. \tag{III.5}$$

It follows that the temperature of the cloud could significantly differ from zero only within the part of the cloud nearest to the Sun, whose mass was of the order of several ten-thousandths of the total mass of the planets.

Inside this domain of the cloud, heated by solar radiation, the dust particles consisted of nonvolatile substances with high melting temperature, from which the inner planets formed. Outside that region, volatile substances were condensing on the dust particles, and thus the total mass of dust was very huge there, and the area gave rise to the giant planets consisting of volatile substances to a considerable degree.

In reality, however, some much greater optical thickness of the cloud had become heated, since the maximal possible flattening corresponding to the beginning of gravitational condensation did not take place simultaneously in different zones of the cloud. The observed ratio of the total masses of inner and outer planets equals one two-hundredth.

In principle, $\tau_z^0=\int_0^H k\rho\, dz$ is a complicated function of R, since both the size of particles, which determines the magnitude of K, and the density ρ depend on R.

However, the assumption that τ_z^0 does not depend on the distance may be supported by the following physical reasoning. Let us evaluate the order of magnitude of τ_z^0:

$$\tau_z^0 \approx k\rho H \approx \frac{\sigma s_1}{m_1} \approx \frac{\sigma}{\rho_1 r_1} .$$

The physical quantity τ_z^0 was approximately constant, since in the domains with a large surface density σ, the sizes of particles were also higher, and their number was probably less in comparison with the areas, where σ was small. This circumstance was connected, first, with the fact that the volatile substances did not form their own nuclei of condensation, but condensed on the already existing particles of dust; second, when the oversaturation of the system was high, even the dust particles of the most refractory substances had to acquire larger sizes, just because of a higher frequency of collisions, accompanied by the vaporization, which resulted in mass redistribution.

To sum up, if the gaseous component of the cloud had the same chemical composition as the stars, then the largest part of the matter of the cloud could be absorbed by its dust particles within the area of outer planets, but only a negligibly small portion could become a part of dust particles close by the Sun. The noncondensing elements (hydrogen, helium, nitrogen, and neon) could form the inner planets, too, but only to a small degree, as a result of insignificant occlusion.

The circumstance that the dust cloud was highly flattened, but the gaseous system was almost spherical, did not prevent the condensation of gas on cold particles, but only made the process of condensation slower.

As soon as a part of the gas which had already existed inside the cloud got absorbed by the dust, fresh portions of gas were immediately supplied by its spherical part, which resulted in its almost perfect exhaustion. There was no gas dissipation from the solar system, so virtually all gases had to become eventually absorbed by dust particles. Naturally, that process had to stop after the planets had been formed. If the mass of the unabsorbed rest of gas was small enough, the gas could be ejected from the solar system, either by selective solar radiation (like the gas flows ejected from the comets), or by the pressure of the corpuscular radiation of the Sun.

4. THE KINETICS OF PLANETARY FORMATION

The process of planetary formation, according to our interpretation, consists of two stages: the formation of secondary condensations from primary ones, and the unification of these into planets. According to Eq. (II.6), the first stage was completed roughly during a thousand revolutions of the cloud around the Sun, that is, during such a brief time, that it is virtually possible to believe that the gravitational condensation immediately leads to the formation of secondary condensations.

The second process developed at a slow rate.

The secondary condensations were becoming stuck together as a result of their mutual perturbations at small distances. The duration of that process was

$$\tau=\frac{1}{nvs},$$

where n is the number of condensations within one volume unit, s is their cross section, and v stands for their relative velocity. The volume of the zone occupied by the condensations is the volume of the ring whose radius is R, the width is $2\epsilon R$, and the thickness is $2iR \approx 2\epsilon R$ (i is the average inclination of the orbits of the condensations). The total number of condensations is $N=M_n/\mu_n$ (the mass of planets), and $\mu_n=0.3M_n\ (M_n/M_\odot)$ is the mass of condensation [Eq. (II.7)]. Consequently,

$$n=\frac{1}{8\pi\times 0.3}\frac{M_\odot}{M_n\epsilon^2R^3} .$$

Further, the cross section of a condensation at the density ρ_1 is

$$s=\pi\left(\frac{1.8}{4\pi}\frac{M_n^2}{\rho_1 M_\odot}\right)^{2/3} .$$

Finally, $v=\epsilon V_0$, where V_0 is the circular velocity. So,

$$\tau=8\frac{\epsilon R^{7/2}\rho^{2/3}}{\gamma^{1/2}M_\odot^{5/6}M_n^{1/3}}, \qquad \text{(III.6)}$$

where γ is the constant of gravitation. Within the zone of Jupiter, when $\rho_1=1$, we have $\tau \approx 10^{15}$ s, that is, around 3×10^7 yr.

Using the same formula for the zone of the Earth, we would have come to a time duration 15 times smaller, whereas for the zone of Pluto the duration would have been 10^4 times more. However, formula (III.6) cannot be actually applied to the latter zones for the reason that they were perturbed by the condensations which were penetrating from the zone of Jupiter, which may be surmised on the basis of the rule of geometric means.

The massive condensations moving from the Jupiter area to other zones affected the process of planetary formation. First, they were able to stop the flattening of the cloud in some zones when the thickness of the cloud had exceeded its critical value. In such cases, the gravitational condensation turned out to become impossible, and the process of dust accretion could take place only on the nuclei which were coming from other zones. Second, the condensations migrating from the Jupiter area affected the bodies within other zones in such a way that their eccentricities increased, which was necessary for them to combine into planets. Thus, the examination of the kinetics of planetary formation should be begun with the formation of Jupiter.

The formation of Jupiter, like the formation of other planets, had begun with the condensation of nonvolatile substances. The total amount of such substances in the interior of Jupiter is not less than their total amount on all the inner planets taken together. Thus, the dust cloud consisting of the nonvolatile substances which eventually formed Jupiter had a mass high enough for its interior to be perfectly shielded from solar radiation, and, consequently, the condensation of volatile substances could take place there. That condensation was increasing the total mass of dust, and the shielding was becoming even greater.

As the matter which the condensations consisted of was being continually fragmented by collisions, the amount of dust which surrounded the formatted condensations had to remain high enough to ensure shielding up to the stage of the complete unification of all the condensations into a planet. This process was quite similar to the modern fragmentation of the small bodies of the solar system—according to Fesenkov's theory,[9] the latter produces the dust which is responsible for the zodiacal light. Then the process was very intensive, but it almost stopped after the majority of smaller bodies had united to form planets. Since the currently existing solar system does not contain a gas cloud anymore, the process of gas condensation had come to an end prior to the unification of secondary condensations into planets. All the other planets were forming under the influence of the already existing condensations of Jupiter.

During the whole process of planetary formation, there were bodies of different size formed by the condensed matter, and so every zone contained bodies whose sizes varied from small dust particles up to secondary condensations. The energy tending to be distributed in a uniform way, the heavier bodies perturbed the smaller ones continually and so the latter's energy was increasing. However, a complete uniform energy distribution could not actually be attained, since the smaller bodies collided more often, a trend which resulted in energy dissipation. Consequently, the process of energy dissipation was developing primarily because the particles of dust continually collided, but its energy source was provided by the kinetic energy of large bodies. Naturally, direct energy exchange between a dust particle and a condensation was much impeded by the enormous difference between their masses. Consequently, such an exchange had to take place via many bodies with intermediate masses.

As a result of that energy exchange, a quasistationary state had to get set up rather quickly at every state of the process. This state was characterized by a balance between the energy gain due to perturbations and its dissipation produced by collisions between dust particles. Because of these perturbations, the velocity dispersion of dust particles could not drop below a certain level, and so the thickness of the dust layer could also not become too small. That means that the gravitational condensation of dust took place either when the surface density σ had reached its critical value (due to a flow of condensing matter from without), or when the perturbatory effects themselves were decreasing due to the continual growth of perturbing bodies, and so the thickness of dust layer reached to its critical value too. The critical thickness evaluated for the region of inner planets on the basis of the formula (I.32) for a quasistatic process,

$$H_{\rm cr}=1.6\times10^{-2}\,\frac{M_n}{M_\odot}\,R,$$

was of the order of several thousand meters. In reality, however, such a small thickness was never reached. It is possible that the bodies which had come in from without served as the nuclei of condensation in that domain.

The principal processes examined in this section were developing so slowly that the stationary distribution always had time to establish itself at every stage, and so the quasistatic modeling done in part I of this work[10] has been justified. The process which took place in reality was considerably different from the scheme suggested in part I of this study. The principal difference is that the cloud was not homogeneous; it consisted of a great many bodies with various masses—from dust particles up to the secondary condensation of the Jupiter zone whose masses were close to the mass of the Earth.

REFERENCES

[1] J. Jeans, *Astronomy and Cosmogony* (Cambridge University Press, Cambridge, 1929).
[2] V. G. Fesenkov, Astron. Zh. **26**, 85 (1949).
[3] O. Yu. Schmidt, *Four Lectures on the Theory of the Origin of the Earth* (Izdatelstro Akad. Nauk SSSR, Moscow) (1949).
[4] V. A. Krat, Dokl. Akad. Nauk SSSR **59**, 455 (1948).
[5] D. Ia. Martynov, Dokl. Akad. Nauk SSSR **60**, 523 (1948).
[6] V. G. Fesenkov, *Priroda* (Moscow) **1948**, (9) 5.
[7] B. Yu. Levin, *Priroda* (Moscow) **1949**, (10) 3.
[8] D. ter Haar, Rev. Mod. Phys. **22**, 119 (1950).
[9] V. G. Fesenkov, *Meteoric Matter in Interstellar Space* (in Russian) (USSR Academy of Sciences Publishing House, Moscow, 1947).
[10] L. E. Gurevich and A. I. Lebedinskii, Izv. Akad. Nauk SSSR Ser. Fiz. **14**, 765 (1950); **14**, 776 (1950).

Cosmogony of the Planetary System and the Evolution of the Sun

B. Yu. Levin
Translated from: Dokl. Akad. Nauk SSSR **91**, 471 (1953)

Examining the process of the evolution of the Sun is a most important, albeit yet unsolved problem of cosmogony. Nowadays, a number of astronomers believe that the Sun emerged as a massive hot star, and later reached its recent state, while moving, because of its mass loss, along the upper part of the main sequence.[1,2]

The very existence of the planetary system in the vicinity of the Sun, which came into being by means of accumulation of sizable bodies in a swarm composed of hard particles, allows the establishment of the upper limit of the change of the Sun's mass in the past.

(1) Let us assume that in the past the Sun had been evolving along the upper part of the main sequence, gradually losing its mass, and also that the formation of planets took place after the Sun had entered the main sequence. If a star is a member of the upper part of the main sequence, its luminosity L is tied to its mass M approximately in accordance with the law

$$L \approx M^4. \tag{1}$$

After the planets—huge bodies rotating around the Sun almost exclusively under the effect of its gravitation—had been formed, the decrease of the Sun's mass had to be accompanied by the increase in the size of planetary orbits according to the law of the conservation of angular momentum:

$$r \approx 1/M. \tag{2}$$

Formula (2) makes it possible to calculate the distance between the planets and the Sun at the moment of planetary formation, if the formation took place when the mass of the Sun had a certain magnitude. Thus we will know the distance between the Sun and the particles which had then formed one or the other of the planets.

The temperature T of the particles which are in thermal equilibrium with solar radiation is determined on the basis of the equality of absorbed and radiated energy. If a is the radius of a spherical particle, and σ is the Stefan–Boltzmann constant, then:

$$4\pi a^2 \sigma T^4 = (\pi a^2 L)/(4\pi r^2).$$

Taking into consideration Eqs. (1) and (2), we will have the equilibrant temperature of the particles with a certain fixed value of angular momentum:

$$T^4 \approx L/r^2 \approx M^6, \quad \text{or} \quad T \approx M^{3/2}.$$

If R is the modern distance between a planet and the Sun measured in astronomical units, and M is the earlier mass of the Sun measured in solar units, then

$$T = (277/\sqrt{R}) M^{3/2}. \tag{3}$$

This formula gives the temperature of the particles which were located at the same distance from the Sun (whose mass was M), as a planet whose modern distance from the Sun (with its modern mass $M=1$) is R.

Formula (3) is based on the assumption that the space was optically transparent, whereas the dust particles were shaped in the form of a very flattened cloud which can produce an enormous opacity. However, we will use the formula for evaluating the upper limit of the possible mass of the Sun in the past. We will be looking for such a mass of the Sun as a star belonging to the main sequence, when no hard particles can definitely exist at the needed distance from it. Then no particles can exist at a closer distance; thus, the Sun's rays will not be intercepted, and formula (3) may be reasonably applied.

Some authors (for example, Krat[3,4]) surmise that the Sun's mass during the period of planetary formation was five- to tenfold more that its recent mass. Formula (3) reveals that such an assumption leads to a contradiction. When $M=10$, the formula gives the temperature $T \approx 9000$ K even for the particles whose specific momentum is the same as of the Earth (to say nothing of Mercury), and no hard bodies could exist under such conditions.

In order that the rocky particles do not melt and evaporate, their temperature should be necessarily less than 1500–1700 K. Demanding this condition to be satisfied for the particles which had formed Mercury, we come to the following conclusion: if the planets emerged during the time when the Sun was evolving along the upper part of the main sequence, then its mass had to be less than 2.5, which means that the Sun belonged to the spectral class A, or a later class.

Should we demand the particles to be capable of condensing from a rarefied gaseous medium, and not only not to evaporate, then the temperature limit will be even lower, and consequently the acceptable mass of the Sun will be less.

(2) Fesenkov, in his talk at the conference on planetary cosmogony, discussed the formation of planets from a disk of dispersed matter which "separated itself from the initial condensation immediately before the formation of the Sun, or even at the same time."[5] Thus two possibilities have been assumed: if the Sun and the protoplanetary cloud happened to emerge simultaneously, then, obviously, the planets them-

selves came into being when the Sun had already become a star, but if the cloud emerged prior to the formation of the Sun, then the planets could arise at a protostellar stage. Let us consider both of the variants.

The assumption that the Sun emerged as a very heavy star with high luminosity could be made consistent with the hypothesis about the simultaneous formation of the Sun and the protoplanetary cloud only if the planets had been emerging gradually, beginning with the distant edge of the cloud. Initially, firm particles could exist only far from the Sun. If the cloud had existed even further, together with the evolution of the Sun, then, as the Sun's mass and luminosity were decreasing, its gases would have condensed into firm bodies, the zone of condensation moving towards the Sun. The formation of planets would have followed the propagation of the zone of condensation. The planets should have been forming under precisely the same temperature condition as that which existed inside the zone of condensation in the inner part of the dust component of the protoplanetary cloud. That could have resulted in a regular and permanent change of the mass and chemical composition of the planets, depending on their distance from the Sun, but that could have never brought about the division of the planets into two groups, with giant Jupiter and small Mars at the junction between them.

This division indicates that all of the planets arose more or less simultaneously. As it has been demonstrated in section (1), such a roughly simultaneous formation could take place in the vicinity of the Sun, which had already become a star belonging to the upper part of the main sequence, only if the mass of the Sun did not exceed 2.5 times its modern mass.

Let us now assume that it will be possible to explain the formation of planets and their above-mentioned division under the conditions which existed in the vicinity of a proto-Sun, that is, the initial condensation of matter which had not yet begun radiating like a star. The upper limit of solar mass may then be calculated as a consequence of the existence of the planet Mercury. Mercury had to survive while also being close to the Sun at the time when the Sun was a bright and heavy star of the main sequence.

Mercury's surface which faced the Sun had to have a higher temperature in comparison with a small particle whose temperature is given by formula (3). At the location where the Sun was in zenith, the surface temperature T_1 was $\sqrt{2}$ times as much:

$$T_1=(392/\sqrt{R})M^{3/2}. \qquad (3')$$

The melting temperature of the rocky substance was some 1500–1700 K, the mean square velocity of the molecules of oxides ejected in the process of its evaporation is more than 1/5 of the parabolic speed on Mercury's surface. Consequently, the melting and evaporation of the surface of Mercury could not be accompanied by the formation of a stable atmosphere composed of the vapors of that substance, which had to be dispersing into space. According to Eq. (3′), such a temperature as 1500–1700 K had to be present when $M\approx2$. However, formula (3′) does not take into account the loss of energy connected with evaporation. A larger solar mass, some $M\approx2.5$, was necessary to support an intensive evaporation of the surface.

Provided that $M=2.5$, Mercury would have been at a distance of 0.16 A.U. from the Sun. Since in such a case $L\approx40$, the "solar constant" for Mercury would have equaled 3000 cal/cm^2 min, and the whole planet would have evaporated in only 10^3–10^4 yr.

Thus the very fact of Mercury's existence at a distance of 0.4 A.U. from the Sun demonstrates that even if the planets were formed at the protostellar stage, the Sun, when entering the upper part of the main sequence, had a mass less than 2.5 times as much as its present mass.

(3) All the stars of the main sequence, even those whose rotational speed achieves the maximum value permitted by the condition of centrifugal stability, have the specific angular momentum of their matter on their equator, which is much less than the specific angular momentum of the distant planets of the solar system. Even at the stage of class B, moreover at the A class stage class and later, no matter separated from the Sun could form in its vicinity a protoplanetary cloud with the necessary angular momentum, which might give birth to the presently existing planetary system.[6] Therefore, the following alternatives consistent with Fesenkov's hypothesis, are conceivable:

(a) If the Sun emerged as a star of the A, or later, spectral class, and if its mass did not exceed 2.5 times its present mass, then the protoplanetary cloud could come into being either at some rather vague preceding stage of its development (by means of capture or otherwise), or already at the star stage (that is, after the Sun had entered the main sequence), and then only by means of capture.

(b) If the Sun emerged as a very heavy star with high luminosity, then the protoplanetary cloud came into existence only afterwards—after the mass of the Sun had become less than 2.5, and necessarily by means of capture.

(4) The motion of small bodies is affected considerably by the hindering action of photon as well as corpuscular radiation. Radzievsky[7] has examined this phenomenon in its general form, under the assumption that the Sun had a variable mass. According to his calculations, if the small bodies now located around the Earth's orbit had existed as independent bodies during the time span when the Sun's mass decreased from $M=1.5$ to $M=1$, then they would have originated from the area of the giant planets, or even from some more remote areas. In order that a body several meters across emerged no further from the Sun than at a distance of the orbit of Jupiter, it is necessary that the Sun's mass experienced less than 1% reduction during its individual life. Unfortunately, nobody knows when that individual life exactly began. Nevertheless, a very rapid evolution of the orbits of small bodies produced by a decrease of the Sun's

mass causes us to believe that the limit noted in (3) is too high.

(5) The above-presented considerations demonstrate that the links between the problems of star and planetary cosmogony are bilateral, not one-sided. Planetary cosmogony should not only take into account a possible evolution of the Sun during the time spanned after the formation of the planets, but also the cosmogony of stars should pay attention to the restrictions put upon the possible former evolution of the Sun due to the fact of the protracted existence of its planetary system and the already firmly established principal features of its formation.

The Geophysical Institute of the USSR Academy of Science,
Delivered 5/7/53

[1] V. G. Fesenkov, Astron. Zh. **26**, 67 (1949).
[2] A. G. Masevich, Astron. Zh. **28**, 26 (1951).
[3] V. A. Krat, Izv. Gl. Astron. Obs. Pulkove **149**, 1 (1952).
[4] V. A. Krat, Vopr. Kosmog. **1**, 34 (1952).
[5] V. G. Fesenkov, Tr. Pervogo Sovetsch. Kosmog. (1951).
[6] V. S. Safronov, Astron. Zh. **28**, 244 (1951).
[7] V. V. Radzievsky, Dokl. Akad. Nauk SSSR **81**, 167 (1951).

THE HYPOTHESIS OF TURBULENCE IN THE PROTOPLANETARY CLOUD

V.S.Safronov & Ye.L.Ruskol
Problems of Cosmogony, Vol.5, (1957), pp. 16-38

The first section of the article critically examines Weizsacker's hypothesis on turbulence in the protoplanetary cloud. The condition of stability of laminar rotational motion for a cloud with a temperature gradient is derived. It is shown that the protoplanetary cloud is stable with respect to small disturbances, and convection cannot exist in it.Turbulence in the cloud could not arise without external perturbations. An estimate of the energy dissipated in turbulent motion shows that the entire gravitational energy of the cloud could be adequate for sustaining turbulence on the large scale proposed by Weizsacker for only 100 revolutions around the sun. This time is several orders less than the time needed for the formation of planets according to Weizsacker's scheme.

In the second section, the redistribution of matter in the protoplanetary cloud at the initial stage of the cloud's evolution is considered, when a brief "turbulent" stage can exist in it. It is shown that in a rotating system at the scale of turbulent mixing of matter comparable with the system's dimensions, the shearing stresses determining the force of friction between the layers depend on the gradient of angular momentum rather than on the angular velocity gradient, as in the case of laminar motion. Since in the protoplanetary clouds the angular velocity decreases with distance from the sun, while the angular momentum increases, the direction of mass and momentum transfer in the cloud proves to be opposite to that found by Weizsacker. Instead of the inner portions of the cloud falling on the sun and its outer parts moving off to infinity, there is a tendency for the cloud to remain in some central region. This result is an additional argument favoring the position which states that if the first stage of the cloud's development exhibits turbulence, then it must very rapidly be damped.

The third section of the study considers the process of flattening of the dust component of the gas-dust cloud in which chaotic macroscopic motion of matter is absent. It is shown that during the time required for dust particles to settle toward the equatorial plane of the cloud, they can increase in size through cohesion resulting from mutual collisions, only to the extent of several centimeters in diameter. Therefore, the process of particle growth in such a cloud does not prevent an increase in density in the dust component of the cloud up to the critical value and the onset of gravitational instability leading to the formation of "primary condensations". However, for this state to be arrived at, the relative velocities of the particles must not exceed 1 cm/sec and in the zone of the earth and 100 cm/second in the zone of the major planets. In the case of higher relative velocities of the particles, the formation of planetary embryos can proceed only through the direct growth of particles.

It can be taken as generally accepted at present, that the planets were formed from a gas-dust cloud rotating around the sun. However, as for the evolution of the protoplanetary cloud and the mode in which the planets were formed, opinions among astronomers vary widely due to differences in their concept of the initial mass of the cloud and of the nature of movement of matter within the cloud. In studies by O. Yu. Schmidt [1], L. E. Gurevich, and A. 1. Lebedinskiy [2], the original mass of the cloud in orders of magnitude is taken as equal to the present mass of the planets. C.Weizsacker [3], G. Kuiper [4], and V. G. Fesenkov [S] believe that the initial mass of the protoplanetary cloud was comparable with the mass of the sun. According to L. E. Gurevich and A. 1. Lebedinsky, the cloud consisted of particles, inelastic collisions between which led to a rapid decrease in their chaotic velocities and to a condensation of the cloud. This caused gravitational instability of the cloud and its breakdown into individual "primary condensations," whose further amalgamation led to the formation of a small number of large bodies -- the planets. G. Kuiper and V. G. Fesenkov believe that gravitational instability existed in the massive gas-dust cloud as a result of which condensations were formed with masses exceeding the masses of the major planets. On the other hand, C. Weizsacker [3], [6], S. Chandrasekhar and D. ter Haar [7], and R. Lust [8] propose that the rotational motion of matter within the protoplanetary cloud was turbulent. The existence of large relative velocities of the volume elements within the cloud must have prevented the onset of gravitational instability. The formation of planetary embryos could have only followed the path of the direct accretion of solid particles. The processes of mass and angular momentum transfer are of very much greater importance in a turbulent cloud than in a laminar one. Weizsacker's cosmogonic concept is built on this contrast. However, this concept has occasioned serious objections. In order to explain the actual character of the evolution of the protoplanetary cloud, it is most important to know whether it existed in a "quiet" state or whether additional macroscopic motions existed in it, besides the over-all revolution around the sun. The purpose of this article is: 1) to examine the stability of laminary rotational motion of the cloud and the possible duration of turbulent motion; 2) to explain the transfer of matter and rotational momentum in the cloud; 3) to examine the initial stage of the growth of planetary embryos in the gas-dust cloud.

1. Stability of Laminar Rotational Motion of Matter Within the Protoplanetary Cloud.

The concept of turbulence in the protoplanetary cloud was first proposed by Weizsacker [3], [6]. He offered the following arguments in its favor:

1) turbulence exists in many interstellar gas clouds;

2) the Reynolds number Re $=\rho l w/\eta$ is large for the protoplanetary cloud. With a scale of turbulence l of the order of the astronomical unit the Reynolds number proves to be the order of 10^{10};

3) the modern law of planetary distances would be well explained by the presence in the past of a regular system of Weizsacker's eddies.

However, these arguments cannot serve as a demonstration that the turbulence existed in the cloud. First of all, the conditions of the motion of matter within the protoplanetary cloud are completely different from conditions in interstellar nebulae, since the gravitational attraction of the sun in this case dominates all other forces; secondly, the large value of the Reynolds number is not adequate for turbulence to have existed in the cloud; it is still further required that this number be greater than the critical value, and the latter is unknown; thirdly, the law of planetary is explained no less adequately from other viewpoints, not appealing to eddies. Thus, according to O. Yu. Schmidt, this law is a consequence of the averaging of the angular momentum of particles amalgamating into the planets. According to L. E. Gurevich and A. I. Lebedinskiy this law reflects also the distribution of the eccentricities of "secondary condensation" in the protoplanetary cloud. V. G. Fesenkov asserts that the law of planetary distances is determined from the condition of tidal stability when planets were formed in the cloud.

In order to decide whether turbulence existed for a long time in the protoplanetary cloud, we must discover whether the conditions for its maintenance prevailed, i.e., we must consider the question of the stability of rotational movement of the gas-dust cloud around the sun. If we assume that no large bodies exerted a marked gravitational influence on the surrounding environment existed in the cloud, and if the mass of the dust was less than the mass of the gas, then in solving the hydrodynamical problems, we must view the cloud as a gas cloud. The dust entrained in the gas does not alter the character of the motion.

As is generally known [9], the condition of stability of laminar motion of a fluid rotating between two coaxial cylinders, for the case of Reynolds number, is of the form:

$$\mu(du/dr) > 0, \qquad (1)$$

where $\mu r = \mu r^2\omega$ -- angular momentum of an element with mass m moving at a distance r from the axis of rotation at an angular velocity ω.

If we neglect the force of gravity of the cloud itself compared to the gravity of the sun (which is valid as long as the cloud does not approach the state of gravitational instability), and also neglect the pressure gradient dp/dr, then the condition (1) when applied to a flat protoplanetary cloud reduces to the condition of stability of circular orbits [10]

$$(\partial r^3/\partial r)(\partial \phi/\partial r) > 0 \qquad (2)$$

where ϕ = potential energy at the point r.

Since the force of attraction of the central body -- $\partial\phi/\partial r$ -- is equal to

-GMr^2, then the conditions (1) and (2) in the protoplanetary cloud are satisfied, which signifies stability of laminar rotational movement of the cloud around the sun. Instability arises only when the central force falls off with increasing distance more rapidly than r^{-3}.

In condition (1), however, we do not take into account the possibility of convection arising within the cloud. Moreover, Weizsacker [3] attempted to substantiate the existence of turbulence in rotating cosmic gas masses, including also within the protoplanetary cloud, by using precisely this condition of the emergence of convection. However, Weizsacker here neglected rotation and did not take into account the condition of stability (1). It is obvious that in order to resolve the question of whether turbulence could arise in the protoplanetary cloud we must consider the condition of stability (1) with account taken of convection. Such an examination has been given by the present author in the article [11].

The gravitational force, centrifugal force, and gas pressure act on the element of the medium of unit mass moving around the sun at velocity v. In the stationary cloud in which motion occurs in circular orbits, these forces are made equal to each other.

$$f = (-GM/r^2) + v^2/r - 1/\rho(dp/dr) = 0 \qquad (3)$$

If this element is shifted to the point r+dr with the same angular momentum v.r, then acting on this element is the force:

$$df = [(2GM/r^3)dr - 3v^2/r^2)dr\,(1/\rho)dp/dr - (1/\rho_{EL})dp/dr_{r+dr}\,] \qquad .....(4)$$

The symbol "EL" designates that this quantity refers to the volume element. The condition of stability of the cloud's motion consists in that $df < 0$ and $dr > 0$. This condition can be written in the form

$$(1/\rho^2)dp/dr[(d\rho/dr)_{EL} - d\rho/dr] < GM/r^2 + (3/\rho r)dp/dr + d/dr(1/\rho)dp/dr \qquad (5)$$

or in the form

$$\mu\, d\mu/dr > (r^2/2r^2)dp/dr[(d\rho/dr)_{EL} - d\rho/dr \qquad (6)$$

The condition of stability (6) differs from (1) by the presence of a right-hand member which takes into account the difference in the density of the displaced element and the surrounding medium. The condition (1) is derived for an uncompressed fluid, while we are considering a gas medium. The condition of absence of convection in a nonrotating cloud used by Weizsacker is obtained from (6) if we place $\mu=0$.

We will assume that the volume element is translated adiabatically, then

$$(d\rho/dr)_{ad} = 1/\gamma\,(\rho/p)dp/dr\,;$$

$$d\rho/dr = \rho/r\; dp/dr - (\rho/T)dT/dr$$

therefore

$$(d\rho/dr)_{ad} - d\rho/dr = \rho/\gamma T[dT/dr-(\gamma-1)(T/p)dp/dr)] \qquad (7)$$

where $\gamma = cp/cv$. The condition of instability here can be written in the form

$$dT/dr > (\gamma-1)T/\rho\; d\rho/dr\{[1+(\gamma/\gamma-1)r^2/(dp/dr)d\rho/dr]$$

$$[GM/r^3 + (3/\rho r)dp/dr + d/dr(1/\rho(dp/dr))]\} \qquad (8)$$

if $dp/dr>0$. If, however, $dp/dr>0$, which is more natural for the protoplanetary cloud, the inequality (8) changes sign. In order for the cloud to be in stable in this case, dT/dr must be negative. The usual condition of convection arising in an immobile medium is obtained from (8) if only unity remains in the bracketed expression on the right. The second member in the bracket refers to stability of the circular orbits.

For an order of magnitude estimation of these members we assume approximately that ρ and ρ are exponential functions of r:

$$\rho \sim r^{-a_1}\,,\quad p \sim r^{-a_2}$$

$$d\rho/dr = -a_1;\; dp/dr = -a_2\, p/r \qquad (9)$$

where a_1 and a_2 = constants of the order of unity. Then the condition of instability takes of the form:

$$dT/dr > (\gamma-1)a_1\; T/r\{1+(\gamma/\gamma-1)(GM/a_1a_2RT^r)$$
$$+(\gamma/\gamma-1)a_2a_1-1)/a_1\} \qquad (10)$$

where R = gas constant. The first and third members in the bracketed expression of (10) are 2-3 orders of magnitude less than the second member, and can be neglected. Then the temperature gradient required for instability must satisfy the inequality:

$$dT/dr > (\gamma/a_2R)GM/r^3 \qquad (11)$$

The temperature gradient in the transparent near-solar equals

$$dT_0/dr = d/dr\,(300^\circ/\sqrt{r_{au}}\,) = -T_0/2r \qquad (12)$$

r_{au} means that r is expressed in astronomical units.

Calculation shows that the value of the gradient required by the inequality (11) exceeds by more than 2 orders of magnitude the temperature gradient in the transparent cloud (12). It is apparent that both the temperature and the temperature gradient in the opaque protoplanetary cloud must be still lower than for the case of the transparent cloud. Therefore, the protoplanetary cloud is stable with respect to small perturbations and convection cannot arise in it.[1]

The arguments that turbulence exists in rotating cosmic gas masses presented by Weizsacker in our opinion are not correct in two points. In the first place, the usual condition of convection arising without any external changes is applied to a revolving cloud. This means that the left hand portion of expression (6) must be equal to zero. Then the cloud revolves around the sun and the velocity is close to the circular (Keplerian); this is completely unacceptable, since the left hand side in (6) is considerably greater than the right hand member. In the second place, in this condition a "turbulent temperature" is used instead of the usual temperature, i.e., the quantity $w^2/3R$ where w= turbulent velocity, considerably surpassing thermal velocity. Therefore, the condition of stability with respect to small perturbations is applied not to the stationary state of the cloud, but to a cloud already perturbed, when intensive turbulent mixing of matter takes place with a mixing distance of the order of the distance of the axis of rotation. In the only concrete example offered by Weizsacker [6] (distribution of density and angular velocity in a two-dimensional model of cosmogonic gas mass) an error exists in the calculations which allows him to conclude that convection is possible.In his cloud model the "turbulence" temperature increases with distance from the axis of rotation, while the density of the matter decreases. Under such conditions convection in reality is impossible.

The possibility of disordered motions arising in the protoplanetary cloud at the very moment of the cloud's formation or as a result of some external perturbation is not precluded. However, to sustain turbulence a constant energy source is required. Without such a source, the dissipation of energy leads to a rapid damping of turbulence. For an order of magnitude estimate of damping rate we can use the expression for the lifetime of turbulent eddies.

$$\tau = (l^{2/3}\, l_m^{\;1/3})/w \qquad (13)$$

According to Kuiper [4], the average size of a turbulent cell is $l=0.27$ r, the maximum size of the cell $l_m=0.62$ r. The average velocity of the chaotic motion w was quite arbitrarily set by Kuiper equal to the average deviation of velocities of the major planets from the circular velocity v_c. Orbital eccentricities of the order of 0.05 lead to $w=(1/80)v_c$. Using these numerical values Kuiper found the lifetime of the eddies to be of the order of 10 revolutions around the sun. Weizsacker [3] assumes that the eddy energy can be replenished from the gravitational energy of the cloud. He believes that turbulent friction leads to the outer regions of the cloud drawing away from the sun and the inner regions approaching the sun. As a whole the potential energy of the cloud falls off, being converted into energy of turbulent motion. However, even if the entire cloud fell in towards the sun, then the potential energy released here would be sufficient to sustain turbulence only for a very brief interval of time. Actually, the dissipation of the energy of turbulent motion equal to $a \sim w^3/l$, per 1 gram of matter per second must be compensated for by the transition of matter from an orbit having a period of revolution t_0 to a smaller orbit with period t_1 in the time

$$t = (\alpha/6\pi k^3)\,(t_0 - t_1) \qquad (14)$$

if the average relative velocity $w = kvc$ and the average size of the turbulence cell $l = \alpha r$. Using Kuiper's values of $k = 1/80$ and $\alpha = 0.3$, this time will be equal to $t = 10^4(t_0 - t_1)$. Therefore, a cloud having turbulent motion of such a scale would have fallen in on the sun in 104 10^5 years if it had occupied the location of the present day planet. For Weizsacker's values of the turbulent velocity ($k \approx 0.1$) the time proves to be an additional two orders of magnitude less. If, however, it is considered that energy is also expended in the external regions of the cloud falling away from the sun, bearing excess rotational momentum, then the time required for the cloud to fall in on the sun proves to be still less. Calculation is carried out for a unit mass of the matter, as the result of which it is practically independent of the initial mass of the cloud. Therefore, for planetary formation the time remains several orders of magnitude less than is required in Weizsacker's scheme (~10 years). This result, derived by ter Haar, shows the internal contradiction of the cosmogonic concept of Weizsacker. However, this contradiction is clearly underestimated by Weizsacker and his supporters. The attempt of ter Haar to overcome this difficulty by an assumption of a regular system of eddies in which the dissipation of energy is several orders of magnitude less, or by an assumption of a relatively great initial mass of the cloud has not proven substantiated.

Thus, we come to the conclusion that the "quiet" protoplanetary cloud is stable with respect to the small perturbations and that turbulence cannot arise in it. If at the first stage of the existence of the protoplanetary cloud a "turbulent" stage was in effect, than evidently was very brief, due to the extremely rapid damping of the turbulence.

2. Redistribution of Matter and Rotational momentum Due to Viscosity in the Protoplanetary Cloud.

The hypothesis of turbulence in the protoplanetary cloud at first glance appears to allow us to proceed to solve the most important problem of planetary cosmogony -- the problem of the distribution of the angular momentum in the solar system. In actual fact, the direction of the development of the revolving cosmic gas masses is determined from the solution of hydrodynamic problems considered by Weizsacker [6] and Lüst [8]. According to Weizsacker, the rotating gravitating gas mass cannot be stable. In a nebula exhibiting central condensation the inner portions rotate more rapidly than the outer. As a result, viscous stresses can arise in the medium, which tend to make the rotation of the solid-body kind. This leads to a retardation of the internal portions and an acceleration of the external portions of the nebula, as the result of which there is no equalizing of the velocities, but an approaching toward the center of the internal portions and a drawing away of the outer. If we neglect gas pressure, then the process of the separation of the gas mass continues until a dense core is formed in the center, rotating as a solid body. Here, the outer gas "shell" separates out "to infinity" bearing with it practically all the rotational momentum. If it is considered that gas pressure is continually "eroding" the outer boundary of the core, then it turns out that the process of separation must continue still further, even though at lesser intensity, embracing ever deeper layers of the central core.

As we will show below, the effect of molecular internal friction within the protoplanetary cloud is very small and cannot lead to any marked redistribution of rotational momentum and matter in cosmogonically relevant periods of time. To explain the transfer of momentum and matter Weizsacker appealed to turbulent friction, believing that it acts similarly to molecular friction. In this appeal, the quantitative effect proved to be 10^{10} times greater, since the coefficient of turbulent viscosity $\eta = \rho l w$ includes the scale of turbulence which is ten orders of magnitude greater than the atomic mean free path.

In the previous section of the article serious objections were raised to the entire concept of turbulence in the protoplanetary cloud: the cloud is stable with respect to small perturbations, and if initially a turbulent state did exist in the cloud, then it must have been very short-lived. However, movements on a large scale could have existed during this brief stage in the cloud (we probably cannot speak here of advanced turbulence in the full sense of the word), when the cloud substance underwent intensive mixing. The rotational momentum here could have been transmitted from some portions of the cloud to others -- along with the matter and through shearing stresses. However, there is reason to believe the direction of transfer as a result of turbulent friction cannot be that which was asserted by Weizsacker, but directly opposite, Therefore, it is interesting to once again consider the mechanism of the redistribution of matter within the protoplanetary cloud with account taken of characteristics of turbulent motion.

Weizsacker [6] and Lust [8] described the rotation of the flattened gaseous disc around the gravitating central mass by the following system of hydrodynamic equations:

$$\rho[\partial v_r/\partial t + v_r(\partial v_r/\partial r) - v_r^2/r] = -\rho(\partial\phi/\partial r) - \partial p/\partial r + T_r \qquad (15)$$

$$\rho[\partial v\varphi/\partial t + vr(\partial v_\varphi/\partial r + v_r v_\varphi/r] = T_\varphi \qquad (16)$$

$$\partial\rho//\partial t = -1/r[\partial/\partial r(r\rho v_r)] \qquad (17)$$

Here, $-\partial \phi/\partial r$ = gravitation of the sun, T_r and T_φ= radial and tangential components of internal friction. These equations, strictly speaking, describe the two dimensional case. However, they can be used also for the three dimensional case if the disc is greatly flattened ($v_x \ll v_\varphi$). It can be assumed, according to Weizsacker and Lust, that the radial velocity vr is low in comparison to the tangential velocity v(p and we can exclude from the equations numbers of a lower order of magnitude than v_r [i.e., $\partial v_r/\partial r$ and $v_r(\partial v_r/\partial r)$]and also the radial friction T_r and the pressure gradient $-\partial p/\partial r$, since they are part of the sum having a considerably higher magnitude. The equations take on the form

$$v_\varphi^2/r = \partial \phi/\partial r = GM/r^2 \qquad (18)$$

$$\rho[\partial v_\varphi/\partial t + v(\partial v_\varphi/\partial r) + v_r v_\varphi/r] = T_\varphi =$$

$$\{(\partial \eta/\partial r)(\partial v_\varphi/\partial r - v_\varphi/r) + \eta[\partial^2 v_\varphi/\partial r^2 + 1/r\,(\partial v_\varphi/\partial r)$$

$$- v_\varphi/r^2\} \qquad (19)$$

$$\partial \rho/\partial t = (-1/r)[\partial/\partial r(r\rho v_r)] \qquad (20)$$

In other words, it is assumed that to the first approximation each volume element moves about a circular keplerian orbit, in which a perturbation caused by the force of tangential friction T operates on this circular movement.

Before we proceed to look at turbulent transfer, we will evaluate the transfer of matter and rotational momentum due to molecular viscosity in a cloud revolving laminarly. The use of equations (18-20) in this case is justified, which cannot be said for the case of turbulent movement.

Substituting in (19) the value of v from (18) and setting $\partial v_\varphi/\partial t = 0$, since the mass of the sun remains practically unchanged, we obtain

$$\rho[v_r(\partial v_\varphi/\partial r) + v_r v_\varphi/r] = T_\varphi = (1/r^{2)}\{\partial/\partial r[\eta r^3(\partial \omega/\partial r)]\} \qquad (21)$$

where $\omega = v_\varphi/r$ -- angular velocity.

Inserting the value of v_φ from (18), we find after uncomplicated transformations

$$\rho v_r = -3\,(\partial \eta/\partial r + \eta/2r) \qquad (22)$$

The radial flow of matter per unit time to a flat layer of unit thickness at adistance r from the center equals

$$J = 2\pi r(\rho v r) = -6\pi[r(\partial \eta/\partial r) + (\eta/2)] \qquad (23)$$

Since molecular viscosity changes slowly with distance

($\eta = 0.04(\sqrt{m_a KT}/r_a^2) \sim \sqrt{T}$, in which change in

temperature T along r is not great), then the second member is dominant in (23). Therefore $J = -3\pi\eta$. The total radial flow of matter per unit time equals J.H, where H = cloud thickness.

If we take H= 10^{13} cm, η= 7(10^{-5}) [See Note 2], then we find that in one billion years, the amount of matter falling in on the sun is of the order of 10^{26} grams, which is low compared to the mass of the entire cloud. We can show that the transfer of momentum associated with the transfer of this amount of matter is also small. Therefore, no marked redistribution of momentum as a consequence of molecular viscosity can occur in the laminar cloud. For the case of the nebular hypothesis leading to rapid revolution of the sun, this mechanism does not provide the transmission of noticeable momentum from the sun to the cloud. For the case of the separate formation of the cloud and the planets, this mechanism does not lead to the transfer by the cloud to the sun of the angular momentum which the sun has at present. Therefore, we can conclude that the role of molecular viscosity in the evolution of the protoplanetary cloud is negligibly small. Since, according to (23), the flow of matter does not depend on the density within the cloud, the previous conclusion is valid for any value of the protoplanetary cloud mass.

The method of approximational solution of hydrodynamic equations presented above was first used by Weizsacker (6), and then by Lust (8) in investigating the evolution of turbulent revolving gaseous masses. Taking $v_r = v_r + v'_r$. $v_\varphi = v_\varphi + v'_\varphi$ where v_r and v_φ= averaged radial and tangential velocities and v'_r and v'_φ= turbulent pulsational velocities in the corresponding coordinates, Weizsacker assumed that v_r and v'_φ entering into the hydrodynamic equations (13-20) must be taken as referring to their averaged values, since the time averaged values of v_r, and v'_φ equal zero. The presence of turbulent velocities appeared only as the force of pressure, which can be neglected, and as the force of turbulent friction. Solution of equations (18-20) for the case of turbulent motion is more involved than for the case of laminar motion, since the coefficient of turbulent viscosity ~ =paw depends on density (the coefficient of molecular viscosity, however depends only on temperature). For the case in which the nebula itself cannot be neglected due to gravity, the equations (18-20) cannot be solved analytically. Weizsacker has undertaken their qualitative examination and concludes that the gaseous mass separates into a nucleus and shell, bearing practically all the momentum. This conclusion of Weizsacker was confirmed by the calculations of Trefftz [12], who used the same method in an application to protogalactic nebulae. For the case of the protoplanetary cloud (the greater point mass in the center) Lust (8) obtained from the equation (18-20) an equation in parabolic type partial derivatives for the cloud density p(r, t), which has a solution of the form

$$\rho(r,t) = \psi\,(r)e\text{-}\beta^{2t}, \qquad (24)$$

in other words, the cloud must scatter with time due to turbulent viscosity. Consideration of this equation shows that the internal portions fall in on the central condensation, while the external fall away lo infinity. This separation is accompanied by the transfer of momentum from the internal portion to the external . The time of dissipation of the cloud is obtained, according to Weizsacker's judgment [13], as a very small quantity -- less than 100 periods of revolution. As already noted above, such a short period of the cloud's evolution is not compatible with the duration of planetary formation.

The conclusion of Weizsacker and his pupils Lust and Trefftz on the direction of transfer of matter and momentum in the protoplanetary cloud appears to us to be unsubstantiated. Strictly speaking, the use of the usual equations of hydrodynamics for a revolving system of eddies in which the length of the mixing path is comparable to the dimensions of the system ($l = \alpha r$, and α = of the order of unity) cannot be taken as valid.

The authors transferred regularities in laminar movement to turbulent without additional investigation. It follows from the usual expression for the viscosity tensor in hydrodynamics that the shearing stresses determinate the force of friction T_φ equal to

$$\rho v'_r v'_\varphi = -\sigma'_{r\varphi} = \eta r(\partial\omega/\partial r) \qquad (25)$$

If the angular velocity gradient in r is negative (as does occur in the solar system), then the right-hand portion of (25) is positive. The force of friction with respect to the inner layers operate in the direction of rotation on an elementary area perpendicular to r. Therefore, the internal layers having greater angular velocity accelerate the internal layers. This is valid with respect to the laminar movement, since in this case there are no grounds to doubt the applicability of the main equations of hydrodynamics and the viscosity tensor. However, their applicability without any changes relevant to turbulent motion are still not as apparent, since turbulent movement differs qualitatively from laminar. If in the rotating medium intense mixing of matter occurs, then it is more natural to expect that as a result of this mixing the equalizing not of angular velocity but of angular momentum will take place.

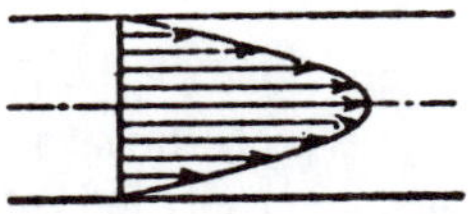

Figure 1. Distribution of velocities in a laminar flow in a tube.

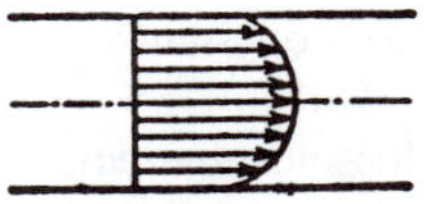

Figure 2. Distribution of velocities in a turbulent flow within a tube.

Let us consider experimental data on the distribution of velocities in laminar and turbulent motion. We can use graphs from the book Aerodinamika (Aerodynamics) by V. F. Durand [14] for a rectilinear flow within a tube. The distribution of velocities in a liquid moving between two rotating cylinders has been studied by Taylor [15] and by Wattendorf [16] and is presented in an article by T. Karman [17].

The qualitative difference between laminar and turbulent movement can be seen already from Figures I and 2. In the laminar rectilinear flow the transverse velocity gradient is practically constant over a great portion of the liquid volume. In the turbulent flow, the velocity is almost constant throughout the entire cross section of the tube and drops sharply along the walls. This difference can be explained by the fact that the coefficient of turbulent viscosity ~=paw is many times greater than the coefficient of molecular viscosity and increases sharply with separation from the walls, while the coefficient of molecular viscosity is constant throughout the entire volume.

Actually, for a rectilinear stationary flow in the tube along the axis x (and this is totally analogous for movement between two parallel planes) we can readily find from the hydrodynamic equations the relationship:

$$\partial^2 v/\partial r^2 = 1/\eta \; (\partial p/\partial x)$$

in which $\partial p/\partial x$ = constant and $\partial p/\partial r = 0$. In the laminar flow η = constant therefore

$$v = v_c(1 - r^2/r_0^2)$$

i.e., we obtain in the cross section a parabola represented in figure 1. In the turbulent stream $\eta \neq$ constant. Far from the walls η is many times larger than near the walls; hence, far from the walls d^2v/dr^2 must be low. This has been confirmed by experiment. The curve v(r) in Figure 2 has a small curvature practically everywhere away from the tube walls. For the case of movement of liquid between two rotating cylinders we can use the condition that the momentum of the forces created by the shearing stresses along the,entire periphery of a disc of any radius r must remain constant:

$$2\pi r^2 \overline{\rho v'_r v'_\varphi} = -2\pi r^2 \sigma'_{r\varphi} = \text{constant} \qquad (26)$$

In the laminar movement, the quantity $\sigma'_{r\varphi}$ is determined by the ratio (25). Therefore, we obtained from (26):

$$r^2\sigma'_{r\varphi} \sim \eta r^3(\partial\omega/\partial r) = \text{constant} \quad (27)$$

This expression does not contradict the distribution of velocities in r for the case in which the external cylinder (Figure 3) is rotated and the motion is stable.

Prandtl has determined for turbulent motion the shearing stresses not in the form of (25) but from the assumption of the transfer of angular momentum -- in the form

$$\rho\overline{v'_r v'_\varphi} = -\sigma'_{r\varphi} = -\eta\ \{1/r[\partial/\partial r(v_\varphi r)]\} \quad (28)$$

Therefore instead of (27) we will have:

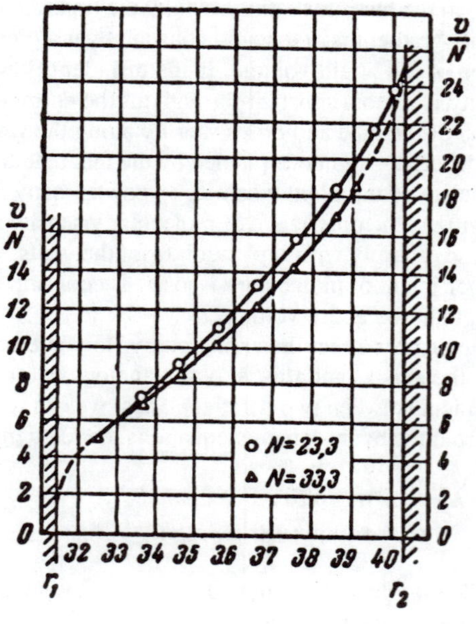

Figure 3. Distribution of velocities in a liquid moving between coaxial cylinders. The external cylinder is moving. The abscissae correspond to distances from the axis of rotation.

$$r^2\sigma_{r\varphi} \sim \eta r\,\partial(v_\varphi r/\partial r) = \text{constant} \quad (29)$$

However, Taylor believes that the expression (28) fits only the very specialized case of turbulent motion. As we have seen above, Weizsacker employed not the relationship (28) but the relationship (25). Which then of these relationships (and, accordingly, of the relationships (27) and (29)) more closely approximates the distribution of velocities in rotational turbulent motion found experimentally and represented in Figure 4. We can assume that as is true for rectilinear flow, the coefficient of turbulent viscosity η rises sharply with distance from cylinder walls. This increase in η must be compensated by a decrease far off from the walls to a very low value of either the quantity $\partial w/\partial r$ or the quantity $\partial(v_\varphi r/\partial r)$, depending on what relationship is valid-- (27) or (29). From figure 4 we can see that in the central region between the cylinders the derivative of the angular momentum is close to zero (solid line) and the derivative of the angular velocity (dotted line) differs significantly from zero. This fact, in our opinion, is a serious argument in favor of the position in which for the case of turbulent rotational motion the relationship (28) is more valid than the relationship (25). In addition, it is precisely this experimentally derived quality of setting equal to zero the derivative of angular momentum within the central region that was considered by Wattendorf and Taylor as the most serious argument against the Prandtl formula (28). However, the precision of the experiment is inadequate to state that $\partial(v_\varphi r)/\partial r$ actually is equal to zero in this region. The velocity was not determined directly, but the pressure, in which the measuring gauge (the pitot tube) introduced large deviations, which were removed only by extrapolation to the diameter of a tube equal to zero. Therefore, we can only say that $\partial(v_\varphi r)/\partial r$ is low, and this agrees just as well with the Prandtl formula. The main objection of Taylor against the relationship (28) was that the angular momentum of the turbulent element, generally speaking, cannot be conserved, since the element must do work against the pressure gradient. However, for the case of turbulent motion in the protoplanetary cloud, this objection disappears, since first of all the gas pressure within the cloud (even when a turbulent cloud model is considered) is low in comparison to the force of attraction of the sun and, secondly, given the circular symmetry of the cloud, the pressure gradient is directed along *r* and cannot alter the angular momentum of the turbulent element. From these objections, of course, it does not follow that the Prandtl formula is wholly correct. However, in our opinion it must never give way in preference to formula (25) derived for the case of very small free path lengths.

Karman (17), in discussing the possibility of using the theory of turbulence in astrophysics, also made use of the Prandtl formula (23) for shearing stresses in a rotating system. Here, he especially emphasized in his text that if the momentum $v_{\varphi r}$ increases with distance from the center, then the correlation between v r and v ~p must be negative, i.e., $v'_r v'_\varphi < 0$. However, Karman did not make any critical comments with respect to Weizsacker's results.

For the protoplanetary cloud ($v_\varphi \sim 1/\sqrt{r}$) the angular momentum, μ is proportional to $\sqrt{r}$ and increases with r, while the angular velocity falls off; therefore, the derivatives $\partial\omega/\partial r$ and $\partial(v_\varphi r/\partial r)$ entering into (25) and into (28) have different signs. This means that the shearing stresses $v'_r v'_\varphi$ have different signs. Using the relationship (25), Weizsacker and Lust came to the conclusion that the outer portions of the cloud were falling away from the sun and that the inner portions were falling in on the sun. In determining the shearing

Figure 4. Distribution of angular momentum in a fluid flow in between coaxial cylinders (solid line). Distribution of angular velocity (dotted line). The inner cylinder is rotating; the motion is turbulent. Distances from the wall of the inner cylinder are plotted along the abcissa axis.

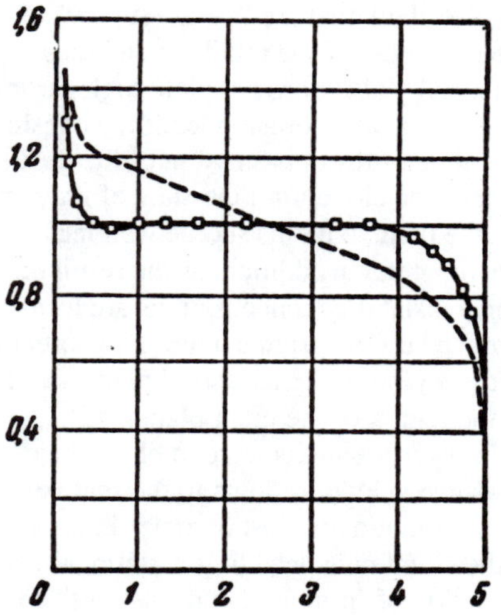

stresses for turbulent motion according to Prandtl as given in (28), which seems more consistent, we obtained the opposite result. The turbulent viscosity began to equalize the momentum and therefore the cloud was held in some central region. The truth is that the movement of matter in the protoplanetary cloud is different from the movement between two cylinders. The absence of the wall makes it possible for some elements of matter to separate from the periphery of the cloud and for others to fall in on the sun. However, this process is not a mechanism capable of sustaining turbulence in the protoplanetary cloud, on the contrary, it promotes its more rapid damping.

Thus, the theory of the evolution of rotating cosmic gases exhibiting turbulent motions constructed by Weizsacker encounters several serious objections both with respect to the possibility of turbulence arising and being sustained, and with respect to the direction in which matter and angular momentum are transferred due to turbulent friction.

3. Growth of Planetary embryos in the Gas-Dust Medium

In the protoplanetary cloud having a mass of the order of the planetary masses, in order to maintain the critical density at which gravitational instability sets in a high degree of condensation is required ($H/r \approx 10^{-4} - 10^{-6}$). In the dust cloud gravitational instability, as is known, is contained due to the condensation of the entire cloud as a whole as a result of the reduced relative velocities of the particles due to their collisions. In the gas-dust cloud, such intense condensation of the gaseous component is possible only at a temperature of several hundreds of fractions of a degree on the absolute scale, which is clearly unrealistic, Therefore, increase in density up to the critical value in the gas-dust cloud can be attained only as a result of the gradual settling of solid particles through the gas to the equatorial plane of the cloud under the action of the z- component of the force of solar gravitation. Such settling is possible in the case in which the gas does not contain macroscopic movements acting on the over-all revolution of the cloud around the sun, and if the cloud does not include large enough bodies capable of exerting a perturbing effect on the medium by virtue of their own gravity.

The time needed for an equal distribution of kinetic energy to be established between atoms and particles is of the order of $\gamma_p\delta_p/\rho_a v_a$, where ρ_a and v_a = density of gas and velocity of atoms, while γ_p and δ_p= particle radius and density. In the protoplanetary cloud $\rho_a \approx 10^{11}$ g/cm3; $v_a \approx 10^5$ cm/sec; $\delta_p \approx$ 1gm/cm3.Therefore, the chaotic velocities of the particles in the cloud drop rapidly to thermal velocities, i.e., to several fractions of a centimeter per second.

The union of readily condensed gas atoms and molecules with the solid particles occurs also quite rapidly, during a period of the order of $r_p\delta_p/\rho_p v_a$ where ρ_p=spatial density of the dust. Therefore, we can see that the particles have settled in an equatorial plane (xy) through uncondensed gas, which is capable of accreting on to the planetary embryos only later, when their gravitation is already significant.

Since the condensation of the cloud must be accompanied by a decrease in temperature in the equatorial plane, most gases acquire the property of condensability. However, we will assume that not only helium but also hydrogen are not condensed, since this would require a temperature less than 3°K, which is of low probability.

Let us compare the time during which sufficient dust will accumulate in the equatorial plane of the cloud and in which gravitational instability will set in with the time required for the direct accretion of solid particles as a result of cohesion up to the dimensions of "primary condensation".

The value of the z-components of the solar gravitation acting on particles of the mass mp equals

$$F_z = -GM_\odot m_p z/(r^2+z^2)^{3/2} = (GM_\odot m_p/r_3)z \qquad (30)$$

since $z^2 \ll r^2$ in virtue of the oblatness of the cloud (r and z = cylindrical coordinates of a particle).

The gravitational attraction of the cloud itself is negligibly small compared to F_z. It becomes comparable with F_z only when the cloud approaches the state of gravitational instability.

The force of resistance of the gaseous medium

$$F_z = (4\pi/3)n_a m_a r^2 p v_a (dz/dt) \qquad (31)$$

(na = number of atoms per unit volume, ma= mass of an atom). The moment of a particle of constant mass is described by the equation

$$d^2z/dt^2 + a_1\ (dz/dt) + a_2 z = 0 \qquad (32)$$

where $a_1 = \rho_a v_a/r_p\delta_p \qquad a_2 = GM_\odot/r^3$

From this equation it follows that particles of small dimensions $[r_p \ll \rho_a v_a/2\delta_p(\sqrt{r^2/GM_\odot})$ settle asymptotically to the plane z=0, in which the coordinate z is reduced down to l/e of its original value during the time

$$t_r = (\rho_a v_a/r_p\delta_p)r^2/GM_\odot \qquad (33)$$

When the gas density is $\rho_a = 10^{-12}$ g/cm^3, characteristic of a protoplanetary cloud with a mass of the order of the planetary masses at a temperature of 300° - 100° K, this time in the region of the terrestrial planets is equal to $1.6(10^3)$ years for particles with $r_t = 10^{-5}$ cm and 1.6 years for particles of size 10^{-2} cm; in the region of the giant planets this time is two orders of magnitude higher.

Large particles with radius $r_p \gg (\rho_a v_a/2\delta_p)\sqrt{r^3/GM_\odot}$ fluctuate relative to the central plane with a period equal to the time of revolution around the sun, and with decreasing amplitude, in which the time of oscillation damping down to l/e of the original value in any portion of the cloud equals

$$t = 2r_p\ \delta_p/p_a v_a \qquad (34)$$

which constitutes an interval of time of one year for one-cm particles up to 10^5 years for particles 10^5 cm in size at a gas density $\rho_a = 10^{-12}$ gm/cm^3. The particles of intermediate size $[r_p = (\rho_a v_a/2\delta p)\sqrt{r^3/GM_\odot}]$ settle toward the equatorial plane during the time of a single revolution around the sun

In the gas-dust cloud, the role of inelastic collisions of particles between each other in the process of cloud thickening is small. Low particle velocities here are attained as a result of their interaction with gases. The settling of particles toward the equatorial plane of the cloud is possible in a period of time not exceeding 10^5 years, and a quite high density is established within the dust component of the cloud for the onset of gravitational instability.

It can be shown that even if each collision of particles leads to their cohesion, their growth during the time of settling toward the equatorial plane is very insignificant. Collisions of solid particles between each other can be caused by the following: 1) a differential settling toward the plane z =0, i.e., differences in dz/dt, and 2) relative velocities in the directions x and y. In spite of the fact that differences in settling velocity can be significant due to the dispersion of particle sizes, their growth by virtue of differences in dz/dt is limited, since the accreting particle can join to itself only matter present in the vertical column which it is passing through. It is easy to show that the increment in particle radius during the time of particle settling toward the central plane equals $\Delta r_p = \sigma/8\delta_p$ where σ =surface density of the cloud. Since the maximum surface density in the planetary system is of the order of 100gm/cm^2, particles can increase only to a diameter several centimeters across.

We can also show that the lateral "supply" of particles due to the smallness of the relative velocities in the directions x and y also does not yield high particle accretion, even if each collision leads to cohesion. During the time of the compression of the dust component of the cloud until the state of gravitational instability the particle radius succeeds in increasing only several centimeters. In addition, as the result of gravitational instability condensations are formed that are very much larger -- from a mass of the order of 10^{15} grams in the region of the terrestrial planets and 10^{22} grams in the region of the giant planets [2].

Thus, if no perturbations were to block the decrease in relative particle velocities down to thermal velocities, then the accumulation of dust particles in the equatorial plane of the cloud must lead to gravitational instability, to the formation of "primary condensations" and to their subsequent aggrandizement. If, however, some macroscopic movements exist in the protoplanetary cloud which are superimposed on the over-all revolution and establish marked relative particle velocities, then the intense flattening of the disc and, therefore, gravitational instability cannot develop. In this case, the planetary embryos can be formed only by their gradual growth in a gas-dust medium.

The thickness of the cloud is associated with the average relative velocity of particles v by the simple relationship (18)

$$H = \sqrt{2\pi r^3 v^2/3GM_\odot}\ ) = \pi/2[r(\bar{v}/v_c)] \qquad (35)$$

where v_c = angular velocity. The density in the equatorial plane of the cloud is equal to $\rho = [Q/(r_2^2 - r_1^2)H] = 2Qv_c/(r_2^2 - r_1^2)\ \pi r v$ (36) where Q = mass of cloud in the zone of the given planct having the boundaries r_1 and r_2 (Q = order of planetary mass). The critical density requisite for the onset of gravitational instability is equal to

$$\rho_{cr} = cM_\odot/r^2 \qquad (37)$$

where the value c is of the order of several units.

Therefore

$$\rho/\rho_{cr} = (2/\pi c)(r^2/r_2^2 - r_1^2)(v_c/v)Q/M_{\odot} \tag{38}$$

Hence, in order for gravitational instability to begin in the cloud, particle velocities are required of the order

$$\bar{v}/v_c \approx (1/c)Q/M_{\odot} \tag{39}$$

(which corresponds to about 1 cm/second for the zone of the earth.

This condition is relatively rigorous. What is required is an ideal symmetry of the cloud and the movement of each volume element strictly in a circular orbit. Such conditions in a purely dust cloud possibly can arise due to the continuous decrease in the relative particle velocities as the result of inelastic conditions. However, in a gas-dust cloud particles interact chiefly with atoms and gas currents are generated. Therefore, even small movements in the gas must prevent the accumulation of dust in a very fine layer and the increase of density to the critical value in the layer. Solar activity must have a perturbing effect on the internal particles of the cloud. Irregular expulsions of corpuscular streams, surges of radiation and the formation of irregular magnetic fields can lead to the emergence of irregular movements of matter within the inner parts of the cloud and prevent gravitational instability from appearing in this region. However, this problem calls for further investigation.

Conclusion

One of the important factors determining the evolution of the protoplanetary cloud is the nature of movement of matter within the protoplanetary cloud.

The existence within the cloud of intensive turbulent movements must have prevented the emergence therein of gravitational instability. The formation of protoplanetary embryos here could have proceeded only by the direct growth of solid particles as they collided with other particles and with gas atoms. However, the hypothesis of turbulence within the protoplanetary cloud suggested by Weizsacker encounters serious objections, since the cloud is usually stable with respect to small perturbations, and since very significant energy sources are necessary to sustain turbulence. If at the first stage of the cloud's existence a turbulent stage prevailed, then it must have been brief. At this stage the redistribution of matter and momentum within the cloud was significantly restricted due to the damping of turbulence. The direction of the transfer of matter and momentum in turbulent motion of matter within the cloud proved to be opposite to the transfer direction found by Weizsacker.

For gravitational instability to have taken hold in the gas-dust cloud having a mass of the order of the present-day planetary masses, it is not only required that there be no turbulence within the cloud, but that the relative particle velocities do not exceed 1 cm/sec at the distance of the earth from the sun and ~10^2 cm/sec in the region of the major planets. This signifies an ideal symmetry of the cloud and the absence of any macroscopic movements within the gas distinct from the revolving circular movement, since the dust is swept along by the gas. In cloud portions close to the sun, it is probable that macroscopic movements could have existed caused by solar activity, which prevented the occurrence of gravitational instability in this area. In cloud portions far from the sun the conditions for gravitational instability were more favorable.

NOTES:

1. We are not here considering the region H II, in which the temperature amounts to 1500°K. According to Kuiper [19, page 361], this region is found at high z-coordinates with a heliocentric width of more than 20°, where the density $\rho < 10^{-20}/cm^3$. The mass of the gas in this region is negligibly small in comparison to the mass of the gas in region HI and cannot have any pronounced effect on it.

2. This corresponds to the viscosity of hydrogen at a temperature of 300°k. The viscosity of heavier gases and dust is still less, therefore for a rough estimate of the upper limit of the flow of matter it is insubstantial.

LITERATURE

1. Schmidt, O. Yu., Doklady AN SSSR, 66, 392, 1945; Chetyre lektsii o proiskhozhdenii Zemli (Four Lectures on the Origin of the Earth, 2nd ed, published by the Academy of Sciences USSR, 1950.
2. Gurevich, L. E., and Lebedinskiy, A. I., Izvestiya AN SSSR, 14, 765, 1950.
3. C. F. von Weizsacker, Zs. f. Astrophys., 22, 319. 1944.
4. Kuiper, G. P., in Astrophysics, Hynek, New York, 1951, page 357.
5. Fesenkov, V. G., Astronomicheskaya zhurnal, 28, 492, 1951; Prois khozhdenive i razvitiye nebesnykh tel po sovremennym dannym, published by the Academy of Sciences USSR, 1953. 6. Weizsacker, C. F. von, Zs. f. Naturforsch., 3a, 524-539, 1948.
7. Chandrasekhar S. and D ter Haar, Astrophys. J., 111, 187, 1950; D. ter Haar, Astrophys. J., 111, 179, 1950.
8. Lust, R., Zs. f. Naturforsch., 7a, No. 1, 87, 1952.
9. Landau, T. D. and Lifshits, Ye. M., Mekhanika sploshnvkh sred (Mechanics of Continuous Media), 2nd ed, Gostekhizdat, 1954, page 134.
10. Chandrasekhar, S., Printsipy zvesdnov dinamiki (Principles of Stellar Dynamics), Foreign Literature Publishing House,1948.
11. Safronov, V. S., Ruskol, Ye. L., Doklady AN SSSR, 108, 413, 1958.
12. Trefftz, E., Zs. f. Naturforsch., 7a, No. 1, 99, 1952.

13. Weizsacker, K., Problemv kosmicheskov aerodinamiki (Problems of Cosmic Aerodynamics), Foreign Literature Publishing House, 1953, page 204.
14. Aerodinamika (Aerodynamics), vol III, ed. by V.F. Durand, Gostekhizdat, 1939, page 143.
15. Taylor, G. J., Proc. Roy. Soc., London, 135, 685, 1932; 151, 494, 1935.
16. Wattendorf, F. L., Proc. Roy. Soc., London, 148, 565, 1935.
17. Karman, T., in the collection: Problemy kosmicheskov aerodinamiki, Foreign Literature Publishing House, 1953, page 157.
18. Safronov, V. S., Astronomicheskiya zhurnal, 31, 499, 1954.
19. Kuiper, G. P., in the collection: La Physique des cometes, Louvain, 1953, page 361.

The Formation of a Protoplanet

Ye. L. Ruskol

Problems of Cosmogony, (Joint Publ. Res. Service, U.S. Dept. Commerce, 1964), Vol. VII, pp. 5–13
Translated from: Vopr. Kosmog. **7** (1960)
Presented at General Meeting of International Astronomical Union, 1958

The author criticizes the hypothesis of the formation of planets from massive protoplanets of "solar" (gaseous) composition, concluding that the mass of such a protoplanetary cloud would be too great.

As is well known, the hypothesis of the formation of the planets from massive protoplanets of cosmic or solar composition is not adequate to explain the chemical composition of the planets, their axial rotation, or certain other observed facts.[1–4] It should be noted in particular that this hypothesis runs into serious difficulties at the start, in connection with the formation of the primary near-solar cloud. Thickenings forming in the cloud could exist and grow only if they maintained themselves by their own gravitational forces against destructive tidal action on the part of the central mass.

The gravitational force of a condensation of mass m along the radius of the condensation r directed to the Sun exceeds the gravitational force of the Sun provided

$$\frac{\alpha G m}{r^3} > \frac{2GM_\odot}{a^3}, \tag{1}$$

where α is a factor depending upon the form of the condensation. If the form is spherical ($\alpha=1$; $m=\frac{4}{3}\pi\rho r^3$) this corresponds to $\rho>2\rho_*$, where $\rho_*=M_\odot/\frac{4}{3}\pi a^3$ is the "smeared out" density of the Sun at distance a. For an oblate spheroid with semiaxes r and h, $h\ll r$ ($\alpha\approx 2$; $m=\frac{4}{3}\pi\rho r^2 h$), this condition gives us

$$\rho>\rho_* \frac{r}{h}. \tag{2}$$

If the condensation was the result of gravitational instability in the nebula, then its minor semiaxis h must be approximately equal to the corresponding half-thickness of the cloud H; while the horizontal radius must be the critical wavelength of instability λ. According to Gurevich and Lebedinskii,[5]

$$\frac{r}{h}\approx 13 \tag{3}$$

must represent the ratio of the axes of the condensations formed in the case of a condensed cloud with differential rotation around the central body. In view of so oblate a form of the condensations, the value of Roche's density is usually assumed to be

$$\rho_R\approx 10\rho_*, \tag{4}$$

and this figure can be regarded as minimal.

A large part of the material in the nebula of solar composition must have been in gaseous form, on account of the predominant excess of uncondensed hydrogen and helium gases. In the case of laminar rotational movement of the nebula, the thermodynamic fluctuations of the density are negligibly small, and the formation of condensations at any point whatever in the equatorial plane is possible only if

$$\rho_0>\rho_R. \tag{5}$$

In the presence of turbulence (and this is quite doubtful for any Reynolds number, as has been pointed out by Jeffreys,[6] and demonstrated in detail by Safronov and the present writer,[7,8] condition (5) must also be observed. In fact, any fluctuation in density arising as a result of turbulence will break down if it does not satisfy Roche's criterion and Jeans's criterion—in other words, if it is not both large and "thick." The scale of the turbulence may not exceed that of the corresponding thickness of the cloud. According to Eq. (3), a self-sustaining fluctuation in density may have the horizontal dimension $\lambda>10H>10l$, that is, it must be greater than 100 cells of the average dimension l. At the same time, Roche's density ρ_R must be present. The probability of achieving such a fluctuation is practically zero if the average density in the equatorial layer of the cloud is lower than ρ_R.

According to Kuiper, protoplanets at the beginning have very large horizontal dimensions which occupy almost the total width of the planetary zones.[9,10] In the case of low temperatures in the cloud such as are assumed by Kuiper, the oblateness of such protoplanets is of the order of 10^2. Consequently, the density close to the equatorial plane of the cloud must in this case by of the order of $10\rho_R$, or $10^2\rho_*$.

Meanwhile, Kuiper assumes the mass of the protoplanetary cloud to be of the order of (0.05–0.1) $M_\odot$, for which the density in the equatorial plane could hardly reach $0.1\rho_R$. Therefore the breakup of such a cloud into stable condensed portions is impossible irrespective of the presence or absence of turbulence.

In two of his works[9,10] Kuiper attempts to support the possibility of the breakup of a near-solar cloud into condensed portions in the presence of a density much lower than ρ_R, by using formulas for the critical wavelength of instability λ:

$$\lambda \geqslant 0.05(\rho_R/\rho)^{1/2}a,$$

$$\lambda \geqslant 0.07(\rho_R/\rho)^{1/2}\tau^{1/2}a, \quad \tau<1.$$

But these formulas were obtained by a formal substitution of ρ_R in Jeans's and Ledoux's criteria for nonrotating media, and they actually do not take Roche's criterion into account. In justifying his derivation of these formulas, Kuiper quotes Chandrasekhar,[11] who shows that Jeans's criterion does not vary in the presence of rotation. However, Chandrasekhar had in mind solid-body rotation of an infinitely extended medium, without tidal forces acting from the central mass; and his result, therefore, just like Kuiper's formulas, cannot be applied to a protoplanetary cloud.

Let us compute the planetary cloud mass necessary for the formation, within it, of gaseous protoplanets. The mass of the cloud is

$$M=\int_{a_0}^{a_n} 2\pi\sigma(a)da. \tag{6}$$

The surface density $\sigma(a)$ in a condensed cloud with given $\rho_0(a)$ may be obtained from the equation of hydrostatic equilibrium with respect to the z coordinate at distance a from the Sun:

$$-\frac{1}{\rho}\frac{dp}{dz}=\frac{GM_\odot z}{(a^2+z^2)^{3/2}}+4\pi G\int_0^z \rho\, dz. \tag{7}$$

Here the z components of the force of gravitation, both of the Sun and of the cloud, are taken into account. Nearly all the material in the cloud was zone HI, and therefore we can consider the cloud to be isothermic with respect to z, and substitute

$$\frac{dp}{dz}=\frac{RT}{\mu}\frac{d\rho}{dz}. \tag{8}$$

Also, we may neglect z^2 in comparison with a^2. Then, differentiating Eq. (7), we arrive at a second-degree differential equation which has a solution for $d\rho/dx$:

$$\frac{d\rho}{dz}=-\rho\sqrt{\frac{8\pi G\mu}{RT}(\rho_0-\rho)+\frac{2GM_\odot\mu}{RTa^3}\ln\frac{\rho_0}{\rho}}. \tag{9}$$

Ledoux[12] has derived an analogous formula for $\rho(z)$ for a compacted nonrotating cloud; Eq. (9) represents a generalization of his method to include a rotating cloud. The surface density in this case is

$$\sigma=2\int_0^\infty \rho\, dz=\sqrt{\frac{2RT\rho_0}{\pi G\mu}}\,I(y^*), \tag{10}$$

where $I(y^*)$ is the integral (determined by numerical means) depending upon the ratio $y^*=\rho_*/\rho_0$:

$$I(y^*)=\frac{1}{2}\int_0^1 \frac{dy}{\sqrt{1-y-\frac{y^*}{3}\ln y}}, \quad y=\frac{\rho}{\rho_0}. \tag{11}$$

In the case studied by Ledoux, $I=1$, with ρ_0 equal to Roche's density ρ_R—which corresponds to $\rho_*/\rho_0=1/10$ and $I=0.9$.

Table 1 presents a comparison of the mass of the protoplanetary cloud, computed for the case of the Roche density [Eq. (4)] in the equatorial plane, with the average molecular weight $\bar{\mu}$ being 2.4 and with the temperature distributions assumed by Kuiper in his basic works.

As is apparent from Table 1, the mass of the protoplanetary cloud must be large. Even for a very low temperature it cannot be less than the solar mass itself. This is an amount 30 times greater than that of the planetary system as we know it, with the addition of the light elements in cosmic proportion. Here is the first hurdle facing the hypothesis of a gaseous protoplanet.

Next, the masses of the condensations formed in the nebula at the temperatures in question must reach the order of 10^{29}–10^{30} g—the order of the masses of the major planets. Along with this, tens, or even hundreds of such condensations must be formed within each planetary zone. It is difficult to imagine that such massive and numerous formations could be converted, for example, into a small number of terrestrial planets of small mass.

We may point out that for the formation of extended condensations almost interlocking in opposition, a mass of the order of $10M_\odot$ would be required, and this is inadmissable.

Finally, for the ejection of enormous masses from the solar system, some very powerful agent is required. It follows from Kuiper's analysis[13] that the only effective mechanism would be solar corpuscular radiation. Assuming Biermann's value[14] for the intensity of corpuscular emission from the Sun, Kuiper arrives at the desired velocity of dissipation of $0.1M_\odot$ from interplanetary space during the life span of the solar system. (Half of this amount must be dissipated at the beginning from the protoplanets; then the latter are ejected from the solar system. Kuiper does not study this process quantitatively). However, the Biermann estimate is 10^4–10^5 times higher than more modern estimates of the intensity of the corpuscular radiation of the Sun (Table 2).

TABLE 1. Mass of a protoplanet cloud expressed in solar masses. Key: (1) temperature of black body $T=300$ K $/\sqrt{\alpha_{ae}}$; (2) $T=300$ K$/\alpha_{ae}$, Kuiper, Ref. 9; (3) $T=$const$=25$ K (Kuiper, approximation, Ref. 10); (4) mass of cloud (Kuiper); (5) mass of planets today.

	(1)	(2)	(3)	(4)	(5)
Within the solar system	2.2	1.4	1.0	0.05–0.1	1.3×10^{-3}
Within the zone of the terrestrial planets	0.4	0.4	0.12	0.22	6×10^{-6}

But even the Biermann value would be insufficient to explain the dissipation of mass actually required for the formation of gaseous protoplanets. In order to eject excesses of matter from the nebula, the Sun itself would have to give up no less an amount of matter. Thus, if we assume that the intensity of corpuscular emission from the Sun during the past was greater than the value assumed by Biermann, we arrive at a mass for the earlier Sun which is far greater than the one known to us (by the amount required for the mass of the protoplanetary cloud). However, such a decrease in mass is not in accordance with modern ideas of the evolution of the Sun.

The difficulties enumerated here, which concern a large mass for the protoplanetary cloud and the condensations, disappear if we assume that only a horizontal layer of solid bodies went into the formation of condensed portions (this possibility has been discussed by Gurevich and Lebedinskiy[5]). Such a view is possible after the solid component of the protoplanetary cloud settles through the gas to the equatorial plane. As Safronov and the author[8] have demonstrated, such settling can take place in a fairly short time.

Such are the supplemental arguments which can be advanced against the hypothesis of massive protoplanets.

TABLE 2. Intensity of solar corpuscular radiation.

Author	Intensity (g/year)
Biermann (Ref. 14)	8×10^{22} ($0.12_{\odot}$/3 000 000 y)
van de Hulst *et al.* (Ref. 15)	5×10^{17}
Vsekhsvyatskiy *et al.* (Ref. 16)	10^{20} [a]
Kiepenheuer (Ref. 17)	1.5×10^{18}
Mustel' (Refs. 18 and 19)	10^{18}
Nikol'skiy (Ref. 20)	9×10^{17}

[a]E. P. Mustel' (Ref. 19) adduces convincing arguments to show that the estimate of Vsekhsvyatskiy *et al.* (Ref. 16) is exaggerated by at least two orders (powers of ten). Moreover, one of the co-authors, Nikol'skiy, gives a much lower estimate in a subsequent article (Ref. 20).

1. H. C. Urey, Astrophys. J. Suppl. Ser. **1**, 147 (1954).
2. H. C. Urey, Astrophys. J. **124**, 623 (1956).
3. I. S. Shklovskiy, Astron. Zh. **28**, 234 (1951).
4. A. I. Lebedinskiy, Tr. Pervogo soveshch. Vopr. Kosmog. (Proceedings of the First Conference on Problems of Cosmogony) (*Publ. House of the Academy of Sciences*, USSR, 1951), pp. 151–167.
5. L. E. Gurevich and A. I. Lebedinskiy, Izv. Akad. Nauk SSSR Ser. Fiz. **14**, 765 (1950).
6. H. Jeffreys, Proc. R. Soc. London Ser. A **214**, 281 (1952).
7. V. S. Safronov and Ye. L. Ruskol, Dokl. Akad. Nauk SSSR **108**, 413 (1956).
8. V. S. Safronov and Ye. L. Ruskol, Vopr. Kosmog. **5**, 22 (1957).
9. G. P. Kuiper, in *Astrophysics*, edited by J. A. Hynek (McGraw-Hill, New York, 1951), pp. 357–424.
10. G. P. Kuiper, Vistas Astron. **2**, 1631 (1956).
11. S. Chandrasekhar, Vistas Astron. **1**, 344 (1955).
12. P. Ledoux, Ann. Astrophys. **14**, 438 (1951).
13. G. P. Kuiper, Mem. Soc. R Sci. Liege Collect. 4 **13**, 361 (1953).
14. L. Biermann, Z. Astrophys. **29**, 274 (1951).
15. H. C. van de Hulst, Bulletin of the Astronomical Institute of the Netherlands **11**, 135 (1950).
16. S. K. Vsekhsvyatskiy, G. M. Nikol'skiy, Ye. A. Ponomarev, and V. I. Cherednichenko, Astron. Zh. **32**, 165 (1955).
17. K. O. Kiepenheuer, *The Sun*, edited by G. P. Kuiper (translation by Publ. House of Foreign Literature, 1957), p. 282.
18. E. R. Mustel', Izv. Krym. Astrofiz. Obs. **16**, 206 (1956).
19. E. R. Mustel', Izv. Krym. Astrofiz. Obs. **19**, 154 (1958).
20. G. M. Nikol'skiy, Astron. Zh. **33**, 588 (1956).

REVIEWS OF MODERN PHYSICS VOLUME 30, NUMBER 3 JULY, 1958

On the Turbulence in the Protoplanetary Cloud

V. S. Safronov

O. J. Schmidt Institute of Earth Physics, Academy of Sciences of the U.S.S.R., B. Grusinskaya, Moscow, U.S.S.R.

The problem of turbulence in the protoplanetary cloud is of importance for planetary cosmogony. Chaotic macroscopic motions probably existed in the cloud during its formation. Further evolution of the cloud depended to a great extent upon whether these original motions damped in a short time, or turbulence supported by some source of energy existed during planet formation. According to Kuiper and Fessenkov's hypotheses, massive protoplanets formed as a result of gravitational instability and turned into planets after the dissipation of light elements. Large-scale turbulent motions with mean velocities exceeding the thermal velocities of atoms and molecules would prevent, however, gravitational instability in the cloud, even if its mass was of the order of the mass of the sun. According to Edgeworth and to Gurevitch and Lebedinsky the planets grew gradually from small condensations formed in a flattened dust disk with a mass equal to that of the present planetary system. But even small scale turbulent motions would prevent extreme flattening of the disk necessary in this case for gravitational instability. The problem of turbulence is also connected with the problem of present distribution of angular momentum between the sun and planets, as large-scale turbulence produces redistribution of matter and of angular momentum in the cloud.

The hypothesis of the presence of large-scale turbulence in the protoplanetary cloud was introduced by von Weizsäcker.[1] But Weizsäcker's arguments do not prove its existence. The Reynolds number is very large for the cloud (about 10^{10}). But for a rotating medium the Reynolds number cannot be considered as a criterion of turbulence. Weizsäcker regards turbulence as a result of convective instability. But he uses the criterion of convection for nonrotating media, which is inapplicable in the case of the rotating cloud. The problem therefore needs further study.

In order to reveal the main features of motions in a flat protoplanetary cloud, one can use the results of investigations of fluid motion between two rotating coaxial cylinders. Rayleigh,[2] Taylor,[3] and Synge[4] proved that such a motion of an incompressible fluid is stable if the angular momentum increases outwards: $d(\omega r^2)/dr>0$. This condition had to be satisfied for the protoplanetary cloud. If we neglect the pressure gradient in the cloud and its own gravitation as compared with the gravitation of the sun, the angular momentum will be proportional to $\sqrt{r}$. Then this condition becomes identical with the condition of stability of circular orbits well known in stellar dynamics. But the condition was obtained for an incompressible fluid and does not take into account the possibility of convection. On the other hand, Weizsäcker, using the criterion of convection, left out of account the condition of stability of circular orbits. These two conditions were combined in the paper by Safronov and E. L. Rouscol.[5] The condition of convection for a flat rotating cloud (cylindrical rotation) was found to be:

$$\omega r^2\frac{d(\omega r^2)}{dr}<\frac{r^3}{2\rho^2}\frac{dp}{dr}\left[\left(\frac{d\rho}{dr}\right)_{ad}-\frac{d\rho}{dr}\right]=\frac{r^3}{2\gamma\rho T}\frac{dp}{dr}\left[\frac{dT}{dr}-(\gamma-1)\frac{T}{\rho}\frac{d\rho}{dr}\right], \quad (1)$$

and $(\omega r^2)^2=GMr+(r^3dp/\rho dr)$. When considering small disturbances, it is possible to approximate smooth functions ρ and T for small intervals of r by power functions

$$\rho\sim r^{-a_1},\quad T\sim r^{-a_2}. \quad (2)$$

The condition of convection is then reduced to

$$2-(\gamma-1/\gamma)(a_1+a_2)>GM/rRT. \quad (3)$$

The protoplanetary cloud being largely an HI region, one can take as maximum value of T on the right-hand side of the inequality (3) the temperature of a blackbody in a transparent cloud, $T_0\approx 300(r_{ae})^{-\frac{1}{2}}$, where r_{ae} is the distance from the sun in a.u. Then

$$(GM/RTr)>350r_{ae}^{-\frac{1}{2}}, \quad (4)$$

and the condition of convection (3) is not satisfied for any acceptable values of a_1 and a_2. Hence, the undisturbed protoplanetary cloud is stable with respect to small disturbances and convection could not arise in it at any admissible values of temperature and of density gradients.

The possibility of large-scale turbulence over a long time scale is open to serious objections from energetic considerations. Solar radiation entering the flat cloud is insufficient to support turbulence. Gravitational energy of the parts of the cloud approaching the sun suffices only for a short time. Weizsäcker's value of the mean turbulent velocity of about one tenth of the orbital velocity leads to a time of disintegration of the cloud of about 10^3 years, while 10^8 years are needed for the planet formation according to Weizsäcker himself.

[1] C. F. von Weizsäcker, Z. Naturforch. **3a**, 524 (1948).
[2] Lord Rayleigh, Proc. Roy. Soc. (London) **A93**, 148 (1916).
[3] G. J. Taylor, Phil. Trans. Roy. Soc. (London) **A223**, 289 (1923); Proc. Roy. Soc. (London) **135** 685 (1932).
[4] J. L. Synge, Trans. Roy. Soc. Canada, **27**, iii, 1 (1933); Proc. Roy. Soc. (London) **167**, 250 (1938).
[5] V. S. Safronov and E. L. Rouscol, Compt. rend. acad. sci. U.R.S.S. **108**, 413 (1956); Problems Cosmogony (Moscow) **5**, 22 (1957).

It seems probable that the ratio of the mean turbulent velocity to the orbital velocity, and the ratio of the mixing length to the distance from the sun, are of the same order of magnitude. Chandrasekhar and ter Haar[6] have obtained $l=0.62r$ from the law of planetary distances and take the value of the turbulent velocity to be slightly higher than one-half of the orbital velocity. Karman's formula[7] for the mixing length in a rotating medium leads to a still higher value, namely, $l=2kr\approx 0.8r$. Under these conditions the time of disintegration of the cloud is less than 10^2 years and formation of the planets is impossible. Large-scale turbulent motions, if such existed at the initial stage of the evolution, had to damp rapidly. According to the energetic considerations only motions of a scale a thousand times less than follows from Karman's formula could exist for a long time.

It is of interest to investigate the problem of the transfer of matter and angular momentum during the existence of turbulence in the cloud. According to Weizsäcker, turbulent friction diminished the angular momentum of the rapidly rotating inner parts of the cloud, which therefore approached the sun. The outer parts acquired the momentum and went away from the sun. Weizsäcker uses shearing stresses depending on the gradient of angular velocity:

$$\tau_{r\varphi}'=\eta r(d\omega/dr). \qquad (5)$$

But this tensor of molecular viscosity stresses is valid, strictly speaking, only for the case of small free paths and is unfit for large-scale turbulent motions. Prandtl found another expression for the stresses as a function of the gradient of angular momentum:

$$\tau_{r\varphi}'=\eta r^{-1}(d/dr)(\omega r^2). \qquad (6)$$

Karman[7] gives the same expression (6) without any comment on Weizsäcker's using expression (5). In the solar system, angular velocity decreases with the distance from the sun, while the angular momentum increases. Hence the direction of the transfer of matter and angular momentum in the cloud according to Prandtl's and Weizsäcker's formulas are opposite.

Taylor[8] believes that the steady value of angular momentum in the central region of turbulent flow (inner cylinder rotating) found experimentally by him and Wattendorf,[9] contradicts Prandtl's formula, as the latter in this case gives zero shearing stresses and would make impossible the transport of angular momentum. However, the equilization of angular momentum in the main part of the flow agrees with Prandtl's expression. The accuracy of the experiment is not sufficient to state that the derivative of angular momentum is exactly zero. We can only say that the derivative is very small, but this conclusion follows just from Prandtl's formula, if the turbulent viscosity is great. The same takes place in the rectilinear flow in tubes. The almost flat velocity profile far from the walls of the tube, and its sharp bending near the walls, can be explained if we suppose that turbulent viscosity is high far from the walls and decreases rapidly when approaching the walls (as the first or the second power of the distance from the walls, for example). A similar suggestion about turbulent viscosity in a rotating flow permits one to explain, by using Prandtl's formula, the almost constant value of angular momentum far from the walls and its sharp fall near the walls. Neither the relation (5) resulting from the molecular viscosity tensor, nor Taylor's suggestion of vorticity conservation explains this peculiarity of turbulent rotational motion.

Probably Prandtl's formula is not quite accurate, because of the semiempirical character of turbulence theory. On the ground of a new interpretation of the mixing length, Wasiutynsky[10] has obtained an expression for stresses in a more general form. For the case of cylindrical rotation, he gives

$$\tau_{r\varphi}'=\frac{\rho K_r^r}{r}\frac{d(\omega r^2)}{dr}-2\rho K_\varphi^\varphi\omega. \qquad (7)$$

When $K_\phi^\phi=0$ (purely radial exchange) one obtains Prandtl's formula; with $K_\phi^\phi=K_r^r$ (isotropy) one obtains a formula of molecular viscous stresses with the exception that turbulent viscosity enters instead of the molecular. He found as condition of nondecreasing turbulence for incompressible ideal fluid

$$\left[2K_\varphi^\varphi\omega r-K_r^r\frac{d(\omega r^2)}{dr}\right]\frac{d\omega}{dr}\leqslant 0. \qquad (8)$$

It is not clear whether this generalization is only formal, or characterizes the turbulent motion more exactly. Nor is it clear which values of the ratio K_ϕ^ϕ/K_r^r are more probable in the actual turbulent flow. One may think that for the rotating system around the gravitating center $K_\varphi^\varphi<K_r^r$. It is well known, for example, that pecular velocities of stars in radial direction are higher than in the direction of rotation. For the solar system ($\omega\sim r^{-\frac{3}{2}}$) the turbulence would decrease according to this formula, if $K_\phi^\phi<\frac{1}{4}K_r^r$. The sign of the stresses is then given by Prandtl's formula and the transfer of matter and of angular momentum is opposite to that found by Weizsäcker. According to energetic considerations it seems probable that this situation occurred for large-scale turbulence. It might be believed that small-scale turbulence would be more isotropic. But small-scale turbulence would be inconsistent with the theoretical value of mixing length found by Karman for a rotating system. It is not clear whether such turbulent motions are possible.

Being only an astronomer the author should like to know the opinions of specialists on turbulence about these questions.

[6] S. Chandrasekhar and D. ter Haar, Astrophys. J. **111**, 187 (1950).
[7] Th. von Karman, "Problems of cosmical aerodynamics," CADO Dayton, Ohio (1951).
[8] G. J. Taylor, Proc. Roy. Soc. (London) **151**, 494 (1935).
[9] F. L. Wattendorf, Proc. Roy. Soc. (London) **148**, 585 (1935).
[10] J. Wasiutynski, "Studies in hydrodynamics and structure of stars and planets," Oslo, p. 32 (1946).

The Development of Planetary Cosmogony

B. Yu. Levin
Defense Technical Information Center, Technical Report (1961) Document No. AD264161
Translated from: Priroda (Moscow) **1959** 10, 19
by Z. Jakubski, Space Technology Laboratories, Los Angeles, Report No. STL-TR-61-5110-37

All contemporary investigators in the field of planetary cosmogony consider that the planets were formed from a cold cloud of gas and dust in the vicinity of the Sun. However, the opinions of scientists differ as to the origin of the cloud itself and the process of formation of planets from it. The majority consider that the principal process in planet formation was the accumulation of cold solid particles. Only the American astronomer G. P. Kuiper and Academician V. G. Fesenkov assume the disintegration of a protoplanetary cloud into large, massive *protoplanets* which subsequently, by decreasing their masses, became the present-day planets. In addition, all scientists except Fesenkov hold that the Earth was formed as a cold body and only after its formation was heated by the accumulation of radioactively generated heat.

There are two opinions about the problem of the origin of the protoplanetary cloud, which is on the boundary between planetary and stellar cosmogony. The first is that the gas-dust substance was captured by the Sun, which was already in existence; the second opinion is that the Sun and the protoplanetary cloud emerged at the same time, in the process of formation of the Sun itself.

The transition to a theory of the gradual growth of the planets by the capturing of surrounding scattered material was initiated by the hypotheses of the German physicist von Weizsäcker and of Academician O. Yu. Schmidt, which appeared almost simultaneously in 1943–44.*

von Weizsäcker considers a disk-shaped gas-dust cloud in the vicinity of the Sun, having a mass of order 0.1 that of the Sun. To explain the regularity of the distances of planets from the Sun, he assumes the existence within the cloud of regular ringlike systems of vortices. Smaller vortices, formed in strictly determined places between the large vortices, created condensations from which subsequently the present-day planets grew.

Mainly because of the extreme artificiality of the assumption of regular systems of vortices, von Weizsäcker's hypothesis was not accepted. It did, however, contain some new ideas which provided the impulse for the work of several western scientists (ter Haar and Kuiper).

Schmidt began his investigation with an analysis of the fundamental regularities of planetary motion, which led him to conclude that the planets were formed from a swarm of bodies moving around the Sun in various elliptical orbits. Later he succeeded in finding a very simple explanation for the origin of the rotation of the planets and showed the way to an explanation of the regularity of their distances from the Sun. In 1949–50, account was taken of the chemical composition of the planets, and then in the work of L. E. Gurevich and A. I. Lebedinskii it was proposed that a swarm of bodies must have been formed in the vicinity of the Sun from a gas-dust cloud.

In order to explain the distribution of angular momentum between the Sun and the planets, Schmidt proposed the hypothesis that the material of the planets was captured by the Sun. Originally it was suggested that a cluster of bodies was captured, and later (after the work of Gurevich and Lebedinskii) that a gas-dust protoplanetary cloud had been captured.

Schmidt paid close attention to the geophysical consequences arising from the new representations of the formation of the Earth. One of the most important of these is the conclusion that the Earth was originally cold. There will be further discussion of the work of Schmidt and his collaborators later in this paper.

In 1949 the American astronomer G. P. Kuiper began his investigation of the origin of the planetary system. Kuiper considered the formation of the planets to be a result of the gradual dispersion of the material of massive condensations of protoplanets into which, in his opinion, the gas-dust cloud in the vicinity of the Sun broke up. He considered that the Sun was formed as a result of the compression of the gas-dust mist, and that the protoplanetary cloud evolved in the course of this process from the same original mist. The protoplanets were formed when the Sun was in the process of formation and not yet radiating, so that the protoplanetary cloud was very cold. After the "heating up" of the Sun in the form of a star, its corpuscular radiation began to "blow off" the external, rarefied layers of the protoplanets, reducing their masses. (According to Kuiper, the mass of the "proto-Jupiter" was three times that of the present-day Jupiter, and the mass of the "proto-Earth" was 120 times greater than the present-day mass of the Earth.) Kuiper does not consider in detail the chemical composition of the Earth, but assumes that the Earth lost not only volatile gases but part of its silicates as well. Meanwhile, many details of the chemical composition of the Earth (e.g., the nitrogen deficit or the still greater deficit of heavy inert gases) contradict its formation from a massive gaseous protoplanet.

In 1958, at the symposium on the origin of the Earth and

*The hypothesis of the German scientist did not become known in the USSR until 1945, from an account in an American journal.

planets which took place in Moscow at the time of the International Astronomical Conference, Kuiper insisted that at least the giant planets were formed from massive protoplanets. However, the proofs he adduced for his point of view were not convincing to the other participants at the symposium. At the same time, Kuiper acknowledged that criticism of his work on the nature of the formation of protoplanets was justified.

In 1951 Academician V. G. Fesenkov, who had up to that time adhered to his own hypothesis of the separation of the planets from the rapidly rotating Sun under the effect of centrifugal force, turned to the investigation of the formation of planets from a gas-dust cloud. In addition, he accepted the point of view of Kuiper of a concurrent formation of the Sun and the cloud, and the disintegration of the cloud into massive protoplanets, one for each planet. It is true, he assumes, that the giant planets retained almost all their original mass, while the planets of the Earth group had protoplanets of masses about 30 times (not hundreds of times, as Kuiper claims) their present masses. Fesenkov considers that the proto-Earth was very much heated in the process of compression; that is, that the incandescent Earth was formed from a cold protoplanetary cloud. The thermal history of the Earth and any other aspects of its development appear in this case to be as they had been predicted earlier on the basis of the hypothesis of J. Jeans and H. Jeffreys.*

In the same year, 1951, an American physical chemist, H. C. Urey, began his investigation of the origin of the planetary system. Unlike the majority of cosmogonists, who are guided by astronomical data concerning the planetary system, Urey proceeds mainly on the basis of data concerning the chemical composition of the planets and meteorites. He has reached the firm conclusion that the principal role in the formation of the Earth was played by the agglomeration of cool solid particles not subjected to heating.

At one time Urey tried to match his representations with the existence of protoplanets similar to those conceived by Kuiper.† But in his most recent work Urey has completely abandoned this attempt, in view of the fact that the chemical composition of the Earth contradicts the possibility of the existence of such a proto-Earth.

The formation of planets by a process of agglomeration of solids is suggested also in the work of Edgeworth (1949), Hoyle (1955), and Gold (1956). This idea has thus had wide recognition in recent years.

The Swedish physicist H. Alfvén has a separate hypothesis, in which it is assumed that during the period of formation of the planets the Sun possessed a very considerable magnetic field, and that the gaseous substance which was the initial material was partly ionized, so that the entire process took place under the influence of an electromagnetic force. However, the actual chemical composition of the planets is not in agreement with the chemical splitting of the protoplanetary cloud which would follow from the basic assumptions of Alfvén's hypothesis. Because of the artificiality of its assumptions, this hypothesis finds no supporters.

At the same time, the idea that the Sun's rotation may have been slowed by the transfer of angular momentum to the partly ionized cloud near the Sun (by the medium of a magnetic field) offers great possibilities for the explanation of the distribution of moments among the Sun and planets. In a qualitative form this idea was used by Kuiper, for instance, and in 1958 during the Moscow symposium on the origin of the Earth and the other planets, Hoyle proposed a new theory, based on this idea, for the origin of the protoplanetary cloud.

According to Hoyle, during the formation of the sun by condensation of a gas, when the solar diameter at the equator was of the order of the diameter of the orbit of Mercury, the rapid rotation caused the separation of a gaseous incandescent mass which did not participate in the later compression of the Sun. Since the gases were partly ionized, the magnetic field rotating with the Sun began to transfer angular momentum to the gases, thus causing them to move farther from the Sun and at the same time slowing its rotation. According to Hoyle's computation, this transfer of momentum from the Sun to the cloud could have occurred with a very moderate solar magnetic field (naturally considerably greater than the present one) and with a very small degree of ionization of the gases.

Hoyle's hypothesis is the first concrete form of the idea of a concurrent formation of the Sun and the protoplanetary cloud. It has not yet been published or subjected to detailed discussion. Nevertheless it can now be said that the hypothesis of Schmidt has a serious rival.

With regard to the question of the formation of the planets themselves, there is much greater unity of opinion. In 1956 the famous geophysicist Gutenberg, concluding a review of the new data on the formation of the Earth, wrote: "There is an ever-increasing number of astrophysicists and geophysicists who consider it probable that the Earth was formed by the gradual agglomeration of cold material, and many geologists express the opinion that the Earth was never completely molten." Since then the number of supporters of this point of view has continued to grow. The development of the idea of the building up of the Earth from cold solid bodies and the idea of its initial cold state, originated in our country by Academician O. Yu. Schmidt, has become today a general line of development of planetary cosmogony throughout the world.

Let us examine at greater length the process of the formation of the Earth and the other planets as it develops from the investigations of Schmidt and his collaborators.

*V. G. Fesenkov refers to an old computation by Jeffreys, according to which the Earth cooled very quickly from an incandescent gaseous cluster. In 1951, however, Jeffreys himself abandoned this conclusion because it was based on a very simplified computation which did not take into account some very important factors.

†See H. C. Urey, *Origin of the Earth*, Collection "Nuclear Geology" (Pub. Foreign Lit., Moscow, 1956).

Let us start with the earliest stages of evolution of the protoplanetary cloud. The gaseous component of the cloud was, because of the high thermal velocities of the molecules, flattened out, but still in a fairly thick ring (see Fig. 1A). The dust particles, on the other hand, which had random velocities, were damped by friction against the gas and by reciprocal collisions and must have settled in the central plane of the cloud, forming a thin disk of higher density (Fig. 1B). At a certain stage in the flattening, the density of the disk reached the critical, or so-called Roche, density, when it became possible for the disk to break up into many condensations in which the internal gravitational forces were greater than the tidal force of the Sun's gravitation (Fig. 1C). The masses of these condensations varied with the distance from the Sun, but were everywhere much smaller than the masses of present-day planets. This was shown in 1950 by Gurevich and Lebedinskii.

Originally only condensations of the dust particles emerged. Later, however, they formed agglomerations tens or hundreds of kilometers in size (Fig. 1D). Further growth of these asteroid-size bodies progressed by a process of capturing material from the surrounding space. First they captured the remnants of "original" particles and later fragments resulting from the breakup of some of the asteroid-size bodies by collision. Those few bodies which outdistanced the others in this process of growth finally became the present-day planets.

The increasing gravitational interactions among the asteroid-type bodies changed the initial circular orbits lying in a common plane to elliptical orbits of varying inclinations (Fig. 2, A and B). During the process of the amalgamation of many small bodies to form the planets, however, the individual characteristics of their motions were averaged, so that the orbits of the planets became nearly circular and nearly coplanar (Fig. 2, C and D).

As was shown by Schmidt (1946), the process of growth by capturing the surrounding material contains within it the mechanism which regulates the distances between the orbits of adjacent growing bodies and leads eventually to the regularity of the distances of the planets from the Sun. The simultaneous application of the laws of conservation of energy and of angular momentum to the process of planet formation by amalgamation provides for the actual straight rotation of the planets.

Examination of the process of planet system formation thus explains the basic astronomical properties of the system. It also explains the division of the planets into two groups, differing by mass and by chemical composition.

Gurevich and Lebedinskii (1950) pointed out that the formation of the disk of dustlike particles must have been accompanied by the emergence of zonal differences in the

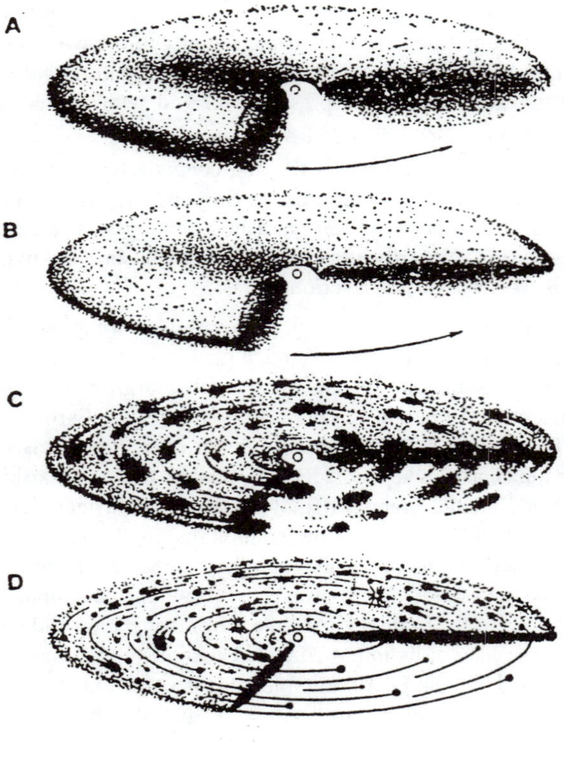

FIG. 1. Suggested mechanism of settling of the dust component of a protoplanetary cloud into a thin disk and its breaking up into multiple clusters which subsequently turn into asteroid-size bodies.

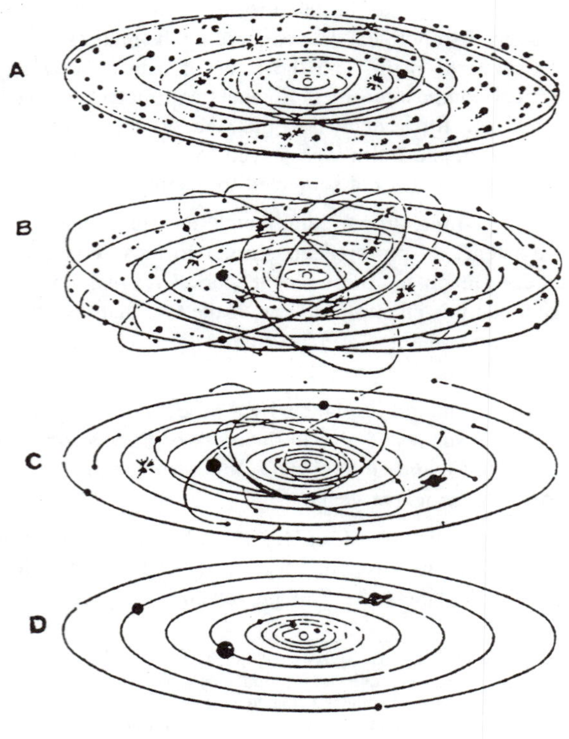

FIG. 2. Suggested mechanism of agglomeration of planets from asteroid-size bodies and fragments.

chemical composition of the general mass of the solid particles. As the dust particles settled to the central plane of the cloud, the tapering dust layer became increasingly transparent and the temperature of this outer portion became lower and lower. Only the narrow inner part of the disk was heated by the Sun, and there the particles must have been of high-melting-point, nonvolatile, substances. In the remaining part of the disk the falling temperature enabled the dust particles to collect frozen molecules of the various volatile compounds present in the gaseous part of the cloud. From the data concerning the relative cosmic abundance of the elements, it follows that in addition to H_2, the compounds present in the greatest abundance must have been H_2O, NH_3, CH_4, and CO_2. As Urey and Dunn showed in 1956, it is apparently necessary to consider the possibility that not only stabilized molecules froze to the dust particles, but radicals as well.

After the material of the disk of dust particles had passed through the stage of clusters of asteroid-size bodies and was amalgamated into planets, these zonal differences in relative abundance and composition of the solid particles were translated into the zonal differences in the masses and composition of the planets. The high-melting-point, rocky particles of the inner zone of the disk formed the small, dense planets of the terrestrial group, while the particles of the outer zone, surrounded by frozen volatile material, formed the giant low-density planets. It can be said as a first approximation that, disregarding the temperature variation during the process of formation, the composition of the planets (as well as comets, asteroids, and satellites) was determined by the temperature conditions prevalent at the time and place of their formation. It is incorrect to suppose that Jupiter contains many light elements simply because it is large and thus prevented their escape; on the contrary, Jupiter is large because it was formed under conditions permitting an abundance of light elements to enter into its composition.

In the event that the protoplanetary cloud was formed according to Hoyle's hypothesis, the explanation of the division of the planets into two categories remains essentially the same.

The planets of the terrestrial group appear to have captured practically all of the solid substance which existed in the region of their orbits (an exception is Mars, which probably did not capture part of the material it should have had, as a result of the large neighboring mass of Jupiter). The problem is different in the case of the giant planets. In their zone the attraction of the Sun was small, so that these planets exerted such a strong attraction even before they reached their final masses that they began to alter sharply the orbits of smaller bodies and particles moving past them. These bodies and particles could then have had orbits taking them beyond the limits of the planetary system; even leaving the Sun permanently. The time came when the quantity of the ejected bodies exceeded that of the annexed bodies. In this way the growth of the giant planets ceased, not because they captured all the scattered matter in the zone of their formation, but because the action of their gravitational fields had ejected all asteroid-size bodies or fragments from that zone.

As the force of attraction decreases rapidly with increased distance, the alteration of orbits was easiest in the region of the most distant planets and hardest in the region of Jupiter. Therefore, Jupiter was able to scatter the bodies in its vicinity only when it had reached a large mass; Saturn, more distant from the Sun, was able to do so with a smaller mass, and Uranus and Neptune had still smaller masses. (Why the mass of Neptune is several times that of Uranus is not yet clear.)

The most complete chemical data are available, naturally, for the Earth (and for meteorites). Analysis of these data confirms the process of formation described above. The very large deficit in the Earth of the heavy inert gases (in comparison with their cosmic abundance), the deficit of chemically inactive nitrogen, and other peculiarities of the chemical composition of the Earth, show that it was formed of substances which were, under the conditions of its formation, in a solid state.

The problem of the nature of the dense core of the Earth is one of great importance. If, according to Ramsey's hypothesis (1949) it is considered that the core of the Earth consists of a silicate, changed under high pressure to a metallic state, then an analysis of the densities of the planets of the terrestrial group and the Moon shows that all, with the exception of Mercury, which is nearest the Sun, have the same composition. This is precisely what would have happened during their amalgamation from the solid particles present in the superheated part of the disk.*

If, however, one retains the previous hypothesis of an iron core of the Earth, as does Urey, for example, then the varying densities of the planets of the terrestrial group must be ascribed to different amounts of metallic iron, and it becomes necessary to look for the process which would have caused these differences. Urey examined several hypotheses, but was obliged to abandon them because of the great difficulties he encountered. The adoption of Ramsey's hypothesis makes the entire investigation unnecessary.

More than 99% of the substance of the Earth is inaccessible to direct investigation. All work concerned with the chemical composition of the Earth as a whole must, therefore, whether explicitly or by implication, make use of cosmogonic concepts of the evolution of the Earth and be based as well on data on the composition of meteorites. Such data were utilized even during the years when meteorites were considered to have come from interstellar space. The motivation for this consideration at that time was simply the fact that both the Earth and meteorites are the results of the evolution of silicate materials. It is now known that meteorites are fragments of asteroids, which are, like the Earth,

*The increased density of Mercury is explained by its formation from the hottest, and therefore densest, particles or even, it may be, from particles which condense at even higher temperatures.

related to the inner zone of the planetary system. The identification of the isotopic composition of a number of elements on the Earth and in meteorites points to a thorough mixing of the protoplanetary matter, at least within the limits of the inner zone. It is true that the asteroids were formed at the junction of the terrestrial planet zone and the giant planet zone and that originally, at the time when the dust-cloud disk was opaque, they were very likely part of the cold zone. The basic variation in the physicochemical conditions in the asteroid band is apparently the cause of the variation in their composition, which is manifested also in the composition of meteorites.

It is obvious that the composition of meteorites reflects the presence on the Earth of only those elements which form substances which are not volatile at temperatures of the order of 0 °C. The origin of the Earth's atmosphere and hydrosphere is probably connected not only with the absorption of gases by solid particles, but also with the fallout on the Earth of bodies of ice similar to the nuclei of comets.

In the present state of knowledge and capabilities it seems most proper to make use of analysis of meteorites to determine the average rock and metal components of meteoritic substances and to consider that the Earth has similar proportions of these substances. With regard to the ratio of rock to metal, this should be determined separately. In 1955 the author, on the basis of some indirect data, determined that the ratio of rock to metal is 6:1; that is, metal (together with the troilite associated with it) constitutes 14.3%. One can proceed from Bullen's hypothesis (1956) to the conclusion that the inner core of the Earth consists, as a result of differentiation, of nickel-rich iron and that in the crust silicate and metal remain fixed. Unfortunately the density and therefore the mass of the inner core are not well known. The mass of the inner core constitutes approximately 8% of the mass of the entire core. The majority of the chemical elements are concentrated in the silicate part of the material, so that various assumptions regarding the ratio of metal tell very little about the proportion in the Earth as a whole. In 1957 we applied this method to the compilation of a table showing the average composition of the Earth. This table includes 78 of the elements.

A comparison with data on the composition of the Earth's crust made it possible to identify those elements which were concentrated in the crust during the formation of the Earth or which, on the contrary, remained principally in the mantle. As a result it appears that there are noticeable concentrations of 25 elements in the Earth's crust: Li, Be, B, N, F, Al, K, Ti, Rb, Sr, Zr, Nb, Cs, Ba, La, Ce, Pr, Nd, Ta, Hg, Tl, Pb, Ra, Th, and U. (On a percentage basis their content in the crust is five times that in the Earth as a whole.) The following 19 elements are in deficit in the Earth's crust: Mg, S, Cr, Fe, Co, Ni, Ge, Se, Ru, Rh, Pd, Cd, Th, W, Re, Os, Ir, Pt, and Au. (Their crust/Earth ratio is 0.2.)

The content of radioactive elements in the Earth is of particular interest. Such knowledge is essential for the study of the thermal history of the Earth, since the interior of the Earth was warmed only gradually and reached its present high temperature as a result of the accumulation of heat generated by the decay of radioactive elements.

The gravitational energy generated during the formation of the Earth was not enough to account for its initial heating. According to the theory of Schmidt, however, as well as according to the concepts of the overwhelming majority of other investigators, the formation of the Earth progressed in such a way that this energy was not used in heating the Earth but was dissipated into space. It was dissipated in this way because of the fact that this energy was not generated in the interior, but at the surface of the growing Earth (by the fall of the bodies and particles forming it) and was generated slowly enough, since the process of formation of the Earth took from 100 to 200 million years.

The concept of an originally cold Earth which gradually grew warmer was in contradiction to the original views of geophysicists, geochemists, and geologists and at first seemed unacceptable to them. Before long, however, they appraised the new possibilities opened up by it for an explanation of the structure and development of the Earth and the Earth's crust.

The content of radioactive elements in meteorites (and their average content in the Earth as well) is very small, so that despite the development of new and excellent laboratory methods, the findings of various authors vary by a factor of more than 10. During the period of formation of the Earth—about 5 billion years ago—radioactive elements (especially actinouranium and ^{40}K) were 10 times as common as today. Taking this into consideration, we can estimate that even with the minimum values of content of radioactive elements, the heat generated was sufficient to account for the present high temperature of the Earth's interior. If the radioactive element content was very small, however, the heating stopped and became a slow cooling process.*

With any appreciable concentration of radioactive elements, it seems most probable that the heat generated in the Earth surpassed the heat losses through the surface, so that the interior would continue to grow warmer.

Since the central portion of the Earth was formed 100–200 million years earlier than the surface and was therefore warmed somewhat longer by the heat of radioactive decay, a temperature gradient arose in the interior, with a resultant flow of heat from the core through the mantle. This creates the possibility of explaining the convective currents in the core, which, as many suppose, are the source of the Earth's magnetic field.

The solid bodies of which the Earth was formed were of all sizes, from asteroid bodies having diameters of tens or hundreds of kilometers to small particles originating from their collisions. A large part of the mass was, however,

*See Priroda (Moscow) **1959** (3), 29.

contributed by large bodies, which played the principal part in the growth of the Earth. Differences in the content and density of these bodies were the preconditions for the nonhomogeneity of the Earth's interior.

One of the most important consequences of the nonhomogeneity of the Earth is the possibility of the shifting of the poles. When it was held that the Earth in an "igneous-liquid" stage separated into concentric layers of decreasing density from center to periphery, the motion of the poles seemed impossible. The nonhomogeneous structure of the interior, however, permits the sinking of heavy volumes with a consequent deformation of the ellipsoid of inertia of the Earth, which inevitably is accompanied by a greater or lesser shifting of the poles. Because the shifting is so slow, the oblateness of the Earth accommodates to the new position of the poles enabling the motion to continue.

The above-presented picture of the cold formation of the Earth, which later grew warm by the accumulation of heat generated by radioactive elements, leads to a concept of a protracted formation of the Earth's crust. Under the condition of radioactive heating of the Earth, when the temperature became high enough, a partial melting of the exterior layers began; that is, there was a melting out from these layers of more compounds having low melting points. When they had accumulated in sufficient amounts, these materials, being less dense, rose to the surface and on the surface. In other words, the crust of the Earth, being from 10 to 40 km thick, was formed as a result of a physicochemical and gravitational differentiation of the Earth's material (essentially to a depth of 700–1000 km).

As was noted above, the average composition of the Earth is close to the average composition of meteorites. Repeated notes have been taken of the possibility of the melting out from meteoritic material of some portion having the composition of the Earth's crust. For example, Goldschmidt (1938), comparing the composition of meteorites and the crust, wrote: "We are thus led to the conclusion that such rocky meteorites could have been analogous to the initial substance of the Earth silicate rocks, in the sense that the lithosphere as a whole may have had a composition similar to chondrites, but in uppermost parts of the lithosphere accessible to us, as a result of fractional crystallization, the residual magmas and maternal magmas of this fractional crystallization appear to be enriched. Such residual magmas as, for example, those of granite and nephelitesyenite rocks have much smaller specific gravities than those of chondrites and therefore, in the course of geological history, should have accumulated chiefly in the uppermost layers of the lithosphere." And further: "results of comparisons of the material composition of the silicate parts of meteorites with Earth rocks, with respect to the rare elements, are also in agreement with the hypothesis that the accessible rocks of the Earth's crust evolved by a process of fractional crystallization from magma of a composition corresponding approximately to that of silicate meteorites."

Only recently, however, Vinogradov (1957–58) confirmed this experimentally by means of zone melting of material of rocky meteorites ground into powder.* One end of the pipe yielded a substance similar to the Earth's crust, and the remaining material proved to be similar to dunite, that is, to ultrabasic subcrust substrate.

A few years ago the idea of the gradual formation of the Earth's crust was proposed by a number of scientists in the USSR and abroad, on the basis of geodesic, geological, and geochemical data.[†,‡] Since that time this point of view has made successful progress, although it has not so far met with universal acceptance. It is confirmed by evidence obtained by measurements of the ages of rocks indicating the gradual growth of the continents. The variation in the thickness of the Earth's crust, which is great under the continents and small beneath the oceans, is obviously connected with the nonhomogeneity of the Earth's interior, as discussed above.

According to the study made by E. N. Lustig, the attribution of a decisive role to the differentiation of plutonic material in tectonic motion of the Earth's crust and in orogenic processes appears to be the most promising avenue for the development of geotectonic theory.

In conclusion, we should note that contemporaneous notions concerning the gradual formation of the atmosphere and hydrosphere by a process of accumulation of gases and vapors escaping from the interior of the Earth are in full agreement with the cosmogonic deduction that the Earth began as a cold agglomeration and was subsequently heated.

*On fractional crystallization and zone melting, see Priroda (Moscow) **1959** (7), 17.

†See V. A. Magnitski; Priroda (Moscow) **1956** (7) 3.

‡See J. T. Wilson; Priroda (Moscow) **1959** (8), 41.

The Formation of Protoplanetary Dust Clouds

V. S. Safronov
Translated from: Vopr. Kosmog. **7** (1960)
In: *Problems of Cosmogony* (Joint Pubs. Res. Service, U.S. Dept. Commerce, 1964), Vol. VII, pp. 120–45

The author derives the relation between density and degree of flatness in condensations formed in the protoplanetary dust cloud. In order that a condensation should not be broken up in connection with the random velocities of differential rotation, we must conclude that the ratio of the density of the condensation to that of the Sun expanded to the distance of the condensation (namely $\rho^*=3M_\odot/4\pi R^3$), where R is the distance from the Sun for condensation, must be 3 to 4 times as great as the ratio of the cloud thickness. The latter ratio turns out to be on the order of 5. When gravitational instability arises in the sheet as the result of strictly symmetrical radial disturbances, with a cloud density of $\geqslant 2.5\rho^*$ and perturbation wavelength ten times the sheet thickness, the formation of contraction rings is possible. With diminution of ring width by a factor of 4, the rotation of the ring around the Sun becomes essentially rigid because of the conservation of angular momentum of the ring material. Such a ring is unstable, and may break up to form condensations which are stable. The masses of the condensations in the latter instance are of the order of 10^{17} at the distance of the Earth's orbit from the Sun, and of 10^{24} at the distance of Jupiter's orbit. After initial contraction, the dimensions of these condensations depend upon angular momentum with respect to the center. The average angular momentum of a condensation in the process of formation, and its alteration as a result of collecting other condensations, are estimated by the author. Condensations which grow by collecting small particles contract very rapidly. Central collisions of the aggregating condensations also lead to rapid contraction. The angular momentum acquired from noncentral collisions is several times the intrinsic momentum of the condensation. The average momentum acquired in this way is small, because of the random direction of the impacts; however, chance deviations from this average are of considerable magnitude, and hinder rapid contraction of the condensation. As a result, there is great diversity in dimensions and densities. The central portions of condensations are readily transformed into solid bodies, but the outer portions may remain diffuse for a longer period. The most massive and densest of the condensations contract most rapidly. They become solid bodies after their masses have been increased by a factor of 10^2–10^3. The time required for this process is of the order of 10^4 years in the case of the Sun–Earth distance, and 10^6 years in the case of the Sun–Jupiter distance. The least massive condensations contract only after central collisions. On the average, the aggregation of these condensations leads to expansion rather than contraction, and so a considerable number of them possibly disintegrate. From the cosmogonical point of view, the entire complex of condensations evolves into a swarm of bodies in a very short time.

Two conclusions which have received recent new support are evidence in favor of the hypothesis of the formation of dust condensations within a protoplanetary cloud.[1,2] First, the cloud lacked the conditions necessary to maintain turbulence for a long period of time.[3,4] This made it possible for the dust to settle to the equatorial plane of the cloud, and also made possible the formation of a layer of increased density at the plane. Second, the initial mass of the protoplanetary cloud as accepted at the present time (not less than $0.1M_\odot$) would not have been sufficient for the gravitational condensation of the gaseous material of the cloud.[5] The density of the gas remained constantly below the critical point, whereas the density of the dust material increased continuously as a result of diminution in the relative velocities of particles upon collision.

However, the conditions for gravitational instability in the dust layer were not everywhere identical. At Jupiter's distance from the Sun, the velocities of particles must be diminished to several hundred meters a second, whereas at the distance of the Earth they must be diminished to several hundred centimeters a second. Therefore, the possibility is not excluded that in the region of the inner planets there was no gravitational instability in the dust layer. Even small deviations from ideal symmetry in the layer, or perturbations, may have prevented the high degree of compression of the layer necessary for achieving the critical density. In such a case we would find the growth of bodies taking place directly from particles upon collision. In one way or another a swarm of bodies was gradually formed in the dust layer. The process of planetary growth as a result of the agglomeration of such bodies has already been discussed.[6–8]

THE FORMATION OF CONDENSATIONS

The question of gravitational instability in a rotating medium is considerably more complex than the corresponding question for a nonrotating medium. Edgeworth,[1] on the basis of Maxwell's familiar study of the stability of the rings of Saturn, proposes that with achievement of density $\rho \approx 0.04\rho^*$ where $\rho^* = 3M_\odot/4\pi R^3$, and R is the distance from the Sun, the cloud ceases to be stable and experiences within itself the rise of eddy motions which in turn cause fluctuations in density. The latter increase, and at $\rho > 3\rho^*$ are transformed into stable condensations. However, the probability of such large random fluctuations is extremely small. At the same time, it is difficult to admit that following the onset of instability, the layer continued to be compacted and that its density almost doubled. Maxwell's result,[9] which in the strict sense is justified only for material points disposed in a ring, is inapplicable to a large number of particles which are colliding and forming an actually continuous medium. The Maxwellian ring disintegrates when the amplitude of oscillation of the material points reaches the mean distance between neighboring points and when collision between them is possible. But collisions between real particles in the layer exert no influence on its stability.*

In a recent article Bel and Schatzman[12] found the condition for gravitational instability in a rotating medium for the case of radial perturbations. In the law for the variation of rotational velocity with distance, as in the case of the solar system ($\omega \sim r^{-3/2}$), instability arises when $\rho = \frac{1}{3}\rho^*$. However, this result was arrived at for the two-dimensional problem: in other words it relates essentially to a system which is infinitely extended in the direction of the axis of rotation. The gravitational force of a flat ring is significantly less than that of an infinite cylinder. Therefore, the critical density necessary for gravitational instability must be considerably less. The computations made by the present writer[13] lead to a minimal value for the critical density of about $2.2\rho^*$, which corresponds to the wavelength of the perturbation, this being equal to 8 times the thickness of the layer. Since strictly symmetrical ring perturbations are evidently the most favorable case, then under real conditions for gravitational instability high density is first of all necessary.

When the critical density in the layer has been achieved, radial perturbations with wavelength 8–10 times the thickness of the layer lead to the formation of rings, which begin to be compressed. The density in the rings increases, and may reach the so-called Roche density (in Kuiper's terminology), which is sufficient for the existence of separate stable condensations, usually assumed to be of the order of 10.

In an article by Gurevich and Lebedinskiy[2] this density is determined on the basis of energy considerations. These authors find what must be the dimensions and density of the spheroidal region, developed in an unperturbed disk, in order that it may maintain itself by its inner forces of gravitation when the surrounding portion of the disk is discarded. The basic condition was determined from the virial theorem, where not only disorderly relative velocities of the particles are taken into account, but also orderly velocities associated with the differential rotation of the cloud around the Sun. It was found that the density of such condensations must be one order (power of ten) higher than the "spread out" density of the Sun ρ^*, and that the condensations originally must have the form of strongly oblate spheroids with axes in the ratio 13:1. The first result supported the estimates of critical density arrived at by other methods; the second result was new, and even somewhat unexpected. For example, Edgeworth, using essentially only Jeans's criterion, concluded that the condensations must have been approximately spherical.

This method gives for mass m_0 and the major axis a_0 of the condensation being formed the following values:

$$m_0 \approx \frac{\sigma^3}{2\rho^{*2}}, \quad a_0 \approx \frac{\sigma}{2\rho^*}, \tag{1}$$

where σ is the surface density of the dust material in the cloud.

The estimate quoted should be regarded as giving only the *order* of the quantities in question. First of all, real condensations may have been formed from a nonspheroidal region. In various directions of the equatorial plane, orderly relative velocities are different with respect to magnitude and are directed differently. Along the line passing through the center of the condensation and turned in the direction of rotation at the angle φ with respect to the radius vector, the angle having the value

$$\varphi = \arctan \sqrt{\frac{dV/dR}{V/R}} \approx 30^\circ$$

moves toward the center of the condensation; and along a line at the same angle φ to the radius vector it moves away from the center. In the first case the differential rotation will make condensation easier; in the second, more difficult. In addition, the balance of all the forms of energy is achieved only in respect to the condensation as a whole. For spherical and nonrotating formations, such an estimate is inadequate. But in strongly compressed condensations conditions in the central region and on the periphery are quite variable, and from the stability of the condensation as a whole (as derived from the virial theorem), the stability of its peripheral portions, which have the highest relative velocities but experience least gravitation, does not, in general, follow. There-

*Jeffreys (Ref. 10), on the basis of Maxwell's criterion, considers that the rings of Saturn are stable and also "single layer." Bobrov (Ref. 11), on the basis of a careful photometric study of the rings, has concluded that the thickness is much greater than the diameter of the component bodies—in other words, that the rings are "multilayer." He also found that, according to Maxwell's criterion, the rings are unstable. However, one cannot conclude from this that "Maxwellian" instability is the cause of thickening of the rings. Another possible cause, for example, could be found in perturbations caused by Saturn's massive satellites.

fore, it would be desirable to make use of another method which does not involve averaging.

As our condition for stability we may take, for example, the stipulation that the condensation must lie completely within the closed surface of Hill's zero velocity, which passes through the point of libration L_1. Let ξ and η characterize the degree of flattening and the degree of consolidation, respectively, of the condensed matter:

$$\xi=\frac{2a}{H}, \quad \eta=\frac{\rho_0}{\rho^*}, \quad \frac{\eta}{\xi}=\frac{a_0}{a}, \tag{2}$$

where $H=\sigma/\rho_0$ is the thickness of the dust layer; ρ_0 is the density of the layer in the equatorial plane; a is the major semiaxis of the condensation. The mass of the flattened condensation may be expressed in the form

$$m\approx\pi a^2\sigma=2\pi\rho^* a^3\,\frac{\eta}{\xi}. \tag{3}$$

The distance from the center of the condensation to the point of libration L_1 is

$$r_{L_1}\approx\sqrt{\frac{m}{3M}}\,R=\sqrt[3]{\frac{m}{4\pi\rho^*}}. \tag{4}$$

From Eqs. (3) and (4) we obtain

$$\frac{\eta}{\xi}\approx\frac{2r_{L_1}^3}{a^3}. \tag{5}$$

Since the relative velocities of particles in the condensation are not equal to zero, $a<r_{L_1}$. Consequently, η must be several times greater than ξ. For a spherical condensation ($\xi=1$), in Eq. (5) in place of 2 we place 3, since $m\approx\frac{4}{3}\pi a^3\eta\rho^*$, and we arrive at

$$\eta\approx\frac{3r_{L_1}^3}{a^3}. \tag{5'}$$

In the direction along the orbit, the width of Hill's region amounts to only 2/3 of r_{L_1}. If for a spherical condensation we take $a<\frac{2}{3}r_{L_1}$, then $\eta>10$.

The values of η and ξ could be estimated with the help of Jacobi's integral; if so, all of the particles of the condensation would have values for Jacobi's constant C greater than for the point of libration C_L. However, this sufficient condition of stability requires too great values of η (of the order of 50). Here we get $a\approx 0.6r_{L_1}$ and according to Eq. (5) the quantity $\xi\approx5$. But for condensations such a condition is obviously not necessary, since particles which are only temporary satellites of the condensation and then fall within it lose their velocity and remain there permanently. It is possible to make an approximate estimate, under the conditions of the two-body problem, of the density of the condensation at which a particle possessing thermal velocity (determined from H) and differential motion will be removed from the condensation to a distance not further than the point of libration. This leads to a value of η of the order of 15–20, and, as before, to a value of ξ of the order of 5. Evidently, this value for the density is more realistic, although not sufficiently accurate. The oblateness of condensations comes out less than half as great as that derived by Gurevich and Lebedinskiy. From Eq. (5) it follows that for a large value of ξ we would need a large value of η—the flatter condensations must be denser.

From Eqs. (2) and (3) we may find an expression for the masses and dimensions of a condensation

$$m\approx m_0\,\frac{\xi^2}{\eta^2}, \quad a\approx a_0\,\frac{\xi}{\eta}, \tag{6}$$

where m_0 and a_0 are these values according to Eq. (1). With $\xi=5$ and $\eta=15$, the mass of the condensation is one order less than the value obtained earlier. But we should keep in mind that $m_0\sim\sigma^3$. Obviously, before the formation of the condensations, compression took place not only with respect to axis z as the result of diminution of the chaotic velocities of the particles, but also in directions parallel to the equatorial plane (as the result of gravitational instability). Along with this, σ grew and became larger than the average value when the mass of the planet is uniformly distributed over its entire zone. Therefore, we may suppose that the masses of the condensations were not less, but instead rather greater, than the values of m_0 obtained from Eq. (1).

Thus we may expect the following differences in properties of forming condensations from the results obtained by the method of Gurevich and Lebedinskiy. First, $\eta=\rho_0/\rho^*$ could be three times greater than $\xi=2a/H$. Before they were nearly identical. Second, the surface density in the forming condensations might exceed the surface density of the dust layer by several times. These factors alter the mass and radius of the condensation in the opposite direction. Therefore, without danger of great error we may limit ourselves to values of m_0 and a_0 according to Eq. (1). For condensations at the distance of the Earth from the Sun, substituting $\sigma=10$ g/cm^2, we get $m_0=3\times10^{15}$ g and $a_0=4\times10^7$ cm. For the zone of the great planets the value of η for the dust layer is difficult to determine. It is still not clear whether hydrogen may freeze onto the particles. It is also uncertain how great is the mass of matter thrown off from a swarm following formation within it of massive bodies which have communicated significant velocities to one another by their gravitational interrelationship. Obviously, σ must be significantly greater than its lower boundary as obtained without allowance for the factors referred to. In the zone of Jupiter, $\sigma_{\min}\approx20$. With $\sigma=50$, we get $m_0=6\times10^{22}$ g.

THE ROTATION OF CONDENSATIONS IN FORMATION

The gravitational force inherent in forming condensations is greater than the external forces present. Therefore a condensation begins to be compressed (flattened) as long as the gravitational force is not equalized by internal pressure and centrifugal force, which as a result of conservation of the angular momentum grows with compression. Diminution of the relative velocities of the particles of the condensation, as a result of nonresilient collisions, is accompanied by compression of the condensation along its axis of rotation. However, frequent external perturbations as the result of approaches and collisions on the part of the condensation prevent an unlimited decrease in the velocities of the particles, and an unlimited flattening of the condensation itself.

The equatorial radius r of the condensation and its angular velocity ω up to the time of compression (subscript 0) and following compression (subscript 1) are associated by the condition of conservation of angular momentum and by the condition of equilibrium:

$$r_0^2\omega_0 = r_1^2\omega_1, \quad r_1^3\omega_1^2 = \lambda Gm, \tag{7}$$

where m is the mass of the condensation and λ is a coefficient which depends upon the form of the condensation. From these relations we have

$$r^1 = \frac{r_0^4\omega_0^2}{\lambda Gm}. \tag{8}$$

In the case of a homogeneous spheroidal condensation rotating like a solid body (Maclaurin's spheroid) with the ratio of axes $c/a=0.6$, we have $\lambda \approx 1/2$; when the ratio $c/a=1/3$, $\lambda \approx 1$; and when $c/a=0.1$, $\lambda \approx 2$. In the second extreme case Roche's model (with practically the entire mass concentrated in the center), $\lambda=1$. Therefore, there is no great error involved in $\lambda=1$.

Thus, the dimensions of the condensation following compression are determined by its rotation at the initial moment. The process of formation of condensations in the rotating layer as the result of the growth of fluctuations in density which arise from some variety of perturbation is very complex, and is still not subject to certain quantitative description. However, one may suppose that the initial rotation of the condensation must in some degree be characterized by the angular momentum with respect to the center of the condensation of that region from which it (the momentum) was formed—that is, a certain volume at every point of which matter is moving with Keplerian circular velocity around the Sun. Such an estimate gives the order of magnitude of the mean angular momentum of the condensation.

Let us adopt a nonrotating system of coordinates with its origin at the center of the condensation, and with axes x and y coinciding, at the moment under discussion, with the direction of the rotation and the radius vector, respectively. Then the specific momentum, with respect to the center of the condensation of the element found at point (x,y) and moving along a circular orbit with velocity $V(R)$, will be[1]

$$x^2\frac{V_0}{R_0} + y^2\left(\frac{dV}{dR}\right)_0 = \omega_c[x^2 - \tfrac{1}{2}(y^2)], \tag{9}$$

where R_0 is the distance of the center of the condensation from the Sun; V_0 and ω_c are its orbital velocity (linear and angular). From this it is apparent that points located on the axis x (on the orbit) have positive momentum with respect to the center of the condensation, while points on axis y (along the radius vector), have negative momentum. The total momentum is positive.

For a region whose dimensions along x and y are identical, integration of Eq. (9) gives the total moment with respect to its center:

$$K = \tfrac{1}{4}(I_0\omega_c), \tag{10}$$

where I_0 is the moment of inertia of the region. Consequently, the average angular velocity of this region with respect to the center of the condensation previous to compression is

$$\omega_0 = \tfrac{1}{4}(\omega_c). \tag{11}$$

The quantity ω_0 in this case can be found considerably more simply if we recall from hydrodynamics that it is half of the eddy

$$\omega_0 = \tfrac{1}{2}(\text{rot } V) = \frac{1}{2R}\frac{d}{dR}(VR) = \tfrac{1}{4}(\omega_c). \tag{11$'$}$$

Actually, the initial dimensions of the region along the y axis may be somewhat less than those along the x axis. For an ellipse in the plane, xy will be

$$\omega_0 = \{(2a^2 - b^2)/[2(a^2+b^2)]\}\omega_c = \alpha\omega_c. \tag{12}$$

The ratio of the axes may be estimated approximately, for example, from the condition of equality at the terminals of the axes of the ratio of kinetic energy of the particles to their potential energy, or from the condition of equality of their total energy. We get, respectively, $b:a=3:4$ or $1:2$. Then the numerical coefficient α will be 0.5 and 0.7.

At the time of compression of the material of this region into a more compact condensation, the initial momentum of any element of volume does not remain constant, since the element lines are not only within the field of gravitation of the internal portions of the condensation, but also within the Sun's gravitational field (the three-body problem). From the general equations of motion, it is possible to find with accuracy within fractions of the second order[14] an expression for the variation in angular momentum as a function of the coordinates (x,y) with respect to the center of the condensation in the rotating system of coordinates:

$$\frac{d}{dt}[\dot{x}y - x\dot{y} + \omega_c(x^2/y^2)] = -\frac{3GM_\odot xy}{R_0^3}. \tag{13}$$

The expression in brackets represents the angular momentum in the nonrotating system of coordinates. Upon integration with respect to the region symmetrical with respect to axis x or axis y, the right-hand member reverts to zero. This means that upon compression of such a region within the Sun's gravitational field, its momentum is conserved. Therefore the angular momentum of a condensation which has been formed may be assumed to be

$$K=\alpha I_0\omega_c, \tag{14}$$

where $\alpha\approx 1/2$. Nonsymmetry of the condensation with respect to the axes x and y may alter this estimate. Thus, if the condensation at first has maximal values in the direction $\varphi=+30°$ to the radius vector at which the material has motion relative to the center of the condensation, then $\overline{xy} > 0$, and in accordance with Eq. (13) the angular momentum of the region upon compression is diminished. This leads to a somewhat smaller value for α.

Substituting in Eq. (8) the values α_0 and m_0 from Eq. (1), and also $\omega_0=\alpha\omega_0$, we arrive at

$$\frac{r_1}{r_0}=\frac{a^2}{\lambda}\approx\frac{1}{4}. \tag{15}$$

If we take our values for α and m on the basis of Eq. (6), then there is still another coefficient, $\xi/\eta<1$. Consequently, the initial angular momentum of the condensation does not diminish the latter's compression by as much as several times.

As was pointed out above, the variation in σ caused by the shift of the material in the equatorial plane may substantially alter the masses and radii of the forming condensation. To this we should add that such displacement of material may reflect substantially on the angular momentum of the condensation. This can be seen by studying the simplest case of radial displacements. Compression of a ring takes place with conservation of the angular momentum of its material with respect to the Sun, and consequently with variation in it orbital velocity. The outer half approaches the Sun and is accelerated; the inner half is moved further away and is decelerated. If the width of the ring is significantly less than its radius R_0, then with reduction of the width by a factor of n_1, the variation in orbital velocity with R will be

$$\frac{dV}{dR}\approx\frac{V_0}{R_0}\left(\frac{n_1}{2}-1\right). \tag{16}$$

If the width is reduced by half, the linear velocity of rotation of all of its parts remains the same, and with reduction to a quarter it rotates as a solid body. Consequently, compression of the ring favors the formation of condensations within it—density grows, and in the first stage there is a decrease in the energy of ordered relative motion of the particles. With solid-body rotation of the ring, the particles are without systematic displacement with respect to their surroundings. To clarify the possibility of disintegration of the ring into separate condensations, we may make use of the condition of instability found by Dibay[15] for an infinite cylinder, which can be written in the form

$$Gm'>\gamma\,\frac{v^2}{2}, \tag{17}$$

where m' is the mass of the cylinder per unit of length and $\gamma=c_p/c_v\approx 1$. The thermal velocities of the particles v are associated with the thickness of the layer, H, and its surface density σ before the formation of the ring by the following relation:

$$v^2=\frac{3\pi G\sigma H}{2I^2}, \tag{18}$$

where I is a quantity close to unity. With an initial breadth of ring $2b_0=cH$, we have

$$m'=2b_0\sigma=c\sigma H. \tag{19}$$

Therefore, the condition of Eq. (17) is transformed into

$$c>\frac{3\pi\gamma}{4I^2}. \tag{20}$$

According to Ref. 13 we may assume $\rho_{cr}/\rho^*=c_1\approx 2.5$, and $c\approx 10$; then $I=0.92$ and condition (20), amounting to $c>3\gamma$, is reliably fulfilled. Consequently, the ring is unstable and will disintegrate. Since the density of a ring rotating as a solid body is equal to $4\rho_{cr}=10\rho^*$, then the condensations forming upon its disintegration must be stable.

The mass of a condensation formed is

$$m=\frac{c^2n\ \sigma^2}{c_l^2 n_1\ \rho^{*2}}, \tag{21}$$

where n is the ratio of the initial dimensions of the condensation along the ring to the width of the ring at the time of the disintegration. With $n=1$, the mass of the condensation exceeds m_0 by one order [as determined from Eq. (1)]. Since the disintegration of rings is possible for various values of n and n_l, the masses, dimensions and particularly the densities of the condensations formed may vary greatly.

SUBSEQUENT EVOLUTION AND GROWTH OF CONDENSATIONS

According to Edgeworth, the subsequent evolution of a condensation is determined by tidal waves of the Sun, which gradually slow down its rotation, permitting the condensation to be compressed gradually with respect to r. However, the compression will take place much more rapidly as the result of collisions and amalgamations. Actually, when two condensations unite which have collided centrally, the mass is doubled, while the specific momentum remains as before. Therefore, the radius of the new condensation, according to Eq. (8) must be diminished by a factor of 2, while its density increases by a factor of 16. With such rapid evolution in condensations, the influence of tidal waves of the Sun upon their rotation is negligibly small.

According to Gurevich and Lebedinskiy, condensations originally moved in nearly circular orbits, since their relative velocities were very small. Only those condensations which were practically on the same orbit could coalesce. In this situation, according to these authors, the union could be effected only with constant specific momentum. Such a process of union, prolonged over a period of time P/e (that is, of the order of 10^5 years in the zone of Jupiter), led to the formation of "secondary condensations" of mass μ, the estimate of which was made on the assumption that during the entire period the eccentricities remained constant. Subsequently, according to these authors, the relative velocities of the condensations were determined by their gravitational interaction as they approached close to one another, and were of the order of

$$v=\sqrt{\frac{Gm}{2r}}. \tag{22}$$

However, one cannot quite agree with even this picture of the evolution of condensations.

In the union of condensations moving along circular orbits, their specific angular momentum with respect to the center is not maintained. Depending upon the magnitude of the initial difference ΔR of their distances from the Sun upon collision, there is achieved either a negative momentum or a positive momentum;* here the variation in momentum may even exceed the momentum of the condensation before the collision. Therefore, it is impossible for us to speak either of the constancy of the specific momentum of the joining condensations or of reverse rotation as the inevitable result of a similar union, to which one is often led in a discussion of Laplace's hypothesis. The increase in relative velocities of condensations as the result of approaches [for example, to a magnitude of the order of Eq. (22)] includes still another mechanism of centralized impacts "left and right." But the process of union does not thereby alter. Therefore, although initially the union of condensations actually takes place in a narrow zone of the orbit, this process is not different qualitatively from the subsequent growth of the condensation. It follows from Eq. (22) that the relative velocities grow in proportion to the masses of condensations. Consequently, the eccentricities of the orbits grow continuously, and the zones of "feed" expand. This expansion of the "feed" zone occurs considerably earlier than the union of all the matter in the narrow zone along the orbit. In view of the continuity of the process of growth in mass and eccentricity, and the absence of qualitative distinctions in their initial stage of growth (from the succeeding), there are no bases for the introduction of the new concepts of "primary" and "secondary" condensations.

If to a condensation of mass m and radius r there is joined another condensation of mass m' and radius r', the latter brings to the former its orbital momentum K_1 with respect to the center of the condensation m and the momentum K_2, associated with proper rotation. The orbital momentum is determined by the relative velocity v up to the time of approach, and by the sighting distance βr:

$$K_1=\beta r v m'. \tag{24}$$

The momentum of the condensation K_2, which is associated with its rotation, on the basis of Eq. (7) is equal to

$$K_2=\tfrac{2}{5}\mu\omega m' r'^2=\tfrac{2}{5}\mu\sqrt{\lambda G m' r'}\,m', \tag{25}$$

where μ is the degree of nonuniformity.

Since the plane of the relative orbit of m' can have any disposition, the direction of the vector of K_1 is also arbitrary. The direction of the vector of K_2 is correlated with the direction of the vector of total momentum of the swarm only at the start. After two or three collisions the direction of K_2 becomes random. Consequently, with union of condensations, the vectors of K_1 and K_2 accumulate as random quantities, and therefore to a considerable degree cancel each other out. However, the total rotational moment K_2 of condensation m will not tend toward zero. This follows from a study of the rotation of planets.

The planetary embryo grows as the result of its collection of bodies and particles, which bring to it a momentum which practically consists only of K_1, since K_2 for such objects is very small. The sum of the vectors K_1 is the present-day rotation of planets, in which are easily delineated the systematic component, or the so-called direct rotation, and the random component, which is characterized by the deviation of the axis of rotation of the planet from the axis of rotation of the swarm. Consequently the average value of β in Eq. (24) is not zero. Having admitted that it does not depend upon upon the mass of the embryo, we may find its approximate numerical value from the present-day rotation of the planets. We will obtain $\beta\approx 0.04$. The random deviations of the momentum from the average value are evidence that the largest bodies falling upon the embryo confer a momentum comparable to that of the embryo itself.

There must be an analogous picture in the case of the union of condensations. Apart from randomly directed components K_1 and K_2, the condensation acquires a momentum K_0, which coincides in direction with the angular momentum of the entire swarm,

$$K_0=\hat{\beta} r v m', \tag{26}$$

where for $\bar{a}$ we may assume a numerical value corresponding to the present-day rotation of the planets.

---

*In view of the smallness of the masses of condensations in comparison with the mass of the Sun, their motion must be regarded as perturbed circular motion. In the first approximation, the condensation is acted upon by a perturbatory force directed along the orbit $f=Gm/r^2$, which sets up a radial velocity $\dot{R}$ and a small excess of tangential velocity over circular velocity, Δv. From the equations of motion, we find

$$\dot{R}=2f/\omega_c, \quad \Delta v=\frac{3}{2}\frac{fR^2}{GM_\odot}\dot{R}. \tag{23}$$

Therefore, if the initial ΔR is such that upon collision it comprises a considerable portion of the radius of the condensation (preserving the original sign), then the maximal positive momentum achieved with this collision is practically equal to $mr\dot{R}$. From Jacobi's integral we find that the initial ΔR necessary for such collisions must be of the order of dimension of the condensation itself.

If we join to the condensation under consideration masses m' which are very small in comparison with m, the random quantities K_1 and K_2 will cancel each other out almost entirely, and the momentum of the condensation is then determined by its average value as obtained by summation of K_0. Taking the relative velocities, as usual, in the form $v = \sqrt{Gm/\theta r}$, we obtain

$$dK_0 = \frac{\bar{\beta}}{\sqrt{\theta}} \sqrt{Gmr}dm = \frac{5\bar{\beta}}{2\mu\sqrt{\lambda\theta}} K \frac{dm}{m},$$

and, consequently,

$$K_0 \sim m^{p_0}, \qquad (27)$$

where $p_0 = 5\bar{\beta}/2\mu\sqrt{\lambda\theta}$. Since $\mu\sqrt{\lambda\theta} \approx 1$, $p_0 \approx 0.1$. With such a small value for the index, the condensation must rapidly be compressed. Actually, in accordance with Eq. (25) the radius of the condensation

$$r \sim \frac{K^2}{m^3} \sim m^{2p_0-3}$$

and the density

$$\rho \sim \frac{m}{r^3} \sim m^{10-6p_0}.$$

Therefore, an increase in the mass of the condensation by one order is sufficient for it to be compressed to the condition of a solid body with $\rho \approx 1$.

An entirely different result is obtained upon the union of condensations of comparable masses. The randomly directed moments K_1 and K_2 brought to every condensation have a magnitude of the order of K of the condensation under discussion and are considerably greater than the average value of K_0, which in this case we may leave out of account. Then the resultant moment is determined by the probable deviation from this average, that is,

$$K^2 = \sum (K_1^2 \cdots K_2^2). \qquad (28)$$

The direction of the vector K will now be arbitrary. Let us estimate the magnitudes of K_1 and K_2. Let l_0 represent the maximal sighting distance at which union of colliding condensations still takes place. The average value of $\overline{l^2}$ in the interval 0 and l_0 will be

$$\overline{l^2} = \frac{1}{\pi l_0^2} \int_0^{l_0} l^2 2\pi l \, dl = l \frac{1}{2} {}_0^2.$$

From the familiar relation between sighting distance l_0 and the distance r_0 at the moment of closest approach, we have

$$l_0^2 v^2 = r_0^2 \left(v^2 + \frac{2G(m-m')}{r_0} \right).$$

With $c^2 = Gm/\theta r$, where θ is of the order of several units, the first term on the left is small in comparison with the second and it may be neglected. We may suppose that with $r_0 > r$, it would be difficult for the condensations to unite since the impacts are almost tangential. Let us assume that $r_0 = \beta' r$, where $\beta' \approx 1$. Then,

$$K_1^2 = \overline{l^2} v^2 m'^2 \approx G(m+m')m'^2 \beta' r = 2p_1 K^2 \frac{m'}{m}, \qquad (29)$$

where

$$p_1 = \frac{25\beta'}{8\lambda\mu^2} \left(1 + \frac{m'}{m}\right) \frac{m'}{m}. \qquad (29')$$

Taking Eq. (25) into consideration, we may write

$$K_2^2 = 2p_2 K^2 \frac{m'}{m}, \qquad (30)$$

where

$$p_2 = \frac{m'^2 r'}{2m^2 r}. \qquad (30')$$

The quantities p_1 and p_2 depend upon the ratio m'/m. If we assume a certain distribution of condensations with respect to mass, then it is possible by simple integration to find the average values $\bar{p}_1$ and $\bar{p}_2$. Then in accordance with Eq. (28), we are able to write the mathematical expectation of the probable increase in the mean square of the angular momentum of the condensation with growth of its mass, in the following manner:

$$\frac{dK^2}{K^2} = 2(\bar{p}_1 + \bar{p}_2) \frac{dm}{m},$$

from which

$$K \sim m^{\bar{p}_1 + \bar{p}_2}. \qquad (31)$$

For the density of the condensation, we will have the expression

$$\rho \sim m^{10-6\bar{p}_1-6\bar{p}_2}.$$

Therefore, for the condensations to be compressed upon joining and not scattered, it is necessary that

$$\bar{p}_1 + \bar{p}_2 < 5/3. \qquad (32)$$

If all of the condensations are identical, this condition is not fulfilled. On the other hand, if the distribution of condensations with respect to mass is such that $\bar{m}' \ll m$, then $\bar{p}_1$ and $\bar{p}_2$ become less than p_0 and K is determined in accordance with Eq. (27). As has already been pointed out, in this case the union of condensations leads to very rapid compression.

Thus, the examination which we have made shows that the results are absolutely different, depending upon the character of the collisions and the ratio of the masses being joined. It should be emphasized that expression (31) gives only the *probable* angular momentum of the condensation. The true picture is bound to be more complicated. Without going into this process in greater detail, we may, on the basis of the results obtained above, draw the following conclusions, which are of a qualitative nature. Near-central impacts confer small momentum, and are accompanied by rapid compression: in the case of union of two identical condensations, the density is increased by a whole order. On the other hand, impacts which are nearly tangential confer large momentum and render compression more difficult. Therefore, from the very beginning there is rapid growth in the dispersion of densities and of dimensions among condensations. Since the most unfavorable peripheral impacts affect the central portion of the condensation comparatively little, we should expect a gradual increase in the concentration of matter towards the center of the condensation. This means that the central portions of the majority of condensations must be transformed into solid bodies at a sufficiently early stage of evolution, whereas the peripheral portions may remain in a diffuse condition for a fairly long period. The transformation into solid bodies proceeds most rapidly in the case of massive and very flat condensations. For them the greatest component in the index (31) is small, since the effective $m' \ll m$. In the first case, it is because m is maximal; in the second, because they (the bodies) readily penetrate the diffuse condensations, receiving from them only the matter which is directly swept up by their cross sections and which confers very little momentum. However, it is sufficient to take $\bar{m}' = \frac{1}{4}m$ in order to fulfill condition (32). The transformation of condensations into solid bodies in this case takes place following increase of the mass of the condensation by 100 times (for the Earth–Sun distance) and by 1000 times (for the Jupiter–Sun distance).

The least massive condensations, probably, are unstable. For them, $m' \approx m$, and in agreement with Eq. (29′) $p_1 \approx 6$. The union of such condensations on the average leads to a decrease rather than an increase in density. However, if upon the formation of condensations not all of the matter of the cloud enters into their composition, but a significant part remains in a diffuse state, then the least massive condensations in acquiring this diffuse material with slight momentum will be effectively compressed, in accordance with Eq. (27).

Assuming a certain average value $p = \bar{p}_1 + \bar{p}_2$ in expression (31) for the angular momentum of the condensation, one may estimate the duration of the evolution of the condensation, using the usual growth formula:

$$\frac{dm}{dt} = \pi r_e^2 \bar{\rho} v.$$

The geometrical cross section of a spheroidal condensation can be represented by $\pi r^2/\xi'$, where $1 < \xi' < d/c$; the mean density of the material $\bar{\rho} \approx \frac{2}{3}\rho_0 = \frac{2}{3}\sigma/H$. Therefore,

$$\frac{dm}{dt} \approx \frac{2\pi}{3\xi'} \frac{c}{H} v r^2 \left(1 + \frac{2Gm}{v^2 r}\right).$$

With $v = \sqrt{Gm/\theta r}$, the coefficient in parentheses is constant. Now let us assume $K \sim m^p$; then, $r \sim m^{2p-3}$, and with allowance for Eq. (15), we arrive at

$$m^{6-4p} dm = \frac{2\pi(1-2\theta)}{3\xi'} \frac{a^4}{\lambda^2} r_0^2 m_0^{6-4p} \frac{\sigma v}{H} dt. \tag{33}$$

The thickness of the layer H is associated with the velocity v given by the formula (18), in which I represents an integral not expressible in terms of elementary functions. At first, when the density of the swarm is close to Roche's density, $H \propto v^2$, so that $I \approx 1$, and H is determined basically by the gravitation of the swarm itself. But immediately upon increase in m, and correspondingly in v, H is significantly increased, and the gravitation of the swarm becomes negligibly small in comparison with the z component of the Sun's gravitation (the process referred to takes place in less than 100 revolutions). Then H becomes practically proportional to v, and, according to Ref. 8,

$$H = \frac{P}{4} v, \tag{34}$$

where P is the period of rotation around the Sun. Then, integrating Eq. (33) and substituting [in accordance with Eq. (1)] $m_0/r_0^2 \approx 2\sigma$, we find the time necessary for the growth of the mass of the condensation from m_1, when relation (34) becomes applicable, to m:

$$t - t_1 = \frac{\lambda^2 \xi'}{4a^4(7-4p)(1-2\theta)} \left[\left(\frac{m}{m_0}\right)^{7-4p} - \left(\frac{m_1}{m_0}\right)^{7-4p}\right] P. \tag{35}$$

Thus, the time of growth of condensations from mass m_0 to mass m turns out to be of the order of $(m/m_0)^{7-4p}P$. Since $p \sim m^{10-6p}$, the time of transformation of condensations into solid bodies must be of the order of

$$T \approx \left(\frac{1}{\rho_1}\right)^{(7-4p)/(10-6p)} P, \tag{36}$$

where ρ_1 is the density of the condensation after its original compression: that is, to a density some 10 times that of Roche's density. In the case cited above, $\bar{m}' = m/4$, we obtained $p \approx 1.2$, and the time of transformation of condensations into solid bodies is of the order of 10^4 years (at the distance of Jupiter). From Eq. (36) it is apparent that for T to become very large, it is sufficient that p increase by a certain amount. However, this would be in error, since $\bar{m}'$ must be several times greater than m. The time of evolution and conversion into solid bodies for individual condensations could vary considerably, but the system on the whole exhibits rapidity. In the region of the planetary group of planets, condensations were converted into solid bodies considerably more rapidly, and also with significantly less mass, than in the region of the great planets.

It should be noted that the process of growth of condensations in the zone of the great planets was still more complicated on account of the presence of gas among them. The resistance of the gas led to lowered relative velocities and rotational momentum of the condensations; and consequently the growth and compression of the condensations was increased.

The usual expression for the resistance of gas is

$$F = ks\rho_g v^2,$$

where s is the cross section of the condensation, while ρ_g is the density of the gas, which leads to an expression for velocity

$$v = v_0 e^{(ks/m)\rho_g l}. \quad (37)$$

Having taken for our characteristic parameter l the length of the free path of the condensations, we find the magnitude of the exponent in Eq. (37) to be

$$\frac{k}{4\sqrt{2}} \frac{\rho_g}{\rho_p} \approx \frac{k}{4\sqrt{2}} \frac{v}{v_g} \frac{\sigma_g}{\sigma_p}.$$

With $v = 0.5$ km/s ($m \approx 10^{25}$ g) and $\sigma_g/\sigma_p = 10$, the index is of the order of unity. In this case, the velocity of the condensation may be diminished over one free path by a factor of 3. With smaller values of m and σ_g/σ_p the retardation is perceptibly smaller.

An analogous expression can be found for diminution of the angular momentum of a condensation. Relative diminution of K shows up more or less as relative diminution of v (the corresponding index of degree is of the order of $0.2\rho_g/\rho_p$).

Thus, it would appear that the retardation of condensations was comparatively small, although the indeterminacy of the quantity ρ_g/ρ_p does not enable us to make a definite quantitative estimate.

In conclusion, we should note that the characteristic feature of the evolution of condensations was the absence of a gradual, even transition from the diffuse to the compact state. Every collision and union of condensations altered sharply the structure and density of the resulting body, and this result was largely a random one, since the angular momentum was altered in a random manner. However, despite this, condensations evolved quickly. A significant diversity in dimensions and densities among the condensations made its appearance. The most dense and massive of them were quickly transformed into solid bodies. The least massive were partially dispersed and partially transformed into a more dense condition. On the whole, the entire system of condensations became a swarm of solid bodies in what was a cosmogonically very short period of time. The formation of numerous bodies accelerated the disintegration of such condensations as they lagged in their evolutionary development. As a consequence of natural concentration, the central portions of the condensations were converted into solid bodies considerably more rapidly than the external ones. On the periphery of a condensation, a large portion of the matter could remain in the diffuse state for a fairly long period without falling into the center as a result of great angular momentum. Such continuously renewed regions of diffuse matter could subsequently exist around the embryos of planets, thus forming the precursors of protosatellite swarms. The question of filling in with matter these peripheral portions, and of their evolution, is a matter of independent interest in connection with the problem of the origin of the satellites.

[1] K. E. Edgeworth, Mon. Not. R. Astron. Soc. **109**, 600 (1949).
[2] L. E. Gurevich and A. I. Lebedinskii, Izv. Akad. Nauk SSSR Ser. Fiz. **14**, 765 (1950).
[3] H. Jeffreys, Proc. R. Soc. London Ser. A **214**, 281 (1952).
[4] V. S. Safronov and Ye. L. Ruskol, Vopr. Kosmog. **5**, 22 (1957).
[5] Ye. L. Ruskol, Vopr. Kosmog. **7**, 8 (1958).
[6] O. Yu. Shmidt, Dokl. Akad. Nauk SSSR **46**, 392 (1945).
[7] V. S. Safronov, Astron. Zh. **31**, 499 (1954).
[8] V. S. Safronov, Vopr. Kosmog. **6**, 63 (1958).
[9] J. C. Maxwell, *Scientific Papers* (Cambridge University Press, Cambridge, 1890), Vol. 1, p. 288
[10] H. Jeffreys, Mon. Not. R. Astron. Soc. **107**, 263 (1947).
[11] M. S. Bobrov, Astron. Zh. **33**, 904 (1956).
[12] N. Bel and E. Schatzman, Rev. Mod. Phys. **30**, 1015 (1958).
[13] V. S. Safronov, Dokl. Akad Nauk SSSR **130**, 53 (1960).
[14] F. Hoyle, Mon. Not. R. Astron. Soc. **106**, 406 (1946).
[15] E. A. Dibay, Astron. Zh. **34**, 954 (1957).

SOVIET ASTRONOMY – AJ VOL. 6, NO. 2 SEPTEMBER-OCTOBER, 1962

ON THE TEMPERATURE OF THE DUST COMPONENT OF THE PROTOPLANETARY CLOUD

V. S. Safronov

Institute of Terrestrial Physics, Academy of Sciences, USSR
Translated from Astronomicheskii Zhurnal, Vol. 39, No. 2, pp. 278-289, March-April, 1962
Original article submitted June 23, 1961

The temperature of an optically thick protoplanetary dust layer heated by solar radiation propagating nearly parallel to its surface and scattered by the dust particles is estimated. It is assumed that the absorption coefficient does not depend on wavelength and that the reradiation of light by the particles is isotropic. The same result would be obtained in the case of pure scattering (conservative case). It is assumed that the equivalent thickness of a uniform layer is proportional to the distance from the sun ($h = \beta R$). The finite size of the sun in the z direction is taken into account. It is shown that the black-body temperature in the dust layer is only very slightly dependent on z. The values of T, computed for different values of β, are given in Table 1. The temperature of the layer decreases with β. Therefore, most of the heating of the layer is due to light scattered in the gas component of the cloud. The black-body temperature in the layer at the distance of Jupiter is found to be 30-35°K and at the distance of Saturn 15-18°K. It is concluded that the condensation of hydrogen was not possible at the distances of Jupiter, Saturn and Uranus and very unlikely at the distance of Neptune.

One of the earliest stages in the evolution of the protoplanetary gas and dust cloud was the separation of the dust from the gas. A thin dust layer was formed in the equatorial plane of the cloud and the formation of protoplanetary condensations and bodies which later amalgamated to produce planets took place in this dust layer. The temperatures in the dust layer must have had a dominant influence on its chemical composition and mass since the condensation of volatile substances depends primarily on the temperature. The chemical composition of the dust layer to a large extent determined the chemical composition of the planets and its mass determined the mass of the condensations formed in it. The division of planets into two groups is associated with the temperature distribution inside the dust cloud. It has been conjectured (for example, by H. Urey, and others) that condensation of hydrogen was possible in the region of the major planets. It is natural to assume that hydrogen, which is the most abundant element in the universe, was also the main component in the protoplanetary cloud. Therefore, in order to study the evolution of the protoplanetary cloud, it is particularly important to settle the question of whether the hydrogen could go into the solid phase or not.

In 1950 L. É. Gurevich and A. I. Lebedinskii [1] determined the temperature distribution in a uniform, two-dimensional optical layer which is infinite in the R direction and has a thickness h in the z direction. The layer is heated at R = 0 and emits radiation in the z direction. The temperature of the layer decreases exponentially along R, the exponent being $-R/4h$, and at distances R from the source much larger than h it is very low. The radiation which diffuses through the layer can easily escape from it in the z direction, so that only a negligible fraction of the radiation penetrates to large distances R.

However, the protoplanetary cloud can be heated by a more efficient method. Because of rotation around the sun under the action of gravitational forces, the layer was not uniform and plane-parallel. Its thickness in the inner region was considerably greater than that away from the sun, while its density decreased with z. The sun was to a large extent outside the layer and its radiation propagating almost parallel to the layer penetrated into the upper rarefied parts of the layer at large distances from the sun and then, after scattering, entered the layer. Although there was not much of this scattered radiation, it was sufficient to keep the temperature of the layer from falling to extremely low values. An attempt on a quantitative analysis of this effect is made in the present paper.

There are two separate aspects to the problem – the first is the determination of the temperature within the layer given the temperature on its boundary and the second is the derivation of the boundary temperature itself. The first problem can be solved comparatively simply inasmuch as the thickness of the layer is much smaller than the distance from the sun. It is shown below that the temperature inside the layer is almost independent of z. At a given R, it is slightly higher in the central plane (z = 0) than on the boundary z_1. However, the relative increase is found to be only of the order of z_1^2/R^2, i.e., less than 10^{-3}. Thus in practice the problem reduces to the determination of the temperature on the boundary of the layer. The density of the layer decreases continuously with increasing z, so that the layer boundary is not a well defined concept. Two requirements, however, must be met. Firstly, in order that the equation used in the section 1 be correct up to a distance z_1, the mean free path of the quanta must be considerably smaller then the half-width h of the layer. This is satisfied when the optical thickness $\tau(z_1)$ measured inwards from the surface of the layer is of the order of 2-3. Secondly, direct solar radiation is present outside the layer and on its boundary, so that the temperature there is high. The temperature falls inside the layer and rapidly approaches its limiting value. The latter, then, is to be taken as $T(z_1)$. In practice, this value is attained at the same value of $\tau(z_1)$ as given above.

It should be emphasized that "temperature" throughout the paper should be understood to mean the temperature of a black body placed at a given point which is unambiguously defined by the density or the intensity J of the integrated (over all wavelengths) radiation at this point. The temperature of actual bodies will in general be different from this. This is so particularly for particles on the surface of the layer which absorb short-wave solar radiation in the visible region of the spectrum and emit in the far infrared region. In the central part of the dust cloud, however, all of the radiation is in the long waves since it has already passed through many stages of absorption and reemission by the particles. Local thermodynamical equilibrium can be expected to hold in this region and the temperature of the particles will coincide with that of a black body.

The determination of T(z) in the second section is made on the assumption of gray absorption in the layer, i.e., it is assumed that total absorption (true absorption plus scattering) is independent of the wavelength. This means that all of the quanta, irrespective of their frequency, have the same mean free path and therefore in the problem of radiation transport it is possible to consider integrated light (summed over the whole of the spectrum) instead of monochromatic radiation. In the case of integrated radiation, but not monochromatic radiation, radiative equilibrium exists since the radiation absorbed by a particle does not disappear but is reradiated in the long-wave region of the spectrum. Consequently, as far as the energy balance is concerned, the propagation of light in a medium with gray absorption takes place in the same way as in the case of pure scattering. Because of this, it is possible to use the results of the theory of diffuse reflection and transmission of light incident on a plane-parallel atmosphere. In the case of isotropic conservative scattering, with increasing depth in the atmosphere the average radiation intensity tends to a definite limit which depends on the intensity of the incident radiation and on the angle of incidence. In our case of gray absorption, the average intensity of the integrated radiation will tend to such a limit.

In the theory of diffuse reflection and transmission, it is assumed that the density of the plane-parallel atmosphere varies only in the z direction, while the radiation incident on it from outside is isotropic. However, the sun is not a source at infinity and the intensity of its radiation changes with distance. The density of the dust layer also depends on R. This can introduce errors into the results. However, both the intensity of the incident radiation and the density of the layer vary relatively slowly with R. Over a distance comparable with the thickness of the layer, these variations are completely negligible. Therefore, in accordance with the results of the first section, we can assume that the inaccuracies introduced by the nonuniformity of the layer and of the incident radiation are negligible.

1. The Temperature Distribution Inside a Dust Layer

Let us consider a plane-parallel dust layer whose plane of symmetry is z = 0 and whose optical properties depend only on the z coordinate. The integrated radiation flux inside the layer obeys the continuity equation, which in cylindrical coordinates (R, z) has the form:

$$\operatorname{div} H = \frac{\partial H_R}{\partial R} + \frac{1}{R} H_R + \frac{\partial H_z}{\partial z} = 0. \quad (1)$$

Here H_R and H_z are the fluxes in the directions R and z. The right-hand side is equal to zero since there is no source within the layer, while the radiation after undergoing "true absorption" is reemitted at other frequencies. Let us consider the case of gray absorption and isotropic reemission. Then the relation between the integrated flux and the integrated average intensity of radiation, $J = 1/4\pi \int I d\omega$, can be found directly from the diffusion equation

$$H_R = -\frac{4\pi}{3\alpha}\frac{\partial J}{\partial R},$$
$$H_z = -\frac{4\pi}{3\alpha}\frac{\partial J}{\partial z}, \quad (2)$$

where $\alpha = \varkappa\rho$ is the coefficient of absorption per unit volume, and ρ the density of the medium. The coefficient of absorption per unit mass, $\varkappa$ = const, is independent of the wavelength.

Let

$$\alpha = \alpha_0 e^{-z/h} \quad (3)$$

and h = const. Then, from (1) and (2) for $z \geq 0$ we obtain

$$\frac{\partial^2 J}{\partial R^2} + \frac{1}{R}\frac{\partial J}{\partial R} + \frac{\partial^2 J}{\partial z^2} + \frac{1}{h}\frac{\partial J}{\partial z} = 0 \quad (4)$$

The boundary conditions will be taken as

$$J = J_0 R^{-p} \text{ when } z = z_1, \; \frac{\partial J}{\partial z} = 0 \text{ when } z = 0. \quad (5)$$

The problem of looking for an exact solution of the second-order partial differential Eq. (4) is fairly complicated and it is more reasonable to look for an approximate solution. Since the uniform half-thickness of the layer is $h \ll R$, then also $z \ll R$. It is therefore reasonable to expect that the value of J inside the layer is not too different from that on boundary (5) and can be written as

$$J = \frac{J_0}{R^p}\left[1 + \frac{u_1(z)}{R^2} + \frac{u_2(z)}{R^4} + \dots\right]. \quad (6)$$

Substituting this expression into (4) and equating to zero the coefficients of terms with the highest powers of R, we find a series of equations for $u_i(z)$. Solving these equations and choosing the integration constant to satisfy boundary conditions (5) we find that

$$u_1(z) = hp^2\,[z_1 - z - h\,(e^{-z/h} - e^{-z_1/h})],$$
$$u_2(z) = h^2p^2\,(p+2)^2\,[(z_1 + h + he^{-z_1/h})\,(z_1 - z) - \frac{1}{2}(z_1^2 - z^2) - h\,(z_1 + h + he^{-z_1/h})\,(e^{-z/h} - e^{-z_1/h}) - h\,(ze^{-z/h} - z_1e^{-z_1/h})]. \quad (7)$$

The quantity z_1 depends on $\varkappa$ and is approximately equal to 2-3 half-widths of the layer <u>h</u>. With $z_1/R \sim 10^{-2}$, we have $u_1(z)\,R^{-2} < 10^{-3}$ and $u_2(z)\,R^{-4} \sim 10^{8}$. Thus the second term in solution (6) is not very important, while the third is negligibly small. However, as the result of their substitution into Eq. (4) the only nonvanishing terms are of order $z_1^4/R^4 \sim 10^{-8}$ This allows us to conclude that the approximate solution (6) is only slightly different from the exact solution and is completely adequate for our purposes. The magnitude of J is very small. The average radiation intensity J increases very slowly from the boundary of the layer towards its central plane. J is related to the black-body temperature T in the layer by a simple relation

$$T^4 = \frac{\pi}{\sigma}J, \quad (8)$$

where σ is the Stefan-Boltzmann constant. Thus, the temperature of the dust layer can be considered to be practically independent of its thickness and a function only of R.

A more complex model of the protoplanetary cloud is used in the next section – the density ρ_0 is a function of R, while $h = \beta R$. It is not difficult to show that the above conclusion about the extremely weak dependence of J on <u>z</u> also holds in the case of this model. In fact, if it is assumed that $\alpha = \alpha_0 R^{-n} e^{-z/h}$, the second term of Eq. (4) now contains a constant factor (1 + n). The approximate solution of this equation can be found by the same method as was used above. This differs from (6) in that $u_1(z)$ now contains an additional factor, $1 - n/p$, which even decreases $u_1(z)$ to some extent. If it is further assumed that $h = \beta R$, then the second term of Eq. (4) will now contain an additional factor $(1 - z/h)$ which is of the order of unity. In this case it is not possible to obtain a simple expression for $u_1(z)$, but from the previous result it can be seen that the order of magnitude of $u_1(z)$ will not change and $u_1(z)\,R^{-2}$ will still be small.

2. Temperature Under the Surface of the Dust Layer

On the basis of the theory of diffuse reflection and transmission of light incident on the boundary of a plane parallel layer developed by V. A. Ambartsumyan, V. V. Sobolev, S. Chandrasekhar [2-4], and others, it is possible to determine the radiation density at a great optical depth as a function of the intensity of the incident light and the angle of incidence. In the case of isotropic, conservative scattering (without absorption) of uniform radiation incident at an angle θ to the inner normal, with increasing optical thickness τ (measured from the surface into the layer) the average intensity $J_\nu(\tau)$ tends to a finite limit:

$$J_\nu(\infty, \mu) = \frac{\sqrt{3}}{4\pi}E_\nu(\mu)\,\varphi(\mu), \quad (9)$$

where $\mu = \cos\theta$, $E_\nu(\mu)$ is the flux of radiation of frequency ν incident per second on one cm² of surface in direction μ, and $\varphi(\mu)$ is a function defined by the integral equation

$$\varphi(\mu) = 1 + \frac{1}{2}\mu\varphi(\mu)\int_0^1 \frac{\varphi(\mu')}{\mu + \mu'}\,d\mu'.$$

Numerical values of this function found by an iterative method are tabulated in [3] and [4].*

Expression (9) is correct for monochromatic radiation of any frequency ν. Inasmuch as the temperature of a black body in the layer is determined by the integrated density of radiation in all wavelengths, it is sufficient for us to determine the integrated average intensity $J(\infty, \mu)$ without determining $J_\nu(\infty, \mu)$. Since the absorption coefficient does not appear in (9), the latter is obviously valid for integrated radiation:

$$J(\infty, \mu) = \frac{\sqrt{3}}{4\pi} E(\mu)\varphi(\mu), \qquad (9')$$

where

$$E = \int_0^\infty E_\nu \, d\nu, \qquad J = \int_0^\infty J_\nu \, d\nu.$$

Although Eq. (9) for $J_\nu(\infty, \mu)$ is valid only in the case of pure scattering, Eq. (9') for $J(\infty, \mu)$ is also correct in the more useful case of gray absorption. In fact, if the total absorption coefficient is independent of the wavelength, the transport equation for integrated radiation is found to be the same as that for monochromatic radiation. Radiative equilibrium also holds for integrated radiation since a stationary case is being considered and there is no transformation of radiation energy into other forms. Both scattering and reradiation are assumed to be isotropic.

The protoplanetary cloud is not a semiinfinite atmosphere, but a flat layer illuminated symmetrically from both sides. However, if the optical thickness of this layer is sufficiently large, then the radiation density on its central plane will be the same as in a semi-infinite atmosphere at large τ with the same intensity of incident radiation, i.e., it will be given by (9'). In both cases the radiation flux through an area parallel to the layer is zero — the amount of radiation reflected by the layer is equal to the amount of incident radiation. The layer illuminated from both sides receives twice as much radiation, but its surface area is also twice as large. Thus, 1 cm^2 of its surface reflects as much radiation as 1 cm^2 of a semiinfinite atmosphere.

The half-thickness h of the dust layer is small by comparison with the distance from the sun R. It is governed by the relative velocities of the particles in the layer and in general is some function of R. The simplest and at the same time most realistic assumption is $h \propto R$. This corresponds to the case in which the relative velocities of the particles are proportional to the circular velocity, while the kinetic temperature is proportional to R^{-1}.

If the layer is very thin and $h \ll R_\odot$, we can neglect all effects due to the departure of the actual model of the cloud and the incident radiation from the ideal model, for which expressions (9) and (9') hold. In this case, the calculations are particularly simple. Since the radius of the sun $R_\odot \ll R$, then $\mu = \cos\theta \le R_\odot / R \ll 1$ and $\varphi(\mu) \approx 1 + R_\odot/R \approx 1$. Therefore, according to (9'), to estimate the temperature of the layer it is sufficient to determine the flux of solar radiation through 1 cm^2 of the layer surface. An element of the solar disc $ds = 2\sqrt{R_\odot^2 - \zeta^2}\,d\zeta$, situated at a height ζ, produces a flux

$$dE = \frac{L}{4\pi R^2}\cos\theta \frac{ds}{\pi R_\odot^2} = \sigma T_e^4 \left(\frac{R_\odot}{R}\right)^2 \frac{\zeta}{R} \frac{2\sqrt{R_\odot^2 - \zeta^2}\,d\zeta}{\pi R_\odot^2},$$

where L is the luminosity of the sun and T_e its effective temperature. The flux from the whole of the solar disc is given by

$$E = \sigma T_e^4 \frac{1}{\pi R^3} \int_0^{R_\odot} 2\sqrt{R_\odot^2 - \zeta^2}\,\zeta\,d\zeta = \frac{2}{3\pi}\left(\frac{R_\odot}{R}\right)^3 \sigma T_e^4. \qquad (10)$$

*It can be seen from the tables that the function $\varphi(\mu)$ is an almost linear function in the case of isotropic scattering. The linear approximation to $\varphi(\mu)$ can be easily found on the basis of a very simple argument. Let us consider a beam of light which is almost parallel to the surface (the flux through the latter must not be zero). After a single isotropic scattering in the optically thin surface region of the layer, one half of the scattered radiation leaves the layer and the other half propagates into the layer uniformly in all directions [$E(\mu)$ = const). Values of $J_\nu(\infty, \mu)$ calculated from (9) for the singly-scattered and incident radiation must be the same. This gives

$$\int_0^1 \varphi(\mu)\,d\mu = 2\varphi(0).$$

The second relation can be obtained from the condition that $J = I$ for isotropically incident radiation (the intensity I = const). Then, since $E(\mu)d\mu = 2\pi I \mu\, d\mu$, Eq. (9) leads to

$$\int_0^1 \varphi(\mu)\,\mu\,d\mu = \frac{2}{\sqrt{3}}.$$

Both expressions are exact. In the linear approximation, we find from them that $\varphi(\mu) \approx 1 + 2\mu$ with the average error of about 2% and a maximum error less than 4%.

According to (8) and (9') with $\varphi(\mu) = 1$, the temperature of the layer will be

$$T_0 = \left(\frac{\sqrt{3}}{4} E\right)^{1/4} = \left(\frac{1}{2\sqrt{3}\pi}\right)^{1/4} \left(\frac{R_\odot}{R}\right)^{3/4} T_e. \quad (11)$$

However, this calculation is incorrect when h is of the order of $R_\odot$ or larger. The radiation incident on the layer propagates almost parallel to its surface and moves an appreciable distance in the other rarefied part of the layer. The fraction of radiation which reaches any given point in this case depends on the density distribution along the ray path. If the layer is not plane-parallel and is not uniform along R, then the calculation of the temperature must be based on a detailed model of the layer. The fact that the sun is not a source at infinity must also be taken into account in the calculations. The optical thickness $\tau(\theta)$ for θ close to $\pi/2$ will be smaller than $\tau(0) \sec \theta = \tau(0)/\mu$ for a source at infinity and will not tend to infinity as $\mu \to 0$. Therefore, $J(\infty, \mu)$ will be larger in this case than the value given by (9'), in which $E(\mu) = \pi F \mu$ tends to zero as $\mu \to 0$ (πF is the flux outside the layer through an area normal to the beam). Rays which are exactly parallel to the surface of the layer ($\mu = 0$) can propagate through its rarefied part. On scattering, they also penetrate into the layer, while in this case according to (9') $E(0) = 0$.

All of these additional factors can be taken into account in the determination of E if we calculate the amount of radiation actually absorbed by particles situated on 1 cm^2 of the layer surface, more accurately, by particles lying within the cylinder of unit cross section whose axis is parallel to the z axis. Let us denote this quantity of singly absorbed (and scattered) radiation by E_1. In the case of a plane-parallel atmosphere illuminated by uniform parallel rays, the incident radiation E is everywhere the same as E_1. In our more complicated case, E_1 provides the best description of the illuminance of a given area of the layer surface. Since $\varphi(\mu) \approx 1$, substituting E_1 into (9) instead of $\int E(\mu) d\mu = E$, we obtain $J(\infty)$ and, according to (8), the temperature inside the layer. There will still be some inaccuracy associated with the nonuniformity of the layer and the radiation. The average radiation intensity $J(z_1)$ on the arbitrary layer boundary z_1 (see section 1) is governed not only by the local value of E_1 above this point, but also by its value at neighboring points since there is radiation mixing inside the layer. The average over the neighborhood of r deviates from the value at the point by a quantity of the order of r^2/R^2 in relative units. Qualitatively, this deviation is of the same character as that found in section (1) for the value of $J(0)$ at the central plane $z = 0$ relative to its value on the boundary, $J(z_1)$. It is true, however, that in the rarefied part of the layer r is greater than h, but the effect is small on the whole and leads to the flattening out of abrupt transitions.

Let us now estimate E_1. It is obvious that

$$E_1 = \int_{z_1}^{z_2} \alpha dz \int_{s_\odot} \frac{I'}{\pi R_\odot^2} e^{-\tau} ds, \quad (12)$$

where I' is the intensity of the solar radiation at the point (R, z) in the absence of absorption, $\alpha = \varkappa\rho$, the coefficient of absorption per unit volume, and τ, the optical thickness of the path from the element of area ds on the solar surface to the point (R, z). The integration with respect to z is in fact limited to a narrow band near the surface of the layer, but it is more convenient to adopt the limits 0 and ∞ for z. This will not alter the result.

For particles of the same size, $\varkappa$ is independent of z and $\alpha \propto \rho$. If, in addition, the kinetic temperature does not vary with z, then $\rho \propto e^{-z^2/h^2}$. However, in the case of particles of different sizes, the smaller ones will move towards the central plane more slowly than the larger ones. Then $\varkappa$ will increase with z (it is assumed that the transverse diameter of the particles is larger than the wavelength), while ρ decreases more slowly than e^{-z^2/h^2}. To simplify the calculation we can assume that $\varkappa\rho \sim e^{-z/h}$. As was mentioned above, the half-thickness of the layer h will be assumed to be proportional to R. Thus, we take

$$\varkappa\rho = \alpha = \alpha_0 e^{-z/h}, \quad h = \beta R. \quad (13)$$

The ray of light reaching the point (R, z) from the point (0, ζ) of the solar surface, situated at a distance ζ from the central plane of the layer, will have at a distance R' from the sun a z coordinate given by

$$z' = \zeta + \frac{R'}{R}(z - \zeta). \quad (14)$$

Since ζ and z are small by comparison with R, the distance between the points (0, ζ) and R, z) is practically equal to R. The optical thickness along the path between these points in view of (13) and (14) is given by

$$\tau(\zeta, z) = \int_0^R \alpha \, dR'$$

$$= e^{-z/h} e^{\zeta/h} \int_0^R \alpha_0 e^{-\zeta/\beta R'} dR' = \tau(\zeta, 0) e^{-z/h} \quad (15)$$

Substituting this value of $\tau(\zeta, z)$ into the integral in (12) and integrating the latter first of all with respect to z with constant ζ and R, taking $\tau(\zeta, z)$ as the independent variable, we obtain from (15)

$$e^{-z/h}\,dz = -\frac{h}{\tau(\zeta,0)}\,d\tau(\zeta,z).$$

Dropping the index on R', we then have

$$\int_{z_1}^{z_2} \alpha_0 e^{-z/h} e^{-\tau(\zeta,z)}\,dz \approx \frac{\alpha_0 h}{\tau(\zeta,0)} \int_0^{\infty} e^{-\tau}\,d\tau$$

$$= \frac{\alpha_0 h}{e^{\zeta/h}\int_0^R \alpha_0 e^{-\zeta/\beta R}\,dR} \tag{16}$$

Next, we introduce the variables u and v instead of R and ζ, where

$$v = \frac{\zeta}{R_\odot}, \qquad u = \frac{\zeta}{h} = \frac{R_\odot}{\beta R}\,v,$$

then

$$\frac{ds}{\pi R_\odot^2} = \frac{2\sqrt{R_\odot^2 - \zeta^2}\,d\zeta}{\pi R_\odot^2} = \frac{2}{\pi}\sqrt{1-v^2}\,dv,$$

$$dR = -R\frac{du}{u}$$

and the integral in (12) becomes

$$E_1 = \frac{2}{\pi} I'\alpha_0 h \int_0^1 \frac{\sqrt{1-v^2}\,dv}{e^u \int_u^{\infty} \alpha_0 R e^{-u}\frac{du}{u}}. \tag{17}$$

The intensity I' of solar radiation at a distance R outside the absorbing layer can be expressed in terms of the effective temperature of the sun, T_e, or in terms of the black-body temperature T' at a distance R from the sun outside the layer:

$$I' = \sigma T_e^4\left(\frac{R_\odot}{R}\right)^2 = 4\sigma T'^4$$

Substituting this value of I' into (17), we obtain on the basis of (8) and (9') the temperature below the surface of the dust layer as

$$T^4 = T'^4\frac{2\sqrt{3}}{\pi}\alpha_0 h \int_0^1 \frac{\sqrt{1-v^2}\,dv}{e^u \int_u^{\infty} \alpha_0 R e^{-u}\frac{du}{u}}. \tag{18}$$

The function $\alpha_0(R)$ is proportional to the density $\rho_0(R)$ of the dust layer at R = 0 if the dimensions of the dust particles are independent of R. It is of interest to consider the following simple cases for distances smaller than that to Jupiter:

1) Layer with constant surface density σ,

$$\sigma = \text{const}, \qquad \alpha_0 \sim R^{-1}.$$

Substituting this value of α_0 into (18), we obtain

$$T^4 = T'^4\frac{2\sqrt{3}}{\pi}\beta\int_0^1 \frac{\sqrt{1-v^2}\,dv}{e^{kv}\,\mathrm{Ei}(kv)}, \tag{19}$$

where $k = R_\odot/\beta R$ and Ei(u) is the tabulated function

$$\mathrm{Ei}(u) = \int_u^{\infty} e^{-u}\frac{du}{u}.$$

2) Layer with constant density in the plane z = 0,

$$\alpha_0 = \text{const}, \qquad \sigma \sim R.$$

In this case (18) becomes

$$T^4 = T'^4\frac{2\sqrt{3}}{\pi}\beta\int_0^1 \frac{\sqrt{1-v^2}\,dv}{1 - kve^{kv}\,\mathrm{Ei}(kv)}. \tag{20}$$

It can be seen from expression (18) that the temperature of the dust layer is independent of the absolute value of the density, but is a function of its gradient along R. T^4 is proportional to the number of scattering particles per unit volume at distance R, i.e., proportional to $\alpha_0(R)$; on the other hand, it decreases with increasing absorption of light along the path length R, i.e., with increasing average values of $\alpha(R)$. Therefore, the more rapid the decrease of density with R, the lower the temperature of the layer. As the thickness of the dust cloud decreases, the difference between the values of T obtained from (19) and (20) becomes less. Thus, with $\beta = 10^{-4}$, the difference is less than half a degree both at the distance of Jupiter and at that of the earth.

Beyond the Jovian zone, the density begins to decrease rapidly with R and the above equations are no longer applicable. For an approximate estimate of the temperature of the layer in the outer region we will adopt the following schematic density distribution which describes the true distribution sufficiently accurately and which leads to comparatively simple calculations:

$$\sigma = \text{const when } R \leqslant R_0, \quad \sigma \sim R^{-2} \text{ when } R > R_0, \tag{21}$$

here R_0 is the distance to Jupiter. Substituting $\alpha_0 \sim \sigma/R$ into (18), we find the following expression for the temperature of the layer for $R > R_0$:

$$T^4 = T'^4\left(\frac{R_0}{R}\right)^2\frac{2\sqrt{3}}{\pi}\beta$$

$$\times\int_0^1 \frac{\sqrt{1-v^2}\,dv}{e^{kv}\,\mathrm{Ei}(k_0v) + (k_0v)^{-2}\left[1 + kv - (1 + k_0v)\,e^{(k-k_0)v}\right]}. \tag{22}$$

TABLE 1.

Planetary zone \ β	10^{-1}	10^{-2}	10^{-3}	10^{-4}	T_0
Mercury	186	136	119	115	115
Venus	130	91	76	72	71
Earth	107	75	61	57	56
Mars	85	58	45	42	41
Jupiter	43	28	20	16.9	16.4
Saturn	23	14.8	10.5	9.5	10.3
Uranus	11.6	7.4	5.9	5.4	6.2
Neptune	7.3	5.0	4.2	3.9	4.6

Values of the temperature T found by a numerical integration of (19) and (22) for R equal to the distances of the planets from the sun and for various values of the parameter β are given in Table 1. In low-temperature regions stellar radiation becomes significant and its density has been taken equivalent to a heating to 3°K. The value $\beta = 10^{-4}$ corresponds to the density of the layer in the Jovian zone, $\rho_0 \approx 3M_\odot / 4\pi R^3$, i.e., to a state which is close to gravitational instability in this region. The value of the temperature T_0 calculated from (11) is given in the last column of the table.

It can be seen from Table 1 that with a constant surface density of the layer (up to the distance of Jupiter) $T \to T_0$ as $\beta \to 0$. However, if the layer is not very thin ($h > R_\odot$) its temperature is found to be considerably higher than T_0. The radiation density in the layer falls faster than R^{-2}, but slower than R^{-3}, as given by (10). As β increases from 0 to 10^{-1}, the exponent falls from 3 to 2.2. The rapid decrease in the surface density in the region of the major planets leads to a more rapid decrease of the temperature with R. However, even with very low values of β the temperatures in this region are also found to be much higher than those given in [1].

3. Radiation Scattered in the Gas Component of the Cloud

Above, we have examined the heating of the dust cloud by solar radiation scattered by particles of the layer itself. However, the dust layer is immersed in a gas cloud and is also heated by radiation scattered by the gas. The methods used above are also applicable in this case. It is true that molecular scattering is not isotropic, but this should not produce significant changes in the numerical results. Thus, the difference between the functions $\varphi(\mu)$ for conservative Rayleigh scattering and isotropic scattering is less than 3%. The thickness H of the gas component of the cloud is considerably larger than that of the dust layer. The magnitude of β for the gas is larger than 10^{-2} and according to Table 1 the temperature of the layer is raised. However, in the derivation of the basic equations it was assumed that the upper limit to the optical thickness $\tau(\zeta, z)$ along R in (16) is infinity. On the other hand, $\tau(\zeta, z)$ in the gas is considerably smaller than in the dust cloud and if it is small on the boundary of the layer z_1, then $\int_0^{\tau(\zeta, z_1)} e^{-\tau}\, d\tau$ will be smaller than unity. In this case, part of the solar radiation transmitted by the gas will be incident on the dust layer. Thus, for a double-layered dust and gas cloud, Eq. (17) will be replaced by

$$E_1 = E_{1g}\,[1 - e^{-\tau(\bar{\zeta}, z_1)}] + E_{1p}\, e^{-\tau(\bar{\zeta}, z_1)}, \qquad (23)$$

where E_{1g} and E_{1p} are of the form (17) for the gas and dust, respectively, and $\bar{\zeta} \approx R_\odot/2$. Since $h_p \leq h_g$, the second term is small and does not compensate the decrease of the first term when $\tau(\zeta, z_1)$ is small. It can be seen from (15) that $\tau(\zeta, z_1) = \tau(\zeta, 0)$ when $z_1 \ll h_g$. Therefore, the correction factor to be applied to (17) is

$$\xi \approx 1 - e^{-\tau(\bar{\zeta}, 0)}. \qquad (23')$$

In the case of Rayleigh scattering, the ratio of the amount of light scattered by one particle (atom or molecule) to the intensity of incident light is given by the expression [5]

$$\sigma_a = \frac{128\pi^5}{3\lambda^4}\left(\frac{n^2 - 1}{4\pi N}\right)^2, \qquad (24)$$

where N is the number of particles per cm^3 and $\underline{n}$ the refractive index. For the polarizability of molecular hydrogen Born [6] obtains $(n^2 - 1)/4\pi N = 8.2 \cdot 10^{-25}$. The same value is obtained if we take $n = 1.0001384$ at STP according to more recent data [5] and calculate the corresponding values of $N = p/kT$. Thus,

$$\varkappa = \frac{\alpha}{\rho} = \frac{\sigma_a N}{\rho} = 2.63 \cdot 10^{-5}\, \lambda^{-4}, \qquad (24')$$

where λ is expressed in microns. Equation (15) for the case σ = const gives

$$\tau(\zeta, 0) = \frac{\varkappa\sigma}{\beta}\, e^{\zeta/\beta R}\, \mathrm{Ei}\left(\frac{\zeta}{\beta R}\right). \qquad (25)$$

The magnitude of $\beta = H/2R$ is governed by the temperature of the gas. The density of the gas in the protoplanetary cloud was always much less than the critical value and its gravitational attraction in the $\underline{z}$ direction was small by comparison with that of the sun. In this case, if T is independent of $\underline{z}$, the density will decrease along $\underline{z}$ as e^{-z^2/h^2} and the uniform thickness of the cloud is [7]

TABLE 2

Planetary zone	Mercury	Venus	Earth	Mars	Jupiter	Saturn	Uranus	Neptune
T	145	100	84	67	35	18.4	9.3	6.1
β	0.019	0.021	0.023	0.024	0.033	0.033	0.033	0.033
$\tau\left(\frac{1}{2}R_{\odot}, 0\right)$	1.35	1.67	1.85	2.1	2.4			
ξ	0.74	0.81	0.84	0.88	0.91			

$$H = P\sqrt{\frac{\Re T}{2\pi\mu}}, \tag{26}$$

where P is the period of rotation around the sun, μ the molecular weight, $\Re$ the gas constant. At the boundary with the dust layer, the gas must have the same temperature as the dust particles, but at large z its temperature was probably considerably higher. As T(z) increases, the decrease in density along z will occur more slowly and the expression $e^{-z/h}$ adopted in our calculations will be qualitatively correct. Here, h is close to H/2, which is calculated according to (26) for the temperature T of the dust layer. Expression (26) and Table 1 give two relations between β and T and thi this allows us to determine both quantities. They are given in Table 2.

If the composition of the planetary cloud is assumed to be the same as that of the sun and its mass taken to be of the order of 0.04 $M_{\odot}$ (Kuiper [8] assumes that it is 0.1 $M_{\odot}$, while Hoyle [9] takes 0.01 $M_{\odot}$), then if the distribution of surface density of the gas, σ, is taken according to (21), we obtain $\sigma \approx 800$ for the region up to the distance of Jupiter. Substituting this value into (25) and taking β from Table 2 and $\bar{\zeta} \approx \frac{1}{2}R_{\odot}$, we find the value of $\tau(\bar{\zeta}, 0)$ given in Table 2. The correction ξ for $\lambda = 1\ \mu$ is given in the last row of the table. Since three quarters of the solar radiation is in the region $\lambda < 1\ \mu$, the correction $(\xi)^{1/4}$ to the temperature of the cloud is not large. It is not important for the region of the major planets and its maximum value (for the Jovian zone) is -4%. For the region of the terrestrial planets it is somewhat larger.

The results in Table 1 have been obtained on the assumption that $H \propto R$, i.e., β = const. It can be seen from Table 2 that this condition is satisfied in the region of the major planets where σ decreases rapidly with R and $T \propto R^{-1}$. In the region of the terrestrial planets, T decreases more slowly, approximately as $R^{0.6}$. Therefore $H \propto R^{1.2}$ and β increases with R. The departure from the condition β = const must lead to higher temperatures than those given in Table 1. However, since this correction is of opposite sign to correction (23'), we will not recalculate T and we will restrict ourselves to the values given in Table 1. Table 1 shows that the estimated temperature is not very sensitive to errors in β.

4. Condensation of Volatile Materials on Particles

A. I. Lebedinskii has shown [10] that solid particles can become heated on account of the random motion of massive protoplanetary bodies. These bodies aquire peculiar velocities as the result of gravitational interactions among themselves. Moving in the dust cloud, they are retarded and transfer energy to the dust particles, so that the latter can acquire temperatures of 5-30°K. On the basis of this result, the author concluded that hydrogen could not have been condensed on dust particles in the region of the major planets. It can be seen from Table 2 that during the earlier stages of the evolution of the protoplanetary cloud, when there were no protoplanetary bodies in it, the temperature of the dust layer was high, so that hydrogen could not condense on particles in the region up to the distance of Neptune. In fact, the condensation of hydrogen gas is related to its density (density of saturated vapor) by [11]

$$\log \rho = -\frac{47.02}{T} - \log T + 0.134 + 0.0363\,T. \tag{27}$$

The actual density of the gas in the central plane of the cloud is $\rho_0 = \sigma g/H$, where H is given by (26).

With the values of σ taken above, the condensation of hydrogen at the distance of Neptune is only possible when the temperature is below 4°K.

It is to be expected that in the presence of appreciable nonuniformities in the cloud screening of radiation must occur, so that the temperature of a region situated immediately behind such a nonuniformity will be lower than the average. However, this possibility, which can occur in a dust layer without a gaseous atmosphere, is excluded in the case of the protoplanetary gas and dust cloud. Here, the heating of the dust layer is mainly due to radiation scattered in the gas component of the cloud and it is difficult to accept the existence of appreciable nonuniformities in the latter.

Thus, hydrogen could enter into the composition of particles only in the form of CH_4, H_2O, NH_3, and other molecules. In the region of the major planets all of these should have been in the solid state. Jupiter and Saturn acquired their hydrogen mainly by accretion during the concluding stages of their growth when their masses became sufficiently large to retain the hydrogen.

Uranus and Neptune grew much more slowly in view of the appreciably lower density in their vicinity [7]. When their masses became sufficiently large for the accretion phase to be possible, there was already very little hydrogen around them. According to A. I. Lebedinskii [10], the gas moved into the zones of Saturn and Jupiter and was absorbed by these planets. F. Hoyle [9], however, considers that the hydrogen in the vicinity of Uranus and Neptune was dissipated into interstellar space. In order for this process to be efficient, the temperature of the gas near Uranus must be of the order of 75°K. However, these two possibilities in fact represent two aspects of one and the same process and cannot be separated from each other. In a rotating system, dissipation is accompanied by a change in the angular momentum. The dissipation of one part of the gas leads to the flow of the remaining gas towards the sun and, conversely, the movement of some of the gas from the outer zone towards the sun must be accompanied by the movement of the other part of the gas away from the sun. The distance from the sun at which the probability of dissipation is equal to the probability of reaching Saturn almost coincides with the distance of Neptune. Consequently, from larger distances hydrogen was mainly dissipated, while from smaller distances it was absorbed by Saturn. The residual gas stayed in the vicinity of Neptune. The dissipation of gas from the distance of Neptune requires a quantitative explanation since it cannot occur easily. Instead of the 75°K found by Hoyle as the critical temperature for dissipation of molecular hydrogen from the distance of Uranus, we find 45-50° for the critical temperature corresponding to the distance of Neptune. The decrease in the temperature required removes some of the difficulties encountered in the explanation of the dissipation of light gases from the solar system, although it is premature to consider that this problem can be solved on the basis of thermal dissipation alone.

LITERATURE CITED

1. L. É. Gurevich and A. I. Lebedinskii, Izv. AN SSSR, Ser. fizich., 14, 765 (1950).
2. V. A. Ambartsumyan, Astron. Zh., 19, 30 (1942).
3. V. V. Sobolev, Radiative Energy Transfer in the Atmospheres of Stars and Planets, Moscow, 1956.
4. S. Chandrasekhar, Radiative Energy Transfer [Russian translation[IIL, Moscow, 1953, p. 95.
5. C. W. Allen, Astrophysical Quantities, London, 1955, p. 87.
6. M. Born, Optics [Russian translation] 1937.
7. V. S. Safronov, Astron. Zh., 31, 499 (1954).
8. G. P. Kuiper, Astrophysics, New York, ed. J. A. Hynek, 1951, 8, p. 357.
9. F. Hoyle, Voprosy Kosmogonii, 7, 15 (1960).
10. A. I. Lebedinskii, Voprosy Kosmogonii, 7, 50 (1960).
11. H. C. Urey, Proc. chem. Soc., March, 67 (1958).

All abbreviations of periodicals in the above bibliography are letter-by-letter transliterations of the abbreviations as given in the original Russian journal. *Some or all of this periodical literature may well be available in English translation.* A complete list of the cover-to-cover English translations appears at the back of this issue.

On the Velocity Dispersion in Rotating Systems of Gravitating Bodies with Inelastic Collisions (abstract)

V. S. Safronov
Vopr. Kosmog. **8**, 168 (1962)

The generation of energy of random motions of gravitating bodies with equal masses m and radius r in the system with differential rotation is estimated by the same method as the heating of viscous medium with the velocity gradient. The expression for the energy output valid for rotating systems with great mean free path is deduced. The higher the relative velocities of the bodies, the more the inelastic collisions decrease their random motions. The condition of equality of generated and absorbed energy of random motions leads to the expression for the relative velocity of bodies showing dependence on m and r similar to that found by L. E. Gurevitch and A. I. Lebedinskii $[v^2 \propto (Gm)/r]$. The factor of proportionality is determined by the degree of inelasticity of collisions. It depends also on the parameter connected with the specificity of energy and momentum transfer in rotating systems with great mean free path. The numerical value of this parameter is not known exactly.

Comparative Analysis of the Internal Constitution and Development of Planets (abstract)

B. Yu. Levin
Mem. R. Soc. Sci. Liege [5] **7**, 39 (1963)

A comparative analysis of the internal constitution and evolution of planets gives important results for planetary physics as well as for geophysics. The most developed branches of such an analysis are (a) the comparison of mean densities of related groups of planets and satellites and (b) a comparative study of thermal histories of terrestrial planets and the Moon.

On the Gravitational Instability and Further Development of Perturbations (abstract)

V. S. Safronov
Vopr. Kosmog. **10**, 181 (1964)

Results of the linearized theory of propagation of small perturbations are critically discussed. Contrary to the conclusion of Simon infinite uniform medium perturbations with wavelength larger than the critical one do not lead to unlimited increase of density. The plane layer with a density $\approx 5\rho_0\lambda^2/\lambda_0^2$ is formed. Only after a repeated perturbation along the layer unlimited compression of the medium and its fragmentation into separate condensations are possible. It is pointed out that there exist no real systems to which the results of the theory of gravitational instability in infinite rotating medium could be applied. The criterion of gravitational instability for flattened rotating systems of finite dimensions is discussed. In linearized theory when the viscosity is not equal to zero (but may be extremely small) the rotation of the medium does not influence the condition of instability. At finite perturbations this result is not valid. A very small viscosity characteristic for the most of astronomical systems does not influence the condition of instability. Also contrary to the conclusion of the linearized theory, a small magnetic viscosity does not prevent the influence of the magnetic field on the gravitational instability.

SOVIET ASTRONOMY — AJ VOL. 10, NO. 4 JANUARY-FEBRUARY, 1967

THE PROTOPLANETARY CLOUD AND ITS EVOLUTION*

V. S. Safronov

Institute of Geophysics, Academy of Sciences of the USSR
Translated from Astronomicheskii Zhurnal, Vol. 43, No. 4, pp. 817-828, July-August, 1966
Original article submitted January 11, 1966

A critical survey is given of current views on the origin and evolution of the protoplanetary cloud. A calculation demonstrates the chief difficulty with Hoyle's hypothesis: solid particles separated from the central disk condensation will not move outward with the gas but will "spiral" inward, and hence cannot spread throughout the solar system. Mass and temperature estimates for the protoplanetary cloud are discussed; the evidence favors an initial mass of $\approx 0.05 M_\odot$.

1. Current Hypotheses on the Origin of the Cloud

It is generally accepted today that the planets were formed from material revolving about the sun in the form of an extended cloud of gas and dust covering the entire space now occupied by the solar system. A theory has been developed for the formation of the planets through gradual accumulation of solid particles into large bodies. Shmidt [1] has given an account of the principles of this theory. However, the problem of the origin of the cloud itself has not yet been solved. Shmidt himself advanced a hypothesis for capture of the cloud by the sun; other variations of this suggestion were worked out by Agekyan [2] and Radzievskii [3]. Alfvén, the founder of cosmic electrodynamics and magnetohydrodynamics, advanced a hypothesis [4] for capture of the cloud by the sun by means of its magnetic field. But no further development of the capture hypothesis has yet been undertaken, and most astronomers have continued to adhere to the classical Laplacian ideas regarding a simultaneous formation of the sun and the cloud, although a firm hypothesis for joint formation that would not conflict with the basic laws of physics has yet to be put forward. The first serious effort in this direction was not made until after Shmidt's death.

In 1948 Hoyle [5] advanced a hypothesis for joint formation of the sun and the cloud from a single rotating nebula. Having shown that no purely hydrodynamic mechanism would be capable of explaining the sun's slow rotation, Hoyle ascribed to magnetic forces the primary role in transferring angular momentum from the sun to the cloud. The protosolar nebula, with initial dimensions of the order of the distance to the nearest stars and an initial angular velocity similar to the angular rotational velocity $\approx 10^{-15}$ sec^{-1} of the Galaxy, would originally have been connected to interstellar clouds by the general galactic magnetic field. In the first stage of slow contraction it would have transferred most of its angular momentum to the interstellar clouds. Then, when the material had acquired the capability of moving easily across the lines of force, the connection would have been broken and free gravitational contraction would have begun, with conservation of angular momentum. When the nebula had contracted to the size of Mercury's orbit it would have become rotationally unstable. In the equatorial region of the nebula a disk (or ring) of mass $0.01 M_\odot$ would have been detached. Between the central condensation (the protosun) and the inner edge of the disk (the protoplanetary cloud) a strong magnetic "coupling" would immediately have been established, making their angular rotational velocities nearly equal and stemming any further flow of material into the disk. Upon acquiring angular momentum, the disk material would have begun to recede from the sun and to disseminate throughout the solar system, while the protosun,

*This paper and the next three were communicated to the Conference on Planetary Physics (see Soviet Astronomy — AJ, Vol. 10, No. 3, 1966).

having lost momentum, would have continued to contract further. The nonvolatile substances in the disk would rapidly have condensed into solid particles. The magnetic field would have exerted no influence on the particles, but they would have been carried along by the gas and would also have been dispersed over the whole solar system. The subsequent process of planet formation would have taken place by combining the solid particles into larger bodies. The rotational energy of the protosun should have been converted into magnetic energy. For this purpose the initial field (≈1 G at detachment of the disk) should have been intensified by a factor of 10^5; that is, 10^5 full turns would have been made because of the small remaining difference in the angular rotational velocity of the disk and the protosun. Only 20% of the energy would have been expended in repelling the disk material from the sun. The remaining 80% of the magnetic energy should have been dissipated inside the central condensation. Perhaps a considerable portion of it would have been dissipated in the form of nonthermodynamic activity at the surface of the protosun. The protosun would have maintained a sufficient degree of ionization, $n^+/n > 10^{-7}$, in the inner part of the disk for no appreciable decay of the magnetic field to have been permitted over the entire course of the process envisioned (≈10^7 years).

Hoyle's hypothesis was favorably received by astronomers and it acquired considerable popularity. But its defects gradually became apparent. In the first place, Hoyle considered that the magnetic field would have been capable of transmitting only to the inner part of the disk. He assumed that momentum transfer to the outer parts of the cloud was accomplished by means of turbulence. But there are serious objections against any possibility that turbulence would exist in a cloud with Keplerian rotation, since it has been shown that such a cloud would be stable against small perturbations. Hence if disordered motions orginally occurred there, they would soon have decayed [6].

Secondly, Hoyle selected the initial parameters for the cloud in the most economical fashion. The mass $0.01M_\odot$ for the cloud rested on the assumption that the composition of Jupiter and Saturn differed little from the solar composition. Other authors' estimates lead to a higher mass for the cloud (see below). In Hoyle's view, the solar nebula would have transferred about 99% of its initial angular momentum to the interstellar medium prior to transition to the gravitational-contraction stage. Cameron [7] considers this assumption unjustified. He gives arguments for a considerably smaller magnetic field in the Galaxy, $3 \cdot 10^{-6}$ G, which could not, in his opinion, have substantially retarded the rotation of the solar nebula. As a result Cameron arrives at a completely different picture for the evolution of the nebula; we shall examine it briefly below.

Thirdly, a major vagueness remains in Hoyle's suggested process for transferring solid matter by means of gas from the inner edge of the disk (from Mercury's distance) outward, and for scattering it over the whole solar system. Hoyle related the withdrawal of gas from the sun to the presence of a tangential (along the orbit) acceleration of the gas by the magnetic field. In Hoyle's opinion the orbital velocity of the gas should then have been greater than the circular velocity by an amount Δv of the same order as the radial velocity v_R. The magnetic field would not have affected the particles and they would have moved with circular velocity. Hence the gas should have imparted to them a tangential acceleration, under whose influence they would have receded from the sun. Hoyle found that the gas would effectively have carried along all bodies with cross sections $d < 1$ m. However, for a purely tangential acceleration f_φ the departure Δv of the gas velocity from the circular velocity would be a small second-order quantity in f_φ (see Eq. (23) of [8]). Correspondingly, the size of the greatest particles carried along by the gas would have been far smaller than the value found by Hoyle. But the force acting on the gas also has a radial component. Hoyle claimed that as the magnetic field was twisted in the disk the direction of the lines of force would gradually become tangential, so that the field would exert a pressure on the gas in an approximately radial direction. However, as Whipple [9] has pointed out, the radial acceleration f_R of the gas would weaken the attraction of the central mass and make the relative velocity of the gas lower than the circular velocity. Hence the gas would not accelerate but decelerate the particles, compelling them to approach the sun. This conclusion, which is so important for the hypothesis, merits closer attention.

If we neglect the pressure gradient of the gas and the viscosity, we may write the equations of axially symmetric motion in the form (see, for example, [10])

$$\frac{\partial v_R}{\partial t} + v_R\frac{\partial v_R}{\partial R} - \frac{v_\varphi^2}{R} = -\frac{GM}{R^2} + f_R, \tag{1}$$

$$\frac{\partial v_\varphi}{\partial t} + v_R\frac{\partial v_\varphi}{\partial R} + \frac{v_R v_\varphi}{R} = f_\varphi. \tag{2}$$

For stationary, nearly circular motion, we have

$$\frac{\partial v_R}{\partial t} = \frac{\partial v_\varphi}{\partial t} = 0, \quad v_\varphi = v_c + \Delta v, \quad v_c = \sqrt{\frac{GM}{R}}. \quad (3)$$

By Eq. (2),

$$v_R = \frac{f_\varphi}{\frac{\partial v_\varphi}{\partial R} + \frac{v_\varphi}{R}} = \frac{2f_\varphi}{\omega}\left[\frac{1}{1 + \frac{2\Delta v}{v_c} + \frac{2}{\omega}\frac{\partial \Delta v}{\partial R}}\right]. \quad (4)$$

By Eqs. (1) and (3),

$$v_\varphi^2 = v_c^2 + 2v_c\Delta v + \Delta v^2 = v_R R \frac{\partial v_R}{\partial R} + \frac{GM}{R} - Rf_R,$$

and

$$\Delta v = -\frac{f_R}{2\omega} + v_R \frac{R\frac{\partial v_R}{\partial R}}{2v_c} - \frac{\Delta v^2}{2v_c}. \quad (5)$$

Retaining only first-order terms, we have

$$v_R \approx \frac{2f_\varphi}{\omega}, \quad \Delta v \approx -\frac{f_R}{2\omega}. \quad (6)$$

Thus Δv will practically always be negative. The gas will move around the sun under the action of the magnetic field at a velocity lower than the Keplerian circular velocity.

The motion of the dust particles can be determined from the same considerations, except that the f_φ and f_R in Eqs. (6) should be replaced by the perturbing accelerations g_φ and g_R exerted on a particle by the gas:

$$g_\varphi = C\omega(\Delta v - \Delta v_p), \quad g_R = C\omega(v_R - v_{pR}),$$

where $C = 2\sigma_g/\pi r\delta$, r and δ are the radius and density of the particle, and σ_g is the surface density of the gas. Thus

$$v_{pR} = \frac{2g_\varphi}{\omega} = 2C(\Delta v - \Delta v_p),$$

$$\Delta v_p = -\frac{g_R}{2\omega} = -\frac{C}{2}(v_R - v_{pR}),$$

and

$$v_{pR} = \frac{C^2}{(1+C^2)\omega}\left(2f_\varphi - \frac{f_R}{C}\right),$$

$$\Delta v_p = -\frac{C^2}{(1+C^2)\omega}\left(\frac{f_\varphi}{C} + \frac{f_R}{2}\right).$$

Small particles, for which $C^2 \gg 1$ and $2Cf_\varphi \gg f_R$, will move practically together with the gas. Particles for which $2Cf_\varphi < f_R$ will approach the sun.

The corresponding condition for the particle size will take the form

$$r > r_0 = \frac{4\sigma_g}{\pi\delta}\cot\theta.$$

Directly after detachment of the disk the magnetic field will not yet be wound up; $\theta \approx 1$, and for $\sigma_g \approx 10^3$, r_0 will be of the order of a few meters. Practically all the particles will be carried along by the gas. But because of the weak field strength the recession from the sun will proceed very slowly. The field strength will grow as the field is twisted. We will have $\theta \to \pi/2$ and $\cot\theta \sim n^{-1}$, where n is the number of turns. When, in Hoyle's argument, n reaches 10^5, r_0 will have become smaller than 10^{-2} cm, so that practically all the particles will move toward the sun under the action of the gas. Meter-sized bodies will grow with roughly ten years. With so short a period the particles will only have been able to recede from the sun by a small fraction of the width of the planetary zone.

This result is a very critical one for Hoyle's hypothesis: if the protoplanetary cloud were detached from the sun at Mercury's distance, solid particles could never have receded from the sun to the distance of the other planets. In order to resolve the contradiction we must reexamine some of the essential propositions of the hypothesis.

Another variant of the joint-formation hypothesis has been proposed by Cameron [7]. According to Cameron the magnetic field did not contribute significantly either in the initial evolutionary stage of the protosolar nebula or in the collapse process. The local angular momentum would have been preserved, and there is no solution that could have yielded a central mass of $M_\odot$ at the end of the collapse (after ionization of H and He). Rotational instability would have set in for $R > 100$ a.u., and almost all the material of the nebula would have passed into the disk. Only after having been twisted in the disk would the magnetic field have transferred momentum outward from within, with the inner parts of the disk moving toward the center, forming the sun. Cameron computes the density distribution in the disk after collapse for two models: a) with polytropic index $n = 1.5$ and mass $4M_\odot$, and b) with $n = 3$ and mass $2M_\odot$. So large a mass for the disk itself raises severe difficulties. First, it is not at all clear how practically all the mass of gas not making up the sun would have been able to dissipate subsequently. Cameron's attempt to pursue an analogy with the mass flow from T Tauri stars at a rate of $1\,M_\odot$ every 10^5 years can hardly be considered satisfactory. Secondly, if one calculates the mass density in the central plane of the disk, then even for the high temperature adopted by Cameron, 1800°K, the value is found to be higher than the Roche density for both models — for the first, everywhere beyond Mercury's orbit, and for the second, inside Jupiter's orbit. This circumstance would make impossible the further evolution of the disk proposed by Cameron — formation of the sun from the inner parts of the disk, withdrawal of the outer parts, and formation of the planets from the remnant condensed solid matter. Instead, there would have formed within the disk thousands of gas condensations of solar composition, each with a mass of the order of Jupiter's mass, and with a combined mass many times as great as the total mass of the planets today. The question of disk evolution essentially has not been treated at all in Cameron's hypothesis.

We see from these considerations that the hypothesis of a joint formation of the sun and the protoplanetary cloud has already met with major difficulties. Its two outstanding variants are very far apart and each of them needs serious revision. In this connection we should mention that the work of Schatzman [11], who has suggested an efficient mechanism for momentum loss by the sun. Schatzman proposes that the electromagnetic activity on the surface of a rotating star results from an interaction between the magnetic field and the convective zone of the star. Material ejected by the star is carried along by its magnetic field and moves with the star's angular velocity out to the distance where $2\rho V\omega = H^2/4\pi R_c$, where ρ and V are the density and velocity of the ejected material, H is the magnetic-field strength, and R_c is the radius of curvature of the magnetic lines of force (it is assumed that $R_c \approx R$). The amount of material ejected is determined under the assumption that about 10^{-2} of the magnetic energy of the activity centers goes into the ejection. Mass and momentum loss through mass flow from the equator and ejection from active regions are also considered. The loss by ejection is small at first, but it increases steadily and becomes greater than the outflow loss. The flow from the equator then ceases, but the rotation continues to be retarded because of the continuing ejection of material from the active regions.

It is possible that the suggestions of Hoyle, Cameron, and Schatzman contain all the basic links in a contradiction-free hypothesis for the joint formation of the sun and the protoplanetary cloud. It would then merely remain to synthesize them, assigning each mechanism its proper place in the evolutionary process of the nebula. But for this purpose a detailed study would first have to be made of the structure of the magnetic field in a rotating, collapsing nebula, particularly at the rotational-instability state.

During the 1960's two other new capture hypotheses have been advanced. According to Lyttleton's hypothesis [12], the capture of matter by the sun would have occurred by means of the previously investigated mechanism of accretion from interstellar clouds having a density of 10^{-23} g/cm^3, a temperature of 3.18°K (an unrealistically low value), and an angular rotational velocity of $\approx 10^{-15}$ sec^{-1}. In order to capture the necessary mass with the necessary momentum, the cloud would have to have a relative velocity $\Delta v \leq 0.2$ km/sec. For the case of a Maxwellian distribution of cloud velocities about a mean value of about 10 km/sec, the fraction of clouds having a velocity ≤ 0.2 km/sec would be 10^{-5}. During the lifetime of the sun prior to its acquisition of the solar system ($\approx$1 billion years) it would have encountered interstellar clouds on about 10^2 occasions. Lyttleton thereby concludes that the probability of capture of protoplanetary material by the sun would have been 10^{-3}. But this estimate is far too high. If we adopt a peculiar velocity of 20 km/sec for the sun, the fraction of clouds having small velocities relative to the sun would be not 10^{-5}, but $4 \cdot 10^{-8}$. We must also recognize that the frequency of encounter is proportional to the value of the relative velocity. This introduces a correction factor of 0.02, so that in place of 10^{-3} we obtain 10^{-7}. Finally, the author does not observe that the mutual attraction of the sun and the cloud would increase their relative velocity during approach. Thus an encounter with $\Delta v \leq 0.2$ km/sec could take place only in exceptional cases (not within the scope of the two-body problem, but under more complicated conditions), so that actually the probability that the sun would have captured interstellar matter by the accretion mechanism is evidently considerably lower even than 10^{-7}.

Wolfson [13] has combined the capture hypothesis with Jeans' separation hypothesis: the sun, it is assumed, experienced a close approach to a star of low mass ($\approx M_\odot/7$) but of very large radius ($\approx$15 a.u.) and captured some of the material ejected from the tidal bulge of the star. His calculation shows that the distance of the star at perihelion should be equal to about three of its radii. A star of such large dimensions may be regarded as still in the gravitational-contraction stage. In this case as well, then, the low probability is a fundamental defect of the hypothesis. To be sure, for both hypotheses as well as for Shmidt's hypothesis, the capture probability would be considerably higher if one were to consider the idea of group formation of stars in large complexes of interstellar clouds [14], assume that the sun did not develop as a single star, and treat the capture problem for the formation stage of the sun itself, when it would still have been in close association with other stars (and nebulae) that had been formed together with the sun. Only after the conditions in the vicinity of the developing sun have been studied and the capture probability estimated will one be able to make a more definite judgment about hypotheses of this type.

2. Mass and Temperature of the Cloud

After the cloud had been formed, most of its mass was concentrated in the region of the major planets. We may evidently accept Hoyle's view that the role of magnetic interaction with the sun was insignificant in this region. As we have seen above, different authors have made very different assumptions about the mass of the cloud: from $10^{-2} M_\odot$ (Hoyle) to $\approx 1 M_\odot$ or more (Cameron [7], Fesenkov [15]). Actually, however, these limits can be narrowed considerably.

A lower limit on the mass of the cloud is usually obtained by diluting planetary material with volatile substances until the solar composition is reached. According to Whipple [9], for reduction to solar composition the mass of a planet of the terrestrial group would have to be increased by a factor of 500, the mass of Jupiter by 10, of Saturn by 30, of Uranus and Neptune by 75 times. This gives a minimum value of $0.028 M_\odot$ for the initial mass of the cloud. Approximately the same result is obtained from calculations of the hydrogen content in Jupiter and Saturn performed by Kozlovskaya [16] with B. I. Davydov's equation of state for hydrogen. During the terminal stage of the growth of a giant planet, its gravitational perturbations would have thrown other bodies beyond the limits of the solar system. Some of these bodies would have remained at the periphery of the system in the form of a comet cloud. Whipple adopts a mass of $0.003 M_\odot$ for the ejected material, with $0.031 M_\odot$ as the total mass of the cloud. As will become clear presently, this estimate should be regarded as a lower bound only. To estimate the mass of the cloud "from above," we may appeal to the following considerations.

a) The chemical composition of the planets indicate that they could not have been formed through the development of gravitational instability in the *gaseous* component of a cloud of solar composition, for no dissipation mechanism would be capable of ensuring a practically complete separation of hydrogen and helium from massive self-gravitating bodies; these constituents originally comprised 99% of the mass [17]. Thus the density ρ_0 in the central plane of the cloud would not have reached the critical density. For local condensations the critical density should be of the same order as the Roche density, that is, $\rho_{cr} \approx 10\rho^*$, where $\rho^* = 3M_\odot/4\pi R^3$. The writer has shown [18] that for the case of annular radial disturbances (such as those propagated from the sun) the value of the critical density would be substantially smaller: $\rho_{cr} \approx 2.1\rho^*$. Hence we may take the value of ρ^* as a reasonable upper limit on ρ_0. This value corresponds to an amount of solid matter in the cloud that is larger than that included in the planets today: for the zone of Jupiter 2.5 times larger, Saturn 6 times, Uranus 12 times, Neptune 6 times. The total mass of the cloud in the major-planet zone alone would then be $0.14 M_\odot$.

b) From a consideration of various mechanisms for dissipation of gas from the solar system, Kuiper [19] obtained $0.1 M_\odot$ for the maximum permissible value of the mass of the protoplanetary cloud. This would correspond to a value of ρ_0 less than ρ^*.

c) The large mass of the cloud implies a need for dissipation of a large amount of solid (not merely gaseous) material from the solar system. The mechanism of gravitational perturbations for ejecting bodies is more efficient than the gas-dissipation mechanism. But it also has more serious consequences, associated with the redistribution of angular momentum. An ejected body would increase in absolute velocity to the parabolic value. As the body approached a planet its relative-velocity vector would rotate without changing in value. The absolute velocity would increase if this rotation took place in the direction of the planet's orbital motion (the orbit may be considered circular). The angular momentum of the body relative to the sun would then increase because of the orbital momentum of the planet. Thus bodies would be ejected preferentially in the direction of motion of the planet. If the total mass of the ejected bodies were comparable with the mass of the planet, the planet would noticeably approach the sun. One could attempt to explain in this way the violation of Bode's law for Neptune: in order for the heliocentric distance of Neptune to decrease from 40 a.u. to the present value of 30 a.u. it would suffice for Neptune to have ejected beyond the solar system a mass of bodies equal to one third of its own mass. This shows how difficult it would be to suppose that by their own perturbations the giant planets had ejected from the solar system a mass comparable to or greater than their own mass. On the other hand, a cloud of solar composition with a mass of $0.14 M_\odot$ would have contained five times as much solid material as the giant planets (without hydrogen and helium). The small mass of Mars and the incompleteness of the process of combining bodies into a planet in the asteroid zone suggest that the distance of Jupiter from the sun at the time of its formation was not much greater than its present distance. Hence the mass of the ejected bodies apparently did not exceed $1/4$-$1/5$ of the mass of the major planets, or 10^2 earth masses. This value corresponds to a total mass of 0.05-$0.06 M_\odot$ for the cloud.

A consideration of the growth rate of the giant planets compels us to prefer the second of the two limiting values 0.03 and 0.05-0.06 $M_\odot$ given above for the mass of the cloud. In 1954 the author showed [20] that for a mean surface density of solid matter in the cloud corresponding to the present mass of the planets, Uranus and Neptune could have grown to their present dimensions only after a period two orders of magnitude longer than the age of the solar system. We can overcome this difficulty and achieve a hundred-fold acceleration in the growth of these planets only by increasing the density of matter in their zones (that is, by increasing the mass of the cloud) and decreasing the relative velocities of the bodies [21, 22]. As the mass of the giant planets grows their zones of supply combine into a single common zone. For a more or less uniform distribution over this zone of nonexhausted matter from the planets themselves and of an additional 10^2 earth masses of solid matter, the surface density in the zone of Neptune could have increased by an order of magnitude. To attain the required growth rate for Neptune, there would have to be another increase by one order in the efficiency of collisions of bodies with Neptune – a drop in the relative velocities of the bodies with a corresponding enhancement of the gravitational focussing. This factor might impose a definite restriction on the distribution function of the bodies with respect to size, a function with which the velocities are closely associated and which has not yet received sufficient study.

The information on one very important physical property of the cloud, its temperature, is quite scanty and inconsistent. Gurevich and Lebedinskii [23] have found that in a flat, homogeneous dust layer of constant thickness H, heated from the edge (along R) and radiating from the surface (in the z direction), the temperature will decline exponentially with the distance R, and will be extremely low for $R \gg H$. Urey [24] proposes that the cloud temperature in the giant-planet region was so low that a considerable portion of the free hydrogen was actually in the solid state. But Whipple [9] considers such low temperatures ($\approx 4°K$) unrealistic. Hoyle [5] assumed that efficient thermal dissipation of hydrogen and then helium took place from the zones of Uranus and Neptune, and found that for this purpose the gas temperature in the zone of Uranus would at first have been $\approx 75°K$, and after the dissipation of hydrogen, $\approx 150°K$.

We have made an estimate for the heating of the dust layer by solar radiation scattered by the gaseous component of the cloud to z coordinates considerably greater than the half-thickness of the dust layer [25]. In the computations the uniform thickness of the cloud (the scattering gas) was assumed proportional to the distance from the sun; this corresponds to a temperature function $\propto R^{-1}$ and is a satisfactory approximation. Table 1 gives the temperature for such a model dust layer, assuming that the sun had its present luminosity.

We find that at Jupiter and Saturn's distances from the sun the temperature would have been appreciably higher than the freezing point of hydrogen. Hydrogen could not have been frozen either in Uranus's zone, or probably even in Neptune's zone. But the volatile hydrogen compounds CH_4 and NH_3 would have been frozen in the giant-planet zone.

This calculation does not allow for the possibility that the outer parts of the cloud could have been screened by the substantially greater thickness of the dust layer in its inner portion. This might have happened, for example, if turbulence or convection were present in the inner part of the cloud and at the same time were absent in the giant-planet region. However, we have already pointed out that a cloud in Keplerian rotation would be stable relative to small perturbations. In order for convection to arise in the cloud a very high temperature gradient in R would be necessary, and this is completely unacceptable for a real cloud [6]. Only if a sharply bounded inner edge were present in the dust layer

TABLE 1

Planet zone	Mercury	Venus	Earth	Mars	Jupiter	Saturn	Uranus	Neptune
Temperature, °K	145	100	84	67	35	18.4	9.3	6.1

could a high T gradient be achieved in the transition region from the part of the cloud illuminated by the sun to the part in shadow, which would have been some 10^{-2} a.u. wide or less. But the existence of so sharp an inner boundary for the dust layer would seem improbable. In this region the Poynting-Robertson effect has a strong influence on dust particles. It acts differently upon particles of different sizes and different density. In the absence of gas, this behavior causes different particles to approach the sun with different velocities, and the inner boundary of the dust layer would have been strongly smeared out. The action of radiative deceleration on particles embedded in a gas is considerably more complicated than for particles in free motion about the sun. The result depends both on the particle size and on the total mass of the particles relative to the gas. Nevertheless, one would expect that in this case too no sharply defined dust-layer boundary would exist, so that the temperature gradient at the inner boundary of the cloud would not be high enough to support convection. The radiative deceleration of the particles would be transmitted to the gas and would induce an outflow from the boundary of the dust layer in the direction of the sun. The gas-pressure gradient dp/dR at the layer boundary would then increase, and might even become positive. In this event convection would not be possible at all. Some perturbations in the dust layer could arise under the action of powerful corpuscular streams ejected by active regions on the sun. It is not yet clear how efficient they might be in transporting solid particles to large z coordinates. Moreover, the activity of individual streams would be local and transient in character. Even if the streams served to screen out, on the average, half the radiation reaching the cloud, the temperature would be lowered only by a factor of $2^{1/4} \approx 1.2$. Thus the estimate we have given for the temperature in the dust layer appears to be correct, in order of magnitude.

3. Evolution of the Protoplanetary Cloud

Because of the stability of a cloud in Keplerian rotation against turbulence and convection [6], primeval disordered macroscopic gas motions would rapidly be damped out. The presence of a magnetic field could only enhance the stability of laminar motion. The succeeding evolution of the cloud would have consisted in separating the dust from the gas component. Very small particles would have settled fairly slowly to the central plane of the cloud: the z coordinate of a particle of constant radius 10^{-4} cm would decrease by a factor of e every $\approx 3 \cdot 10^5$ revolutions about the sun. But the agglomeration of particles would considerably accelerate the settling process. If while a particle is sinking, all the other particles (assumed stationary) that it encounters along the way adhere to it, such as by cold welding [26], then even for an initial radius $r_0 = 10^{-5}$ cm the settling time to the central plane for a particle would be shortened to 10^3 revolutions about the sun. The particle radius would increase to about 1 cm during the settling period.

Thus, with the decay of turbulence, solid particles would sink to the equatorial plane of the cloud, forming a flat dust layer of enhanced density.[1] When the density of the layer reached a critical value, gravitational instability would have arisen there and numerous condensations would have been formed. The fragmentation process of the dust layer into condensations was described almost concurrently by Edgeworth [27] and more fully by Gurevich and Lebedinskii [23]. The energy considerations presented in [23] show that the condensations would have been strongly flattened (spheroid axial ratio 13 : 1), and their mass would have been of order 10^{16} g at the earth's distance from the sun, and 10^{22} g at Jupiter's distance.

A treatment of gravitational instability in a plane rotating layer has shown [18] that for radial, axially symmetric perturbations along the layer the critical density would be several times smaller than the "Roche density" required for the existence of local self-gravitating condensations. Hence even before the Roche density is attained the dust layer could begin to break up into rings, segments of rings, or other structures elongated in one dimension, depending on the character of the perturbations. As these structures contracted their density would have risen and they would have broken up into individual local condensations.

The initial condensation of these agglomerations would have exceeded their density relative to the dust-layer density by only an order of magnitude. Their rotation would have impeded further contraction. But as the clumps combined they would

[1]We have in mind here the sequence of events occurring at a given distance from the sun. The duration of the stage of turbulence damping, settling of the dust layer, and formation of condensations in the layer would have been considerably shorter than the contraction time of the protosun ($\approx 10^7$ years, according to Fowler and Hoyle) and the formation time of the protoplanetary cloud. At different distances from the center of the system all these processes would have been completed at different times, and there would not have been a unified disk with simultaneous evolution in every part of the structure.

have become denser: after two individual condensations had joined with conservation of specific momentum (a central collision), the radius of the resulting condensation would have been reduced by a factor of two, and the density would have risen by 16 times. The entire system of condensations would have evolved rapidly into a body. To convert the condensations into compact bodies, it would on the average have been sufficient for their mass to have increased by 1.5-2 orders [8]. This process would have taken about 10^4 years at the earth's distance from the sun, and 10^6 years at Jupiter's distance. Thus a populous swarm would have developed, consisting of a large number of bodies, and through a complicated evolutionary process they would have accumulated to form planets. A description of this process falls beyond the scope of this paper.

We should emphasize that in order for critical density to be attained and gravitational instability to set in in the dust layer a very high degree of quiescence and flattening would be required. For $\rho_{cr} = 3\rho^*$, the relative particle velocities should be equal to $2 \cdot 10^2$ cm/sec at Jupiter's distance from the sun, 10 cm/sec at the earth's distance, and 0.5 cm/sec at Mercury's distance, while the thickness H of the layer should be 10^{-4} R, $2 \cdot 10^{-6}$ R, and $5 \cdot 10^{-8}$ R respectively. For $\rho_{cr} = 10\rho^*$, the particle velocities should be even smaller, by a factor of about $3^{1/2}$, and the thickness H three times smaller. Thus in the parts of the cloud close to the sun the conditions required for gravitational instability would have been exceptionally severe. But it is in these inner parts that macroscopic gas motions could have arisen, associated with solar activity and with the ejection of corpuscular streams, and could have exerted a significant disturbing influence on the dust layer, preventing its condensation to the state of gravitational instability. Thus we cannot exclude the possibility that in the zone of Mercury (and perhaps of Venus) there was no gravitational instability in the dust layer. Dust condensations would then not have formed in this region, but the particles would gradually have grown into bodies through aggregation during collisions. A consideration of the velocity dispersion of the bodies in the protoplanetary swarm shows [28] that until the size of the largest bodies exceeds several kilometers, the relative velocities of the bodies due to mutual gravitational perturbations would remain less than 1 m/sec. Collisions at these velocities should occur practically without fragmentation. In the inner zone, bodies with cross sections of several kilometers should accumulate within 10^5 years. As a result, the absence of gravitational instability could not have inhibited the formation process of large bodies and their evolution into planets.

It still remains uncertain then whether gravitational instability occurred in the portion of the dust layer close to the sun. If not, and if the bodies were formed through direct growth of the particles, then the question arises whether this circumstance would have led to any distinctions in the dynamical or physical and chemical properties of the planets formed in this zone.

Dissipation of gas from the solar system would have proceeded in parallel with the evolution of the dust component of the protoplanetary cloud. Since the earth grew in $\approx 10^8$ years [21] the absence of substantial amounts of hydrogen and helium on the earth would suggest that these gases evidently dissipated from the vicinity of the terrestrial-planet group in a period of roughly one-third of the age of the group, that is, $10^{7.5}$ years. Uranus and Neptune also do not contain sizable amounts of helium or free hydrogen. By the time that they had become sufficiently massive the gas had already dissipated from this region. The dissipation time probably did not exceed $10^{8.5}$ years. Jupiter and Saturn reached full growth considerably earlier than Uranus and Neptune. They consumed all the gas remaining in the vicinity of these planets (by accretion).

We have already pointed out that Hoyle explains the loss of gases from the vicinity of Uranus and Neptune by thermal dissipation. But this process would require a fairly high temperature. Gold [29] considers thermal dissipation an inefficient mechanism. But if the energy were nonuniformly distributed over the entire system, concentrated in small regions (in solar flares, in particle streams, perhaps even in external interstellar "winds"), then for dissipation of the same amount of gas the temperature could be considerably lower. Evidence against thermal dissipation of a large amount of gas from the solar system is also provided by a consideration of the redistribution of angular momentum, as given above in estimating the mass of solid matter ejected by the planets. Thermal gas dissipation, as well as the ejection of solid bodies, proceeds preferentially in the direction of rotation of the solar system. The remaining gas thereby loses its angular momentum, moves closer to the sun, and should be absorbed by Saturn and Jupiter. Hence in the absence of other dissipation mechanisms the thermal dissipation mechanism can serve to explain the loss of only a small mass of gas, no greater than the mass of Jupiter and Saturn, that is, $\approx 10^{-3} M_{\odot}$, or 2-3% of the required amount. Next to the problem of angular-momentum redistribution in the solar nebula, which has not yet been satisfactorily solved under current hypotheses for the origin of the protoplanetary cloud, the problem of gas dissipation from the cloud is the most difficult one.

Until recently the greatest difficulties in planetary cosmogony have arisen from the fact that to solve this complex problem only empirical data referring to the structure of the solar system have been applied. Now the situation has changed. During the 1940's Shmidt called attention to the importance of applying a variety of data from the earth sciences to cosmogony. In the 1950's Urey followed a new route and utilized for cosmogony a wealth of information from physicochemical research on the composition and structure of meteorites and terrestrial rocks. Urey independently concluded [24, 30] that during the epoch of meteorite formation the material of the planets consisted of solid bodies of asteroidal dimensions. Now in the 1960's one further new trend has appeared — a study of the nuclear evolution of protoplanetary material, enabling one to utilize the extensive data on the abundances and isotopic composition of different elements in meteorites, the earth, and the terrestrial atmosphere. The remarkable, if controversial, work of Fowler, Greenstein, and Hoyle [31] has shown how these data can be analyzed on the basis of a unified cosmogonical concept, and what valuable information they can yield for an understanding of the conditions prevailing during the evolution of the protoplanetary cloud. In particular, these authors have found that when the material of the earth and the meteorites was being bombarded by high-energy particles emitted by a magnetoactive sun in the process of formation, the material was in the form of small, cold, solid bodies. Finally, a real opportunity has opened up for using the abundant data of stellar cosmogony to interpret the origin of the protoplanetary cloud. The solution of this problem now depends mainly on progress in developing a theory for star formation.

Thus planetary cosmogony has been placed on the firm foundation of a variety of empirical data acquired from all related scientific fields.

LITERATURE CITED

1. O. Yu. Shmidt, Four Lectures on the Theory of the Earth's Origin [in Russian] (Moscow, Izd-vo AN SSSR, 1957), 3rd ed.
2. T. A. Agekyan, DAN SSSR, 69, 515 (1949); Uchen. Zap. Leningr. Gos. Univ., No. 136, 33 (1950).
3. V. V. Radzievskii, DAN SSSR, 72, 861 (1950).
4. H. Alfvén, On the Origin of the Solar System, Oxford Univ. Press (1954).
5. F. Hoyle, Voprosy Kosmogonii, 7, 15 (1960).
6. V. S. Safronov and E. L. Ruskol, DAN SSSR, 108, 413 (1956); Voprosy Kosmogonii, 5, 22 (1957).
7. A. G. W. Cameron, Icarus, 1, 13 (1962).
8. V. S. Safronov, Voprosy Kosmogonii, 7, 121 (1960).
9. F. L. Whipple, Proc. Natl. Acad. Sci., 52, 565 (1964).
10. L. D. Landau and E. M. Lifshits, Mechanics of Continuous Media [in Russian] (1933), p. 70.
11. E. Schatzman, Ann. Astrophys. Ser. B, 25, 18 (1962).
12. R. A. Lyttleton, Monthly Notices Roy. Astron. Soc., 122, 399 (1961).
13. M. M. Wolfson, Proc. Roy. Soc., A282, 485 (1964).
14. W. A. Fowler and F. Hoyle, Roy. Observ. Bull., No. 67, 301 (1963)
15. V. G. Fesenkov, Astron. zh., 28, 6, 492 (1951); The Origin and Evolution of Celestial Bodies according to Recent Research [in Russian] (Moscow-Leningrad, Izd-vo AN SSSR, 1953).
16. S. V. Kozlovskaya, DAN SSSR, 108, 409 (1956).
17. I. S. Shklovskii, Astron. zh., 28, 234 (1951).
18. V. S. Safronov, DAN SSSR, 130, 53 (1960) [Soviet Physics — Doklady, Vol. 5, p. 13].
19. G. P. Kuiper, Colloq. Intl. Astrophys., 4, Mem. Soc. Roy. Sci. Liège, (4)13, 361 (1953).
20. V. S. Safronov, Astron. zh., 31, 499 (1954).
21. V. S. Safronov, Voprosy Kosmogonii, 6, 63 (1958).
22. B. Yu. Levin, Voprosy Kosmogonii, 7, 55 (1960).
23. L. É. Gurevich and A. I. Lebedinskii, Izv. Akad. Nauk SSSR, Ser. fiz., 14, 765 (1950).
24. H. C. Urey, Proc. Amer. Chem. Soc. (March, 1958), p. 67; Space Science, New York-London (1963), p. 123.
25. V. S. Safronov, Astron. zh., 39, 278 (1962) [Soviet Astronomy — AJ, Vol. 6, p. 217].
26. B. Yu. Levin, Accumulation in the Solar Nebula, International Dictionary of Geophysics, Pergamon Press (1966).
27. K. E. Edgeworth, Monthly Notices Roy. Astron. Soc., 109, 600 (1949).
28. V. S. Safronov, Voprosy Kosmogonii, 8, 168 (1962).
29. T. Gold, Origin of the Solar System, Ed. by R. Jastrow and A. Cameron, Academic Press (1963), p. 171.
30. H. C. Urey, J. Geophys. Res., 61, 394 (1956); Astrophys. J., 124, 623 (1956); Geochim. Cosmochim. Acta, 26, 1 (1962).
31. W. A. Fowler, J. L. Greenstein, and F. Hoyle, Geophys. J. Roy. Astron. Soc., 6, 148 (1962).

SOVIET ASTRONOMY – AJ VOL. 15, NO. 5 MARCH-APRIL, 1972

MASS DISTRIBUTION OF PROTOPLANETARY BODIES

E. V. Zvyagina and V. S. Safronov

Institute of Terrestrial Physics, Academy of Sciences of the USSR
Translated from Astronomicheskii Zhurnal, Vol. 48, No. 5,
pp. 1023-1032, September-October, 1971
Original article submitted October 26, 1970

The mass distribution of protoplanetary bodies can be established by the general techniques of coagulation theory. The distribution function is usually described by an inverse power law $n(m) = cm^{-q}$. A qualitative study of the coagulation equation enables the sign of q to be determined for various ranges of variation in the mass m and for different forms of the coagulation coefficient as a function of m. If the bodies in the system suffer no fragmentation and the coagulation coefficient is proportional to the sum of the masses of the colliding bodies, then for all values of m except those close to the upper and lower limits of the n(m) distribution, the exponent will asymptotically approach the value $q_0 = 1.5$. This result agrees with the analytic solution obtained previously by one of the authors [4]. The distribution is found to be stable. For values of m approaching the upper limit M of the distribution, the exponent q_0 is sharply lower; thus the distribution function for $m \gtrsim 0.1$ M cannot be represented by a power law. As m approaches the lower limit of the distribution, q_0 diminishes only slightly (to 1.43). The coagulation coefficient in the protoplanetary cloud would have been a more complicated function of the masses of the colliding bodies. In the absence of fragmentation the characteristic expression for this coefficient for small bodies (as determined by their geometric cross section) implies that $q_0 = 1.54$. For large bodies, whose gravitational attraction significantly enlarges their collisional cross section, q_0 diminishes from 1.7 to 0.5 as the mass increases. Except for a small mass range including the very largest bodies, the mass distribution can be approximated by a power law with exponent $q_0 \approx 1.55 \pm 0.15$. The form of the asymptotic solution is independent of the form of the initial distribution function. Any distribution with $q < 2$ will approach an asymptotic law with exponent q_0 in the time required for the masses of the bodie. ɔ grow by several times.

1. INTRODUCTION

At some stage in the formation process of the planets, circumsolar space was filled with bodies having various masses. As they moved around the sun in different orbits the bodies would have collied with one another, and depending on their relative velocities they could either have amalgamated or have suffered partial or even complete disruption. The mass distribution of the bodies would then have continually been changing. So long as the masses of the largest bodies were small, their velocities would not have been great ($\ll$ 1 km/sec) [1], and they would have tended to consolidate through collisions. The range in mass would have extended toward larger bodies, and the number of bodies would have diminished. This process would have resulted in the formation and rapid growth of embryonic planets. As the masses of the protoplanets increased the relative velocities of the other bodies would have become larger, and collisions would often have culminated in fragmentation. All the small bodies of the solar system evidently have repeatedly suffered fragmentation. It is therefore difficult to tell from the observed distribution of the asteroids, meteorites, lunar craters, and other objects in the solar system what the mass distribution of the bodies was while the planets were being formed. Yet it is important to know this distribution if we wish to study the planetary accumulation process, estimate the primordial temperature of the planets, and identify primordial inhomogeneities in their interiors, which may have significantly influenced the subsequent evolution of the planets.

The process whereby bodies and particles grow through collisions without fragmentation is analogous to the processes considered in colloidal chemistry, and it can be studied by the general methods of coagulation theory. In this case the mass distribution function of the bodies will be determined by the standard coagulation equation, which describes the time variation in the average number n(m, t)dm of bodies per unit volume within the small mass interval (m, m + dm).

The coagulation equation is usually written in the form

$$\frac{\partial n(m,t)}{\partial t} = \frac{1}{2}\int_0^m A(m', m-m')n(m',t)n(m-m',t)\,dm'$$

$$- n(m,t)\int_0^\infty A(m,m')n(m',t)\,dm', \qquad (1)$$

where A(m, m') is called the coagulation coefficient. The quantity A(m, m')n(m, t)n(m', t)dm dm' represents the number of collisions per unit volume and unit time between bodies with masses in the interval (m, m + dm) and bodies with masses in the interval (m', m' + dm'). This equation assumes the condition of conservation of total mass per unit volume,

$$\rho = \int_0^\infty mn(m,t)\,dm = \text{const},$$

and the condition that the coagulation coefficient A(m, m') is independent of the coordinates of the bodies m and m'. The relative velocities of the bodies in the protoplanetary cloud (and thereby the coagulation coefficient) are determined by the mass of the embryonic planets. Thus A(m, m') and n(m, t) would not have varied over the supply zone of a given protoplanet, but they might have varied from zone to zone. If n(m, t)dm ≫ 1, it would be close to the number of bodies per unit volume in the mass interval dm, but if n(m, t)dm is small it would merely represent the mathematical expectation of that number.

If the function n(m, t) were known, the mean expected values for the mass m_k of the bodies (in decreasing order: m_0, m_1, ...) contained in some volume V could be determined from the condition

$$m_k(t) = V\int_{M_{k+1}}^{M_k} mn(m,t)\,dm, \qquad (2)$$

where $M_0 = \infty$, and the other limits of integration M_k are so chosen that for all values k = 0, 1, 2, ..., the mathematical expectation of the number of bodies in the interval (M_{k+1}, M_k) within the given volume V will be unity:

$$V\int_{M_{k+1}}^{M_k} n(m,t)\,dm = 1. \qquad (3)$$

If the increments ∂n in Eq. (1) represent infinitesimal rather than integer increments Δn, the description of the mass growth process of the bodies will be distorted. The distortion will increase with the interval $\Delta m = M_k - M_{k+1}$ containing a single body. In fact, the probability P_k for collision[1] between a body of mass m and a body of mass m_k in unit time is equal to A(m, m_k). But if the mass m_k of the body is distributed continuously over the interval (M_{k+1}, M_k) in accordance with Eqs. (2) and (3), then the probability $\bar{P}_k$ for collision between m and this "distributed" body m_k will be

$$\bar{P}_k = \overline{A(m,m_k)} = V\int_{M_{k+1}}^{M_k} A(m,m')n(m')\,dm'. \qquad (4)$$

For a coagulation coefficient proportional to the sum of the p-th powers of the masses of the interacting bodies [A(m, m') ∝ $m^p + m'^p$], the ratio of the probabilities will be

$$\chi = \frac{\bar{P}_k}{P_k} = \frac{m^p + \overline{m_k^p}}{m^p + m_k^p}$$

where $\overline{m_k^p}$ is the mean value of the p-th power of the mass of the k-th body in the mass interval (M_{k+1}, M_k). If m ≪ m_k, then

$$\chi \approx \overline{m_k^p}/m_k^p.$$

Only if p = 1 (a coagulation coefficient proportional to the sum of the masses of the colliding bodies) will χ = 1. For p > 1 the probability ratio χ > 1;

[1] It would be more accurate to consider not the collision probability but the mathematical expectation of the collision frequency. However, if the unit of time is taken to be short enough and if $P_k \ll 1$, the two concepts will coincide.

TABLE 1

q	p			
	4/3	1	2/3	1/3
3/2	1.04	1	0.98	0.98
5/3	1.07	1	0.98	0.97
11/6	1.08	1	0.97	0.96

Eq. (1) will then yield too large a falloff in the number of bodies of mass m through interaction with large bodies. For $p < 1$ the opposite situation will obtain. For example, with a power-law mass distribution $n(m) = cm^{-q}$, we will have for $k = 0$ (the largest body) and $M = \rho V$

$$\chi = \frac{2-q}{1+p-q}\cdot\frac{1-(2-q)^{(p-q+1)/(q-1)}}{[1-(2-q)^{(2-q)/(q-1)}]^{p}}.$$

Table 1 gives the values of χ for selected p and q. For large bodies $p \approx {}^4/_3$. In the range of q values that is of practical interest, χ differs from unity by less than 10%. Only if q is very close to two will the result be appreciably distorted by using a continuous function n(m) in the range of large m values, where the number of bodies is small.

Equations of the type (1) can be solved analytically only for a few special forms of the coagulation coefficient (Smolukhovskii [2], Melzak [3], Safronov [4]). For this reason one usually seeks approximate solutions for such equations [5-7]. But the approximations are suitable only over short time intervals. A more satisfactory approximation for describing the accumulation process of protoplanetary bodies can be obtained by taking the coagulation coefficient to be proportional to the sum of the masses of the colliding bodies:

$$A(m, m') = a(m + m'). \tag{5}$$

If the initial mass distribution is described over the entire infinite range of m variation by the exponential function

$$n(m, 0) = c(0)e^{-b(0)m}, \tag{6}$$

the solution of Eq. (1) will take the form [4, 8]

$$n(m,\tau) = \frac{N_0(1-\tau)}{m\sqrt{\tau}}\, e^{-(1+\tau)bm} I_1(2bm\sqrt{\tau}), \tag{7}$$

where $\tau = 1 - e^{-apt}$ and $I_1(x)$ is the modified Bessel function.

For $x \gg 1$, $I_1(x) \approx e^x(2\pi x)^{-1/2}$; thus for $m \gg 1/(2b\sqrt{\tau})$, or practically throughout the entire range of mass variation (except for very small m), the function n(m, t) will represent the product of a power function by an exponential:

$$n(m, t) \approx c(t)m^{-3/2}e^{-(1-\sqrt{\tau})^2 bm}. \tag{8}$$

Its departure from an inverse power function will increase with the value of m, and for given m the departure will be greater when earlier times are considered. The exponent $-{}^3/_2$ is independent of time.

On the basis of these results Marcus [9, 10] has endeavored to use an inverse power law to describe the distribution of bodies in the protoplanetary cloud:

$$n(m) = cm^{-q}. \tag{9}$$

However, as we have shown elsewhere [11], absurd conclusions are obtained if this law is applied to the largest bodies, including the protoplanets themselves. Attempts to use a power law are to a large extent justified by the simplicity of operating with power functions in theoretical calculations. In this paper we shall examine the possibility of approximating the distribution of protoplanetary bodies by power functions having different values for the exponent q.

Piotrowsky [12] and one of the present authors (Safronov [1]) have begun a qualitative investigation of the coagulation equation. It enables one to follow the changes in the exponent q with time for different intervals in the mass m and for coagulation coefficients of different types, and to study the character of the mass distribution of the bodies during the accumulation process of the planets.

2. THE DISTRIBUTION FUNCTION IN THE MASS INTERVAL $m_m \ll m \ll M$

It is natural to test the validity of our proposed method for qualitatively investigating the coagulation equation by first applying it to the case for which the analytic solution (7) has already been found. For a coagulation coefficient proportional to the sum of the masses of the colliding bodies, $A(m, m') = a(m + m')$, the coagulation equation may conveniently be written in the form

$$\frac{\dot{n}}{an} = m\int_0^{m/2} n(m')\frac{n(m-m')}{n(m)}\,dm' - m\int_0^{\infty} n(m')\,dm' - \rho. \tag{10}$$

Let us assume that at initial time t_0 the mass distribution of the bodies is described by an inverse power law in a restricted mass interval:

$$n(m, t_0) = \begin{cases} c(t_0)\, m^{-q(t_0)} & 0 < m_m < m < M \\ 0 & m \leqslant m_m \text{ and } m \geqslant M, \end{cases} \tag{9'}$$

where M has some finite value. For convenience one usually takes $M = \infty$, but the passage to an infinite limit is not correct. Although for $1 < q < 2$ it would hardly affect the number of large bodies, their probable mass would become infinite. The last integral in Eq. (1) would diverge, and we would obtain $\rho = \infty$ in Eq. (10). When $q > 2$ the value $m_m = 0$ would also lead to divergent integrals.

According to Eqs. (10) and (9'), the relative increment in the function $n(m, t_0)$ per unit time for arbitrary m in the interval $2m_m < m < M$ will be

$$\frac{\dot n}{n} = ac\left[m^{2-q}\int_\varepsilon^{1/2} x^{-q}(1-x)^{-q}dx - m\frac{M^{1-q} - m_m^{1-q}}{1-q} - \frac{\rho}{c}\right] \tag{11}$$

where $\varepsilon = m_m/m \leq 1/2$ is a small quantity.

If the power law $n = cm^{-q}$ remains in force as time passes, with parameters c and q depending only on time, not on m, then the relation

$$\left[\frac{\dot n}{n}\right]_1 = \frac{\dot c}{c} - \dot q \ln m \tag{12}$$

should be satisfied. The mass enters only logarithmically into this expression, and for $\dot q \neq 0$ it cannot be reconciled with Eq. (11), which was derived from the coagulation equation. Thus the power law breaks down for $t > t_0$. For the new mass distribution of the bodies, as determined by the coagulation equation, for $t = t_0 + dt$ the parameters c(t) and q(t) are found to differ for different m. The direction and rate of variation of c and q can be determined at any point m from the conditions that the curves (11) and (12) representing the function $\dot n/n = f(m)$ at that point should have the common tangent

$$\frac{\dot n}{n} = \left[\frac{\dot n}{n}\right]_1,$$
$$\frac{\partial}{\partial m}\left(\frac{\dot n}{n}\right) = \frac{\partial}{\partial m}\left[\frac{\dot n}{n}\right]_1. \tag{13}$$

The first of these conditions determines the relative variation in c with time at the point m as a function of $\dot q$:

$$\frac{\dot c}{c} = \dot q \ln m + ac\left[m^{2-q}\int_\varepsilon^{1/2} x^{-q}(1-x)^{-q}dx - \frac{m(M^{1-q} - m_m^{1-q})}{1-q} - \frac{\rho}{c}\right]. \tag{14}$$

The second condition determines the variation in the quantity q:

$$\frac{\dot q}{acm^{2-q}} = (q-2)\int_\varepsilon^{1/2} x^{-q}(1-x)^{-q}dx - \varepsilon^{1-q}\left[(1-\varepsilon)^{-q} + \frac{1}{1-q}\right] + \frac{E^{1-q}}{1-q} \tag{15}$$

where $E = M/m \geq 1$. Evidently, then, the sign of $\dot q$ depends in an essential way on the initial value of q, on ε, and on E. To evaluate $\dot q$ we shall expand the integral and the second term of Eq. (15) in powers of ε. After some straightforward operations we obtain

$$\frac{\dot q}{acm^{2-q}} = 2^{q-1}(q-2)S_1(q) - \varepsilon^{3-q}q(q+1)S_2(q,\varepsilon) + \frac{E^{1-q}}{1-q}, \tag{15'}$$

where $S_1(q)$, $S_2(q, \varepsilon)$ are the convergent series

$$S_1(q) = \frac{1}{1-q} + \sum_{n=1}^{\infty}\frac{q(q+1)(q+2)\ldots(q+n-1)}{n!(1+n-q)2^n}$$
$$S_2(q,\varepsilon) = \frac{1}{2(3-q)} + \sum_{n=1}^{\infty}\frac{(n+1)(q+2)(q+3)\ldots(q+n+1)}{(3+n-q)(n+2)!}\varepsilon^n \tag{16}$$

Let us first consider the range of values of m satisfying the condition

$$m_m \ll m \ll M \tag{17}$$

so that

$$\varepsilon \ll 1, \quad E \gg 1. \tag{17'}$$

Then for $q < 1$ the last term in Eq. (15) will predominate, and $\dot q > 0$. For $q > 3$ the second term will predominate, and we will again have $\dot q > 0$ because of the negative first term in $S_2(q, \varepsilon)$. In the range $1 < q < 3$ the main term is the first one.

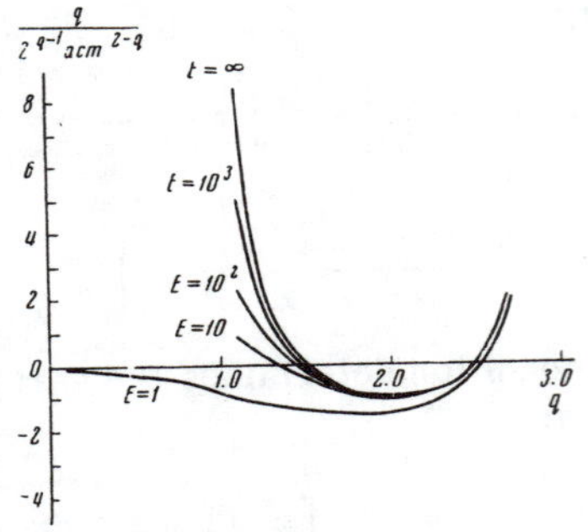

Fig. 1. Dependence of $\dot{q}$ on q for the case $\varepsilon = m_m/m = 0$.

In the limiting case

$$\varepsilon = 0 \text{ and } E = \infty \tag{18}$$

a case which is not actually reached, the quantity $\dot{q}$ will vanish for the values $q_0 = 1.5$ and $q_0 = 2.5$, the roots of the equation $(q - 2)S_1(q) = 0$. The value $q = 2$ is not a root because of the factor $2 - q$ in the denominator of the second term in $S_1(q)$. The upper curve in Fig. 1 shows that in this case $\dot{q} < 0$ for $1.5 < q < 2.5$. Thus if we initially have $q < 2.5$ the variation will take place in the direction of the root $q_0 = 1.5$. But if initially $q > 2.5$, then q will increase with time. It does not follow, of course, that in this case q can in fact become arbitrarily large, because our analysis remains valid only for a mass distribution approximated by a power law, and does not hold if the distribution departs strongly from such a law. For $q = q_0$ we will have $\dot{q} = 0$ for all m, so that q will remain constant. Hence the distribution function formally described by the inverse power law $n(m) = cm^{-q_0}$ will be an asymptotic solution of the coagulation equation throughout the infinite mass interval $0 < m < \infty$. The value $q_0 = 1.5$ yields a stable solution, since q will approach q_0 for all $q < 2.5$. The power law with exponent $q_0 = 2.5$ is an unstable solution: a small departure of q from q_0 arising accidentally will increase with time.

For $1 < q < 2.5$ the time required for q to approach $q_0 = 1.5$ will be

$$\Delta t = \int_q \frac{dq}{\dot{q}} = \int_q \frac{dq}{acm^{2-q}(q-2)2^{q-1}S_1(q)}\,.$$

This integral diverges at the upper limit of integration, so that q should approach $q_0 = 1.5$ asymptotically over an infinitely long time interval.

For the case where the distribution function is specified on an infinite interval of mass variation [Eqs. (18)] and where it has the form (6) at initial time, the analytic solution (7) of the coagulation equation is known for all epochs. This solution asymptotically approaches the power function with exponent $q = 1.5$; $\tau \to 1$ as $t \to \infty$, and according to Eq. (8), n(m) approaches $m^{-3/2}$ throughout the entire range of masses.

We have found above that $q \to 1.5$ for a wide range of initial values of q. This circumstance implies that the form of the asymptotic solution does not depend on the form of the initial distribution function. (The case $q > 2.5$ will be considered in the next section).

In the more general case where the limits m_m and M are finite, there exists no value of q_0 such that $\dot{q}$ would approach zero in Eq. (15') for all values of m simultaneously. But in the range of m values satisfying the condition (17) the equation $\dot{q} = 0$ holds for values of $q_0(m)$ that differ little for different m.

Although in this case the power law (9) with exponent q_0 is not actually an exact solution of the coagulation equation, it may nevertheless be regarded as an approximate expression for an asymptotic solution of the equation.

It is of interest to compare the rate of change in q with the mass growth rate of the bodies. The increase in the mass m due to aggregation of smaller-sized bodies is given by

$$\dot{m} = \int_{m_m}^{m} A(m, m')n(m')m'dm'$$
$$= acm^{3-q}\left[\frac{1-\varepsilon^{2-q}}{2-q} + \frac{1-\varepsilon^{3-q}}{3-q}\right]. \tag{19}$$

As the characteristic time scale for approach of q to q_0 we may take the quantity $\tau_0 = (q_0 - q)/\dot{q}$, and as the characteristic time for growth of m, the quantity $\tau_m = m/\dot{m}$. For $\varepsilon \ll 1$ and $E \gg 1$ the ratio of these time scales is

$$\frac{\tau_q}{\tau_m} = \begin{cases} \dfrac{(q_0 - q)(5-2q)(1-q)}{(2-q)(3-q)} E^{q-1}, & \text{if } q < 1, \\[2ex] \dfrac{(5-2q)(q-q_0)\,2^{1-q}}{(3-q)(2-q)^2 S_1(q)}, & \text{if } 1 < q < 2, \\[2ex] \dfrac{(q_0 - q)\,2^{1-q}}{(2-q)^2 S_1(q)} \varepsilon^{2-q}, & \text{if } q > 2. \end{cases} \tag{20}$$

TABLE 2

ε	q			
	2.1	2.2	2.3	2.4
10^{-1}	3.5	2.6	2.9	4.8
10^{-2}	4.4	4.1	5.8	12.0
10^{-3}	5.6	6.5	11.6	30.0

TABLE 3

E	q			
	0.2	0.4	0.6	0.8
10	0.15	0.17	0.16	0.11
10^2	0.02	0.04	0.06	0.07
10^3	0.00	0.01	0.02	0.05

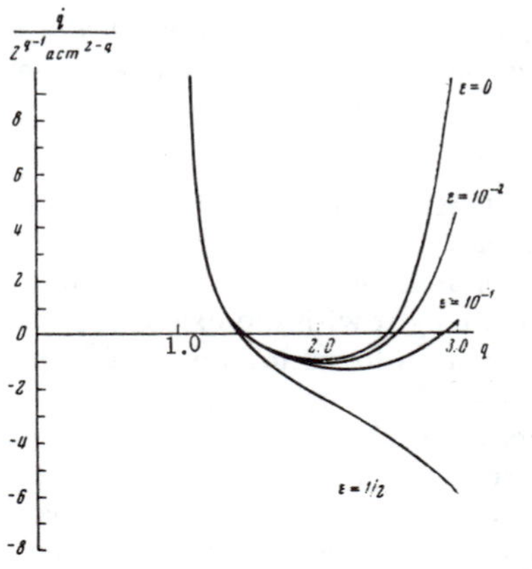

Fig. 2. Dependence of $\dot{q}$ on q for the case $E = M/m = \infty$.

If $q > 2$ the ratio τ_q/τ_m will depend strongly on ε; the masses of the bodies will experience considerable change before q approaches a stable value ($\tau_q \gg \tau_m$). But for smaller bodies (with $m_m/m = \varepsilon \gtrsim 10^{-2}$) the time required for variation in q will be comparable to the mass growth time of the bodies. Table 2 gives the values of τ_q/τ_m for several q and ε. For initial values in the range $1 < q < 2$, the ratio τ_q/τ_m is nearly independent of the mass m of the bodies, and the time τ_q is comparable to the characteristic mass growth time τ_m of the bodies:

q	1.2	1.4	1.6	1.8
τ_q/τ_m	0.13	0.31	0.52	1.2

If $q < 1$, the ratio τ_q/τ_m will be even smaller; it is given in Table 3 for selected q and E.

We have found, then, that for any initial value less than two, q will approach $q_0 = 1.5$ within a time during which the masses of the bodies will change by less than a factor of two to three. If q is initially in the range $2 < q < 2.5$, the approach to q_0 will proceed considerably more slowly.

3. MASS DISTRIBUTION OF THE LARGEST AND SMALLEST BODIES

Let us consider values of m for which the conditions (17) are not satisfied. For large bodies with $m \approx M$, Eq. (15') indicates that if $q < 3$ the quantity $\dot{q}$ will vanish provided that

$$2^{q-1}(q-2)S_1(q)+\frac{F^{1-q}}{1-q}=0. \qquad (21)$$

It is apparent from Fig. 1 that the roots of this equation are values $q_0(E) \leq q_0 = 1.5$, which determine a stable solution of the coagulation equation, and values $q_0(E) \geq 2.5$, which determine an unstable solution. The roots yielding the stable solution are as follows:

$\lg E$	0	0.2	0.5	1.0	1.5	2.0	3.0	4.0
$q_0(E)$	0	0.63	1.10	1.32	1.42	1.46	1.49	1.50

The larger the body, the smaller the local value of the exponent q in the asymptotic solution. Accordingly, an inverse power law with exponent $q_0 \approx 1.5$ would not be suitable for describing the mass distribution of the largest bodies with $m \gtrsim 10^{-1}$ M.

In the range of small values of m, the root of $q_0(\varepsilon)$ yielding a stable solution shifts slightly toward the left as ε increases, reaching a value of 1.43 for the smallest bodies (see Fig. 2). The root corresponding to the unstable solution, on the other hand, shifts strongly toward the right. Hence if $q > 2.5$, then for the smallest bodies $\dot{q} < 0$ and $q \to q_0(\varepsilon)$. In particular, for $q = 2.58$ this will be the case for bodies with $m < 10^2 m_m$, and for $q = 2.9$, bodies with $m < 10\, m_m$. This is a highly important property of the distribution function for small m, since if $q > 2$ the small-sized bodies will contain the bulk of the mass. For example, if $q = 2.9$ the mass of the bodies for which q is increasing will comprise less than 13% of the mass of all the bodies, and if $q = 2.58$ it will be less than 7% of the mass. Thus even for initial values $q > 2.5$, the local value of q for the bulk of the mass of the bodies will approach a value $q_0(\varepsilon)$ differing little from $q_0 = 1.5$.

We conclude that in a system of bodies not subject to fragmentation, a stable mass distribution will be established quite rapidly (after an increase by several times in the mass of the bodies), and it can be described approximately by an inverse power law with an exponent $q \approx q_0$ for all $m \leqslant 0.1$ M. The distribution of the largest bodies deviates considerably from a power law.

4. ASYMPTOTIC SOLUTIONS FOR OTHER FORMS OF THE COAGULATION COEFFICIENT

We can verify that the method described above for seeking an asymptotic solution to Eq. (1) is correct by comparing it with the exact analytic solution of the equation found for the case of a coagulation coefficient of the form (5). Hence we have every reason to apply our method to an equation with a different coagulation coefficient, for which no exact solution is available. If an aggregation of bodies takes place in every collision, then allowing for their gravitation [1] we may write

$$A(m, m') = \pi \bar{V} (r + r')^2 \left[1 + \frac{2G(m + m')}{\bar{V}^2 (r + r')}\right], \quad (22)$$

where $\bar{V}$ is the mean velocity of the bodies of mass m' relative to the bodies of mass m away from the epochs of encounter. The velocity $\bar{V}$ depends weakly on the mass of the bodies [1] and may be regarded as approximately constant. For small bodies the second term in brackets, which arises from the gravitational attraction of the bodies, will be small, and the coagulation coefficient will be determined by their geometrical cross sections:

$$A \approx A_1 \propto (m^{1/3} + m'^{1/3})^2. \quad (23)$$

For large bodies, on the other hand, the second term will predominate, and

$$A \approx A_2 \propto (m + m')(m^{1/3} + m'^{1/3}). \quad (24)$$

The expression $A \propto m + m'$ adopted above represents the simplest dependence of A on m and m', and is intermediate between the expressions (23) and (24) for small and large bodies.

According to Eqs. (12) and (13), the exponent q in the required power-law asymptotic solution of Eq. (1) should be a root of the equation

$$\frac{\partial}{\partial m}\left(\frac{\dot{n}}{n}\right) = -\frac{\dot{q}}{m} = 0, \quad (25)$$

where $\dot{n}/n$ is determined by the coagulation equation. Table 4 presents the values of q_0 found in this manner, corresponding to stable solutions and computed for several forms of the coagulation coefficient in the limiting case $\varepsilon = 0$, $E = \infty$.

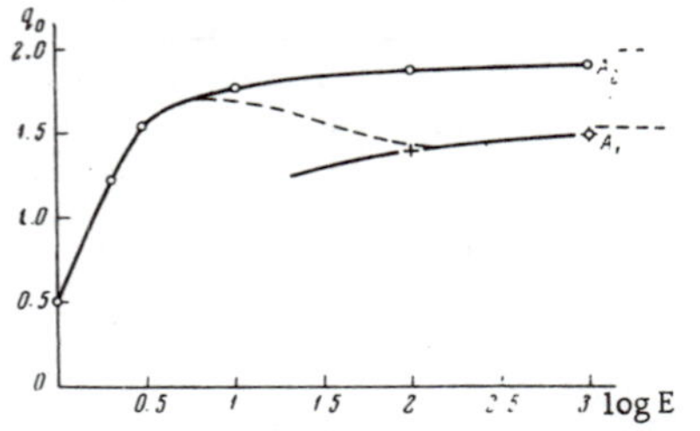

Fig. 3. Dependence of q_0 on E = M/m for the coagulation coefficients A_1 and A_2. The dashed lines on the right represent the limiting values of q_0 as $E \to \infty$. The curve A_1 gives the values of q_0 for small bodies (large E) the curve A_2 for the largest bodies and the dashed curve, for the intermediate range.

In all cases the exponent q_0 falls within the range $1 < q_0 < 2$.

However, the assumption $E = \infty$ does not provide us with the mass distribution of larger bodies. In Fig. 3 the values of q_0 are shown as a function of E for $\varepsilon = 0$ and coagulation coefficients of the form (23) and (24). For small bodies Eq. (23) yields $q_0 = 1.54$ if $E = \infty$, and $q_0 = 1.4$ if $E = 100$. Equation (24) is applicable only for large bodies with $E \leqslant 10$; it yields values of q_0 diminishing from 1.76 to 0.5 as E drops from ten to unity.

Since the coagulation coefficient enters linearly into Eq. (1), the condition (25) will be satisfied for $A = A_1 + A_2$ if it is satisfied for A_1 and for A_2 individually. Thus for equal values of the exponent q_0 in the power-law asymptotic solution for the coagulation coefficients A_1 and A_2, the asymptotic solution for the coagulation coefficients A_1 and A_2 and the asymptotic solution for $A = A_1 + A_2$ will have the same exponent q_0. If, however, the corresponding exponents q_{01} and q_{02} for A_1 and A_2 are not equal, the exponent for A should have an intermediate value between q_{01} and q_{02}. Hence the exponent q_0 corresponding to the general expression for the coagulation coefficient (22) will lie between the values of the exponents corresponding to Eqs. (23) and (24). Its dependence on E is illustrated in Fig. 3 by a dashed curve.

Thus in an actual system of gravitating bodies coalescing through collisions and not suffering frag-

TABLE 4

A/const	$(m + m')$	$(m^{2/3} + m'^{2/3})$	$(m^{1/3} + m'^{1/3})^2$	$(m^{4/3} + m'^{4/3})$	$(m^{2/3} + m'^{2/3})^2$
q_0	1.50	1.31	1.54	1.67	1.80

mentation, the mass distribution of all the bodies except the largest ones (E > 3) may be described approximately by an inverse power law with an exponent $q_0 \approx 1.5$-1.6. The degree of approximation provided by this law will be characterized by a difference in the values of q_0 for different m no greater than 0.3. The mass distribution of the largest bodies will deviate sharply from a power law. The abrupt decline in the local value of q_0 as the upper limit of the distribution is approached is a distinctive consequence of an effect we have found previously [11, 13]: the large discontinuity that develops between the mass of the largest body (the embryonic planet) and the masses of the other large bodies. In order to improve our understanding of the masses of the very largest bodies, further investigation will be needed.

LITERATURE CITED

1. V. S. Safronov, The Evolution of the Protoplanetary Cloud and the Formation of the Earth and Planets [in Russian], Nauka, Moscow (1969).
2. M. Smoluchowski, in: The Brownian Motion [in Russian] (1936), p. 332; The Coagulation of Colloids [in Russian] (1936), p. 7.
3. Z. A. Melzak, Quart. Appl. Math., 11, 231 (1953).
4. V. S. Safronov, Dokl. Akad. Nauk SSSR, 147, 64 (1962) [Sov. Phys.–Dokl., 7, 967 (1963)].
5. T. E. W. Schumann, Quart. J. Roy. Meteorol. Soc., 66, 195 (1940).
6. S. V. Pshenai-Severin, Dokl. Akad. Nauk SSSR, 94, 865 (1954).
7. P. Welander, Tellus, 11, 297 (1959).
8. A. M. Golovin, Izv. Akad. Nauk SSSR, Ser. Geofiz. (1963), No. 5, p. 783.
9. A. H. Marcus, Icarus, 4, 267 (1965).
10. A. H. Marcus, Astron. J., 72, 814 (1967).
11. V. S. Safronov and E. V. Zvyagina, Icarus, 10, 109 (1969).
12. S. Piotrowsky, Acta Astron., A5, 115 (1953).
13. V. S. Safronov, Astron. Zh., 42, 1270 (1965) [Sov. Astron.–AJ, 9, 987 (1966)].

Qualitative solution of the coagulation equation with allowance for fragmentation

E. V. Zvyagina, G. V. Pechernikova, and V. S. Safronov

Institute of Terrestrial Physics,
Academy of Sciences of the USSR
(Submitted March 14, 1973)
Astron. Zh., 50, 1261-1273 (November-December 1973)

A qualitative analysis is given of the coagulation equation describing the accumulation of protoplanetary bodies, with allowance for fragmentation through collisions between bodies of comparable size. A condition is derived for the disintegration of a gravitating body of mass m_2 in collision with another body of smaller mass m_1. If the bodies are not too massive this condition may be approximated by a relation of the form $\alpha = m_1/m^* \propto m^{*\lambda}$, where $m^* = m_1 + m_2$. In accordance with experimental evidence, the mass distribution of the fragmentation products is assumed to follow an inverse power law with exponent $q_1 \approx 1.8$. Asymptotic solutions of the coagulation equation including fragmentation are derived by a method developed previously for the case with no fragmentation, and their stability is investigated. The solutions can satisfactorily be approximated by an inverse power law with exponent $q \approx 1.8$ over the entire range of m values, except for the few largest bodies. A study is made of how the exponent in the power-law asymptotic distribution depends on various parameters, such as the coagulation coefficient, the fragmentation process, and the size of the largest fragments. The asymptotic solutions obtained by Dohnanyi, Hellyer, and Bandermann for the asteroid mass distribution are here derived as a special case, where there is no accumulation but only fragmentation. These authors' solution with $q = 11/6$ can be obtained only by assuming that the fragmentation condition is independent of the mass of the bodies ($\lambda \to 0$). Evidence is given that actually $\lambda \approx 2/3$. The agreement with the results of others lends support to our approach. The method offers promise for analyzing the complicated equations that simultaneously describe the accumulation and fragmentation of bodies.

In a previous paper[1] we have investigated the coagulation equation for protoplanetary bodies that have suffered no fragmentation. We here give a similar analysis which does allow for the fragmentation of the colliding bodies. The procedure we shall follow represents a development of a method employed by one of us[2] and originally introduced by Piotrowski[3] for studying the distribution function of asteroid sizes (fragmentation without accumulation).

1. FRAGMENTATION IN HIGH-VELOCITY COLLISIONS

The mass distribution of protoplanetary bodies must have depended significantly on the relative velocity of their mutual collisions. So long as the relative collisional velocities remained low, the bodies would merely have amalgamated, resulting in a rapid growth of the embryonic planets, a strengthening of gravitational perturbations, and accordingly to an increase in the velocity dispersion of the bodies in the protoplanetary cloud. But at velocities $V \gtrsim 1$ km/sec collisions between bodies might no longer have led to amalgamation, but rather to a partial or even complete disruption. Asteroids, which have experienced multiple fragmentation, testify to the important role of breakup in the formative processes of the bodies comprising the solar system.

The process of disruption of bodies through collisions at high velocities has not received very much study. Experiments on the partial fragmentation of basalt and quartz objects when bombarded by small particles at velocities of 5-20 km/sec have been described by Gault et al.[4] The target body m_2 was found to lose a mass m_e two or three orders of magnitude greater than the mass m_1 of the missile (for $m_1 \ll m_2$). The ratio $m_e/m_1 = \Gamma$ depends strongly on the mechanical properties of the target m_2, but only weakly on the properties of the missile m_1.

From an analysis of work by Gault et al.,[4-6] Marcus has found[7] that Γ may be regarded as proportional to the square of the relative collisional velocity: $\Gamma = CV_i^2$, where $C \approx 5$ (km/sec)$^{-2}$ for basalt and is three times larger for weakly consolidated quartz sand. It follows that if the velocities V_i are high enough the target m_2 should break up completely in a collision with a much smaller body having $m_1 = m_2/\Gamma$.

However, a confined body m_2 differs in an essential way from an unconfined target. In the latter case a shock wave will be reflected from the opposite surface and will induce a further disruption. Hence the complete disruption of the body m_2 can actually take place through collisions with even smaller missiles, having $m_1 \approx m_2/50\,\Gamma$ (see Dohnanyi's discussion[8] based on a private communication from Gault).

Recent impact experiments have shown[9] that in the velocity range 0.2-0.5 km/sec microparticles of iron will rebound from targets of aluminum, copper, tungsten, and other materials. At high velocities they will adhere to the target. Up to velocities of ≈ 13 km/sec about 70-100% of the mass of the particle will remain in the crater. Similar experiments by Kerridge and Vedder[10] in the velocity range $1.5 < V_i < 9.5$ km/sec have demonstrated that silicates behave differently. For silicates the proportion of the particle substance that adheres to the target is generally small, and the size of the crater (at least for velocities above 2 km/sec) is greater than the size of the impacting particle. On the basis of these distinctions attempts have been made to explain the fractionation of substances during the formation of the planets.

Gravitation will play a major role for large bodies, and collisions will become less catastrophic. Although gravitation does increase the collisional velocity, at the same time it hinders the dispersal of the fragmented material, causing it to return to m_2. Using laboratory data for the velocity distribution function of the fragments following a collision in the case where $m_1 \ll m_2$ and a complete disruption of m_2 does not occur, Marcus[7] has estimated the mass of particles having a velocity higher than the parabolic velocity at the surface of m_2 and leaving m_2. He finds that the largest of the bodies m_2 lose less mass through the infall of small bodies m_1 than they acquire, so that there is a net growth in their mass (accretion). This will be the case, for example, with asteroids more than 180 km across if the relative velocities of the bodies ("meteoroids") prior to their encounter with m_2 are no higher than 2 km/sec.

When bodies of comparable mass collide, the energy of impact may prove sufficient for a complete dispersal of their material if the target m_2 is not too massive. It is usually assumed, following Piotrowski[3] (for example, Hellyer[11]), that nongravitating bodies m_1 and m_2 will break up completely in a collision if $m_2/m_1 < 1/\alpha' \approx 125$. But for gravitating bodies the quantity α' should depend on their mass. It can be estimated as follows.

In a coordinate system referred to the center of gravity of the bodies m_1 and m_2, their kinetic energy T_0 prior to the encounter will be given by

$$T_0 = \frac{m_1 m_2}{2(m_1 + m_2)} V^2, \tag{1}$$

where V is the average relative velocity of the bodies before the encounter. The conservation of the total energy of the system in three states – before the encounter, immediately after the encounter (when the bodies m_1 and m_2 may, to some approximation, be regarded as a single fragmented body with a mass $m_{12} = m_1 + m_2$ and a radius r_{12}), and following dispersal of the fragments to large distances – may be written in the form

$$T_0 + U_1 + U_2 = T + Q + U_{12} = T' + Q + \beta U_{12}, \tag{2}$$

where T_0, T, T' denote the kinetic energy for each of the three states; U_1, U_2, U_{12}, βU_{12} denote the potential energy of the bodies m_1, m_2, m_{12}, and their fragments; and Q represents the energy converted into heat as well as that expended in fragmentation of the bodies and in phase transitions.

Suppose that T comprises the fraction χ of the total T+Q that is released in the collision. The condition $T' \geq 0$ may be regarded as a condition for complete disintegration of the system; it may be written in the form

$$\chi(T_0 + U_1 + U_2) \geqslant (\chi + \beta - 1) U_{12}. \tag{3}$$

Substituting $U = -3Gm^2/5r$ and $V^2 = 2GM/\theta R$, where M and R are the mass and radius of the largest body in the zone,[2] we obtain after a calculation the following criterion for disintegration:

$$\frac{m_1}{m_1 + m_2} \geqslant \frac{1-\chi-\beta}{\chi}\left[\frac{5}{3\theta}\frac{m_2}{m_1+m_2}\left(\frac{M}{m_1+m_2}\right)^{2/3} - \left(\frac{m_1}{m_1+m_2}\right)^{2/3} - \frac{m_2}{m_1}\left(\frac{m_2}{m_1+m_2}\right)^{2/3}\right]^{-1} = \alpha. \tag{4}$$

For a power-law mass distribution of the fragments with an exponent $-q_1$ (see below) and with an upper limit on the distribution of m_{12}, the coefficient $\beta = (2 - q_1)/(8/3 - q_1)$. If $q_1 = 1.8$ the coefficient $\beta = 0.23$.

We are unaware of any estimates of χ for completely disrupted bodies. For shallow explosions Dokuchaev et al.[12] have found $\chi \approx 0.15$. For ejections of aluminum particles in collisions with a basalt target (Gault and Heitowit[13]), and also for a target of quartz sand (Braslau[14]) with a velocity of 6.37 km/sec, $\chi \approx 0.4$-0.5. In Table 1

TABLE 1

m_2/m_1	α	M/m_2	
		$\chi = 1/3$	$\chi = 1/2$
1	0.5	160	75
10	0.09	550	290
100	0.01	13400	7300
$n \gg 1$	$1/(n+1)$	$13 \cdot n^{3/2}$	$7.1 \cdot n^{3/2}$

we give values for the ratio M/m_2 and α corresponding to the criterion (4) as a function of the ratio m_2/m_1 (with $m_2 > m_1$), for $\theta = 4$, $\beta = 0.23$, and $\chi = 1/3$ and $1/2$.

Bodies with masses $m_2 > M/160$ for $\chi = 1/3$ or $m_2 > M/75$ for $\chi = 1/2$ will not disintegrate in collisions with any bodies of mass $m_1 \leq m_2$, although considerable fragmentation will take place.

Very little information is available on the mass distribution function of the fragments formed through two-body collisions. Usually an inverse power law $n_1(m) = c_1 m^{-q_1}$, the simplest and most convenient for mathematical analysis, is adopted, with $q_1 \approx 1.8$, in accordance with the findings of Gault et al.[4] Impact experiments have led Hartmann[15] to conclude that q_1 lies within the range $1.5 \leq q_1 \leq 2.0$. Hawkins' experiments and observations,[16] as well as Anders' study[17] of the distribution of asteroids in the Hirayama families, yield the value $q_1 = 5/3$.

We shall adopt a power-law mass distribution of the combined fragments of both the colliding bodies. Then the constant c_1 will be determined by the mass-conservation condition

$$m_1 + m_2 \equiv m'' = \int_{m_0}^{M_1} m n_1(m)\, dm. \tag{5}$$

We find

$$c_1 = (2 - q_1) M_1^{q_1 - 2} m'' [1 - (m_0/M_1)^{2-q_1}]^{-1}, \tag{6}$$

where M_1 represents the upper limit of the mass distribution function of the fragments. On the basis of the experiments by Gault et al.,[4] Dohnanyi[8] considers that in a partial fragmentation, when $m_1 \ll m_2$, the quantity M_1 will be proportional to the mass of the smaller fragmenting body m_1, and will not depend on the mass of the larger body m_2: $M_1 = \Lambda m_1$, where $\Lambda \approx 10$-100. However, in a collision of bodies of comparable size, an inverse relation between M_1 and m_1 will be more probable: The larger m_1 is, the greater will be the collisional energy per unit mass of m_2, and the smaller the fragments: $M \propto (m_1 + m_2)m_2/m_1$.

In view of the lack of experimental data we may adopt the simplest assumption, that the upper limit on the mass distribution function of the fragments (not the mass of the largest fragment) is proportional to the combined mass of the two colliding bodies:

$$M_1 \approx \xi (m_1 + m_2). \tag{7}$$

If $\xi = 1$ the expected mass M_b of the largest fragment [Safronov and Zvyagina,[18] Eqs. (8) and (9)] will be

$$M_b = [1 - (2 - q_1)^{(2-q_1)/(q_1-1)}](m_1 + m_2), \tag{8}$$

which is equal to about 1/3 the combined mass of the bodies for $q_1 \approx 1.8$.

2. TIME VARIATION OF THE MASS DISTRIBUTION FUNCTION

Suppose that the mass distribution n(m, t) of the bodies is a continuous function of m and t for $m_0 \leq m \leq M$. The number of bodies in any mass interval will vary only through collisions of the bodies. The collision probability $A(m_1, m_2)$ for the bodies m_1 and m_2 will depend on their masses. If the bodies are small the probability will be proportional to their geometrical collision cross section:

$$A \propto (m_1^{1/3} + m_2^{1/3})^2 \propto (r_1 + r_2)^2. \tag{9}$$

For the largest bodies in the system, the coagulation coefficient, which characterizes the collision probability, will be determined by the gravitational cross section:[2]

$$A \propto (m_1^{1/3} + m_2^{1/3})(m_1 + m_2). \tag{10}$$

In a swarm of protoplanetary bodies consisting of both large ("gravitating") and small objects, one may as a first approximation adopt a coagulation coefficient representing a suitable average[2] between the expressions (9) and (10):

$$A \propto m_1 + m_2. \tag{11}$$

All three of the expressions (9)-(11) for $A(m_1, m_2)$ may be written in the form

$$A(m_1, m_2) = A_1 \sum_{i \leqslant k} b_i m_1^i m_2^{k-i} = A_1 m_1^k \sum_{i \leqslant k} b_i \left(\frac{m_2}{m_1}\right)^{k-i}, \tag{12}$$

where i assumes certain fractional values from 0 to k.

We shall assume that collisions of bodies terminate either in an amalgamation or in a complete disruption if the disintegration condition (4) is satisfied. Partial mass loss by a body through crater formation (erosion) will not be considered here. According to Bandermann,[19] catastrophic collisions will have a far stronger influence on the mass distribution than erosion activity. The contribution of the two collision processes will be determined by the parameters Γ and Γ', respectively; their variation will have a similar effect on the distribution. It is for this reason that Hellyer's fragmentation equation,[11] which does not take erosion into account, yields results similar to those of Bandermann. Accordingly, in our case as well erosion may be included simply by making a slight change in the disintegration condition (the parameter α) for catastrophic collisions.

The expression (4) is too cumbersome to be useful for further analysis. We shall consider a simpler disintegration condition which, although it differs from the criterion (4), still reflects the basic manner in which α depends on the masses of the colliding bodies and on the mass M of the largest body:

$$\frac{m_1}{m''} \geqslant \alpha(m'') = a\left(\frac{m''}{2M}\right)^{\lambda}. \tag{13}$$

For small bodies (of mass less than 10^{-3} M), the condition (4) reduces to the expression (13) with $\lambda = 2/3$. In the case of larger bodies λ has a considerably higher value. For $\lambda = 0$ the quantity α no longer depends on m". If

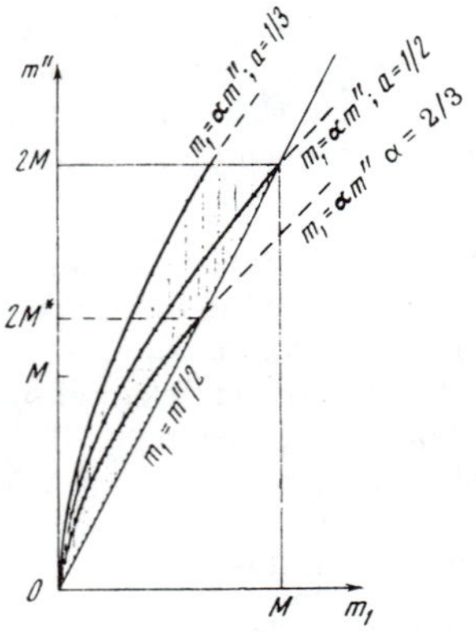

Fig. 1. Shaded area, range of mass for bodies disrupted by collisions, according to the condition (13). Abscissa, mass m_1 of the smaller body; ordinate, combined mass $m'' = m_1 + m_2$.

$a = 0$ any collision will lead to a breakup, while if $a = 1/2$ the largest body M will not disintegrate as a result of collisions with any bodies.

In the case $\lambda = 2/3$ one finds that bodies m_2 of mass larger than 10^{-2} M will not be disrupted (see Table 1) if $a \approx 10$.

Figure 1 serves to illustrate the fragmentation condition (13) for colliding masses m_1 and m_2. The abscissa is the value of m_1; the ordinate is $m'' = m_1 + m_2$. The fragmentation region is bounded on the right by the line $m_1 = m''/2$ and on the left by the curve $m_1 = \alpha m''$ (in the figure α is taken for $\lambda = 2/3$ and for three different values of a). The line intersects the curves at the values $m'' = 2M/(2a)^{1/\lambda}$. Furthermore, it is evident that $m'' < 2M$. Thus an upper limit on the combined mass m" of the two bodies experiencing fragmentation will be 2M*, where

$$M^* = \begin{cases} M & \text{for } a \leqslant 1/2 \\ M/(2a)^{1/\lambda} & \text{for } a > 1/2. \end{cases}$$

Let us assume that the mass distribution of the fragmentation products conforms to an inverse power law with exponent q_1:

$$n_1(m, m'') = c_1(m'') m^{-q_1}, \tag{14}$$

where $1.5 < q_1 < 2$, $m_0 \leq m \leq M_1$, and $c_1(m'')$ is given by Eq. (6) with the value $M_1 = \xi m''$ (for ξ = const). The time variation of the number n(m, t) of bodies within unit volume and unit mass interval may be described by the equation

$$\frac{\partial n(m, t)}{\partial t} = A - B + C, \tag{15}$$

where A denotes the number of bodies of mass m formed through the amalgamation of bodies m_1 with bodies of mass $m - m_1$, B represents the decrease in the number of bodies of mass m due to collisions with any other objects, and C is the number of bodies of mass m formed through the fragmentation of larger bodies. The terms A and B in this equation may be written in the same form as in our previous paper,[1] except that in the first integral the upper limit of integration should be restricted by the onset of fragmentation according to the condition (13):

$$A - B = \int_{m_0}^{\alpha^*(m)\cdot m} A(m_1, m - m_1)\, n(m_1) n(m - m_1)\, dm_1 - \int_{m_0}^{M} A(m, m_1)\, n(m)\, n(m_1)\, dm_1, \quad (16)$$

where

$$\alpha^*(m) = \begin{cases} \alpha(m) & \text{for } \alpha(m) \leqslant 1/2 \\ 1/2 & \text{for } \alpha(m) > 1/2. \end{cases}$$

The number of pairs of bodies of combined mass m″ breaking up per unit time is given by

$$N(m'') = \int_{\alpha^*(m'')m''}^{m''/2} A(m_1, m'' - m_1)\, n(m_1)\, n(m'' - m_1)\, dm_1. \quad (17)$$

The number of fragments of mass m formed per unit time through collisions of bodies with a combined mass m″ will evidently be $n_1(m, m'')$ N(m″). We can obtain the term C by taking into account all collisions yielding fragments of mass m. The value of m″ should then be restricted to the range $m/\xi < m'' < 2M^*$;

$$C = \int_{m/\xi}^{2M^*} n_1(m, m'')\, N(m'')\, dm''$$

$$= \int_{m/\xi}^{2M^*} n_1(m, m'') \int_{\alpha(m'')m''}^{m''/2} A(m_1, m'' - m_1)\, n(m_1)\, n(m'' - m_1)\, dm_1\, dm''. \quad (18)$$

We shall seek a solution of Eq. (15) in the form of an inverse power law,

$$n(m, t) = c(t)\, m^{-q(t)}. \quad (19)$$

To determine the exponent q we shall apply the method of qualitative analysis that we have developed[1] for the coagulation equation without fragmentation. As time passes the exponent q will approach an asymptotic value given by the condition

$$\frac{\partial}{\partial m}\left(\frac{\dot n}{n}\right) = -\frac{\dot q}{m} = 0. \quad (20)$$

Substituting Eqs. (6), (12), (14), (16), (18), and (19) into the initial equation (15), we obtain

$$\dot n(m, t) = A_1 c^2 \int_{m_0}^{\alpha^*(m)m} m^k \sum_{i\leqslant k} b_i \left(1 - \frac{m_1}{m}\right)^{k-i} \left(\frac{m_1}{m}\right)^i m_1^{-q}$$

$$\times (m - m_1)^{-q}\, dm_1 - A_1 c^2 \int_{m_0}^{M} m^k \sum_{i\leqslant k} b_i \left(\frac{m_1}{m}\right)^{k-i} m^{-q} m_1^{-q}\, dm_1$$

$$+ A_1 c^2 \int_{m/\xi}^{2M^*} (2 - q_1) \frac{m''^{q_1 - 1} m^{-q_1}}{\xi^{2-q_1} - \left(\frac{m_0}{m''}\right)^{2-q_1}}$$

$$\times \int_{\alpha(m'')m''}^{m''/2} m''^k \sum_{i\leqslant k} b_i \left(1 - \frac{m_1}{m''}\right)^{k-i}$$

$$\times \left(\frac{m_1}{m''}\right)^i m_1^{-q} (m'' - m_1)^{-q}\, dm_1\, dm''. \quad (21)$$

Next we divide the left- and right-hand members of Eq. (21) by $A_1 c n(m) = A_1 c^2 m^{-q}$, introduce the new variables $x = m_1/m$ into the A term and $x = m_1/m''$ into the C term, and evaluate the derivative with respect to m:

$$\frac{\partial}{\partial m}\left(\frac{\dot n}{n}\right)\frac{1}{A_1 c} = (k - q + 1)\, m^{k-q} \int_{\varepsilon}^{\alpha^*(m)} x^{-q}(1 - x)^{-q} P(x)\, dx$$

$$+ m^{k-q+1}\left[\alpha^{*-q}(m)(1 - \alpha^*(m))^{-q} P(\alpha^*(m)) \frac{\partial \alpha^*(m)}{\partial m}\right.$$

$$\left. - \varepsilon^{-q}(1 - \varepsilon)^{-q} P(\varepsilon) \frac{\partial \varepsilon}{\partial m}\right] - \sum_{i\leqslant k} b_i i m^{i-1} \int_{m_0}^{M} m_1^{k-i-q}\, dm_1$$

$$+ (2 - q_1)(q - q_1)\, m^{q-q_1-1} \int_{m/\xi}^{2M^*} \frac{m''^{k+q_1-2q}}{\xi^{2-q_1} - \left(\frac{m_0}{m''}\right)^{2-q_1}}$$

$$\times \int_{\alpha(m'')}^{1/2} P(x)\, x^{-q}(1 - x)^{-q}\, dx\, dm''$$

$$- \frac{2 - q_1}{\xi^{k+q_1-2q+1}} \frac{m^{k-q}}{\xi^{2-q_1} - \left(\frac{m_0}{m}\right)^{2-q_1}} \int_{\alpha(m/\xi)}^{1/2} P(x)\, x^{-q}(1 - x)^{-q}\, dx, \quad (22)$$

where we have set

$$\sum_{i\leqslant k} b_i (1 - x)^{k-i} x^i = P(x), \quad (23)$$

$$m_0/m = \varepsilon < 1, \quad M/m = E > 1. \quad (24)$$

Combining terms with the same powers of m, we obtain after straightforward transformations a final expression for $\dot q$, provided $q_1 < 2$:

$$-\frac{\dot q}{A_1 c m^{1+k-q}} = m \alpha^{*-q}(m)\, P(\alpha^*(m))$$

$$\times (1 - \alpha^*(m))^{-q} \frac{\partial \alpha^*(m)}{\partial m} + \alpha^{1-q}(m)$$

$$\times \left[Q(\alpha(m))(1 + k - q)\right.$$

$$\left. + \frac{(2 - q_1)\, Q(\alpha(m/\xi)) + (2 - q_1)(q - q_1)\, S(\alpha(m/\xi))}{\xi^{k-2q+3+\lambda(1-q)}}\right]$$

$$+ (2E)^{k+q_1-2q+1} \frac{(2 - q_1)(q - q_1)}{\xi^{2-q_1}}$$

$$\times \left[\frac{(1/2)^{1-q} Q(1/2)}{k + q_1 - 2q + 1} - \alpha^{1-q}(2M^*)\, S(\alpha(2M^*))\right]$$

$$\times \left(\frac{M^*}{M}\right)^{1+k+q_1-2q} - \sum_{i\leqslant k} b_i i \frac{E^{k-q+1-i}}{k - q + 1 - i}$$

$$- (2 - q_1)\left(\frac{1}{2}\right)^{1-q} Q(1/2)$$

$$\times \frac{k + 1 - q}{k + 1 + q_1 - 2q} \xi^{2q-k-3}$$

$$+ \varepsilon^{1-q}\left[P(\varepsilon)(1 - \varepsilon)^{1-q} - Q(\varepsilon)(k - q + 1)\right.$$

$$\left. + \sum_{i\leqslant k} b_i i \frac{\varepsilon^{k-i}}{k - q + 1 - i}\right], \quad (25)$$

where the integrals

$$\int_y^z P(x)\, x^{-q}(1 - x)^{-q}\, dx = z^{1-q} Q(z) - y^{1-q} Q(y) \quad (26)$$

and

$$\int_y^z Q(\alpha(x))\, x^{k+q_1-2q+\lambda(1-q)}\, dx$$
$$= z^{k+q_1-2q+\lambda(1-q)+1} S(\alpha(z)) - y^{k+q_1-2q+\lambda(1-q)+1} S(\alpha(y)), \qquad (27)$$

with $Q(x)$ and $S[\alpha(x)]$ given by the convergent series

$$Q(x) = \sum_{i \leqslant k} b_i x^i \left[\frac{1}{1-q+i} + \sum_{n=1}^{\infty} \frac{(q+i-k)(q+i-k+1)\dots(q+i-k+n-1)}{n!\,(i-q+n+1)} x^n\right], \qquad (28)$$

$$S(\alpha(x)) = \sum_{i \leqslant k} b_i \alpha^i(x) \times \left[\frac{1}{(1-q+i)[k+q_1-2q+\lambda(1-q+i)+1]} + \sum_{n=1}^{\infty} \frac{(q+i-k)(q+i-k+1)\dots(q+i-k+n-1)}{n!\,(i-q+n+1)[k+q_1-2q+\lambda(1-q+i+n)+1]} \alpha^n(x)\right]. \qquad (29)$$

Upon substituting Eqs. (23) and (28) for $P(\varepsilon)$ and $Q(\varepsilon)$ into the last term of the expression (25) and expanding $(1-\varepsilon)^{k-i-q+1}$ in powers of ε, we may reduce this term to the form

$$\varepsilon^{2-q} \sum_{i \leqslant k} \varepsilon^i b_i \left[\frac{(k-i-q)(i+1-k)}{2-q+i} + O(\varepsilon)\right]. \qquad (30)$$

For $q < 2$ the mass of the smallest body, $m_0 = m\varepsilon$, may be taken equal to zero; the expression (30) then will also vanish. In the case $k = 1$ this result holds for all values $q < 3$. Thus if

$$q_1 < 2, \quad q < 2, \quad \varepsilon \ll 1, \qquad (31)$$

Eq. (25) may be written in the form

$$-\frac{\dot{q}}{A_1 c m^{1+k-q}} \approx (2E)^{\lambda(q-1)} \Big[a^{1-q} F_1(a, \lambda, k, q_1, q) + E^{(2+\lambda)(q_m - q)} F_2(a, \lambda, k, q_1, q) - \sum_{0<i\leqslant k} E^{(1+\lambda)(q_i - q)} F_{3i}(k, q, i)\Big] - F_4(k, q_1, q) = F(q, \dots), \qquad (32)$$

where

$$q_m = \frac{1+k+\lambda+q_1}{2+\lambda}, \quad q_i = \frac{1+k+\lambda-i}{1+\lambda} \qquad (33)$$

and the functions F_1, F_2, F_3, F_4 may be represented by convergent series.

3. ASYMPTOTIC POWER-LAW SOLUTIONS OF THE COAGULATION EQUATION

For $a > 1/2$, bodies more massive than M^* will not break up in any collisions. We have previously discussed the variation in the mass distribution of bodies when fragmentation is absent.[1] Thus we shall here consider only bodies smaller than M^*, corresponding to values $E > (2a)^{1/\lambda}$. In this event $\alpha^* = \alpha$. We exclude the case $a = 1/2$ for $\lambda = 0$, which corresponds to complete coagulation.

Equation (32), obtained under the conditions (31), enables us to tell the direction in which the exponent q will vary with time. This behavior can readily be ascertained for comparatively small bodies, with

$$E \gg 1. \qquad (34)$$

The term with the largest positive power of E will then predominate in Eq. (32). The first term will predominate for values of q satisfying the conditions

$$q_m < q < 2 \text{ and } q > \frac{1+k+\lambda}{1+\lambda} \qquad (35)$$

(we shall call these ranges of q region I). Then

$$\frac{\dot{q}}{A_1 c m^{1+k-q}} \approx -(2E)^{\lambda(q-1)} a^{1-q} F_1(a, \lambda, k, q_1, q). \qquad (36)$$

Asymptotic power-law solutions of the coagulation equation will be determined by the equation $\dot{q} = 0$, that is, by a relation

$$F(q) = 0. \qquad (32')$$

If the conditions (31) and (34) hold, the roots of Eq. (32′) will practically coincide with the roots of the equation

$$F_1(a, \dots) = Q(\alpha(m))(1+k-q) + P(\alpha(m))(1-\alpha(m))^{-q\lambda} + (2-q_1)[Q(\alpha(m/\xi)) + (q-q_1)S(\alpha(m/\xi))]\xi^{2q-k-3+\lambda(q-1)} = 0. \qquad (37)$$

Let us consider the function F_1 more carefully. Terms depending on α enter into it with a positive power. They therefore contain E to a negative power and are negligibly small. Accordingly,

$$F_1(a, \dots) \approx b_0 \frac{(q-q_{01})(q-q_{02})(1+\lambda)}{(q-1)(q-q_m)} \xi^{(2+\lambda)(q-q_{01})}, \qquad (38)$$

where q_{01}, q_{02} are roots of Eq. (37):

$$q_{01} = \frac{k+3+\lambda}{2+\lambda}, \quad q_{02} = \frac{k+1+\lambda}{1+\lambda}. \qquad (39)$$

It is of interest to observe that both of these roots are independent not only of the exponent q_1 in the fragment mass distribution but also of the upper limit of that distribution. The roots q_{01}, q_{02} will represent asymptotic solutions if they fall within the region (35). This will be the case if the following relations hold between the parameters k, λ, q_1, and i:

$$\text{for the root } q_{01}\text{:}\quad k < 1+\lambda, \quad q_1 < 2;$$
$$\text{for the root } q_{02}\text{:}\quad k < 1+\lambda, \quad q_1 < 1 + \frac{k}{1+\lambda}. \qquad (40)$$

The condition for stability of any root q_0 implies the inequality

$$(2q - q_{01} - q_{02})_{q=q_{01},\, q=q_{02}} > 0. \qquad (41)$$

For $k < 1 + \lambda$ we have $q_{01} > q_{02}$. Hence the root q_{01} will be stable; the root q_{02}, unstable.

The second term in Eq. (32), the term containing F_2, will predominate for values of q within the following region II:

$$q < \min\left(q_m, \frac{1+k+q_1}{2}, q_1 + i\right), \qquad (42)$$

ere $i > 0$. In this region the asymptotic solution should represented by the root q_{03} of the equation

$$F_2 = \frac{(2-q_1)(q-q_1)}{\xi^{2-q_1}} \times \left[\frac{(2^{q-1}Q(^1/_2)}{1+k+q_1-2q} - \alpha^{1-q}(2M^*)S(\alpha(2M^*))\right] \times \left(\frac{M^*}{M}\right)^{1+k+q_1-2q} = 0. \quad (43)$$

e root $q_{03} = q_1$ will satisfy the condition (42) if

$$q_1 < \frac{1+\lambda+k}{1+\lambda} = q_{02}. \quad (44)$$

order for the root q_{03} to be stable we must have

$$\left(\frac{\partial}{\partial q}\dot{q}\right)_{q=q_{03}} = \alpha^{1-q_1}(2M^*)S(\alpha(2M^*)) + \frac{2^{q_1-1}Q(^1/_2)}{q_1-1-k} < 0. \quad (45)$$

e condition (44) imposed on the root q_1 may be written the form $\lambda < (k+1-q_1)/(q_1-1)$. For these values of we obtain from Eqs. (26) and (27):

$$\frac{2^{q_1-1}Q(^1/_2)}{k+1-q_1} - \alpha^{1-q_1}(2M)^* S(\alpha(2M^*)) = \int_0^{(2a)^{-1/\lambda}} x^{k-q_1} \int_{a(x)}^{^1/_2} P(y)y^{-q_1}(1-y)^{-q_1}dy\,dx > 0. \quad (46)$$

us the root q_1 will be stable throughout the region II, ere it represents a solution of Eq. (32').

According to Eqs. (25) and (33), the function $F_{3i} = /(k+1-q-i)$ will not yield asymptotic solutions of coagulation equation, because the third term in Eq. ') predominates in the region $q_1 + i < q < 1 + (k-i)\cdot + \lambda)^{-1}$, that is, only for unrealistically small values q_1.

All the roots of Eq. (32') that we have found are inde- ıdent of the parameter a [see Eqs. (39)]. From a com- rison of the relations (4) and (13), which define this rameter, we see that a depends on the mean relative ocity of the bodies in the system through the quantity Hence the mean relative velocity of the bodies (as- ned to be the same for all masses) in the range $m_0 \ll$ $\ll$ M will not affect their ultimate mass distribution. ly the rate at which an asymptotic distribution is estab- hed will depend on the mean relative velocity.

The parameter λ in Eq. (13) specifies how the frag- ntation of the bodies depends on their combined mass. r $\lambda = 0$, the mass ratio α will be independent of m. In this event $q_{01} \approx 2$, and a substantial portion of the mass of the system will consist of small bodies. As λ increases, the fragmentation of small bodies compared to large ones will increase sharply. Small bodies will be intensively disrupted, leaving fine-scale fragmentation products behind in the system, which subsequently will be accreted by the large bodies. A return mass flow will take place for large bodies and the exponent q_{01} in the power-law mass distribution (for $E = \infty$ and $m_0 = 0$) will decrease.

The dependence of the roots q_{01}, q_{02}, q_{03} on λ for certain values of k and for $q_1 = 1.8$ is illustrated in Figs. 2a-c. The solid curves represent the values of the stable roots of Eqs. (37) and (43) in the regions of q values defined by the conditions (35) and (42), respectively, where these roots are also roots of Eq. (32'). The dashed curves represent the unstable roots in these regions, while in Fig. 2c the curves to the left of the value $\lambda_0 = k - 1$ indicate roots outside the regions I and II [not roots of Eq. (32')]. The roots q_{01} and q_{02} decrease more rapidly with increasing λ as k becomes larger, that is, as m increases.

If $k = 2/3$ (Fig. 2a), the asymptotic solution is determined solely by the root q_{01} (real roots q_{02}, q_{03} could exist only for $q_1 < 5/3$). The value $k = 2/3$ corresponds to the geometrical collisional cross section of the bodies when their masses still remain small and gravitation is unimportant. For bodies that are not too large the condition (4) gives $\lambda \approx 2/3$, whereby $q_{01} \approx 1.65$. But in all investigations of asteroid fragmentation[8,11,19,20] the disruption of the bodies is assumed to be governed only by their mass ratio, and α is regarded as constant. Then $\lambda = 0$, and we obtain the unique stable solution

$$q_{01} = \frac{11}{6}. \quad (47)$$

However, there are in fact no grounds for accepting that $\lambda = 0$. According to the relation (4), λ could be smaller than 2/3 if θ decreases with increasing m. But $\theta(m)$ has been found[2] to have so weak a dependence that λ remains practically unchanged.

If $k = 1$ (Fig. 2b), then for small enough λ (less than 1/4) two stable roots will exist simultaneously as well as one unstable root q_{02} located between them. If the initial value of q is greater than q_{02}, the asymptotic solution of the equation will yield the root q_{01} (q will "move away" from the unstable root q_{02}). For an initial value $q < q_{02}$, the asymptotic solution will be determined by the root q_1, so that the mass distribution in the system will coincide with that of the fragments of the disrupted bodies. For $\lambda > 1/4$, only the single stable root q_{01} remains; at $\lambda = 2/3$ we have $q_{01} = 1.75$.

The value $k = 4/3$ (Fig. 2c) refers to large bodies,

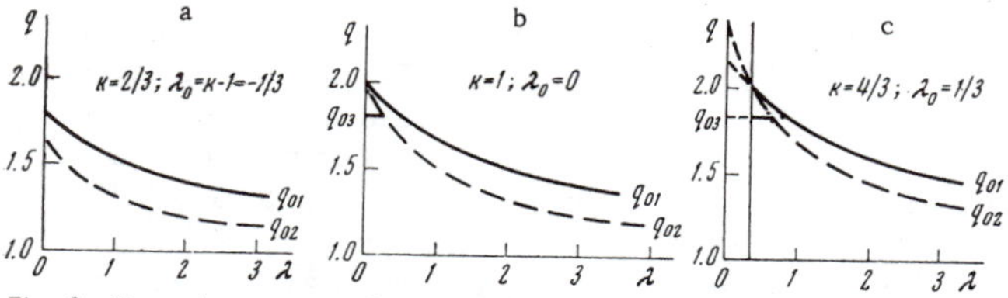

Fig. 2. Dependence on λ of the roots q_0 of Eq. (32'), which for $E \gg 1$ represent the exponent in the inverse power law for the asymptotic solution of the coagulation equation.

where gravitation renders the effective collisional cross section considerably larger than the geometrical cross section. For $\lambda < 1/3$ there is only one stable asymptotic solution q_1. In the range $1/3 < \lambda < 2/3$ there exist two stable roots q_1 and q_{01}, as well as one unstable root q_{02}, representing in a sense a "watershed," with the mass distribution varying with time in the sense of an increasing departure of the exponent q from the value q_{02}.

These results may be interpreted in the following manner. As k increases, collisions of large bodies will become more frequent, while as λ increases there will be a more intensive fragmentation of small bodies. If k is large and λ is small, fragmentation products of large-sized bodies will predominate in the system; hence the mass distribution will asymptotically approach the distribution of fragments of the disrupting bodies, that is, an inverse power law with exponent q_1. The larger the value of k, the wider the range of λ values for which this behavior will take place.

Thus the approximation of an asymptotic mass distribution for protoplanetary bodies, as established through an accumulation process together with fragmentation, leads to the following values for the exponent q_0 in the inverse power law: In the range of small bodies, $11/6 = 1.83$ for $\lambda = 0$ or 1.65 for $\lambda = 2/3$; in the range of medium-sized bodies, 1.75; and in the range of large bodies, 1.8. These slight differences between the values of q_0 show that the mass distribution may be described perfectly well by a power law with an exponent $q_0 \approx 1.8$.

However, the assumption $E \gg 1$ used for this estimate is not applicable to the largest bodies in the system. For these objects $E \approx 1$, and all terms must be taken into account when solving Eq. (32'). A numerical solution of this equation for the case $k = 1$, $\lambda = 1$, and $a = 1/2$ yields the relation between the stable root q_{01} and E shown in Fig. 3. For small m (that is, for $E = M/m \gg 1$), the root $q_{01} = 5/3$ (compare with Fig. 2b). As the mass of the bodies increases, the root q_{01} at first diminishes very gradually; but in the region of large bodies it falls off quite rapidly: from 1.56 to 1.26 as E decreases from 10 to 1. Thus just as in a system without fragmentation,[1] the distribution of the largest bodies deviates in a major way from a power law. The question of the distribution of bodies in this mass range warrants a special investigation which also takes into account the fact that an asymptotic relation might not be reached for the most massive objects. Strictly speaking, the departure from a power law in the region of the largest bodies should cause some change in the estimate of q_0 for all the other bodies of the system; but this distorting effect would appear to be small.

4. MASS DISTRIBUTION OF THE ASTEROIDS

Recently the mass distribution of the asteroids has been studied in some detail. Steady-state solutions have been found for the distribution function in systems where fragmentation occurs without accumulation. Dohnanyi[8] allows for two types of collisions: catastrophic (complete disruption of a body) and erosion (partial mass loss). In both cases the distribution of the fragments is assumed to follow an inverse power law with an exponent $q_1 = 1.8$. Taking the effective collisional cross section to be equal to the geometrical cross section and adopting the disintegration condition α = const for bodies in the mass range $m_0 \ll m \ll M$, Dohnanyi has obtained a steady-state solution in the form of an inverse power law with an exponent $q \approx 11/6$. This value agrees with our solution for the case $k = 2/3$ (geometrical cross section for collision) and $\lambda = 0$. As pointed out above, a unique solution $q_{01} = 11/6$ exists in this case.

Subsequently Hellyer[11] solved an analogous problem, but only with catastrophic fragmentation; no allowance was made for erosion or accumulation. Piotrowski's disintegration condition was adopted: $\alpha = 1/125$. Adopting a power-law fragment distribution for each of the colliding bodies and a geometrical collisional cross section, Hellyer has also obtained a steady-state solution for $m_0 \ll m \ll M$ with an exponent $q = 11/6$. Furthermore, by solving the fragmentation equation for large masses ($m \approx m_{max}$) and neglecting the fragments (the "restricted" problem), Hellyer obtains a simple solution with $q = 5/3$, agreeing with the solution derived by Piotrowski. A numerical solution of the same equation[20] has shown that the distribution of small masses actually does conform to a stationary power law, and the exponent $q = 1.825$ practically coincides with the value found previously by Hellyer analytically. However, the exponent in the distribution of large masses ($m \approx m_{max}$) increases with time and does not level off at the predicted value $q = 5/3$, even though it fails to reach the value 11/6 applicable to small bodies. From these calculations Hellyer concludes that the restricted problem has an inadmissible solution. It is worth noting that in Fig. 2a, for $\lambda = 0$ and $k = 2/3$, there is also a root $q_{02} = 5/3$ of the equation $F_1 = 0$ which agrees with that found by Hellyer. But it does not satisfy the condition (35), and is not a root of the full equation (32'). Accordingly, our result supports Hellyer's conclusion.

A more detailed study of the fragmentation equation (again neglecting accumulation), with initial assumptions similar to those made by Dohnanyi, has been carried out by Bandermann.[19] His basic findings are in accord with ours. In particular, the conclusion is reached that after a sufficiently long time a distribution independent of the initial distribution will be established. The steady-state solution is also independent of the mass distribution of the fragments and of its upper limit.

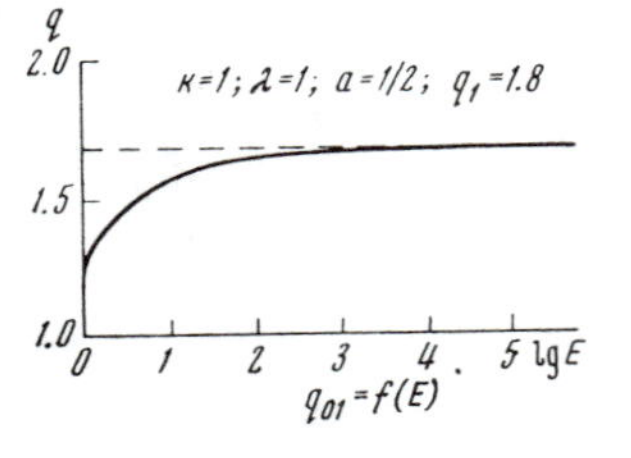

Fig. 3. Dependence of the root q_{01} on $E = M/m$, where M is the upper limit of the mass distribution in the system.

5. CONCLUDING REMARKS

The general analysis we have presented of the coagulation equation with allowance for fragmentation has enabled us to find asymptotic solutions for various forms of the coagulation coefficient and for various fragmentation conditions (as expressed by the parameters k and λ). For bodies of all masses other than the largest, the solution can satisfactorily be approximated by an inverse power law with an exponent q close to 1.8, which is independent of the exponent for the mass distribution of the fragments. An investigation shows that only some of the solutions obtained are stable.

Upon comparing the general solution for the particular case where no accumulation takes place in the system, but only fragmentation, with distributions found by other authors from analyses of asteroid fragmentation, we observe that the results are in agreement, lending support to the theory developed here.

The case $\lambda = 0$, the one which these authors have considered, does not appear adequately to represent the actual circumstances of asteroid fragmentation. The dependence of the basic root q_{01} on λ that we have obtained above indicates that a more careful analysis will be needed of the conditions for the disintegration of colliding bodies.

[1]E. V. Zvyagina and V. S. Safronov, Astron. Zh., 48, 1023 (1971) [Sov. Astron., 15, 810 (1972)].
[2]V. S. Safronov, The Evolution of the Protoplanetary Cloud and the Formation of the Earth and Planets [in Russian], Nauka, Moscow (1969).
[3]S. Piotrowski, Acta Astron., Sér. a, 5, 115 (1953).
[4]D. E. Gault, E. M. Shoemaker, and H. J. Moore, NASA Tech. Note D-1767 (1963).
[5]D. E. Gault, E. D. Heitowit, and H. J. Moore, in: The Lunar Surface Layer (proc. Lunar Surface Materials Conf., Boston, 1963), Academic Press (1964), p. 151.
[6]D. E. Gault, W. L. Quaide, V. R. Oberbeck, and H. J. Moore, Science, 153, 985 (1966).
[7]A. H. Marcus, Icarus, 11, 76 (1969).
[8]J. S. Dohnanyi, J. Geophys. Res., 74, 2531 (1969).
[9]H. Dietzel, G. Neukum, and P. Rauser, J. Geophys. Res., 77, 1375 (1972).
[10]J. F. Kerridge and J. F. Vedder, Science, 177, 161 (1972).
[11]B. Hellyer, Mon. Not. Roy. Astron. Soc., 148, 383 (1970).
[12]M. M. Dokuchaev, V. N. Rodionov, and A. N. Romashev, Ejective Explosions [in Russian], USSR Acad. Sci. Press, Moscow (1963).
[13]D. E. Gault and E. D. Heitowit, Proc. 6th Sympos. Hypervelocity Impact, 2, Part 2, 419 (1963).
[14]D. Braslau, J. Geophys. Res., 75, 3987 (1970).
[15]W. K. Hartmann, Icarus, 10, 201 (1969).
[16]G. S. Hawkins, Proc. Lunar and Planetary Exploration Colloq. (North American Aviation, Inc.), 2, No. 4 (1961), p. 5.
[17]E. Anders, Icarus, 4, 399 (1965).
[18]V. S. Safronov and E. V. Zvyagina, Icarus, 10, 109 (1969).
[19]L. W. Bandermann, Mon. Not. Roy. Astron. Soc., 160, 321 (1972).
[20]B. Hellyer, Mon. Not. Roy. Astron. Soc., 154, 279 (1971).

Mass distribution of protoplanetary bodies. I. Initial data for numerical solution

G. V. Pechernikova

Institute of Earth Physics, Academy of Sciences of the USSR

(Submitted May 20, 1974)
Astron. Zh., 51, 1305-1315 (November-December 1974)

A brief review is given of a qualitative study and numerical solution of the equation of accumulation of the protoplanetary bodies and the equation of fragmentation of the asteroids. The equation describing the variation in the distribution function of the protoplanetary bodies in the process of accumulation with fragmentation is examined. A method is worked out for the numerical solution of this equation. Some particular results of this solution are given.

1. INTRODUCTION

At some stage in the development of the solar system the space near the sun was filled with bodies of different masses. The bodies collided with each other in their motion around sun and depending on their relative velocities they either united or underwent partial or even total destruction. Hence the mass distribution of the protoplanetary bodies constantly varied.

Several authors have attempted to describe this process mathematically.

The variation in the mass distribution function n(m, t) of the bodies in the process of their growth during collisions without fragmentation is described by the well-known coagulation equation. An analytic solution of this equation is possible only for a few particular forms of the coagulation coefficient which characterizes the collision frequency of two bodies.[1-5] A satisfactory approximation can be obtained to describe the process of growth of the protoplanetary bodies by taking the coagulation coefficient as proportional to the sum of the masses of these bodies:

$$A(m_1, m_2)=A_1(m_1+m_2). \tag{1}$$

Then with an initial mass distribution described by the exponential function

$$n(m, 0)=c(0)e^{-b(0)m}, \tag{2}$$

the solution found[3,4] makes it possible for practically any range of variation of mass to represent the function n(m, t) in the form of the product of a power function times an exponential function

$$n(m, t)\approx c(t)m^{-3/2}e^{-b(t)m}. \tag{3}$$

The role of the exponential function decreases with time at the same values of m since b(t) decreases monotonically. Thus, in a system of nonfragmenting bodies the mass distribution approaches asymptotically with time to the stable distribution $n(m) = cm^{-q}$ with q = 1.5 for all values of m except for those close to the ends of the range.[4]

However, in the protoplanetary cloud the coagulation coefficient was a more complicated function of the masses of the colliding bodies. E. V. Zvyagina and V. S. Safronov[6] made a qualitative study of the coagulation equation for protoplanetary bodies not undergoing fragmentation. The method they used made it possible to follow the direction of the change in the exponent q with time for different intervals of masses m and with different forms of the coagulation coefficient.

In the collisions of bodies with velocities $V \gtrsim 1$ km/sec partial loss of mass (erosion) or even complete destruction is possible. The asteroid belt can probably serve as an example of such a system. S. Petrovskii[7] first used the qualitative method of studying the fragmentation equation in a study of the size distribution function of the asteroids. Considering only the large asteroids and neglecting the fragments of previous collisions he found an asymptotic distribution of particle radii equivalent to a mass distribution of the type $n(m) \propto m^{-5/3}$.

Dohnanyi,[8] examining the fragmentation of asteroids in more detail, considered two types of collisions, catastrophic (total destruction of the bodies) and erosion (partial loss of mass), adopting a mass distribution function of the fragments in the form of an inverse power function with the exponent q_1 = 1.8. Following Piotrowsky, Dohnanyi assumes that the condition of fragmentation of the bodies is independent of their masses and that the probability of collision of the bodies is proportional to their geometrical cross sections. With allowance for the contribution of the fragmented material, for bodies in the region of masses $m_0 \ll m \ll M$ Dohnanyi obtained a stationary asymptotic mass distribution of the asteroids in the form $n(m) \propto m^{-1.837}$.

Hellyer[9] has studied the effect of catastrophic collisional fragmentation on the mass distribution function (ignoring erosion and accumulation). He proposes a distribution of fragments according to an inverse power function with an exponent q_1 for each of the bodies which have collided. With the condition of destruction of α = 1/125 and a geometrical collision cross section Hellyer also finds an asymptotic power solution of the fragmentation equation for the region of masses $m_0 \ll m \ll M$ with the exponent q = 11/6 which does not depend on the exponent q_1 of the distribution of fragments if vaporization does not occur during the fragmentation of the bodies and the total mass of the fragments is equal to the mass of the destroyed bodies. Solving the "restricted problem" for the region of large masses (ignoring fragments), Hellyer obtained a stationary solution with the exponent q = 5/3, confirming the result of Piotrowsky. Thus, asymptotic solutions have been found also for the distribution function in systems having fragmentation without accumulation.

A qualitative analysis of fragmentation and accumulation equations enables one to find only asymptotic solutions as $t \to \infty$. The stationary mass distribution function of the bodies found only makes it possible to predict the state which should be reached by a system of colliding bodies after a sufficiently long time period. Numerical integration of the equation is necessary in order to study the variation of the mass distribution with time.

Hellyer has transformed a fragmentation equation which he obtained earlier[9] into a form suitable for machine computation and has solved it numerically.[10] His calculations for the region of smaller asteroid masses showed that within the lifetime of the solar system the distribution arrives at a stable state with an exponent $q = 1.825$ from the initial value $q(0) = 5/3$ independently of the value of the exponent q_1 of the mass distribution of the fragments. The solution obtained practically coincides with that found analytically ($q = 11/6$). However, in the region of large masses the exponent of the distribution does not remain at the predicted value of $q = 5/3$ but increases with time, not arriving at a stationary value in the lifetime of the solar system. It must be concluded from this that the fragments play a considerable role even in the region of large bodies. The solution obtained for small masses agrees well with observations. But in the region of large asteroids a surplus of bodies is observed relative to the power function, which is only partially explained by the decrease in q for large masses.

With initial assumptions analogous to those of Dohnanyi, Bandermann[11] made a more detailed numerical study of the fragmentation equation. Confirming the conclusions of the previous works, the author showed that with a sufficiently long time the mass distribution becomes independent of the initial distribution, catastrophic collisions affect the mass distribution considerably more strongly than erosion, and that the rate of evolution, shape, and average slope of the distribution curve as a function of time depend weakly on q_1. However, the calculations showed that the evolution of the distribution depends critically on the ratio of limiting masses present in the system and with time can lead to a power, bimodal, or nonstationary type. The author considers that the asteroids, for which the range of masses is more than 20 orders of magnitude, are subject to the first type of evolution, whereas the interplanetary dust should give a bimodal distribution. This conclusion evidently agrees with observations,[12] which for small particles gives exponents q from 1.4 to 2.4.

In order to trace the rate of evolution of the mass distribution n(m, t) from the presumed initial distribution of the asteroids and to detect the stochastic fluctuations of the exponent q, Dodd, Napier, and Preece[13] conducted direct modelling of the collisional processes by the Monte Carlo method. The authors studied separately the accreting systems and systems with fragmentation on the assumption that the collision probability of the bodies is proportional to their geometrical cross section. In the model of a system of bodies which are combining together in random collisions a mass spectrum arises which at the lower end of the masses is satisfactorily described by the power function $n(m) \propto m^{-q}$, where q decreases with time. But in the region of massive bodies a deficit is obtained relative to this power function, while in reality an excess of large bodies is observed in the asteroid system.

In the models of systems having fragmentation with an initial power distribution these authors assume that each of two colliding bodies is broken into fragments distributed according to the power function $n_1(m) \propto m^{-q_1}$, where q_1 is chosen randomly from the uniformly distributed interval (1.5, 2). Power distributions of the bodies in the region of small masses and a surplus of large bodies were obtained as a result. The exponent q, undergoing considerable fluctuations, increases with time from the initial value $q_0 = 1.6$ and reaches the value ~ 2.4, which differs markedly from the observed value of 1.8 for asteroids.

In a report[14] on the problem of the origin of the asteroids Napier and Dodd again turned to numerical modelling of the collision processes in order to evaluate the correctness of the separate hypotheses. In the case of accretion (in a special model with a constant number of bodies and an increasing total mass) a stable solution is obtained with $q = 1.5$. In the model of fragmentation of many bodies with an initial distribution $n(m, 0) \propto m^{-1.8}$ the distribution is stable for a certain time, but after 580 collisions q begins to increase, possibly similar to Hellyer's result for large masses. In studying the hypothesis of the fragmentation of several large initial bodies the authors obtained a stable distribution for small masses with $q = 1.8$ and an excess of large bodies over the power function, i.e., the large bodies have a tendency toward preservation.

2. VARIATION IN DISTRIBUTION FUNCTION OF PROTOPLANETARY BODIES IN PROCESS OF ACCUMULATION WITH FRAGMENTATIONS

The mass distributions of bodies in collisional processes of fragmentation or accumulation are studied in the works given above. However, in a protoplanetary cloud at a certain stage with a large dispersion of masses and velocities these processes could proceed in parallel, one overlapping the other under certain conditions.

An equation describing the variation with time in the mass distribution function of the protoplanetary bodies in the process of their accumulation with allowance for catastrophic fragmentation was obtained by V. S. Safr .ov.[15]

This equation for bodies in the region of masses

$$m_0 \leq m \leq M \tag{4}$$

is written in general form in the following way[16]

$$\frac{\partial n(m,t)}{\partial t} = \int_{m_0}^{\alpha(m)m} A(m_1, m-m_1)\, n(m_1,t)\, n(m-m_1,t)\, dm_1 - n(m,t)\int_{m_0}^{M} A(m,m_1)\, n(m_1,t)\, dm_1$$
$$+ \int_{m^*}^{M^*} n_1(m,m'') \int_{\alpha(m'')m''}^{m''/2} A(m_1, m''-m_1)\, n(m_1,t)\, n(m''-m_1,t)\, dm_1\, dm'', \tag{5}$$

where $A(m, m_1)$ is the coagulation coefficient; $n_1(m, m'')$ is the distribution by mass m of the fragments formed in the fragmentation of two colliding bodies having a total mass m''; m^* and M^* are the lower and upper limits, respectively, of the sum of the masses of the fragmenting bodies providing to the system a supply of fragments of mass m. The condition of fragmentation or joining together of the colliding bodies m_1 and m_2 is determined by the parameter α:

$$\text{when} \quad \frac{m_1}{m_1+m_2} \leqslant \alpha \quad \text{the bodies are joined together,}$$

$$\text{when} \quad \frac{m_1}{m_1+m_2} > \alpha \quad \text{complete fragmentation occurs.} \qquad (6)$$

A study of this equation is presented in the report of E. V. Zvyagina, G. V. Pechernikova, and V. S. Safronov.[16] First of all, the form of the functions entering into the equation is determined. The condition of destruction of a gravitating body of mass m_2 during a collision with another body of smaller mass m_1 is obtained from energetic considerations. For bodies which are not very large this condition can be approximated by the expression

$$\frac{m_1}{m_1+m_2} = \frac{m_1}{m''} > \alpha(m'') = a\left(\frac{m''}{2M}\right)^{\lambda}. \qquad (7)$$

The mass distribution of the collection of fragments of both colliding bodies is taken in accordance with experimental data in the form of an inverse power function with an exponent $q_1 \approx 1.8$

$$n_1(m, m'') = c_1 m^{-q_1} \quad \text{for} \quad m_0 \leqslant m \leqslant M_1, \qquad (8)$$

where c_1 is determined from the condition of conservation of mass

$$c_1 = (2-q_1) M_1^{q_1-2} m'' [1-(m_0/M_1)^{2-q_1}]^{-1}. \qquad (9)$$

The upper limit M_1 of the mass distribution function of the fragments is taken as proportional to the total mass of the destroyed bodies

$$M_1 = \xi m'', \quad \xi \leqslant 1. \qquad (10)$$

The collision probability $A(m_1, m_2)$ is taken in the form

$$A(m_1, m_2) = A_1 \sum_{i \leqslant k} b_i m_1^{i} m_2^{k-i}, \qquad (11)$$

where i takes certain whole or fractional values from 0 to k. The parameter k determines the collision cross section. For small bodies $k = 2/3$ characterizes the geometrical cross section, for large bodies $k = 4/3$ characterizes the gravitational cross section, while for $k = 1$ and integers i we have $A \propto m_1 + m_2$.

Taking into account Eqs. (6)-(11), the solution of Eq. (5) was postulated in the form of the inverse power function

$$n(m, t) = c(t) m^{-q(t)}. \qquad (12)$$

A qualitative study was made of the equation to determine the exponent q. Stable asymptotic values of the exponent q, depending on the coagulation coefficient, the nature of the fragmentation of the bodies, and the sizes of the largest fragments, were obtained for bodies in the region of masses $m_0 \ll m \ll M$.

In order to trace the variation in the mass spectrum of the colliding bodies with time it is necessary to apply numerical methods of solution. Analytically found solutions are obtained only with a number of simplifying assumptions, in particular concerning the coagulation coefficient. Numerical integration is the sole method of solving the problem if we wish to study the models closest to reality and not simplified, idealized schemes.

Because of the very large n in Eq. (5), for small m the first two integrals in the right-hand part represent the difference between large values. Therefore it is more convenient for a numerical solution to take as the unknown function not n(m, t) but the function $\mu(m, t) = mn(m, t)$ which determines the total mass incorporated in bodies with masses in the range from m to m + dm at the time t and per unit volume.[15] The function $\mu(m, t)$ provides a graphic picture of the mass distribution in the system and subsequently simplifies the control of the conservation of mass. Now the equation is written in the form

$$\frac{\partial \mu(m,t)}{\partial t} = m \int_{m_0}^{m\alpha(m)} \frac{A(m_1, m-m_1)\mu(m_1,t)\mu(m-m_1,t)}{m_1(m-m_1)} dm_1$$
$$-\mu(m,t) \int_{m_0}^{M} \frac{A(m, m_1)\mu(m_1,t)}{m_1} dm_1 + \int_{m^*}^{M^*} \mu_1(m, m'')$$
$$\times \int_{m''\alpha(m'')}^{m''/2} \frac{A(m_1, m''-m_1)\mu(m_1,t)\mu(m''-m_1,t)}{m_1(m-m_1)} dm_1\, dm''. \qquad (13)$$

At each stage in time in the process of the numerical solution of this equation it is necessary to calculate four integrals in the right-hand part, with the inner integral in the last term having to be calculated repeatedly. This greatly increases the computation time. Therefore let us convert this equation to a form more economical for calculation:[15]

$$\frac{\partial \mu(m,t)}{\partial t} = m \int_{m_0}^{m/2} \frac{\mu(m_1,t)}{m_1} \left[\frac{A(m_1 m-m_1)\mu(m-m_1,t)}{m-m_1} \right.$$
$$\left. - \frac{A(m_1, m)\mu(m,t)}{m} \right] dm_1 - \mu(m,t) \int_{m/2}^{M} \frac{A(m, m_1)\mu(m_1,t)}{m_1} dm_1$$
$$-m \int_{m\alpha(m)}^{m/2} \frac{A(m_1, m-m_1)\mu(m_1,t)\mu(m-m_1,t)}{m_1(m-m_1)} dm_1 + \int_{m^*}^{M^*} \mu_1(m, m'')$$
$$\times \int_{m''\alpha(m'')}^{m''/2} \frac{A(m_1, m''-m_1)\mu(m_1,t)\mu(m''-m_1,t)}{m_1(m''-m_1)} dm_1\, dm''. \qquad (14)$$

By combining part of the second integral with the first and separating from the first integral the part containing fragmentation in the form of a third integral, we observe that it completely coincides in form with the inner integral in the fourth term. In calculating the right-hand part we can simplify the calculation of the fourth integral by retaining in the machine memory the values of the third integral for all m.

In Eq. (14) $\mu(m, t)$ is the unknown function and $\mu_1(m, m'') = mn_1(m, m'')$, where $n_1(m, m'')$ is determined according to (8) with allowance for (9) and (10).

3. CONDITION OF FRAGMENTATION AND COAGULATION COEFFICIENT

In the simplest case, taking α = const, we will have

at $\alpha = 1/2$ the absence of fragmentation in the entire range of masses, while at $\alpha = 0$ the collision of two bodies of any sizes results in their destruction.

The condition of destruction of two colliding bodies as a function of their masses m_1 and m_2, obtained in ref. 16, has the form

$$\frac{m_1}{m_1+m_2} \geqslant \frac{1-\chi-\beta}{\chi}\left[\frac{5}{3\theta}\frac{m_2}{m_1+m_2}\left(\frac{M_0}{m_1+m_2}\right)^{2/3} - \left(\frac{m_1}{m_1+m_2}\right)^{2/3} - \frac{m_2}{m_1}\left(\frac{m_2}{m_1+m_2}\right)^{2/3}\right]^{-1} = \alpha, \tag{15}$$

where χ is the fraction of the kinetic energy T out of all the energy T + Q liberated in the collision, and β is the fraction of the potential energy of the fragmented bodies having a total mass $m_{12} = m_1 + m_2$ which goes into the potential energy of the fragments obtained as a result of the collision of m_1 and m_2. M_0 and R are the mass and radius of the largest body in this system which determines the relative velocities of the bodies[15] according to the equation

$$v = \sqrt{GM_0/\theta R} \tag{16}$$

(θ is a dimensionless parameter depending on the properties of the system).

In the approximation of the complicated condition of destruction of the bodies (15) by the simpler function (7) the mass M_0 of the largest body was identified with the upper limit of the mass distribution of bodies in the system. The factor 2 in the denominator of (7) allows for the possibility of the collision of bodies of comparable masses in the region of large bodies.

However, such an approach is undesirable in the numerical solution of the equation, when the time dependence of the process of accumulation and fragmentation of bodies during collisions is studied for any initial mass distribution. The condition of fragmentation of the bodies depends on their relative velocity, which in turn is determined by the mass of the largest body in the system at the given moment. Therefore at each successive step in time in the numerical solution M in the condition of destruction (7) is taken as equal to the mass of the largest body, estimated in the preceding step.

The qualitative study of Eq. (4) was conducted for intermediate masses m remote enough from the upper and lower limits of the distribution so that the first term in the brackets in (15) is much larger than the other terms and they can be neglected in the simplified description of the condition of fragmentation (7). In the numerical solution of the equation the interaction of bodies in the entire range of masses is taken into account and consequently all the terms of (15) must be kept in mind. In this case from the numerical solution obtained for the inequality (15) $\alpha(m'')$ is equal to

$$\alpha(m'') = \frac{1-\beta}{\chi}\frac{3\theta}{5.5}\left(\frac{m''}{M_0}\right)^{2/3}\left[1-1.1(3\theta)^2\frac{m''^{4/3}}{M_0^{4/3}}\right]^{-1}. \tag{17}$$

which is correct as long as $\alpha \leq 0.5$.

With allowance for the gravitational attraction of the bodies[15] their collision frequency is equal to

$$A(m_1, m) = \pi(r+r_1)^2[1+2G(m+m_1)/V^2(r+r_1)]V, \tag{18}$$

where r and r_1 are the radii of bodies with masses m and m_1, respectively, and V is the velocity of body m relative to body m_1 before their approach. In the case of small bodies $A(m, m_1)$ is determined by their geometrical cross section $\pi(r + r_1)^2$, while in the case of large bodies the collision cross section increases because of attraction and $A(m_1, m)$ is approximately proportional to $(m + m_1)(r + r_1)$.

Converting from radii to masses and taking V = v, according to (16) we obtain

$$A(m_1, m) = A_2[M_0^{1/3}(m^{1/3}+m_1^{1/3})^2 + 2\theta M_0^{-1/3}(m^{1/3}+m_1^{1/3})(m+m_1)], \tag{19}$$

where $A_2 = \sqrt{3\pi G/4\theta\delta}$ and δ is the density of the bodies.

4. METHOD OF CALCULATION

The solution of the problem on an electronic computer requires a program which provides the following 1) the numerical solution of the intergrodifferential equation (14) with a given accuracy; 2) assignment of the initial function $\mu(m, 0)$ in analytical form or in tabular form; 3) obtaining the solution in the form of the parameters of a given analytical function which approximates the mass distribution of the bodies, or in tabular form; 4) control of the conservation of mass in the system; 5) automatic regulation of the step in mass and the step in time in accordance with the assigned criteria; 6) calculation of the mass of the largest body in the system for the subsequent calculation of the function determining the condition of fragmentation and the coagulation coefficient.

Numerical integration requires the transition from continuous functions to discrete functions. For this the entire region of masses determined by the inequality (4) is divided into n equal intervals of Δm, m_i designates the value $i \cdot \Delta m$, where $i = 1, 2, \ldots, n$, and $\mu_i(t)$ is the mass incorporated in the bodies m_i at the time t. The variation in the total mass of the bodies m_i with time is determined by the equation

$$\partial\mu_i(t)/\partial t = \varphi_i(t), \tag{20}$$

where $\varphi_i(t)$ designates the right-hand side of Eq. (14) for $m = m_i$. Thus, Eq. (14) is separated into a system of n first-order differential equations whose right-hand sides are sums of integrals. The extrapolation method of Adams[17-19] is used for the integration of Eq. (20) with retention of differences up to the third order inclusively. Of all methods of approximately the same accuracy this method is the simplest.

Simpson's method[20,21] is used to calculate the integrals in the computation of the right-hand side. One peculiarity of the present problem is the movable right-hand limit of the function $\mu(m, t)$. During fragmentation the number of large bodies decreases and the upper limit of $\mu(m, t)$ moves to the left with time, while if accumulation prevails the mass flows over to the region of large bodies and the boundary moves to the right. The magnitude of n, up to which the calculation of the functions $\mu_i(t)$ is carried, varies as a function of the intensity of the process. And the size Δt_{k+1} of the step in time is regulated as a function of the size of the increment $\Delta n_k = n_k - n_{k-1}$ at the k-th step in time.

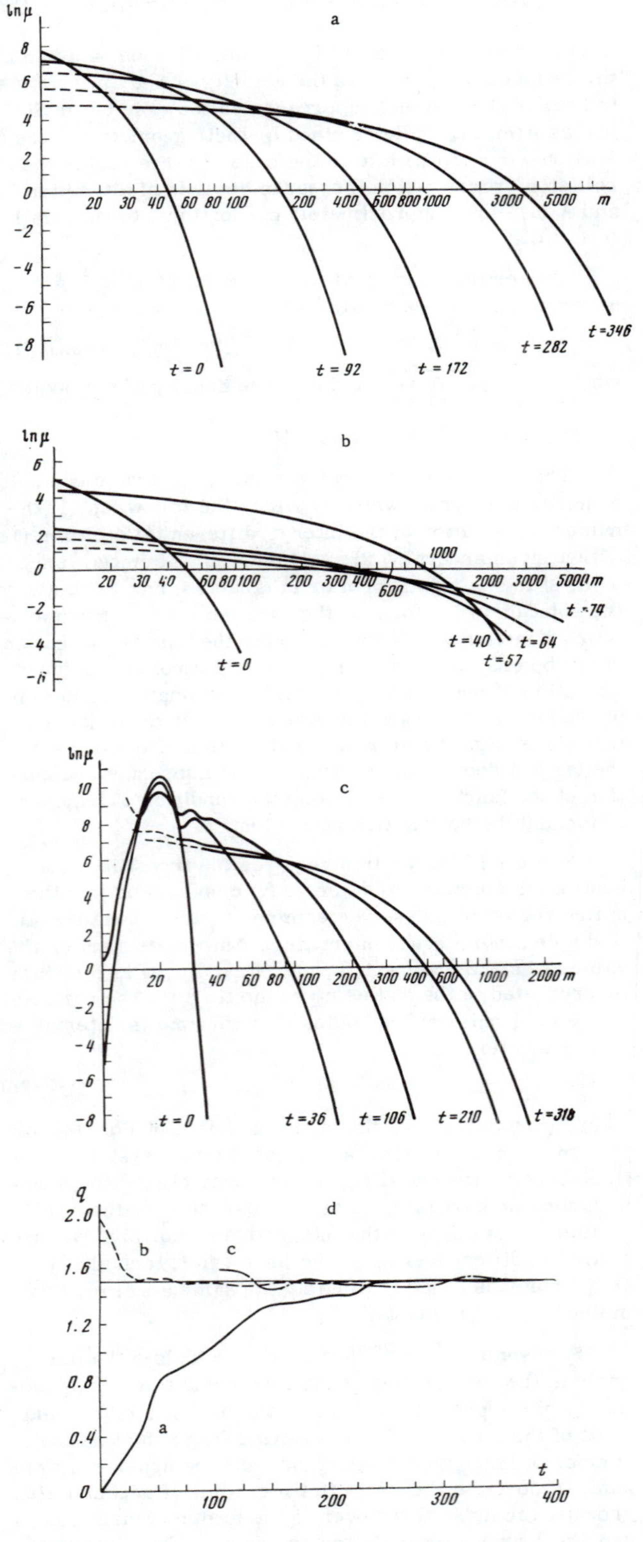

Fig. 1. Evolution of mass distribution of bodies with coagulation coefficient $A \propto m_1 + m_2$ and α = const from an initial distribution given in the form: a) $\mu(m, 0) = a_0 m e^{-m}$ without fragmentation ($\alpha = 0.5$); b) $\mu(m, 0) = a_0 m^{-1} e^{-0.01m}$ without fragmentation ($\alpha = 0.5$); c) $\mu(m, 0) = a_0 m e^{-b(m-m)^2}$ with weak fragmentation ($\alpha = 0.48$); d) variation with time of exponent q of inverse power function of distribution.

The masses of the bodies increase in the process of accumulation and consequently the range of the mass distribution of bodies in the system increases. When the range of masses reaches twice the value of the initial range $n_0 (n = 2n_0)$ a twofold decrease is made in the scale, i.e., the initial size of the step in mass is doubled $\Delta m = 2\Delta m_0$ and the number n of steps is cut in half ($n/2 \approx n_0$). For the control of the conservation of mass of the system the following integral is calculated at each step in time:

$$\int_{m_0}^{M} \mu(m, t)\, dm = \rho. \tag{21}$$

The mass of the largest body is calculated in two stages. First, that value of m_1 beyond which only one body remains is determined from the condition

$$\int_{m_1}^{M} \frac{\mu(m)}{m}\, dm = 1$$

for the values of the function $\mu(m)$ obtained at the given step in time. Then the mass of this body is calculated:

$$M_0 = \int_{m_1}^{M} \mu(m)\, dm.$$

By analogy with the analytical solution of (3) for the equation of coagulation without fragmentation, as well as for convenience in the analysis of the mass distribution function, $\mu(m)$ is approximated by the analytical function

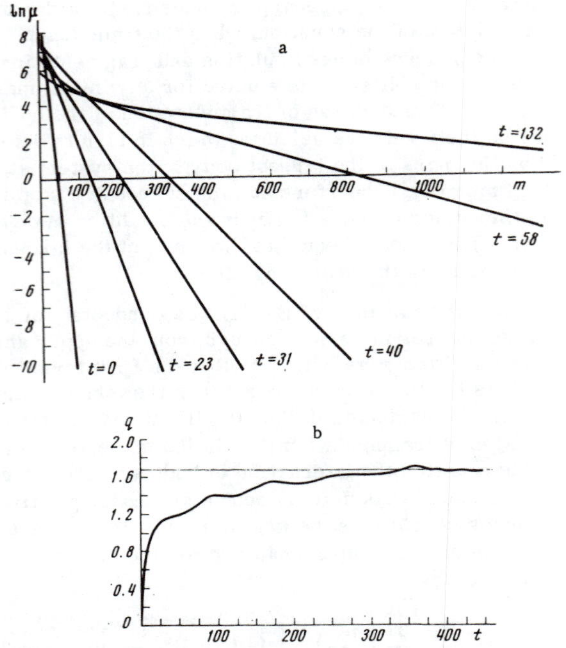

Fig. 2. a) Variation with time of function $\mu(m, t)$ from initial distribution $\mu(m, 0) = a_0 m e^{-m}$ in process of accumulation with fragmentation at $\alpha = 0.4$ and $A \propto m_1 + m_2$; b) variation with time of exponent g of approximating function $\tilde{\mu}(m, t)$ from initial value $q_0 = 0$ for the case $\alpha = 0.4$ and $A \propto m_1 + m_2$.

$\tilde{\mu} = a m^{1-q} e^{-bm}$, for which the parameters q, a, and b are calculated for the different mass ranges. The variation with time of the values q, a, and b reflects the evolution of the mass distribution of the system. A comparison of the values obtained in the process of the numerical integration of Eq. (14) with the values calculated from the parameters q, a, and b makes it possible to estimate the accuracy of such an approximation.

Figure 1, a–c illustrates the variation of the function μ(m, t) with time for different initial distributions μ(m, 0) for the cases of accumulation without fragmentation and with very weak fragmentation. As seen from Fig. 1, d all three distributions converge to an asymptotic power solution with the exponent q = 1.5.

An example of the evolution of the distribution function μ(m, t) is shown in Fig. 2 for the case when fragmentation plays an important role in the accumulation process. In this specific case the exponent of the power distribution approaches the value q = 1.66. The waves in the curve of Fig. 2, b are connected with the specifics of the calculation: The section of the curve μ(m, t) for which q is estimated changes upon doubling of the step in mass. A detailed analysis of the results will be given in the next section of the report.

[1]M. Smolukhovskii, in: Brownian Motion [in Russian], (1936).
[2]Z. A. Melzak, Quart. Appl. Math., 11, 231 (1952).
[3]V. S. Safronov, Dokl. Akad. Nauk SSSR, 147, 64 (1962) [Sov. Phys. – Dokl., 7, 967 (1963)].
[4]A. M. Golovin, Izv. Akad. Nauk SSSR, Ser. Geofiz., No. 5, 783 (1963).
[5]B. A. Trubnikov, Dokl. Akad. Nauk SSSR, 196, 13 (1971).
[6]E. V. Zvyagina and V. S. Safronov, Astron. Zh., 48, 1023 (1971) [Sov. Astron., 15, 810 (1972)].
[7]S. Piotrowsky, Acta Astron., Ser. A, 5, 115 (1953).
[8]J. S. Dohnanyi, J. Geophys. Res., 74, 2531 (1969).
[9]B. Hellyer, Mon. Not. Roy. Astron. Soc., 148, 383 (1970).
[10]B. Hellyer, Mon. Not. Roy. Astron. Soc., 154, 279 (1971).
[11]L. W. Bandermann, Mon. Not. Roy. Astron. Soc., 160, 321 (1972).
[12]P. M. Millman, The Moon, 8, 228 (1973).
[13]R. J. Dodd, W. M. Napier, and A. A. Preece, Astrophys. Space Sci., 18, 196 (1972).
[14]W. M. Napier and R. J. Dodd, Mon. Not. Roy. Astron. Soc., 166, 469 (1974).
[15]V. S. Safronov, Evolution of the Protoplanetary Cloud and Formation of the Earth and Planets [in Russian], Nauka, Moscow (1969).
[16]E. V. Zvyagina, G. V. Pechernikova, and V. S. Safronov, Astron. Zh., 50, 1261 (1973) [Sov. Astron., 17, 793 (1974)].
[17]W. E. Milne, Numerical Calculus, Princeton (1949).
[18]E. Kamke, Handbook on Simultaneous Differential Equations [Russian translation], Nauka, Moscow (1971).
[19]L. Collatz, Numerical Treatment of Differential Equations, Springer (1966).
[20]I. S. Berezin and N. P. Zhidkov, Methods of Computation [in Russian], Nauka, Moscow (1966).
[21]B. P. Demidovich and I. A. Maron, Fundamentals of Computational Mathematics [in Russian], Nauka, Moscow (1966).

Gravitational instability in rotating systems with radial perturbations

I. L. Genkin and V. S. Safronov

Astrophysical Institute, Academy of Sciences of the Kazakh SSR, Alma-Ata
Institute of Earth Physics, Academy of Sciences of the USSR, Moscow
(Submitted January 4, 1974)
Astron. Zh., 52, 306-315 (March-April 1975)

Conditions for the development of gravitational instability through radial perturbations are established for four model rotating system: an unbounded homogeneous medium, an infinitely thin disk, and a layer of finite thickness with solid-body or Keplerian rotation. The dispersion relations for the finite layer and the thin disk are quite different. The minimum density for onset of instability is $\rho_{cr} \approx 2\rho^*$, where $\rho^* = 3M/4\pi R^3$ (here M denotes the mass within a sphere of radius R), and the critical wavelength $\lambda_{cr} \approx 8H$, where H represents the characteristic thickness of the disk. Explanations are given for the substantial differences in the values of ρ_{cr} estimated by various authors.

Much attention has lately been devoted to the gravitational stability of rotating systems. One reason is the prospect for obtaining some insight into the origin of such galactic structures as bars, spiral arms, and star clouds. Although there has been definite progress in this direction, the problem as a whole is far from solved. The models generally considered are highly schematic and often bear little resemblance to real objects. Some investigations contain outright mistakes or arbitrary assumptions. Many of the results obtained even for simple model stellar systems therefore need to be reconsidered.

In this paper we shall examine four models for a rotating system: a homogeneous mass distribution, an infinitely thin disk, and disks of finite thickness having either solid-body or Keplerian rotation. These models in a sense represent limiting cases, and the properties of actual stellar systems can to some extent be obtained simply by interpolation.

We shall here adopt a hydrodynamic description of the models. The more accurate kinetic treatment is substantially more complicated, and furthermore in most cases it would yield no new information. An exception is provided by models with specially selected (as a rule, singular or nonmonotonic) distribution functions, wherein distinctive instabilities appear that have no analog in hydrodynamics. Kinetic instabilities will not, however, be considered in this paper.

1. Bel and Schatzman[1] have discussed the stability of an infinite homogeneous medium with nonuniform rotation. They have obtained a condition for instability of the medium against cylindrically symmetric perturbations (although they incorrectly regard their results as applicable to plane systems). Because of the approximate character of the method the condition contains a small extra term. But the problem can be solved exactly.

With the customary assumption that the perturbations are time-dependent according to the $e^{i\omega t}$ harmonic law (that is, $\partial/\partial t = i\omega$), Bel and Schatzman's initial linearized equations may be written in the form

$$i\omega u_1 - 2\Omega v_1 = \frac{\partial\Phi_1}{\partial R} - c^2\frac{\partial s}{\partial R},$$

$$i\omega v_1 + \frac{\varkappa^2}{2\Omega}u_1 = 0,$$

$$i\omega s + \frac{1}{R}\cdot\frac{\partial}{\partial R}(Ru_1) = 0,$$

$$\frac{1}{R}\cdot\frac{\partial}{\partial R}\left(R\frac{\partial\Phi_1}{\partial R}\right) = -4\pi G\rho s, \quad (1)$$

where $s = \rho_1/\rho$ is the relative density perturbation, u and v are the radial and tangential velocity components, Φ is the potential (the subscript 1 designates small perturbations), Ω is the angular rotational velocity of the system, ω is the angular frequency of harmonic oscillations, and $\varkappa^2 = 4\Omega^2[1 + (R/2\Omega)d\Omega/dR]$ represents the square of the epicyclic frequency.

Multiplying the first equation by R, differentiating the result with respect to R, and expressing the quantities v_1, $\partial(Ru_1)/\partial R$, and $\partial(R\partial\Phi_1/\partial R)/\partial R$ in terms of s by means of the other three equations (this can be done uniquely in all cases), we obtain the following equation:[1)]

$$\frac{c^2}{R}\cdot\frac{\partial}{\partial R}\left(R\frac{\partial s}{\partial R}\right) + s(\omega^2 + 4\pi G\rho - \varkappa^2) = 0.$$

Its solution, which represents a diverging wave, is given by a Hankel function, and with the time dependence we have adopted it takes the form

$$s = AH_0^{(2)}(qR)\exp i\omega t.$$

Within a bounded cylinder a standing wave will develop:

$$s = BJ_0(qR)\exp i\omega t.$$

The quantity q is nearly equal to the radial wave number $k = 2\pi/\lambda$, and is determined by the equation (which is, in fact, the dispersion relation)

$$q^2c^2 = \omega^2 + 4\pi G\rho - \varkappa^2. \quad (2)$$

The instability condition $\omega^2 < 0$ may also be written as

$$4\pi G\rho > q^2 c_{is}^2 + \varkappa^2 \approx \frac{4\pi^2 c_{is}^2}{\lambda^2} + \frac{2\Omega}{R}\cdot\frac{d}{dR}(\Omega R^2). \quad (3)$$

The analogous condition of Bel and Schatzman contains the extra term $c^2/4R^2$, which enters because these authors sought an unimportant sine-wave solution for a certain auxiliary quantity F.

We should also point out that the model considered above is not self-consistent.[2] For a self-consistent system with ρ = const, one must either include additional masses in the form of stars, so as to provide for a specified $\Omega(R)$, or assume that $\Omega^2 = 2\pi G\rho$ = const. In this event $\varkappa^2 = 4\Omega^2$, and Eq. (2) will agree exactly with Chandrasekhar's dispersion relation[3] for plane waves propagated perpendicular to the rotation axis, whereas Bel and Schatzman's equation will still contain an extra term. For cylindrical waves, unlike plane waves, the amplitude will decrease with distance R, while the wavelength λ will only asymptotically approach the value $2\pi/q$.

If one allows for heat conduction (Kato and Kumar;[4] see also a recent discussion by one of us[5]), then in the expressions for the critical quantities the ordinary adiabatic sound velocity c will be replaced by the isothermal value $c_{is} = (kT/m)^{1/2}$, which appears in Eq. (3). We shall henceforth regard the subscript "is" as understood and omit it; that is, c will be interpreted as the rms thermal velocity along a single coordinate. We may equivalently use the astronomical notation for the velocity dispersion σ (= c_{is}).

2. Infinitely thin layer. The velocity dispersion (like the sound velocity) may depend on direction. Let us consider the idealized case, where $c_z \ll c_R$. If c_z does not vanish entirely, the layer will have a certain (if very small) equivalent thickness H related to the surface density μ by $\mu = H\rho(0)$, where $\rho(0)$ is the density at z = 0. The density in the layer may be expressed mathematically as $\rho(R, z) = \mu(R)\delta(z)$.

The system of basic equations governing the development of perturbations in such a layer will agree with the system (1) if s is interpreted as the relative perturbation μ_1/μ in the surface density. Only the Poisson equation will change, as it will now include a dependence of the potential on the z coordinate:

$$\frac{1}{R}\cdot\frac{\partial}{\partial R}\left(R\frac{\partial\Phi_1}{\partial R}\right)+\frac{\partial^2\Phi_1}{\partial z^2}=-4\pi G\mu_1(R)\delta(z).$$

Polyachenko and Fridman[6] have shown that this equation has the solution

$$\Phi_1=\frac{2\pi G\mu_1}{q}\exp(-q|z|).$$

The method of solution, as these authors themselves point out, is applicable only for disturbances of short wavelength. In Appendix I we shall demonstrate that the solution is suitable for any q. The dispersion relation accordingly has the form

$$\omega^2-\varkappa^2+2\pi G\mu q-q^2c_R^2=0. \qquad (4)$$

Goldreich and Lynden-Bell[7] have obtained the same equation in a different way.

We see that the transition from a distribution homogeneous in z (ρ = const, $H \to \infty$) to a disk ($H \to 0$) involves only a replacement of the $4\pi G\rho$ term by $2\pi G\mu q$, or if plane waves are considered, by $2\pi G\mu k$. The reason is that perturbations in a plane ring will induce an attraction smaller by a factor of $2\rho/\mu q \approx \lambda/\pi H$ than in a cylinder of the same density infinitely long in the z direction. The question of the self-consistency of the model will not be examined here; we do not know whether a $\varkappa^2$ exists such that μ = const throughout all space. The formal validity of the dispersion relation does not depend on the answer. But we may consider a disk of finite size $R \le R_m$ for which a self-consistent solution does exist.

A model with $c_R \ne c_z$ is admissible in collisionless systems. But in systems with interaction, and a fortiori in a continuous medium, we should take $c_R \approx c_z$. Thus if the disk is infinitely thin it should be "cold." We will then have $c_R = 0$ in Eq. (4), and the disk will be unstable against disturbances of wavelength $\lambda < \lambda_{cr}' = 4\pi^2 G\mu/\varkappa^2$. However, an arbitrarily small velocity dispersion σ_R will stabilize disturbances of the shortest wavelength. If, on the contrary, $\sigma_R > \pi G\mu/\varkappa$, then by Eq. (4) we will have $\omega^2 > 0$, and the disk will be stable against any radial perturbations. This criterion is analogous to the stability condition[8] $\sigma_R > 3.36\, G\mu/\varkappa$ established from kinetic considerations. A boundary curve was derived[8] for the entire wavelength range from 0 to ∞, so that the criterion does not apply only to short-wavelength disturbances. Yet its domain of applicability is nevertheless restricted. In the first place, it is localized in the sense that the perturbations are assumed to have the form exp iqR. Secondly, it refers only to a Gaussian distribution function, which is by no means compatible with all density distributions $\mu(R)$. For the distribution function of R. Graham, the corresponding criterion[9] is $\sigma_R > 2.99\, G\mu/\varkappa$. Thus our hydrodynamic approach yields an intermediate value.

3. Disk of finite thickness (layer): $c_z = c_R \ne 0$. In this more realistic case the dispersion relation differs from Eq. (4). Nevertheless, Eq. (4) is often applied to a layer of finite thickness, perhaps because of the shortcomings of the customary normal-mode method. Waves propagated in such a layer cannot be either plane harmonic or cylindrical waves. The density in the layer, and thereby the wave propagation velocity, will depend on the distance from the plane of symmetry: $v_p = \omega/k = [c^2 + f(\rho, \varkappa, k)]^{1/2}$. In general the wave front should therefore be distorted.

A situation can occur where conditions suitable for gravitational instability arise in the plane of symmetry, while far from that plane the medium is almost stable (λ_{cr} is very large). We are evidently justified in regarding the entire layer as stable if it is stable at z = 0. Otherwise we could easily specify an instability condition at z = 0 analogous to the relation (3) by estimating the radial force in the plane of symmetry from a disturbance of a given scale λ. This force is proportional to $4\pi G\rho(0)\cdot f^*(\lambda/H)$, where the correction factor f^* approaches unity as $H \to \infty$, while $f^* \to kH/2 = \pi H/\lambda$ as $H \to 0$. The function f^* is almost independent of the boundary conditions, but under various assumptions regarding the z distribution of the density the estimates of f^* may depart by as much as 10% from their mean value. For arbitrary H it is convenient to adopt the simple interpolation equation

$$f^*\left(\frac{\lambda}{H}\right)=\left(1+\frac{\lambda}{\pi H}\right)^{-1}=\left(1+\frac{2}{kH}\right)^{-1}. \qquad (5)$$

Despite its simplicity, this expression is quite accurate. It yields, for example, a stability condition for a nonrotating isothermal layer ($\varkappa$ = 0). The dispersion relation for z = 0 then takes the form

$$\omega^2=k^2c^2-4\pi G\rho(0)\left(1+\frac{2}{kH}\right)^{-1}. \qquad (6)$$

We can find the critical value k_{cr} by setting $\omega^2 = 0$ and $c^2 = \sigma^2$. Then

$$k_{cr}=-\frac{1}{H}+\sqrt{\frac{1}{H^2}+\frac{4\pi G\rho(0)}{\sigma^2}}$$

or

$$\lambda_{cr} = \frac{\sigma^2 + \sqrt{\sigma^4 + 4\pi G\rho(0)\sigma^2 H^2}}{2G\rho(0)H}.$$

Since $H = 2\sigma^2/\pi G\mu = \sigma[2/\pi G\rho(0)]^{1/2}$, we find that

$$\lambda_{cr} = \sigma\sqrt{\frac{2\pi}{G\rho(0)}} = \frac{2\sigma^2}{G\mu} = \pi H,$$

which agrees with Ledoux's result.[10] Thus Eq. (6) gives accurate values not only for λ/H approaching zero or infinity, but also for $\lambda/H = \pi$. Satisfactory agreement may be expected for other values of λ/H as well. As long ago as 1960, one of us (Safronov[11]) calculated a quantity f analogous to f^* by using a model for a bounded homogeneous ring. Throughout the range of their definition the quantities f and f^* differ by no more than 10%, and they become equal near λ_{cr}.

The dispersion relation for a rotating layer with $c_z = c_R$ is given similarly by

$$\omega^2 = \varkappa^2 + q^2c^2 - 4\pi G\rho(0)\left(1 + \frac{2}{qH}\right)^{-1}. \qquad (6')$$

If we adopt $\lambda_{cr} = 2\pi/q_{cr}$, we obtain for λ_{cr} [omitting the argument in $\rho(0)$] the cubic equation

$$\lambda_{cr}^3\frac{\varkappa^2}{\pi H} + \lambda_{cr}^2(\varkappa^2 - 4\pi G\rho) + \lambda_{cr}\frac{4\pi\sigma^2}{H} + 4\pi^2\sigma^2 = 0.$$

One root of this equation is negative and is not physically meaningful. The other two roots may be either real and positive or complex, depending on the sign of the discriminant D. For $D < 0$ we have two positive roots λ_{cr} and λ'_{cr}. The interval between them corresponds to unstable perturbations. The values of $\varkappa$ are less than $\varkappa_{cr}$ when $D < 0$. Instability will therefore set in if the system is rotating slowly. The critical value of the discriminant is determined by the equation

$$(\varkappa_{cr}^2 - 4\pi G\rho)^3 + 8\frac{\varkappa_{cr}\sigma^2}{H^2} + 16\frac{\varkappa_{cr}^2\sigma^4}{H^4} + 80\frac{\varkappa_{cr}^2\sigma^2}{H^2}\pi G\rho - 16\frac{\sigma^2}{H^2}(\pi G\rho)^2 = 0.$$

In order to evaluate $\varkappa_{cr}$ the model must be made more specific. Consider a layer rotating around a central body of mass M at the Keplerian velocity $\Omega = (GM/R^3)^{1/2}$. Assume that $\rho(0) = \text{const}$ in its plane of symmetry. Unfortunately, even so simple a model evidently cannot be described analytically; in particular, we are not able to find the density distribution along the z coordinate. But we need only know the equivalent thickness H of the system as a function of $\rho(0)$ and σ^2. The corresponding calculations are set forth in Appendix II. They yield

$$H = \sqrt{\frac{2\sigma^2}{\pi G\rho(0)}}J = H_0 J,$$

where H_0 is the thickness of the "Ledoux layer," and the correction factor J differs significantly from unity only at short distances from the central body (where the rotation may be considered rapid and the layer stable).

Taking $J = 0.89$ as a first approximation, we obtain

$$\varkappa_{cr}^2 = \Omega_{cr}^2 = 0.74\pi G\rho(0).$$

For instability we must have $\Omega < \Omega_{cr}$; that is,

$$R^3 > \frac{M}{0.74\pi\rho(0)}$$

or equivalently

$$\rho(0) > \rho_{cr} = 1.8\rho^*.$$

A second approximation is unnecessary, because for $\rho/\rho^* = 1.8$ the value of J is 0.89. From our expression for $\varkappa_{cr}^2$ we find

$$\lambda_{cr} = \lambda_{cr}' = 7.8H = 6.9H_0.$$

Previously,[11] with another procedure,

$$\rho_{cr} = 2.1\rho^* \text{ and } \lambda_{cr} = \lambda_{cr}' = 8.2H = 7.3H_0$$

were obtained. These values differ slightly because ρ_0 was estimated[11] so as to ensure the instability of a homogeneous layer of thickness H with $\bar{\rho} \approx 0.9\rho_0$, whereas we have here considered instability in the plane $z = 0$ (as was done by Ledoux). In practice it suffices to adopt the values[2)] $\rho_{0,cr} \approx 2\rho^*$ and $\lambda_{cr} \approx 8H$.

Thus in a gaseous disk at a finite temperature, instability will appear at distances where the density in the plane of symmetry is twice the spread-out density of the central body. The critical length will be eight times the equivalent thickness of the layer at a given point in the system. As the distance from the central body increases, λ_{cr} will decrease, approaching the Ledoux value $\lambda_{cr} = \pi H_0$ as $R \to \infty$ (provided the layer with $\rho_0 = \text{const}$ is infinite). Meanwhile $\lambda'_{cr} \to \infty$. A whole zone of instability will therefore exist here.

When $\lambda = \lambda_{cr}$ or $\lambda = \lambda'_{cr}$, the growth rate, defined as $(-\omega^2)^{1/2}$, will vanish, so that disturbances on that scale will not in fact increase with time. The disturbances with the largest growth rate will have an intermediate wavelength, which can readily be determined by differentiating ω^2 in the dispersion relation with respect to λ and equating the derivative to zero. At very large distances, where $J = 1$, disturbances with $\lambda = 6.7H_0$ will have the maximum growth rate. Accordingly, in the event of radial perturbations a Keplerian system will be divided, beginning at a certain distance, into rings of approximately the same width. The outer rings, whose growth rate is

$$\sqrt{4\pi G\rho(0)\left(1 + \frac{\lambda}{\pi H}\right)^{-1}} \approx \sqrt{1.3\pi G\rho(0)},$$

will become separated most rapidly. One will recall that in a homogeneous system the Jeans dispersion relation indicates that disturbances of sufficiently large scale will grow equally rapidly at the rate $(4\pi G\rho)^{1/2}$.

A criterion for stability of the system may be written in yet another form. We multiply the condition $\varkappa^2 > 0.74 \cdot \pi G\rho(0)$ by the equation $\sigma^2 = \pi G\rho(0)H^2/2J^2$. Inasmuch as $\mu = H\rho(0)$, we obtain

$$\sigma^2 > \frac{0.74\pi^2}{2J^2}\left(\frac{G\mu}{\varkappa}\right)^2$$

or

$$\sigma > 2.15\frac{G\mu}{\varkappa}.$$

4. We turn now to another limiting case: a homogeneous disk of finite thickness in solid-body rotation (a highly flattened Maclaurin spheroid). B. Lindblad had been concerned with problems of this nature in his research on the origin of spiral structure. His results nevertheless are not fully comparable to ours, as Lindblad did not consider Maclaurin spheroids but approximately spheroidal stellar systems with an exponential density distribution along the z coordinate.

Maclaurin ellipsoids have been analyzed from the kinetic standpoint by Polyachenko and Shukhman.[13,14] But the Fridman velocity distribution function used by them is quite artificial, a circumstance that becomes especially noticeable for systems of small oblateness. In particular, homogeneous spheres with arbitrary elliptical particle or star orbits are stable, whereas the same spheres with the Fridman distribution function are unstable.

Goldreich and Lynden-Bell[7] have considered the stability of a uniformly rotating layer with a specified density distribution along the z coordinate – in particular, with ρ = const for $z < z_e$ and $\rho = 0$ for $z > z_e$. The layer is regarded as homogeneous in other directions, despite the rotation. Unless z_e becomes infinite, such a model will not be self-consistent. The centrifugal force is therefore assumed to be compensated by a gravitational field exterior to the layer. In effect, Goldreich and Lynden-Bell examine the influence of Coriolis forces on the stability of this external field. We have already mentioned that a violation of self-consistency will not affect the form of the dispersion relation; hence these authors' results are compatible with ours for a self-consistent model. The coefficients differ somewhat merely because of the distinction in the laws adopted for the density distribution and in the velocity dispersion along the z direction. The Maclaurin spheroid has a density ρ = const out to a boundary determined by the equation

$$\frac{R^2}{a^2}+\frac{z^2}{b^2}=1.$$

According to the hydrostatic-equilibrium conditions, the velocity dispersion in a homogeneous spheroid should be equal to

$$\sigma^2(R, z)=\sigma^2(0,0)\left(1-\frac{b^2R^2+a^2z^2}{a^2b^2}\right).$$

In particular, this relation implies that

$$\overline{\sigma^2}={}^2/_5\sigma^2(0,\ 0).$$

For highly flattened Maclaurin spheroids we have the expansion

$$\frac{\Omega^2}{2\pi G\rho}=\frac{\pi}{2}\cdot\frac{b}{a}-3\left(\frac{b}{a}\right)^2+\ldots$$

Using the virial theorem

$$M\overline{v^2}+I\Omega^2+U=0,$$

where $v^2 = 3\sigma^2$, I is the moment of inertia, and U is the potential energy, and omitting terms of order b^3/a^3 or higher, we obtain from these expressions

$$b^2=\frac{\sigma^2(0,0)}{2\pi G\rho},$$

whence the thickness of the spheroid is given by

$$H(R)=\sqrt{\frac{2\sigma^2(R,0)}{\pi G\rho}}.$$

This equation is analogous to Ledoux's expression for an isothermal layer, except that here H is not the effective but instead the actual R-dependent thickness of the homogeneous spheroid.

The dispersion relation and the equations for the critical wavelength agree now with the equations obtained for a layer with a massive central body if in those equations we put J = 1 and $\varkappa^2 = 4\Omega^2$. We then readily find

$$\varkappa_{cr}^2=4\Omega_{cr}^2=0.84\pi G\rho$$

and

$$\lambda_{cr}(R)=\lambda_{cr}'(R)=6.9H(R).$$

The critical value $\varkappa_{cr}^2$ corresponds to an axial ratio

$$\left(\frac{b}{a}\right)_{cr}=0.0783\approx\frac{1}{13},$$

or an eccentricity e_{cr} = 0.9969.

The stability condition may also be written in the form

$$\sigma>2.03\frac{G\mu}{\varkappa}=2.14\frac{G\mu}{\varkappa_c},$$

which is nearly the same as the condition (7). Thus the stability criteria expressed in terms of the "circular" epicyclic frequency $\varkappa_c$ (as defined in Appendix III) practically coincide for the two limiting cases of a homogeneous spheroid and a layer with a massive central nucleus. The estimates for the other models obtained by Goldreich and Lynden-Bell[7] also are nearly the same. In their calculations these authors use the mean density $\bar{\rho}$ of the layer and the thickness T defined by the relation $\mu = T\bar{\rho}$. Converting to our notation, we obtain for a layer with $\gamma = 1$

$$\varkappa_{cr}^2=0{,}91\pi G\rho(0),\quad \sigma_{cr}=2.12\frac{G\mu}{\varkappa_c},\quad \lambda_{cr}=6.7H.$$

The limiting cases considered above cover a substantial portion of the range of conditions encountered in rapidly rotating gaseous and stellar systems. As the results do not differ very much (even if one includes an incompressible medium, which was also investigated by Goldreich and Lynden-Bell), it is appropriate to accept mean values. A system may be regarded as stable against radial perturbations if the epicyclic frequency satisfies the condition

$$\varkappa^2\geqslant0.8\pi G\rho(0)$$

or if the velocity dispersion exceeds the critical minimum value

$$\sigma_{R,\min} \approx 2.1 \frac{G\mu}{\varkappa_c}.$$

The two conditions are equivalent, with one implying the other. If we compare this last result with the corresponding criterion for an infinitely thin disk, we find that in a layer of finite thickness perturbations can be stabilized with a velocity dispersion 1.5 times lower, and with 1.5 times less energy of random motion (allowing for the difference in the number of degrees of freedom).

Returning to a spheroid, we observe that at the critical oblateness $\lambda_{cr}(0)$ is approximately half the radius a. The properties of the system (such as its thickness H) will change significantly over such distances, and a local analysis based on the dispersion relation is not fully justified. Hence estimates of the critical spheroid parameters, in particular the value of e_{cr} itself, should be regarded as highly approximate. The same is true of Toomre's estimates.[8]

If $e > e_{cr}$, then $\lambda'_{cr} > \lambda_{cr}$, and a whole spectrum of unstable perturbations will exist. On the other hand, if $e < e_{cr}$ the spheroids will be stable against radial perturbations. But out to some other critical value of the eccentricity they will still be unstable with respect to nonradial perturbations, such as azimuthal Lindblad density waves or changes in shape.

APPENDICES

I. It is natural to assume that if μ_0 = const and $H \to 0$, as in the case ρ_0 = const, Bessel functions for the perturbation μ_1 should provide a solution of the linearized equations. That is, μ_1 would satisfy the Bessel equation

$$\frac{1}{R} \cdot \frac{\partial}{\partial R} \left(R \frac{\partial \mu_1}{\partial R} \right) + q^2 \mu_1 = 0.$$

The procedure discussed above for solving the system (1) indicates that such a result will be obtained only if at z = 0 we have

$$\frac{1}{R} \cdot \frac{\partial}{\partial R} \left(R \frac{\partial \Phi_1}{\partial R} \right) = -\alpha s = -\alpha \frac{\mu_1}{\mu_0};$$

where α is a constant not yet specified. Upon comparing these two equations we readily see that

$$\Phi_1 = \frac{\alpha \mu_1}{q^2 \mu_0} f(z), \quad \text{with} \quad f(0) = 1.$$

If we substitute this expression into the Poisson equation we find

$$-\alpha \frac{\mu_1}{\mu_0} f(z) + \frac{\alpha \mu_1}{q^2 \mu_0} f''(z) = -4\pi G \mu_1 \delta(z)$$

or

$$\alpha f''(z) - \alpha q^2 f(z) + 4\pi G \mu_0 q^2 \delta(z) = 0.$$

The solution of this equation is

$$\alpha f(z) = 2\pi G q \mu_0 e^{-q|z|}.$$

Since $f(0) = 1$, we obtain

$$\alpha = 2\pi G \mu_0 q,$$

$$f(z) = \exp(-q|z|).$$

We now have for Φ_1 the expression

$$\Phi_1 = \frac{2\pi G \mu_1}{q} \exp(-q|z|)$$

for any q.

II. Suppose that an isothermal layer with $\rho(0)$ = const is rotating about a massive body M. Assume further that in the radial direction the rotation is fully balanced by the gravitational force. For the z direction we may write the ordinary hydrostatic-equilibrium condition

$$-\frac{kT}{m} \cdot \frac{1}{\rho} \cdot \frac{d\rho}{dz} = \frac{GMz}{(R^2+z^2)^{3/2}} + 4\pi G \int_0^z \rho \, dz.$$

If z is small we may neglect the value of z^2 compared to R^2 as well as the variation of ρ in the integrand. Then we have

$$-\frac{kT}{m} \cdot \frac{1}{\rho} \cdot \frac{d\rho}{dz} = \frac{GM}{R^3} z + 4\pi G \rho(0) z,$$

whence we obtain for the density distribution the expression

$$\rho(z) = \rho(0) \exp\left(-\frac{z^2}{z_e^2} \right),$$

where

$$z_e^2 = \frac{2\sigma^2}{4\pi G \rho(0) + GM/R^3}.$$

By introducing a conventional averaged density $\rho^* = 3M \cdot (4\pi R^3)^{-1}$ we may write the preceding expression in the form

$$z_e^2 \approx \frac{\sigma^2}{2\pi G \rho(0)} J^2$$

where

$$J^2 = \frac{3\rho/\rho^*}{1+3\rho/\rho^*}.$$

More accurate calculations of J with allowance for the mass distribution at large z have been given elsewhere.[11,15] We find, then, that if $\rho(0)$ = const the equivalent thickness of the system and the surface density depend on R (or on ρ^*). According to Ruskol,[15] we have

$$H = \sqrt{\frac{2\sigma^2}{\pi G \rho(0)}} J = H_0 J,$$

where H_0 is the equivalent thickness of the nonrotating isothermal layer. As $R \to \infty$ we will have $J \to 1$. For $\rho(0)/\rho^* = 2$, which corresponds approximately to the distance at which instability sets in, the value of J is 0.89.

III. If the velocity dispersion is different from zero, the system will rotate at less than the circular velocity:

$$\Omega^2 = \Omega_c^2 + \frac{1}{R\rho} \cdot \frac{\partial}{\partial R} (\rho \sigma_R^2) + \frac{\sigma_R^2 - \sigma_\theta^2}{R^2}.$$

In the case we have considered, ρ = const and $\sigma_R = \sigma_\theta$, so that

$$\Omega^2 = \Omega_c^2 + \frac{1}{R} \cdot \frac{\partial}{\partial R} \sigma_R^2 = \Omega_c^2 - 4 \frac{b^2}{a^2} \pi G \rho.$$

Using the critical values obtained for b/a and $\pi G\rho$, we find

$$\Omega = 0.946\, \Omega_c.$$

For an infinitely thin disk ($\sigma_z \equiv 0$) we have the analogous relation

$$\Omega^2 = \Omega_c^2 + \frac{1}{R\mu} \cdot \frac{\partial}{\partial R} (\mu \sigma_R^2),$$

where now $\mu \neq$ const. With the hydrodynamic stability criterion $\sigma_R = \pi G\mu/\varkappa$ derived in the text, we obtain for the disk $\Omega = 0.876\Omega_c$. If we replace Ω by Ω_c in the expression for $\varkappa$, we arrive at the "circular" epicyclic frequency

$$\varkappa_c^2 = 4\Omega_c^2 \left(1 + \frac{R}{2\Omega_c} \cdot \frac{d\Omega_c}{dR}\right),$$

which ordinarily will exceed the actual epicyclic frequency. In particular, $\varkappa = 0.946\varkappa_c$ for the critical Maclaurin ellipsoid, while for a layer with a massive central body we will essentially have $\varkappa = \varkappa_c$.

1) Strictly speaking, this expression is obtained for $\varkappa^2$ = const, that is, for $\Omega^2(R) = C_1R^{-4} + C_2$. However, the inexactness introduced by using a more general expression for $\Omega(R)$ probably should be of no consequence.

2) These estimates differ considerably from the results of Ginzburg et al.,[12] who obtain an instability criterion in the form

$$\frac{M}{m} < 4\pi \frac{R}{H},$$

where $m = \pi R^2 \rho H$. This condition corresponds to

$$\rho_{cr} = \frac{1}{3\pi} \rho^* \approx 0.1\rho^*.$$

An inconsistency in the treatment of the problem accounts for the small values of m and ρ_{cr}. The dispersion relation is derived for a cold disk [it agrees with Eq. (4) for c ≡ 0], but at the same time the layer is considered to have a finite thickness H, corresponding to c ≠ 0.

1 N. Bel and E. Schatzman, Rev. Mod. Phys., 30, 1015 (1958).
2 V. S. Safronov, in: Problems of Magnetohydrodynamics and Cosmical Gas Dynamics [in Russian], Voprosy Kosmog., 10, 181 (1964).
3 S. Chandrasekhar, Vistas in Astron., 1, 344 (1955).
4 S. Kato and S. S. Kumar, Publ. Astron. Soc. Japan, 12, 290 (1960).
5 I. L. Genkin, in: Dynamics of Galaxies and Star Clusters [in Russian], ed. T. B. Omarov, Astrophys. Inst. Kazakh Acad. Sci., Alma-Ata (1973), p. 199.
6 V. L. Polyachenko and A. M. Fridman, "Multiarm spirals in a model disk galaxy" [in Russian], Preprint Novosibirsk. Inst. Yadernoi [Nuclear] Fiz., No. 50-70 (1970).
7 P. Goldreich and D. Lynden-Bell, Mon. Not. Roy. Astron. Soc., 130, 97 (1965).
8 A. Toomre, Astrophys. J., 139, 1217 (1964).
9 W. H. Julian, Astrophys. J., 155, 117 (1971).
10 P. Ledoux, Ann. Astrophys., 14, 438 (1951).
11 V. S. Safronov, Dokl. Akad. Nauk SSSR, 130, 53 (1960) [Sov. Phys. – Dokl., 5, 13 (1960)]; Ann. Astrophys., 23, 979 (1960).
12 I. F. Ginzburg, V. L. Polyachenko, and A. M. Fridman, Astron. Zh., 48, 815 (1971) [Sov. Astron., 15, 643 (1972)].
13 V. L. Polyachenko and I. G. Shukhman, Astron. Zh., 50, 97, 721 (1973) [Sov. Astron., 17, 62, 460 (1973-1974)].
14 A. G. Morozov, V. L. Polyachenko, and I. G. Shukhman, in: Dynamics of Galaxies and Star Clusters [in Russian], Astrophys. Inst. Kazakh Acad. Sci., Alma-Ata (1973), p. 172.
15 E. L. Ruskol, Voprosy Kosmog., 7, 8 (1960).

Mass distribution of protoplanetary bodies. II. Numerical solution of generalized coagulation equation

G. V. Pechernikova, V. S. Safronov, and E. V. Zvyagina

Institute of Earth Physics, USSR Academy of Sciences

(Submitted May 29, 1975)

Astron. Zh. **53**, 612–619 (May–June 1976)

An analysis is given of the main results of a numerical solution of the generalized coagulation equation describing the time variation in the mass distribution function of the protoplanetary bodies in the process of accumulation with fragmentation. It is shown that the mass spectrum in the region of large bodies can be satisfactorily approximated by the product of an inverse power function (asymptotic solution) times an exponential function. The dependence of the exponent of the power-law asymptotic distribution on the initial distribution and the nature of the fragmentation of the bodies is studied. The advantage of not including the largest body (the planetary embryo) in the general distribution function is discussed. The proper allowance for the relative velocities of the bodies in the fragmentation parameter allows one to study the protoplanetary swarm at different stages of its evolution.

PACS numbers: 95.10.+b

The time variation in the mass distribution function of protoplanetary bodies in the process of their combination and destruction in random collisions is described by a generalized coagulation equation.[1] A study of this equation is made in ref. 2. Stable asymptotic solutions of the power-law type are obtained for bodies in the range of masses sufficiently far from the upper and lower boundaries of the distribution, in which upon the collision of two bodies with masses m_1 and m_2 not only their combination but also their destruction are possible. Since the analytically found solutions are obtained only for certain simple cases, the use of numerical methods of solution was required in the further study of the generalized coagulation equation.

After the replacement of the distribution function $n(m, t)$ by the function $\mu(m, t) = mn(m, t)$ the initial coagulation equation was reduced to a form suitable for its numerical integration:

$$\frac{\partial\mu(m,t)}{\partial t} = m\int_{m_0}^{m/2}\frac{\mu(m_1,t)}{m_1}\left[\frac{A(m_1,m-m_1)\mu(m-m_1,t)}{m-m_1} - \frac{A(m_1,m)\mu(m,t)}{m}\right]dm_1 - \mu(m,t)\int_{m_0}\frac{A(m,m_1)\mu(m_1,t)}{m_1}dm_1$$
$$- m\int_{m\alpha(m)}^{m/2}\frac{A(m_1,m-m_1)\mu(m_1,t)\mu(m-m_1,t)}{m_1(m-m_1)}dm_1$$
$$+ \int_{m^*}^{M^*}\mu_1(m,m'')\int_{m''\alpha(m'')}^{m''/2}\frac{A(m_1,m''-m_1)\mu(m_1,t)\mu(m''-m_1,t)}{m_1(m''-m_1)}dm_1\,dm'', \quad (1)$$

where $A(m, m_1)$ is the coagulation coefficient, proportional to the probability of collision of bodies with masses m and m_1; the unknown function $\mu(m, t)$ determines the total mass included in bodies with masses in the interval from m to $m + dm$ at the time t per unit volume; $\mu_1(m, m'')$ is the mass distribution function of the collection of fragments of the bodies m_1 and $m'' - m_1$ which have collided; M and m_0 are the upper and lower limits of the masses of bodies in the system; (M^*, m^*) is the region of values of m'' for which fragments with a mass m are possible; $\alpha(m)$ is the fragmentation parameter, determining the critical mass ratio at which combination of the bodies is replaced by fragmentation: Bodies m_1 and m_2 fragment upon collision if

$$\frac{m_1}{m_1+m_2} > \alpha. \quad (2)$$

In ref. 3 the initial data are determined and a method is developed for the numerical solution of Eq. (1) on a computer. Calculations performed by this method made it possible to obtain the mass distribution of the bodies for different initial distributions, different types of coagulation coefficient, and different fragmentation conditions. For convenience of analysis of the solutions the mass distributions $\mu_k(m)$ calculated for successive times t_k were approximated by a function of the type

$$\tilde{\mu} = c_k m^{1-q_k} e^{-b_k m}. \quad (3)$$

The values of c_k, q_k, and b_k were calculated from three points for different sections of the curve $\mu_k(m)$. The results of the calculations are represented in the form of graphs of two types – functions $\mu(m)$ or $\ln\mu(m)$ for a given time t and dependences of the parameters c,

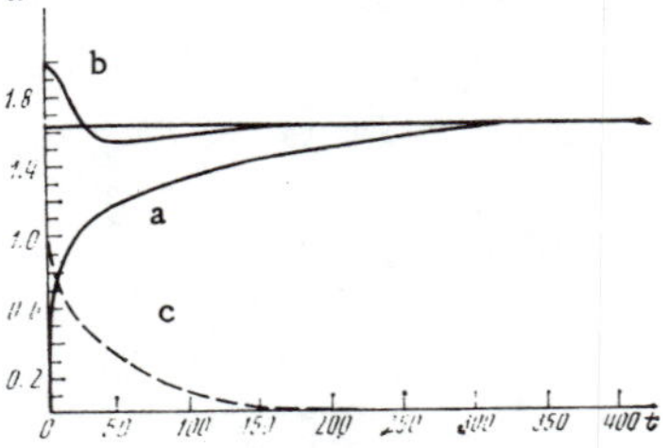

Fig. 1. Evolution of mass distribution of bodies – variation in exponent q – from the initial distribution, assigned in the form $n(m, 0) = c_0 m^{-q(0)} \cdot e^{-b(0)m}$, with a coagulation coefficient $A \propto m_1 + m_2$ and a fragmentation parameter $\alpha = 0.4$: a) $q(0) = 0$, $b(0) = 1$; b) $q(0) = 2$, $b(0) = 0.01$; c) time variation in the parameter b for an exponential function corresponding to case a.

q, and b of the approximating function $\tilde{\mu}(m)$ on time.

1. DEPENDENCE OF SOLUTION ON INITIAL DISTRIBUTION

Two series of calculations with a coagulation coefficient proportional to the sum of the masses of the colliding bodies were performed to study the dependence of the solution on the initial distribution.

The time variation in the function $\mu(m)$ and the parameters b and q of the distribution in a system without fragmentation were calculated with three different initial mass distributions n(m, 0) of the bodies [remember that $\mu(m) \equiv mn(m)$]:

a) an exponential function $n(m, 0) = c_0 e^{-m}$;

b) the product of a power-law function times an exponential function $(m, 0) = c_0 m^{-2} e^{-0.01 m}$;

c) a Gaussian law $n(m, 0) = c_0 e^{-b(m-\bar{m})^2}$.

The calculations showed[3] that independent of the initial mass spectrum assigned the process of combination of the bodies during collisions leads to a power-law asymptotic distribution of the type $n(m) = cm^{-3/2}$, which coincides with the exact solution obtained for a system without fragmentation.[4]

The dependence of the asymptotic solution on the initial distribution of the bodies was also studied for the equation of accumulation with fragmentation. Two solutions of Eq. (1) for $\alpha = 0.4$ [colliding bodies are fragmented when their mass ratio is $m_1/m_2 > 2/3$ and for a power-law distribution of fragments $n_1(m) = c_1 m^{-q_1}$ with an exponent $q_1 = 1.75$] are presented in Fig. 1. Curve a reflects the time variation in the exponent q from the initial value $q_0 = 0$ [distribution of bodies $n(m, 0) = c_0 e^{-m}$]; curve b reflects the variation in q from the initial value $q_0 = 2$ [i.e., $n(m, 0) = c_0 m^{-2} e^{-0.01 m}$]. In both cases q approaches the value $q = 1.66$ with time. The time variation in the parameter b from the initial value $b_0 = 1$, corresponding to curve a for q, is shown by a dashed line in the same figure.

The calculation of the parameters q and b was carried out in the mass interval of $5\Delta m$-$60\Delta m$ (see ref. 3), if m is expressed in units of Δm. With an increase in the mass of the bodies the mass interval is increased, and therefore the mass step Δm was regularly doubled. Thus, at the initial time the indicated interval $(5\Delta m, 60\Delta m)$ corresponded to masses of $(0.76$-$9.2) \cdot 10^{20}$ g, at the 150th step in time the parameters q and b were calculated for the distribution of bodies in the mass region of $(6.0$-$73) \cdot 10^{20}$ g, and at t = 400 steps in time after sixfold doubling of Δm this interval now corresponded to $(49$-$588) \cdot 10^{20}$ g, i.e., the mass region for which the calculated parameters of the distribution are valid is expanded in the direction of large bodies. In the same time b decreases from 1 to 0.003 (see Fig. 1), i.e., faster than Δm increases. The product $b\Delta m$ decreases and the role of the exponential function accordingly decreases in the region of masses for which the parameters q and b are estimated.

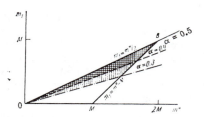

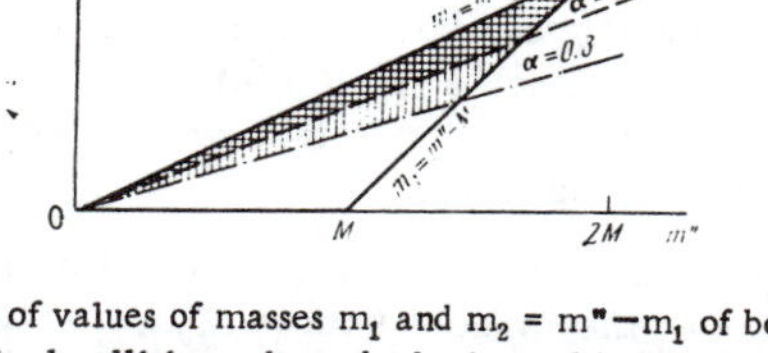

Fig. 2. Region of values of masses m_1 and $m_2 = m'' - m_1$ of bodies fragmenting in mutual collisions, shown by horizontal hatching for $\alpha = 0.4$ and by vertical hatching for $\alpha = 0.3$.

Thus, the calculations performed showed that with the elapse of a long enough time the mass distribution in a system of colliding bodies approaches an asymptotic distribution which is described by an inverse power law with an exponent q and does not depend on the initial distribution but is determined by the conditions of collision and fragmentation of the bodies.

2. DEPENDENCE OF SOLUTION ON FRAGMENTATION PARAMETER

a) The case α = const. A series of calculations with an initial distribution α of bodies and a coagulation coefficient $A(m, m_1) = A_1(m + m_1)$ are performed for different (constant) values of α. The corresponding regions of fragmentation are shown in Fig. 2: values of the total mass $m'' = m_1 + m_2$ of two bodies which have collided are laid out along the abscissa and values of the mass m_1 of the smaller of them are laid out along the ordinate; M is the upper limit of the mass distribution in the system; the triangle OBM bounds the region of possible values of m_1 as a function of m''. The region of fragmentation in the collision of bodies m_1 and m_2 is distinguished by vertical hatching for $\alpha = 0.3$ and by horizontal hatching for $\alpha = 0.4$; if $\alpha = 0.5$ there is no fragmentation and the bodies combine in the collisions.

The time variation in the exponent q of the power-law distribution found with the indicated values of α and with an exponent $q_1 = 1.75$ of the distribution of fragments is shown in Fig. 3. It follows from the results obtained that the exponent of the power-law asymptotic distribution increases with a decrease in the parameter α, i.e., with an increase in the fragmentability of the colliding bodies.

b) The case

$$\alpha(m'') = a(m''/M)^{\lambda}. \tag{4}$$

In this expression for α each of the three parameters *a*, M, and λ was varied separately.

The influence of the choice of M on the asymptotic value of q was estimated from a comparison of two solutions of Eq. (1) — one was obtained with a value of M equal to the upper limit of the mass distribution of the bodies and the other with a value of M equal to the mass M_0 of the largest body in the system at the given time, with the other conditions being equal. The calculations showed that the asymptotic solution is reached rather quickly: M_0 is able to increase by only two to four times and still remains several orders of magnitude smaller

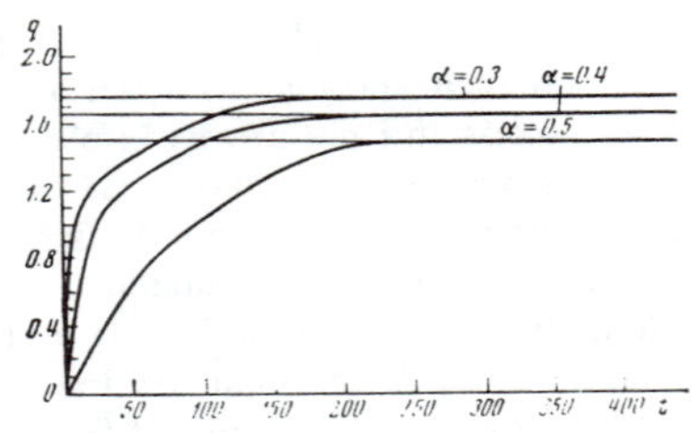

Fig. 3. Time variation in the exponent q for a coagulation coefficient $A \propto m_1 + m_2$ and different values of the fragmentation parameter α.

than the total mass of the system. Under these conditions the upper limit of the distribution is close to the mass of the largest body in the system. The asymptotic values of q obtained numerically almost coincided.

This result agrees with the data of ref. 5, in which the dependence of the upper limit M of the power-law mass distribution on the mass M_0 is estimated for several values of the exponent q.

The exponent λ in the function $\alpha(m'')$ characterizes the dependence of the fragmentability of the colliding bodies on their total mass. When $\lambda = 0$ the condition (4) is reduced to the previous case of α = const. With an increase in λ the parameter α is almost unchanged in the region of large bodies and increases strongly in the region of small bodies. It is seen from the condition (2) that the smaller α, the wider the interval of mass ratios m_1/m_2 in which fragmentation of the bodies occurs. With an increase in λ the fragmentability of small- and medium-sized bodies grows considerably whereas the very largest bodies in the system are subject to destruction only by moderately large bodies. The number of the latter decreases because of their easier fragmentation, and the probability of survival of the largest bodies in the system thereby is relatively increased and their accumulation is hastened. Consequently, the predominant strengthening of the fragmentation of small bodies which occurs upon an increase in λ leads to a redistribution of mass in the system such that a large portion of it is concentrated in the region of large bodies and the exponent q decreases. By a numerical solution of Eq. (1) one obtains asymptotic values of q = 2.2 and 1.9 for $\lambda = 1/3$ and $2/3$, respectively (a was taken as 1/2). Hence it is seen that the decrease in q with an increase in λ, obtained for masses sufficiently far from the boundaries of the distribution,[2] also extends to the region of large bodies (the case of $\lambda = 1/3$ is evidently physically unrealistic).

Power-law asymptotic solutions of the generalized coagulation equation which do not depend on the coefficient a in the expression for the fragmentation parameter α were obtained in ref. 2. This is valid only under those assumptions for which the qualitative analysis of the equations was performed and for the mass region $m_0 \ll m < M^*$. We note that M^*, the upper limit of the mass region where fragmentation is possible, is itself determined by the parameter a. Thus, the question of the dependence of the mass distribution of the bodies on a in the entire mass range $m_0 > m > M$ and not only in the region with fragmentation requires additional study. For this purpose Eq. (1) was solved numerically with α taken in the form (4), with an initial distribution $n(m, 0) = c_0 e^{-m}$, and with different values of the factor a. With the indicated form of the function α the upper limit of masses m'' of the fragmentation region equals[2]

$$M^* = \frac{M_0}{(2a)^{1/\lambda}}. \tag{5}$$

With a decrease in a the region of masses of bodies fragmented in collisions is increased. The ratios M^*/M for different a with $\lambda = 2/3$ are presented in Table 1. The values of the exponent q of the asymptotic power-law mass distribution of the bodies (3), which are found through the numerical solution of Eq. (1) with the same values of a and λ, are presented in the last row of Table 1.

TABLE 1

a	2.5	1	0.5
M^*/M	0.07–0.09	0.27–0.30	0.75–0.80
q	1.56	1.68	1.9

Thus, the exponent q of the power-law asymptotic distribution grows with a decrease in a and accordingly with a widening of the region of masses of bodies subject to fragmentation.

3. STUDY OF THE MASS SPECTRUM IN A SYSTEM OF BODIES CONTAINING A MASSIVE PLANETARY EMBRYO

The condition for the destruction of two colliding bodies as a function of their masses is presented in ref. 3, Eq. (15). A comparison of this complicated expression with the simpler function $\alpha(m'') = a(m''/M_0)^\lambda$ which approximates it leads to an expression for a:

$$a \approx \frac{1-\chi-\beta}{\chi} \cdot \frac{3\theta}{5} (1-2), \tag{6}$$

where χ and β are physical characteristics of the solid bodies under consideration (χ is the fraction of the collision energy which goes into mechanical energy and β is connected with the mass distribution of the fragments); θ is a dynamic characteristic of the entire system of bodies (it is determined by their average relative velocities). By varying the numerical value of a we can thereby analyze different systems of bodies.

It is shown in ref. 1 that the considerable increase in the effective cross section for the collision of bodies with a planetary embryo because of its gravitation leads to its more rapid growth. As a result, the ratio of the mass of the embryo to the mass of the body next in size in its zone increases, approaching a limiting ratio on the order of 10^2–10^3. Such a mass distribution can also be formally described by a power law with the appropriate choice of the parameters.

Expressions are obtained in ref. 5 for the probable mass ratios of the largest bodies $M_0, M_1, M_2, \ldots$ as a function of the upper limit of the distribution and the mass of the planetary embryo when they have a power-

TABLE 2

M_0	0.05	0.1	0.3	0.5	0.9	M_0	0.05	0.1	0.3	0.5	0.9
1	2	3	4	5	6	1	2	3	4	5	6
M_1/M_0	0.905	0.818	0.538	0.333	0.053	M_0'	0.048	0.091	0.231	0.333	0.477
M_2/M_1	0.909	0.833	0.625	0.5	0.357	M_1'/M_0'	0.909	0.833	0.625	0.500	0.357
M_3/M_2	0.913	0.846	0.684	0.6	0.513	M_2'/M_1'	0.913	0.846	0.684	0.600	0.513
M	0.053	0.111	0.429	1.0	9.00	M'	0.050	0.100	0.300	0.500	0.912

law distribution. Successive mass ratios of the bodies, calculated with $q = 3/2$, are presented in Table 2; the mass of the entire system is taken as unity.

It is seen from the data of the third column that as long as the mass M_0 of the largest body does not exceed 0.1 of the mass of the entire system the successive mass ratios of the bodies (with a power-law distribution) are close to each other and the upper limit of the distribution almost coincides with the mass of the largest body. But if M_0 comprises more than half the mass of the entire system then the upper limit of the distribution must be considerably increased, which in turn entails a proportional increase in the mass step $\Delta m = M/100$. In this case the form of the distribution cannot be brought out by the numerical methods used in the work, since almost all the bodies of the system except for the very largest are within the same first mass step Δm.

Let us exclude the planetary embryo from the system described. The remaining bodies form a system of bodies with the same power-law distribution and a total mass equal to $1-M_0$. We will denote the masses of bodies belonging to the system without the embryo by a prime. Then the body M_1 second in size in the former system becomes the largest body M_0' in the new system and its mass will comprise a fraction $M_1/(1-M_0)$ of the total mass of the new remaining system. Using the same method as above, we obtain the mass ratios of the largest bodies and the upper limit of the distribution for the remaining system as a function of the mass M_0 of the embryo (see the right side of Table 2).

It is seen from a comparison of these ratios that by excluding the embryo M_0 from the total system of bodies we obtain a system with a closer mass ratio, with a distribution limit close to the mass M_0' of the largest body, and consequently with a smaller mass step $\Delta m' = M'/100$.

In the region of large bodies the real distribution departs from a power-law distribution and the planetary embryo considerably outstrips in mass the second largest body long before it collects a large part of the mass. For such a distribution the exclusion of the body M_0 from the total system and the decrease in the upper limit of the distribution in the remaining system prove to be even more important than for a power law.

The fragmentation parameter for the remaining system of bodies can be written in the same form (4) as for the total system:

$$\alpha = a(m''/M_0)^\lambda = a\left(\frac{M_0'}{M_0}\right)^\lambda \left(\frac{m''}{M_0'}\right)^\lambda = a'(m''/M_0')^\lambda, \qquad (7)$$

where $a' = a(M_0'/M_0)^\lambda < a$. Since the separation of the embryo from the other bodies does not occur at once, M_0/M_0' increases gradually and a' decreases accordingly. Therefore, by taking different values of a' we can study the distribution of bodies in the system at different stages of its evolution.

The essence of the difference between systems with different a' consists in the fact that the fragmentation of the bodies increases with an increase in the relative velocities, which are determined by the mass M_0 of the planetary embryo, whereas the "survivability" of the bodies in collisions depends on their proper masses. The greater the separation in mass of the embryo from the other bodies (i.e., the smaller a'), the more the first factor dominates over the second and the more intensively the bodies are destroyed.

The results of numerical calculations with different values of the parameter a are presented in Fig. 4. The time is laid out along the abscissa: The lower scale is the number of time steps and the upper scale is the time in years; the scale of the mass M_0 in grams is laid out on the left ordinate and the parameters q and b on the right ordinate. Curve 1, obtained with $a = 2.5$ and $\lambda = 2/3$ and with allowance for the gravitational collision cross section [ref. 3, Eq. (19)], shows the growth of the largest body M_0 in the system from a mass of 10^{21} g to $2 \cdot 10^{22}$ g in a time $\Delta t = 1.9 \cdot 10^5$ years. It is close to curve 2, which reflects the growth, in a medium with the same density, of a planetary embryo which combines with all the material incident on it.[1] Consequently, with the parameter $a = 2.5$ (meaning at large values of this parameter) the growth of the largest body in the system is not slowed by fragmentation.

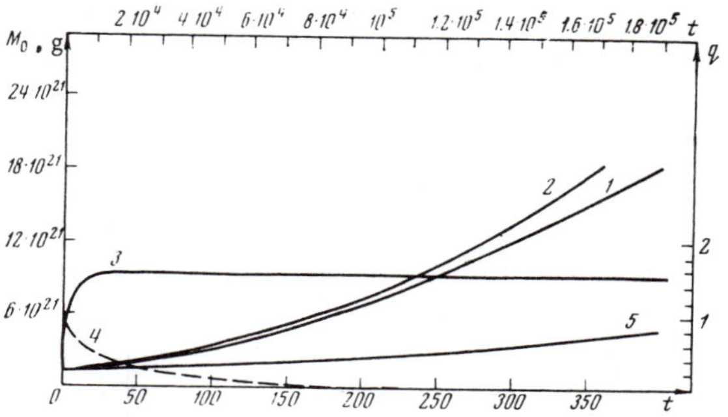

Fig. 4. Time variation in parameters of the distribution of bodies in the system: 1, 3, 4) growth of mass M_0 of largest body and variation in parameters q and b of the distribution function calculated with $a = 2.5$ and a coagulation coefficient which allows for the gravitation of the bodies; 2) growth of mass of a planetary embryo not subject to fragmentation; 5) growth of mass M_0' of largest body in secondary system, in which the velocities of the bodies are determined by the planetary embryo which is not included in the distribution function and correspond to $a = 0.5$.

Curves 3 and 4 show the time variation in the parameters q and b of the approximating distribution function $\tilde{\mu}$ in the mass interval of $15\Delta m$–$60\Delta m$. It follows from the calculations that the exponent q reaches the asymptotic value $q \approx 1.56$ rather rapidly (in the time of one to two doublings of Δm) while the parameter b in the same mass interval decreases smoothly from 1 to 0.02 in the calculated time period.

Curve 5 in Fig. 4 illustrates the growth of the largest body in the system with the same conditions as for curve 1 but with the parameter $a = a' = 0.5$. In this case the mass of the largest body grows much slower – it increased by only four times in the time of the calculation (and not by 20 times, as when $a = 2.5$). If $a = 2.5$ corresponds to a system of bodies in which M_0 is included in the total continuous distribution function then according to Eq. (7) $a' = 0.5$ corresponds to a secondary system of bodies, from which the planetary embryo M_0, which outweighs the second body $M_1 = M_0'$ in size by 10–12 times, is excluded. The slowing of the growth of body M_0' in this system is due to the decrease in effective collision cross sections of the bodies of the secondary system owing to the increase in their relative velocities, which are determined by the massive planetary embryo M_0.

Being the largest body in the system, the embryo grows faster than the other bodies since the effective collision cross section is $\propto r^4$ owing to gravitational focusing.[3] With an increase in the separation of the planetary embryo M_0 in mass from the second body in size M_1 the value of a' in the secondary system (without M_0) decreases [see (7)] and the region of fragmentation widens accordingly. At some critical ratio M_0/M_1 the fragmentation encompasses the entire secondary system and begins to considerably retard the growth of body M_1, leading to a still larger separation between the embryo M_0 and M_1.

Thus, not only the rate but also the direction of evolution of the mass distribution of the protoplanetary bodies depend on the quantity a'. Calculations with different values of a corresponding to real physical conditions allow one to trace this evolution.

4. CONCLUSION

The numerical study of the generalized coagulation equation showed that the mass distribution in the region of large bodies can be approximated by the product of an inverse power-law function times an exponential function. The first is analogous to an asymptotic distribution in the region of small bodies while the second depends on time and reflects the fact that in the region of large bodies an asymptotic solution has not yet succeeded in being established. The asymptotic dependence found does not depend on the form of the intial distribution.

The exponent q of the asymptotic distribution is less than 2. This means that a large part of the mass of the system is concentrated in the large bodies. The formation of a planetary embryo having a mass which considerably exceeds the mass of the other large bodies in its zone and determines their relative velocities makes it necessary to exclude it from the mass distribution under study and change to a secondary system with a smaller value of a.

The investigation of the evolution of the mass distribution of bodies with different values of the factor a in the expression for the fragmentation parameter makes it possible to study the distribution of the bodies at different stages of the evolution of the swarm, as well as under more complicated conditions, when the relative velocities of the bodies in the system are determined by external perturbations.

[1] V. S. Safronov, Evolution of the Protoplanetary Cloud and Formation of the Earth and Planets [in Russian], Nauka, Moscow (1969).
[2] E. V. Zvyagina, G. V. Pechernikova, and V. S. Safronov, Astron. Zh. 50, 1261 (1973) [Sov. Astron. 17, 793 (1974)].
[3] G. V. Pechernikova, Astron. Zh. 51, 1305 (1974) [Sov. Astron. 18, 778 (1975)].
[4] V. S. Safronov, Dokl. Akad. Nauk SSSR 147, 64 (1962) [Sov. Phys. Dokl. 7, 967 (1963)].
[5] V. S. Safronov and E. V. Zvyagina (Zvjagina), Icarus 10, 109 (1969).

Translated by Edward U. Oldham

Effect on eccentricity of a planet's orbit of its encounters with bodies of the swarm

I. N. Ziglina

O. Yu. Shmidt Institute of Earth Physics, Academy of Sciences of the USSR

(Submitted December 29, 1975)

Astron. Zh. **53**, 1288–1294 (November–December 1976)

An examination is made of the problem of the eccentricity of the orbit of a forming planet which is varying due to the impacts and close encounters of preplanetary bodies. The mass distribution of these bodies is taken in the form of an inverse power law. It is assumed that the eccentricities of the orbits of bodies of the swarm have been determined by their interaction with one another and with the planet and have increased in the course of the planet's growth. It is shown that in both the plane and three-dimensional cases the contribution to the eccentricity of the planet's orbit from encounters is on the same order as that from impacts of the bodies. For the majority of the planets of the solar system the calculated initial eccentricities of the orbits lie within the limits given by the Lagrange theory of secular perturbations. It is concluded that the current eccentricities of the planetary orbits are determined to a greater extent by their gravitational interaction than by the process of accumulation of the planets itself.

PACS numbers: 96.40.–z

In ref. 1 we examined the problem of the eccentricity of the orbit of a planet which is accumulating from many bodies for which the mass and velocity distributions are given. Here we will examine in addition the effect of encounters of bodies of the swarm with the growing planet on the eccentricity e of the planet's orbit, retaining the restrictions and assumptions of ref. 1. Let us given them briefly.

1. The plane problem is considered and rotation of the planet is ignored.

2. The accretion of gas by Jupiter and Saturn are ignored.

3. The mean square velocity of bodies of the swarm at a given point relative to the Keplerian circular velocity $\mathbf{v}_c$ at the same point is determined by their interaction with the planet and with each other and equals[2]

$$v'=\sqrt{Gm/\theta r}, \tag{1}$$

where m and r are the current mass and radius of the planet, θ is a parameter on the order of several units, characterizing the properties of the system of bodies, and G is the gravitational constant. Equation (1) is valid up to the time when the planet's mass has increased so much that the ejection of bodies from the solar system by the planet begins, after which v' ceases to grow and is determined from[3]

$$v'^2 \approx \frac{1}{12} v_c^2. \tag{2}$$

4. The number of bodies of the swarm with masses of from m' to m' + dm' per unit area at a distance R from the sun equals

$$n(m', R, t)\,dm'=C(R, t)\,m'^{-q}dm'. \quad m'\leqslant m_1(t),$$

where C(R, t) is a factor which does not depend on the mass, the exponent[4,5] $q \approx 11/6$, and $m_1/m \approx$ const (m_1 is the upper limit of the distribution, approximately equal to the mass of the largest body in the swarm, not counting the embryo).

5. The velocity distribution of the bodies of the swarm does not depend on their masses.

6. The eccentricities e' of the orbits of bodies of the swarm are taken as identical and dependent on the mass of the planetary embryo, while the distribution by semimajor axes of the number of bodies per unit area is taken in the form of an inverse power law

$$n_1(a', t)\,da'=C_1(t)\,a'^{-p}da', \quad R(1-e')\leqslant a'\leqslant R(1+e'), \tag{3}$$

where the constant p is on the order of several units. Such a distribution can be considered as acceptable since it leads to a fully natural law of decline of density in the swarm – a power law.

As was shown in ref. 1, the following distribution by true anomalies φ' ($0 \le \varphi' \le 2\pi$) of the number of bodies per unit area results from the distribution (3):

$$n_2(\varphi', R, t)\,d\varphi'=C_2(t)R^{-p+1}|\sin\varphi'|\,(1-pe'\cos\varphi'+O_1(e'^2))\,d\varphi', \tag{4}$$

as well as

$$e'^2=\frac{4}{3}\frac{v'^2}{v_c^2}+O_2(e'^3). \tag{5}$$

The ejection of bodies is practically impossible in the zone of planets of the earth group. In this zone, in accordance with (5) and (1),

$$e'^2 \approx \frac{4}{3}\frac{Gm}{\theta r v_c^2}. \tag{6}$$

For the giant planets the mass m_E of the planet beginning with which the planet efficiently ejects bodies from the solar system is considerably less than the mass M of the planet which has formed. In accordance with (2) and (5) one can assume that for this zone

$$e'^2\approx 1/9=\text{const}. \tag{7}$$

Like Chandrasekhar,[6] we will analyze each encounter as an independent two-body problem. In this case one must consider only those encounters which took place within the sphere of influence of the growing planet. As the radius S of the sphere of influence of the planet one can take the distance to the libration point L_1. In the plane problem under consideration, however, it is enough to use the fact that $S \gg l$, where l is the "effective radius" of the planet, i.e., the limiting distance at which a passing body touches the surface of the planet:

$$l=r\sqrt{1+v_e^2/V^2},$$

where V is the velocity of the body relative to the planet before the counter and $v_e = \sqrt{2Gm/r}$ is the parabolic velocity at the surface of the planet.

Suppose that in a time Δt, not small in comparison with the period of revolution of the planet, k bodies have fallen on it and n ecounters with bodies of the swarm have occurred. Suppose the numbers k and n are large enough that $\Delta e^2 \approx \mathscr{M}(e^2)$, $\Delta m \approx \mathscr{M}(\Delta m)$. Then

$$\Delta e^2=\sum_{i=1}^{k}\Delta e_{ai}^2+\sum_{i=1}^{n}\Delta e_{gi}^2\approx k\mathscr{M}(\Delta e_a^2)+n\mathscr{M}(\Delta e_g^2), \quad (8)$$
$$\Delta m\approx k\mathscr{M}(m'),$$

where Δe_a^2 is the change in the quantity e^2 due to the falling of a body and Δe_g^2 is the change due to an encounter with a body of the swarm.

In ref. 1 it was found that

$$\mathscr{M}(\Delta e_a^2)=-\frac{2e^2\mathscr{M}(m')}{m}+\frac{e'^2\mathscr{M}(m'^2)}{m^2}=-\frac{2e^2\mathscr{M}(m')}{m}+\frac{Ie'^2\mathscr{M}(m')}{m}, \quad (9)$$

where

$$I=\frac{\mathscr{M}(m'^2)}{m\mathscr{M}(m')}=\frac{2-q}{3-q}\frac{m_1}{m}=\frac{1}{7}\frac{m_1}{m}\approx \text{const}.$$

Let us find Δe_a^2 and then $\mathscr{M}(\Delta e_g^2)$. For a body moving along an ellipse it is easy to find

$$\mathrm{v}=\{v_r, v_\varphi\}=v_c\{e\sin\varphi+O_3(e^2),\ {}^1\!/_2 e\cos\varphi+O_4(e^2)\}, \quad (10)$$

where v_r and v_φ are the radial and transverse components of the velocity vector of the body relative to circular velocity. Assuming that the radius of the sphere of influence of the planet is small in comparison with the distance to the sun, one can write the relative velocity of the body and planet before the encounter in projection onto the radius vector of the planet and transverse to it as follows[1)]:

$$\mathbf{V}=v_c\{(e'\sin\varphi'+O_3(e'^2))(1+O_5(S/R))+({}^1\!/_2 e'\cos\varphi' +O_4(e'^2))O_6(S/R)+O_6(S/R)-e\sin\varphi-O_3(e^2),\ {}^1\!/_2(e'\cos\varphi'-e\cos\varphi)+\ldots\}\ . \quad (11)$$

As a result of an encounter with a body of mass m' the planet receives a velocity increment

$$\Delta\mathrm{v}=-\frac{m'\Delta\mathbf{V}}{m+m'}\approx-\frac{m'\Delta\mathbf{V}}{m},$$

since $m' \ll m$. As is known from the two-body problem,[6] the direction of the relative velocity V of the bodies at the end of the encounter is deflected by an angle $\pm 2\psi$, determined from

$$\sin\psi=\frac{1}{\sqrt{1+D^2V^4/\mu^2}}, \quad (12)$$

where D is the impact parameter and $\mu = G(m + m')$. It is convenient to take the quantity D with a plus sign if the deflection angle is positive and a minus sign if the deflection angle is negative.

Thus

$$\Delta\mathrm{v}=\frac{m'}{m}v_c\{2\sin^2\psi[(e'\sin\varphi'+O_3(e'^2))(1+O_5(S/R)) +O_6(S/R)-e\sin\varphi-O_3(e^2)+\ldots]\pm\sin2\psi(e'\cos\varphi'-e\cos\varphi+\ldots), \mp\sin2\psi(e'\sin\varphi'-e\sin\varphi+\ldots)+\sin^2\psi(e'\cos\varphi'-e\cos\varphi+\ldots)\}. \quad (13)$$

From Eq. (10) we get

$$e^2\approx(v_r/v_c)^2+4(v_\varphi/v_c)^2.$$

Hence

$$\Delta e^2\approx\frac{1}{v_c^2}[2v_r\Delta v_r+8v_\varphi\Delta v_\varphi+(\Delta v_r)^2+4(\Delta v_\varphi)^2].$$

By substituting into this formula the expressions (10) for v_r and v_φ and (13) for the change in the velocity components Δv_r and Δv_φ resulting from the encounter, we obtain Δe_g^2:

$$\Delta e_g^2\approx\frac{m'}{m}\Big\{4\sin^2\psi\Big[-e^2-2e\sin\varphi O_3(e^2)+e'\sin\varphi' O_3(e^2) +e\sin\varphi O_3(e'^2)+ee'\cos(\varphi-\varphi')\Big(1+O_5\Big(\frac{S}{R}\Big)\Big)+e\cos\varphi O_5\Big(\frac{S}{R}\Big) +e\sin\varphi O_6\Big(\frac{S}{R}\Big)+\ldots\Big]\pm\sin2\psi e\Big(e'\cos\varphi'\sin\varphi-4e'\sin\varphi'\cos\varphi +\frac{3}{2}e\sin2\varphi+\ldots\Big)\Big\}+\Big(\frac{m'}{m}\Big)^2\Big\{4\sin^4\psi\Big(e'\Big(1+2O_5\Big(\frac{S}{R}\Big)\Big) +e^2-2e'\sin\varphi' O_6\Big(\frac{S}{R}\Big)+2e'\cos\varphi' O_5\Big(\frac{S}{R}\Big)+e'\sin\varphi' O_3(e'^2) -2ee'\cos(\varphi-\varphi')\Big]+\sin^2\psi\cos^2\psi[(\cos^2\varphi'+16\sin^2\varphi')e'^2 +(\cos^2\varphi+16\sin^2\varphi)e^2-2ee'(\cos\varphi\cos\varphi'+16\sin\varphi\sin\varphi')] \mp12\sin^3\psi\cos\psi(e'\cos\varphi'-e\cos\varphi)(e'\sin\varphi'-e\sin\varphi)+\ldots\Big\}. \quad (14)$$

The probability density $f(\varphi, \varphi', D/e, m)$ for an approach (φ, φ', D) when the eccentricity of the planet's orbit is e and the mass is m must obviously be taken as equal to

$$f(\varphi,\varphi',D/e,m)=\frac{n_3(\varphi,\varphi',D,e,m)}{\int_{-S}^{-l}\int_0^{2\pi}\int_0^{2\pi}n_3\,d\varphi\,d\varphi'\,dD-\int_l^S\int_0^{2\pi}\int_0^{2\pi}n_3\,d\varphi\,d\varphi'\,dD},$$

where

$$n_3=n_2(\varphi',R',e'(m),t)V\frac{dt}{d\varphi},$$

R' being the distance from the sun to the point on the sphere S at which the body is found at the start of the encounter with the planet, $|R - R'| \leq S$. Taking n_2 in accordance with (4), V in accordance with (11), $dt/d\varphi \propto R^2$, and assuming that the quantities e and S/R are small in comparison with e', we find

$$f(\varphi, \varphi', D/e, m) = \frac{1+O_7(l/S)+O_8(S^2/R^2e'^2)}{4\pi S \int\limits_0^{2\pi} |\sin\varphi'| \sqrt{1-{}^3/_4\cos^2\varphi'}\, d\varphi'}$$
$$\times[1+(p-3)e\cos\varphi+\ldots]\,|\sin\varphi'[1-pe'\cos\varphi'+O_1(e'^2)]\sqrt{1-{}^3/_4\cos^2\varphi'}$$
$$\times\Big[1-\frac{e(\sin\varphi\sin\varphi'+{}^1/_4\cos\varphi\cos\varphi')}{e'(1-{}^3/_4\cos^2\varphi')}\Big(1+O_5\Big(\frac{S}{R}\Big)\Big)$$
$$+\frac{O_6^2(S/R)+O_5^2(S/R)}{2e'^2(1-{}^3/_4\cos^2\varphi)}+O_5\Big(\frac{S}{R}\Big)-e\cos\varphi\, O_5\Big(\frac{S}{R}\Big)+\ldots\Big],$$
$$\int\limits_0^{2\pi} |\sin\varphi'| \sqrt{1-{}^3/_4\cos^2\varphi'}\, d\varphi' \approx 3.42. \qquad (15)$$

Let us take the mathematical expectation of the expression (14) for Δe_g^2. Then we will have

$$\mathscr{M}(\Delta e_g^2) = -6e^2\frac{\mathscr{M}(m')}{m}\mathscr{M}(\sin^2\psi)\Big[1+O_9(e'^2)+O_{10}\Big(\frac{S^2}{R^2e'^2}\Big)\Big]$$
$$+\frac{\mathscr{M}(m'^2)}{m^2}e'^2\mathscr{M}[4\sin^4\psi+\sin^2\psi\cos^2\psi(\cos^2\varphi'+16\sin^2\varphi')]$$
$$\times\Big[1+O_{11}(e'^2)+O_{12}\Big(\frac{S^2}{R^2e'^2}\Big)\Big]. \qquad (16)$$

Substituting (9) and (16) into (8), after elementary transformations we obtain

$$\Delta e^2 \approx -\frac{2e^2\Delta m}{m} - \frac{6e^2\mathscr{M}(\sin^2\psi)}{m}\frac{n}{k}\Delta m + \frac{Ie'^2\Delta m}{m}$$
$$+\frac{Ie'^2\Delta m}{m}\frac{n}{k}\mathscr{M}(16\sin^2\psi-12\sin^4\psi$$
$$-15\sin^2\psi\cos^2\varphi'+15\sin^4\psi\cos^2\varphi'). \qquad (17)$$

It should be noted that the expression (17) gives the principal part of Δe^2 not only with the expression (4) for $n_2(\varphi')$ but with any distribution of the bodies by φ' or by semimajor axes, which is the same.

By dividing (17) by Δm we obtain a linear differential equation for $e^2(m)$:

$$\frac{de^2}{dm} = \Big[-\frac{2}{m}-\frac{6\mathscr{M}(\sin^2\psi)}{m}\frac{n}{k}\Big]e^2$$
$$+\frac{Ie'^2}{m}\Big[1+\frac{n}{k}\mathscr{M}(16\sin^2\psi-12\sin^4\psi$$
$$-15\sin^2\psi\cos^2\varphi'+15\sin^4\psi\cos^2\varphi')\Big]. \qquad (18)$$

Without the allowance for encounters of the planet with bodies the equation for e_a^2 has the form

$$\frac{de_a^2}{dm} = -\frac{2}{m}e_a^2+\frac{Ie'^2}{m}. \qquad (19)$$

In ref. 1 it was found from this equation that for planets of the earth group $e_{a1}^2 \approx (1/4)(I/\theta)(v_e/v_c)^2$, while for the giant planets $e_{a2}^2 \approx (I/18)$.

Clearly, n/k should be taken as equal to

$$\frac{n}{k} = \frac{\int\limits_{-S}^{-l}\int\limits_0^{2\pi}\int\limits_0^{2\pi} n_3\, d\varphi\, d\varphi'\, dD + \int\limits_l^S\int\limits_0^{2\pi}\int\limits_0^{2\pi} n_3\, d\varphi\, d\varphi'\, dD}{\int\limits_{-l}^{l}\int\limits_0^{2\pi}\int\limits_0^{2\pi} n_3\, d\varphi\, d\varphi'\, dD}$$

Hence

$$\frac{n}{k} \approx \frac{S\int\limits_0^{2\pi} n_2(\varphi', R, t)\, V\, d\varphi'}{rv_e\int\limits_0^{2\pi} n_2(\varphi', R, t)\, d\varphi'} \approx \frac{0.86e'v_cS}{v_er}. \qquad (20)$$

Retaining only the dominant terms, from (12) and (15) we have:

a) for planets of the earth group with[2] $\theta = 3$

$$\mathscr{M}(\sin^2\psi) = \frac{1.37\mu}{e'^2v_c^2S},\quad \mathscr{M}(\sin^4\psi) = \frac{1}{2}\mathscr{M}(\sin^2\psi)$$
$$-\frac{0.36rv_e}{Se'^2v_c},\quad \mathscr{M}(\sin^2\psi\cos^2\varphi') = \frac{0.55\mu}{Se'^2v_c^2},$$
$$\mathscr{M}(\sin^4\psi\cos\varphi') = \frac{1}{2}\mathscr{M}(\sin^2\psi\cos^2\varphi') - \frac{0.13v_er}{Se'v_c},$$

b) for the giant planets

$$\mathscr{M}(\sin^2\psi) = \frac{2.2\mu}{Se'^2v_c^2},\quad \mathscr{M}(\sin^4\psi) = \frac{1}{2}\mathscr{M}(\sin^2\psi),$$
$$\mathscr{M}(\sin^2\psi\cos^2\varphi') = \frac{0.86\mu}{Se'^2v_c^2},$$
$$\mathscr{M}(\sin^4\psi\cos^2\varphi') = \frac{1}{2}\mathscr{M}(\sin^2\psi\cos^2\varphi').$$

Substituting these quantities into (18) and using (20), we obtain

a) $$\frac{de_1^2}{dm} = -\frac{9.5e_1^2}{m}+\frac{11.7Ie'^2}{m}.$$

Hence

$$e_1^2 = \int\limits_0^M \frac{4}{3}\frac{11.7GI}{\theta v}\Big(\frac{4\pi\delta}{3m}\Big)^{1/3}\Big(\frac{m}{M}\Big)^{9.5} dm = \frac{0.77}{\theta}I\Big(\frac{v_e}{v_c}\Big)^2 = 3.06e_{a1}^2,$$
$$e_1 = 0.19\sqrt{\frac{m_1}{m}}\frac{v_e}{v_c} = 1.8e_{a1}. \qquad (21)$$

(Here δ is the average density of the planet.)

b) $$\frac{de_2^2}{dm} = -\frac{2e_2}{m}-\frac{17e_2^2}{m}\frac{v_e}{v_c}+\frac{Ie'^2}{m}\Big(1+20\frac{v_e}{v_c}\Big)$$
$$= -\frac{2e_2^2}{m}-\frac{\alpha e_2^2}{m^{2/3}}+\frac{I}{9}\Big(\frac{1}{m}+\frac{\beta}{m^{2/3}}\Big).$$

Hence

$$e_2^2 = \int\limits_0^M \frac{I}{9}\Big(\frac{1}{m}+\frac{\beta}{m^{2/3}}\Big)\exp\Big[\int\limits_m^M\Big(-\frac{2}{\tau}-\frac{\alpha}{\tau^{2/3}}\Big)d\tau\Big]dm$$
$$\approx \frac{I}{9}\frac{\beta}{\alpha} = 2.35e_{a2}^2,$$
$$e_2 = 0.14\sqrt{m_1/m} = 1.5e_{a2}. \qquad (22)$$

TABLE 1

Planet	m_1/m	e_{obs}	e_{min}	e_{max}	$\bar{e}$	e_c
Mercury	$1.8 \cdot 10^{-3}$	0.205	0.109	0.241	0.18	0.00073
Venus	$1.8 \cdot 10^{-3}$	0.0068	–	0.074	0.037	0.0023
Earth	$1.8 \cdot 10^{-3}$	0.017	–	0.067	0.034	0.003
Mars	$3.5 \cdot 10^{-3}$	0.093	0.004	0.141	0.073	0.0025
Jupiter	$5.3 \cdot 10^{-4}$	0.048	0.027	0.062	0.045	0.003
Saturn	$7.0 \cdot 10^{-2}$	0.055	0.012	0.086	0.049	0.035
Uranus	$12.3 \cdot 10^{-2}$	0.047	–	0.067	0.034	0.046
Neptune	$12.3 \cdot 10^{-3}$	0.0085	0.005	0.013	0.009	0.015

Consequently, the allowance for the effect of passing bodies on the eccentricity of the planet's orbit, carried out for the plane case, does not lead to its significant increase.

In the case of the three-dimensional problem the equation for $e^2(m)$ should be analogous to Eq. (18). The factor $\mathcal{M}(\sin^4\psi)$ becomes relatively small and the terms which include it can be neglected. In the three-dimensional case the quantity $(n/k)\mathcal{M}(\sin^2\psi) = \nu$ depends little on the mass of the planetary embryo, and one can set $\nu = \mathrm{const} \gg 1$. Then Eq. (18) takes the form

$$\frac{de^2}{dm} = -\frac{2e^2}{m}(1+3\nu) + \frac{Ie'^2}{m}(1+16\nu-15\nu\zeta),$$

where $\zeta = \overline{\cos^2\varphi'}$.

For planets of the earth group and for the giant planets, respectively, its solution gives

$$e_1^2 = \frac{1+\nu(16-15\zeta)}{2\frac{2}{3}+6\nu} 2\frac{2}{3} e_{a1}^2 \approx \frac{4}{9}(16-15\zeta)e_{a1}^2,$$

$$e_2^2 = \frac{1+\nu(16-15\zeta)}{1+3\nu} e_{a2}^2 \approx \frac{1}{3}(16-15\zeta)e_{a2}^2.$$

With $\zeta = 1/2$ we have

$$e_1 = 1.9e_{a1}, \quad e_2 = 1.7e_{a2}.$$

Thus, in the three-dimensional case also no significant increase in the eccentricity of the planet's orbit occurs as a result of encounters of bodies of the swarm with the planet.

The values of e_c calculated from Eqs. (21) and (22) are presented in Table 1. The quantities m_1/m are taken from ref. 2, where they were determined from the inclinations of the axes of rotation of the planets.[2)] The current values of the eccentricities e_{obs} of the planetary orbits, the minimum and maximum eccentricities e_{min} and e_{max} calculated in accordance with the Lagrange theory of secular perturbations,[7] and the "mean" eccentricities $\bar{e}$, taken simply as the arithmetic mean of the respective e_{min} and e_{max}, are given for comparison.

As seen from Table 1, the calculated eccentricities of the orbits of the only planets which formed generally fall in the interval (e_{min}, e_{max}). Mercury, Mars, and Jupiter are the only exceptions. The growth of the eccentricity of Jupiter's orbit can evidently be explained by the process of capture into resonance with Saturn. The velocities of the bodies of the swarm in the zone of Mars could increase by fully two times owing to the effect of the massive Jupiter, and then e_c of Mars would also lie in the indicated interval. The large eccentricity of Mercury's orbit may be connected with its closeness to the sun or with some peculiarities of its accumulation.

In comparing e_c with $\bar{e}$ one can conclude that the eccentricities of the planetary orbits are determined mainly by their gravitational interaction and not by the process of accumulation of the planets itself.

The author thanks V. S. Safronov for support and attention to the work.

1)In order not to encumber the equations, in (11), (13), (14), and (15) only some typical terms are written out of those which later prove to be unimportant.

2)The accretion of gas by Jupiter and Saturn was not taken into account in the calculation. When accretion is allowed for the equation for the eccentricity of the orbit (and the inclination of the axis of rotation) includes a factor less than unity which, however, is exactly compensated by an increase in m_1/m in this case.

[1]I. N. Ziglina and V. S. Safronov, Astron. Zh. 53, 429 (1976) [Sov. Astron. 20, 244 (1976)].

[2]V. S. Safronov, Evolution of the Preplanetary Cloud and Formation of the Earth and Planets [in Russian], Nauka, Moscow (1969).

[3]V. S. Safronov, The Motion, Evolution of Orbits and Origin of Comets, Proceedings of IAU Symposium No. 45, Leningrad, 1970, G. A. Chebotarev et al., Eds. (1972), p. 329.

[4]J. S. Dohnanyi, J. Geophys. Res. 74, 2531 (1969).

[5]E. V. Zvyagina, G. V. Pechernikova, and V. S. Safronov, Astron. Zh. 17, 793 (1974).

[6]S. Chandrasekhar, Principles of Stellar Dynamics, Univ. Chicago Press, Chicago (1942).

[7]D. Brouwer and G. M. Clemence, in: Planets and Satellites [Russian translation], collection of translations, Inostr. Lit., Moscow (1963), p. 43.

Translated by Edward U. Oldham

Averaging of the orbital eccentricities of bodies which are accumulating into a planet

I. N. Ziglina and V. S. Safronov

O. Yu. Shmidt Institute of Earth Physics, USSR Academy of Sciences

(Submitted August 15, 1975)

Astron. Zh. **53**, 429–435 (March–April 1976)

The averaging of the orbital eccentricities of protoplanetary bodies during their accumulation into a planet is studied in the report. The mathematical expectation of the orbital eccentricity of the planetary nucleus in the course of its accumulation, which is considered as a random process, is estimated. It is assumed that the orbital eccentricities of the protoplanetary bodies were determined by gravitational perturbations on the part of the growing planet and that they increased with time. The perturbations of the planet during the encounters with the bodies are not taken into account. The mass distribution of the protoplanetary bodies is taken in the form of an inverse power law. The largest bodies made the principal contribution to the orbital eccentricity of the planet. For the three outer planets the calculated eccentricities proved to be comparable with the observed values, while for the remaining planets they were considerably smaller than the observed eccentricities. Evidently the orbital eccentricities of the majority of the planets are mainly determined not by the process of their accumulation but by the mutual gravitational perturbations of the planets during the entire time of their existence.

PACS numbers: 96.10.+i

According to modern concepts, in the course of the evolution of the protoplanetary cloud a swarm of solid bodies was formed which rotated about the sun in orbits with relatively small eccentricities and inclinations. The gravitational interaction of the bodies increased and their relative velocities grew in proportion to the increase in the masses of the bodies owing to their accumulation. An approximate estimate of the velocities was first derived by L. É. Gurevich and A. I. Lebedinskii.[1] V. S. Safronov[2] made a more detailed analysis. The largest bodies grew faster during the collisions and became planetary nuclei. Some of them, having encountered the most favorable conditions, were converted into planets.

One of the principal regularities of the planetary system is the almost circular orbits of the planets, lying practically in a single plane. A qualitative explanation of this regularity has already been given by O. Yu. Shmidt: such orbits are "a consequence of the natural statistical averaging of the motions of the separate bodies, from the accumulation of which the planets arise."[3] A quantitative estimate of the effect of averaging of the orbital eccentricities of many bodies during their accumulation into a single planet is made in the present article. The gravitational perturbations of the planet on the part of the bodies flying past it are not taken into account. From general considerations one can expect that these perturbations should lead to eccentricities of the same order as would the impacts of the falling bodies. The problem is considered as a plane problem and the rotation of the growing planet is not taken into account.

Suppose m_i', e_i', and φ_i' are the mass of the i-th body falling on the planet, the eccentricity of its orbit, and its angular distance from perihelion (the true anomaly) at the moment of impact and m_{i-1}, e_{i-1}, and φ_i are the same quantities for the growing planet immediately before the i-th impact ($i = 1, 2, \ldots, n$). The initial mass and orbital eccentricity of the planetary nucleus are m_0 and e_0, respectively. It is obvious that

$$m_i = m_0 + \sum_{j=1}^{i} m_j'.$$

We will assume that $m_n = M \gg m_0$, where M is the mass of the planet which has formed.

For a body moving along an ellipse, confining ourselves to terms of first order with respect to the eccentricity, we obtain

$$v_\varphi=\frac{1}{2}v_c e\cos\varphi,\quad v_r=v_c e\sin\varphi, \tag{1}$$

where v_r and v_φ are the radial and transversal components of the relative velocity vector (relative to the circular Keplerian velocity v_c).

From these equations it follows that

$$e^2=\left(\frac{v_r}{v_c}\right)^2+4\left(\frac{v_\varphi}{v_c}\right)^2. \tag{2}$$

We can express e_i through e_{i-1}. We represent the velocities of the nucleus and the incident body before impact in the form (1). To determine the velocities $v_{\varphi i}$ and v_{ri} of the nucleus after impact we use the condition of conservation of total momentum:

$$v_{\varphi i}=\frac{v_c}{2m_i}(m_{i-1}e_{i-1}\cos\varphi_i+m_i'e_i'\cos\varphi_i'),$$
$$v_{ri}=\frac{v_c}{m_i}(m_{i-1}e_{i-1}\sin\varphi_i+m_i'e_i'\sin\varphi_i'). \tag{3}$$

Substituting these expressions into Eq. (2), we find e_i^2:

$$e_i^2=\frac{1}{m_i^2}[(m_{i-1}e_{i-1})^2+(m_i'e_i')^2$$
$$+2e_i'm_i'm_{i-1}e_{i-1}(\cos\varphi_i\cos\varphi_i'+\sin\varphi_i\sin\varphi_i')]. \tag{4}$$

The process of formation of the planet can be considered as a random process and the values φ_i, φ_i', e_i ($i = 1, 2, \ldots, n$) as random values. The masses and orbital eccentricities of the incident bodies will be considered as given.

Since the number of impacts n is large, $e_n^2\approx\mathcal{M}(e_n^2)$ and $e_n\approx\sqrt{\mathcal{M}(e_n^2)}$, and therefore we take

$$e_{acc}=\sqrt{\mathcal{M}(e_n^2)}.$$

[Obviously, $e_n\approx\mathcal{M}(e_n)$ simultaneously, but as seen from Eq. (4), it is more convenient to find $\mathcal{M}(e_n^2)$ rather than $\mathcal{M}(e_n)$.]

Taking the mathematical expectation of the left and right sides of Eq. (4) (in doing this we average the expression over all φ_j and φ_j', where $j \leq i$), we obtain

$$\mathcal{M}(e_i^2)-\mathcal{M}(e_{i-1}^2)=\frac{1}{m_{i-1}^2}\{-2m_i'm_{i-1}\mathcal{M}(e_{i-1}^2)$$
$$+(m_i'e_i')+2e_i'm_i'm_{i-1}[\mathcal{M}(e_{i-1}\cos\varphi_i\cos\varphi_i')$$
$$+\mathcal{M}(e_{i-1}\sin\varphi_i\sin\varphi_i')]\}. \tag{5}$$

At first we take the orbital eccentricities of all the bodies in the swarm as the same (depending on the mass of the nucleus). We assume that the distribution by semimajor axes of the number of bodies per unit area (we are considering the plane problem) at some distance R from the sun is described by an inverse power law:

$$n(a',t)\,da'=C(t)\,a'^{-p}da'. \tag{6}$$

In this case the values of the semiaxes clearly lie within the limits

$$R(1-e')\leqslant a'\leqslant R(1+e'). \tag{7}$$

By integrating (6) over a' within the limits (7) we obtain the total number of bodies per unit area as a function of the distance to the sun, also in the form of a power law:

$$N(R,t)=C_1(t)R^{-p+1}. \tag{8}$$

The distribution of the number of bodies with respect to R which is obtained seems quite natural and reasonable. This can be considered as an argument in favor of the adopted distribution (6) over the semimajor axes.

An elementary consequence of the presumed axial symmetry of the problem is the fact that the number of bodies with true anomaly φ' per unit area is equal to the number of bodies with true anomaly $2\pi-\varphi'$. Then, using the relation $a'\approx R(1+e'\cos\varphi')$ between a' and φ' at a given distance R, from the distribution (6) over the semiaxes of the number of bodies per unit area we obtain the distribution of the number of bodies by true anomalies:

$$n_1(\varphi',R,t)\,d\varphi'=\frac{C_1(t)}{4}R^{-p+1}|\sin\varphi'|(1+e'\cos\varphi')^{-p}\,d\varphi'. \tag{9}$$

The same expressions for N and n_1 are also obtained with a more general expression for $n(a', t)$, when the constant in (6) is a power-law function of R.

The number of bodies with a true anomaly in the interval $[\varphi', \varphi'+d\varphi']$ which fall on the planet in the small time Δt is equal to

$$n_1(\varphi',R,t)\,d\varphi' SV\Delta t,$$

where S is the collision cross section and V is the velocity of the falling bodies relative to the planet.

In the plane problem SV in the first approximation does not depend on V, which means it also does not depend on φ'. Therefore, for a body chosen randomly from the bodies which fell on the planet in the time Δt the probability that the true anomaly lies in the interval $[\varphi', \varphi'+d\varphi']$, with the condition that the orbital eccentricity of the nucleus equals e and its true anomaly equals φ, is equal to

$$f(\varphi'/e,\varphi)\,d\varphi'=\frac{n_1(\varphi',R,t)SV\Delta t\,d\varphi'}{\int^{2\pi} n_1(\varphi',R,t)SV\Delta t\,d\varphi'}$$
$$=\frac{n_1(\varphi',R,t)\,d\varphi'}{N(R,t)}\approx\frac{1}{4}|\sin\varphi'|(1-pe'\cos\varphi')\,d\varphi'. \tag{10}$$

The expression (10) obtained does not depend on e and φ, and consequently the values of φ' and of e and φ are independent. Therefore,

$$\mathcal{M}(e_{i-1}\cos\varphi_i\cos\varphi_i')=\mathcal{M}(e_{i-1}\cos\varphi_i)\mathcal{M}(\cos\varphi_i'),$$
$$\mathcal{M}(e_{i-1}\sin\varphi_i\sin\varphi_i')=\mathcal{M}(e_{i-1}\sin\varphi_i)\mathcal{M}(\sin\varphi_i'). \quad (11)$$

It follows from (10) that

$$\mathcal{M}(\cos\varphi_i')=-\frac{e_i'p}{3}, \quad \mathcal{M}(\sin\varphi_i')=0. \quad (12)$$

The quantity $\mathcal{M}(e_{i-1}\cos\varphi_i)$ can be represented in the form

$$\mathcal{M}(e_{i-1}\cos\varphi_i)=\mathcal{M}\left[e_{i-1}\int_0^{2\pi}\cos\varphi_i f(\varphi_i/e_{i-1})\,d\varphi_i\right], \quad (13)$$

where $f(\varphi_i/e_{i-1})d\varphi_i$ is the probability that body i falls on the planet when the true anomaly of the planet lies in the limits from φ_i to $\varphi_i + d\varphi_i$, with the condition that the eccentricity of the planet's orbit equals e_{i-1}. For conciseness we again omit the indices.

It is obvious that $f(\varphi/e)d\varphi$ can be taken as equal to

$$f(\varphi/e)\,d\varphi=\frac{N(R(\varphi),t)SV\frac{dt}{d\varphi}d\varphi}{\int_0^{2\pi}N(R(\varphi),t)SV\frac{dt}{d\varphi}d\varphi},$$

where N(R, t) is the above-mentioned number of bodies per unit area at the distance R from the sun.

Taking N(T, t) in accordance with (8), taking $dt/d\varphi \propto R^2$, and assuming that the variation in $C_1(t)$ during one revolution can be neglected, we obtain

$$f(\varphi/e)=\frac{1}{2\pi}\left[1+\left(-2-\frac{\partial N}{\partial R}\bigg|_a\cdot\frac{a}{N(a,t)}\right)e\cos\varphi\right]$$
$$=\frac{1}{2\pi}[1+(p-3)e\cos\varphi]. \quad (14)$$

Consequently,

$$\int_0^{2\pi}\cos\varphi_i f(\varphi_i/e_{i-1})\,d\varphi_i=\frac{p-3}{2}e_{i-1}. \quad (15)$$

Substituting (15) into (13), we obtain

$$\mathcal{M}(e_{i-1}\cos\varphi_i)=\frac{p-3}{2}\mathcal{M}(e_{i-1}^2). \quad (16)$$

Finally, substituting (16) and (12) into (11) and then substituting (11) into (5), we obtain

$$\mathcal{M}(e_i^2)-\mathcal{M}(e_{i-1}^2)$$
$$=\frac{1}{m_{i-1}^2}\left\{m_i'm_{i-1}\left[-2-\frac{p(p-3)}{3}e_i'^2\right]\mathcal{M}(e_{i-1}^2)+(m_i'e_i')^2\right\}. \quad (17)$$

If the number i is large then $\mathcal{M}(e_{i-1}^2)$, depending on m_{i-1}, is approximately equal to $e^2(m)$, where e and m are the current orbital eccentricity and mass of the planet. Suppose that in a time Δt (not small compared with the period of revolution of the planet) k bodies have fallen on the planet, where k is large enough that it can be assumed that $\mathcal{M}(e_{j+k}^2-e_j^2)\approx\Delta e^2$, but at the same time the mass $\Delta m=\sum_{i=j}^{j+k}m_i'$ which has fallen is small.

We sum Eqs. (17) over i within the limits of $j\le i\le j+k$. Changing from the mathematical expectations to the quantities themselves, as was done above, we obtain

$$\Delta e^2=\frac{1}{m^2}\left\{\Delta m\cdot m\,[-2-be'^2(m)]e^2+e'^2(m)\sum_{i=j}^{j+k}m_i'^2\right\}, \quad (18)$$

where

$$b=\frac{p(p-3)}{3}. \quad (19)$$

Let us estimate the sun in the last term of (18). Suppose the number of bodies with masses in the interval $[m', m'+\Delta m']$ which fell on the planet in the time Δr is equal to $n(m', t)\Delta m'\Delta t$, where[1] $n(m', t)=C(t)m'^{-q}$ $(0<q<2)$.

Then

$$\sum_{i=j}^{j+k}m_i'^2=\int_0^{m_1}n(m',t)\Delta t\,m'^2\,dm'=Im\,\Delta m, \quad (20)$$

where

$$I=\frac{\int_0^{m_1}n(m',t)m'^2\,dm'}{m\int_0^{m_1}n(m',t)m'\,dm'}=\frac{2-q}{3-q}\cdot\frac{m_1}{m},$$

and m_1 is the upper limit of the mass distribution of the bodies.

As the planet's mass m(t) grows $m_1(t)$ also grows. Approximately, one can take $m_1/m=\text{const}$ and $I=\text{const}$.

Substituting (20) into (18) and dividing by Δm, we obtain the differential equation

$$\frac{de^2}{dm}=-\frac{1}{m}(2+be'^2(m))e^2+I\frac{e'^2}{m}=-a_1(m)e^2+a_2(m). \quad (21)$$

Up to now we have assumed that the orbital eccentricities e' of all the bodies falling on the planet were the same. It is not hard to show that the expression (21) obtained is also valid with more general assumptions, when the orbital eccentricities of the bodies are distributed by some law which does not depend on the masses of the bodies. In this case it is only necessary to take as e' on the right side of (21) its mean-square value.

As is known, the general solution of this equation has the form

$$e^2=e_0^2\exp\left[-\int_{m_0}^{M}a_1(m)\,dm\right]+\int_{m_0}^{M}a_2(m)\exp\left[-\int_{m}^{M}a_1(\tau)\,d\tau\right]dm. \quad (22)$$

From the distribution (9) of the number of bodies by true anomalies per unit area one can find with allowance for (1) the connection between $e'(m)$ and the mean-square velocity $v(m)$ of the bodies relative to the circular velocity:

$$e'^2 = \frac{4}{3}\frac{v^2}{v_c^2}. \tag{23}$$

The gravitational interactions of the bodies of the swarm with the planetary nucleus and with one another lead to an increase in their relative velocities. Conversely, inelastic collisions decrease them. As a result, some equilibrium value v is established which depends primarily on the mass m and radius r of the largest body in the zone and somewhat less strongly on the mass distribution of the bodies. It is usually written in the form[2]

$$v^2 = \frac{Gm}{\theta r}, \tag{24}$$

where G is the gravitational constant and θ is a dimensionless parameter on the order of several units which characterizes a property of the system of colliding bodies and which remains approximately constant in the course of the planet's growth.

When the relative velocity vector **v** of a body is parallel to the vector $\mathbf{v}_c$ of the Keplerian circular velocity then the absolute heliocentric velocity of the body is maximal. If v reaches the value $(\sqrt{2}-1)v_c$ in this case then the body leaves the solar system. In the zone of planets of the earth group v is small in comparison with v_c and ejection of bodies is practically impossible. But in the zone of the giant planets ejection begins at a mass m_E of the planetary nucleus which is much less than the modern mass M of the planet. The perturbations change the direction of the vector **v** faster than its magnitude. Therefore the highest velocities v_m of the bodies cannot significantly exceed the minimum velocity necessary for ejection in the direction of the orbital motion of the planet. With the growth in the mass of the planet the further increase in the velocities v which follows from (24) ceases. For approximate estimates in accordance with ref. 4 we take the average velocity of the ejected bodies as equal to $v_E = (1/2)v_c$ and the average velocity of all the remaining bodies as equal to

$$v^2 = \frac{1}{3}v_E^2 = \frac{1}{12}v_c^2. \tag{25}$$

From the last three expressions we accordingly have for the planets of the earth group

$$e'^2 = \frac{4}{3}\frac{Gm}{\theta r v_c^2} \tag{26}$$

and for the giant planets

$$e'^2 \approx \frac{1}{9}. \tag{27}$$

For the planets of the earth group $e'^2 \ll 1$. Therefore one can neglect the value be'^2 in comparison with 2 in Eq. (21).

From (26) and (22) for the modern masses M of these planets we have

$$e^2 = e_0^2\left(\frac{m_0}{M}\right)^2 + \int_{m_0}^{M}\frac{4}{3}\frac{GI}{\theta v_c^2}\left(\frac{4\pi\delta}{3m}\right)^{1/3}\left(\frac{m}{M}\right)^2 dm$$
$$= e_0^2\left(\frac{m_0}{M}\right)^2 + \frac{1}{4}\frac{I}{\theta}\left(\frac{v_e}{v_c}\right)^2\Bigg|_{m_0}^{M},$$

where δ is the mean density of the planet and v_e is the parabolic velocity at its surface.

Since $m_0 \ll M$ one can take $m_0 = 0$. Then

$$e \approx \frac{1}{2}\sqrt{\frac{I}{\theta}}\frac{v_e}{v_c}. \tag{28}$$

For the giant planets $m_E \ll M$ and therefore, taking $m_E = 0$ [one can show that the relative error in this case is on the order of $(m_E/M)^2$] and $m_0 = 0$, from (27) and (22) we obtain

$$e \approx \sqrt{\frac{I}{18+b}}. \tag{29}$$

It is seen from Eqs. (28) and (29) that for planets of the earth group e grows in proportion to r while for the giant planets e, having reached the value (29), remains constant.

In the derivation of Eq. (21) for de^2/dm and Eq. (23) for e'^2 we assumed that the orbital eccentricities of the falling bodies were small. For the bodies of the zone of giant planets this is evidently not so and the final equation (29) is inaccurate. It gives values for the orbital eccentricities of the giant planets which are correct in order of magnitude – the relative error does not exceed 30-40%.

The values of e_{acc} calculated from Eqs. (28) and (29) with $\theta = 3$ (ref. 2), $q = 11/6$ (refs. 5 and 6), and with $p = 4$, which corresponds to a variation proportional to R^{-3} in the surface density in the swarm,[3] are presented in the third column of Table 1. The orbital eccentricities of the bodies in the final stage of accumulation are presented in the second column and the modern orbital eccentricities of the planets in the last column for comparison. The values of m_1/m needed for the determination of I are taken from ref. 2, where they were estimated from the inclinations of the axes of rotation of the planets. For Mercury and Venus m_1/m is unknown, since their axial rotation has been braked by the tidal interaction with the sun. For them we took m_1/m as the same as for the earth.

The estimate of e_{acc} for Jupiter and Saturn is not fully correct, since the accretion of gas by these planets was not taken into account. Allowance for accretion should

TABLE 1

Planet	m_1/m	e_{max}	e_{acc}	e_{obs}
Mercury	$1.8\cdot10^{-3}$	0.042	0.00042	0.206
Venus	$1.8\cdot10^{-3}$	0.14	0.0013	0.0068
Earth	$1.8\cdot10^{-3}$	0.18	0.0017	0.017
Mars	$3.5\cdot10^{-3}$	0.10	0.0014	0.093
Jupiter	$5.3\cdot10^{-4}$	0.33	0.002	0.048
Saturn	$7.0\cdot10^{-2}$	0.33	0.023	0.056
Uranus	$12.3\cdot10^{-2}$	0.33	0.03	0.047
Neptune	$12.3\cdot10^{-3}$	0.33	0.0096	0.0086

lead to a considerable increase in m_1/m (i.e., in the sizes of the largest bodies). But e_{acc} changes to a considerably lesser extent.

It is seen from Table 1 that during the accumulation of bodies into a planet the effective averaging of their orbital eccentricities actually occurs. The orbital eccentricity e_{acc} of the just-formed planet proves to be one to two orders of magnitude less than the orbital eccentricities e'_{max} of the bodies in the final stage of its growth. This is also valid for the region of the outer planets, where the bodies had strongly elongated orbits.

It is further seen from Table 1 that for the outer three planets the modern orbital eccentricities e_{obs} are close to the calculated values e_{acc} whereas for all the other planets they are considerably larger. One might have tried to explain this disagreement by gravitational perturbations of the growing planets by the large bodies, which were not taken into account in the present article. This question requires special analysis. But the approximate preliminary estimates show that the increase in the orbital eccentricity of a planet owing to such perturbations is relatively small. One can therefore assume that the modern orbital eccentricities of the planets are determined to a greater extent by their prolonged gravitational interaction during the entire course of their existence (and particularly by the appearance of resonance motions) than directly by the process of formation of the planets.

The authors wish to thank E. V. Zvyagina for critical remarks and helpful advice.

[1]L. É. Gurevich and A. I. Lebedinskii, Izv. Akad. Nauk SSSR Ser. Fiz. 14, 765 (1950).
[2]V. S. Safronov, Evolution of the Protoplanetary Cloud and Formation of the Earth and Planets [in Russian], Nauka, Moscow (1969).
[3]O. Yu. Shmidt, Four Lectures on the Theory of the Origin of the Earth [in Russian], Izd. Akad. Nauk SSSR, Moscow (1957).
[4]V. S. Safronov, "The motion, evolution of orbits, and origin of comets," G. A. Chebotarev (ed.), Proceedings of LAU Symposium No. 45, Leningrad, 1970 (1972), p. 329.
[5]J. S. Dohnanyi, J. Geophys. Res. 74, 2531 (1969).
[6]E. V. Zvyagina, G. V. Pechernikova, and V. S. Safronov, Astron. Zh. 17, 793 (1974).

Translated by Edward U. Oldham

Some problems concerning the accumulation of planets

B. Yu. Levin

Astronomical Council, Academy of Sciences of the USSR Moscow

(Submitted September 15, 1977)

Pis'ma Astron. Zh. 4, 102–107 (February 1978)

The random velocities of the planetesimals in the protoplanetary swarm are shown to have been much lower than hitherto believed on the basis of V. S. Safronov's analysis. Some modification should therefore be made in O. Yu. Shmidt's scheme for the growth of the planets from material in the zones feeding them. The present author's 1972 arguments for a protoplanetary nebula with a massive, extended periphery are no longer tenable.

PACS numbers: 96.10.+i, 95.10.Ce

1. INTRODUCTION

Practically all students of planetary cosmogony in the past few decades have assumed that the planets of the terrestrial group were formed by accumulation – by a gradual agglomeration of dispersed solid material around an originally small nucleation center. And most investigators believe the giant planets developed by a similar process: an aggregation of solid material accompanied, to varying degrees, by accretion of gases. Unfortunately, many aspects of the accumulation process remain unclear, quite apart from the problem of the physical and chemical evolution of the protoplanetary medium.

One of the most important reasons for the difficulties that have been encountered is that the process whereby the protoplanetary cloud developed near the sun remains completely unknown. But whatever sort of process this may have been, its duration should have been such that planet formation commenced, and perhaps even progressed significantly, while the cloud was still being formed. Furthermore, planet formation probably did not begin simultaneously at different distances from the sun, and it undoubtedly proceeded at differing rates.

The accumulation of the solid component of protoplanetary material may conveniently be divided into two stages: 1) the formation of planetesimals from the dust component of the cloud; 2) further gravitational accumulation of such planetesimals and their fragments. Calculations show that the first stage ought to have been very rapid, whereas the second would have been considerably more protracted. We shall examine below several topics connected with this second, more lengthy phase of accumulation.

2. RANDOM VELOCITIES OF PLANETESIMALS AND THEIR AGGREGATION INTO PLANETS

One of the chief parameters governing the growth rate of an embryonic planet is the mean velocity of random motion of the planetesimals, for the ratio of the effective cross section of the embryo to its geometrical cross section will depend on this velocity.

In a swarm of planetesimals revolving about the sun, gravitational interaction among them during close encounters would inevitably have induced chaotic motions superimposed on the general rotation of the swarm. The mean velocity of these relative (random) motions can be expressed in terms of another quantity, which characterizes the gravitational activity of the planetesimals and has dimensions of velocity: the velocity of escape from the surface of a planetesimal.

In their analysis of a swarm of identical planetesimals, Gurevich and Lebedinskii[1] assumed that the mean relative (random) velocity of these objects could be expressed by

$$v = \sqrt{Gm/2r} = v_e/2. \quad (1)$$

Here G is the gravitational constant, m and r are the mass and radius of a planetesimal, and v_e is the escape velocity at its surface. A similar assumption ($v \approx v_e$) has been adopted recently by Goldreich and Ward.[2] To determine the exact value of the corresponding effective distance, a complete analysis of the consequences of encounters should be made. If the planetesimals in the swarm differ in mass, an average should be taken over the masses.

Applying an equation of the type (1) to a collection of planetesimals differing in mass, Safronov[3,4] has taken m and r to mean the mass and radius of the largest planetesimal, and he has introduced a dimensionless parameter θ whose numerical value should express the result of averaging with respect to individual relative velocities, impact parameters, and masses:

$$v = \sqrt{Gm/\theta r} = v_e/\sqrt{2\theta}. \quad (2)$$

For several alternative mass distributions of planetesimals, Safronov finds[3,4] that $\theta \approx 3$-5. Equation (2) with these numerical values for θ has been applied in all calculations entering into the analysis of the accumulation process for the earth, as well as for other planets and the moon. However, Safronov himself has shown (Sec. 23 of his book[4]) that if the largest body in a given zone of the protoplanetary swarm (for instance, an embryonic planet) is much more massive than all the others, then if one takes $\theta \approx 3$-5 in Eq. (2), one should not adopt the mass and radius of this individual body, but the mass m_1 and radius r_1 of the second largest body, that is, the largest of all the remaining objects that make up the bulk of the population in that zone. In other words, if $m \gg m_1$, then one should either replace Eq. (2) by

$$v = \sqrt{Gm_1/\theta r_1} \quad (3)$$

0360-0327/78/0401-0054$02.00

and continue to regard θ as about 3-5, or retain Eq. (2) but replace the θ in it by

$$\theta' = \theta \frac{mr_1}{m_1 r} = \theta \frac{r^2}{r_1^2} \quad (4)$$

(keeping $\theta \approx$ 3-5, as before).

Safronov believes this situation will arise only in the terminal phase of the accumulation process. He does not take into account the result of considering the case $m \gg m_1$, either in other chapters of his book[4] or in later papers. Yet this change is a necessary one for analyzing the main part of the accumulation.

If a large planetesimal is present, standing out in mass from the host of diminutive ones, then the mean random velocity of these small planetesimals will be determined not by their rare encounters with the lone specially massive body, but by their frequent encounters with one another. This circumstance has escaped the attention of all those who have made use of Safronov's results, including the present author,[5,5] Ruskol,[7] Kaula,[8] and others.

Even during the first phase of the accumulation, planetesimals ought to form that differ considerably from one another in mass. As the gravitational agglomeration proceeds, these differences should be enhanced rapidly. When the growth rate depends on the effective rather than the geometrical radius of the growing body, the larger bodies will very swiftly leave the smaller ones behind. The planets will therefore acquire the bulk of their mass after their embryos have already become substantially larger than the other planetesimals, that is, under conditions whereby Eq. (3) should be applied or θ should be replaced by θ' according to Eq. (4).

Even under circumstances such that Eq. (2) is applicable, we find that compared with the largest planetesimals, the mean relative velocity at infinity for the other, small planetesimals will be two or three times lower than the escape velocity at the surface of those largest planetesimals. As a consequence, the material of most of the smaller bodies colliding with planetesimals will adhere to them almost completely.

Matters improve even more in the case of an embryonic planet that comes forward in the course of its growth. In this situation, if we substitute into Eq. (2) the θ' of Eq. (4), we find that the mean relative velocity of the planetesimals is tens of hundreds of times lower than the escape velocity at the surface of the embryo.

The smaller the planetesimal, the lower the escape velocity at its surface will be; hence the same collision velocities that can result in amalgamation when the collision involves a planetesimal or even an embryo will serve instead to eject a substantial amount of material and disperse it into space when smaller bodies are involved. This enhanced "vitality" of the largest planetesimals, particularly embryonic planets that come to the fore, will ensure their preferential growth and will ultimately lead to the creation of a planetary system.

Making improper use of Eq. (2), Safronov and Zvyagina[9,10] (see also Sec. 26 of Safronov's book[4]) have concluded that the mass m of the embryo cannot surpass the mass m_1 of the second largest body by more than a factor of $(2\theta)^3$. For $\theta \approx$ 3-5 we would then have $m \approx (0.2\text{-}1) \cdot 10^3 m_1$. However, in the main part of the accumulation process, with $m \gg m_1$, Eq. (3) or (2) should be used in conjunction with Eq. (4); it then turns out that an upper limit for the ratio m/m_1 will not exist.

3. MODIFICATION OF O. Yu. SHMIDT'S SCHEME

The substantial distances between the orbits of successive planets ensure the prolonged survival of the planetary system, and they also imply that each planet will absorb material from a broad supply or feeding zone. The concept of feeding zones of planets was introduced by Shmidt[11] and underlies the scheme he used to estimate the length of time the earth required to grow. According to this scheme, an embryonic planet would travel along an approximately circular orbit near the middle of its feeding zone, and solid material not yet used up would be distributed uniformly over the whole width of the zone throughout the entire accumulation process. In order for accretion to encompass the full width of the feeding zone, the orbital eccentricties of the planetesimals should be close to 0.2, which in turn implies that their random velocities should amound to $\approx$ 0.2 of the circular velocity, or $\approx$ 6 km/sec in the case of the earth. When Eq. (2) was erroneously applied in the past, such velocities seemed possible, at least in the terminal growth phase. However, the embryonic planet should have a radius larger by an order of magnitude or more than the other largest planetesimals, and as we have shown in Sec. 2, the mean random velocity of the planetesimals should then be at least an order of magnitude smaller than hitherto believed. But if the planetesimals have low random velocities, then an embryo traveling along a practically invariant circular orbit could not sweep up a broad feeding zone, but only a narrow ring along its orbit.

As indicated above, the larger the mass of the growing body the faster it will grow. Hence even at an early stage of the accumulation some of the planetesimals should come forward and occupy a dominant position. A big planetesimal of this kind will quickly sweep up the immediate vicinity of its orbit and become the embryo of a future planet even before less sizable planetesimals have succeeded in going through an analogous agglometation elsewhere in the broad feeding zone. Further growth of the embryo will proceed at an increasingly slow rate, as planetesimals that have acquired random velocities through mutual gravitational perturbations, and that have orbital eccentricities considerably larger than average, fly into the narrow zone of the embryo.

Apart from the diminishing reserve of planetesimals not yet swept up, another important factor retarding the growth of the embryo will be the gradual expansion along its orbit of the zone already swept up. Thus during the planet formation phase there should have been an era when in all or nearly all the gaps between the planets' orbits, rings of asteroidal bodies existed, similar to the asteroid belt today. These rings have now completely disappeared between the orbits of the terrestrial planets, but in the giant-planet zone, and especially in its outer part, remnants of these rings probably still survive, although inacessible to observation. Bombardment of the terrestrial planets by bodies from these primordial asteroid belts would have been responsible for the tail in the accumulation process during which craters were formed on the surfaces of these planets—structures that have now managed to become covered with a crust consisting of the magmatic differentiation products of the planets' interiors.

Shmidt's scheme, which presupposes that the bodies accreted would have a uniform distribution over the whole wide feeding zone throughout the entire accumulation process, permits a simple mathematical analysis of the length of the planet formation phase. Adopting Shmidt's formulation and using Eq. (2) with a parameter $\theta \approx 3$-5, Safronov[3,4] concluded that the duration of the formative period for the earth, defined conventionally as the time the earth required to accumulate 97% of its mass, was $\approx 10^8$ yr. But since the radius of the embryo during most of the accumulation process was an order of magnitude or more larger than the radius of the biggest planetesimals, one should take in Eq. (2) the value θ' given by Eq. (4). The numerical value of θ' will in fact be two or three orders of magnitude larger than θ. Since the length of the accumulation phase in Shmidt's scheme is approximately in inverse proportion to θ, the formal introduction of more legitimate values for this parameter would imply that the accumulation phase of the earth lasted only 10^5-10^6 yr (and there would be a corresponding rise in its initial temperature as governed by impacting planetesimals). But such an estimate would be an improper one, because if the planetesimals have low random velocities Shmidt's scheme should be modified as outlined above. A mathematical treatment of such a modification has not yet been worked out.

4. FEEDING AND ACCUMULATION ZONES OF THE GIANT PLANETS

For a parameter $\theta \approx 3$-5, calculations indicate that early in the growth phase of the giant planets, when their embryos were still much less massive than the earth today, the concept of isolated feeding zones should have been deprived of meaning, because the rise in the random velocities should have resulted in a complete overlapping of the feeding zones (Safronov,[4] Sec. 31). If we recognize, however, that the random velocities were at least an order of magnitude lower than assumed by Safronov, then we find that overlap of the feeding zones of the giant planets should have begun to develop when the radii of the embryos were an order of magnitude greater, that is, for masses 10^3 times as large as those given by Safronov[4] (Table 13). In other words, the feeding zones should have begun to overlap when the giant planets were nearly as massive as they are now. This result is consistent with our idea[12] that the growth of the giant planets halted not through complete exhaustion of the material in their feeding zones, as was the case for the planets of the terrestrial group, but because of ejection from the feeding zones of material not yet swept up, followed by its ejection from the entire planetary system. If Safronov's estimates of random velocities were valid, the ejection process would not even have permitted the planets distant from the sun to grow to their present mass.

Several years ago the author, using Eq. (2) and, with Safronov, taking $\theta \approx 3$-5, came to the conclusion that the accumulation of Uranus and Neptune over an admissible period shorter than the age of the solar system would require that the protoplanetary cloud have had a massive outer zone extending well beyond the present limits of the planetary system.[6,13] This situation would have implied a large primeval mass for the protoplanetary cloud, of the order of the solar mass, and a large primordial angular momentum. But if, as we have explained, values of θ some 10^2 times as great should be taken, then the need for a massive protoplanetary cloud will be avoided.

During the initial growth phase of the embryo, before its formation appreciably lowered the (surface) density of the accreting solid material, the radius of the embryo will have increased at the rate[6,13]

$$\frac{dr}{dt} = \frac{1+2\theta}{\delta} \frac{\sigma_0}{P}. \tag{5}$$

Here P denotes the revolution period around the sun and δ is the density of the embryo, which may be taken as 1 g/cm^3 for ice substances. The minimum value of the initial surface density σ_0 of the solid matter in the Uranus – Neptune zone, based on the spread of 80-90% in the present masses of these planets and not allowing for the ejection of solids out of that zone, would be ≈ 0.3 g/cm^2.

Taking $1 + 2\theta \approx 10$, the author previously found that $dr/dt \approx (2\text{-}4) \cdot 10^{-2}$ cm/yr. It appeared that over the billions of years the solar system has been in existence, only small embryos of Uranus and Neptune about 1000-2000 km in radius could have been formed. Now, however, we can take $1 + 2\theta \approx 1000$ and can conclude that such embryos could have developed in just 10^7 yr. Moreover, there is no question but that ejection of solid material from the formation zone of Uranus and Neptune actually occurred,[14] so that the true value of σ_0 would have been greater than the value used above. The embryos of Uranus and Neptune might accordingly have been formed in just 10^6 yr. In any event, the requirement of a massive protoplanetary cloud is removed.

Any estimate of the total accumulation period of the giant planets is impeded not only by uncertainty as to the stage when the gases accumulated but by a circumstance already intimated above in connection with the modification of Shmidt's ideas: for small random velocities an embryo would not be able to sweep up a broad feeding zone, and instead of a single large planet in each zone, several small ones should have accumulated initially. Probably they would have consisted wholly of solid matter—rocks and ices, and only after they amalgamated into large planets would thay have acquired hydrogen and helium, in huge abundance for Jupiter and Saturn but in a very moderate amount for Uranus and Neptune. The low hydrogen and helium abundance in Uranus and Neptune might possibly indicate that their formation consumed more time and that by the epoch when it ended the gases had largely been able to disperse. As for Pluto, it should be regarded as one of the primordial little planets in Neptune's zone, having escaped becoming attached to Neptune thanks to its acquisition of a resonance orbit which precludes such encounters.

[1]L. É. Gurevich and A. I. Lebedinskii, "Gravitational condensation of a dust cloud," "The properties of the clouds from which the planets of the

solar system were formed," "Laws of planetary distances and planetary revolution" [in Russian], Dokl. Akad. Nauk SSSR 74, 673-676, 905-907, 1061-64 (1950); "The formation of the planets I-III" [in Russian], Izv. Akad. Nauk SSSR, Ser. Fiz. 14, 765-799 (1950).

[2]P. Goldreich and W. R. Ward, "The formation of planetesimals," Astrophys. J. 183, 1051-61 (1973).

[3]V. S. Safronov, "On the growth of the terrestrial planets" [in Russian], Voprosy Kosmogonii, No. 6, USSR Acad. Sci. Press (1958) pp. 63-77.

[4]V. S. Safronov, "Evolution of the Protoplanetary Cloud and the Formation of the Earth and Planets," Nauka, Moscow (1969) [NASA TT F-677 (1972)].

[5]B. Yu. Levin, "The origin of the earth," Izv. Akad. Nauk SSSR, Fiz. Zemli (1972) No. 7, 5-21 [Izv. Acad. Sci. USSR, Phys. Solid Earth (1972) No. 7, 425-434; Tectonophysics 13, 7-29 (1972)].

[6]B. Yu. Levin, "Revision of the initial size, mass, and angular momentum of the solar nebula, and the problem of its origin," in: Sympos. on Origin of Solar System (Nice, April 1972), Cen. Natl. Rech. Sci., Paris (1972), pp. 341-360.

[7]E. L. Ruskol, "Formation of the moon from a cluster of particles encircling the earth," Izv. Akad. Nauk SSSR, Fiz. Zemli (1972) No. 7, 99-108 [Izv. Acad. Sci. USSR, Phys. Solid Earth (1972) No. 7, 483-488].

[8]W. M. Kaula, in: Cosmochemistry of the Moon and Planets (Proc. Soviet–Americal conf.), Nauka, Moscow (1975), p. 630.

[9]V. S. Safronov and E. V. Zvyagina, "Relative sizes of the largest bodies during the accumulation of the planets," Icarus 10, 109-115 (1969).

[10]V. S. Safronov and E. V. Zvyagina, in: Physics of the Moon and Planets [in Russian], Nauka, Moscow (1972), p. 219.

[11]O. Yu. Shmidt, "The astronomical age of the earth," Dokl. Akad. Nauk SSSR 46, 392-395 (1945).

[12]B. Yu. Levin, "On the distances and masses of the planets" [in Russian], Voprosy Kosmogonii, No. 7, USSR Acad. Sci. Press (1960), pp. 55-58.

[13]B. Yu. Levin, in: "The origin of the Solar System" [in Russian], Mir, Moscow (1976), p. 505.

[14]E. J. Opik, "Comets and planets: their interrelated origin," Irish Astron. J. 12, 1-48 (1975).

 0360-0327/78/0401-0057$2.00

Mass of the largest bodies and the velocity dispersion during the accumulation of the planets

G. V. Pechernikova and A. V. Vityazev

Shmidt Institute of Earth Physics, Academy of Sciences of the USSR, Moscow

(Submitted June 21, 1978)

Pis'ma Astron. Zh. **5**, 54–59 (January 1979)

A model is proposed for the growth of embryonic planets that have expanding and overlapping feeding zones. The mass ratio m/m_1 of the largest bodies in each zone is comparable with the mass ratio of successive planets today. The relative velocities of the bodies would have been determined by the largest ones, which would have incorporated the bulk of the matter in their zones. The embryonic planets would have had orbital eccentricities similar to those of the present planets.

PACS numbers: 96.10. + i, 95.10.Eg, 95.10.Ce

1. INTRODUCTION

According to current views, when the protoplanetary cloud was evolving a swarm of solid bodies developed, traveling around the sun in orbits with relatively small eccentricities and inclinations. As the bodies amalgamated and grew in mass, their gravitational interaction became stronger and their relative velocities increased. In his book on the subject, Safronov[1] assumed a Maxwellian velocity distribution and derived the distribution function n(m, t) of the bodies with respect to mass, as well as an expression for their mean relative velocity v. For a function $n(m) \propto m^{-q}$, with $1.5 \leqslant q < 2$, the quantity v turns out to be of the same order as the escape velocity v_e determined for the largest body in a given distribution:

$$v = \frac{v_e}{\sqrt{2\theta}} = \sqrt{\frac{Gm}{\theta r}}, \quad \theta \simeq 3\text{–}5, \tag{1}$$

where G is the gravitational constant, and m and r denote the mass and radius of the largest body.

For the body m the effective collision cross section is $\pi r^2(1 + 2\theta)$, while for the next largest body m_1 it is $\pi r_1^2(1 + 2\theta r_1^2/r^2)$. Hence the break in mass between m and m_1 will grow with time, approaching,[1] for constant $\theta > 1/2$, a maximum value $m/m_1 = (2\theta)^3$. If the gap is substantial, the mass of the body m_1 will become the effective upper limit of the distribution function, and, according to Safronov, v will now have to be taken in the form

$$v = \sqrt{\frac{Gm}{\theta' r}\frac{r_1}{r}} = \sqrt{\frac{Gm_1}{\theta' r_1}}, \tag{2}$$

where the values of θ' are somewhat smaller than the values given for θ.

Confining our attention to the relative growth of the two largest bodies in a single zone, we can find the time dependence of the ratio r/r_1 corresponding to Eqs. (1) and (2). The growth rates of the largest body and the next in size are given by

$$\frac{dr}{dt} = \frac{(1 + r_e^2/r^2)\,\sigma(t)}{P\delta}, \quad \frac{dr_1}{dt} = \frac{(1 + r_{e1}^2/r^2)\,\sigma(t)}{P_1\delta_1}, \tag{3}$$

where P, P_1 denote the periods of revolution of the bodies around the sun; δ, δ_1 are their densities; v_e, v_{e1} are the escape velocities; and $\sigma(t)$ is the surface density of the material in the given zone. If we here set $P = P_1$, $\delta = \delta_1$, and take the value (1) for v, then we find, upon integrating Eqs. (3), that

$$\varkappa^*(t) = \frac{r(t)}{r_1(t)} = \frac{1 + 2\theta(r/r_0)^b(\varkappa_0^* - 1)/(2\theta - \varkappa_0^*)}{1 + (r/r_0)^b(\varkappa_0^* - 1)/(2\theta - \varkappa_0^*)},$$

$$b = \frac{2\theta - 1}{2\theta + 1}, \quad \theta > 1/2,$$

$$r(t) = \frac{1 + 2\theta}{P\delta}\int_{t_0}^{t}\sigma(t)\,dt + r_0, \tag{4}$$

where r_0, r_{10}, $\varkappa_0^* = r_0/r_{10}$ represent the values of r, r_1, $\varkappa^*$ at time t_0. Equation (4) implies that $\varkappa^* \to 2\theta$ as $r/r_0 \to \infty$.

If instead we adopt the value (2) for v, then for $P = P_1$, $\delta = \delta_1$, Eqs. (3) give

$$\varkappa(t) = \frac{r(t)}{r_1(t)} = \frac{1 - (r_1/r_{10})^{b_1}(\varkappa_0 - 1)/(2\theta'\varkappa_0 - 1)}{1 - 2\theta'(r_1/r_{10})^{b_1}(\varkappa_0 - 1)/(2\theta'\varkappa_0 - 1)},$$

$$b_1 = \frac{2\theta' - 1}{2\theta' + 1}, \quad \theta' > 1/2,$$

$$r_1(t) = \frac{1 + 2\theta'}{P\delta}\int_{t_0}^{t}\sigma(t)\,dt + r_{10}. \tag{5}$$

As r_1 approaches the value $r_1 = r_{10}[(2\theta'\varkappa_0 - 1)/2\theta'(\varkappa_0 - 1)]^{1/b_1}$, $\varkappa$ rapidly becomes infinite.

It is this possibility, the development of a gap with catastrophic speed and the corresponding rise in $\theta = \theta' r^2/r_1^2$, which Levin[2] had in mind in a recent letter. In his view, this effect would have caused a more gradual growth (by a factor $\varkappa$) of random velocities, with the progenitors of the present planets having been formed at an early stage when the remaining bodies had been far outdistanced. The asymptotic values of $\varkappa^*$ and $\varkappa$ given by Eqs. (4) and (5) will evidently be reached only if we suppose primordially broad feeding zones, comparable with the ones today. Allowance should be made for the depletion of material in the feeding zone, the amalgamation of the largest bodies during the growth process, and the combining of their zones, effects which will produce considerably smaller values of $\varkappa$ and θ. We shall now analyze a simple model for the agglomeration of the largest bodies in the accumulation process, and shall discuss the constraints on the value of the velocity dispersion.

 0360-0327/79/01 0031-04 $02.00

2. A MODEL FOR THE BREAK IN THE ACCUMULATION PROCESS

According to geo- and cosmochemical evidence, gas would have played an insignificant role during the late stages of the accumulation in the region of the terrestrial planets. Thus in considering the mechanism for development of a break in this region we may neglect the influence of gas on the random velocities. The drift of the embryonic planets with respect to R may also be neglected. Since the mean eccentricities achieved in this region are relatively small ($\bar{e} \lessapprox 0.2$), we shall limit the treatment to the first approximation with respect to $\bar{e}$.

From coagulation theory we can estimate the time variation of the mass spectrum in a system of many bodies that amalgamate and fragment by collisions. The main result derived from this theory is that most of the mass will be concentrated in a relatively small number of large bodies. In order to describe the agglomeration of a small number of bodies we may conveniently use Eq. (3), remembering that it is permissible to do so only over times shorter than the interval between collisions of m and m_1 with bodies of comparable mass. Neglecting such collisions here, we shall apply Eqs. (3)-(5) up to the time when the bodies m, m_1 collide with each other.

a. Allowance for depletion of material. Assume that m and m_1 grow in the same feeding zone, that is, in an annular region whose width $2\Delta R$ will be defined in terms of the mean orbital eccentricity $\bar{e}$ of the preponderance of bodies:

$$\Delta R \approx \bar{e} R \approx \sqrt{2}\, \frac{v}{v_c} R, \tag{6}$$

where v_c denotes the Keplerian velocity of the body m at distance R from the sun. Then the mass of material in the growing feeding zone will be expressed by

$$Q(R,\, t) = 2\pi \int\limits_{R-\Delta R(t)}^{R+\Delta R(t)} \sigma(x,\, 0)\, x\, dx. \tag{7}$$

We now write the surface density $\sigma(t)$ for this zone, allowing for the depletion of material by the bodies m and m_1, and taking $\sigma(R, 0) = \sigma_0 = \text{const}$:

$$\sigma(t) = \sigma_0 \left(1 - \frac{m(t) + m_1(t)}{Q(t)}\right) = \sigma_0 (1 - B\varkappa^3 r_1^2 - B r_1^2),$$
$$B = \frac{1}{2\sigma_0} \sqrt{\frac{\theta' \delta M_\odot}{6\pi R^5}}\,. \tag{8}$$

Here v is determined by Eq. (2). The bodies m, m_1 will cease to grow when $\sigma(t)$ has dropped to zero. We thereby obtain an expression for the value of $\varkappa$ such that the feeding zone is exhausted:

$$\varkappa^3 = \frac{1}{B r_1^2} - 1. \tag{9}$$

A simultaneous solution of Eqs. (5) and (9) yields the maximum possible values $\varkappa_{max}$, on the condition that the body m_1 survives until the end of the accumulation process.

To estimate $\varkappa_{max}$ we must specify values for θ', r_{10}, and $\varkappa_0$. Safronov[1] (Sec. 23) has evaluated θ' for the earth's zone in the absence of gas. As the expression for $\theta'(r, r_1, q)$ is rather cumbersome, we shall not reproduce it here but shall use the numerical values in Safronov's Table 8. The quantity θ' falls off with increasing r, r_1 more weakly than logarithmically, so it may be regarded as constant in obtaining the expression (5) from Eqs. (3). According to Safronov's analysis of gravitational instability in a sufficiently flat protoplanetary disk,[1] we may take r_0, $r_{10} \approx 10^5$ cm. One can readily show that with these values of r_0, r_{10}, the boundaries of the zone swept out by the gravitational cross section of the growing body m (for $m \approx 10^{-4}$–$10^{-5}\, m_\oplus$) will reach the original boundaries of the feeding zone as specified in terms of r_0 and r_{10} according to Eq. (6).

For r_0, $r_{10} \approx 10^6$ cm and $\varkappa_0 \approx 1.1$-2, the body m_1 will begin to cross the Hill sphere of the body m when r has increased to just $(2\text{-}6) \cdot 10^7$ cm; then $\varkappa \approx (2\text{-}3)\varkappa_0$. Assuming that m_1 nevertheless survives to the end of the accumulation process, remaining the second largest body, we find for $\theta' = 3$ and $\varkappa_0 = 1.1$-1.4 that $\varkappa_{max} = 3$-10; for $\varkappa_0 = 2$ we obtain $\varkappa_{max} = 30$, and the final mass of the largest body becomes $m_{max} \approx 10^{-2}\, m_\oplus$.

The large values of $\varkappa_0$ adopted above, which give a substantial gap in mass, correspond to rare occurrences of collision and amalgamation of the major bodies. Coagulation theory for m, $m_1 \ll Q$ yields values for $\varkappa_0$ practically equal to unity.

b. Agglomeration of bodies with amalgamation of zones. As the bodies m and m_1 grow their feeding zones will expand, embracing the adjacent regions in which other large bodies with masses of order m had been growing independently. We shall now show that the principal agglomerating bodies in the combined zone will be the largest bodies of the adjacent zones, and we shall estimate the ratio $r(n, t)/r(n+1, t)$ of their radii, invoking the simple idea of a doubling of the zones.

Suppose that at time t_0 the zones of adjacent embryonic planets are contiguous. As the bodies grow the zones will begin to overlap. When r_1 has increased by a factor of two its supply zone will have doubled ($\Delta R \propto r_1$), and two adjacent embryos will be within it. Because of differences in P(R) and σ(R) ($\sigma_0 \approx R^{-\nu}$, with $|\nu| \lessapprox 2$), the size ratio of the embryonic planets in adjacent zones[3] will be $r_n \cdot (r_{n+1})^{-1} \approx 1 + (3 + 2\nu)\bar{e}$. Equation (5) may be used to estimate the value of $\varkappa(t') = r(t')/r_1(t')$ by doubling time t'. Estimates show that, on the average $r(n, t')/r(n+1, t') < \varkappa(t')$, so the two embryos will become the largest bodies in the amalgamated zone from this time onward. The doubling process will be repeated so long as there is enough material in the zone for the radius of the largest bodies to double. Using Eq. (5) as a recursion relation, where $\varkappa_0 = 1 + (3 + 2\nu)\bar{e}$, we can find $\varkappa(t)$; by the time of the last doubling of the zones, $\varkappa(t)$ will remain $\lessapprox 2$.

c. Final stage of accumulation. When the mass of the embryo has become comparable with the total mass of all the other bodies in its zone, Eq. (2) will no longer be correct, and we must use Safronov's more general expression[1] for θ:

$$\theta = \frac{\theta' \varkappa^2}{(1+\chi)^{1/2}}\,, \tag{10}$$

$$\chi = f \varkappa^3 \frac{m}{Q-m}\,, \quad f \approx 2\text{—}4. \tag{10a}$$

Here χ denotes the ratio of the relaxation time T_D, as defined according to Chandrasekhar[4] by the perturbations of all bodies except m, to the relaxation time T_D^* determined by the perturbations of the embryo. The expression we have used for χ is analogous to that obtained by Safronov[1] (Chap. 7), differing only in the factor of ≈ 0.4-0.6 that appears when we allow for the embryo's own orbital eccentricity and the boundedness of its zone of influence. The zone of influence will enlarge as m/m_1 increases; at an intermediate stage, when m ≈ 0.1Q, it will measure approximately 10 Hill radii of the body m, according to estimates based on the three-body problem.[5]

Using the ratio

$$m(t)/Q(t) = r^2(t)\, B\sqrt{\theta/\theta'}, \tag{11}$$

inferred from Eq. (8), we obtain from Eq. (10a) the expression

$$\chi = \frac{Bf\varkappa^{1}r^2}{(1+\chi)^{1/4} - B\varkappa r^2}\,. \tag{12}$$

For m(t) ≪ Q(t) we find from Eq. (10a) that $\chi \ll 1$, and Eq. (10) now yields Eq. (2), that is, $\theta = \theta'\varkappa^2$. But as $m + m_1 \to Q$ the denominator $(1+\chi)^{1/2}$ will increase and must be taken into account. For the earth's zone, $B/\sqrt{\theta'} = 1.25 \cdot 10^{-18}$, and if $r \gtrsim 0.1\, r_\oplus$ and $\chi \gtrsim 1$ the velocity dispersion will here be determined mainly by the embryonic planet. The quantity $\varkappa/(1+\chi)^{1/4}$ will diminish, and by the end of the accumulation process θ will have dropped to 1.2-1.5.

The mean orbital eccentricity $\bar{e}$ of the preponderance of bodies in the zone will increase in proportion to $r/\sqrt{\theta}$. We can estimate the variation in the orbital eccentricity e of the embryo by generalizing somewhat the method of Ziglina[6] to the case of variable $\varkappa$ and Q. At the stage when the feeding zones combine, $\varkappa$ will vary only weakly; e will rise in proportion to $\bar{e}$ without exceeding values in the range $(0.2\text{-}0.3)\bar{e}$. The largest bodies in the zone (the former embryos) will have eccentricities larger than e, so it is sufficient for the width of the zone to enlarge by a factor of 3-4 in order for these bodies to begin to cross the orbit of the embryo and ultimately to collide with it; by this time they will have $\varkappa \gtrsim 3$. In the final stage of the accumulation, when m > 0.1Q, as the largest bodies fall onto the embryo the role of the body m_1 will be transferred to bodies of increasingly small mass, and $\varkappa$ will effectively increase. The growth rate of the orbital eccentricity of the embryo will decline, and by the close of the accumulation we will have $e \approx 0.1\bar{e}$, corresponding to $e \approx 10^{-2}$ in the case of the earth.

We have neglected in the discussion above the fragmentation and erosion of the large bodies. The breakup of small bodies is taken into account, however, by the exponent q in the mass distribution function of the bodies. We can estimate the influence of erosion on the growth rate of the largest bodies by introducing an additional factor of the form $[1 - \alpha(1 + v^2/v_e^2)]$ into the right-hand member of Eq. (3). According to O'Keefe and Ahrens,[7] the proportion of energy going into ejection will be small: $\alpha \approx 10^{-2}$, so for $\varkappa \lesssim 4$ the influence of erosion will be unimportant. Thus far the process whereby the largest bodies are disrupted by tidal forces in close encounters with the embryonic planet is not entirely clear.

3. CONCLUSIONS

Unless we postulate at the outset highly irregular physical conditions in the accumulation zones [in particular, sharp fluctuations in the surface density $\sigma(t_0, R)$], the independent growth of the largest bodies in adjacent zones and the agglomeration of those bodies during subsequent overlap of the zones will seriously restrict the role of any possible mass fluctuations. The mass ratio of the chief bodies in each zone turns out to be of the same order as the ratio of the present masses of successive planets in the corresponding region of the solar system. Initially, so long as m < 0.1W, the relative velocities will be determined by the largest bodies, which incorporate the bulk of the material in the zone, and the orbital eccentricity of the embryo will differ little from the mean eccentricity of the preponderance of the bodies. When m > 0.1Q and the embryo has an eccentricity comparable with the value today, the relative velocities of the bodies will be determined mainly by the embryo ($\chi > 1$), and will be adequate to exhaust the material within the zone. By this time the increase in the eccentricity of the embryo will have practically ceased, and mutual perturbations of embryonic planets will no longer be sufficient to alter the planetary orbits very much.

With the improved expressions for $v(\varkappa)$ obtained in this analysis we can reconcile the computed masses, distances, orbital eccentricities, and inclinations of the rotation axes of the planets with the values observed.

a. The results of Vityazev et al.[3] indicate that for the terrestrial planets the observed masses and distances would agree with the theoretical values if θ = 1.5-3 during the closing phase of the accumulation process.

b. In order to arrive at acceptable growth times for Uranus and Neptune following the dissipation of the gas in the outer parts of the protoplanetary cloud, we would need θ = 15-30 (e ≈ 0.3).

If the different zones have comparable values of $\varkappa$, it would be hard to satisfy concurrently these two conditions imposed on θ, because Eq. (2) implies that $\theta \propto \varkappa^2$. However, by including in Eq. (10) the factor $[1 + \chi(\varkappa, R)]^{-1/2}$, which compensates for $\varkappa^2$ in the accumulation zone of the terrestrial planets, we remove the apparent contradiction.

According to Safronov,[1] the inclinations of the rotation axes of the planets were determined by the size of the infalling bodies. Solving the converse problem, Safronov obtained for the largest bodies m' incident on m the ratio $m/m' = 10^2\text{-}10^3$, which seemed to confirm the asymptotic relation obtained by considering the relative growth rate $m/m' = (2\theta)^3$ for θ = 3-5. As we have shown in this letter, such an asymptotic value is not achieved for the largest bodies; indeed, by the end of the accumulation process $\theta < 3$ in the earth's zone. We therefore have to explain the large m/m' values obtained from consideration of the inclinations of the planets' rotation axes.

We have estimated the inclinations on the basis of a spherically symmetric accumulation process for three cases of $m'/m \propto m^n$, with n = −1, 0, 1. By allowing for the flattening of the cloud and the smaller than average orbital inclinations for the largest bodies, we can lower the m/m' values given above by a factor of 1.5-2. In using one of the three cases for m'/m to cover the entire accumulation, we fail to take into account the change in the functional dependence m'(m) with time. At the stage of growing and overlapping zones, $m' \sim m$. For any body falling onto the planet, the following conditions should hold in the zones that have formed: $m' \le m$ for $m \le Q/2$, and $m' < Q - m$ for $m > Q/2$.

If, in solving the converse problem, we attempt to evaluate m'/m, say in the form $m'/m = \varepsilon(1 - m/Q)$, then for constant ε the m/m' values given above will diminish by another factor of 4-6. This result is understandable, as collisions of bodies of comparable mass early in the formation of a planet will have little effect on the inclination of the rotation axis observed today. For the early stages, then, Safronov's estimates for the mass of the largest infalling bodies (and Ziglina's estimates[6] for the orbital eccentricities of the embryonic planets, as based on these mass estimates) are evidently too low by an order of magnitude.

The authors wish to thank V. S. Safronov for discussions of this subject and B. Yu. Levin for helpful comments on the manuscript.

[1]V. S. Safronov, Evolution of the Protoplanetary Cloud and Formation of the Earth and the Planets, Nauka, Moscow (1969) [English transl., NASA TT F-677 (1972)].
[2]B. Yu. Levin, "Some problems concerning the accumulation of planets," Pis'ma Astron. Zh. 4, 102-107 (1978) [Sov. Astron. Lett. 4, 54-57 (1978)].
[3]A. V. Vityazev, G. V. Pechernikova, and V. S. Safronov, "Limiting masses, distances, and times for the accumulation of planets of the terrestrial group," Astron. Zh. 55, 107-112 (1978) [Sov. Astron. 22, 60-63 (1978)].
[4]S. Chandrasekhar, Principles of Stellar Dynamics, Univ. Chicago Press (1942).
[5]C. Hayashi, K. Nakazawa, and I. Adachi, "Long-term behavior of planetesimals and the formation of the planets," Publ. Astron. Soc. Jpn. 29, 163-196 (1977).
[6]I. N. Ziglina, "The effect on a planet's orbital eccentricity of its encounters with bodies of the protoplanetary swarm," Astron. Zh. 53, 1288-94 (1976) [Sov. Astron. 20, 730-733 (1977)].
[7]J. D. O'Keefe and T. J. Ahrens, "Meteorite impact ejecta: dependence of mass and energy lost on planetary escape velocity," Science 198, 1249-51 (1977).

Evolution of orbital eccentricities of the planets in the process of their formation

G. V. Pechernikova and A. V. Vityazev

O. Yu. Shmidt Institute of Earth Physics, USSR Academy of Sciences

(Submitted May 15, 1979)

Astron. Zh. **57**, 799–811 (July–August 1980)

In the theory of the accumulation of the planets it is shown that, as a result of mutual gravitational perturbations, the eccentricities and inclinations of the orbits of bodies in the "supply zones" of the planets grew with the growth in the masses of the largest bodies. In a number of accumulation problems it is necessary to allow for the nonzero eccentricities of the planets in the process of growth of their masses also. In the report we give the derivation of equations connecting the orbital eccentricity e of a planet with the mean orbital eccentricity $\bar{e}$ and with the mass distribution function of the bodies encountering and falling onto the planet. It is shown that in the early stage of accumulation the orbital eccentricities of the planets were comparable with $\bar{e}$ and increased with an increase in $\bar{e}$, and then decreased. By the end of the accumulation the calculated eccentricities lie within the limits given by D. Brouwer and G. M. Clemence for the present-day planetary system.

PACS numbers: 96.10. + i, 95.10.Ce, 95.10.Eg

According to modern concepts, in the course of the evolution of the protoplanetary cloud a swarm of solid bodies formed, revolving about the sun along orbits with relatively small eccentricities and inclinations. With the increase in the masses of the bodies owing to their amalgamation, the gravitational interaction of the bodies was strengthened and their relative velocities grew. The growth of the planets under such conditions has been analyzed by Shmidt,[1] Safronov,[2] and recently by the authors.[3,4] In contrast to the earlier work, we introduced the concept of variable zones of supply of the planets, growing and overlapping with the growth of the masses of the latter, and we obtained asymptotic estimates for the mass spectrum of the planets and their mutual distances as a function of the initial conditions in the protoplanetary cloud. Refined expressions for the relative velocities were obtained in Ref. 4, and the question of the size of the largest bodies falling onto the planets was reexamined. It was noted that within the framework of the new approach one can reconcile the calculated masses, distances, eccentricities, inclinations of the axes of rotation, and velocities of rotation of the planets with the observed ones. The evolution of the orbital eccentricities of the planets in the process of their growth is discussed in the present report.

One of the properties of the planetary system is the almost circular orbits of the planets, lying in practically the same plane. Shmidt first gave the qualitative explanation for this property: such orbits are "the consequence of the natural statistical averaging of the motions of the separate bodies, from the combining of which the planets arise."[1] The first attempt to estimate this effect quantitatively was made in the reports of Safronov and Ziglina (Refs. 5, 6). The calculated eccentricities, especially for planets of the terrestrial group, proved to be considerably less than the observed ones. The conclusion was drawn that the orbital eccentricities of the planets are determined mainly by their gravitational interaction, rather than by the process of accumulation of the planets itself. We note that according to the theory, the accumulation lasted 10^8 years, and if the mean orbital eccentricities of the planets of the terrestrial group were less than the present ones by an order of magnitude,[5,6] then the cited conclusion is equivalent to the assertion of the existence of gravitational driving with a characteristic time of 10^9 years. This conclusion has not yet been tested by numerical calculations, although there is no doubt of the importance of such a calculation in connection with the problem of the stability of the planetary system. On the other hand, in Ref. 4 we pointed out that an understated value was used in Refs. 5 and 6 for the upper limit to the mass distribution function of the bodies. Thus, a more careful examination of the problem of the variation in the eccentricities of the planets in the course of accumulation is required. In part 1 of the present article we give considerations involving the functional dependence of the upper limit of the mass distribution of the bodies encountering and falling onto a planet on the mass of the planet: $M_1(m)$. It is concluded that M_1 and m are comparable over a considerable part of the evolution. The fair size of the quantity M_1/m at these stages compels one to allow for higher degrees in expansions with respect to M_1/m when determining the influence of close passes and the falling of bodies on the eccentricity of the planetary orbit. This is done in Parts 2 and 3. In Part 4 the results obtained in the work are discussed and it is concluded that by the end of accumulation, the orbital eccentricities of the planets were close to the present-day "mean" for the planets according to Brouwer and Clemence.[7] Furthermore, if it is not specifically mentioned, we are talking about the zone of planets of the terrestrial group, where the accumulation process was not complicated by gas accretion and the ejection of bodies.

1. MASSES OF THE LARGEST BODIES ENCOUNTERING AND FALLING ONTO A PLANET

Following Refs. 2–6, we will assume that the mass distribution of the colliding, fragmenting, and amalgamating bodies is given in the first approximation by the coagulation theory,

$$n(m,t)=n_0(t)\,m^{-q}, \qquad (1)$$

where, in accordance with crater statistics and the observed distribution in the asteroid belt, $q \approx 1.8$.

For a small number of large bodies the random deviations from the cited distribution can be large, and they are usually treated separately. The point under discussion here is the difference in mass, which separates the planetary embryo from the other large bodies, and its variation in the process of accumulation. In Refs. 2, 5, and 6 it was assumed that $m_1/m \approx 10^{-2}$–10^{-3} (m_1 is the mass of the largest body, not counting the planet itself). Hoyle[8] and Marcus[9] assumed that the bodies which fell onto the planets were considerably larger. The quantity m_1/m plays an important role in many aspects of accumulation theory: in the problem of the orbital eccentricities and inclinations, the rotation of the planets, their heating during impacts, etc.

Back in 1944 Shmidt pointed out that the inclinations of the axes of rotation of the planets were caused by the impacts of the largest bodies. For a model of spherically symmetric accumulation of bodies onto a planet Safronov[2] obtained the following equation connecting the torque moment K, the inclination ε of the axis of rotation, and the parameters characterizing the mass distribution function of the bodies:

$$K^2\sin^2\varepsilon=\frac{2}{3}G\frac{2-q}{3-q}\int_0^m m_1\,mr\,dm. \qquad (2)$$

Here G is the gravitational constant, r is the radius of the planet, and $m_1 \approx M_1$. Assuming that as the planet's mass m(t) grows, $m_1(t)$ also grows, i.e., $m_1/m = \text{const}$, Safronov obtains $m_1/m \approx 10^{-2}$–10^{-3}. Just these values were used in Refs. 5 and 6 in calculating the eccentricities. However, we can cite simple arguments in favor of considerable variations in m_1/m in the course of accumulation.

a) By definition, $m_1 \le m$ and $m_1 \le Q - m$, where Q is the total mass of the bodies in the supply zone:

$$Q=2\pi\int_{R_1}^{R_2}\sigma(R)\,R\,dR. \qquad (3)$$

Here R_1 and R_2 are the inner and outer boundaries of the supply zone and $\sigma(R)$ is the surface density of material. At the start of accumulation, $m_1, m \ll Q$ and, in accordance with the coagulation theory,[10] $m_1/m \approx 1$. In Fig. 1 the segments 1A and A1 cut out the region of definition of the function $m_1(m)$ in the variables $m_1/m = y$ and $m/Q = (m/M)^{2/3} = x$. The corresponding values of the dimensionless (normalized to the present-day) radius of the planet, r/r_{max}, the half-width of the supply zone $\Delta R\cdot(\Delta R_{max})^{-1}$, and the mass of the planet, m/M, are inserted below for convenience. We note that the actual behavior of m_1/m in the region of $m/Q < 10^{-2}$, the initial condition for the accumulation problems under consideration by us, is unimportant, since the corresponding contribution of this stage to the final values of the eccentricities, inclinations, etc. is negligibly small.

b) In the first stage of growing and overlapping supply zones, when the largest of the neighboring competing embryos is taken as the body of mass m while the smaller is taken as the body of mass m_1, the ratio m_1/m can be obtained from a comparison of their growth times,[3,4]

$$m_1/m\simeq(1+{}^3/_2\bar{e})^{-3}, \qquad (4)$$

where $\bar{e}$ is the mean eccentricity of the main mass of bodies,[2]

$$\bar{e}=2\sqrt{mR/3\theta r m_\odot}, \qquad (5)$$

R being the distance from the sun and $m_\odot$ the mass of the sun. θ depends weakly on m and m_1, and here we neglect the variations of θ in the individual stages, taking θ = 2–3 for the zone of planets of the terrestrial group and θ ≈ 20–30 for the outer zones.[2,4] In the model of growing and overlapping zones the final combining of zones sets in when $m/Q \lesssim 0.25$ (i.e., $r/r_{max} \lesssim 1/2$), after which the separation in mass between m and the next largest body in the zone, m_1 (before it falls or is destroyed in a close pass), is described by the equations[4]

$$\frac{m_1}{m}=\left[\frac{1+(m/m_0)^{b/3}(\varkappa_0^*-1)/(2\theta-\varkappa_0^*)}{1+2\theta(m/m_0)^{b/3}(\varkappa_0^*-1)/(2\theta-\varkappa_0^*)}\right]^3,$$

$$b=\frac{2\theta-1}{2\theta+1},\quad r(t)=\frac{1+2\theta}{P\delta}\int_{t_0}^{t}\sigma(t)\,dt+r_0, \qquad (6)$$

where r_0, r_{10}, m_0, and $\varkappa_0^* = r_0/r_{10}$ are the values of r, r_1, m, and $\varkappa^*$ at the time t_0 of the final overlapping of the zones, P is the Keplerian period of revolution, and δ is the average density of the planet. From (4) and (6) we obtain an approximate expression for the parameters of the terrestrial group (see Fig. 1, curve 1):

$$m_1/m\approx 1-0.62(m/Q)^{0.3}. \qquad (7)$$

c) In the second stage ($m/Q \gtrsim 0.25$), after the fall or destruction of m_1, m_2 becomes the largest body (transition from curve 1 to curve 2), and after the falling of m_2 there is a transition to curve 3, etc. The exact times of fall of m_1 and the subsequent large bodies cannot be determined. One can only speak of the probability of a fall in the interval of $0.25 \lesssim x \lesssim 1$, adopting a reasonable stochastic model. The dashed curve BD1 in Fig. 1 represents the result of a calculation for the limiting case when the embryo grows through the successive fall of the largest bodies, starting with m_1. It is approximately described by the equation

$$y=\frac{(1-x)^{1/(2-q)}}{x}f(x,q),\quad f(x,q\simeq 1.8)\approx 0.65. \qquad (8)$$

This function bounds the region of possible values of m_1/m from below. This region is bounded on the other side by the curves BC and C1. If the jumps in the function m_1/m are neglected, then one can assume that it is a monotonically declining, concave function which varies from a value of ~0.6 at $m/Q = 0.25$ to zero at $m/Q = 1$. The sizes of the largest craters on Mars and Mercury show that at the end of accumulation ($m/Q \approx 0.99$) bodies with masses of ~ 10^{-6} m fell onto the planets. And as one

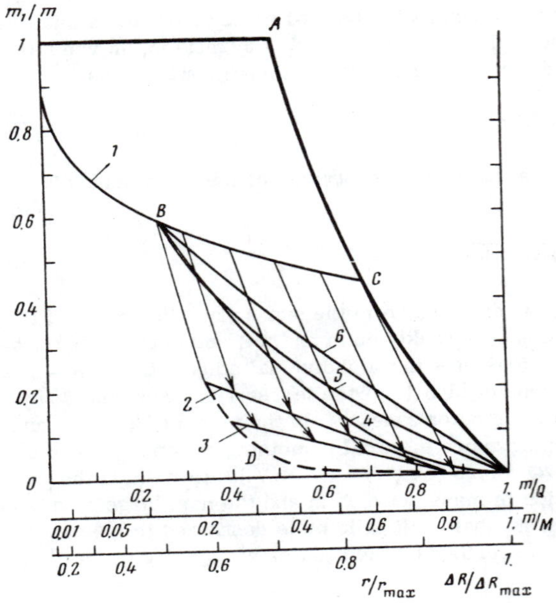

FIG. 1. Region of definition and model distributions for the function m_1/m: 1) $m_1/m = 1 - 0.62(m/Q)^{0.3}$; 2) m_2/m; 3) m_3/m; 4) $y = \alpha(1-x)^2$; 5) $y = -\alpha \ln x$; 6) $1 - x^{\alpha}$.

would expect, the function (8) given above leads to considerably lower values. For preliminary estimates one can assign model functions, determining the required parameters from the indicated reference points.

We consider several simple functions satisfying the restrictions obtained above:

one-parameter

$$y=e^{-\alpha x}, \quad y=1-x^{\alpha},$$
$$y=1-\alpha x, \quad y=\alpha \ln 1/x; \tag{9}$$

two-parameter

$$y=\alpha x^{-\beta}, \quad y=1-\alpha x^{\beta},$$
$$y=\beta(1-x)^{\alpha}, \quad y=\alpha x^{-\beta} \ln 1/x.$$

Some of these functions are shown in Fig. 1. Having in mind their future use in the analytical theory, it is reasonable to be confined to those which permit the successive taking of quadratures in problems on the proper rotation and orbital eccentricities of the planets. In the present work we used the function $y = -\alpha \ln x$ (α = 0.3, 0.4) in the calculations for the second stage. It is shown in Ref. 11 that when this function is used the rotation velocities of the planets and the inclinations of the axes of rotation given by the theory are close to those observed in the solar system.

Comment 1. Above we identified M_1, the upper limit of the distribution function (1), with the mass of the largest body, m_1, after the planet in the supply zone. This does not lead to a large error, since $m_1 \ll Q - m$ in the entire range of values of m/Q. It can be shown[10] that

$$m_1=(Q-m)\left[1-\left(1+\frac{q-1}{2-q}\frac{M_1}{Q-m}\right)^{(2-q)/(1-q)}\right].$$

For $m_1/m = \alpha \ln x$ with $q = 11/6$, in particular,

$$\frac{m_1}{m}=\frac{1-x}{x}\left[1-\left(1-5\alpha\frac{x\ln x}{1-x}\right)^{-1/5}\right]. \tag{10}$$

It is seen from Table I that the transition from M_1/m to m_1/m in the second stage ($m/Q \geq 0.25$) corresponds to a decrease in the factor α by 1.5–2 times. In the first stage (of growing and overlapping zones) the difference between M_1/m and m_1/m is still smaller.

Comment 2. We also did not distinguish between the upper limit to the distribution function of the bodies falling onto the planet and the upper limit to the distribution function of the bodies in the supply zone, which is equivalent to the assumption of a low probability of the breakup and subsequent dispersion of the largest bodies during their close passes near the planet before their fall. Allowance for this effect is an interesting problem for future investigations. Since the influence of impacts (see Parts 2 and 3) on the orbital eccentricity is not dominant, however, the corresponding corrections will be small.

2. VARIATION OF ORBITAL ECCENTRICITY OF AN EMBRYO DURING IMPACTS OF FALLING BODIES

Let us consider the variation in the orbital eccentricity of a growing planet caused by the impacts of growing bodies. We consider the plane problem: we neglect the inclinations of the orbits and the collision cross section is $s = \pi r^2(1 + 2\theta)$.

Let m_i', e_i', and φ_i' be the mass of the i-th body falling onto the planet, the eccentricity of its orbit, and its true anomaly at the moment of impact and let m_{i-1}, e_{i-1}, and φ_i be the same quantities for the growing planet immediately before the i-th impact. By definition,

$$m_i=m_0+\sum_{j=1}^{i} m_j',$$

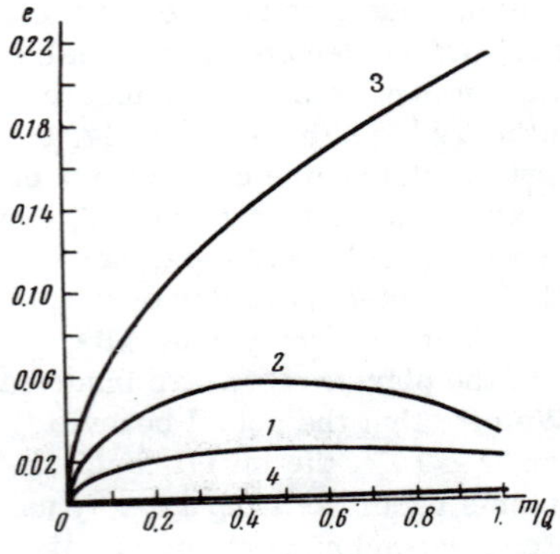

FIG. 2. 1) Orbital eccentricity of a growing planet, produced by collisions with bodies; 2) eccentricity produced by collisions and close passes of bodies; 3) mean eccentricity of orbits of main mass of bodies of zone; 4) orbital eccentricity of growing planet according to earlier estimate of Ref. 6.

TABLE I

x	0.1	0.2	0.3	0.4	0.5	0.6	0.7	0.8	0.9	0.99
M_1/m	0.689	0.617	0.518	0.394	0.298	0.220	0.153	0.096	0.045	0.004
m_1/m	0.565	0.432	0.324	0.232	0.167	0.118	0.079	0.048	0.022	0.002

where m_0 is the initial value of the mass of the planetary embryo.

For the orbital eccentricity of the embryo after the i-th impact one finds,[5] from the condition of conservation of the total momentum,

$$e_i^2 = \frac{1}{m_i^2}[(m_{i-1}e_{i-1})^2 + (m_i'e_i')^2 + 2e_i'm_i'e_{i-1}m_{i-1}(\cos\varphi_i\cos\varphi_i' + \sin\varphi_i\sin\varphi_i')]. \quad (11)$$

Since $m_i = m_{i-1} + m_i' = m_{i-1}(1 + m_i'/m_{i-1})$, by changing from m_i to m_{i-1} on the right side of (11) and subtracting e_{i-1}^2 from the right and left sides, we obtain the increment in the square of the orbital eccentricity of the body m_i when the body m_i' falls onto it:

$$e_i^2 - e_{i-1}^2 = \left[\frac{1}{(1+m_i'/m_{i-1})^2} - 1\right]e_{i-1}^2 + \frac{(m_i'/m_{i-1})^2}{(1+m_i'/m_{i-1})^2}e_i'^2 + 2\frac{m_i'/m_{i-1}}{(1+m_i'/m_{i-1})^2}e_i'e_{i-1}(\cos\varphi_i\cos\varphi_i' + \sin\varphi_i\sin\varphi_i'). \quad (12)$$

The formation of the planet is treated as a random process, and the quantities φ_i, φ_i', e_i (i = 1, 2, ...) are treated as random. The orbital eccentricities of the falling bodies will be taken as known from the accumulation theory, and we take the mass distribution of the bodies in the form (1).

Taking the mathematical expectations of the left and right sides of (12), we obtain

$$\mathcal{M}(e_i^2) - \mathcal{M}(e_{i-1}^2) = \left\{-2\mathcal{M}\left[\frac{m_i'/m_{i-1}}{(1+m_i'/m_{i-1})^2}\right] - \mathcal{M}\left[\left(\frac{m_i'/m_{i-1}}{1+m_i'/m_{i-1}}\right)^2\right]\right\}\mathcal{M}(e_{i-1}^2) + \mathcal{M}\left[\left(\frac{m_i'/m_{i-1}}{1+m_i'/m_{i-1}}\right)^2\right]e_i'^2 + 2e_i'\mathcal{M}\left[\frac{m_i'/m_{i-1}}{(1+m_i'/m_{i-1})^2}\right][\mathcal{M}(e_{i-1}\cos\varphi_i\cos\varphi_i') + \mathcal{M}(e_{i-1}\sin\varphi_i\sin\varphi_i')]. \quad (13)$$

For a distribution function of the number of bodies per unit area with respect to semimajor axes a' of $n(a', t)da' = c(t)a'^{-p}da'$ the distribution of the number of bodies with respect to true anomaly was obtained in Ref. 5 in the form

$$n_1(\varphi', R, t)d\varphi' = {}^1/_4 c_1(t)R^{-p+1}|\sin\varphi'|(1+e'\cos\varphi')^{-p}d\varphi'. \quad (14)$$

The number of bodies with a true anomaly in the interval of $(\varphi', \varphi' + d\varphi')$ falling onto the planet in a small time Δt is

$$n_1(\varphi', R, t)d\varphi' sV\Delta t,$$

where V is the velocity at an incident body relative to the planet. For a body chosen randomly from the bodies which fell onto the planet in the time Δt the probability of having a true anomaly in the interval of $(\varphi', \varphi' + d\varphi')$ with the condition that the orbital eccentricity of the embryo is e and its true anomaly is φ is

$$f(\varphi'/e, \varphi)d\varphi' = n_1(\varphi', R, t)sV\Delta t\,d\varphi' \Big/ \int_0^{2\pi} n_1(\varphi', R, t)sV\Delta t\,d\varphi'. \quad (15)$$

The plane problem, in which the product sV was assumed to be independent of V, and hence of φ', φ, and e, was analyzed in Ref. 5. Therefore, φ' and e and φ were assumed to be independent. Actually, $sV \propto 1/V$, and the probability density of a collision for (φ, φ') will also depend on the velocity of the body m' relative to the planet, and hence on φ and e also. Then in place of (15) we have

$$f_1(\varphi', \varphi/e', e)d\varphi'd\varphi = n_1(\varphi', R, t)sV\frac{dt}{d\varphi}d\varphi'd\varphi \Big/ \int_0^{2\pi}\int_0^{2\pi} n_1 sV\frac{dt}{d\varphi}d\varphi'd\varphi, \quad (16)$$

where $d\varphi/dt \propto R^{-2}(\varphi)$,

$$V = {}^1/_2 V_c\sqrt{(e'\cos\varphi' - e\cos\varphi)^2 + 4(e'\sin\varphi' - e\sin\varphi)^2}, \quad (17)$$

and V_c is the Keplerian circular velocity.

Taking $s \approx \pi r^2 v_e^2/V^2$ and $n_1(\varphi', R, t)$ in the form (14), we obtain

$$f_1(\varphi', \varphi/e', e)d\varphi'd\varphi \approx |\sin\varphi'|(1 - pe'\cos\varphi') \times [1-(3-p)e\cos\varphi]d\varphi'd\varphi\frac{8\pi^2}{3\sqrt{3e'}} \times [\sqrt{(e'\cos\varphi' - e\cos\varphi)^2 + 4(e'\sin\varphi' - e\sin\varphi)^2}]^{-1}. \quad (18)$$

We calculate the mathematical expectations in the last term in (13):

$$\mathcal{M}(e_{i-1}\cos\varphi_i\cos\varphi_i') = \mathcal{M}\left[e_{i-1}\int_0^{2\pi}\int_0^{2\pi}\cos\varphi_i\cos\varphi_i' f_1(\varphi_i, \varphi_i')d\varphi_i d\varphi_i'\right] \simeq \frac{1}{3}\mathcal{M}(e_{i-1}^2)e_i'[1 + 0.6p(3-p)], \quad (19)$$

$$\mathcal{M}(e_{i-1}\sin\varphi_i\sin\varphi_i') \approx \mathcal{M}(e_{i-1}^2)/3e'. \quad (20)$$

Substituting (19) and (20) into (13), we have

$$\mathcal{M}(e_i^2) - \mathcal{M}(e_{i-1}^2) = -2\mathcal{M}\left[\frac{m_i'/m_{i-1}}{(1+m_i'/m_{i-1})^2}\right]\mathcal{M}(e_{i-1}^2) - \mathcal{M}\left[\left(\frac{m_i'/m_{i-1}}{1+m_i'/m_{i-1}}\right)^2\right]\mathcal{M}(e_{i-1}^2) + \mathcal{M}\left[\left(\frac{m_i'/m_{i-1}}{1+m_i'/m_{i-1}}\right)^2\right]e_i'^2 + \frac{2}{3}\mathcal{M}\left[\frac{m_i'/m_{i-1}}{(1+m_i'/m_{i-1})^2}\right][\mathcal{M}(e_{i-1}^2)e_i'^2(1+0.6p(3-p)) + \mathcal{M}(e_{i-1}^2)]. \quad (21)$$

When i ≫ 1 we have $\mathcal{M}(e_{i-1}^2) \approx e^2(m)$, where e and m are the current eccentricity and mass of the planet. Suppose that in a time Δt (not small compared with the period of revolution of the planet) k bodies fell onto the planet, where k is large enough that we can take $\mathcal{M}(e_{j+k}^2 - e_j^2) \approx \Delta e_k^2$, but at the same time, the mass Δm which fell is sufficiently small. Then

$$\mathcal{M}(\Delta m) = \mathcal{M}\left(\sum_{i=j}^{j+k} m_i'\right) = \int_0^{M_1} m' n(m', t)\, \Delta t\, dm' \approx \Delta m. \tag{22}$$

As in Ref. 5, we sum (21) over i in the limits of $j \le i \le j+k$. Then, using (22) and (1), we have

$$\mathcal{M}\left[\sum_{i=j}^{j+k} \frac{m_i'/m_{i-1}}{(1+m_i'/m_{i-1})^2}\right] = \int_0^{M_1} \frac{m' n(m')\, \Delta t\, dm'}{m(1+m'/m)^2}$$

$$= (2-q)\, \mu_1^{q-2} \frac{\Delta m}{m} \int_0^{\mu_1} \frac{x^{1-q}\, dx}{(1+x)^2}, \tag{23}$$

where $x = m'/m$ and $\mu_1 = M_1/m < 1$ is the ratio of the upper limit in the mass distribution function of the bodies to the mass of the embryo. When $q < 2$ we have

$$\int_0^{\mu_1} \frac{x^{1-q}\, dx}{(1+x)^2} = \frac{\mu_1^{2-q}}{2-q}\, {}_2F_1(2, 2-q, 3-q, -\mu_1), \tag{24}$$

where ${}_2F_1(\alpha, \beta, \gamma, z)$ is a Gaussian hypergeometric function. Substituting (24) into (23), we have

$$\mathcal{M}\left[\sum_{i=j}^{j+k} \frac{m_i'/m_{i-1}}{(1+m_i'/m_{i-1})^2}\right]$$

$$= \frac{\Delta m}{m}\left(1 - 2\frac{2-q}{3-q}\mu_1 + 3\frac{2-q}{4-q}\mu_1^2 - 4\frac{2-q}{5-q}\mu_1^3 + \ldots\right). \tag{25}$$

Similarly, we find

$$\mathcal{M}\left[\sum_{i=j}^{j+k}\left(\frac{m_i'/m_{i-1}}{1+m_i'/m_{i-1}}\right)^2\right] = \int_0^{M} \frac{(m'/m)^2 n(m')\, \Delta t\, dm'}{(1+m'/m)^2}$$

$$= (2-q)\, \mu_1^{q-2} \frac{\Delta m}{m} \int_0^{\mu_1} \frac{x^{2-q}\, dx}{(1+x)^2},$$

where

$$\int_0^{\mu_1} \frac{x^{2-q}\, dx}{(1+x)^2} = \frac{\mu_1^{3-q}}{3-q}\, {}_2F_1(2, 3-q, 4-q, -\mu_1).$$

Consequently,

$$\mathcal{M}\left[\sum_{i=j}^{j+k}\left(\frac{m_i'/m_{i-1}}{1+m_i'/m_{i-1}}\right)^2\right]$$

$$= \frac{\Delta m}{m}\frac{2-q}{3-q}\mu_1\left(1 - 2\frac{3-q}{4-q}\mu_1 + 3\frac{3-q}{5-q}\mu_1^2 - \ldots\right). \tag{26}$$

Substituting (25) and (26) into (21), we obtain

$$\Delta e^2 = -2\frac{\Delta m}{m}\left(1 - 2\frac{2-q}{3-q}\mu_1 + 3\frac{2-q}{4-q}\mu_1^2 - \ldots\right) e^2$$
$$- \frac{\Delta m}{m}\frac{2-q}{3-q}\mu_1\left(1 - 2\frac{3-q}{4-q}\mu_1 + 3\frac{3-q}{5-q}\mu_1^2 - \ldots\right) e^2$$
$$+ \frac{2}{3}\frac{\Delta m}{m}\left(1 - 2\frac{2-q}{3-q}\mu_1 + 3\frac{2-q}{4-q}\mu_1^2 - \ldots\right) e^2$$
$$+ \frac{\Delta m}{m}\frac{2-q}{3-q}\mu_1\left(1 - 2\frac{3-q}{4-q}\mu_1 + 3\frac{3-q}{5-q}\mu_1^2 - \ldots\right) e'^2$$
$$+ \frac{2}{3}\frac{\Delta m}{m}\left(1 - 2\frac{2-q}{3-q}\mu_1 + 3\frac{2-q}{4-q}\mu_1^2 \ldots\right)[1 + 0.6p(3-p)]\, e^2 e'^2. \tag{27}$$

The last term in (27) can be neglected, since $e^2 e'^2 \ll e^2$, e'^2.

Dividing the right and left sides by Δm, we arrive at the differential equation for the square of the planet's orbital eccentricity. In particular, neglecting terms containing μ_1 to more than first power and taking $e' = \bar{e}$, we obtain

$$\frac{de^2}{dm} = -\frac{1}{m}\left(\frac{4}{3} - \frac{5}{3}\frac{2-q}{3-q}\mu_1\right) e^2 + \frac{1}{m}\frac{2-q}{3-q}\mu_1 \bar{e}^2. \tag{28}$$

We designate e^2 as y_1 and m/M as x_1 in (28), and then $\bar{e}^2 = \zeta x_1^2$ from (5) and in place of (28) we have

$$\frac{dy_1}{dx_1} = \frac{1}{x_1}\left(-\frac{4}{3} + \frac{5}{3}\frac{2-q}{3-q}\mu_1\right) y_1 + \frac{2-q}{3-q}\zeta \mu_1 x_1^{-1/3}, \tag{29}$$

where $\mu_1 = \mu_1(x_1)$. This equation has a general solution of the type

$$y(x) = \exp\left(-\int_{x_0}^{x} f(s)\, ds\right)\left[y_0 + \int_{x_0}^{x} g(s) \exp\left(\int_{x_0}^{s} f(t)\, dt\right) ds\right], \tag{30}$$

where

$$f(x_1) = \frac{1}{x_1}\left(\frac{4}{3} - \frac{5}{3}\frac{2-q}{3-q}\mu_1\right), \quad g(x_1) = \frac{2-q}{3-q}\zeta \mu_1 x_1^{-1/3}.$$

In accordance with (7) of Part 1, for the first stage of growth of the embryo we take $\mu_1(x_1) = 1 - 0.62 x_1^{0.2}$. Then for q = 11/6 we obtain from (30)

$$e^2\left(\frac{m}{Q}\right) \approx 0.0811\bar{e}^2\left\{\left[1 - 0.0752\left(\frac{m}{Q}\right)^{0.3} + 0.0051\left(\frac{m}{Q}\right)^{0.6} - 0.0003\left(\frac{m}{Q}\right)^{0.9}\right] - 0.5566\left(\frac{m}{Q}\right)^{0.3}\left[1 - 0.0683\left(\frac{m}{Q}\right)^{0.3} + 0.0043\left(\frac{m}{Q}\right)^{0.6} - 0.0002\left(\frac{m}{Q}\right)^{0.9}\right]\right\}. \tag{31}$$

Taking μ_1 for the second stage of growth of the planet in the form $\mu_1(x_1) = \alpha \ln(Q/m) = -{}^2/_3 \alpha \ln x_1$, where $\alpha = 0.4$, from (30) we obtain

$$e^2\left(\frac{m}{Q}\right) \approx \left(\frac{m_0}{Q_0}\frac{Q}{m}\right)^2 \left(\frac{m_0}{Q_0}\right)^{0.077 \ln(m_0/Q_0)} \left(\frac{m}{Q}\right)^{-0.077 \ln(m/Q)} e_0^2$$
$$+ 0.0205\bar{e}^2\left[\left(\frac{m_0}{Q_0}\frac{Q}{m}\right)^3\left(\frac{3}{2}\ln\frac{m_0}{Q_0} - \frac{1}{2}\right) - \frac{3}{2}\ln\frac{m}{Q} + \frac{1}{2}\right], \tag{32}$$

where m_0/Q_0 is the boundary between the first and second stages of growth and $e_0 = e(m_0/Q_0)$.

Curve 1 in Fig. 2 shows e(m/Q) calculated in accordance with (31) and (32) for the parameters corresponding to the terrestrial zone with $m_0/Q_0 = 0.25$. We note that $e\left(\frac{m}{Q}=1\right)= 0.022$, whereas the final value obtained in Ref. 5 was e = 0.0017.

3. INFLUENCE OF CLOSE PASSES OF BODIES OF THE SWARM ON ORBITAL ECCENTRICITY OF PLANET

Let us consider the variation in the orbital eccentricity e of a growing planet during encounters of bodies of the swarm with it, by analogy with Ref. 6 but with allowance for the fair size of the quantities M_1/m and e/e'.

Suppose that k bodies fell onto the planet and there were n encounters with bodies of the swarm over a time Δt. Let the numbers k and n be large enough that $\Delta e^2 \approx \mathscr{M}(\Delta e^2)$ and $\Delta m \approx \mathscr{M}(\Delta m)$. Then

$$\Delta e^2=\sum_{i=1}^{k}\Delta e_{ai}{}^2+\sum_{i=1}^{n}\Delta e_{gi}{}^2\approx k\mathscr{M}(\Delta e_{ai}{}^2)+n\mathscr{M}(\Delta e_{gi}{}^2), \tag{33}$$

where Δe^2_{ai} is the change in e^2 due to the fall of a body and Δe^2_{gi} is that due to an encounter with a body of the swarm.

In Part 2 of the present report we obtained the expression (27) for $\Delta e_a{}^2=k\mathscr{M}(\Delta e_{ai}{}^2)$, where $\Delta m\approx k\mathscr{M}(m')$ is determined by (22).

In Ref. 6 the following expression was obtained for Δe^2_g, the square of the increment in the orbital eccentricity of an embryo during the close flight past it of a body with a mass m':

$$\begin{aligned}\Delta e_g{}^2 \approx \frac{m'}{m}\Big\{&4\sin^2\psi[-e^2+e'e\cos(\varphi-\varphi')]\\ &\pm e\sin 2\psi\left(c'\cos\varphi'\sin\varphi-4e'\sin\varphi'\cos\varphi+\frac{3}{2}e\sin 2\varphi+\ldots\right)\Big\}\\ &+\left(\frac{m'}{m}\right)^2\{4\sin^4\psi[e'^2+e^2-2e'e\cos(\varphi'-\varphi)]\\ &+\sin^2\psi\cos^2\psi[(\cos^2\varphi'+16\sin^2\varphi')e'^2+(\cos^2\varphi+16\sin^2\varphi)e^2\\ &-2e'e(\cos\varphi\cos\varphi'+16\sin\varphi\sin\varphi')]\\ &\mp 12\sin^3\psi\cos\psi(e'\cos\varphi'-e\cos\varphi)(e'\sin\varphi'-e\sin\varphi)+\ldots\},\end{aligned} \tag{34}$$

where we took $m'/(m + m') \approx m'/m$, since it was assumed that $m' \ll m$; 2ψ is the angle by which the direction of the relative velocity V of the bodies is deviated at the end of the encounter, determined from

$$\sin\psi=1/\sqrt{1+D^2V^4/\mu^2}, \tag{35}$$

where D is the impact parameter and $\mu = G(m + m')$.

Retaining m' in the denominators of (34) and the terms containing e'e and e^2, we write the mathematical expectation of (34) as

$$\begin{aligned}\mathscr{M}(\Delta e_g{}^2)=&\mathscr{M}\left(\frac{m'/m}{1+m'/m}\right)\{4\mathscr{M}[\sin^2\psi(-e^2+e'e\cos\varphi'\cos\varphi\\ &+e'e\sin\varphi'\sin\varphi)]\}+\mathscr{M}\left[\left(\frac{m'/m}{1+m'/m}\right)^2\right]\{4\mathscr{M}[\sin^4\psi(e'^2+e^2-\\ &-2e'e\cos\varphi'\cos\varphi-2e'e\sin\varphi'\sin\varphi)]\\ &+\mathscr{M}(\sin^2\psi\cos^2\psi[(\cos^2\varphi'+16\sin^2\varphi')e'^2+(\cos^2\varphi+16\sin^2\varphi)e^2\\ &-2(\cos\varphi'\cos\varphi+16\sin\varphi'\sin\varphi)e'e])\}.\end{aligned} \tag{36}$$

The probability density $f_2(\varphi, \varphi', D/e, m)$ for the encounter (φ, φ', D), when the orbital eccentricity of the planet is e and the mass is m, was taken in Ref. 6 as

$$f_2(\varphi,\varphi',D/e,m)=n_2(\varphi,\varphi',D,e,m)\Big/\left(\int_{-S}^{S}\int_0^{2\pi}\int_0^{2\pi}n_2\,d\varphi\,d\varphi'\,dD-\int_{-l}^{l}\int_0^{2\pi}\int_0^{2\pi}n_2\,d\varphi\,d\varphi'\,dD\right),$$

where $n_2 = n_1(\varphi', R, e', t)Vdt/d\varphi$ and S is the radius of the of the sphere of influence of the planet. Taking n_1 in accordance with (14), V in accordance with (17), $dt/d\varphi\propto R^2(\varphi)$, $l=r\sqrt{1+v_e{}^2/V^2}$ and considering that the quantities S/R and l/S are small in comparison with e', we find

$$\begin{aligned}f_2(\varphi,\varphi',D/e,m)=&\frac{\sqrt{3}\,|\sin\varphi'|(1-pe'\cos\varphi')[1-(3-p)e\cos\varphi]}{8\pi S(1+4\pi/3\sqrt{3})(1+0.44e^2/e'^2)}\\ &\times\sqrt{\frac{4}{3}-\cos^2\varphi'}\left[1+\left(\frac{2}{3}\frac{e^2}{e'^2}-\frac{e^2}{2e'^2}\cos^2\varphi-\frac{4e}{3e'}\sin\varphi'\sin\varphi\right.\right.\\ &\left.\left.-\frac{e}{3e'}\cos\varphi'\cos\varphi\right)\Big/\left(\frac{4}{3}-\cos^2\varphi'\right)\right].\end{aligned} \tag{37}$$

Let us determine the corresponding mathematical expectations of the quantities from (36):

$$\begin{aligned}\mathscr{M}(\sin^2\psi)&=2\int_l^S\int_0^{2\pi}\int_0^{2\pi}\sin^2\psi f_2(\varphi,\varphi',D/e,m)\,dD\,d\varphi'\,d\varphi\\ &=2.22\frac{\mu}{SV_c{}^2e'^2}(1-0.96e^2/e'^2),\\ \mathscr{M}[\sin^2\psi(\cos\varphi'\cos\varphi+\sin\varphi'\sin\varphi)]&=1.11\frac{\mu e}{SV_c{}^2e'^3}(1-0.44e^2/e'^2),\\ \mathscr{M}(\sin^4\psi)&=0.14\frac{\mu}{SV_c{}^2e'^2}(1-0.96e^2/e'^2),\\ \mathscr{M}[\sin^4\psi\cos(\varphi'-\varphi)]&=0.07\frac{\mu e}{SV_c{}^2e'^3}(1-0.44e^2/e'^2),\\ \mathscr{M}\{\sin^2\psi\cos^2\psi[(\cos^2\varphi'+16\sin^2\varphi')e'^2&+(\cos^2\varphi+16\sin^2\varphi)e^2\\ -2(\cos\varphi'\cos\varphi+16\sin\varphi'\sin\varphi)e'e]\}&=21.10\frac{\mu}{SV_c{}^2}(1-1.27e^2/e'^2).\end{aligned} \tag{38}$$

By analogy with what was done in Part 2 [see (25) and (26)], we find

$$\begin{aligned}\mathscr{M}\left(\frac{m'/m}{1+m'/m}\right)&=\frac{\Delta m}{m}(2-q)\mu_1^{q-2}\int_0^{\mu_1}\frac{x^{1-q}\,dx}{1+x}\\ &=\frac{\Delta m}{m}{}_2F_1(1,2-q,3-q,-\mu_1)=\frac{\Delta m}{m}\left(1-\frac{2-q}{3-q}\mu_1+\frac{2-q}{4-q}\mu_1^2-\ldots\right).\end{aligned} \tag{39}$$

Substituting (26), (38), and (39) into (36), after some transformations we obtain

$$\Delta e_g{}^2=-\frac{\Delta m}{m}\left[4.44\left(1-\frac{2-q}{3-q}\mu_1+\ldots\right)+\right.$$

$$+26.73\frac{2-q}{3-q}\mu_1\left(1-2\frac{3-q}{4-q}\mu_1+\ldots\right)\Bigg]$$

$$\times\frac{e^2}{e'^2}\frac{\mu}{SV_c^2}+\frac{\Delta m}{m}21.66\frac{2-q}{3-q}\mu_1\left(1-2\frac{3-q}{4-q}\mu_1+\ldots\right)\frac{\mu}{SV_c^2}. \qquad (40)$$

It follows from (33) that the contribution to Δe^2 of close passes of bodies grows with an increase in n/k, where, according to Ref. 6,

$$\frac{n}{k}=\Bigg(\int_{-S}^{S}\int_{0}^{2\pi}\int_{0}^{2\pi} n_2\,d\varphi\,d\varphi'\,dD$$

$$-\int_{-l}^{l}\int_{0}^{2\pi}\int_{0}^{2\pi} n_2\,d\varphi\,d\varphi'\,dD\Bigg)\Bigg/\int_{-l}^{l}\int_{0}^{2\pi}\int_{0}^{2\pi} n_2\,d\varphi\,d\varphi'\,dD$$

$$\simeq\frac{0.86e'V_cS}{v_cr}.$$

Changing from v_er to V, we obtain

$$n/k=0.43\sqrt{3\theta/2}V_c^2Se'^2/\mu. \qquad (41)$$

We substitute Eqs. (27) and (40) into (33), multiply by n/k, divide the right and left sides by Δm, and, being confined to terms containing μ_1 to no higher than first power, we obtain a differential equation describing the variation in the the orbital eccentricity of a growing planet under the action of encounters and the impacts of falling bodies:

$$\frac{de^2}{dm}=-\frac{1}{m}\left(5.38+18.67\frac{2-q}{3-q}\mu_1\right)e^2+\frac{20.76}{m}\frac{2-q}{3-q}\mu_1e'^2. \qquad (42)$$

This equation has a general solution in the form (30). Taking $\mu_1 = 1 - 0.62(m/Q)^{0.3}$ in the first stage of growth of the embryo, with q = 11/6 we obtain from (42)

$$e^2(m/Q)\simeq0.341[1-0.4202(m/Q)^{0.3}-0.0764(m/Q)^{0.6}$$
$$-0.0136(m/Q)^{0.9}]\bar{e}^2. \qquad (43)$$

For the second stage $\mu_1 = -\alpha\ln(m/Q)$; substituting this expression into (42) and (30), we obtain

$$e^2(m/Q)\approx\left(\frac{m_0}{Q_0}\frac{Q}{m}\right)^8e_0^2$$

$$+0.1405\left[\left(\frac{m_0}{Q_0}\frac{Q}{m}\right)^9\left(\frac{3}{2}\ln\frac{m_0}{Q_0}-0.1654\right)-\frac{3}{2}\ln\frac{m}{Q}+0.1654\right]\bar{e}^2, \qquad (44)$$

where m_0/Q_0 is the boundary between the first and second stages of growth and $e_0 = e(m_0/Q_0)$.

The variation in the orbital eccentricity of an embryo in the terrestrial zone under the influence of close passes and impacts of bodies of the swarm, calculated in accordance with (43) and (44) with $m_0/Q_0 = 0.25$, is shown by curve 2 in Fig. 2. Curve 3 shows the variation in the mean eccentricity of the orbits of the main mass of bodies in the supply zone of the embryo with the growth in its mass. Curve 4 shows the orbital eccentricity of the embryo to the earlier estimate of Ref. 6.

4. CONCLUSION

The main purpose of the present work consisted in a more rigorous derivation of the equations describing the evolution of the "mean" eccentricities of the growing planets, with allowance for possible variations in the mass distribution function of the bodies encountering and falling onto the planets. The resulting equations (28) and (42) differ from those given in Refs. 5 and 6 both in the coefficients and in additional terms appearing as a consequence of the allowance for the fair size of M_1/m. These equations can be used in the solution of those problems of accumulation theory where it is important to allow for the nonzero orbital eccentricities of the growing planets. A general concept of the character of the variation in eccentricities can be obtained by using a power function for the mass distribution of the bodies with a variable upper limit. The value of the latter at individual stages of accumulation is not determined very reliably, but it can be refined subsequently by using data on the inclinations of the axes of rotation and on the rotation rates of the planets. The results of preliminary calculations for the parameters characterizing the terrestrial zone are given in Fig. 2. It is seen that the final value of the eccentricity exceeds the estimate obtained in Refs. 5 and 6 by an order of magnitude. A similar result can also be obtained for other planets of the terrestrial group. With the exception of the orbital eccentricity of Mercury, the orbital eccentricities of the newly formed planets lie in the middle of the range of values calculated by Brouwer and Clemence[7] in accordance with the theory of secular perturbations. Thus, the accumulation theory does not provide a basis for assuming that the "mean" orbital eccentricities could be strongly altered even after completion of the process of accumulation of the planets. A new result is the presence of a maximum of the function e(m/Q). When model functions are used, $e_{max} \simeq \bar{e}/2$ when m/Q = 0.3-0.6.

Whereas in the zone of planets of the terrestrial group the ejection of bodies beyond the limits of the zone is negligibly small, in the zone of the giant planets ejection is efficient and begins with embryo masses much smaller than the masses of the present-day planets. In this case the velocities of the ejected bodies are close to $(\sqrt{2}-1)V_c$. If, following Refs. 5 and 6, one takes the mean eccentricity of the remaining bodies as one third, then it follows from (42) that the orbital eccentricities of the giant planets by the end of accumulation were ~0.05. This value is close to the mean value calculated from the data of Brouwer and Clemence[7] for Jupiter, Saturn, and Uranus (0.045; 0.049; 0.034) while differing markedly from the mean of Nepture, 0.01.

[1]O. Yu. Shmidt, Selected Works on Geophysics and Cosmogony [in Russian], Izd. Akad. Nauk SSSR, Moscow (1960).
[2]V. S. Safronov, Evolution of the Protoplanetary Cloud and Formation of the Earth and Planets [in Russian], Nauka, Moscow (1969).
[3]A. V. Vityazev, G. V. Pechernikova, and V. S. Safronov, Astron. Zh. 55, 107 (1978) [Sov. Astron. 22, 60 (1978)].
[4]G. V. Pechernikova and A. V. Vityazev, Pis'ma Astron. Zh. 5, 54 (1979). [Sov. Astron. Lett. 5, 31 (1979)].
[5]I. N. Ziglina and V. S. Safronov, Astron. Zh. 53, 429 (1976) [Sov. Astron. 20, 244 (1976)].
[6]I. N. Ziglina, Astron. Zh. 53, 1288 (1976) [Sov. Astron. 20, 730 (1976)].

[7]D. Brouwer and G. M. Clemence, in: Planets and Satellites (collection of translations) [Russian translation], Inostr. Lit., Moscow (1963), p. 43.
[8]F. Hoyle, Vopr. Kosmog. 7, 15 (1960).
[9]A. H. Marcus, Icarus 7, 283 (1967).
[10]V. S. Safronov and E. V. Zvyagina (Zvjagina), Icarus 10, 109 (1969).
[11]A. V. Vityazev and G. V. Pechernikova, Preprint No. 10, Inst. Fiz. Zemli Akad. Nauk SSSR (1980).

Translated by Edward U. Oldham

 0038-5301/80/04 0467-03 $02.70

The Settling of Grains in a Contracting Protoplanetary Cloud (abstract)

V. S. Kessel'man
Translated from: Astron. Zh. **58**, 58 (1981)
Sov. Astron. **25**, 33 (1981)

In regions of strongly turbulent gas motion in a contracting interstellar cloud, if solid particles have grown much larger than the average grains size in the interstellar medium, they will sink down to the equatorial plane of the cloud. The time scale of this settling process is calculated, and a possible mechanism is discussed for the accumulation of solid particles during the gravitational settling phase.

Formation of Planets (abstract)

E. L. Ruskol
In *The Solar System and its Exploration*, Proceedings of Alpbach Summer School, 1981 (European Space Agency, Paris, 1981), pp. 107–113

The evolution of a protoplanetary gas-dust cloud of mass (0.05–0.1) $M_{\odot}$, laminarly rotating around the Sun in the boundaries of the present planetary system, results in the formation of solid planetesimals grouped in concentric zones of growing embryo planets. The growth of terrestrial planets takes about 10^8 yr, several times longer than the period of dissipation of gas from the inner parts of the protoplanetary cloud. The formation of giant planets is a two-stage process: the formation of nuclei from the solid material with the mass of 2–3 Earth's masses in 10^7–10^8 yr and then the gaseous accretion on nuclei in 10^5–10^6 yr. The origin of satellites is a process accompanying the growth of planets in the most active stage, via circumplanetary swarms for terrestrial planets and accretional disks for giants.

Solid Particles at an Early Stage of Evolution of a Protoplanetary Cloud (abstract)

V. S. Kessel'man
Translated from: Astron. Zh. **59**, 810 (1982)
Sov. Astron. **26**, 489 (1982)

The sizes of solid particles grown in the process of coagulation in a protoplanetary nebula are calculated. The influence of the variation of the average sizes of solid particles on the process of formation of massive solid bodies in a protoplanetary nebula is discussed.

THE DEVELOPMENT OF SOVIET PLANETARY COSMOGONY

V.S.Safronov

O.Yu. Schmidt i sovetskayaia geofizika 80-kh,godov
pp.41-57. Moscow: Izdatelstvo "Nauka" (1983).

1. The model explaining the formation of planets in terms of the accumulation of solid matter is the most wide-spread and well recognized one nowadays. This idea, as well as the opposite hypothesis of the planets arising from the condensation of gaseous clouds, was first suggested some time ago. However, its recent revival and the beginning of its systematic development grounded on modern knowledge are directly connected with the name of Academician O.Yu. Schmidt. Schmidt laid the cornerstones of the modern theory of planetary formation, and he also created a small research team (the Department of the Evolution of the Earth) at the Institute of the Physics of the Earth of the USSR Academy of Sciences (in those days, the Geophysical Institute) whose members began systematic studies of the problem. O.Yu. Schmidt [1-2] outlined the methods for explanation of the basic regularities of a planetary system arising from the gradual combination of a great number of bodies of a pre-planetary swarm revolving around the Sun. He also suggested some promising, albeit still qualitative, ideas about the formation of the planetary satellites and asteroids. Slightly later, and quite independently, Harold Urey [3] came to the same conclusion about the formation of the planets from solid bodies (approximately the size of the Moon); his theory was based on the physico-chemical analysis of the composition and structures of the meteorites. O.Yu. Schmidt also suggested a hypothesis about the formation of the pre-planetary cloud by means of its capture by the Sun which had been then traveling through an interstellar nebula; he believed this model permitted getting around the problem of the distribution of the angular momentum between the Sun and the planets in the easiest way. The hypothesis was submitted to intense discussion. The possibility of the realization of gravitational capture (in the three-body problem), as well as some other forms of capture, was demonstrated in the works by O.Yu. Schmidt, G.F. Khilmy, T.A. Agekian, V.V. Radzievsky, etc. (see references, for example, in [2]). However, the problem of the reality of such a capture still remains open, and other variants of this capture have occasionally been proposed. Nevertheless, most specialists in the field share the concept of the simultaneous formation of the Sun and the pre-planetary cloud from the same nebula. This idea has recently been supported by a number of new studies performed by different scientists.

L.E. Gurevich and A.I. Lebedinsky [4] examined the evolution of a dust cloud (containing no gas) revolving around the Sun. They came to the conclusion that the particles ' relative velocities were decreased because of inelastic collisions between them, and the cloud therefore was flattened. If the density of these particles is high enough, the disintegration of the lens-shaped cloud into isolated clumps becomes energetically advantageous. If their velocities were disordered and they were in the presence of the tidal forces exerted by the Sun, the authors estimated the parameters of a spheroidal condensation able to withstand the disintegration due to the mutual gravitational attraction amongst its particles. The density of such a condensation should be of the order of magnitude of the so-called "Roche's critical density," and the dimensions of the equatorial axis and the axis which is orthogonal to it should be related to each other as 13 to 1. These condensations were contracting in the course of their collisions accompanied by mutual agglomeration, and, as a result, the swarm of solid bodies, which was regarded by Schmidt as the initial substance to form the planets, had come into existence.

The direction of the research initiated by O.Yu. Schmidt has proved to be promising, and the development of the new theory has not stopped even after 1956 when he passed away. Step by step, the main stages of the process of planetary formation were examined at Schmidt's department. The almost completely "ready" Sun and the dust-gas cloud of the same chemical composition revolving around the Sun with the Keplerian circular velocity were taken as the initial state of the system. The description of the development of that system first was made in qualitative terms [5], but gradually the qualitative approach gave way to a more detailed quantitative description. The evolution of the gas-dust cloud was studied (V.S. Safronov, E.L. Ruskol [6]). The analysis of the stability of the revolving cloud with respect to the radial convection of its particles revealed that the temperature gradient inside the cloud along its radial dimension had been at least two orders of magnitude smaller than the amount necessary to secure this kind of convection, and, consequently, the initial turbulence inside the cloud could not have been secured by that mechanism for the duration of more than 10-100 revolutions around the Sun without some additional energy sources. After the turbulence had faded, the dust sank down through the gas towards its central plane, where a dust layer eventually came into existence, much denser than the gas. In such a way one more step into the past was made; that is, it was demonstrated that the dust lens surmised by Gurevich and Lebedinsky as the initial state for their model, had been a natural outcome of the evolution of the primordial gas-dust cloud revolving around the Sun. The development of an instability inside a flattened dust layer with a differential rotation in the presence of radial (ring-like) disturbances was also examined [7-9]. It was shown that the critical density causing the disintegration of the layer into isolated rings was minimal when the disturbancy wavelength was equal to the thickness of the disk increased eightfold and five to seven times as small as the Roche density ($\rho_R = 2\rho^x$, where $\rho^x = (3M)/(4\Pi R^3)$). As a compressing ring becomes more and more dense, and as soon as its density approaches the Roche limit $\rho_R \sim (10$ to $15)\rho^x$ it disintegrates into isolated condensations.

Further, the evolution of dust condensation has been studied. As has been demonstrated, the condensations were compressing and transforming into solid bodies during the process of agglomeration. Inside the zone of the terrestrial planets the stage of condensations was only short-lived, and so its role in the process of the accumulation of the Earth was rather modest. It took more time in the zones of such outer planets as Uranus and Neptune, but even there its duration was roughly ten times as small as the time of their accumulation [10]. An interest in this state has been recently revived in connection with the works by T.M. Eneev and N.N. Kozlov [11-12] on the numerical modeling of the process of planetary accumulation wherein the principal role was ascribed just to the stage of condensations.

The formation of planets inside the swarm of protoplanetary bodies has been studied in more detail. The evolution of the swarm was governed by the temporary change of its principal parameters, that is, its mass and velocity distributions. The velocities were evaluated on the basis of the energy balance of the relative motions of the bodies and it was surmised that the masses were spatially distributed in accordance with an inverse power law; the energy was assumed to be acquired as a result of the reciprocal gravitational disturbances in the course of drawing the bodies near to each other and to be lost as a result to subsequent collisions between them [8]. As has been discovered, the velocities only weakly depended on the masses of the bodies within the zone of their interaction. They were increasing in the process of accumulation in proportion to the linear measures of the latter. The mass distribution was found from the coagulation equation. A qualitative examination of this equation, taking into account the splintering of colliding bodies, permitted finding its asymptotic solutions when $t \to \infty$ (E.V. Zviagina et al., [13]). A numerical solution of the equation was obtained by G.V. Pechernikova on a computer [14-15]. It has been found that almost over the whole range of masses (with the exception of the largest ones) they are distributed according to an inverse power law with the index $q<2$, which means that the mass of matter is mainly concentrated inside the largest bodies. One characteristic feature of the process of accumulation has been identified: the largest bodies tend to grow ahead of all the others, due to their much greater gravitational cross-section [8]. This type of growth resulted in the formation of a great number of the "embryos" of planets with their own ring-like feeding zones . These embryos were growing in the process of the absorption of the other bodies, and their feeding zones were broadening and overlapping. In this way, a smaller embryo could find itself within the zone of a larger one; then its motion experienced gravitational disturbances from the side of its larger counterpart, it was forced to leave its former quasi circular orbit, and finally it either closely approached one of the larger embryos and became destroyed by its tidal forces, or its existence was terminated in the course of a direct collison with such a body. This process of scooping out the neighboring bodies by the embryos, accompanied by the parallel reduction of their number, lasted 10^8 years, and eventually it brought about the currently existing planetary system whose members are located at some special distances from each other (to be more specific, from the Sun) determined by the condition of the stability of the system over the billions of years of its existence. It was first described in [8], and then in more detail in [16-18]. The tidal destruction of the bodies has been considered by I.N. Ziglina [19], the evolution of the orbital eccentricities of the embryos has been studied by I.N. Ziglina and V.S. Safronov in [20-21] and G.V. Pechernikova and A.V. Vitazev in [22].

In the 1970's, the promising prospects of our approach were recognized abroad. Strong research groups arose one after another in the United States, Japan, Italy, and West Germany, which became intensively engaged in working out the theory of the formation of planets from solid particles and bodies. The basic difference amongst them is that every team made its own decisions concerning that which factors in this complicated process should be regarded as the most important ones. We keep in regular touch with these scientists.

A number of parallel processes accompanied the basic process of the formation of planets. These processes, albeit collateral, were nevertheless of importance since only they determined the existing variety of the bodies of the Solar system. The monotonic growth of the proportion of volatile substances (ices) in the solid component of the cloud resulted in the division of the planets into two groups. On the boundary between these groups no planet could arise, and so the belt of small planets (asteroids) came into being there. This process has been examined in [23]. Two different stages of the formation of giant planets (Jupiter and Saturn) have been identified and studied by V.S. Safronov and E.L. Ruskol [24]. In the first stage, the solid cores of these planets were formed in the same way as the planets of the inner group. As soon as their masses had reached some critical magnitude, the accretion of the neighboring gas onto the cores began. When the giant planets had acquired sufficiently high masses, they turned out to be able to eject (by means of gravitational interactions) other bodies from the Solar system and towards its distant periphery [25], where the so-called Oort's cometary cloud consequently arose. The quantitative aspects of this process are examined in [26]. The evolution of the smaller bodies of the Solar system and interconnections among them have been considered by B.Yu. Levin [27]. The possibility of chemical fractionation in the cloud during the stage of the growth of its small particles was examined by A.B. Makalklin [28].

Protosatellite swarms came into being in the vicinity of planets simultaneously with the planet's growth. The latter process took place because of the effect of capturing solid bodies and particles by the planets' gravitation when they collided in their vicinity. On the basis of this idea, suggested by O.Yu. Schmidt, the formation of such swarms and their transformation into satellite swarms surrounding Jupiter and Saturn arose from gas-dust accretion disks.

Whereas the solid substance of the primordial cloud was amalgamating and forming planets, its gaseous component was gradually dissipating away from the

Solar system. The principal mechanism of such a dissipation is commonly believed to have been a process of blowing the gases away by the solar wind during the stage when the young Sun had not yet reached the phase of membership in the "main sequence" of stars belonging to the T Taurus type. However, the duration of the T Taurus stage is certainly less than 10^7 years, whereas the heavy cores of Jupiter and Saturn that were able to switch on the process of gas accretion required a minimum of 10^8 years to form. Hence a good proportion of gaseous matter still remained inside the Solar system even after the T Taurus stage had been completed. That means that another mechanism of gas dissipation whose lifetime exceeded 10^8 years had to exist as well. G.V. Pechernikova and A.V. Vityazev [31] have demonstrated that the thermal dissipation of gases from a hot "exosphere" heated by the ultraviolet radiation of the young Sun could provide such a mechanism.

2. The above-listed studies belong, according to Schmidt's terminology, to the central problem of planetary cosmogony, that is, to the area of the examination of the evolution of the protoplanetary cloud and the formation of planets. The most important results have been achieved just there. Still, cooperation with neighboring sciences has been broadening also. O.Yu. Schmidt paid special attention to the necessity of using copious data provided by geosciences for boosting the progress of cosmogonic studies. In order to reach this aim, it was necessary to build a bridge between cosmogony and geosciences, that is, to make geophysical, geological, geochemical, etc. deductions on the basis of cosmogonic notions. It appeared natural that it would be an easier task to begin the examination of the progeologic stage of the evolution of the Earth "along the direction of the time arrow," that is, on the basis of some initial state which looked plausible from the cosmological point of view. Earlier works of that kind were related to the examination of the thermal evolution of the Earth. They were performed first by E.A. Liubimova [32], and further by S.V. Mayeva and B.Yu. Levin [33-34]. All of these authors surmised that initially the Earth had been comparatively cool and homogeneous. This conclusion followed from the concept of the formation of the Earth from small particles and bodies. A comparative analysis of different models describing the internal structure of the Earth and the other planets of the inner group, which resulted in an evaluation of the differences of their chemical composition, has also been made (S.V. Kozlovskaia [35-36]). The demonstration of an important role played by sizable bodies in the process of planetary formation [37] has brought about a considerable change in our ideas about the initial stage of the Earth. It has been proved that the larger bodies had been heating our planet to a greater extent than the smaller ones, because their kinetic energy was released in the course of impacts not upon the surface, but at the depth of about two diameters of a falling body. Qualitative evaluations could not be made prior to developing a theory of crater formation and the shock mixing of matter after the fall of very sizable bodies (when the gravitational field of the planet disturbs the "geometric similarity" radically, and the picture of a strike looks like a "camouflet explosion" [8]). The calculations revealed that the temperature of the central zone of the Earth had been relatively low (some results were about 1000 K, that is, several times smaller than the melting point). The peak of heating corresponded to a depth of about 300 kilometers. The upper layer located between the depth levels of 100 and 1500 kilometers was on averaged heated up to the temperature of the beginning of melting [38]. Despite the very high energy of the impacts, further heating is impossible because as soon as the decreasing viscosity (due to heating) of the matter had reached the level of 10^{16} - 10^{17} poises, the process of effective convection was switched on inside the heated zone. This convection transported to the outside all the heat released due to impacts. This means that only some hundredths of the whole mass of the Earth experienced melting in the process of its accumulation [39].

The emergence of large-scale local heterogeneties in the Earth's mantle, whose horizontal size exceeded one thousand kilometers, was the second important consequence of the impact of large bodies [40]. The proportion of melted substance was higher in these areas and the differences in chemical composition also could be maximal there. These domains happened to play an important role in the development of the differentiation of the Earth's substance conditioned by its density at an earlier stage of the evolution of our planet. This differentiation began at the end of the process of accumulation within the most heated external layers [41-42], but not in the center of the Earth as many authors believed. Under one kind of condition, the differentiation could become cascade-like, under other conditions it could be accompanied by thermal instability (A.V. Vityazev and S.V. Maeva [43]) when the released energy greatly accelerated the further development of the process of accumulation. Approximate evaluations, together with some data related to the isotopes, demonstrated that the main mass of the Earth's matter could complete its differentiation within several hundred million years with the resulting formation being the planetary core whose size was comparable with the size of the recent core. The whole process of differentiation is possibly not finished even now, although it has been dying down gradually. Hence, planetary cosmogony supplies geoscientists with sufficiently definite concepts about the initial state of the Earth, which should be taken into consideration when the evolution of our planet is studied.

Some authors, while indicating certain problems related to the process of differentiation, made attempts to reject the whole idea. One group tried to explain the stratification of the Earth in terms of the phase transition of its matter into a metallic state, whereas the partisans of another approach suggested a hypothesis of the "heterogeneous" accumulation of the Earth, according to which the metallic core had been formed first, and only later was it augmented by a shell composed of silicates. Modern cosmogonic data witness against a similar hypothesis on a large scale, although some special details of the Earth's structure can possibly be connected with the phase transitions; equally, a modest radial heterogenity in the initial chemical compostion of the Earth may not be excluded from consideration. One more way of eliminating the problems of differentiation was also suggested: some authors inferred an initially hot Earth by means of an unjustified reduction of the duration of its accumulation by a factor of one thousand and possibly more (A. Ringwood's earlier models etc.). However, a prolonged discussion on the duration of the process has been ended by the confirmation of our time scale (10^8 years). A higher initial temperature of the upper mantle (prior to the beginning of melting) can be explained by the impacts from the side of large bodies falling onto the Earth.

3. Up to this point we have only been discussing the problem of planetary formation inside the preexisting gas-dust cloud surrounding the Sun; the question of the origin of the cloud itself has not been touched upon at all. The latter is not less complicated than the former, since it is related to a much earlier stage of the process for which the available data are considerably scarcer; moreover, it belongs more to the domain of the (rather than planetary) cosmogony. No generally recognized theory of star formation existed during Schmidt's time, and some serious doubts concerning the very possibility of developing the correct theory of planetary formation were voiced. However, Schmidt insisted that the problem of planetary formation inside the preplanetary cloud was sufficiently independent on the problem of the formation of the cloud itself, and thus that it had to be studied without waiting for a solution of the latter question. Due to such an approach, our studies in that area began more than ten years earlier than foreign ones.

In the 1970's, studies of the origin of stars based upon massive observations of stars and interstellar media made over the whole spectrum of wavelengths became considerably more active. As V.A. Ambartsumian had demonstrated as early as the end of the 1940s [45], stars are presently being formed; they are coming into existence not one by one, but in groups located inside the domains which he called associations; these domains exist in the vicinity of (or inside) the giant concentrations of interstellar gas-dust clouds. In turn, these concentrations (complexes) of clouds are located along the spiral arms of the Galaxy. The well known Orion Nebula studied by astrophysicists in detail is the most striking example of such a complex. The formation of such complexes, whose masses amount to 10^5 solar masses and whose average size is fifty parsecs, is usually interpreted in terms of the Rayleigh-Taylor-Parker instability in the gas disk of the Galaxy. The thermal instability results in rising hot and cool areas there. When the spiral density waves advance through these areas, the matter becomes considerably denser (almost tenfold) and so the conditions are more favorable for star formations to arise. The gas becomes cooler while interacting with the molecules of CO and H_2O, which brings about a thermochemical instability, and the latter results in the formation of cool and very dense (10^4 - 10^6 atoms per cubic centimeter) "molecular" clouds whose masses average 10 - 100 solar masses. These clouds experience contraction and thus they disintegrate into isolated fragments, which then collapse and transform into stars. One can easily identify several groups of young stars existing within the Orion Nebula, the difference between their lifetimes amounting to several million years. The molecular clouds wherein no stars have had time to arise yet are the youngest ones. One can reasonably believe that the process of star formation spreads sequentially from one area to another, like a "forest-fire" or "infection" (H. Reeves,[46]). The "wind" propagating the "flame" inside the Orion Nebula blows from North-West to South-East. This process can initiate density shock waves, for example, spreading from newly flashing very bright stars in the form of ionization fronts, or generated by the exploding supernovas (the latter waves are especially powerful). The heaviest stars (whose masses reach or exceed twenty solar masses) complete their evolution in several million years, and after burning down all of their fuel, they exploded and ejected a good part of their mass in the form of fastly spreading shells. Accepting the Sun to have been born within an Orion-type association, and having evaluated the plausible frequency of supernovas arising in that nebula, H.Reeves comes to the conclusion that "the sun came into being amidst a firework of supernovas." Some dozen supernovas had to flash in the vicinity of the Solar system during the period of its formation. Nevertheless, there are no sufficient reasons to believe that a supernova and nothing else had initiated the collapse of the solar nebula. There were several triggering mechanisms, and most probably, the collapse had been generated by their mutual action.

The cosmochemical aspect of the participation of supernovas in the formation of the Solar system appears more important. The fact that some isotopic anomalies of chemical elements have been discovered in the meteorites, especially in the refractory intrusions of Allende (^{16}O, ^{26}Mg) directly indicates that some newly synthesized chemical elements, including even some short-lived isotopes such as ^{26}Al, whose half-life is less than one million years, had been injected into the solar nebula. Such a synthesis would be most naturally switched on by an exploding supernova. A number of different ideas concerning the role of supernovas in the cosmogonic process have been suggested, including even an extreme variant - which is that the Sun itself had exploded as a supernova some time in the remote past, and the protoplanetary cloud was nothing but a remnant of its shell. However, this hypothesis is certainly unacceptable from the point of view of its dynamical aspect, since it faces some insuperable difficulties, particularly in connection with the problem of angular momentum.

In the framework of our research interests, the question of supernovas looked important physically as well as dynamically. Concerning its physical side, it seems very essential to learn what was the effect of heating of the bodies of the Solar system by short-lived radioactive isotopes like ^{26}Al, etc. The sources of their earliest heating are unidentified as yet. In the case of the refractory intrusions being confined inside the protoplanetary bodies (meteorites) not later than some 10^6 years after their formation (in particular, if these intrusions condensed within the area of the solar nebula polluted by the products of an exploded supernova), then the heating effects produced by the radioactive ^{26}Al could be considerable. However, if those intrusions should be regarded, as many people believe, as "interstellar" (meaning they condensed inside the shell of a supernova and survived the process of injection into the solar nebula) particles, then the time period between the synthesis of ^{26}Al and the penetration of the particles into the protoplanetary bodies cannot be dated, and so the problems of the thermal contribution of the short lived elements and of the switching on the collapse of the solar nebula by exploding supernovas remain as yet unsolved.

The dynamic influence of a supernova on the collapse of the solar nebula is weaker, and so it may be excluded from consideration at the first degree of approximation. The same "effect of forgetting the past," which Schmidt had in mind while noting that the process of planetary formation only weakly had depended on the mechanism of the emergence of the protoplanetary cloud, takes place there, too. The most important regularities of the collapse and of the subsequent evolution of the solar nebula were determined primarily by the initial mass and angular momentum of the nebula. Its earlier history had been so strongly obliterated that only a very careful analysis of new astrophysical and cosmochemical data may allow restoration of it. Some geochemists believe that some already-available data on isotopic anomalies and meteorites indicate the chemical, physical, and dynamic heterogeneity of the solar nebula. In their opinion, it was so strong that one should reject the currently existing models of initially uniform nebula and begin working out inhomogeneous models. For the time being, however, such suggestions hardly may be regarded opportune. First, the astrophysicists have not properly understood the principal regularities in the uniform models yet, and secondly, cosmochemistry has not produced concrete data on the character of the heterogeneties. Their ad hoc introduction into the dynamic models of the nebula would bring about so many arbitrary conjectures that our chances to grasp the truth would be nullified. One thing is certainly true: further studies in that area should be performed on the basis of close cooperation between such neighboring sciences as astrophysics, cosmochemistry, and planetary and stellar cosmogony.

4. The numerical calculations of the collapse of the cool interstellar clouds which have approximately a mass of one solar mass which have been recently performed permit us to begin an analysis of the principal regularities of the evolution of the solar nebula. The initial density of such clouds was assumed to equal 10^{-19} - 10^{-18} g/cm^3, in accordance with the Jeans criterion of gravitational instability. At the first stage of contraction the nebula is transparent and it experiences isothermic contraction, its temperature being about 10 K. A dense nucleus then arises in the center. When $\rho \geq 10^{-12}$ g/cm^3, it becomes opaque and is quickly heated, and its contraction stops. Gas continues to fall down from the shell to the nucleus, thus the latter's mass grows. A nonrotating nebula transforms into a solitary star, which, naturally, has no protoplanetary cloud. The contraction of fastly rotating nebulas took place in another way. The majority of calculations were performed for the case when $\omega \sim 10^{-15}$ sec^{-1} (the Galaxy's angular speed), which corresponds to the nebula's angular momentum $\sim 10^{54}$ g cm^2/sec. As it turns out, even prior to the formation of an opaque nucleus, a dense ring arises under the conditions of reduced density in the nebula's central area. The ring is unstable and disintegrates into two or more fragments. As a result, a binary or multiple star emerges. On the basis of these considerations we have come to the conclusion that the Solar system arose from a nebula whose angular momentum was much smaller [47]. It has been discovered that no binary star can emerge if the initial momentum of the nebula is smaller than 10^{53} g cm^2/sec [48]. It has also been demonstrated that if we equate the amount of volatile substances contained in the modern planets to their rich concentration on the Sun and simultaneously take into account the matter ejected from the Solar system in the course of the accumulation of the giant planets, the initial mass of the protoplanetary could will be about 0.04 $M_\odot$, and, correspondingly, the initial momentum of the cloud will become (1 to 2)10^{52} g cm^2/sec. I should note that in the majority of Camerone's models of the solar nebula, including the most popular ones, this momentum has been taken = 10^{54} g cm^2/sec.

The problem of the initial momentum of the nebula is far from being of secondary importance. The dynamic and chemical evolution of the nebula depended considerably on its magnitude. When it is $\sim 10^{53}$ g cm^2/sec, the size of the central area ("core"), wherein the matter from the shell falls, exceeds the size of the orbit of Jupiter. Therefore the core remains sufficiently cool, and the primoridal interstellar particles do not vaporize. It is still unclear how such a core can lose its momentum and experience transformation into the Sun. When its momentum $\approx 10^{52}$ g cm^2/sec, the core is hot, and its diameter is such that it lies within the orbit of Mercury. The particles falling into this zone vaporize. In this case, the formation of the Sun offers no great difficulty. However, some sufficiently effective mechanism providing the transfer of the angular momentum is needed to explain the transportation of matter from the area of the inner planets outward, towards the most remote planets. When astrophysicists study the evolution of the disks existing in the vicinity of stars, they usually regard turbulent viscosity as such a mechanism. When the energy which initiates the turbulence is large enough, the speed of the process is of the same order as the speed of sound, and the length of mixing is of the order of the disk's thickness. Some authors formally accept the same magnitude of viscosity for the solar nebula, and then the duration of the essential evolution of the protoplanetary disk turns out to have been less than 10^5 years. However, the problem of the sources of turbulence in the protoplanetary disk is quite vague yet, and so such an approach may not be regarded as well justified. Instead of it, D.N.C. Lin and some other scholars [52] hypothesized the idea of the convection spreading along the Z-axis (that is, perpendicularly to the plane of the disk) as a possible source supporting the turbulent movements. In turn, the dissipation of the energy of these movements is believed to stabilize the gradient of temperature along the Z-axis which is necessary to secure continuous convection. It is not completely clear yet whether the physical conditions needed to guarantee the existence of stationary turbulent convection (such as the sufficiently fast growth of the opacity index of the medium when its temperature increases, etc.) could exist in the disk.

There could be a more reliable source of turbulence in the disk, namely the energy of the falling gas which experienced accretion during the collapse. P.M. Cassen et al. have made an approximate analytic examination of the problem of the expansion of the disk in the course of accretion on the assumption that the viscosity of the disk's substance was uniform [53]. The dependence of viscosity on the energy of the falling matter was taken into consideration by T.V. Ruzmaikina [49] and by the author of this review [18]. As has been shown, this energy theoretically suffices for the stabilization of the turbulent viscosity, able to guarantee the expansion of the protoplanetary disk up to the size of the currently existing planetary system during the time of the collapse of the nebula. However, some functional links between the effectiveness of that mechanism and such factors as the density fluctuations within the nebula, the shape of the disk, etc. have also been established. These links indicate the necessity to study the whole process in much more detail. The problem of the redistribution of gas within the disk at the end of the process of accretion and immediately afterward towards some quasi-stationary state also seems to deserve attention (A.V.Vityazev, G.V. Pechernikova [50]).

One can find a more complete review of the Soviet and foreign studies on the problems of planetary cosmogony mentioned in this review, as well as on a number of other important questions which unfortunately could not be touched upon here, in [51].

The theory of the formation of planets from accreting solid material, whose steady and consistent elaboration was first begun by O.Yu. Schmidt, has become very promising. This theory has successfully passed through different obstacles, and nowadays it is developing in different countries.

The current stage in the development of planetary cosmogony is characterized by the permanent expansion of its ties with such neighboring sciences as stellar cosmogony, astrophysics, cosmochemistry, and geosciences. The results which may be achieved at the junction of different sciences are especially important, since they eliminate arbitrary and undefined concepts and lead to the theory-building on the basis of all available factual data.

REFERENCES

[All of the following publications are in Russian except for refs. 3, 17, 23, 24, 26, 28, 31, 39, 43, 44, 52, 53]

1. Schmidt, O.Yu. "A meteoric theory of the origin of the Earth and planets. "Doklady Akad. Nauk SSSR, 1944, 45, No.6, p.245.
2. Schmidt, O.Yu. Four Lectures on the Theory of the Origin of the Earth. Moscow: Izd-vo Akademii Nauk SSSR, 1957, 144 pp.
3. Urey, H.C. The Planets. "Their" Origin and Development. New Haven: Yale University Press, 1952.
4. Gurevich, L.E., and Lebedinsky, A.I. "On the formation of the planets". Izv. Akad. Nauk SSSR, Ser. Fizich., 14, No. 6, p.765.
5. Levin, B.Yu. The Origin of the Earth and Planets. Moscow: "Nauka," 1964.
6. Safronov, V.S., and Ruskol, E.L. "On the hypothesis of turbulence in the protoplanetary cloud. Voprosy Kosmogonii, 1957, 5, p.22.
7. Safronov, V.S. "On the gravitational instability in flattened rotating systemswith axial symmetry. Doklady Akad. Nauk SSSR, 1960, 130, No. 1 p.53.
8. Safronov, V.S. Evolution of the Protoplanetary Cloud and Formation of the Earth and Planets. Moscow: "Nauka," 1969, 244pp.
9. Genkin, I.L., and Safronov, V.S. "Gravitational instability in rotating systems with radial perturbations." Astron. Zhur., 1975, 52, No. 2 p.36-315.
10. Safronov, V.S. "The duration of the formation of the Earth and the planets and its role in their geological evolution" In the book Kosmokhimiya luny i Planet [Comochemistry of the Moon and Planets]. Moscow: "Nauka," 1975, p. 624.
11. Eneev, T.M., and Kozlov, N.N. "Model of the accumulation process in the formation of planetary system. 1. Numerical experiments." Astron. Vestn., 1981, 15, No. 2, p.80.
12. Eneev, T.M., and Kozlov, N.N. "Model of the accumulation process in the formation of the planetary system. 2. Rotation of the planets and the relation of the model to the theory of gravitational instability." Astron. Vestn., 1981, 15, No. 3, p.80.
13. Zvyagina, E.V., Pechernikova, G.V., and Safronov, V.C. "Qualitative solution of the coagulation equation with allowance for fragmentation." Astron. Zhurn., 1973, 50, No.6, p.1261.
14. Pechernikova, G.V. "Mass distribution of proto-planetary bodies". Astron. Zhurn., 1976, 51, p.1305.
15. Pechernikova, G.V., Safronov, V.S., Zvyagina, E.V. Mass distribution of protoplanetary bodies II. Numerical solution of generalized coagulation equation." Astron. Zhurn., 1976, 53, No. 3, p.612.
16. Vityazev, A.V., Pechernikova, G.V., Safronov, V.S. "Limiting masses, distances, and times for the accumulation of planets of the terrestrial group." Astron. Zhurn., 1978, 55, No. 1, p.107.
17. Safronov, V.S. "Accumulation of the protoplanetary bodies." In: Early solar system processes and the present solar system." LXIII Corso Soc. Italiana di fisica. Bologna, Italy, 1980, p.58.
18. Safronov, V.S. "State-of-the-Art for the Theory of the Origin of the Earth." Fizika Zemly, 1982, No.6, p.5.
19. Ziglina, I.N. "Tidal destruction of bodies near a planet". Fizika Zemly, 1978, No.7, p.3.
20. Ziglina, I.N., Safronov, V.S. "Averaging of the orbital eccentricitites of bodies which are accumu-lating into a planet." Astron. Zhurn., 1976, 53, No. 2, p.429.

21. Ziglina, I.N. "Effect on eccentricity of a planet's orbit of its encounters with bodies of the swarm." Astron. Zhurn., 1976, 53, No.6, p.1288.
22. Pechernikova, G.V., Vityazev, A.V. "Evolution of orbital eccentricities of the planets in the process of their formation." Astron. Zhurn., 1980, 57, No.4, p.799.
23. Safronov, V.S. "On the origins of asteroids." In: Asteroids, edited by T.Gehrels. Tucson: University of Arizona Press, 1979, p.975.
24. Safronov, V.S., Ruskol, E.L. "On the origin and initial temperature of Jupiter and Saturn." Icarus, 1982, 49, No. 2, p.284.
25. Levin, B.Yu. "On the origin of comets". Voprosy Kosmogonii, 1963, No.9, p.215.
26. Safronov, V.S. "Ejection of bodies from the solar system in the course of the accumulation of giant planets and the formation of the Oort cometary cloud." In: The Motion, Evolution of Orbits, and Origin of Comets.Edited by G. A. Chebotarev, E.I. Polskaya & B. G. Marsden. Dordrecht: Reidel, 1972, p.329.
27. Levin, B.Yu. "Asteroids, comets, meteoritic matter - their role and place in the cosmogony of solar system". Fizika Zemli, 1982, No. 6, p.25.
28. Makalkin, A.V. "Possibility of formation of an originally unhomogeneous Earth." Phys. Earth and Planet. Int. 1980, No.22, p.302.
29. Ruskol, E.L. Origin of the Moon. Moscow:"Nauka," 1975, 188pp.
30. Safronov, V.S., Ruskol, E.L. "The accumulation of satellites." Astron. Zhurn., 1977, 54, No. 2, p.378.
31. Pechernikova, G.V. and Vityazev, A.V. "Thermal dissipation of gas from the protoplanetary cloud." Adv. Space Res. 1981, No.1, p.55.
32. Liubimova, E.A. Thermal Processes of the Earth and the Moon. Moscow: Nauka, 1968, 279pp.
33. Levin, B.Yu. and Mayeva, S.V. "Thermal history of the Earth." Izv. Akad. Nauk SSSR, Ser. Geofiz., 1960, No. 2, p.3.
34. Mayeva, S.V. "Thermal history of an iron core Earth." Fizika Zemili, 1971, 43, No. 5, p.1081.
35. Kozlovskaya, S.V. "Models of the internal structure of the Earth, Venus, and Mars". Astron. Zhurn., 1966, 43, No.5, p.1081.
36. Kozlovskaya, S.V. "The internal structure of Venus and the content of iron in the planets of the terrestrial group". Astron. Vestn., 1982, 16, No. 1, p. 3.
37. Safronov, V.S. "Sizes of the largest bodies falling on the planets during their formation". Astrron. Zhur., 1965, 42, p. 1270.
38. Safronov, V.S. and Kozlovskaya, S.V. "Heating of the Earth by the impact of accreted bodies". Izv. Akad Nauk SSSR, Fizika Zemli, 1977, No. 10, p. 3.
39. Safronov, V.S. "Initial state of the Earth and its early evolution." In: Evolution of the Earth., edited by R.J. O'Connell & W. S. Fyfe. Geodynamics Series, Vol. 5. Washington, DC: American Geophysical Union, 1981, p. 249.
40. Safronov, V.S. "On the initial nonuniformities of the Earth's mantle". Izv. Akad. Nauk SSSR, Ser. Geofiz.1965, No. 7, p. 1.
41. Safronov, V.S. "The initial state of the Earth and certain features of its evolution". Fizika Zemli,1972, No. 7, p.35.
42. Safronov, V.S., Vityazev, A.V. and Mayeva, S.V. "Problems of the initial state and early evolution of the Earth". Geokhimiya, 1978, No. 12.
43. Vityazev, A.V. "Heat generation and heat mass transfer in the early evolution of the Earth." Phys. Earth and Planet. Int. 1980, 22, p.289.
44. Vityazev, A.V. and Mayeva, S.V. "Simulation of the Earth core and mantle formation." Phys. Earth and Planet. Int. 1980, 22, p.296.
45. Ambartsumian, V.A. Star Evolution and Astrophysics. Erevan , 1947.
46. Reeves, H. "The 'Bing Bang' theory of the Origin of the Solar System" In: Protosvezdy i planety (Protostars and Planets). Moscow: "Mir," 1982 p.45
47. Safronov, V.S. and Ruzmaikina, T.V. "General problems for planetary and stellar cosmogony." In Rannie stadii evolutsii svezd (Early stages of stellar evolution). Kiev: "Naukova Dymka," 1977,p.147.
48. "On the transfer of angular momentum and the accumulation of solid bodies in the solar nebula" In Protosvezdy i Planety (Protostars and Planets),ed. T. Gehrels. Moscow: "Mir," 1982, p.623.
49. Ruzmaikina, T.V. Ursprung des Sonnensystems (diskussionforum). Mitt. Aastron. Ges., 1982, No. 57, 49.
50. Vityazev, A.V. and Pechernikova, G.V. "Models of the protoplanetary disks neighbouring stars of classes F-G". Pisma v Astronomicheskii Zhurnal, 1982, 8, No. 6, p. 371.
51. Safronov, V.S. and Vityazev, A.V. "Origin of the Solar System". In Astrofizika vysokikh energii i kosmogoniya (High-energy astrophysics and cosmogony). Itogi nauki i tekhniki. Ser. Astronomiya, 1983, Vol. 24. Moscow: VINITI, p. 5.
52. Lin, D.N.C. "Convective accretion disk model for the primordial solar nebula." Astrophys. J. 1981, 246, p.972.
53. Cassen, P.M. and Moosman, A. "On the formation of protostellar disks." Icarus, 1981, 48, No. 3, p.353.

Current Problems of the Cosmogony of the Solar System (abstract)

V. S. Safronov
Translated from: Astron. Vestn. **18**, 322 (1984)
Sol. Syst. Res. **18**, 208 (1984)

The possibility of the formation of the Sun and preplanetary nebula from a protosolar nebula with an initial mass of $\sim 1.05 M_{\odot}$ and initial angular momentum of $\sim 10^{52}$ g cm^2/s is discussed. Turbulence was generated during the collapse of the nebula, at the stage of gas accretion onto the nucleus and surrounding nascent disk, and the gas viscosity could have been sufficient for the expansion of the disk to the size of the planetary system. The formation of a dust layer in the disk, its disintegration into clumps, and their transformation into solid bodies are discussed. The evolution of the swarm of preplanetary bodies, the formation of the planetesimals (the number of which initially reached 100 in the zone of the Earth), and their accumulation into the planets are examined. It is shown that the formation of the small bodies of the solar system—the satellites, asteroids, and comets—was a natural side effect of the process of the accumulation of the planets. The initial temperature of the Earth, the inhomogeneity of its interior, and the possibility of its differentiation at the final stage of accumulation are discussed.

The Magnetic Field in the Collapsing Protosolar Nebula (abstract)

T. V. Ruzmaikina
Translated from: Astron. Vestn. **19**, 101 (1985)
Solar Syst. Res. **19**, 65 (1985)

The behavior of the magnetic field in the contracting protosolar nebula with a relatively small angular momentum of $\sim 10^{52}$ g cm^2/s and a mass $M \approx 1 M_{\odot}$ is discussed in the paper. In the process of collapse, following a short stage in which the magnetic field is frozen into the matter, the growth of the magnetic field limits the dissipation processes—first ambipolar diffusion and then ohmic dissipation. Before the start of the stage of hydrogen dissociation, when the temperature increases to 1600 K, dust is vaporized, and alkali metals are ionized; the magnetic field again becomes frozen in. For an ionization rate of 10^{-17}–10^{-21} s^{-1} in the protosolar nebula, the magnetic field strength in the starlike solar embryo is 3×10^3–10 G. The magnetic field, strengthened by differential rotation, is capable of redistributing angular momentum over a time less than the time of evolution of the core. This leads to the formation of a differentially rotating disk from the outer equatorial layers of the core.

On the Evaporation [of] Interstellar Dust during the Preplanetary Disk Formation (Abstract)

A. V. Vitjazev and G. V. Pechernikova
Lun. Plan. Sci. **16**, 885 (1985)

The hypothesis on the formation of meteoritic chondrules and Ca, Al rich inclusion from interstellar 1-mm grains by heating by aerodynamic drag in the nebular gases was suggested by Wood.[1] The main heating of the grains (at a radial distance $r \gtrsim 1$ A.U.) occurs during their deceleration behind the shock-wave discontinuity of interstellar gas accreting to the preplanetary disk. It seems to us that this model of the formation of the chondrules and CAI's has many advantages over other models. However, a number of questions arise. (1) Wood's assumption concerning the great abundance of the interstellar particles of about 1 mm. According to well-known data[2] the bulk mass of interstellar dust is contained in particles with size $\sim 10^{-5}$ cm. (2) Wood's estimations for physical characteristics behind the shock front of velocity u_2, density ρ_2, pressure P_2, and temperature of the gas T_2 as functions of the corresponding preshock parameters u_1, ρ_1, P_1, T_1 are maximum values insofar as the normal density discontinuity is only a particular case. In the general case, the slanting density discontinuity should be considered, where the preshock normal component velocity $u_{n1} = \xi u_1$ and $\xi < 1$. (3) There are difficulties with the explanation of element abundances (e.g., deficit Fe), isotopic anomalies (oxygen, magnesium, etc.), and mineralogical composition. Here we will not discuss these questions and further we will consider briefly some interesting effects connected with the vaporization of the dust grains with a dominant particle size of $r_p \sim 10^{-5}$ cm.

It is easy to show that for parameters which may be expected in the collapse models of presolar nebula with a moderate ($\sim 1.1 M_\odot$) mass and specific angular momentum ($\lesssim 10^{20}$ cm^2/c), the physical preshock (subscript 1) and postshock (subscript 2) properties of a gas are related by the following expressions:

$$\rho_2 = \frac{\mu}{R}(2\sigma)^{1/4}\rho_1^{3/4}(\xi u_1)^{5/4}, \quad P_2 = \rho_1(\xi u_1)^2, \quad T_2 = (\xi^3 u_1^3 \rho_1 / 2\sigma)^{1/4},$$

$$u_2 = \sqrt{(R/\mu)^2(\rho_1/2\sigma\xi u_1)^{1/2} + (1-\xi^2)u_1^2}, \quad u_{n2} = R(\rho_1/2\sigma\xi u_1)^{1/4}/\mu, \tag{1}$$

where μ is the mean molecular weight, σ is the Stefan–Boltzmann constant, R is the gas constant, $\xi = \sin\varphi$, φ is the angle between the direction of the accreting gas flux and the surface of the shock front, $\varphi < \varphi_{\text{crit}}$. Values of ρ_2, P_2, T_2, and u_2 calculated according to Eq. (1) for $\xi = 1$ and $\mu = 2$ as functions of radial distance from the forming star, ρ_1 and u_1 are given in Table 1. The quantities ρ_1 and u_1 were taken as in Ref. 1. For comparison the similar results obtained by Wood[1] by means of an iteration procedure are shown.

Heating of dust particles during aerodynamic drag in the gas is described by known equations of Whipple.[3] Heating is determined by three characteristic times: (1) the characteristic time of changing of the particle velocity v due to the gas drag is $\tau_v = 4\delta r_p/3\rho_2 v$, where δ is the particle density; (2) the characteristic time of heating on ΔT is $\tau_+ = 8c_p\delta\Delta T r_p/3\alpha\rho_2 v^3$, where α is the accommodation coefficient, c_p is specific heat; (3) the characteristic time of the cooling, when the particle temperature

TABLE 1.

r	2 A.U. (Ref. 1)	2 A.U.	5 A.U.	10 A.U.
ρ_1 (g/cm^3)	1.7×10^{-15}	1.7×10^{-15}	1.3×10^{-15}	1.1×10^{-15}
μ_1 (cm/c)	2.4×10^{6}	2.4×10^{6}	1.5×10^{6}	9.3×10^{5}
ρ_2 (g/cm^3)	2.0×10^{-12}	2.0×10^{-12}	8.7×10^{-13}	4.3×10^{-13}
P_2 (g/cm c^2)	9.6×10^{-3}	9.9×10^{-3}	2.8×10^{-3}	9.5×10^{-4}
T_2 (K)	121	120	78	53
u_2 (cm/c)	2.0×10^{3}	2.1×10^{3}	2.2×10^{3}	2.4×10^{3}

is T_p, is $\tau_- = \delta c_p r_p / 3\sigma T_p^3$. In our case we have $\tau_v \gg \tau_+, \tau_-$ so T_p can be found simply from the equality $\tau_+ = \tau_-$, where $v = \bar{u}_1 - \bar{u}_2 = u_{n1} - u_{n2} \approx \xi u_1$:

$$T_p = \left(\frac{\alpha\mu(2\sigma)^{1/4}}{8\sigma R} \right)^{1/4} \left(\frac{GM_\odot}{\text{A.U.}} \frac{M_*(t)}{M_\odot r_{\text{a.u.}}} \right)^{17/32} \rho_1^{3/16} = 3.24 \times 10^5 \rho_1^{3/16} \left(\frac{M_*(t)}{M_\odot r_{\text{a.u.}}} \right)^{17/32}.$$

The effective evaporation of the ice mantles of the dust grains requires the temperature $T_e \approx 180$ to 200 K. These temperatures are reached at $r \leqslant r_{\max} = (3.24 \times 10/T_e)^{32/17} \rho_1^{6/17} M_*(t)/M_\odot$. With $\rho_1 = 1.5 \times 10^{-15}$ g/cm we have $r_{\max} = (6.4$ to $7.9) M_*/M_\odot$ A.U. Evaporation of the ices causes the increase of the postshock density ρ_2 and accordingly the pressure P_2. In these conditions the shock front would move off with a rate $w \approx \xi\zeta' u_1 \rho_1/\rho_2 = \xi\zeta' R(2\sigma)^{-1/4}(\rho_1/u_1)^{1/4}/\mu$, where ζ' is the fraction of the evaporated ices. With $\zeta' \sim 10^{-2}$ we have $w \simeq \xi(1$ to $10)$ cm/c.

The ice vaporization starts earlier than the melting and evaporation of silicates. Before examining the process of the silicate evaporation it is necessary to investigate the effect of slow spreading of the shock front. It is necessary to investigate the stability of such a density discontinuity also. The solution of these questions is important for the examination of the origin and growth of the primary disk as well as for estimation of the generation efficiency of the turbulent velocities in the disk by the accreting gas.

[1] J. A. Wood, Earth Plan. Sci. Lett. **70**, 11 (1984)
[2] *Protostars and Planets*, edited by T. Gehrels (University of Arizona Press, Tucson, Arizona, 1978).
[3] F. L. Whipple, Proc. Natl. Acad. Sci. USA **86**, 687 (1950).

Eccentricities and inclinations of the orbits of growing planets

I. N. Ziglina

Institute of Earth Physics, USSR Academy of Sciences

(Submitted September 13, 1983)

Astron. Zh. **62**, 141–152 (January–February 1985)

The evolution of the eccentricity e and inclination i of the orbit of a planet of the terrestrial group as a result of collisions and encounters with bodies of the swarm during the process of accumulation is discussed in the paper. Encounters, being more numerous than collisions, had the main influence. Differential equations are derived describing the variation of the expected average values of e^2 and i^2 with the growth of the planet's mass. The rms eccentricity and inclination of the planet's orbit prove to be smaller than the corresponding quantities for bodies of the swarm by about $\sqrt{\bar{m}'/m}$ times, where $\bar{m}'$ is the average mass of bodies in the swarm and m is the planet's mass. The Fokker-Planck equation for the distribution density of the eccentricity and inclination of the orbit of a growing planet is solved. The problem of comparing the theoretical results with observations is discussed.

1. INTRODUCTION

One of the most important problems of planetary cosmogony is the calculation of the parameters of the planetary orbits which can be expected on the basis of the Shmidt–Safronov scheme. The present paper is devoted to an investigation of the formation of the eccentricities and inclinations of the orbits of planets of the terrestrial group. The influence of collisions and encounters with bodies of the swarm on the orbital eccentricity of a growing planet was estimated in a two-dimensional formulation in the papers of Ziglina and Safronov,[1] Ziglina,[2] and Pechernikova and Vityazev.[3] In Safronov's model[4] the half-thickness of the swarm of preplanetary bodies exceeds the radius of the sphere of influence of the planetary embryo over the entire duration of the process, so that the effective cross section of the interaction of bodies with the planet has the form of a circle rather than a segment, as in the two-dimensional problem. In the present work we consider the more realistic three-dimensional case, which differs considerably from the two-dimensional case.

The following model of the accumulation of planets of the terrestrial group is postulated. In the course of evo-

 0038-5301/85/01 0081-07 $04.10

lution of the preplanetary swarm several bodies are formed considerably exceeding the remaining bodies in mass.[4] These bodies can be treated as potential planetary embryos. The mass distribution of the remaining bodies (the quantities characterizing them will be marked by primes) is described by an inverse power law,[5]

$$n(m')=Cm'^{-q}, \quad m_{\min}\leqslant m'\leqslant m_1, \tag{1}$$

where $q \approx 1.8$ and $m_1 \gg m_{\min}$. For simplicity we shall assume that the masses and velocities of the preplanetary bodies are distributed independently. We adopt a Maxwellian distribution of the random velocities of the bodies.

$$f_0(\mathbf{v}')=(3/2\pi j^2)^{3/2}\exp(-3v'^2/2j^2), \tag{2}$$

where $\mathbf{v}'$ is the velocity of a body relative to the average velocity of the bodies, equal to $\mathbf{v}_K$, the Keplerian circular velocity in the central plane to the swarm. In the feeding zone of the planetary embryo under consideration the velocity dispersion of the bodies can be expressed in the form[4]

$$j^2=Gm/\theta r, \tag{3}$$

where m and r are the mass and radius of the embryo, G is the gravitational constant, and θ is a parameter having the order of several units ($\theta \approx 1$-5).

Only "close" encounters of preplanetary bodies with the planetary embryo, taking place within the sphere of influence of the embryo, are taken into account in the present work. When the relative velocities are not very low, each such encounter can be treated as a two-body problem. Over a time Δt let the embryo undergo n "close" encounters and k collisions with bodies of $m' > m_{\min}$, with its mass increase Δm being small compared with m. From the properties of mathematical expectation we get

$$\begin{gathered}\Delta\langle e^2\rangle=\langle\Delta e^2\rangle=k\langle\Delta_s e^2\rangle+n\langle\Delta_g e^2\rangle,\\ \Delta\langle i^2\rangle=\langle\Delta i^2\rangle=k\langle\Delta_s i^2\rangle+n\langle\Delta_g i^2\rangle,\end{gathered} \tag{4}$$

where e and i are the eccentricity and inclination of the embryo's orbit, $\langle\Delta_s x\rangle$ is the average change (the mathematical expectation of the change) in the value of x as a result of a collision with one body, and $\langle\Delta_g x\rangle$ is the average change in x as a result of an encounter with one body. Assuming that $m' \ll m$, we have

$$\Delta m\approx k\langle m'\rangle, \tag{5}$$

where $\langle m'\rangle$ is the average value of m' with a weight function n(m'). Below we shall estimate the values of $\Delta\langle e^2\rangle/\Delta m$ and $\Delta\langle i^2\rangle/\Delta m$ and find the rms values of the eccentricity and inclination of an embryo's orbit.

2. INFLUENCE OF COLLISIONS OF PREPLANETARY BODIES WITH THE EMBRYO ON THE ECCENTRICITY AND INCLINATION OF ITS ORBIT

We designate as $\mathbf{v}$ the velocity of the planetary embryo relative to the $\mathbf{v}_K$ corresponding to the point of the embryo's orbit under consideration. It is easy to obtain the following equations for the r, φ, and z projections of the vectors $\mathbf{v}$ in a cylindrical coordinate system with the center at the sun and the z axis perpendicular to the central plane of the swarm:

$$\begin{gathered}v_r=v_K[e\sin\varphi+O(e^2,\ i^2)],\\ v_\varphi=v_K[e\cos\varphi/2+O(e^2,\ i^2)],\\ v_z=v_K[i\cos(\omega+\varphi)+O(e^2,\ i^2)],\end{gathered} \tag{6}$$

where φ is the true anomaly of the embryo while ω is the argument of the latitude of the pericenter. Hence it follows that for small e and i

$$e^2=(v_r^2+4v_\varphi^2)/v_K^2. \tag{7}$$

Moreover, we have

$$z=R\sin i\sin(\omega+\varphi), \tag{8}$$

where R is the distance from the sun. From (8) and (6) with small e and i we find

$$i^2=v_z^2/v_K^2+z^2/R^2. \tag{9}$$

For the velocity distribution law (2),(3), the time of encounter with the planet is short compared with its period of revolution around the sun for the majority of bodies. Then the variation of the spatial coordinates of the planet during the encounter can be neglected. Designating the increments in the components of the vector $\mathbf{v}$ as a result of an encounter (ending with a collision or not) as Δv_r, Δv_φ, and Δv_z, from (7) and (9) we obtain

$$\begin{gathered}\Delta e^2=[2v_r\Delta v_r+8v_\varphi\Delta v_\varphi+(\Delta v_r)^2+4(\Delta v_\varphi)^2]v_K^{-2},\\ \Delta i^2=[2v_z\Delta v_z+(\Delta v_z)^2]v_K^{-2}.\end{gathered} \tag{10}$$

From the law of conservation of total momentum it follows that the change in the velocity of an embryo m in its collision with a body m' is

$$\Delta\mathbf{v}=(m'/m+m')(\mathbf{v}'-\mathbf{v})\approx(m'/m)(\mathbf{v}'-\mathbf{v}). \tag{11}$$

Substituting (11) into Eq. (10) and taking the mathematical expectation, we obtain

$$\begin{gathered}\langle\Delta_s e^2\rangle=2\langle m'\rangle m^{-1}\langle(-v_r^2-4v_\varphi^2+v_rv_r'+4v_\varphi v_\varphi')v_K^{-2}\rangle\\ +\langle m'^2\rangle m^{-2}\{\langle(v_r'^2+4v_\varphi'^2)v_K^{-2}\rangle+\langle(2v_rv_r'+8v_\varphi v_\varphi'+v_r^2+4v_\varphi^2)v_K^{-2}\rangle\},\\ \langle\Delta_s i^2\rangle=2\langle m'\rangle m^{-1}\langle(-v_z^2+v_zv_z')v_K^{-2}\rangle\\ +\langle m'^2\rangle m^{-2}\{\langle v_z'^2v_K^{-2}\rangle+\langle(2v_zv_z'+v_z^2)v_K^{-2}\rangle\}.\end{gathered} \tag{12}$$

We retain only the first terms inside the curly brackets in Eqs. (12), assuming that $\langle v^2\rangle\ll\langle v'^2\rangle$. The results of Sec. 3 confirm this assumption.

Just like e, i, and m', we shall treat φ and $\mathbf{v}'$ as random quantities characterizing a collision. For the probability density of a collision (e, i, φ, $\mathbf{v}'$) with a planet of mass m we have

$$f_s(e,\ i,\ \varphi,\ \mathbf{v}';\ m)=f(e,\ i;\ m)f_s(\varphi;\ m/e,\ i)f_s(\mathbf{v}';\ m/e,\ i,\ \varphi), \tag{13}$$

where f(e, i; m) is the distribution function of the quantities e and i; $f_s(\varphi;\ m/e,\ i)$ is the probability density of a collision at the point of the embryo's orbit with a true anomaly φ for given e and i; $f_s(\mathbf{v}';\ m/e,\ i,\ \varphi)$ is the probability density of a collision with a body moving at a velocity $\mathbf{v}'$ for given e, i, and φ. Thus, successive averaging can be carried out over $\mathbf{v}'$, φ, e, and i with the corresponding weight functions. As follows from the definition,

$$f_s(\mathbf{v}';m/e,i,\varphi)=N(\mathbf{R})\sigma Vf_0(\mathbf{v}')\Big/\int N(\mathbf{R})\sigma Vf_0(\mathbf{v}')dv'^3, \tag{14}$$

$$f_s(\varphi;m/e,i)=N(\mathbf{R})(dt/d\varphi)\int\sigma Vf_0(\mathbf{v}')dv'^3\Big/\iint N(\mathbf{R})(dt/d\varphi)\sigma Vf_0(\mathbf{v}')dv'^3d\varphi. \tag{15}$$

Here N(**R**) is the number of bodies per unit volume at the point with the radius vector **R**; $\sigma=\pi l^2=\pi r^2(1+v_e^2/V^2)$ is the collisional cross section of the embryo, l is the gravitational radius of the embryo, $v_e = \sqrt{2Gm/r}$ is the escape velocity at its surface, and $\mathbf{V} = \mathbf{v}' - \mathbf{v}$ is the velocity of m' relative to m before the encounter. From (15) we get

$$f_s(\varphi;\ m/e,\ i) = (2\pi)^{-1}[1+O(e,\ i)]. \tag{16}$$

For simplicity we set

$$\sigma = \pi r^2 v_e^2/V^2. \tag{17}$$

We thereby roughly estimate the influence of bodies with velocities $v' \gtrsim v_e$. However, for a Maxwellian velocity distribution (2) and for $j^2 \ll v_e^2$ there are few such bodies in the swarm and their contribution to the variation of the quantities e and i is negligibly small. Substituting (17) into (14), we have

$$f_s(\mathbf{v}';\ m/e, i, \varphi) = f_0(\mathbf{v}')V^{-1} \Big/ \int f_0(\mathbf{v}')V^{-1}\,dv'^3. \tag{18}$$

Upon integration over $\mathbf{v}'$ we obtain

$$\int f_0(\mathbf{v}')V^{-1}\,dv'^3 = 2(3/2\pi j^2)^{1/2}[1+O(v^2/j^2)],$$

$$\int v_x v_x' f_0(\mathbf{v}')V^{-1}\,dv'^3 = (1/3)v_x^2 2(3/2\pi j^2)^{1/2}[1+O(v^2/j^2)], \qquad x = r, \varphi, z.$$

Thus,

$$\int v_x v_x' f_s(\mathbf{v}';\ m/e, i, \varphi)\,dv'^3 \approx (1/3)v_x^2, \qquad x = r, \varphi, z.$$

From this, with allowance for (7) and (13), we obtain

$$\langle(-v_r^2 - 4v_\varphi^2 + v_r v_r' + 4v_\varphi v_\varphi')v_K^{-2}\rangle = -(2/3)\langle e^2\rangle. \tag{19}$$

Taking into account (16) and the last equation in (6), we have

$$\langle(-v_z^2 + v_z v_z')v_K^{-2}\rangle = -(2/3)\iint\Big[\int_0^{2\pi} f_s(\varphi;\ m/e, i)v_z^2 v_K^{-2}\,d\varphi\Big] f(e, i;\ m)\,de\,di$$

$$= -(2/3)\iint\Big[\int_0^{2\pi} i^2\cos^2(\omega+\varphi)(2\pi)^{-1}\,d\varphi\Big] f(e, i;\ m)\,de\,di = -(1/3)\langle i^2\rangle. \tag{20}$$

From (13), (16), and (18) we get approximately

$$\langle v_r'^2 v_K^{-2}\rangle = \langle v_\varphi'^2 v_K^{-2}\rangle = \langle v_z'^2 v_K^{-2}\rangle = \frac{1}{3}\int \frac{v'^2 3e^{-\frac{3}{2}\frac{v'^2}{j^2}}\,dv'^3}{v_K^2 4\pi j^2 v'} = \frac{2}{9}\frac{j^2}{v_K^2}. \tag{21}$$

The substitution of (19), (20), and (21) into (12) gives

$$\langle\Delta_s e^2\rangle = -(4/3)\langle m'\rangle m^{-1}\langle e^2\rangle + (10/9)\langle m'^2\rangle m^{-2} j^2 v_K^{-2},$$
$$\langle\Delta_s i^2\rangle = -(2/3)\langle m'\rangle m^{-1}\langle i^2\rangle + (2/9)\langle m'^2\rangle m^{-2} j^2 v_K^{-2}. \tag{22}$$

The quantity $\langle m'^2\rangle$ can be represented in the form

$$\langle m'^2\rangle = \overline{m'}\langle m'\rangle, \tag{23}$$

where $\overline{m'}$ is the average value of the masses of the bodies with a weight function m'n(m'), i.e.,

$$\overline{m'} = \int_{m_{min}}^{m_1} m'^2 n(m')\,dm' \Big/ \int_{m_{min}}^{m_1} m' n(m')\,dm'. \tag{24}$$

For the mass distribution law (1) of the bodies we have

$$\overline{m'} = m_1(2-q)/(3-q). \tag{25}$$

Using (22), (23), and (5), we obtain

$$\Delta_s\langle e^2\rangle/\Delta m = k\langle\Delta_s e^2\rangle/\Delta m = -(4/3)m^{-1}\langle e^2\rangle + (10/9)\overline{m'}m^{-2}j^2 v_K^{-2},$$
$$\Delta_s\langle i^2\rangle/\Delta m = k\langle\Delta_s i^2\rangle/\Delta m = -(2/3)m^{-1}\langle i^2\rangle + (2/9)\overline{m'}m^{-2}j^2 v_K^{-2}. \tag{26}$$

A comparison of (26) with Eq. (21) from Ref. 1 of Ziglina and Safronov shows that the influence of collisions on the orbital eccentricity of a planet is about the same in the plane and three-dimensional cases. The unimportant difference in the coefficients is connected with the fact that σV = const in the plane problem and $\sigma V \propto V^{-1}$ in the three-dimensional problem.

3. INFLUENCE OF ENCOUNTERS OF PREPLANETARY BODIES WITH THE EMBRYO ON THE ECCENTRICITY AND INCLINATION OF ITS ORBIT

From the condition of conservation of total momentum we find that, as a consequence of an encounter of a body m' with an embryo m,

$$\Delta\mathbf{v} = -(m'/m+m')\Delta\mathbf{V} \approx -(m'/m)\Delta\mathbf{V}.$$

We substitute this expression into Eq. (10) by components and take the mathematical expectation; then

$$\langle\Delta_g e^2\rangle = -2\langle m'\rangle m^{-1}\langle(v_r\Delta V_r + 4v_\varphi\Delta V_\varphi)v_K^{-2}\rangle$$
$$+\langle m'^2\rangle m^{-2}\langle[(\Delta V_r)^2 + 4(\Delta V_\varphi)^2]v_K^{-2}\rangle,$$
$$\langle\Delta_g i^2\rangle = -2\langle m'\rangle m^{-1}\langle v_z\Delta V_z v_K^{-2}\rangle + \langle m'^2\rangle m^{-2}\langle(\Delta V_z)^2 v_K^{-2}\rangle. \tag{27}$$

Within the framework of the two-body problem the projections of the vector $\mathbf{V}$ can be expressed as

$$\Delta V_r = -2\sin^2\psi V_r - \frac{VV_\varphi\sin 2\psi\cos\eta}{\sqrt{V_r^2+V_\varphi^2}} - \frac{V_z V_r\sin 2\psi\sin\eta}{\sqrt{V_r^2+V_\varphi^2}},$$
$$\Delta V_\varphi = -2\sin^2\psi V_\varphi + \frac{VV_r\sin 2\psi\cos\eta}{\sqrt{V_r^2+V_\varphi^2}} - \frac{V_z V_\varphi\sin 2\psi\sin\eta}{\sqrt{V_r^2+V_\varphi^2}}$$
$$\Delta V_z = -2\sin^2\psi V_z + \sqrt{V_r^2+V_\varphi^2}\sin 2\psi\sin\eta, \tag{28}$$

where 2ψ is the angle of rotation of the vector $\mathbf{V}$ and is determined by the relation

$$\sin^2\psi = (1+D^2V^4/\mu^2)^{-1}, \tag{29}$$

where $\mu = G(m+m') \approx Gm$. The quantities D and η can be defined as the polar coordinates of the body on a target passing through the center of the planet and orthogonal to $\mathbf{V}$. The azimuthal angle η is reckoned from a line parallel to the central plane.

It is natural to assume that the bodies are uniformly distributed over the area of the target for a fixed $\mathbf{V}$. Then the probability density of the encounter (e, i, φ, $\mathbf{v}'$, D, η) for a given planetary mass m is

$$f_g(e, i, \varphi, \mathbf{v}', D, \eta;\ m) = f(e,\ i;\ m')f_g(\varphi;\ m/e,\ i)$$
$$\times f_g(\mathbf{v}';\ m/e,\ i,\ \varphi)(2\pi)^{-1}2D/(S^2-l^2), \tag{30}$$

where S is the radius of the planet's sphere of influence. We take

$$S = R_{L_1} \approx \sqrt[3]{m/3M_\odot}\,R,$$

where R_{L_1} is the distance from the planet to the libration point L_1. The expressions for $f_g(\varphi;\ m/e,\ i)$ and $f_g(\mathbf{v}';\ m/e,\ i,\ \varphi)$ for encounters are analogous to Eqs. (14) and (15) for collisions, where σ must be set equal to πS^2. We have

$$f_g(\varphi;\ m/e,\ i) = (2\pi)^{-1}[1+O(e,\ i)], \tag{31}$$

$$f_g(\mathbf{v}';\ m/e, i, \varphi) = f_0(\mathbf{v}')V \Big/ \int f_0(\mathbf{v}')V\,dv'^3. \tag{32}$$

Let us estimate the quantities $\langle v_r\Delta V_r v_K^{-2}\rangle$, $\langle v_\varphi\Delta V_\varphi v_K^{-2}\rangle$ and $\langle v_z\Delta V_z v_K^{-2}\rangle$. Taking ΔV_r, ΔV_φ, and ΔV_z in accordance with (28), after averaging over η we obtain

$$\langle v_x\Delta V_x v_K^{-2}\rangle = 2\langle\sin^2\psi v_x V_x v_K^{-2}\rangle, \qquad x = r,\ \varphi,\ z. \tag{33}$$

The averaging of $\sin^2\psi$ over the impact parameter D yields

$$\langle\sin^2\psi\rangle^D = \int_l^S (1+D^2V^4/\mu^2)^{-1}\,dD^2/(S^2-l^2) = \ln(1+S^2V^4/\mu^2)\,\mu^2/S^2V^4, \tag{34}$$

since $l^2 \ll S^2$. With allowance for (34), Eqs. (33) take the form

$$\langle v_x\Delta V_x v_K^{-2}\rangle = -2q^{-2}\langle v_x V_x v_K^{-2}V^{-4}\ln(1+q^2V^4)\rangle,\quad x=r,\ \varphi,\ z, \tag{35}$$

where $q^2 = S^2/\mu^2$.

The next step is averaging over **v**' for a fixed **v** with the weight function (32). We designate

$$\mathbf{A} = \int \mathbf{V}V^{-4}\ln(1+q^2V^4)f_0(\mathbf{v}')V\,d\mathbf{v}'^3 \Big/ \int f_0(\mathbf{v}')V\,d\mathbf{v}'^3. \tag{36}$$

It is easy to calculate that

$$\int f_0(\mathbf{v}')V\,d\mathbf{v}'^3 = 4(j^2/6\pi)^{1/2}[1+O(v^2/j^2)]. \tag{37}$$

The integral in the numerator of Eq.(36) is designated as **B**. By virtue of the spherical symmetry of the distribution $f_0(\mathbf{v}')$, the vector **B** has no component orthogonal to **v**, so that

$$\mathbf{B} = \boldsymbol{\gamma}_0\int \mathbf{V}\boldsymbol{\gamma}_0 V^{-3}\ln(1+q^2V^4)f_0(\mathbf{v}')\,d\mathbf{v}'^3$$
$$= \boldsymbol{\gamma}_0\int_0^\infty f_0(\mathbf{v}')v'^2\left[\int_0^\pi \mathbf{V}\boldsymbol{\gamma}_0 V^{-3}\ln(1+q^2V^4)\,2\pi\sin\lambda\,d\lambda\right]dv', \tag{38}$$

where $\boldsymbol{\gamma}_0 = \mathbf{v}/v$; λ is the angle between the vectors **v** and **v**'. Let us evaluate the inner integral in (38). We have

$$\mathbf{V}\boldsymbol{\gamma}_0 = v'\cos\lambda - v,\quad V = (v^2+v'^2-2vv'\cos\lambda)^{1/2}.$$

Let us consider the case of $v' < v$. We expand the cofactors inside the integral in a series with respect to v'/v and, retaining the first terms, we obtain

$$\int_0^\pi (v'\cos\lambda - v)v^{-3}[1+3\cos\lambda v'/v + O(v'^2/v^2)][\ln(1+q^2v^4)$$
$$-4q^2v^4(1+q^2v^4)^{-1}\cos\lambda v'/v + O(v'^2/v^2)]\,2\pi\sin\lambda\,d\lambda$$
$$= -4\pi\ln(1+q^2v^4)v^{-2}[1+O(v'^2/v^2)]. \tag{39}$$

In the case of $v' > v$ we have

$$\int_0^\pi (v'\cos\lambda - v)v'^{-3}[1+3\cos\lambda v'/v + O(v^2/v'^2)][\ln(1+q^2v'^4)$$
$$-4q^2v'^4(1+q^2v'^4)^{-1}v/v' + O(v^2/v'^2)]\,2\pi\sin\lambda\,d\lambda$$
$$= -(16/3)\pi q^2v'v(1+q^2v'^4)^{-1}[1+O(v^2/v'^2)]. \tag{40}$$

Substituting (39) and (40) into (38), we obtain

$$\mathbf{B} \approx -\mathbf{v}(3/2\pi j^2)^{3/2}\left[\int_0^v 4\pi\ln(1+q^2v^4)v^{-3}v'^2\exp(-3v'^2/2j^2)\,dv'\right.$$
$$\left.+\int_v^\infty (16/3)\pi q^2v'^2(1+q^2v'^4)^{-1}\exp(-3v'^2/2j^2)\,dv'\right]$$
$$= -v(3/2\pi j^2)^{3/2}$$
$$\left[\int_0^\infty (16/3)\pi q^2v'^2(1+q^2v'^4)^{-1}\exp(-3v'^2/2j^2)\,dv' + O(v^2/j^2)\right]. \tag{41}$$

Combining Eqs. (41), (37), and (36), we find

$$\mathbf{A} = -3J\mathbf{v}/2j^2,$$

where

$$J = \int_0^\infty \exp(-\alpha x^{1/2})(1+x)^{-1}\,dx,\quad \alpha = 3/2\,qj^2.$$

Let us evaluate J:

$$J = \int_0^\infty \exp t\,2t(t^2+\alpha^2)^{-1}\,dt = 2\,\mathrm{Re}\int_0^\infty \exp t(t+i\alpha)\,dt.$$

We make the substitution $y = t+i\alpha$. Then

$$J = 2\,\mathrm{Re}\int_{i\alpha}^{-\infty} y^{-1}\exp y\exp(-i\alpha)\,dy = -2\,\mathrm{Re}[(\cos\alpha - i\sin\alpha)\,\mathrm{Ei}(i\alpha)]. \tag{42}$$

For $z \neq 0$ the integral exponential function Ei(z) admits the expansion[6]

$$\mathrm{Ei}(z) = -\int_z^\infty t^{-1}\exp t\,dt = C - \pi i + \ln z + \sum_{n=1}^\infty z^n/(nn!), \tag{43}$$

where C = 0.5772 ... is Euler's constant. Substituting (43) into (42), we find

$$J = -2\ln\gamma\alpha + \alpha(\pi-2) + O(\alpha^2),\quad \gamma = 1{,}781\ldots \tag{44}$$

We substitute the components of the vector A into (35). Then

$$\langle v_x\Delta V_x v_K^{-2}\rangle = (3J/q^2j^2)\langle v_x^2/v_K^2\rangle,\quad x=r,\ \varphi,\ z. \tag{45}$$

From this, with allowance for (7), we obtain

$$\langle(v_r\Delta V_r + 4v_\varphi\Delta V_\varphi)v_K^{-2}\rangle = (3J/q^2j^2)\langle e^2\rangle. \tag{46}$$

Replacing x by z in (45), using (6) and (39) we obtain

$$\langle v_z\Delta V_z v_K^{-2}\rangle = (3J/2q^2j^2)\langle i^2\rangle. \tag{47}$$

To find the quantities $\langle(\Delta V_r)^2v_K^{-2}\rangle$, $\langle(\Delta V_\varphi)^2v_K^{-2}\rangle$, and $\langle(\Delta V_z)^2v_K^{-2}\rangle$, appearing in (27), we square Eqs. (28) and average over η:

$$\langle(\Delta V_r)^2\rangle^\eta = 2[\sin^2\psi + O(\sin^4\psi)][V_\varphi^2+V_z^2],$$
$$\langle(\Delta V_\varphi)^2\rangle^\eta = 2[\sin^2\psi + O(\sin^4\psi)][V_r^2+V_z^2],$$
$$\langle(\Delta V_z)^2\rangle^\eta = 2[\sin^2\psi + O(\sin^4\psi)][V_r^2+V_\varphi^2].$$

Then averaging over D, v', and φ and neglecting small quantities, we find

$$\langle(\Delta V_x)^2v_K^{-2}\rangle \approx (4/3)\langle v'^2\sin^2\psi v_K^{-2}\rangle^{D,v',\varphi} \approx (4/3)v_K^{-2}$$
$$\times\int_0^\infty v'^2\ln(1+q^2v'^4)q^{-2}v'^4(9/8\pi j^2)\exp(-3v'^2/2j^2)4\pi v'^2\,dv'$$
$$= 2v_K^{-2}q^{-2}j^{-2}\int_0^\infty \exp(-t)\ln(1+\alpha^2t^2)\,dt = 2v_K^{-2}q^{-2}j^{-2}J,\quad x=r,\varphi,z, \tag{48}$$

since

$$\int_0^\infty \exp(-t)\ln(1+\alpha^{-2}t^2)\,dt = -\exp(-t)\ln(1+\alpha^{-2}t^2)\Big|_0^\infty$$
$$+\int_0^\infty 2t(t^2+\alpha^2)^{-1}\exp(-t)\,dt = J.$$

Substituting (48), (47), (46), and (23) into (27), we obtain

$$\langle\Delta_g e^2\rangle = -6\langle m'\rangle m^{-1}q^{-2}j^{-4}J\langle e^2\rangle + 10\langle m'\rangle\overline{m'}m^{-1}q^{-2}j^{-2}Jv_K^{-2},$$
$$\langle\Delta_g i^2\rangle = -3\langle m'\rangle m^{-1}q^{-2}j^{-4}J\langle i^2\rangle + 2\langle m'\rangle\overline{m'}m^{-1}q^{-2}j^{-2}Jv_K^{-2}. \tag{49}$$

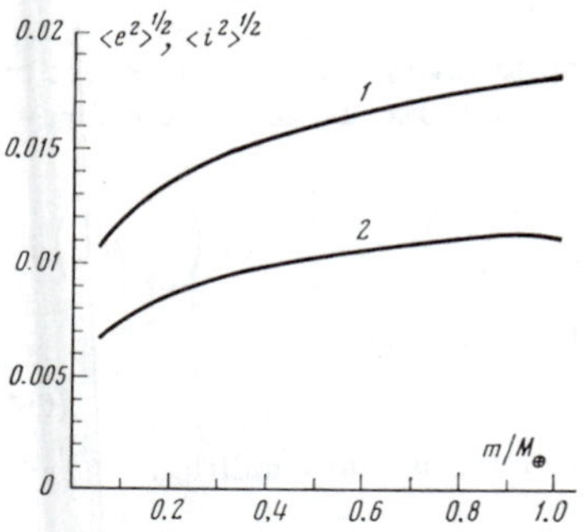

FIG. 1. The rms eccentricity (1) and inclination (2) of the orbit of the growing protoearth as a function of its mass.

Obviously, the ratio n/k equals

$$n/k=\iint N(R)\pi S^2 V f_0(\mathbf{v}')\,(dt/d\varphi)\,d\varphi\, dv'' \Big/ \iint N(R)\sigma^2 V f_0(\mathbf{v}')$$
$$\times(dt/d\varphi)\,d\varphi\, dv''\approx S^2\int v' f_0(\mathbf{v}')\,dv''/r^2 v_e^2\int f_0(\mathbf{v}')\,v'^{-1}\,dv''$$
$$=(^2/_3)\,j^2 v_e^{-2}S^2 r^{-2}. \tag{50}$$

We substitute (49) and (50) into (4) and divide by Δm. Taking j^2 in accordance with Eq. (2) and going to the limit as $\Delta m \to 0$, we find

$$d\langle e^2\rangle/dm=-(^4/_3)\,m^{-1}\langle e^2\rangle+(^{10}/_9)\,\overline{m'}m^{-1}G\theta^{-1}r^{-1}v_K^{-2}$$
$$-2\theta J m^{-1}\langle e^2\rangle+(^{10}/_3)\,\overline{m'}m^{-1}GJr^{-1}v_K^{-2},$$
$$d\langle i^2\rangle/dm=-(^2/_3)\,m^{-1}\langle i^2\rangle+(^2/_9)\,\overline{m'}m^{-1}G\theta^{-1}r^{-1}v_K^{-2}$$
$$-\theta J m^{-1}\langle i^2\rangle+(^2/_3)\,\overline{m'}m^{-1}GJr^{-1}v_K^{-2}, \tag{51}$$

where

$$J\approx-2\ln(\gamma\alpha)=2\ln(2S/3\gamma\theta r).$$

For planets of the terrestrial group $J\approx 5$–8. The influence of encounters proves to be dominant, and the first two terms in Eqs. (51), connected with collisions, can be neglected. A comparison of the first equation with the analogous linear equation in the plane problem[2] shows that the corresponding coefficients are 1.5 orders of magnitude larger in the three-dimensional case. In the case of $m'\ll m$ that we are considering, the variation of the planet's semimajor axis over the time of its growth is small. Therefore, we can set $v_K = \text{const}$ in the equations. Then the solution has the form

$$\langle e^2\rangle=e_0^2\exp\Big[-\int_{m_0}^{m}2\theta J\tau^{-1}d\tau\Big]+\int_{m_0}^{m}(^{10}/_3)\,\overline{m'}GJv_K^{-2}(4\pi\rho/3)^{1/3}x^{-4/3}$$
$$\times\exp\Big[-\int_{x}^{m}2\theta J\tau^{-1}d\tau\Big]\,dx,$$
$$\langle i^2\rangle=i_0^2\exp\Big[-\int_{m_0}^{m}\theta J\tau^{-1}d\tau\Big]+\int_{m_0}^{m}(^2/_3)\,\overline{m'}GJv_K^{-2}(4\pi\rho/3)^{1/3}x^{-4/3}$$
$$\times\exp\Big[-\int_{x}^{m}\theta J\tau^{-1}d\tau\Big]\,dx, \tag{52}$$

where m_0, e_0, and i_0 are the initial values of the planet's mass and the eccentricity and inclination of its orbit and ρ is the average density of the planet.

For further progress we must assign the function $\theta(m)$ and $\overline{m'}(m)$. In the simplest case of $\theta = \text{const}$ and $\overline{m'}/m = I = \text{const}$ we have

$$\langle e^2\rangle=e_0^2(m_0/m)^{2\theta J}+(^{10}/_3)\,GJ(2\theta J+{}^2/_3)^{-1}v_K^{-2}I\,[mr^{-1}-m_0r_0^{-1}(m_0/m)^{2\theta J}],$$
$$\langle i^2\rangle=i_0^2(m_0/m)^{\theta J}+(^2/_3)\,GJ(\theta J+{}^2/_3)^{-1}v_K^{-2}I\,[mr^{-1}-m_0r_0^{-1}(m_0/m)^{\theta J}]. \tag{53}$$

The dependence on the initial conditions is rapidly wiped out. Therefore,

$$\langle e^2\rangle\approx(^5/_3)\,(Gm/\theta r)\,v_K^{-2}I=\langle e'^2\rangle\,(\overline{m'}/m),$$
$$\langle i^2\rangle\approx(^2/_3)\,(Gm/\theta r)\,v_K^{-2}I=\langle i'^2\rangle\,(\overline{m'}/m), \tag{54}$$

since from Eqs. (7), (8), and (9), valid for any body moving along an ellipse, it follows that the average values of e'^2 and i'^2 are

$$\langle e'^2\rangle=(^5/_3)\,j^2v_K^{-2},\quad \langle i'^2\rangle=(^2/_3)\,j^2v_K^{-2}.$$

Equations (54) evidently signify the equidistribution of the energy of random motion between the planetary embryo and the bodies of mass $\overline{m'}$. These equations are also approximately correct when I depends on m. It is easy to obtain the following estimate for the relative error:

$$(\langle e^2\rangle-I\langle e'^2\rangle)/I\langle e'^2\rangle\approx-I'(m^*)\,m/I(2\theta J+{}^5/_3),\qquad m_0<m^*<m,$$
$$(\langle i^2\rangle-I\langle i'^2\rangle)/I\langle i'^2\rangle\approx-I'(m^{**})\,m/I(2\theta J+{}^5/_3),\qquad m_0<m^{**}<m,$$

where $I' = dI/dm$.

Further, it follows from (54) that the rms values of the eccentricity and inclination of the embryo's orbit are connected by the relation

$$\langle e^2\rangle^{1/2}\approx1.6\langle i^2\rangle^{1/2}.$$

An example of the evolution of $\langle e^2\rangle^{1/2}$ and $\langle i^2\rangle^{1/2}$ for the growing protoearth is shown in Fig. 1. At the initial time $r_0 = 1000$ km and $m_0/m_{10} = 4$. We also assumed that the upper limit in the mass distribution varies by the same law by which the mass of the largest body after the planetary embryo would vary if it was preserved up to the end of the accumulation. Then, according to Ref. 7,

$$m_1/m=[(1+(m/m_0)^{b/3}\varkappa)/(1+2\theta(m/m_0))]^3,$$

where $b = (2\theta-1)/(2\theta+1)$, $\varkappa = (r_0/r_{10}-1)/(2\theta-r_0/r_{10})$. It was assumed that the densities of bodies m_1 and m are the same and that $\theta = \text{const} = 3$.

4. PROBABILITY DENSITIES OF THE ECCENTRICITY AND INCLINATION OF THE ORBIT OF A GROWING PLANET

In this section we find the distribution function f(e, i; m) itself, i.e., we find a more general result than the result of the preceding sections, where we estimated only the second moments of the quantities e and i. We use the Fokker–Planck equation to find the distribution function. This can be done, since a time scale Δt exists for which the quantities e and i undergo a set of fluctuations, but yet their rms increments over this time are small. In the problem under consideration the Fokker–Planck equation has the form

$$\partial f/\partial m=-\partial/\partial e\,(f\langle\Delta e\rangle/\Delta m)$$
$$-\partial/\partial i\,(f\langle\Delta i\rangle/\Delta m)$$
$$+\partial^2/\partial e^2\,(f\langle(\Delta e)^2\rangle/2\Delta m)$$
$$+\partial^2/\partial i^2\,(f\langle(\Delta i)^2\rangle/2\Delta m)$$
$$+\partial^2/\partial e\partial i\,(f\langle\Delta e\Delta i\rangle/\Delta m), \tag{55}$$

where m plays the role of the independent variable instead

of time. As before, we neglect the variation of the semi-major axis of the planet's orbit.

Extracting the root of Eqs. (7) and (8) and expanding in a series with respect to Δv_r, Δv_φ, and Δv_z, we obtain

$$\Delta e=(v_r\Delta v_r+4v_\varphi\Delta v_\varphi)v_K^{-2}e^{-1}+2[v_\varphi^2(\Delta v_r)^2+v_r^2(\Delta v_\varphi)^2 -2v_rv_\varphi\Delta v_r\Delta v_\varphi]v_K^{-4}e^{-3}+\ldots,$$

$$\Delta i=v_z\Delta v_z v_K^{-2}i^{-1}+(1/2)z^2R^{-2}(\Delta v_z)^2v_K^{-2}i^{-3}+\ldots, \qquad (56)$$

from which

$$(\Delta e)^2=[v_r^2(\Delta v_r)^2+16v_\varphi^2(\Delta v_\varphi)^2+8v_rv_\varphi\Delta v_r\Delta v_\varphi]v_K^{-4}e^{-2}+\ldots,$$

$$(\Delta i)^2=v_z^2(\Delta v_z)^2v_K^{-4}i^{-2}+\ldots,$$

$$\Delta e\Delta i=(v_rv_z\Delta v_r\Delta v_z+4v_\varphi v_z\Delta v_\varphi\Delta v_z)v_K^{-4}e^{-1}i^{-1}+\ldots \qquad (57)$$

Then we find the average values of the quantities (56) and (57) over a time interval during which the planet's mass increases by the small amount Δm, analogous to what was done above. Then setting $\Delta m\approx\langle\Delta m\rangle$, we obtain

$$\langle\Delta e\rangle/\Delta m\approx-(2/3)m^{-1}e-\theta Jm^{-1}e+(5/18)\overline{m'}m^{-2}j^2v_K^{-2}e^{-1} +(5/6)\overline{m'}m^{-1}GJr^{-1}v_K^{-2}e^{-1},$$

$$\langle(\Delta e)^2\rangle/2\Delta m\approx(5/18)\overline{m'}m^{-2}j^2v_K^{-2}+(5/6)\overline{m'}m^{-1}GJr^{-1}v_K^{-2},$$

$$\langle\Delta i\rangle/\Delta m\approx-(1/3)m^{-1}i+(1/2)\theta Jm^{-1}i+(1/18)\overline{m'}m^{-2}j^2v_K^{-2}i^{-1} +(1/6)\overline{m'}m^{-1}GJr^{-1}v_K^{-2}i^{-1},$$

$$\langle(\Delta i)^2\rangle/2\Delta m\approx(1/18)\overline{m'}m^{-2}j^2v_K^{-2}i^{-1}+(1/6)\overline{m'}m^{-1}GJr^{-1}v_K^{-2}i^{-1},$$

$$\langle\Delta e\Delta i\rangle\approx 0. \qquad (58)$$

With allowance for (58), Eq. (55) takes the form

$$\partial f/\partial m=\beta_e(m)\partial/\partial e(fe)-p_e(m)\partial/\partial e(fe^{-1}) +p_e(m)\partial^2f/\partial e^2+\beta_i(m)\partial/\partial i(fi) -p_i(m)\partial/\partial i(fi^{-1})+p_i(m)\partial^2f/\partial i^2, \qquad (59)$$

where

$$\beta_e(m)=\theta Jm^{-1},\quad p_e(m)=(5/6)\overline{m'}m^{-1}GJr^{-1}v_K^{-2},$$

$$\beta_i(m)=(1/2)\theta Jm^{-1},\quad p_i(m)=(1/6)\overline{m'}m^{-1}GJr^{-1}v_K^{-2},$$

if we neglect terms connected with collisions between the planet and bodies.

As is seen from Eq. (59), the variables e and i have been separated. Therefore,

$$f(e,i;m)=f_e(e,m)f_i(i,m),$$

where $f_x(x, m)$ satisfies the equation

$$\partial f_x/\partial m=\beta_x(m)\partial/\partial x(f_xx)-p_x(m)\partial/\partial x(f_xx^{-1})+p_x(m)\partial^2f/\partial x^2, \quad x=e,i. \qquad (60)$$

To simplify the notation we omit the index x below.

Equation (60) can be interpreted as a diffusion equation, where the flux through the point x equals

$$F(x)=-\beta fx+pfx^{-1}-p\partial f/\partial x.$$

The flux through the point $x = 0$ must equal zero, since $e\geq 0$ and $i\geq 0$. Therefore, the boundary condition has the form

$$F(0)=(-\beta fx+pfx^{-1}-p\partial f/\partial x)|_{x=0}=0.$$

We shall seek a fundamental solution, i.e., a solution with the initial condition

$$f(x,m_0)=\delta(x-x_0),$$

where $\delta(x)$ is the Dirac delta function.

In Eq. (60) we make the following successive substitutions for the independent variables x and m and the function f:

$$y=x\exp\left(\int_{m_0}^{m}\beta d\tau\right),\quad n=f\exp\left(-\int_{m_0}^{m}\beta d\tau\right),\quad t=\int_{m_0}^{m}p\exp\left(\int_{m_0}^{\tau}2\beta d\xi\right)d\tau.$$

As a result, we arrive at the equation

$$\partial n/\partial t=-\partial/\partial y(ny^{-1})+\partial^2n/\partial y^2, \qquad (60')$$

where $n = n(y, t)$ must satisfy the initial condition

$$n(y,0)=\delta(y-x_0)$$

and the boundary condition

$$\left[-\beta ny+pny^{-1}\exp\left(\int_{m_0}^{m}2\beta d\xi\right)-p(\partial n/\partial y)\exp\left(\int_{m_0}^{m}2\beta d\xi\right)\right]\Big|_{y=0}=0.$$

Equation (60') can be solved by the Fourier method. The solution has the form

$$n(y,t)=(y/2t)\exp[(-y^2+x_0^2)/4t]I_0(yx_0/2t),$$

where $I_0(z)$ is a zeroth-order Bessel function of an imaginary argument. Returning to the variables x and m and the function f, we obtain

$$f(x,m) =\frac{x\exp\left[-\left(x-x_0e^{-\int_{m_0}^{m}\beta d\tau}\right)^2\Big/4\int_{m_0}^{m}pe^{-\int_{\tau}^{m}2\beta d\xi}d\tau\right]}{2\int_{m_0}^{m}pe^{-\int_{\tau}^{m}2\beta d\xi}d\tau}\exp(-z)I_0(z),$$

where

$$I_0(z)=\sum_{n=0}^{\infty}(z/2)^{2n}(n!)^{-2}=1+z^2/4+\ldots,$$

$$z=xx_0\exp\left(\int_{m_0}^{m}\beta d\tau\right)\Big/2\int_{m_0}^{m}p\exp\left(\int_{m_0}^{\tau}2\beta d\xi\right)d\tau.$$

For a test, we used the distribution function (61) to determine the quantities $\langle e^2\rangle$ and $\langle i^2\rangle$, which coincided with those found earlier.

The dependence of the distribution function on the initial condition dies out rapidly through the action of dynamic friction. After a sufficient time all the solutions prove to be close to the solution

$$\tilde{f}(x,m)=\frac{x\exp\left[-x^2/4\int_{m_0}^{m}p\exp\left(-\int_{\tau}^{m}2\beta d\xi\right)d\tau\right]}{2\int_{m_0}^{m}p\exp\left(-\int_{\tau}^{m}2\beta d\xi\right)d\tau}. \qquad (62)$$

A graph of this function has the form of a column, the width of which is comparable with the rms value of x, equal to

$$\langle x^2\rangle^{1/2}=2\left[\int_{m_0}^{m}p\exp\left(-\int_{\tau}^{m}2\beta d\xi\right)d\tau\right]^{1/2}.$$

With the highest probability the quantities e and i take values of the order of their rms values or less.

5. CONCLUSION

The problem of the evolution of the eccentricity and inclination of the orbit of a growing planetary embryo under the influence of collisions and encounters with bodies of the swarm in the three-dimensional case was discussed in the paper. The velocity distribution of bodies of the swarm was taken as Maxwellian with a dispersion dependent on the mass of the embryo and independent of the masses of the bodies themselves. The results of encounters were described within the framework of the two-body problem, and only encounters with bodies lying within the Hill sphere of the planetary embryo were taken into account.

The rms value of the eccentricity and inclination of the embryo's orbit were found under these simplifying assumptions. The distribution function f(e, i; m) was found using the Fokker–Planck equation. As a consequence of dynamic friction the distribution approaches an equilibrium distribution of the Gaussian type.

The gravitational interaction of the planets with each other, leading to secular oscillations of the eccentricities and inclination of their orbits with periods $T \approx 10^5$–10^6 yr (Ref. 8), was not taken into account in the development of the theory. Jupiter plays the major role in the process of secular oscillatiions. If Jupiter could have formed earlier than the planets of the terrestrial group and in an orbit close to its present orbit, then it influenced the formation of the eccentricities and inclinations of the orbits of planets of the terrestrial group even during the very accumulation process. It is proposed to estimate this influence in the future.

The calculations made above on the basis of the modern accumulation model of Ref. 4 give rms values of e and i for the earth and Venus comparable with the average eccentricities and inclinations for these planets over the period of the secular oscillations. The large eccentricity and inclination of Mercury's orbit definitely could not have been acquired through its interaction with bodies of considerably smaller mass. There are attempts to explain them as the result of the passage of Mercury through secular resonances with Venus connected with deceleration of the rotation of the young sun.[9] The orbit of Mars was formed under the strong influence of perturbations from Jupiter, both gravitational and through bombardment by bodies of its zone flying into the zone of Mars. Other factors influencing the evolution of the orbit of Mars are also possible. Heppenheimer[10] assumed, e.g., that the increase in the eccentricity and inclination of the orbit of Mars occurred during its passage through secular resonances with the Jupiter–sun system during the dissipation of the preplanetary nebula. Further development of cosmogonic theory is needed to reveal the most important mechanisms of formation of the eccentricities and inclination of the planetary orbits.

The author thanks V. S. Safronov, V. N. Zharkov, and A. B. Makalkin, for a discussion and comments during the preparation of the paper, as well as V. V. Leont'ev, for help in the work.

[1]I. N. Ziglina and V. S. Safronov, Astron. Zh. 53, 429 (1976) [Sov. Astron. 20, 244 (1976)].
[2]I. N. Ziglina, Astron. Zh. 53, 1288 (1976) [Sov. Astron. 20, 730 (1976)].
[3]G. V. Pechernikova and A. V. Vityazev, Astron. Zh. 57, 799 (1980) [Sov. Astron. 24, 460 (1980)].
[4]V. S. Safronov, Evolution of the Preplanetary Cloud and Formation of the Earth and Planets [in Russian], Nauka, Moscow (1969).
[5]E. V. Zvyagina, G. V. Pechernikova, and V. S. Safronov, Astron. Zh. 50, 1261 (1973) [Sov. Astron. 17, 793 (1974)].
[6]E. Yanke, F. Emde, and F. Lesh, Special Functions [in Russian], Nauka, Moscow (1968).
[7]G. V. Pechernikova and A. V. Vityazev, Pis'ma Astron. Zh. 5, 54 (1979) [Sov. Astron. Lett. 5, 31 (1979)].
[8]D. Brouwer and G. M. Clemence, in: Planets and Satellites, G. P. Kuiper and B. M. Middlehurst, (eds.), Univ. Chicago Press, Chicago (1961).
[9]W. R. Ward, G. Colombo, and F. A. Franklin, Icarus 28, 441 (1976).
[10]T. A. Heppenheimer, Icarus 41, 76 (1980).

Translated by Edward U. Oldham

Process of Formation of the Protoplanetary Disk (abstract)

T. V. Ruzmaikina and S. V. Maeva
Translated from: Astron. Vestn. **20**, 212 (1986)
Sol. Syst. Res. **20**, 132 (1986)

A numerical investigation is made of the evolution of the preplanetary disk to the accretion stage of collapse of the protosolar nebula, with an angular momentum of $(1–2)\times10^{52}$ g cm^2/s, with allowance for the accretion of material onto the disk. It is shown that during moderate subsonic accretion in the disk, $\nu\lesssim(0.3–1)\times10^{16}$ cm^2/s, its radius grows over the accretion time from $R_d\lesssim0.3$ A.U. to the present size of the solar system. Accretion onto the disk results in an increase in the disk's mass and a weak dependence of the mass on the boundary condition at the core and it also causes a change in the direction of flow of material in the disk as one moves from the core toward the edge. Near the core, material flows out from the disk into the core, $u_R<0$, at radii 0.3–0.5 A.U.$<R<0.6R_d$, $u_d>0$, then $u_r<0$ up to $R>(0.93-0.95)R_d$, and at the very edge ($R\gtrsim0.95R_d$) it becomes positive again, making the disk grow. In the region of planets of the terrestrial group and the asteroid belt, $u_R>0$ during most of the accretion. This explains the enrichment of meteoritic material with products that were subjected to the action of high temperature near the core.

Premature Particles in the Solar Nebula (abstract)

T. V. Ruzmaikina and V. S. Safronov
Lun. Planet Sci. **17**, 720 (1986)

Structural and isotopic features of meteorites imply that the most presolar condensed matter was vaporized during the formation of the solar system while a part of the matter remained almost unchanged.[1] We explain here this fact in the model of formation of the solar nebula (SN) as a result of collapse of a presolar nebula (PN) possessing relatively small angular momentum $J\sim10^{52}$ g cm^2 s^{-1}.[2–4]

According to this model a circumsolar disk (SN) is formed simultaneously with the proto-Sun during 10^5–10^6 yr. The disk expands to the size of the present planetary system from the size of a few solar radii due to the angular momentum transfer outward by turbulent and magnetic stresses. The turbulence can be supported by the shear flow of a gas accreted onto the disk with a relatively moderate value of a viscosity $\nu\sim10^{16}$ cm^2 s^{-1}, which in terms of "α disk" ($\nu\sim c_s h$, where c_s is the sound speed and h is the height scale in the disk) corresponds to $\alpha\sim10^{-2}$. Most of the PN envelope material falls onto the proto-Sun and onto the inner part of the disk inside a radius $R_k=0.08(J/10^{52})^2(M_c/M_\odot)^{1/3}$ A.U., where M_c is the mass of the growing proto-Sun.[3] The gas moving near the equatorial plane joins the disk and flows along its surface inward until $R\sim R_k$ in a layer (a) which has a thickness $\Delta z\sim(\nu/\Omega)^{1/2}$; Ω is the angular velocity of the Keplerian rotation. In the underlayer (b) along the vertical coordinate z the turbulent viscosity forces the gas to flow outward in the outer part of the disk $R>R_k$ and inward in the inner part.

Premature solid particles falling with the gas inside evaporate in a hot inner part of the envelope and disk. We consider a stage of accretion when a significant part of the PN mass is already in the proto-Sun while the envelope is not exhausted yet ($0.3<M_c/M_\odot<1$). The envelope is heated by radiation produced in the accretion shock and by the turbulent energy dissipated in the disk. The luminosity L is about $25L_\odot$ when the core radius is $R_c=5R_\odot$ and the rate of accretion is $M=10^{-5}M_\odot$ yr^{-1}. The radius of the destruction front for silicate particles with typical dimensions $r\sim10^{-5}$ cm is $R_{ev}=0.4(L/25L_\odot)^{1/2}$ A.U. and $R_{ev}=0.2(L/25L_\odot)^{1/2}$ A.U. when $r\sim1$ mm. The temperatures at these distances equal $T_{ev}=1500$ and 1900 K correspondingly. Graphite grains of $r\sim10^{-5}$ cm evaporate at temperature 2000 K.[5] Behind the shock front at the disk surface the infalling particles are heated additionally by the UV radiation[6] and by the aerodynamic drag.[7] Silicate particles of $r=10^{-5}$ cm are heated here up to $T_{gr}=(3\times10^{-21}nv^3/r+T^5)^{1/5}$, and those of $r\sim1$ mm up to $T_{gr}=(10^{-19}nv^3+T^4)^{1/4}$, where n is the number density of the preshock gas and v is the normal (to the front) component of the velocity of the infalling gas. The silicate particles of $r\sim10^{-5}$ cm and $r\sim1$ mm evaporate in the shock within $R\lesssim0.5$ A.U. and $R\lesssim0.3$ A.U. correspondingly for $L=25L_\odot$ and

$v \lesssim (GM/R)^{1/2}$. The region where the dust is evaporated should be more extended in the equatorial plane of the disk.[8] The solid particles moving in the layer (a) to R_k are vaporized as they reach R_{ev} ($R_{ev} > R_k$ when $J < 3\times 10^{52}$ g cm^2 s^{-1}). However, a fraction ξ of the matter (gas and dust) accreted by the disk diffuses into the deeper z layer (b) and does not reach R_{ev}. One can find that $\xi \simeq 1 - \mathrm{erf}\, x_0$, where $x_0^2 = -\ln\{2\rho_d\sqrt{\pi v \Omega}[R^{1/2}/(R_{ev}^{1/2} - R_k^{1/2})]/\dot{\sigma}_a\}$, ρ_d is the gas density in the layer considered, and $\dot{\sigma}_a$ is the mass accreted on the unit square per unit time. The expression for ξ (<1) can be reduced to $\xi \sim \alpha^{1/2} v/c_s$ in the case $R \sim R_{ev} R_k$ by taking $\rho_d = (\rho_{\mathrm{envelope}}) v^2/c^2$. At the disk equator ξ may approach to 1. However, the normal component of the velocity $v \sim c_s$ in the region $R_k \ll R \ll R_D$ hence $\xi \sim 10^{-1}$. For particles of $r \sim 1$ mm ξ should be larger than estimated. These particles speed through the postshock gas before stopping at a distance greater than the thickness of layer (a) Δz and penetrate directly into layer (b). When $R > R_k$ they are transferred here outward by the turbulent diffusion. Thus a significant fraction of the accreted particles should survive in the greater part of the SN. The particles accreted within 0.3–0.5 A.U. are heated to $T \gtrsim 1500$ K. One can believe[7] that these conditions are favorable to their melting and transformation into chondrules. Unfortunately, nothing is known about the fraction of the mass of premature solid material which can be contained in particles of 1 mm in size.

The particles of $r \sim 10^{-5}$ cm in the envelope could grow less than 10^{-4} cm during the collapse of the PN. The relative velocities of small particles stirred by turbulence are also very small inside the disk. The particles of different sizes collide with each other more frequently due to the differences in their settling velocities to the central plane $\Delta v_z \sim \delta\Omega^2 z \Delta r / c_s \rho$, where δ is the density of a particle. For the usual power law mass distribution with exponent $\simeq 1.8$ the larger particles grow exponentially $[r \simeq r_0 \exp(t/\tau)]$ with a characteristic time $\tau \simeq c_s/(4\zeta q \Omega^2 z)$, where ζ is the efficiency of sticking in one collision, $q \sim 10^{-2}$ is the mass ratio of solid and gaseous matter and $\bar{z} \sim h/2$. For $\zeta \sim 1$ the solid particles grow to 1 mm in size in several hundred years.

In conclusion, we note that the particles with the hot and cold past should be mixed throughout the disk in our model of the formation of the solar nebula from the PN with $J \sim 10^{52}$ g cm^2 s^{-1}. Their relative abundances are varied in time and over the distance from the proto-Sun. The fraction of material which was strongly heated increases inward. A characteristic radius of the region with more abundant recondensed solids grows in time and becomes remarkably larger than R_{ev}. Finally, we note an alternative model.[9]

[1] J. T. Wasson, "Meteorites: Their Record of Solar System History," preprint (1983).
[2] T. V. Ruzmaikina (1981) Adv. Space Res. **1**, COSPAR, 49 (1981).
[3] P. Cassen and A. Moosman, Icarus **48**, 353–376 (1981).
[4] V. S. Safronov and T. V. Ruzmaikina in *Protostars and Planets II*, edited by D. Black (in press, 1984).
[5] S. W. Stahler, F. H. Shu, and R. E. Taam, Astrophys. J. **248**, 727 (1981).
[6] D. Hollenbach and C. F. McKee, Astrophys. J. Suppl. **41**, 555 (1979).
[7] J. A. Wood, Earth Plan. Sci. Lett. **70**, 11 (1984).
[8] A. B. Makalkin, V. A. Dorofeyeva, and S. P. Borunov (Lunar and Planetary Science, Volume 17).
[9] G. E. Morfill, Icarus **53**, 41 (1983).

Changes in the Semimajor Axis of a Planet's Orbit in the Accumulation Process (abstract)

I. N. Ziglina
Translated from: Astron. Vestn. **20**, 328 (1986)
Sol. Syst. Res. **20**, 194 (1986)

A regular variation in the semimajor axis of a planetary orbit from its initial value and a mean square deviation Δa_{ms} resulting from interactions with the swarm's bodies are evaluated. It is shown that the former quantity is small in comparison to the latter. The value of Δa_{ms} for the Earth is found to be from 0.01 to 0.3 A.U., depending on the ratio $\bar{m}'/m$ of the average mass of planetesimals to the mass of the planet during the accumulation.

Thermal Conditions of the Protoplanetary Disk (abstract)

A. B. Makalkin
Translated from: Astron. Vestn. **21**, 324 (1987)
Sol. Syst. Res. **21**, 209 (1988)

The problem of the thermal conditions in the protoplanetary gas-dust disk is important for the understanding of the physical and chemical evolution of preplanetary material. The kinetics of chemical reactions, the condensation and evaporation of material, the efficiency of dust particle fusion upon collision, and the thermal conditions within the planetesimals depend greatly on the temperature.

Evolution of the Dusty Component of the Circumsolar Protoplanetary Disk (abstract)

V. S. Safronov
Translated from: Astron. Vestn. **21**, 216 (1987)
Sol. Syst. Res. **21**, 135 (1987)

We examine the settling of dust particles toward the central laminarly rotating plane of the circumsolar gas and dust disk and the origin in the dust subdisk of a gravitational instability. We obtain a dispersion relation for radial perturbations and find critical values of the density, the wavelength of the perturbation λ_c, and the masses of the largest condensations (10^{22} and 10^{24} g in the zones of Jupiter and Neptune, respectively). The damping effect of the gas on the thermal velocities of the particles renders unstable all modes with $\lambda<\lambda_c$. Short-wavelength perturbations grow faster than long-wavelength ones. In particular, prior formation of small condensation, which escaped being gravitationally connected to the massive condensations, eliminates the difficulties with forming kilometer-sized bodies in the zone of the giant planets. These bodies are necessary for the creation of the Oort comet cloud.

Evolution times for disks of planetesimals

S. I. Ipatov

Institute of Applied Mathematics, USSR Academy of Sciences

(Submitted September 4, 1986; resubmitted July 5, 1987)
Astron. Zh. **65**, 1075–1085 (September–October 1988)

We describe the basic design principles behind an algorithm for computer modeling of the evolution of disks consisting of a large ($\gtrsim 10^4$) number of planetesimals, making it possible to investigate the main stages of solid-body accumulation of planets or their cores. The accumulation times for planets are estimated on the basis of computer models of the evolution of disks consisting of hundreds of bodies, and analytical investigations of a series of models for the evolution of disks consisting of a considerably larger number of bodies. As in the work of Safronov, Vityazev, and Pechernikova, the accumulation time for the main mass of the earth is $\sim 10^8$ years, while for Neptune it is $\sim 10^9$ years. In the zone of Neptune certain planetesimals, moving in eccentric and inclined orbits, could be preserved up to the present.

1. INTRODUCTION

Many authorities now maintain[1] that the planets of the terrestrial group and the cores of the giant planets were formed from a disk of solid bodies or planetesimals, initially moving in almost circular orbits. According to Refs.1-3, the total mass of the initial bodies in the zones of the giant planets exceeded by severalfold the mass of solid material that went into these planets. Having reached ~ 1-3 $m_\oplus$ ($m_\oplus$ is the mass of the earth) in $3\cdot 10^7$ and $3\cdot 10^8$ years, the embryos of Jupiter and Saturn, respectively, began to accrete gas.[4] Other models of the formation of the planets also exist.[1,5-7] The process of solid-body accumulation has been investigated both analytically[1-2,8] and by computer modeling.[3,6-7,9-19] The results of computer modeling of the evolution of disks initially consisting of hundreds of bodies have been used to study the final stages of solid-body accumulation of the planets.[3,6,10-12,14-18] The evolution of swarms that initially consisted of a large number ($\sim 10^{11}$-10^{12}) of planetesimal bodies has been studied by a computer investigation of the variations of the distribution functions of the bodies with respect to masses and distances from the sun.[9,13,19] Upon the appearance of large bodies (planetary embryos) in the swarm, these calculations were ended, since the "particle in a box" method used in Refs. 9 and 19 became inapplicable.

The actual process of accumulation of the planets depended in a complex way on many factors. The study of relatively simple models, however, makes it possible to draw a number of important conclusions about the process of accumulatioin of the planets. In the present paper we consider the evolution of disks of bodies moving around a massive central body (the sun). Models in which the bodies combine in

 0038-5301/88/05 0560-07 $04.10

collisions are predominantly studied, but we also discuss the influence of fragmentation of bodies and gas drag on the accumulation process. The mutual gravitational influence of the bodies is taken into account by the method of spheres (mainly spheres of action),[14,20] i.e., inside a sphere the relative motion of bodies is treated within the framework of the two-body problem, while outside the sphere the bodies move around the sun in unperturbed Keplerian orbits.

In Sec. 2 we consider certain general questions in the construction of an algorithm for computer modeling of the evolution of disks consisting of a large number ($\gtrsim 10^4$) of bodies. As in Refs. 9, 13, and 19, the bodies of the disk are divided into groups (bins), but the interactions of the bodies are taken into account differently. In the limit, when each group consists of one body, this algorithm, not yet implemented on a computer, is the same as the algorithm used to investigate the evolution of disks consisting of hundreds of bodies.[18,20] The formulas that we use to determine the number of encounters between bodies up to the radius r_c of the sphere used and the number of collisions between bodies in a time Δt (Sec. 3) were obtained for a more complex model than the "particle in a box" model considered in Refs. 9 and 19, and they enable us to take into account not only the relative velocity of the bodies but also the orbital elements of proximate bodies. The time of the fall of small bodies into a larger body is determined (Sec. 4) for models differing from the models considered by Safronov et al.[1-2,8] In Sec. 5 we give the evolution time for disks initially consisting of hundreds of bodies. The evolution of such disks was investigated by computer modeling. On the basis of the results given in Secs. 3 and 5, the evolution time for disks consisting of a large number of original bodies is estimated for a series of models of Secs. 6 and 7. Although the evolution of the disks of planetesimals was three-dimensional in nature,[2] a plane model was also considered in the paper along with the three-dimensional model. The investigation of the plane model enables us to estimate the minimum changes in the orbits of gravitationally interacting planetesimals.

2. BASIC DESIGN PRINCIPLES BEHIND AN ALGORITHM FOR COMPUTER MODELING OF THE EVOLUTION OF DISKS CONSISTING OF A LARGE NUMBER OF PLANETESIMALS

The group of bodies examined usually includes bodies for which the masses m and the semimajor axes a (as well as the eccentricities $\bar{e}$ and inclinations i when a sufficiently powerful computer is available) of the orbits lie within certain fairly narrow limits. In the algorithm under consideration, each (k-th) group is characterized by the number $N^{(k)}$ of bodies included in it (instead of $N^{(k)}$ we may consider the total mass $M_{\Sigma}^{(k)}$ of all the bodies of the group) and the mean values of m, a, e, and i designated as $\bar{m}^{(k)}$, $\bar{a}^{(k)}$, $\bar{e}^{(k)}$, and $i^{(k)}$, respectively. Moreover, in the three-dimensional case each group can also be characterized by the mean values $\bar{\Omega}^{(k)}$ and $\bar{\omega}^{(k)}$ of the longitude Ω of the ascending node and the argument of perihelion ω.

As was done in Refs. 9, 13, and 19, we break down the entire process into a series of successive steps Δt in time. The number of encounters $N^{(k,\ell)}$ between bodies up to a distance r_c and the number of collisions $N_{col}^{(k,\ell)}$ which occur between bodies of the k-th and ℓ-th groups, as well as the changes in the quantities $N^{(k)}$, $\bar{m}^{(k)}$, $\bar{e}^{(k)}$, $\bar{i}^{(k)}$ ($\bar{\Omega}^{(k)}$ and $\bar{\omega}^{(k)}$) in a time Δt can be determined differently for different models. Below we shall consider, in particular, a relatively simple model, which we call the model of group-averaged bodies. The more complex models can be investigated on the basis of this model. In the model of group-averaged bodies we shall assume that in the interaction of bodies of two groups, for all $N^{(k)}$ bodies of one (the k-th) group the masses m and the orbital elements, a, e, i, (Ω and ω) are the same and equal to $m^{(k)}$, $a^{(k)}$, $e^{(k)}$, $i^{(k)}$, (Ω and $\omega^{(k)}$), respectively. Such bodies will be called averaged bodies. To improve conservation of the integrals of motion, it is better to determine the values of $\bar{a}^{(k)}$, $\bar{e}^{(k)}$, $\bar{i}^{(k)}$ and $\bar{\Omega}^{(k)}$ not as the mean values of the corresponding orbital elements (as was done in Refs. 9 and 19) but so that the mechanical energy $h^{(k)}/2$ and the angular momentum $\mathbf{c}^{(k)}=\{c_x^{(k)}, c_y^{(k)}, c_z^{(k)}\}$ of all the averaged bodies of the group are the same as for all the actual (different) bodies of the group. For each averaged body of the k-th group the values of c and h are $\bar{\mathbf{c}}^{(k)}=\{\bar{c}_x^{(k)}, \bar{c}_y^{(k)}, \bar{c}_z^{(k)}\} = \mathbf{c}^{(k)}$ and $\bar{h}^{(k)} = h^{(k)}/N^{(k)}$ (remember that in the method of spheres, bodies outside the spheres do not interact with each other). Using the laws of conservation of energy and angular momentum and dropping the group (k) we have

$$\bar{a}=-\bar{m}\bar{\mu}/\bar{h},\quad \bar{e}=\sqrt{1-[(\bar{c}_x)^2+(\bar{c}_y)^2+(\bar{c}_z)^2]/\bar{\mu}\bar{m}\bar{a}},$$
$$\cos\bar{i}=\bar{c}_z/\bar{m}\sqrt{\bar{\mu}\bar{p}},\quad \sin\bar{i}=\sqrt{(\bar{c}_x)^2+(\bar{c}_y)^2}/\bar{m}\sqrt{\bar{\mu}\bar{p}},$$

$$\cos\bar{\Omega}=-\bar{c}_y/\bar{m}\sin\bar{i}\sqrt{\bar{\mu}\bar{p}},\quad \sin\bar{\Omega}=\bar{c}_x/\bar{m}\sin\bar{i}\sqrt{\bar{\mu}\bar{p}}, \tag{1}$$

where $\bar{p} = (1-\bar{e}^2)a$, $\mu = G(M_\odot + \bar{m})$, $M_\odot$ is the mass of the sun, G is the gravitational constant. In accounting for the mutual gravitational influence of the bodies by the method of spheres, h and c do not depend on ω or the true anomaly ν. Therefore, if ω is treated as one of the properties of the group, we can calculate $\bar{\omega}$ as the mean value of ω_j for all the bodies of the k-th group.

Witout dwelling in detail here on modeling the gravitational interactions and collisions of bodies, we note that at each step Δt of the algorithm one must know the values of $N_c^{(k,\ell)}$ and $N_{col}^{(k,\ell)}$ for $k=1,\ldots,N_g$ and $\ell = 1,\ldots, N_g$, where N_g is the number of groups in the disk. The values of Δt are chosen, in particular, so that there is no more than one collision per body of a group, on the average, in time Δt. If the number of bodies in the groups is large, then the results of interactions between bodies in the time Δt can be determined from computer modeling of a relatively small number of characteristic encounters and collisions.

3. NUMBERS OF ENCOUNTERS AND COLLISIONS BETWEEN BODIES IN TIME Δt

For $N_c^{(k,\ell)}$ and $N_{col}^{(k,\ell)}$ we shall take the expectation values of these quantites. In the stochastic approach under consideration,

$$N_c^{(k,l)}=N^{(k)}N_e^{(l,k)}\Delta t/\bar{\tau}^{(k,l)}. \tag{2}$$

where $N^{(k)}N_e^{(\ell,k)}$ is the number of pairs of bodies (body j_1 belongs to the k-th group while body j_2

belongs to the ℓ-th group) for which the values of the aphelion and perihelion distances (r_α and r_π) allow an encounter between the j_1-th and j_2-th bodies (the minimum distance between the segments $[r_\pi(j_1), r_\alpha(j_2)]$ does not exceed r_c); $1/\bar{\tau}^{(k,\ell)}$ is the mean values of the quantities $1/\tau(j_1, j_2)$, where $\tau(j_1, j_2)$ is the time between encounters (up to r_c) of the j_1-th and j_2-th bodies. We can represent $N_e(\ell, k)$ in the form

$$N_e^{(l,k)}=N^{(l)}\bar{\varkappa}^{(l,k)}, \text{ where } \bar{\varkappa}^{(l,k)}\leqslant 1. \quad (3)$$

In the general case $\bar{\kappa}(\ell, k)$ can be calculated numerically. In the interaction between bodies of the same group (for $k = \ell$), instead of $n^{(k)}N_e^{(\ell,k)}$ in Eq. (2) we have $N^{(k)}(N^{(k)}-1)\bar{\kappa}^{(k,k)}/2$.

First let us consider the plane model. Let $\Delta\phi$ be the sum of the angles (in radians), with the apex at the sun, within which the distance between the orbits of the j_1-th and j_2-th bodies along the central ray (with apex at the sun), is less than r_c, while $T(j)$ is the period of revolution of the j-th body around the sun. Bodies that lie on the same central ray will, after a time close to (equal to, for circular orbits) the synodic period of revolution $T_s = T(j_1)T(j_2)/|T(j_1)-T(j_2)|$, again lie on the same central ray. If as an approximation we take the direction of this ray to be random, then the ratio $\Delta\phi/2\pi$ is the probability that the distance between bodies lying on this ray is less than r_c. Also allowing for the fact that the mean value of the initial angle (with apex at the sun) between the directions toward the bodies is π for a random location of the bodies, while after an encounter it is 2π, we find that for $\Delta t \le T_s/2$ the expectation value of the number of encounters (up to r_c) between the j_1-th and j_2-th bodies in the time Δt is $\Delta t/\tau(j_1, j_2) \approx \Delta t\Delta\phi/\pi T_s$, while for $\Delta t > T_s/2$ it is $\Delta t/\tau(j_1, j_2) \approx (0.5 + \Delta t/T_s)\Delta\phi/2\pi$. In the study of the process of solid-body accumulation it is proposed to take $\Delta t \gg T_s$. In this case, therefore, designating the mean value of $\Delta\phi$ for different pairs of bodies of the k-th and ℓ-th groups as $\overline{\Delta\phi}$ and assuming that the values of T_s for these pairs differ only slightly, we obtain

$$\bar{\tau}^{(k,l)}\approx 2\pi\bar{T}_s/\overline{\Delta\varphi}, \quad (4)$$

where $\bar{T}_s$ is the mean value of T_s. For eccentric orbits ($e_m = \max\{\bar{e}^{(k)}, \bar{e}^{(\ell)}\} < r_c/R$) we consider the the quantity $k_\phi = \overline{\Delta\phi}/r_c^*$, where $r_c^* = r_c/R$ and R is the distance of the bodies from the sun. The values of k_ϕ are smaller for larger eccentricities. It is found numerically, for example, that for $m = 0.01\, m_\oplus$ we have $k_\phi \approx 30$ for $e_m = 0.1$, while $k_\phi \approx 10$ for $e_m = 0.3$ (different mutual orientations of the orbits of bodies of the different groups were considered for these estimates of k_ϕ). Using Eqs. (2)-(4), we can determine the value of $N_c^{(k,\ell)}$ for $k \ne \ell$.

In an investigation of interactions between bodies of the same group, the bodies can be divided into two subgroups in a certain way and the same Eqs. (2)-(4) can be applied. Moreover, in that case the values of $N_c^{(k,k)}$ can also be determined from Eq. (2), assuming that

$$\bar{\tau}^{(k,k)}\approx k_T(\bar{a}^{(k)}/a_\oplus)^{3/2}T_\oplus/r_c^\star\approx 2\pi k_T(\bar{a}^{(k)})^{3/2}/r_c^\star\sqrt{GM_\odot}, \quad (5)$$

where $k_T \approx 3.5$, $a_\oplus$ is the semimajor axis of the earth's orbit, and $T_\oplus$ is the period of its revolution around the sun. Equation (5) was obtained on the basis of computer modeling of the evolution of a series of disks consisting of 100 identical particles of mass m_0. The initial eccentricities e_0 of the particle orbits were the same. In such modeling it was assumed that the ratio of the maximum to the minimum values of the semimajor axes of the orbits of bodies of the initial disk was $a_{max}/a_{min} = 2$, $10^{-8}\, m_\oplus \le m_0 \le m_\oplus$, and $0.01 \le e_0 \le 0.3$, and it was found that $\tau(k, k)$ hardly depended on $\bar{e} \approx e_0$.

Now let us consider the calculation of $N_c^{(k,\ell)}$ for the three-dimensional model. We designate the angle between the orbital planes of two encountering bodies as Δi, while u' is the angle between one of the two rays formed by the line of intersection of the orbital planes and the direction toward one of these bodies, which we call the second. Encounters between bodies up to r_c can occur only for those values of u' for which $h(u') = R_2 \sin u' \sin \Delta i \le r_c$, where R_2 is the distance from the sun to the second body. We designate the sum of the angles, with apex at the sun, within which this inequality is satisfied as $\Delta u'_\Sigma$. Taking $R_2 \approx R$, $\sin u' \approx u'$, $\sin \Delta i \approx \Delta i$, and $\max\{h(u')\} \ge r_c$ (i.e., $\Delta i \ge r_c^*$), we obtain $\Delta u'_\Sigma \approx 4r_c^*/\Delta i$, and the probability $p(\Delta i)$ of an encounter between two bodies up to r_c in a time Δt is smaller in the three-dimensional case than in the plane case by a factor of $k_i \approx (2\pi/\Delta u'_\Sigma)(r_c/r_c') \approx \pi R\, \Delta i/2r_c'$, where r_c' is the mean value (in the region $\Delta u'_\Sigma$) of the projection of r_c onto the orbital plane of the first particle. Since

$$r_c'=\Bigg[\int\limits_{h(u')\leqslant r_c}\sqrt{(r_c)^2-(h(u'))^2}\,du'\Bigg]\Bigg/\int\limits_{h(u')\leqslant r_c}du'\approx r_c/2,$$

$p(\Delta i)$ and N_c are smaller by about a factor of $k_i = \pi\Delta i/r_c^*$ for $\Delta i \ge r_c^*$ than for $\Delta i = 0$ (for the same values of m, e, and a). In the case when Δi assumes values from 0 to $\Delta i^*_{max} - r_c^*$ with equal probability, taking $k_i \approx 2$ (where $1/\bar{k}_i$ is the mean value of $1/k_i$ for different Δi) for $0 \le \Delta i \le r_c^*$, for $0 \le \Delta i \le \Delta i^*_{max}$ we obtain

$$1/\bar{k}_i=\Big[\int(1/k_i)\,d(\Delta i)\Big]\Big/\int d(\Delta i)=r_c^\star\eta_i/\pi\Delta i^\star_{max}, \quad (6)$$

where $\eta_i = 0.5 + \ln(\Delta i^*_{max}) - \ln(r_c^*)$. For i uniformly distributed from 0 to $i^{(k)}_{max}$, the probability that $\Delta i \approx \Delta i_{max} = i^{(k)}_{max} + i^{(\ell)}_{max}$ is lower than it is for $\Delta i \approx 0$. Therefore, $\Delta i^*_{max} < \Delta i_{max}$. We shall take approximately $\Delta i^*_{max} \approx \bar{i}^{(k)} + \bar{i}^{(\ell)}$. For example, if r_c is the radius of the sphere of action and $\Delta i^*_{max} = 0.15$, we have $\eta_i \approx 9$ for $m = 10^{-6}\, m_\oplus$ and $\eta_i \approx 4$ for $m = m_\oplus$. Equation (6) was obtained for the case when the bodies are not divided into different groups according to i. But if the values of i for bodies of k-th group range from $i^{(k)}_{min}$ to $i^{(k)}_{max}$, while the minimum distance Δi_{min} between the segments $[i^{(k)}_{min}, i^{(k)}_{max}]$ and $[i^{(\ell)}_{min}, i^{(\ell)}_{max}]$ is greater than r_c^*, then, integrating k_i from Δi_{min} to Δi^*_{max}, by analogy with Eq. (6), we obtain

$$\bar{k}_i\approx\pi(\Delta i^\star_{max}-\Delta i_{min})/r_c^\star \ln(\Delta i^\star_{max}/\Delta i_{min}).$$

Using Eq. (2), we obtain

$$N_{col}^{(k,l)}=N_c^{(k,l)}/\bar{N}_c^{(k,l)}=N^{(k)}\Delta t/k_N T_\varphi^{(k,l)}, \quad (7)$$

where $N_c^{(k,\ell)}$ is the mean number of encounters up to r_c leading to one collision and

$$T_\varphi^{(k,l)}=\bar\tau^{(k,l)}\bar N_c^{(k,l)}/k_N N_e^{(l,k)} \quad (k_N=1 \text{ for } k\neq l \text{ and } k_N=2 \text{ for } k=l). \tag{8}$$

Usually in the algorithm under consideration $\Delta t < T_\phi^{(k,\ell)}$, but formally we find from (7) that in a time $T_\phi^{(k,\ell)}$ each body of the k-th group takes part in one collision with some body of the ℓ-th group, on the average. We assume below that bodies collide (combine or are destroyed) when the distance between their centers of mass becomes equal to the sum r_Σ of their radii. With allowance for the additional capture or destruction of some of the bodies that have entered the Roche zone,[1,13] it must be borne in mind that the accumulation time is $T \propto (r_3)^{-1}$ for $\Delta i = 0$ and $T \propto (r_3)^{-2}$ for $\Delta i\ r_c^*$, where r_3 is the capture radius. According to Refs. 21 and 23, the additional capture is minor.

Using the concept of the effective radius r_e of a body, in the plane case for averaged bodies we have

$$\bar N_{c\,pl}^{(k,l)} \approx r_c/r_e \approx r_c/r_\Sigma\sqrt{1+2\theta'}, \tag{9}$$

where $\theta' = 0.5(v_{par}/v_{rel})^2$, $v_{par} = \sqrt{2G(\bar m^{(k)}+\bar m^{(\ell)})/r_\Sigma}$, and v_{rel} is the relative velocity of bodies encountering each other at a distance r_c. According to Refs. 1 and 2, the rms velocity of bodies relative to a circular orbit of a planet of mass M and radius r is $v_{rel}^M = \sqrt{GM/\theta r}$, where θ is Safronov's parameter. If $m^{(k)} = M \geq m^{(\ell)} = \gamma M$, then $\theta' = \theta(1+\gamma)/(1+\gamma^{1/3})$, where $3/4 \leq (1+\gamma)/(1+\gamma^{1/3}) \leq 1$. We designate $\xi = v_{rel}/v_{orb} = k_\xi e$, where v_{orb} is the velocity of a body in an orbit of radius $(\bar a^{(k)} + \bar a^{(\ell)})/2$. On the basis of Eq. (9) and the values of $\bar N_c$ obtained in computer modeling of the evolution of plane disk consisting of hundreds of bodies, for a disk consisting of one group we have $k_\xi = 1$. If $\Delta i > r_c^*$, then $\bar N_c \approx (\bar N_{c\,pl}^*)^2$, where $N_{c\,pl}^*$ is the value of $\bar N_c$ for a plane model with the same orbital eccentricities as in the three-dimensional case. For $\bar i = \bar e/\sqrt{2}$ (Ref. 13) the values of ξ and k_ξ are larger by a factor of about $\sqrt{3}/2$ than for $\bar i = 0$ with the same values of $\bar e$.

In the course of evolution of a disk of planetesimals, θ varies. In the zone of the giant planes, for example, $\theta \gg 1$ for a gasless model disk of approximately equal bodies.[23] We represent r_c in the form $r_c = R(m/M_\odot)^\alpha(1+\delta)^{\alpha/2}$. When spheres of action are used, $\alpha = 0.4$, while in Ref. 10 $\delta = m_2/m_1$, where m_1 and m_2 are the masses of the bodies encountering each other ($m_2 \leq m_1 = m$). If $N^{(k)} \gg 1$, using Eqs. (5), (6), (8), and (9) and omitting the group numbers, in the interaction between bodies (of density ρ) of the same group and for $i_{max} \approx 2\bar i$ we have:

a) for $i+2\theta' \approx 1$,

$$\bar N_{c\,pl} \approx \bar a\rho^{1/3}\bar m^{\alpha-1/3}(\pi/6)^{1/3}(1+\delta)^{\alpha/2}/(M_\odot)^\alpha,$$
$$T_{\phi\,pl} \approx \bar a^{5/2}\rho^{1/3}k_T(\pi/6)^{1/3}T_\oplus/\bar m^{1/3}N\bar\varkappa(a_\oplus)^{3/2} \propto \bar m^{-1/3}/M_\Sigma,$$
$$T_{\phi 3d} \approx \bar a^{7/2}\rho^{2/3}\bar i k_T\pi(\pi/6)^{2/3}T_\oplus/\bar m^{2/3}N\bar\varkappa\eta_i(a_\oplus)^{3/2} \propto \bar m^{1/3}/M_\Sigma, \tag{10}$$

b) for $1+2\theta' \approx 2\theta'$ (for $\theta' \gg 1$),

$$\bar N_{c\,pl} \approx \bar a^{1/2}\rho^{1/6}\bar m^{\alpha-2/3}\xi(M_\odot)^{1/2-\alpha}(1+\delta)^{\alpha/2}(\pi/6)^{1/6}/2 \propto k_\xi\bar e,$$
$$T_{\phi\,pl} \approx \bar a^2\rho^{1/6}(\xi/\bar\varkappa)k_T(M_\odot)^{1/2}(\pi/6)^{1/6}T_\oplus/2\bar m^{2/3}N(a_\oplus)^{3/2} \propto \bar m^{1/3}/M_\Sigma.$$
$$T_{\phi 3d} \approx \bar a^{5/2}\rho^{1/3}\xi^2\bar i k_T\pi(\pi/6)^{1/3}M_\odot T_\oplus/4\bar m^{4/3}N\bar\varkappa\eta_i(a_\oplus)^{3/2} \propto \bar m^{-1/3}/M_\Sigma. \tag{11}$$

For the "particle in a box" model,[9,19] $N_{col}^{(k,\ell)} \approx N^{(k)}N^{(\ell)}v_{rel}\pi(r_e)^2\Delta t/V$, where V is the volume of space while v_{rel} is determined through $\bar e$ and $\bar i$. The values of N_c and N_{col}, determined using Eqs. (2)-(9), also depend on $\Delta\phi$, T_s, and κ, which are determined, in turn, by the orbital elements of the bodies encountering each other. The different values of the semimajor axis of the orbits of the bodies are thereby taken into account, in contrast to Refs. 9 and 19. Usng the equations obtained above, we can analyze, in particular, the cases of $i = 0$ and $e = 0$. The algorithm whose basic principles were discussed above enables us to investigate not only the initial stage, as in Refs. 9 and 19, but also subsequent stages of solid-body accumulation of the planets (or of their cores). If each group consists of only one body, then such an algorithm is analogous to the algorithm used for computer modeling of the evolution of disks consisting of hundreds of gravitating bodies.[17-18,20] In studying the interaction between single large bodies, instead of using the stochastic approach one can determine the times until encounters between pairs of bodies and subsequently model (with allowance for encounters that have already occurred) the encounters between bodies occurring in the time interval Δt under consideration.

4. TIMES OF FALL OF SMALL BODIES INTO A LARGER BODY (PLANET)

When the k-th group consists of one large body (planet) of mass M and the ℓ-th group consists of N bodies of mass m, in a time Δt about $N\Delta t/T_\phi^{\ell,k}$ bodies fall onto body M. Using the results of Sec. 3 in the case of $\bar e = \bar e^{(\ell)} \gg \bar e^{(k)}$ and $\bar i = \bar i^{(\ell)} \gg \bar i^{(k)}$ we have

$$T_{\phi\,pl}^{(l,k)} \propto M^{-1/3}/\bar\varkappa, \quad T_{\phi 3d}^{(l,k)} \propto \bar i M^{-2/3}/\bar\varkappa \quad \text{for} \quad 1+2\theta' \approx 1 \tag{12}$$

and

$$T_{\phi pl} \propto \bar e M^{-2/3}/\bar\varkappa, \quad T_{\phi 3d} \propto \bar i\bar e^2 M^{-4/3}/\bar\varkappa \text{ for } 1+2\theta' \approx 2\theta'. \tag{13}$$

From Eqs. (12) and (13) it is seen, in particular, that, other conditions being equal, larger bodies grow faster. If the number of bodies m that can encounter the body M is proportional to e, then $\kappa \propto \bar e$ in Eqs. (12) and (13). In the case when all the bodies m can encounter body M (for $N_e^{(\ell,k)} = 1$), the mean time for a body m to fall onto body M is $T = T_\phi^{(\ell,k)}N_c$. If the inclinations of the orbits of the bodies m to the orbital plane of body M are the same and equal to Δi, then, with allowance or the results of Sec. 3, for eccentric orbits of the bodies m we find that in the plane case $\bar T_{pl} \approx \bar N_{c\,pl}2\pi T_s/k_\phi r_c^* \approx 2\pi T_s R/r_\Sigma k_\phi\sqrt{1+2\theta'}$, while for $\Delta i > r_c^*$, $\bar T_{3d} \approx \bar N_{c\,3d}2\pi^2 T_s\Delta i/k_\phi(r_c^*)^2 \approx 2\pi^2 T_s\Delta i(R/r_\Sigma)^2/k_\phi(1+2\theta')$.

Analytic formulas for the time of growth of a planet as it exhibits the bodies from its feeding zone are given in Refs. 2 and 8. Here the thickness $H = 2R\sin i$ of the zone and the eccentricities $\bar e$ of the orbits (the same for all the bodies) are determined by the mass M of the planet, which varies greatly in the course of evolution. In obtaining these formulas, it is assumed that θ does not vary with time, and that the surface density of material near the planet's orbit is the same as in the entire feeding zone (continuous mixing of bodies in the zone occurs). The formulas given in Sec. 4 were obtained for the case when M varies little over the in vestigated time interval, while e and $\sin i$ (or Δi) cannot depend on each other or on M.

Let us consider the fall of small bodies m_0 into a large body, a planet, of mass M ($M \gg m_0$) and radius r in the case when the feeding zone of body M has the form of a torus, formed by rotation about the Oz axis of part of a circular ring, bounded by two arcs of circles of radii $R_1 = R(1 - \delta_1)$ and $R_2 = R(1 + \delta_2)$ and by two segments of rays with the apex at the sun. We take the angle between these rays to be 2i and we assume that the bisector of this of this angle is the Oy axis, perpendicular to Oz. The volume of this zone is $V_T \approx 4\pi[(R_2)^3 - (R_1)^3] \cdot \sin i/3$. We designate the total mass of the bodies m_0 lying inside such a torus as m^0_Σ. We shall assume that the ratio λ of the mass of the bodies thrown into hyperbolic orbits or falling into other planetary embryos to the mass $m^!_\Sigma$ of bodies falling into body M is constant in the course of evolution. We assume that near the orbit of body M the spatial density of bodies of the swarm is $\rho_0 = \kappa' \rho_{cs}$, where $\rho_{cs} = [m^0_\Sigma - m^!_\Sigma(1 + \lambda)]/V_T$ and $\kappa' = \text{const}$. Then, using the well-known[2,8] relation $dM/dt = \pi(r_e)^2 \rho_0 v^M_{rel}$, for $i > r_e/R$ and nearly constant values of M, $\sin i$, and v^M_{rel} (and hence of $\bar{e}$ and ξ) in the course of evolution, we find that the time in which bodies m_0 with a total mass $k_s m^0_\Sigma(1 + \lambda)$ fall into body M is $k_s m_\Sigma^0/(1+\lambda)$, $T(k_s) \approx -2(\delta_1+\delta_2)\sin i (R/r)^2 P \ln|1-(1+\lambda)k_s|/\pi(1+2\theta')\xi\kappa$, where P is the period of revolution of body M around the sun.

The formulas given in this section can be used to determine the total mass of bodies m that fell into body M (not necessarily a planetary embryo) in a time Δt, as well as to determine the time of existence of bodies in the vicinity of a nearly formed planet. While bodies are falling into a planet $\theta' \approx \theta = (4\pi/3)^{1/3} \rho^{1/3} M^{2/3} a/\xi^2 M_\odot$ (Ref. 23) and $\delta_1 \approx \delta_2 \approx e$, where ρ is the density of the planet and a is the semimajor axis of its orbit. If $\bar{e} = \sqrt{2}\xi$ (Ref. 23), $\lambda = 0$, and $k_s = 0.97$, then for the earth $T(k_s) \approx 3 \cdot 10^9 \sin i/(1 + 2\theta)$ yr, while for Neptune $T(k_c) \approx 3 \cdot 10^3 \sin i(1 + 2\theta)$ yr. In this case, in particular, for $\bar{e} = 0.2$ in the terrestrial zone $T(k_s) = 5 \cdot 10^8 \sin i$ yr, while for $\bar{e} = 0.4$ in the zone of Neptune $T(k_s) = 2 \cdot 10^{11} \sin i$ yr. If $T_s = 5P$, then for these planets and values of $\bar{e}$ we have $\bar{T}_{3d} = 4 \cdot 10^8 \Delta i$ (for $k_\phi = 20$) and $\bar{T}_{3d} = 2 \cdot 10^{11} \Delta i$ (for $k_\phi = 10$), resectively. The estimates given above indicate that in the zone of Neptune individual planetesimals may still exist, moving in inclined and eccentric orbits. The extended gaseous envelopes possessed earlier by the giant planets[12] could contribute to greater efficiency of capture of planetesimals (smaller values of $\bar{T}$ and $T(k_s)$).

5. TIMES OF EVOLUTION OF DISKS CONSISTING OF HUNDREDS OF BODIES

The distribution of bodies of an evolving disk as a function of their masses and orbital elements was investigated earlier[3,14-18] by computer modeling of the evolution of disks initially consisting of hundreds of bodies. The mutual gravitational influence of the bodies was taken into account mainly by the method of spheres of action, and it was assumed that colliding bodies combine. In this case the bodies were not divided up into groups, pair-wise interactions of all the bodies were taken into account, and the algorithm described in Refs. 14 and 20 was used. In this section we give the time T of evolution of these disks (up to the last collision between bodies). The value of T can vary by an order of magnitude with variation of the pseudorandom numbers used to determine the positions in their orbits of the bodies encountering each other.[20] Besides T, therefore, we also consider the values of $T^{(25)}$, the time of evolution until 25 bodies remain in the disk.

For the three-dimensional model (with $\bar{i} = \bar{e}/2$) in the terrestrial zone, with a mass M_Σ of solid material in the feeding zone of the planet equal to $m_\oplus$, we have $T^{(25)}_{3d} \approx 10^7$ yr and $T_{3d} \approx 5 \cdot 10^7\text{-}5 \cdot 10^8$ yr, while in the zone of Uranus and Neptune with $M_\Sigma = 200\ m_\oplus$, we have $T^{(25)}_{3d} \approx 5 \cdot 10^7$ yr and $T_{3d} \approx 3 \cdot 10^8\text{-}10^{11}$ yr. Here, as below, the numerical estimates are made for the case when the density of bodies in the terrestrial zone is ~5.6 g/cm³, while in the zone of Neptune it is ~2 g/cm³. The results obtained, like the data of Ref. 11, indicate that the time it takes for the individual bodies from its feeding zone to fall into the earth may reach $5 \cdot 10^8$ years. In a computer-aided investigation of the evolution of three-dimensional disks initially consiting of nearly formed planets and several hundred bodies in the zone of Uranus and Neptune,[3] the time it took for most of the bodies to fall into these planets was $T_{UN} \approx 10^9$ yr. Although the evolution of the disk of planetesimals had a three-dimensional character,[2] for comparison we note that $T_{UN} \approx 10^7$ yr in the plane case. When the initial mass of the plane disk of approximately equal bodies wss close to the mass of solid material in the corresponding planets, for the terrestrial zone $T^{(25)}_{3d} \approx 10^4$ yr and $T_{pl} \approx 2 \cdot 10^4\text{-}10^5$ yr (here not one but several small planets are formed[15]), while for the zone of Neptune $T^{(25)}_{3d} \approx 10^6$ yr and $T_{pl} \approx 3 \cdot 10^6\text{-}3 \cdot 10^7$ yr.

6. TIME OF EVOLUTION OF A DISK CONSISTING OF APPROXIMATELY EQUAL BODIES

Using the formulas given in Sec. 3, we can make certain analytic estimates of the time of evolution of disks for a number of the simplestmodels. In Sec. 6 we consider an auxiliary model in which the masses m of the bodies of the evolving disk are always approximately equal to each other (although they change in time), and in which the bodies combine in collisions. In that case we can treat the entire feeding zone of a planet as one group and use Eqs. (10) and (11), taking $\kappa \propto e$. Such a model does not occur in reality, but we can use it to investigate more complex models, treating the approximately equal bodies of the disk as one of the groups. For a model of equal bodies, in a time T_ϕ the bodies combine in pairs and the masses of the bodies double.

As can be seen from Eqs. (10) and (11), $T_{\phi pl}$ grows with an increase in m. At the end of evolution $\theta > 1$ (Refs. 2 and 23) and $T_{\phi pl}$ does not depend on e for $\bar{\kappa} \propto eT_{\phi pl}$. Therefore, most of the time of evolution of a plane disk is spent in the end stages of evolution. In the terrestrial zone this statement is also valid for a three-dimensional model, since in this case $T_{\phi 3d}$ is larger for larger m. For a disk of different bodes it follows from this, in particular, that in the terrestrial zone the majority of the initial planetesimals were not preserved up to the final stage of formation of this planet. In the case when the influence of gravitational interactions of the bodies on the variation of $\bar{e}$ and $\bar{i}$ is greater than that of the gas, for the three-dimensional model in the zone of the giant planets e and i can reach a maximum rather rapidly, while, apart from the initial stage of evolution, $\theta > 1$ (Refs. 1 and 25). Therefore (see Eq. (11)), in this case $T_{\phi 3d}$ can decrease in the couse of evolution. For example, if in the zone of Neptune the mass of the disk is $M_\Sigma = 100\ m_\oplus$, $\bar{e} = 0.2$, $\bar{i} \approx \bar{e}/\sqrt{2}$, and $\rho = 2$ g/cm^3, the maximum of $T_{\phi 3d} \approx 4\cdot10^9$ yr is reached for $m \approx 10^{-4}$-$0.01\ m_\oplus$, while for $m = m_\oplus$ and $m = 10^{-6}\ m_\oplus$ the values of $T_{\phi 3d}$ are lower than this maximum by a factor of two to three. For $\bar{e} = 0.4$, $T_{\phi 3d}$ is about four times larger than for $\bar{e} = 0.2$. Therefore, in the evolution of a disk of different bodies combining in collisions, bodies with masses of $\sim10^{-4}$-$0.01\ m_\oplus$ could exist in the final stages of the accumulation of Neptune.

Using Eqs. (7), (10), and (11), we can compare the rate of growth of bodies in the feeding zones of different planets. If the ratio of the total masses of the planetesimals in the zones of Jupiter and Earth is $M_\Sigma{}^R \geq 5^{5/2}$, while $\theta \gg 1$, then in the case of a thre-dimensional gasless model with equal values of m and e, the values of $T_{\phi 3d}$ are smaller for the zone of Jupiter, and hence the rate of growth is greater, than in the terrestrial zone. Since $T_{\phi 3d} \propto m^{-1/3}/M_\Sigma$ for $\theta \gg 1$, in this case allowance for the larger bodies in the zone of Jupiter than in the terrestrial zone only strengthens this statement. If $1 + 2\theta \approx 1$, then with $M_\Sigma{}^R \leq 5^{7/2}$ for the three-dimensional model and with $M_\Sigma{}^R \leq 5^{5/2}$ for the plane model, the values of T_ϕ in the zone of Jupiter are larger (for the same m and $\bar{e}$). The drag of the gas decreases $\bar{e}$ and i (Ref. 7). According to Ref. 13, $\bar{e} \approx \bar{i}\sqrt{2}$. Therefore, with allowance for Eqs. (19) and (11), we find that (except for the plane case with $1 + 2\theta \approx 1$) gas drag decreases T_ϕ and hence the time of evolution. Gas probably disappeared earlier from the terrestrial zone than from the zone of Jupiter. Therefore, allowance for the influence of gas only accelerates the relatively faster (in comparison with the terrestrial zone) growth of bodies in the zone of Jupiter. This also results from allowance for the fact that θ may be larger in the zone of Jupiter than in the terrestrial zone. The results obtained indicate that the embryo of Jupiter with a mass of $\sim$2-3 $m_\oplus$ (capable of accreting gas) could have formed before the accumulation of the earth ended. The maximum value of e for bodies moving in a gas are estimated in Refs. 18 and 23.

7. TIMES OF EVOLUTION OF DISKS CONSISTING OF DIFFERENT BODIES

Let us consider a gasless model of the evolution of a disk consisting of large bodies m and smaller bodies m'. We designate their total masses as m_Σ and $m_\Sigma^!$, respectively, and their mean eccentricities as e(m) and e(m'). We shall assume that for each time under consideration, the masses of the bodies m are approximately equal to each other. Let $T_{m,m'}$ be the mean time between collisions of two bodies m and m', while $k' = T_{m,m'}/T_{m,m}$. Then for a model in which bodies combine in collisions, the rate of growth of the bodies m is proportional to $[m_\Sigma 2\bar{e}(m) + m_\Sigma'(\bar{e}(m) + \bar{e}(m'))/k']/T_{m,m}$. If the mean eccentricities and inclinations of the orbits of bodies in the variants being compared are about the same for bodies m and m' (this condition can be satisfied for $m_\Sigma^! < m_\Sigma$), as well as for different variants, then $k' > 1$ and the time in which the masses of the bodies m double is $T_* = T_\phi(m_\Sigma, m)k'm_\Sigma/(k'm_\Sigma + m_\Sigma') \approx T_\phi(M_\Sigma, m)c_k$, where $c_k = k'/[1 + (k'-1)m_\Sigma/M_\Sigma]$, $M_\Sigma = m_\Sigma + m_\Sigma$, while $T_\phi(m_\Sigma, m)$ is the value of T_ϕ for the case when the disk consists of bodies m whose total mass is m_Σ. For such a model $k' \leq 2$ in the plane case and $k' \leq 4$ in the three-dimensional case and for M_Σ = const, T_* is of the same order as for a model disk of bodies that are always approximately equal.

Let us consider the model of evolution of a disk, consisting of bodies m and m', in which bodies m colliding with each other are converted into bodies m', and bodies m' colliding with bodies m always combine with them. Then, under the assumption that the mean eccentricities and inclinations of the orbits of bodies m and m' are approximately the same, and in the absence of the ejection of bodies into hyperbolic orbits, we can show that the disk evolves to a state in which $m_\Sigma^! = k'm_\Sigma$. If this relation is satisfied, then for the same values of M_Σ, the bodies m grow more slowly by a factor of $M_\Sigma/M_\Sigma = k' + 1$ than in the case when the disk consists only of equal bodies m.

The results given above indicate that in the evolution of a disk of different bodies, for a model allowing for fragmentation of the bodies and gas drag, the times of growth of the largest bodies of the disk do not exceed (by more than a factor of five, in any case) the times of growth of bodies for the model of approximately equal bodies if the masses of the disks are the same in both cases. Bodies with larger masses grow faster, Therefore, allowance for the differentiation of the bodies by mass only strengthens the statement made above. Although planetary embryos may be destroyed in collisions with large bodies, breaking up and partially vaporizing, in the opinion of Cox and Lewis[10] a considerable part of the fragments can be collected into one body again under the influence of gravity. Even while losing some mass in collisions with large bodies, however, planetary embryos could grow, on the whole, due to the accumulation of small bodies.

8. CONCLUSION

An algorithm for computer modeling of the evolution of a disk initially consisting of a large ($\geq 10^4$) number of bodies, the basic design of which was discussed in the present paper, makes it possible to study only the initial stage, as in Refs. 9, 13, and 19, but also subsequent stages of solid-body accumulation of planets (or their cores).

We obtained the analytic dependence of the time of evolution of a disk on the number, masses, and mean eccentricities and inclinations of the orbits of the bodies comprising. On the basis of these functions, as well as computer modeling of the evolution of disks of hundreds of bodies, the characteristic times of evolution of disks consisting of a considerably larger number of bodies were studied for a number of models. The times for the falling of small bodies into large bodies and planetary embryos were studied. The estimates obtained for the time of accumulation of the main mass of the planets are close to the results of Safronov, Vityazev, and Pechernikova[1,2,8] ($\sim10^8$ years for Earth and $\sim10^8$ years for Neptune). In the zone of Neptune certain planetsimals, moving in inclined and eccentric orbits, might be preserved up to the present, and some bodies with masses of $\sim10^{-4}$-$0.01\ m_\oplus$ could survive up to the final stages of formation of the planet.

[1]V. S. Safronov and A. V. Vityazev, Itogi Nauki Tekh., Astron. 24, 5 (1983).
[2]V. S. Safronov, Evolution of the Protoplanetary Cloud and the Formation of the Earth and Planets [in Russian], Nauka, Moscow (1969) [NASA Tech. transl. F-677, Washington, D.C. (1972)].
[3]S. I. Ipatov, "Numerical studies of the evolution of three-dimensional rings of gravitating bodies corresponding to the feeding zones of the giant planets," Preprint No. 1, Inst. Prikl. Mat., Acad. Nauk SSSR, Moscow (1984).
[4]V. S. Safronov and E. L. Ruskol (Ruscol), Icarus 49, 284 (1982).
[5]T. M. Eneev and N. N. Kozlov, Pis'ma Astron. Zh. 5, 470 (1979) [Sov. Astron. Lett. 5, 252 (1979)].
[6]G. W. Wetherill, Ann. Rev. Astron. Astophys. 18, 77 (1980).
[7]Protostars and Planets II, D. C. Black and M. S. Matthews (eds.), Univ. Arizona Press, Tucson (1985).
[8]A. V. Vityazev, G. V. Pechernikova, and V. S. Safronov, Astron. Zh. 55, 107 (1978) [Sov. Astron. 22, 60 (1978)].
[9]R. Greenberg, J. F. Wacker, W. K. Hartman, and C. R. Chapman, Icarus 35, 1 (1978).
[10]L. P. Cox and J. S. Lewis, Icarus 44, 706 (1980).
[11]G. W. Wetherill, Geol. Assoc. Can., Spec. Pap., No. 20, 3 (1980).
[12]J. A. Fernandez and W.-H. Ip, Icarus 58, 109 (1984).
[13]Y. Nakagawa, C. Hayashi, and K. Nakazawa, Icarus 54, 361 (1983).
[14]S. I. Ipatov, Astron. Zh. 58, 1085 (1981) [Sov. Astron. 25, 617 (1981)].
[15]S. I. Ipatov, "Numerical studies of the accumulation of planets of the terrestrial group," Preprint No. 144, Inst. Prikl. Mat., Akad. Nauk SSSR, Moscow (1982).
[16]S. I. Ipatov, "Numerical studies of a plane model of the accumulation of cores of the giant planets," Preprint No. 117, Inst. Prikl. Mat., Akad. Nauk SSSR, Moscow (1983).
[17]S. I. Ipatov, Astron. Vestn. 21, 207 (1987).
[18]S. I. Ipatov, Earth, Moon, Planets 39, 101 (1987).
[19]R. Greenberg, S. J. Weidenschilling, C. R. Chapman, and D. R. Davis, Icarus 59, 87 (1984).
[20]S. I. Ipatov, "Evolution of a plane ring of gravitating particles," Preprint No. 2, Inst. Prikl. Mat., Akad. Nauk SSSR, Moscow (1978).
[21]H. R. Aggarwal and V. P. Oberbeck, Astrophys. J. 191, 577 (1974).
[22]H. Mizuno and A. P. Boss, Icarus 63, 109 (1985).
[22]S. I. Ipatov, "Evolution of orbital eccentricities in the initial stage of solid-body accumulation of the planes," Preprint No. 4, Inst. Prikl. Mat., Akad. Nauk SSSR, Moscow (1985).

Translated by Edward U. Oldham

Properties of the Drop Model of a Protoplanetary Disk (abstract)

E. M. Levin
Translated from: Astron. Zh. **65**, 73 (1988)
Sov. Astron. **32**, 38 (1988)

The evolution of the drop model of a protoplanetary disk—a swarm of spherical bodies moving in the same plane in circular orbits around an attracting center and combining in collision—is analyzed on the basis of the ideas and results of T. M. Éneev and N. N. Kozlov within the framework of coagulation theory. The coagulation equation is derived and its analytic solution is given, which reproduces the following dynamic features of the evolution: (1) the generation of annular standing waves of density as a form of development of instability of the differentially rotating disk of accumulating bodies; (2) arrangement of the boundaries of the zones of compression and rarefaction in a geometrical progression; (3) a close connection between the radial redistribution of mass and the accumulation of angular momentum of the proper rotation of the bodies; (4) the predominance of direct proper rotations; (5) proportionality of the specific angular momenta of the bodies to their masses to the 2/3 power, which results in isochronism of the rotations.

Evolution of dust clusters in the preplanetary disk

G. V. Pechernikova and A. V. Vityazev

O. Yu. Shmidt Institute of Earth Physics, USSR Academy of Sciences

(Submitted April 4, 1986)

Astron. Zh. **65**, 58–72 (January–February 1988)

Following the settling of dust in the circumsolar preplanetary disk and gravitational instability in the dust subdisk, dust clusters were formed. The relation between the size, mass, density, and angular momentum of the clusters is investigated. The time of their growth and conversion into compact bodies is estimated. The largest of them could reach nearly planetary masses.

1. INTRODUCTION

Models of the formation and evolution of the circumsolar gas–dust disk and the formation of planets in it are attracting the interest of an ever greater number of investigators. Models of a disk with a moderate mass ($\lesssim 0.1\ M_\odot$), momentum ($\lesssim 3 \cdot 10^{52}$ g·cm·sec^{-1}), and turbulence are receiving the greatest recognition today.[1,2] A description of models of the preplanetary disk that formed around the young sun and the distribuitons of density, pressure, temperature, and other parameters in it can be found in Ref. 3. It has been shown in a number of papers (see, e.g., the review Ref. 3) that, following the damping of turbulence and the settling of dust toward the central plane of the disk, the resulting dust subdisk, upon reaching a certain critical density ρ_{cr}, breaks up into dust clusters with masses $m_0 \approx 8\sigma_d^3(R)/\rho_{cr}^2$, where $\sigma_d(R)$ is the surface density in the dust subdisk at the given distance R from the sun, $\rho_{cr} \approx 2\rho^*$, and $\rho^* = 3M_\odot/4\pi R^3$ is the "smeared out density of the sun."[4,5] Naturally, the problem arose of their further evolution and the mechanisms and time of conversion into solid bodies. In early work it was assumed that compression of the clusters occurred due to slow rotation by the solar tide. The duration of such evolution is very long, however, because of the small sizes of the clusters. According to Ref. 4, the combining of clusters in collisions led to their compression. According to the estimates obtained in Ref. 5, the time of conversion of the clusters into solid bodies is $\sim 10^4$ years for the terrestrial zone and $\sim 10^6$ years at the distance of Jupiter, while the corresponding increases in mass are $\sim 10^2$ and $\sim 10^3$ from the initial values. It was concluded that, although the time of evolution and conversion into bodies could differ considerably for individual clusters, the entire system of clusters as a whole was converted into a swarm of solid bodies in a cosmogonically short time.

A model of the formation of the planets from rarefied planetesimals, the material of which was assumed to have occupied the largest possible volume – the Hill sphere, was analyzed in Refs. 6 and 7. The mechanism capable of maintaining the protoplanets in the rarefied state during the entire time of accumulation was not specified. Because of the overstated value of the effective collision cross section adopted by the authors, a short time scale of formation of the planets ($\sim 10^3$–10^4 years), inconsistent with isotopic data,[3] was obtained by numerical modeling. Additional difficulties arise in such a model with the explanation of the internal structure of the primordial planets and the formation of satellites and other minor bodies of the solar system.

In the review Ref. 3, it was noted that noncentral collisions were the only mechanism capable of maintaining the planetesimals in a rarefied state. The combining of clusters leads to very efficient compression, on the average. Thus, in the case of the combining of two clusters of comparable masses colliding centrally, the mass practically doubles while the angular momentum remains as before: the radius of the cluster decreases twofold while the density increases 16-fold. With an increase in the masses of the clusters, the eccentricities and inclinations of their orbits increase, i.e., their relative velocities grow. As a result, cluster can acquire both positive and negative momentum in collisions, and the change in momentum can even exceed the momentum before the collision.[5] In this case, the clusters can collect sufficiently large masses for them to be converted into solid bodies. Before proceeding to a detailed examination of this process, let us dwell briefly on the mechanisms capable of the efficient compression of clusters in the earliest stages of their evolution.

2. PRECOLLISIONAL EVOLUTION OF CLUSTERS

The internal gravitational force of a cluster that has formed is greater than the external forces. Therefore, it starts to contract, until gravitation is balanced by centrifugal force, increasing during contraction.[5] The equatorial radius r of a cluster and its angular velocity ω of rotation before contraction (index 0) and after contraction (index 1) are connected by the condition of conservation of angular momentum:

$$r_0^2\omega_0 = r_1^2\omega_1, \quad r_0 \simeq \sigma_d/\rho_{cr}. \tag{1}$$

So long as the densities δ of the clusters are low ($\delta \ll 1$ g/cm^3), their velocities of rotation are close to the Keplerian velocity,

$$\omega^2 = \xi \frac{Gm}{r^3}, \tag{2}$$

where ξ is a coefficient that depends on the shape of a cluster and G is the gravitational constant.

If a region, lying inside the Hill sphere of the mass contained in it, contracts in the gravitational field of the sun, then its angular momentum is conserved and the average rotation proves to be forward, with

$$\omega_0 = \frac{1}{2}\,\mathrm{curl}\ V_c = \frac{1}{4}\,\omega_c, \tag{3}$$

where V_c and ω_c are the linear and angular ve-

 0038-5301/88/01 0031-08 $04.10

locities of revolution around the sun. From (1)-(3) it follows that the initial contraction of the cluster leads to a decrease in its initial radius by a factor of three to four and to an increase in density by an order of magnitude or more.[5] Gravitational instability in the dust layer can start before the complete settling of dust to the central plane; in addition, some of the dust may not go into the clusters that formed and may remain in a dispersed state.

The presence of a dispersed dust component after the formation of the clusters and its absorption by the clusters lead to acceleration of their growth in density and to faster conversion into solid bodies. We designate the fraction of the dust fallen to the central plane by the start of gravitational instability as ε_1, the fraction of the dust falling from large z coordinates with a characteristic sedimentation time $\tau_{sed} \gg \tau_{gr.inst.}$ as $(1 - \varepsilon_1)$, the fraction of dust that went into the dust clusters as ε_2, and the fraction of the dust remaining in the dispersed state by the end of the formation of clusters as $(1 - \varepsilon_2)$. Only the case of $\varepsilon_1 = 1$ and $\varepsilon_2 = 1$ was considered in Refs. 4-8.

Let $\varepsilon_1 \approx 1$ and $\varepsilon_2 \approx 1$. With a considerable fraction of the dust not having gone into the initial clusters, their contraction is determined by the absorption of this dust and the decrease in angular momentum due to the drag of the medium.

Let us consider the evolution of an individual cluster over times less than the characteristic time of collisions between clusters:

$$\tau_s = \frac{\lambda}{v} = \frac{1}{vn\pi r^2} = \frac{m}{v\varepsilon_2\rho_d\pi r^2} \approx \frac{mP}{4\sigma_d\varepsilon_2\pi r^2}. \tag{4}$$

Here λ is the mean free path, v is the relative velocity, n is the number density, respectively, of the clusters, and ρ_d is the density in the dust subdisk $\left(\rho_d = \frac{4\sigma_d}{Pv}\right.$, where P is the period of revolution around the sun). The characteristic time τ_f of variation of the angular momentum due to the drag of the medium can be estimated from the equation

$$\frac{d}{dt}\frac{2}{5}m\omega r^2 = -8\pi\eta\omega r^3, \tag{5}$$

from which we have, in order of magnitude,

$$\tau_f \approx \frac{m}{20\pi r\eta}, \tag{6}$$

where the viscosity coefficient n in the case of a nonturbulent gas is $\eta_g \approx 10^{-3}$ P, but it may be two to three orders of magnitude greater with moderate turbulence, maintained, e.g., by mixing of the gas by randomly moving clusters with an average relative velocity $v \sim \sqrt{\frac{Gm}{2r}}$. In the case of a pronounced dispersion of the velocities v_d of dust particles, the drag exerted by the dispersed dust component becomes important. It has the order of magnitude $\eta_d \approx a_d\delta_d v_d$, and a_d is the characteristic size, δ_d is the density, and v_d is the relative velocity of the grains. If we take $a_d \approx 0.1$ cm, $v_d \approx 10$ cm/sec, and $\delta_d \approx 1$ g/cm^3, then $\eta_d \approx 10^3\eta_g \approx 1$ P, and for the initial clusters in the terrestrial zone, we have

$$\tau_{f_0} = \frac{m_0}{20\pi r_0\eta} \approx P\left[\frac{1\text{P}}{\eta}\right]\left[\frac{\rho^*}{\rho_{cr}}\right]. \tag{7}$$

The characteristic time τ_m of doubling of the mass of a cluster due to the absorption of dispersed dust can be estimated from the growth equation

$$\frac{dm}{dt} = \pi r^2(1-\varepsilon_2)\rho_d v, \tag{8}$$

from which

$$\tau_{m_0} \simeq \frac{m_0}{\pi r_0^2(1-\varepsilon_2)\rho_d v} \approx \tau_{s0}\frac{\varepsilon_2}{1-\varepsilon_2} \propto \frac{P}{1-\varepsilon_2}. \tag{9}$$

Equations (5) and (8), with allowance for (1) and (2) with the given ε_2 and η, form a closed system of equations for the unknowns m(t), r(t), $\varepsilon_2(t)$, and $\omega(t)$, having solutions in quadratures. These solutions can be used only at times $t \lesssim \tau_s$, however. An analysis shows that in the case of $\varepsilon_2 \lesssim 0.1$ and $\eta \gtrsim 1$ P, the initial clusters are converted in a time $\tau \lesssim \tau_s$ into slowly rotating, sufficiently dense bodies ($\delta \approx 0.1$-1 g/cm^3) with masses exceeding the masses of the initial clusters by a factor of ε_2^{-1}.

In the case of $\varepsilon_2 \approx 1$ and $\varepsilon_1 \ll 1$, the influence of the dispersed dust component is determined by the intensity of its entry into the system of clusters, and the increase in the mass of the clusters due to the absorption of the settling dust will be

$$\frac{dm}{dt} \approx \frac{\pi r^2(1-\varepsilon_1)\sigma_d}{\tau_{sed}}. \tag{10}$$

For $\varepsilon_2 \approx 1$, we have $\tau_s \approx \tau_m$ and we must allow for the consequences of the collisions between clusters, the growth of the average mass of a cluster, and the variation of the radius, density, and velocity of rotation. Below we consider the evolution of the clusters under the assumption $\varepsilon_1 = \varepsilon_2 = 1$; the influence of small ε_1 and ε_2 is discussed at the end of the paper.

3. COLLISIONAL EVOLUTION OF CLUSTERS

The momentum of a cluster associated with its rotation is

$$K = \frac{2}{5}\mu(\xi Gmr)^{1/2}m, \tag{11}$$

where μ is the coefficient of nonuniformity. We assume, in accordance with Ref. 5, that the momentum of a cluster is determined only by its mass,

$$K \propto m^p, \tag{12}$$

where p is a quantity dependent on the parameters of the cluster [shape, degree of nonuniformity $\delta(r)$, etc.]. Then from (11) we obtain

$$r \propto \frac{K^2}{m^3} \propto m^{2p-3} \tag{13}$$

and, eliminating r from (11), we write the square of the angular momentum of a cluster of mass m in the form

$$K^2(m) = \left(\frac{2\mu}{5}\right)^2 \xi G m_0^3 r_0 \left(\frac{m}{m_0}\right)^{2p}, \tag{14}$$

where m_0 and r_0 are the initial mass and radius of the cluster.

Let us estimate the variation of the angular momentum K of a cluster of mass m in the process of its growth by the attaching of small clusters of

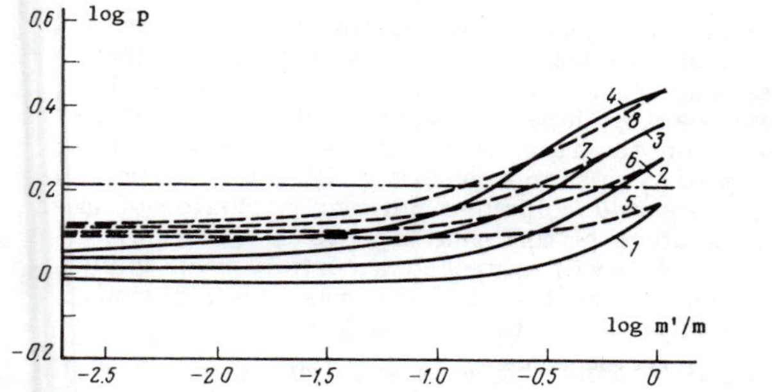

FIG. 1. Dependence of the parameter p on m'/m. p(m'/m) for massive clusters are shown by solid lines: 1) B_2 = 1.56; 2) 3.12; 3) 6.25; 4) 12.5; p(m'/m) for low-mass clusters are shown by dashed ines: 5) B_1B_2 = 1.56; 6) 3.12; 7) 6.25; 8) 12.5. The dash-dot line corresponds to the value p = 5/3.

mass m'. A cluster of mass m', in falling onto a cluster of mass m, imparts to it an orbital angular momentum $\mathbf{K}_1$ relative to the center of the cluster of mass m and a momentum $\mathbf{K}_2$, associated with its own rotation. The orbital momentum is determined by the relative velocity V before the encounter and by the impact parameter ℓ:

$$K_1 = lVm'. \tag{15}$$

Let ℓ_0 be the maximum impact parameter for which the colliding clusters can combine. Then the mean value of $\bar{\ell}^2$ is

$$\overline{l^2} = \frac{1}{\pi l_0^2}\int_0^{l_0} l^2 2\pi l\, dl = \frac{l_0^2}{2}. \tag{16}$$

Within the framework of the two-body problem, the cross section for the collision of a body of mass m with a body of mass m', having a relative velocity V before encounter, is written in the form

$$s = \pi l_0^2 = \pi d^2\left[1 + \frac{2G(m+m')}{V^2 d}\right], \tag{17}$$

where ℓ_0 is the impact parameter corresponding to a grazing collision with a distance d = r + r' between the centers of the bodies at the time of their closest approach. Since it is difficult for clusters to combine in nearly grazing impacts, we assume that clusters can combine for $d_0 = \beta_0(r + r')$, where r and r' are the radii of the colliding clusters and $\beta_0 < 1$. Then from (16), with allowance for (17), for ℓ_2 we have

$$\overline{l^2} = \frac{1}{2}\beta_0^2(r+r')^2\left[1 + \frac{2G(m+m')}{\beta_0(r+r')V^2}\right] \tag{18}$$

and from (15) we obtain

$$K_1^2(m, m') = \frac{1}{2}\beta_0^2(r+r')^2\left[1 + \frac{2G(m+m')}{\beta_0(r+r')V^2}\right]V^2 m'^2. \tag{19}$$

Let the square of the mean relative velocity of the two clusters be

$$V^2 \approx 2\overline{v^2} = \frac{Gm^*}{\theta r^*}, \tag{20}$$

where m* and r* are the effective mass and radius of clusters, determining the relative velocities in the zone, and θ is a parameter of the order of several units.[5] Then clusters for which the condition

$$\frac{2G(m+m')}{\beta_0(r+r')V^2} \leqslant 1 \tag{21}$$

is satisfied, i.e., the collision cross section can be taken as the geometrical cross section, will be called small (low-mass) clusters. Using (12), (13), and (21), we convert this condition to the form

$$\frac{\beta_0}{\theta}\frac{r}{m}\frac{m^*}{r^*} = B_1 \geqslant 1. \tag{22}$$

For small clusters, for which gravitational focusing is unimportant, from (19) with allowance for (22) we obtain

$$K_1^2(m, m') \approx \frac{1}{2}\beta_0^2(r+r')^2 m'^2 V^2, \tag{23}$$

and for the largest clusters, the velocity of escape from which exceeds V, we can write

$$K_1^2(m, m') \approx \beta_0 G(m+m')(r+r')m'^2. \tag{24}$$

The intrinsic angular momentum of a cluster of mass m' is, from (14),

$$K_2^2(m') = \left(\frac{2\mu}{5}\right)^2 \xi G m_0^3 r_0 \left(\frac{m'}{m_0}\right)^{2p} \tag{25}$$

(under the assumption that ξ = const, μ = const, and p does not depend on the mass of a cluster). For random directions of impacts of clusters of mass m' incident on a cluster of mass m and random directions of the momenta $\mathbf{K}(m)$ and $\mathbf{K}_2(m')$, the imparted momenta are added by the law of random quantities: Their squares are summed. Then the change in the momentum of the cluster of mass m when a cluster of mass m' falls on it is determined by the expression

$$\Delta K^2 = K^2(m+m') - K^2(m) = K_1^2(m, m') + K_2^2(m'). \tag{26}$$

Substitution (14), (24), and (25) into (26), with allowance for (13) we obtain the equation for p,

$$\left(1+\frac{m'}{m}\right)^{2p} - \left(\frac{m'}{m}\right)^{2p} = B_2\left(1+\frac{m'}{m}\right)\left(\frac{m'}{m}\right)^2\left[1+\left(\frac{m'}{m}\right)^{2p-3}\right] + 1, \tag{27}$$

where B_2 is a factor that depends on the parameters

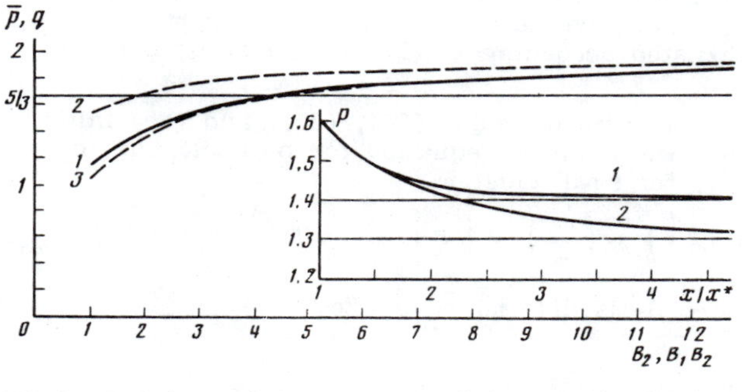

FIG. 2. Dependence of the exponent q of the mass distribution of clusters and of the parameter $\bar{p}$ on B_1 and B_2. The values of B_2 for massive clusters and of B_1B_2 for low-mass clusters are laid out along the abscissa axis. Curve 1 shows $\bar{p}(B_1B_2)$ = $q(B_2)$ for massive clusters; curve 2 shows $\bar{p}(B_1B_2)$ and curve 3 shows $q(B_1B_2)$ for small clusters. The inset shows the decrease in the parameter p in the process of growth of the largest cluster for the case of $p(x^*)$ = 1.6 at the time t* [$x^* = m(t^*)/m_0$]. Curve 1 was obtained from (47) with $\bar{p}$ = 1.4 while curve 2 was obtained with $\bar{p}$ = 1.3.

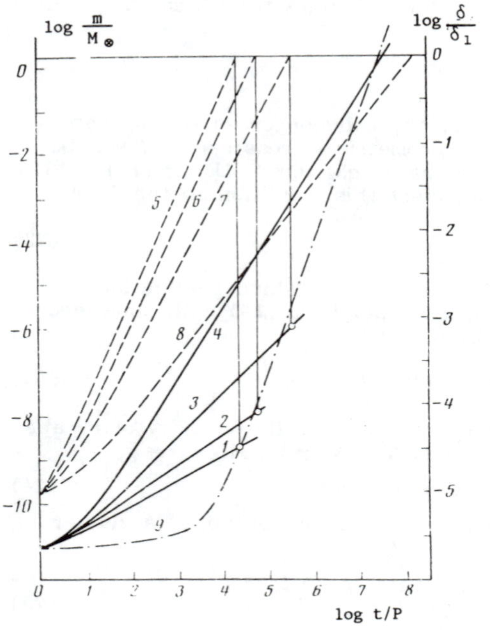

FIG. 3. Growth of the mass m(t) of a cluster and of its density δ(t) with time as functions of $\bar{p}$. Solid lines show m(t) for: 1) $\bar{p}$ = 1.3; 2) 1.4; 3) 1.5; 4) 1.6; dashed lines show δ(t) for: 5) $\bar{p}$ = 1.3; 6) 1.4; 7) 1.5; 8) 1.6. The growth of the mass of a solid body with a density δ = 1 g/cm³ is shown by a dash-dot line 9.

of a cluster:

$$B_2 = \frac{25\beta_0}{4\mu^2\xi}.$$

In Table I we give values of B_2 for $\beta_0 = {}^1/_2$ and certain values of μ and ξ.

The value of p determined by Eq. (27) gives the value of the exponent in (12) if a growing cluster of mass m is increasing its axial angular momentum K through the relative orbital momentum K_1 and intrinsic angular momentum K_2 of the joined cluster of mass m', with the exponent p being the same for all three clusters (m, m', and m + m'). The same thing pertains to the parameter B_2. In Fig. 1 we give the dependence of p on m'/m, calculated according to (27) for certain values of B_2.

Substituting Eqs. (14), (23), and (25) into (26), we obtain an equation for p, analogous to (27), for small clusters,

$$\left(1+\frac{m'}{m}\right)^{2p} - \left(\frac{m'}{m}\right)^{2p} = B_1 B_2 \left[1+\left(\frac{m'}{m}\right)^{2p-3}\right]^2 \left(\frac{m'}{m}\right)^2 + 1, \tag{28}$$

where B_1 is determined by Eq. (22).

The dependence p(m'/m) for small clusters with certain values of B_1B_2 is shown by dashed curves in Fig. 1. It is seen from the figure that the exponent p depends rather weakly – logarithmically – on the ratio m'/m, increasing somwhat for clusters of comparable masses. It can be expected that in repeated collisions between a cluster of mass m and clusters of different masses m' (m' ≲ m), the value of p will approach a certain mean value $\bar{p}_m$, which can be found if the mass distribution function n(m, t) of the clusters is known.

We determine the mass spectrum n(m, t) of the clusters using coagulation theory. Let the mass m of a cluster grow through joining with other clusters of smaller mass m' in collisions. We shall ignore the breakup of clusters or partial mass loss in nearly grazing collisions or in close passages of other clusters. Then the variation of the mass distribution function of the clusters with time can be described by an equation of the type

$$\frac{\partial n(m,t)}{\partial t} = \frac{1}{2}\int_{m_0}^{m} A(m-m', m')\,n(m-m', t)\,n(m', t)\,dm' - \int_{m_0}^{m_1} A(m, m')\,n(m, t)\,n(m', t)\,dm', \tag{29}$$

where A(m, m') is the coagulation coefficient and m_0 and m_1 are the lower and upper limits of the distribution. The coagulation coefficient A(m, m') is proportional to the frequency of collisions between bodies. We write it in the form

$$A(m, m') = A_0(m^\alpha + m'^\alpha). \tag{30}$$

then in the case of a geometrical collision cross section $s \propto r^2$ [cf. (18)], for solid bodies we have α = 2/3, while for clusters, using (13) and taking some mean value $\bar{p}$, we obtain $s_1 \propto r^2 \propto m^{2(2\bar{p}-3)}$, i.e., $\alpha_1 = 2(2\bar{p}-3)$. For the largest bodies, with allowance for their gravitation, the collision cross section is $s \propto mr$, i.e., α = 4/3 for solid bodies and hence $\alpha_2 = 2\bar{p} - 2$ for clusters.

An investigation of Eq. (29) showed that it has a stable, asymptotic, power-law solution of the type

$$n(m, t) \propto m^{-q}, \tag{31}$$

where the exponent q is determined by the form of the coagulation coefficient A(m, m') and does not depend on time. For a coagulation coefficient of type (30) with α < 2,

$$q = 1 + \frac{\alpha}{2} \tag{32}$$

was obtained.[8]

Substituting the value of α_1 into (32), for low-

TABLE I

ξ	1		3/2		2	
μ	1	1/2	1	1/2	1	1/2
B_2	3.125	12.5	2.083	8.333	1.562	6.25

TABLE II

Zone	$\bar{p}$	1.3	1.4	1.5	1.6
Terrestrial	$\log t/P_\oplus$	4.1	4.4	5.0	(7.5)
	$\log m/m_0$	2.3	3.1	5.0	(12.5)
Of Jupiter	$\log t/P$	5.7	6.1	7.0	(10.5)
	$\log t/P_\oplus$	6.8	7.2	8.1	(11.6)
	$\log m/m_0$	3.2	4.4	7.0	(17.5)
Of Uranus	$\log t/P$	7.4	7.9	(9.0)	(13.5)
	$\log t/P_\oplus$	9.3	9.8	(10.9)	(15.4)
	$\log m/m_0$	4.1	5.6	(9.0)	(22.5)
	$\log t_{cr}/P$	(7.7)	7.1	6.3	5.0

mass clusters we obtain

$$q_1=2\bar{p}-2, \tag{33}$$

whereas for massive clusters, the value of α_2 leads to

$$q_2=\bar{p}. \tag{34}$$

It was noted above that to find $\bar{p}_m$, one must know the mass distribution function $n(m, t)$ of the clusters. At the same time, to determine $n(m, t)$ in the solution of the coagulation Eqs. (29), one must know the dependence $p(m)$, appearing in the coagulation coefficient through α. For simplicity, we used only the mean value of the parameter p inside the integrals in (29). This does not lead to significant errors, since p depends relatively weakly on the ratio m'/m (see Fig. 1). And the dependence of p on m, manifested in the difference between q_1 and q_2, is also minor.

Now let us estimate $\bar{p}$ for growing clusters. With an increase in the mass m of the cluster by Δm, the increase in its momentum due to the angular momenta of all the possible clusters of mass m' [see (25)] will be

$$\Delta K_2^2=\left(\frac{2\mu}{5}\right)^2 \xi G m_0^3 r_0 \frac{\int\limits_{m_0}^{m_1}\left(\frac{m'}{m_0}\right)^{2\bar{p}} n(m',t)\,dm'}{\int\limits_{m_0}^{m_1} m' n(m',t)\,dm'}\Delta m.$$

For the power-law distribution (31), we obtain

$$\Delta K_2^2=\left(\frac{2\mu}{5}\right)^2 \xi G m_0^{3-2\bar{p}} r_0 \frac{2-q}{2\bar{p}-q+1} m_1^{2\bar{p}-1}\Delta m, \tag{35}$$

where m_1 is the upper limit of the mass distribution of the incident clusters.

Similarly, for the orbital momenta of clusters of mass m' acquired by a massive cluster, according to (24), we have

$$\Delta K_1^2=\beta_0 G m r \frac{\int\limits_{m_0}^{m_1}\left(1+\frac{m'}{m}\right)\left(1+\frac{r'}{r}\right) m'^2 n(m',t)\,dm'}{\int\limits_{m_0}^{m_1} m' n(m',t)\,dm'}\Delta m$$

$$=\beta_0 G m m_1 r\left[\frac{1}{3-q}+\frac{1}{4-q}\frac{m_1}{m}+\frac{1}{2\bar{p}-q}\left(\frac{m_1}{m}\right)^{2\bar{p}-3}+\frac{1}{2\bar{p}+1-q}\left(\frac{m_1}{m}\right)^{2\bar{p}-2}\right](2-q)\Delta m, \tag{36}$$

whereas in the case of low-mass clusters, in accordance with (23), we obtain

$$\Delta K_1^2=\frac{\beta_0^2 v^2}{2} r^2 \frac{\int\limits_{0}^{m_1}\left(1+\frac{r'}{r}\right)^2 m'^2 n(m',t)\,dm'}{\int\limits_{0}^{m_1} m' n(m',t)\,dm'}\Delta m$$

$$=\frac{\beta_0^2 v^2}{2} r^2 m_1\left[\frac{1}{3-q}+\frac{2}{2\bar{p}-q}\left(\frac{m_1}{m}\right)^{2\bar{p}-3}+\frac{1}{4\bar{p}-3-q}\left(\frac{m_1}{m}\right)^{4\bar{p}-6}\right](2-q)\Delta m. \tag{37}$$

We write the total increment in the square of the angular momentum of the cluster of mass m when its mass increases by Δm in the form

$$\Delta K^2=\Delta K_1^2+\Delta K_2^2. \tag{38}$$

For massive clusters, we substitute Eqs. (35) and (36) into (38) and integrate over m. For $m_1 = m$, we obtain

$$K^2(m)=\left(\frac{2\mu}{5}\right)^2 \xi G m_0^3 r_0\left(\frac{m}{m_0}\right)^{2\bar{p}}=\left[\left(\frac{2\mu}{5}\right)^2 \xi G\frac{2-q}{2\bar{p}-q+1}+\beta_0 G(2-q)\left(\frac{1}{3-q}+\frac{1}{4-q}+\frac{1}{2\bar{p}-q}+\frac{1}{2\bar{p}+1-q}\right)\right] m_0^3 r_0\left(\frac{m}{m_0}\right)^{2\bar{p}}\frac{1}{2\bar{p}},$$

from which, after simple transformations, we obtain the equation for $\bar{p} = q$:

$$2\bar{p}-\frac{2-\bar{p}}{\bar{p}+1}-(2-\bar{p})\left(\frac{1}{3-\bar{p}}+\frac{1}{4-\bar{p}}+\frac{1}{\bar{p}}+\frac{1}{\bar{p}+1}\right)B_2=0. \tag{39}$$

The dependence of $\bar{p}$ (and hence of q) on the value of B_2 for massive clusters is shown in Fig. 2.

By a similar substitution of (22), (35), and (37) into (38), for low-mass clusters we obtain

$$2\bar{p}-\frac{2-q}{2\bar{p}+1-q}-\frac{2-q}{\bar{p}-1}\left(\frac{1}{3-q}+\frac{2}{2\bar{p}-q}+\frac{1}{4\bar{p}-3-q}\right)\frac{B_1B_2}{2}\bar{p}=0. \tag{40}$$

In this case, the values of $\bar{p}$ and q are found from (33) and (40). The dependence of $\bar{p}$ and q on the value of B_1B_2 is shown in Fig. 2. One can see that $\bar{p}$ for massive clusters is determined only by the parameters of the clusters, whereas for small clusters, $\bar{p}$ also depends (although weakly) on the mass (the factor B_1), and the smaller the ratio m/m^*, the closer it is to two.

The parameters μ, ξ, and β_0 of clusters that appear in Eqs. (39) and (40) (the factor B_2) are not free parameters for individual clusters. They are connected with one another, as well as with

$\bar{p}$ and q, changing in each collision. Confining ourselves to the consideration of spherical clusters for simplicity, we can estimate the values of μ as a function of the density distribution $\delta(r)$ of a cluster, under the assumption that the angular velocity of rotation $\omega(r)$ is Keplerian. Larger values of μ (a considerable part of the mass is in the outer region of the cluster), other conditions being equal, correspond to larger values of β_0. A more detailed investigation of the functional relations between μ, ξ, β_0, p, and q is desirable.

The parameters μ, ξ, and β_0 do not appear separately in Eqs. (39) and (40) but in the complex $B_2 = 25\beta_0/4\mu^2\xi$, so that the quantity B_2 can be considered as some generalized parameter which, on the average, varies in a relatively smaller intervals than each of the constituent parameters individually. Values of B_2 that seem reasonable, at least at the start of the evolution of clusters, evidently lie in the interval $2 \lesssim B_2 \lesssim 5$ ($\mu \approx 1$, $\xi \approx 1$, $\beta_0 \approx 1/2$). The corresponding values of $\bar{p}$ for massive clusters lie in the interval of 1.4-1.7, according to Fig. 2.

With a known mass distribution (31) of clusters, we can use the condition (22) to estimate the fraction of the mass of the system incorporated into the massive clusters,

$$\zeta = \int_{m_2}^{m_1} mn_2(m)\,dm \Big/ \left[\int_{m_0}^{m_2} mn_1(m)\,dm + \int_{m_2}^{m_1} mn_2(m)\,dm\right].$$

where m_2 is determined from the condition $\frac{\theta}{\beta_0}\frac{m_2}{r_2}\frac{r^*}{m^*} = 1$, while $n_1(m)$ and $n_2(m)$ are the distribution functions of the small and massive clusters, respectively. Taking $m^* = m_1$, for $q_1 \approx q_2 \approx q$ we obtain

$$\zeta = \left[1 - \left(\frac{\beta_0}{\theta}\right)^{(2-q)/(4-\bar{p})}\right] \Big/ \left[1 - \left(\frac{m_0}{m_1}\right)^{2-q}\right].$$

For $\beta_0 \approx 1/2$, $\theta \gtrsim 2$, and $q \approx \bar{p} \approx 1.6$-$1.8$, we find $\zeta \gtrsim 0.5$. Henceforth we shall confine ourselves to consideration of the evolution of massive clusters.

We can use $\bar{p}$ to estimate the time duration of the evolution of the clusters.[5] Integrating the expression

$$\frac{dm}{dt} = \pi l_0^2 \bar{\rho}_d v \simeq \frac{8\pi}{3}\frac{(1+2\theta)\sigma_d}{P}\frac{\alpha'}{\xi^2} r_0^2 \left(\frac{m_0}{m}\right)^{6-4\bar{p}}, \quad \frac{\alpha'}{\xi^2} \approx 10^{-1}, \tag{41}$$

for the rate of growth of a cluster, we obtain the time of growth of the cluster from the mass m_0 to m:

$$t_1 \approx \left(\frac{m}{m_0}\right)^{7-4\bar{p}} P. \tag{42}$$

Since $\delta \propto m^{10-6\bar{p}}$, the time of conversion of clusters into solid bodies will be of the order of

$$t_2 \approx \left(\frac{\delta_0}{\delta_1}\right)^{-\frac{7-4p}{10-6p}} P, \tag{43}$$

where $\delta_1 \approx 1$ g/cm³ and δ_0 is the density of a cluster after its initial contraction: $\delta_0 \gtrsim 10\rho^*$. In Ref. 5 it was assumed that $\bar{m}' = m/4$, which gave $\bar{p} = 1.2$ and a time of $\sim 10^4$ years for the conversion of clusters into bodies at the earth's distance from the sun. In the process, the masses of the initial clusters increase by two orders of magnitude.

We set $\delta_0 = 10^{-5}$ g/cm³ for the terrestrial zone, and from (43) we find the time of conversion of clusters into bodies for the values of $\bar{p}$ determined for massive clusters. Then, taking $t_1 = t_2$, we estimate the corresponding increases $\tilde{m}/m_0$ in their masses. In Table II we give estimates of the duration of the evolution of clusters and the corresponding increase in their mass, obtained for the terrestrial zone as well as for the zone of Jupiter with $\delta_0 = 10^{-7}$ g/cm³ and the zone of Uranus with $\delta_0 = 10^{-9}$ g/cm³. For the zones of Jupiter and Uranus, the time is given in the corresponding orbital periods (log t/P) and in years (log $t/P_\oplus$).

It must be noted that Eq. (41) for the growth rate is valid so long as the masses of the largest clusters are much less than the mass M of the future planet in the given zone, and we can neglect depletion of the material, taking $\sigma_d = \sigma_{d0}\left[1 - \left(\frac{m}{M}\right)^{2/3}\right] \approx \sigma_{d0}$.

As can be seen from Table II, depletion effects must be taken into account starting with $\bar{p} \gtrsim 1.5$, so that estimates for which the inequality $\frac{m}{M} \ll 1$ is not satisfied have a purely formal character and are given in parentheses. Moreover, in the region of the giant planets, when the clusters reach a certain critical mass m_{cr}, the average relative velocity v of the clusters becomes comparable with the velocity $V_e(R)$ of escape from the system, and the ejection of material (bodies, clusters) in hyperbolic orbits becomes important. The value of $m_{cr} = \frac{1}{18\sqrt{3\pi\delta}}\left(\frac{\theta M_\odot}{R}\right)^{3/2}$ is found from the equality $v(m_{cr}) = V_c/3$. With the onset of ejection, one must take into account not only the variation of the surface density σ_d but also the increase in the parameter θ with time. Therefore, the application of (41) to zones of the giant planets is correct only up to m_{cr}. The times of the growth of clusters to m_{cr} as a function of $\bar{p}$ for the zone of Uranus are given in the last row of Table II. One can see that rapid growth of the surviving clusters occurs for $\bar{p} \approx 1.5$-1.6 (corresponding to $B_2 \approx 3$), and m_{cr} is reached in a time $\sim(10^5$-$10^6)$P, whereas for the bodie this time is $\sim 10^8$P. This effect is important for the solution of the problem of the growth of the distant planets.[9,10]

Curves of the growth of clusters in the terrestrial zone, calculated for several values of $\bar{p}$ (solid lines), and of their increase in density (dashed lines), are shown in Fig. 3. The masses $\tilde{m}(p)$ of the clusters at which their density becomes $\sim$1 g/cm³ are marked by circles. The curve of the growth of a solid body with a density $\delta = 1$ g/cm³ is shown by a dash-dot line for comparison: The times of growth from m_0 to $\tilde{m}$ for a cluster and a solid body prove to be of the same order.

It is seen from Table II and Fig. 3 that the larger the value of $\bar{p}$, the faster a cluster grows and the greater the mass it can gather. It is obvious that in a system of clusters with a certain average $\bar{p}$, the accelerated growth of individual clusters with $p > \bar{p}$ is possible. We shall show that in this case there is a mechanism preventing a considerable separation in masses.

4. GROWTH OF THE LARGEST CLUSTERS

For a cluster with a mass $m > m_1$, the upper

limit of the mass distribution of the remaining clusters in its zone, a collision with one like itself is unlikely, and Eq. (39), derived under the assumption $m_1 = m$, is inapplicable. Let us consider the trend of the further evolution of such a cluster on the example of outstripping in growth for $p > \bar{p}$. From a comparison of the growth rates of clusters with $m(p)$ and $m_1(\bar{p})$, we can obtain

$$\frac{m_1(t)}{m(t)} \approx \left(\frac{7-4\bar{p}}{7-4p}\right)^{1/(7-4\bar{p})} \left(\frac{m(t)}{m_0}\right)^{-4(p-\bar{p})/(7-4\bar{p})} = Ax^{-b}, \tag{44}$$

where $x = m/m_0$. Substituting (44) into (35) and (36) and allowing for the fact that $\bar{p} = q$ in accordance with (34), we obtain the increment in the square of the momentum of a cluster of mass m upon an increase in its relative mass by dx,

$$\begin{aligned}\frac{dK^2}{dx} &= \left(\frac{2\mu}{5}\right)^2 \xi G m_0^3 r_0 \frac{2-\bar{p}}{\bar{p}+1} A^{2\bar{p}-1} x^{(1-b)(2\bar{p}-1)} \\ &+ (2-\bar{p})\beta_0 G m_0^3 r_0 A x^{2p-1-b} \left[\frac{1}{3-\bar{p}} + \frac{A}{4-\bar{p}} x^{-b}\right. \\ &\left. + \frac{A^{2\bar{p}-3}}{\bar{p}} x^{-b(2\bar{p}-3)} + \frac{A^{2\bar{p}-2}}{\bar{p}+1} x^{-b(2\bar{p}-2)}\right],\end{aligned} \tag{45}$$

where $\bar{p}$ is determined from Eq. (39). On the other hand, from (14) we obtain

$$\frac{dK^2}{dx} = 2\left(\frac{2\mu}{5}\right)^2 \xi G m_0^3 r_0 x^{2p-1}\left(p + x \ln x \frac{dp}{dx}\right). \tag{46}$$

Equating (45) and (46), after certain transformations we arrive at the equation

$$\begin{aligned}\frac{dp}{dx} = \frac{1}{2x\ln x}\Bigg\{&\frac{2-\bar{p}}{\bar{p}+1}\left(A^{2\bar{p}-1}x^{-2,5b}-1\right) + (2-\bar{p})B_2\left[\frac{1}{3-\bar{p}}\left(Ax^{-b}-1\right)\right. \\ &+ \frac{1}{4-\bar{p}}\left(A^2x^{-2b}-1\right) + \frac{1}{\bar{p}}\left(A^{2\bar{p}-2}x^{-b(2\bar{p}-2)}-1\right) \\ &\left. + \frac{1}{\bar{p}+1}\left(A^{2\bar{p}-1}x^{-b(2\bar{p}-1)}-1\right)\right] + 2(\bar{p}-p)\Bigg\}.\end{aligned} \tag{47}$$

For $p = \bar{p}$, we obtain $A = \left(\frac{7-4\bar{p}}{7-4p}\right)^{1/(7-4\bar{p})} = 1$, $b = \frac{4(p-\bar{p})}{7-4\bar{p}} = 0$ and $\frac{dp}{dx} = 0$, while for $p > \bar{p}$ and $x > 1$, we obtain $\frac{dp}{dx} < 0$. This means that a cluster with $p > \bar{p}$ outstrips the others in mass, but as soon as its mass becomes appreciably greater than m_1, p starts to decrease, tending toward $\bar{p}$, which leads to a corresponding slowing in the growth of the cluster and its more rapid compaction. The variation of p with an increase in x from the initial value $p(x^*) = 1.6$, calculated from (47), is shown in the inset to Fig. 2 for $\bar{p} = 1.4$ and $\bar{p} = 1.3$. One can see that an increase in the mass of the growing cluster by a factor of two to four is sufficient for the value of the exponent p to decrease almost to its mean value $\bar{p}$. The separation between the mass m of the cluster and those of the other clusters in the zone does not decrease in the process and will even increase further owing to the larger gravitational cross section for collisions. Since the condition $m_1/m < 1$ is retained, p will accordingly decrease even further. This leads to more rapid growth of the cluster's density and a decrease in its size, i.e., a decrease in the collision cross section and slowing of the growth.

The decrease in the specific momentum of the largest growing cluster (i.e., the decrease in p) is due to the fact that its mass increases through the joining of small clusters, which (see Fig. 1) contribute a relatively small momentum. Thus, negative feedback exists in the system of clusters, leading to a slowing, on the average, in the growth of clusters that have considerably outstripped the remaining clusters in mass.

5. DISCUSSION OF RESULTS

We have considered processes in the system on the average. Fluctuations must play an important role, however, by virtue of the very character of the random process of collisions. Thus, as a result of the collision of clusters of comparable masses and sizes, the mass of a cluster doubles, while the value of the parameter p may reach or even exceed two (see Fig. 1). In this case, the massive cluster can be pulled apart into "leaders." In the region of small clusters, where collisions between comparable clusters are more likely, and $\bar{p}$ is also somewhat larger (see Fig. 2), such collisions should lead to the clusters becoming less dense. From $\delta \propto m^{10-6p}$, it follows that this occurs for $p > 5/3$. Some of the mass of the dispersed material becomes comparable with the mass of gas in the given zone, it is efficiently depleted by the remaining clusters, which leads to a decrease in $\bar{p}$ (see Sec. 2). This feedback mechanism is evidence in favor of $\bar{p}_{max} \simeq 5/3$ (the value $p = 5/3$ is shown in Fig. 2).

In Sec. 2 it was noted that the presence of the dispersed dust component in the system promotes the efficient compression of the clusters as they grow. Therefore, the estimate of the time for the conversion of the system of clusters into a system of solid bodies (see Table II) for a given $\bar{p}$ and $\varepsilon_1 = \varepsilon_2 = 1$ is an upper estimate. It is still premature to draw final conclusions, but it seems reasonable that the conversion of clsuters into bodies in the zone of planets of the terrestrial group occurred in a time of $\sim 10^5$-10^6 years. The largest bodies could evidently reach 1000-km sizes. Thus, by the end of the stage of clusters, a swarm of bodies is formed with a wide mass spectrum, rather than equal 1-km bodies. In the zone of the outer planets, the stage of clusters could be more prolonged, while efficient compaction occurs for far larger masses.

In models describing the evolution of the preplanetary disk from the formation of clusters to the formation of planets, one must reject a one-to-one correspondence between the mass and radius of a body and allow for the variation of the specific momenta of bodies in the process of their interaction.

The model of the collisional evolution of dust clusters is also interesting in the cosmochemical aspect. In the collision of rotating clusters with a relative collision velocity of $\sim 2v$, some of the particles may have an impact velocity of up to 4v. The high-velocity collisions of small particles should have led to more efficient heating, melting, and degassing of the material than in collisions of large dense bodies.

A similar analysis may be of interest in the investigation of the dynamics of rotating interstellar clouds.

The authors thank V. S. Safronov for a useful discussion during the preparation of the paper.

[1]T. Gehrels (ed.), Protostars and Planets, Univ. Arizona Press, Tucson (1978).
[2]D. C. Black and M. S. Matthews (eds.), Protostars and Planets II, Univ. Arizona Press, Tucson (1985).
[3]V. S. Safronov and A. V. Vityazev, Itogi Nauki Tekh., Astron. 24, 5 (1983).
[4]L. E. Gurevich and A. I. Lebedinskii, Izv. Akad. Nauk SSSR, Ser. Fiz., 14, 765 (1950).
[5]V. S. Safronov, Evolution of the Protoplanetary Cloud and the Formation of the Earth and Planets [in Russian], Nauka, Moscow (1969) [NASA Tech. Transl. F-677, Washington, D.C. (1972)].
[6]T. M. Eneev and N. N. Kozlov, Astron. Vestn. 15, 80 (1981).
[7]T. M. Éneev and N. N. Kozlov, Astron. Vestn. 15, 131 (1981).
[8]G. V. Pechernikova, Kinemat. Fiz. Nebesn. Tel. 3, No. 5, 85 (1987).
[9]A. V. Vityazev, "Some problems of the origin and early evolution of the earth," Author's abstract of candidate's dissertation, Inst. Fiz. Zemli, Akad. Nauk SSSR, Moscow (1974).
[10]V. S. Safronov, in: Cosmic Chemistry of the Moon and Planets [in Russian], Nauka, Moscow (1975), p. 624.

Translated by Edward U. Oldham

 0038-5301/88/01 0038-07 $04.10

Analytical Model of the Evolution of the Protoplanetary Accretion Disk (abstract)

A. D. Grechinskii
Translated from: Astron. Vestn. **23**, 125 (1989)
Sol. Syst. Res. **23**, 77 (1989)

An analytical model of the evolution of the protoplanetary disk with small angular momentum $J \approx 10^{52}$ g cm^2/s and the condition of direct matter accretion on it has been developed. It is shown that the resulting evolutionary scenario is qualitatively different from that obtained in the works by Cassen and Moosman (1981) and Cassen and Summers (1983), in which the accretion of matter on the disk was not considered.

Evolution of Self-Gravitating Clumps of a Gas-Dust Nebula Participating in the Accumulation of Planetary Bodies (abstract)

V. P. Myasnikov and V. I. Titarenko
Translated from: Astron. Vestn. **23**, 14 (1989)
Sol. Syst. Res. **23**, 7 (1989)

As a result of the application of the method of two-scale expansions to the system of hydrodynamic and thermodynamic equations describing the behavior of self-gravitating gas-dust clumps, a mathematical model of processes of evolution of clumps is constructed which comes down to finding solutions of the boundary-value problem for the Lane–Emden equations, the Cauchy problem for the first-order, nonlinear differential equation of the evolution of entropy, allowing for the energy source due to the decay of radioactive admixtures, and radiation transfer in the diffusional approximation. Numerical calculations made for clumps with a wide range of variation of the masses and values of the characteristic density allowed us to choose the probable initial distributions of density, temperature, and pressure for each clump. Numerical modeling was performed and the main stages of the process of evolution were investigated for a gas clump ($\gamma = 1.57$) having a mass equivalent to the earth's mass, a characteristic density of 0.4 g/cm^3, and a characteristic heat capacity at constant pressure of 1.5×10^8 erg/(g K) with a concentration $c_R \approx 10^{-3}$ of admixtures of short-lived isotopes of ^{26}Al in its material; the time of evolution of the clump to the start of condensation is estimated.

Evolution of a Self-Gravitating Gas-Dust Clump with Allowance for Radiative Transfer in a Diffusional Approximation (abstract)

V. P. Myasnikov and V. I. Titarenko
Translated from: Astron. Vestn. **23**, 207 (1989)
Sol. Syst. Res. **23**, 126 (1990)

The basic equations of a mathematical model of the problem of the evolution of a self-gravitating gas-dust clump are investigated. Using group methods, solutions of the boundary-value problem for the Lane–Emden equation are obtained from the corresponding solutions of the Cauchy problem for each specific polytropic index, and the corresponding solutions of the Cauchy problem and the boundary-value problem for this equation are analyzed for different γ. The probable initial distributions of the main thermodynamic characteristics of clumps (temperature, pressure, and density) and their limiting radii are chosen from their masses and characteristic densities, varied over a wide range, calculated for different polytropic indices γ. A complete investigation is made of the homogeneous Neumann problem for an ordinary, linear, second-order differential equation for the radiation energy density with singularities at both ends of the integration interval, representing radiative transfer in the differential approximation. A supplementary integral equation is found that enables one to obtain a unique solution to this problem, and an algorithm is developed for its solution as a function of the similarity parameter α, equal to the ratio of the characteristic clump radius to the photon mean free path. The time dependence of the evolution of a clump up to the onset of condensation is studied as a function of the optical properties of the material, as well as of the type and concentration of short-lived radioactive isotopes appearing as impurities in the composition of the material. It is established that the condition of contraction imposes limits on the abundance of radioactive impurities responsible for heat release.

Kinetic analysis of an evolving protoplanetary disk: gravitational interactions

I. N. Ziglina

Shmidt Institute of Earth Physics, USSR Academy of Sciences, Moscow

(Submitted December 1, 1988)

Pis'ma Astron. Zh. **15**, 661–666 (July 1989)

Transport processes in a particulate protoplanetary disk can be assessed by the Goldreich–Tremaine method of analyzing the dynamics of an ensemble of inelastically colliding particles (as would be true of Saturn's rings) if allowance is made for another important effect: gravitational interactions between particles experiencing close encounters. Single-integral expressions are obtained for the corresponding rates of change of the principal stress-tensor components—quantities that occur in the Goldreich–Tremaine equations.

In 1978 Goldreich and Tremaine[1] chose a kinetic approach to study the dynamics of rings composed of particles revolving about a central body and colliding inelastically with one another. The postulated that, to a first approximation, the velocity distribution function of the particles might be represented by a triaxial Gaussian in velocity-space, with different velocity dispersions along the principal axes.

If all the terms in the Boltzman equation for the distribution function of objects in phase space $(\mathbf{r}, \mathbf{v})$ (where $\mathbf{r}$ is the radius vector and $\mathbf{v}$ is the velocity) are multiplied in turn by 1, v_i, v_iv_j and then integrated over $\mathbf{v}$, one will obtain the hydrodynamic equations: the equation of continuitv, the Eulerian equations of motion, the viscous stress equations.

Neglecting gravitational interactions among the particles, Goldreich and Tremaine calculated the rates of change of the principal components of the stress tensor due to inelastic collisions in an ensemble of individual particles. By solving the viscous stress equations numerically, they were able to evaluate the rms particle velocities σ_1, σ_2, σ_3 along the principal directions for the rings of Saturn as well as the orientation of the principal axes of the velocity ellipsoid, all as functions of the optical depth of the rings. In addition the kinematic viscosity coefficient

$$\nu = \frac{\sigma_1^2 - \sigma_2^2}{3\Omega} \sin 2\delta \tag{1}$$

was determined; here Ω denotes the angular velocity of Keplerian motion around the central body, the unit vectors e_1, e_2 along the principal axes of the velocity ellipsoid lie in the central ring plane, and δ ($|\delta| \le \pi/4$) is the angle that e_1 froms with the radial direction.

This method has successfully been applied[2] to analyze the dynamics of a particulate ring in orbital resonance with a satellite, and in the same context allowance has been made[3] not only for interparticle collisions but also for the particles' axial rotation and finite size. One other application of the Goldreich–Tremaine approach would be to investigate the dynamics of a protoplanetary disk: the conventional equations of gas-kinetic theory will not suffice, because the bodies' mean free path during the primeval accumulation phase will exceed the radius of the system.[4] Unlike the case with Saturn's rings, however, in a protoplanetary swarm the gravitational interactions between bodies experiencing close encounters will play a decisive role.

Our aim in this letter is to evaluate the gravitational collision terms.

If we assume that the particles have independent mass and velocity distributions, the components of the stress tensor will be given by

$$p_{ii} = \int m v_i^2 f(\mathbf{v}) \, n(m) \, d\mathbf{v} \, dm, \quad i = 1, 2, 3. \tag{2}$$

Here v denotes a particles's velocity relative to the mean velocity at a certain point (to a first approximation the mean velocity will equal the Keplerian circular velocity in the central plane of the cloud); $n(m)\,dm$ represents the number of bodies per unit volume with a mass in the interval $(m, m + dm)$, and the function

$$f(\mathbf{v}) = \frac{1}{(2\pi)^{3/2} \sigma_1\sigma_2\sigma_3} \exp\left(-\sum_{j=1}^{3} \frac{v_j^2}{2\sigma_j^2}\right). \tag{3}$$

In a close encounter between two bodies, the gravitational interaction will change the components p_{ii} at a rate which may be expressed as

$$\left.\frac{\partial p_{ii}}{\partial t}\right|_{\text{grav}} = \frac{1}{2}\int (m_1 v_{1i}'^2 + m_2 v_{2i}'^2 - m_1 v_{1i}^2 - m_2 v_{2i}^2) \times n(m_1)\, n(m_2)\, f(\mathbf{v}_1)\, f(\mathbf{v}_2) \mathbf{V} D\, dD\, d\eta\, d\mathbf{v}_1\, d\mathbf{v}_2\, dm_1\, dm_2, \tag{4}$$

where $\mathbf{V} = \mathbf{v}_2 - \mathbf{v}_1$ is the relative velocity of bodies m_1, m_2 prior to the encounter, D is the impact parameter, and η is the angle between the plane of the bodies' relative orbit and some fixed plane passing through $\mathbf{V}$. We shall consider only two-body interactions and shall assume that the distance S within which an interaction would be significant is comparable with the disk half-thickness but is much shorter than the characteristic length for appreciable modifications in the bodies' mass and velocity distributions. The primes in Eq. (4) refer to the velocities after the encounter.

To evaluate the quantities $\partial p_{ii}/\partial t|_{\text{grav}}$ we must perform successive integrations. It is convenient to transform from the variables v_1, v_2 to the relative velocity V and the velocity $\mathbf{v}_c = (m_1\mathbf{v}_1 + m_2\mathbf{v}_2)/(m_1 + m_2)$ of the center of mass of m_1, m_2. This transformation has unit Jacobian, and the parenthesized factor in Eq. (4) takes the form $(V_i'^2 - V_i^2) m_1 m_2/(m_1 + m_2)$.

We shall describe encounters in terms of the two-body problem, regarding the bodies as being infinitely far apart before the encounter and as becoming infinitely separated afterward. An encounter will turn the relative velocity vector **V** by an angle 2ψ such that

$$\tan\psi = \frac{\mu}{DV^2}, \tag{5}$$

where $\mu = G(m_1 + m_2)$, with G denoting the gravitation constant. The quantites D and η characterizing an encounter may be defined as the polar coordinates of body m_2 in the plane passing through m_1 and orthogonal to **V**. Let the angle η be measured from the intersection line between the target plane and a plane passing through m_1 orthogonal to the i-th principal direction; then

$$V_i' = V_i - 2(\sin^2\psi)V_i + \sqrt{V^2 - V_i^2}\sin 2\psi\sin\eta.$$

After some straightforward operations we now find that

$$\int_0^{2\pi}(V_i'^2 - V_i^2)\,d\eta = \pi\sin^2 2\psi\,(V^2 - 3V_i^2). \tag{6}$$

Using Eq. (5), we next integrate $\sin^2 2\psi$ over the impact parameter D to obtain

$$\int_l^S \sin^2 2\psi D\,dD = \frac{2\mu^2}{V^4}\ln\left(1 + \frac{D^2V^4}{\mu^2}\right)\Bigg|_l^S + \frac{2\mu^2}{V^4}\frac{1}{\left(1+\frac{D^2V^4}{\mu^2}\right)}\Bigg|_l^S, \tag{7}$$

where $l = (r_1 + r_2)\sqrt{1 + \frac{2G(m_1+m_2)}{(r_1+r_2)V^2}}$ is the maximum impact parameter for which the bodies are still regarded as being in collision. The half-thickness of the protoplanetary swarm will be much greater than ℓ, so the dominant term on the right in Eq. (7) will be the one containing $\ln(1 + S^2V^4/\mu^2) \approx 10$. Henceforth the other terms will be neglected.

Upon substituting the formulas (3) for $f(\mathbf{v}_1)$, $f(\mathbf{v}_2)$ into Eq. (4) and expressing $\mathbf{v}_1$, $\mathbf{v}_2$ in terms of **V** and $\mathbf{v}_c$, we can integrate the expression (4) over $\mathbf{v}_c$ without any difficulty. After having integrated over $\mathbf{v}_c$, η, and D we arrive at the expression

$$\frac{\partial p_{ii}}{\partial t}\Bigg|_{\text{grav}} = \int \pi G^2 m_1 m_2 (m_1+m_2)\,n(m_1)\,n(m_2)$$
$$\times \ln(1 + S^2V^4/\mu^2)V^{-3}(V^2 - 3V_i^2)\frac{1}{2^3\pi^{3/2}\sigma_1\sigma_2\sigma_3}$$
$$\times \exp\left(-\sum_{j=1}^{3}\frac{V_j^2}{4\sigma_j^2}\right)d\mathbf{V}\,dm_1\,dm_2. \tag{8}$$

In **V**-space we introduce the spherical coordinates V, ϕ, θ:

$$V_i = V\cos\theta,\ V_j = V\sin\theta\cos\varphi,\ V_k = V\sin\theta\sin\varphi;$$
$$d\mathbf{V} = V^2\sin\theta\,dV\,d\varphi\,d\theta.$$

We begin by integrating over V. To do so we have to evaluate the integral

$$J_i = \int_0^\infty \ln(1 + q^2V^4)\,V\exp(-\alpha_i^2V^2)\,dV,$$

where

$$q^2 = \frac{S^2}{\mu^2},\quad \alpha_i^2 = \frac{\cos^2\theta}{4\sigma_i^2} + \frac{\sin^2\theta\cos^2\varphi}{4\sigma_j^2} + \frac{\sin^2\theta\sin^2\varphi}{4\sigma_k^2}.$$

We introduce the new variable $t = \alpha_i^2V^2$; then

$$J_i = \frac{1}{2\alpha_i^2}\int_0^\infty \exp(-t)\ln\left(1 + \frac{q^2t^2}{\alpha_i^4}\right)dt$$
$$= \frac{1}{\alpha_i^2}\int_0^\infty \exp(-t)\frac{t\,dt}{t^2 + \beta_i^2},$$

where $\beta_i^2 = \alpha^4/q^2$. In a protoplanetary swarm the quantity $\beta^2 \ll 1$. In that event, as the author has shown previously,[5] one may use the approximation

$$J_i \approx \frac{1}{\alpha_i^2}\ln\frac{q}{\gamma\alpha_i^2}, \tag{9}$$

where $\gamma = 1.781\ldots$ is Euler's constant.

Next we evaluate the integral

$$J_{1i} = \int_0^{2\pi}(a_i^2\cos^2\varphi + b_i^2\sin^2\varphi)^{-1}$$
$$\times \ln\left[\frac{4q}{\gamma(a_i^2\cos^2\varphi + b_i^2\sin^2\varphi)}\right]d\varphi, \tag{10}$$

where

$$a_i = \sqrt{\frac{\cos^2\theta}{\sigma_i^2} + \frac{\sin^2\theta}{\sigma_j^2}},$$
$$b_i = \sqrt{\frac{\cos^2\theta}{\sigma_i^2} + \frac{\sin^2\theta}{\sigma_k^2}}. \tag{11}$$

We transform from the angle ϕ to an angle χ, related to it by

$$\cos\chi = \frac{a_i\cos\varphi}{\sqrt{a_i^2\cos^2\varphi + b_i^2\sin^2\varphi}},$$
$$\sin\chi = \frac{b_i\sin\varphi}{\sqrt{a_i^2\cos^2\varphi + b_i^2\sin^2\varphi}}.$$

Then

$$J_{1i} = \frac{1}{a_ib_i}\int_0^{2\pi}\ln\left[\frac{4q}{\gamma}\left(\frac{\cos^2\chi}{a_i^2} + \frac{\sin^2\chi}{b_i^2}\right)\right]d\chi.$$

Substituting $z = e^{i\chi}$ (in this instance $i = \sqrt{-1}$), we proceed to integrate over the unit circle in the complex plane centered at the origin. Evaluating the appropriate residues we obtain

$$J_{1i} = \frac{2\pi}{a_ib_i}\ln\left[\frac{q}{\gamma}\left(\frac{1}{a_i^2} + \frac{1}{b_i^2}\right)^2\right]. \tag{12}$$

If we now combine Eqs. (8), (9), (12) we find that

$$\frac{\partial p_{ii}}{\partial t}\Bigg|_{\text{grav}} = \frac{G^2\pi^{1/2}}{\sigma_1\sigma_2\sigma_3}\int m_1m_2(m_1+m_2)\,n(m_1)\,n(m_2)$$
$$(1 - 3\cos^2\theta)\frac{\sin\theta}{a_ib_i}\ln\left[\frac{S}{\gamma G(m_1+m_2)}\left(\frac{1}{a_i} + \frac{1}{b_i}\right)^2\right]dm_1\,dm_2\,d\theta. \tag{13}$$

To a certain approximation we can replace the

sum $m_1 + m_2$ in the bracketed expression by the mean value 2m. We then arrive at our final expression, integrable by computer:

$$\left.\frac{\partial p_{ii}}{\partial t}\right|_{grav} = \frac{2\pi^{1/2}\rho^2\bar{m}G^2}{\sigma_1\sigma_2\sigma_3}\int_0^\pi \frac{(1-3\cos^2\theta)\sin\theta}{a_i b_i} \times \ln\left[\frac{S}{2\gamma G\bar{m}}\left(\frac{1}{a_i}+\frac{1}{b_i}\right)^2\right]d\theta, \quad (14)$$

where ρ denotes the combined mass of particles per unit volume and m is the average mass of the individual bodies comprising the bulk of the swarm:

$$\bar{m} = \int m^2 n(m)\,dm \Big/ \int mn(m)\,dm. \quad (15)$$

The quantities $\partial p_{ii}/\partial t|_{grav}$ (Eq. (14)) are to be added to the right-hand sides of the Goldreich–Tremaine equations for the viscous stresses:

$$\frac{\partial p_{11}}{\partial t} - \frac{3\Omega}{2}(\sin 2\delta)\,p_{11} = \left.\frac{\partial p_{11}}{\partial t}\right|_{coll} + \left.\frac{\partial p_{11}}{\partial t}\right|_{grav},$$
$$\frac{\partial p_{22}}{\partial t} + \frac{3\Omega}{2}(\sin 2\delta)\,p_{22} = \left.\frac{\partial p_{22}}{\partial t}\right|_{coll} + \left.\frac{\partial p_{22}}{\partial t}\right|_{grav},$$
$$p_{11}(1+3\sin^2\delta) - p_{22}(1+3\cos^2\delta) = 0,$$
$$\frac{\partial p_{33}}{\partial t} = \left.\frac{\partial p_{33}}{\partial t}\right|_{coll} + \left.\frac{\partial p_{33}}{\partial t}\right|_{grav}. \quad (16)$$

The $\partial p_{ii}/\partial t_{coll}$ terms, describing the inelastic collision effects, are expressed in terms of elliptic integrals. A numerical solution of Eqs. (16) on the premise that the $\partial p_{ii}/\partial t \equiv 0$ would yield the unknowns σ_1, σ_2, σ_3 and the angle δ. By using Eq. (1) to express the effective viscosity in terms of these quantities, one would be able to estimate the transport of mass and angular momentum in a protoplanetary cloud—processes that will have a major impact on the growth of planets.

I wish to thank V. S. Safronov for valuable criticisms.

[1] P. Goldreich and S. D. Tremaine, Icarus 34, 227 (1978).

[2] N. Borderies, P. Goldreich, and S. D. Tremaine, Icarus 55, 84 (1983).

[3] I. G. Shukhman, Astron. Zh. 61, 985 (1984) [Sov. Astron. 28, 574 (1985)].

[4] V. S. Safronov, Evolution of the Protoplanetary Cloud and the Formation of the Earth and Planets, Nauka, Moscow (1969) [NASA TT F-677 (1972)].

[5] I. N. Ziglina, Astron. Zh. 62, 141 (1985) [Sov. Astron. 29, 81 (1985)].

Translated by R. B. Rodman

Dissipative Instability of the Protoplanetary Disk and the Law of Planetary Distances (abstract)

N. N. Gor'kavyi, V. L. Polyachenko, and A. M. Fridman
Translated from: Pisma Astron. Zh. **16**, 183 (1990)
Sov. Astron. Lett. **16**, 79 (1990)

The stability of a viscous, differentially rotating protoplanetary disk is examined. It is shown that besides the negative diffusion instability discovered by Lin and Bodenheimer (1981) and Ward (1981), there is a quasisecular instability in the disk with a positive diffusion coefficient. The dissipative instabilities in the disk form a series of rings around the Sun. The conditions under which these rings are localized at the planetary orbits are determined. These conditions conform to current cosmogonic concepts.

Dust Particle Transport in Protoplanetary Accretion Disks (abstract)

A. D. Grechinskii
Translated from: Astron. Vestn. **24**, 134 (1990)
Sol. Syst. Res. **24**, 85 (1990)

An approximate theory of transport of dust particles in a protoplanetary accretion disk has been constructed. The possibility of conservation of the organic mantle of some particles in molecular clouds in the process of their accretion on the disk as well as in the disk itself is shown.

Kuiper Prize Lecture: Some Problems in the Formation of the Planets (abstract)

V. S. Safronov
Icarus **94**, 260 (1991)

I consider several problems of the evolution of the solar nebula, the precipitation of dust particles toward its central plane, and the possibility that a gravitational instability developed in the dust subdisk. A solution of the dispersion equation is given for the development of an instability in a two-component (gas-dust) rotating disk under the assumption that there is no instability in the gaseous component. This solution is compared to one found earlier for one-component disks. I emphasize that a monodisperse "initial" state with equal kilometer-sized bodies, as traditionally assumed in many numerical simulations, never existed. I note that turbulence induced by the gradient of the rotation velocity $d\omega/dz$ and differential radial drift of particles having different sizes in the gas does not exclude the possibility of a gravitational instability in a thin equatorial layer of the subdisk which contains a small fraction of the total dust mass. A runaway scenario for the accumulation of preplanetary bodies is discussed. I find that a transition to higher relative velocities of bodies and a slower growth of planet embryos began already before the sum of masses of the embryos reached one-tenth the mass of all the other bodies. Mechanisms for removal of almost all the initial mass of solids from the asteroidal zone are considered. A reliable mechanism for the sweeping out of this zone by bodies which penetrate from Jupiter's zone could work efficiently at values of the velocity parameter $\theta \lesssim 30-50$, once Jupiter's embryo reached a mass of $\sim(5-10)\ M_{\oplus}$. Also at such moderate values of θ the outer planets ejected planetesimals into the cometary cloud.

Influx of Interstellar Material onto the Protoplanetary Disk (abstract)

I. N. Ziglina and T. V. Ruzmaikina
Translated from: Astron. Vestn. **25**, 53 (1991)
Sol. Syst. Res. **25**, 40 (1991)

The influx of interstellar material onto the circumsolar disk during the collapse of the slowly rotating nebula is examined. The dependence of the disk parameters on the heliocentric distance is assumed to be a power law for the surface density and a proportionality for the thickness. It is found that the ratio of the mass flux onto the peripheral parts (between the radius a significantly greater than the centrifugal radius for the accreting material and the outer radius a_2 of the disk) of the lateral surface of the disk of mass M_d to the flux at the edge is the order of $(M_d/M)(a_2/a)^{1/2}$, where M is the mass of the proto-Sun. Organic compounds and ices could have survived the process of material accretion onto the disk, in spite of heating in the shock wave.

IV. Rotation of Planets

On the Problem of Planet Rotation (abstract)

V. S. Safronov
Translated from: Vopr. Kosmog. **8**, 150 (1962)

O. Yu. Schmidt's explanation of planet rotation based on his theory of planet accumulation from solid particles and bodies is considered. Analysis of equations for constancy of total energy and momentum has shown that Schmidt's simplified model of circular orbits of accumulating bodies is insufficient.

The present velocity of the Earth's rotation being assumed, the condition found by Schmidt for direct rotation of planets is not fulfilled in this model. To fulfill this condition the Earth's rotation should be 10^4 times faster than it is now.

The energy and angular momentum equations for a more general case of the motion of bodies in elliptical orbits are deduced. It is shown that in this new model the contradictions mentioned above are removed.

The relation between the planet rotation and the heat losses during the accumulation process is analyzed. Taking into account the energy of rotation of the planet the previous opinion on this relation must be changed. It is shown that the faster the planet rotates, the lesser are the heat losses.

It follows from the expressions obtained that the simplest supposition about the velocity of rotation derived from dimensional considerations leads to the proportionality between the rotational energy increase and the potential energy on the surface of the growing planet. This leads in its turn to the approximate constancy of angular velocity during the process of growth in agreement with the results obtained earlier on the assumption of the asymmetry of shocks of falling bodies and with the fact that differences of periods of rotation of the planets are comparatively small.

SOVIET PHYSICS–ASTRONOMY VOL. 9, NO. 1 JULY-AUGUST, 1965

THE ORIGIN OF THE AXIAL ROTATION OF PLANETS

A. V. Artem'ev and V. V. Radzievskii

State Pedagogical Institute, Yaroslavl'
Translated from Astronomicheskii Zhurnal, Vol. 42, No. 1, pp. 124-128, January-February, 1965
Original article submitted July 15, 1962

It is shown that contrary to the opinion held since the time of Laplace, the Keplerian velocity distribution of particles moving around the sun and falling on a planet gives rise to a rotation in a direct and not the reverse direction. Quantitative computations lead to periods of axial rotation which for the main planets of the solar system coincide with the actual periods.

From the time that Laplace's hypothesis was published until very recently it has been generally assumed that the Keplerian motion of particles forming protoplanetary matter is incompatible with direct axial rotations of the planets condensing from it. It is for this reason that numerous investigators were forced in attempting to explain the direct rotation of planets either to assume that the protoplanetary cloud or part of it rotated like a rigid body (which always gave rise to objections in view of the low density and, hence, low viscous cohesion between the particles of the cloud), or to draw upon special artificial schemes for the agglomeration of the planets.

According to O. Yu. Shmidt's hypothesis [1], direct rotation of the planets is considered as a purely formal result due to the loss of energy of particles falling on a planet together with the conservation of their angular momentum. However, the actual mechanism for imparting rotation to the growing planet was not investigated. This approach to the solution of the problem shows little promise. Indeed, the particles approaching the planet "do not know" whether they will have to lose energy or not. And, of course, their positive angular momentum relative to the planet's axis does not arise at the instant of energy loss or because of it. Approaching the planet, these particles must have a positive angular momentum for reasons of mechanics, completely independently of the subsequent energy loss. These mechanical factors should be investigated.

O. Yu. Shmidt's hypothesis also contains an appreciable error which was pointed out by V. S. Safronov [2]. Let us also note another fundamental error. The fact is that in forming the energy and angular momentum balance, O. Yu. Shmidt erroneously assumed that in the system associated with the sun the laws of conservation of kinetic energy and angular momentum are valid, while these laws only hold in inertial frames of reference. The frame of reference tied to the sun is inertial up to the time of formation of the planets, while the center of mass of the cloud coincides with the center of the sun, but ceases to be inertial after the planets are formed and the sun begins to have orbital motion. It would not be difficult to show that the neglected effects of solar orbital motion are comparable in magnitude to the effects due to the axial motion of the planets.

Without rejecting the desirability of studying the other reasons facilitating the initiation of direct rotation, we nevertheless consider that the latter is the direct consequence of the Keplerian distribution of velocities of the particles forming the protoplanetary cloud.

As early as 1946 Edgeworth [3] and more recently V. S. Safronov [2] have shown that the total angular momentum of particles moving in Keplerian orbits and situated at a distance r from an arbitrarily chosen center is positive. We will develop these ideas further and, in particular, we will show that particles moving in Keplerian orbits and falling on a planet impart to it a positive and not a negative angular momentum, as has been considered up to the present time.

We begin by considering an idealized case in which all of the particles move in circular Keplerian orbits in planes that are inclined to the orbital plane of the planet at such small angles that in the vicinity of the latter the particle orbits may be considered to be coplanar. Suppose that a planet of mass m and radius r_0 is also moving in a circular orbit at a distance R_0 from the sun, whose mass we denote by M. The neglect of the dynamic interaction between the planet and particles which underlies this idealized scheme is not the rough approximation that it appears to be at first sight. Indeed, since we are interested in planetocentric angular momenta of the particles, it is necessary to recall that the gravitational attraction of the planet has no direct influence on the magnitude of these angular momenta.

Let us consider the plane which forms a cross section of the planet parallel to its orbit (see Fig. 1). The particles in zone A move faster than the planet's center and, overtaking the planet, fall onto it in sector aa'. The particles in zone B move slower than the planet and fall on it in sector bb'. However, both groups of particles will accelerate the rotation of the planet in an counter-clockwise direction, even if this appears paradoxical. This is accounted for by the fact that the rotation of the velocity vector v plays a more important role than the decrease of the magnitude of velocity with increasing distance R.

Let us determine the magnitudes of the transverse, v_t, and radial, v_r, components of the velocity of the particle situated at the point C on the planet's surface with longitude λ measured from the midnight meridian ob (the planet's axis is assumed to be perpendicular to its orbital plane). It is clear that

$$v_t = v\cos\lambda' - v_0\cos\lambda, \quad v_r = v\sin\lambda' - v_0\sin\lambda. \tag{1}$$

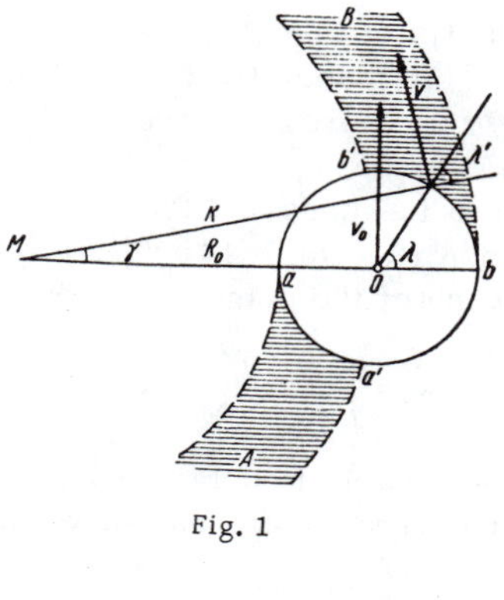

Fig. 1

Because

$$v = \sqrt{GM/R}, \tag{2}$$

where G is the gravitational constant, and

$$R^2 = R_0^2 + r^2 + 2rR_0\cos\lambda, \quad \lambda' = \lambda - \gamma,$$

and $v_0 = \omega R_0$, where ω is the angular velocity of the planet's rotation, we easily find after a transformation of (1) that correct to first-order terms in r/R_0

$$v_t = \omega r(1 - {}^3/_2\cos^2\lambda), \tag{3}$$

$$v_r = -\omega r\,{}^3/_2\cos\lambda\sin\lambda. \tag{4}$$

As can be seen from (3), particles falling on the planet impart a negative angular momentum when $0° < \lambda < 35°$ and $180° < \lambda < 215°$ and a positive angular momentum when $35° < \lambda < 90°$ and $215° < \lambda < 270°$. An averaging over λ with a uniform distribution of particles yields the average value of their specific angular momentum:

$$k = {}^1/_4\omega r^2, \tag{5}$$

which is found to be positive. It is necessary to take into account, however, that the number of particles reaching the planet at different longitudes will not be the same. It is proportional to the radial component of the velocity and, as can be seen from (4), will be a maximum at $\lambda = 45°$ and 235° (i.e., at 3 a.m. and 3 p.m. local time).

If it is assumed that the density of particles ρ is everywhere constant, then the mass of all particles falling on the planet and the angular momentum imparted by them in each of the sectors ($0° < \lambda < 90°$ and $180° < \lambda < 270°$) will be given by the following integrals:

$$\frac{dm}{dt} = -\int_{-\pi/2}^{\pi/2}\int_0^{\pi/2} \rho v_r r r_0 \cos\varphi\, d\lambda\, d\varphi, \tag{6}$$

$$\frac{dK}{dt} = -\int_{-\pi/2}^{\pi/2}\int_0^{\pi/2} \rho v_r v_t r^2 r_0 \cos\varphi\, d\lambda\, d\varphi, \tag{7}$$

where r_0 is the radius of the planet and r the distance from the axis of rotation at latitude φ.

Substituting into (6) and (7) the values of v_t and v_r and taking into account that

$$r = r_0\cos\varphi, \tag{8}$$

we find after integration that

$$\frac{dm}{dt} = \rho\omega r_0^3, \tag{9}$$

$$\frac{dK}{dt} = \frac{1}{5}\rho\omega^2 r_0^5, \tag{10}$$

from which the average specific angular momentum acquired by the planet in the idealized case of circular Keplerian motion of all particles will be

$$k = {}^1/_5 \omega r_0^2. \tag{11}$$

Result (11) has only a qualitative significance, but it shows convincingly that the Keplerian particles falling on the planet will impart a direct rotation to it. However, the magnitude of the specific angular momentum (1) is too small for it to be possible to explain the observed rotation of the planets within the framework of the above simplified scheme. Thus, for example, the angular momentum (1) would be equal to the actual specific angular momentum of the earth if the radius of our planet were 160,000 km. Therefore, let us proceed to the next stage of elaboration of our model by taking into account the gravitational field of the planet. However, even here we will be forced to use some simplifications in view of the absence of a sufficient general solution of the three-body problem.

In accordance with Laplace [4], we will assume that the particles move along unperturbed heliocentric Keplerian orbits, although, having fallen into the zone of influence of the planet, they may be captured by the latter. The zone of influence of the planet is approximately spherical in shape with radius

$$r_0 = R_0 \left(\frac{m}{M}\right)^{2/5}. \tag{12}$$

As we now know, the capture of the particle requires a change in its Jacobi constant, which, in particular, may change as the result of collisions between particles according to the scheme considered by E. L. Ruskol [5]. Taking this factor into account, we can represent the physical picture of the capture in the following terms. The particles move in elliptical orbits and only on the average do they have circular orbits. Any collision of particles, on the one hand, decreases their velocities and consequently changes their Jacobi constant in a direction favorable to capture and, on the other hand, makes their orbits more circular. If the collisions occur within the zone of influence of the planet, then the particles may be captured.

Obviously, the probability of capture will be higher if the particles are closer to the planet. Therefore, the average distance at which particle capture occurs will be smaller than the radius of the zone of influence of the planet given by (12) and will be a certain fraction $q < 1$ of the latter. The coefficient q will also tend to decrease because a considerable number of the particles will fall on the planet directly without a preliminary collision or capture. According to the above arguments, these particles also impart a positive angular momentum given by formula (11), with r_0 equal to the radius of the planet.

Thus, let us assume that the capture of particles in one way or another takes place on the average at a distance

$$r_0 = qR_0 \left(\frac{m}{M}\right)^{2/5}, \tag{13}$$

the particles that enter this sphere carrying the angular momentum (11), which, together with the mass of the particles themselves, will sooner or later be imparted to the planet as the captured matter falls on it.

It is not difficult to see that most of the particles fall on the earth immediately after capture. Indeed, the condition for a particle reaching the planet's surface is the possession by it of a specific angular momentum given by $k < (Gmr_p)^{1/2}$. The momentum of the captured particle according to (3) is $k < \omega r_0^2$ and, on the average, is ${}^1/_5\, \omega r_0^2$, where r_0 is given by expression (13). With the value of q that will be adopted below, the condition for fallout is satisfied for most particles. However, it is necessary to emphasize that the average momentum of the captured particles remains practically unchanged even if they are not captured immediately, in view of the fact that the perturbing action of the sun varies in sign. Inasmuch as both the angular momentum and the mass of the planet increase by the addition of the same matter, we can write

$$dK = kdm.$$

Substituting (11) and (13) into this expression and integrating, we get

$$K = \frac{1}{9}\, q^2 \omega R_0{}^2 \cdot \frac{m^{9/5}}{M^{4/5}}. \tag{14}$$

Let us substitute into (14) $\omega = (GM/R_0^3)^{1/2}$, where $G = 6.7 \cdot 10^{-8}$ and $M = 2 \cdot 10^{33}$, as well as $K = \alpha m r_p^2 (2\pi/\tau)$, where α is the coefficient of gyration of the planet, r_p is the radius of the planet, related to its mass and the average density δ by the relation $r_p^2 = (3m/4\pi\delta)^{2/3}$, and τ is the period of rotation of the planet. We then find that

$$\tau = \frac{8.5 \cdot 10^{14}\, \alpha}{q^2 R_0{}^{1/2} \delta^{2/3} m^{2/15}}. \tag{15}$$

In order to be able to obtain from (15) the correct period of rotation of the earth, we must take

$q = 0.24$. With this value of q the periods of axial rotation of all the planets in hours are

Mercury	−60	Jupiter	− 9.5
Venus	−30	Saturn	−12.6
Earth	−24	Uranus	− 6.8
Mars	−28.8	Neptune	− 4.8

The insignificant deviations of the true periods from those calculated for most of the planets do not require any explanations since the mechanism considered by us does not claim high rigor or accuracy. Thus, for example, the deviations could be due to the fact that the true orbits of the particles are not circular and the mean eccentricity may vary with position in the solar system. As regards Mercury, Uranus, and Neptune, we can assume that their axial rotations have been affected by some extraneous factors. Thus, for example, Mercury may have been affected by tidal friction; the rotation of Uranus may have been changed by the fallout of a retrograde semipolar satellite; the retardation of Neptune's rotation may have been caused by a close approach to Pluto.

In conclusion, it should be noted that the mechanism of particle capture considered by us may also be extended to large asteroidal bodies and since in the sector of the planet with $0° < \lambda < 35°$, $145° < \lambda < 215°$, and $325° < \lambda < 360°$ matter with a negative angular momentum is captured, we have the prospect of being able to interpret the phenomenon of retrograde satellites.

LITERATURE CITED

1. O. Yu. Shmidt, Izv. AN SSSR, Ser. Fiz., 14, 1, 29 (1950).
2. V. S. Safronov, Voprosy Kosmogonii, 8, 150 (1962).
3. K. E. Edgeworth, Monthly Notices Roy. Astron. Soc., 106, 470 (1946).
4. P. Laplace, Traité de mecanique celeste, 4, 218, Paris (1805).
5. E. L. Ruskol, Astron. Zh., 37, 690 (1960) [Soviet Astronomy – AJ, Vol. 4, p. 657].

All abbreviations of periodicals in the above bibliography are letter-by-letter transliterations of the abbreviations as given in the original Russian journal. *Some or all of this periodical literature may well be available in English translation.* A complete list of the cover-to-cover English translations appears at the back of this issue.

SOVIET ASTRONOMY—AJ VOL. 9, NO. 6, MAY-JUNE, 1966

SIZES OF THE LARGEST BODIES FALLING ONTO THE PLANETS DURING THEIR FORMATION

V. S. Safronov

Geophysics Institute, Academy of Sciences of the USSR
Translated from Astronomicheskii Zhurnal, Vol. 42, No. 6,
pp. 1270-1276, November-December, 1965
Original article submitted March 25, 1965

Application of coagulation theory to the process of accumulation of the planets from solid matter leads to the conclusion that this matter was in the form of particles and bodies of different sizes. Falling onto the planets, the bodies imparted to them a rotational moment consisting of two components of different nature: a regular component ("direct" rotation), related to rotation of the system as a whole, and a random component, related to the random direction of velocity of the falling bodies relative to the planet and manifested in the inclinations of the axes of rotation of the planets. The largest bodies made the principal contribution to the random component of rotation. This article gives the derivation of expressions relating the values of the random component of rotation to the masses m_1 of the largest bodies falling onto a planet of mass m on the assumption of an exponential distribution function of the sizes of the bodies. Table 1 gives the values m_1/m determined from a comparison of the theoretically computed angles of inclination of the axes of rotation of the planets and the observed values. The largest bodies falling onto the earth had masses of about 10^{-3} of the earth's mass, that is, they were of the size of the largest asteroids. This same mechanism makes it possible to explain the anomalous rotation of Uranus if it is assumed that the random component of the rotation of Uranus was greater than the systematic component. The mass of the largest body falling onto the surface of Uranus in this case would have to be 0.05 of the mass of that planet.

An estimate of the size of the largest bodies falling onto the planets during their formation is important for determining the principal laws of the process of accumulation of the planets. It also is of considerable interest for geophysics because it is necessary for determining the earth's initial temperature and makes it possible to judge the scale of the initial inhomogeneities of the earth's mantle [1] which could exert an influence on the entire subsequent development of the earth. We already have noted [2] that the maximum size of the bodies from which the planets were formed can be determined from the inclinations of the axes of rotation of the planets. It was determined roughly that in the final stage of growth of a planet the masses of bodies falling onto it were less than 10^{-2} of the mass of the planet itself. This problem is considered in greater detail in this article, taking into account the probable distribution function of the protoplanetary bodies.

The observed rotation of the planets can be broken down into two components — systematic (regular) with the moment $\mathbf{K}_0$ (directed perpendicular to the central plane of the planetary system), characterizing direct rotation, and the random component with $\mathbf{K}_1$, manifested in the inclinations of the axes of rotation of the planets. The latter is related to the discreteness of the process of formation of the planets. It shows that a considerable part of the mass fell onto the planet in the form of individual bodies having randomly directed relative motion at the time of the impact. An identical order of magnitude of the angles of inclination of the axes of most of the planets is a characteristic property of a planetary system which has not yet been given due attention and which indicates a definite regularity in the growth process and a regularity in the size distribution of the bodies.

Assume m and r are the mass and the radius of the growing planet ("nucleus") and m'_i is the mass of the bodies falling onto it. First as a clarification we will consider the case when all the falling bodies have identical masses $m'_i = m'$ and move in the plane Oxy relative to the planet m, whose center is situated at O. Assume v is the velocity of the body relative to the planet prior to its approach to the planet. Then the momentum imparted to the planet by the mass m'_i

$$\Delta K_{1i} = m'vl_i \tag{1}$$

is directed along the z axis and represents a random value since the impact parameter l_i of the falling body is a random value with a constant density probability distribution in the interval $(-l_0, +l_0)$. The mathematical expectation l (mean value $\bar{l}$) is equal to zero, but the mathematical expectation l^2 (l dispersion) is not equal to zero:

$$\mathrm{M}l = \bar{l} = 0, \qquad \mathrm{D}l = \overline{l^2} = \frac{1}{2l}\int_{-l_0}^{+l_0} l^2\,dl = \frac{1}{3}l_0^2. \tag{2}$$

The value l_0 is the maximum impact parameter leading to a collision of m' with m and is related to the radii r and r' by the known relation

$$l_0^2v^2 = (r+r')^2\left[v^2 + \frac{2G(m+m')}{r+r'}\right], \tag{3}$$

being an elementary corollary of the laws of conservation of energy and moment of momentum in a two-body system.

With the falling of several bodies m' onto m, in accordance with the theorem of addition of dispersion as the sum of the independent random values [3], when m'v = const we have

$$\mathrm{D}\sum_{i=1}^{n}\Delta K_{1i} = (m'v)^2\,\mathrm{D}\sum_{i=1}^{n} l_i$$

$$= (m'v)^2\sum_{i=1}^{n}\mathrm{D}l_i = (m'v)^2\frac{nl_0^2}{3}. \tag{4}$$

Therefore, the mean value of the square of momentum, imparted by n bodies m' with the total mass $\Delta m = nm'$, is equal to

$$\overline{\Delta K_1^2} = (m'v)^2\frac{nl_0^2}{3} = (vl_0)^2\frac{m'}{3}\Delta m. \tag{5}$$

Since $\overline{\Delta K_1} \propto \bar{l} = 0$, the random component of the moment of momentum ΔK_1 imparted to the planet by the falling bodies obviously is determined by its mean-square deviation, related to m' by expression (5). Expression (5) shows that the imparted momentum is the greater the larger the body m'. Small particles make virtually no contribution to ΔK_1.

In the more general case of motion of bodies in all possible directions an estimate of the imparted momentum can be made in the following way. Assume one third of all the bodies (n/3) move parallel to the x axis, a third parallel to the y axis and a third parallel to the z axis. This method is applied in the kinetic theory of gases.

We will consider bodies moving toward the surface of a planet in the direction of the z axis. Upon falling onto the planet they will impart to it the momentum components K_{1ix} and K_{1iy} along the x and y axes respectively. Obviously

$$K_{1ix} = m'vl\sin\varphi, \quad K_{1iy} = m'vl\cos\varphi, \tag{6}$$

where φ is the angle between the plane Oxz and the orbital plane of the body relative to the planet. The dispersion of the random value K_{1ix} is equal to

$$\mathrm{D}K_{1ix} = \overline{K_{1ix}^2} = (m'v)^2\frac{\int_0^{l_0}\int_0^{2\pi}(l\sin\varphi)^2\,l\,dl\,d\varphi}{\int_0^{l_0}\int_0^{2\pi}l\,dl\,d\varphi} = \frac{(m'vl_0)^2}{4}. \tag{7}$$

Similarly

$$\mathrm{D}K_{1iy} = \frac{(m'vl_0)^2}{4}.$$

The component of momentum along the x axis is also introduced by bodies moving parallel to the y axis toward the surface of a planet; in this case the dispersion $\mathrm{D}K_{1ix}$ is determined by expression (7). The dispersion of the sum of the random values K_{1ix} is equal to the sum of the dispersions of the terms

$$\mathrm{D}\sum_{i=1}^{n}K_{1ix} = 2\,\frac{n}{3}\,\mathrm{D}K_{1ix} = \frac{n}{6}(m'vl_0)^2. \tag{8}$$

The dispersion of the components of momentum along the y and z axes will be the same. According to (8), the mathematical expectation of the square of the component of momentum along the x axis is

$$\overline{\Delta K_{1x}^2} = \frac{v^2l_0^2m'\Delta m}{6}. \tag{9}$$

Therefore,

$$\Delta K_1^2 = \Delta K_{1x}^2 + \Delta K_{1y}^2 + \Delta K_{1z}^2 = \frac{1}{2} v^2 l_0^2 m' \Delta m. \quad (10)$$

We substitute here $v^2 l_0^2$ from (3), on the right-hand side of the latter, assuming $v^2 = Gm/\theta r$, where θ is of the order of unity, and omitting the terms m' and r', which, as will be demonstrated below, are small in comparison with m and r. Then

$$\Delta K_1^2 = \left(1 + \frac{1}{2\theta}\right) Gmrm'\Delta m = \left(1 + \frac{1}{2\theta}\right) Gmrnm'^2. \quad (11)$$

The imparted specific momentum is inversely proportional to the root of n:

$$\Delta K_1 / \Delta m = \sqrt{(1 + 1/2\theta) Gmr / n}. \quad (11')$$

On the basis of the rule of addition of dispersion it is easy to obtain an expression for ΔK_1^2 in the more general case when the masses m'_j of falling bodies are different. This requires that expression (11) be summed for all m'_j. Assume n(m') is the mass distribution of bodies falling onto the planet; these bodies have the total mass

$$\Delta m = \int_0^{m_1} m' n(m') dm'. \quad (12)$$

Integrating (11) for all m' and substituting Δm from (12), we obtain

$$\Delta K_1^2 \approx \left(1 + \frac{1}{2\theta}\right) Gmr \frac{\int_0^{m_1} n(m') m'^2 dm'}{\int_0^{m_1} n(m') m' dm'} \Delta m, \quad (13)$$

where m_1 is the mass of the largest body, not counting the planet itself. Obviously, this relation makes sense when $m_1 \ll \Delta m \ll m$.

In general, the expression

$$J(m, m_1) = \frac{\int_0^{m_1} n(m') m'^2 dm'}{m \int_0^{m_1} n(m') m' dm'} \quad (14)$$

is a function of planetary mass m, since n(m') is dependent on time. If

$$n(m', t) = c(t) m'^{-q}, \quad (15)$$

then, when $q < 2$,

$$J(m, m_1) = \frac{\int_0^{m_1} m'^{2-q} dm'}{m \int_0^{m_1} m'^{1-q} dm'} = \frac{2-q}{3-q} \frac{m_1}{m}. \quad (16)$$

The masses m' of the falling bodies increase parallel with the growth of the planet; therefore, in the first approximation it can be assumed that m_1/m is constant. Then J = const. Assuming the density of the planet to be constant and integrating (13) for m, we find the value of the square of the random component of the rotational moment of the planet:

$$K_1^2 = \sum \Delta K_1^2 = \left(1 + \frac{1}{2\theta}\right) GJ \int_0^m m^2 r \, dm$$

$$\approx \left(1 + \frac{1}{2\theta}\right) GJ \frac{3}{10} m^3 r$$

and

$$K_1 = m \sqrt{\frac{3}{10}\left(1 + \frac{1}{2\theta}\right) JGmr}. \quad (17)$$

Allowance for an increase of density of the planet with m exerts virtually no influence on the results: the right-hand side of (17) increases only by a value of about 1%. It is possible that m_1/m increased in the final stage of growth of m. Then the masses of the largest falling bodies could be several tens of percent greater than the values determined below on the basis of (20) on the assumption J = const.

It was assumed in (13) that Δm is the total increment of the earth's mass, since we assume that virtually all bodies falling onto the earth imparted both regular and random components of rotation. According to the A. V. Artem'ev-V. V. Radzievskii hypothesis [4], the regular component of rotation was imparted by bodies not falling directly onto the planet but trapped by it as a result of their inelastic collisions in its zone of attraction, that is, by essentially the same mechanism which according to E. L. Ruskol [5] led to the formation of a satellite swarm around the planet. If the largest bodies were not present in the matter trapped in this way, or if these bodies were greatly broken down during collisions, the random rotational component which they imparted to the planet was small. Then (if the authors' hypothesis is accepted) these bodies should not be included in Δm. However, according to the authors' own estimate, the part of the matter falling onto the planet by such a "two-stage" method should be only several percent. The corresponding correction to

m_1 therefore falls within the limits of accuracy of our estimate.

The vector $\mathbf{K}_1$ has a random direction in space. Assume the angle between the systematic component of momentum $\mathbf{K}_0$ directed perpendicular to the orbital plane and $\mathbf{K}_1$ is equal to δ, and the angle between $\mathbf{K}_0$ and the vector of the total moment of momentum of the planet $\mathbf{K} = \mathbf{K}_0 + \mathbf{K}_1$ (inclination of the axis of rotation) is equal to ε. Then the component of momentum perpendicular to $\mathbf{K}_0$ is equal to

$$K_1 \sin\vartheta = K \sin\varepsilon. \tag{18}$$

The right-hand side of (18) is known from observations. On the left-hand side K_1 represents the relation (17) and the angle ϑ can have any value between 0 and π. As the probable value $\sin\vartheta$ in (18) it is natural to use its mean value. In the case of a uniform distribution of vectors K_1 over the sphere,

$$\overline{\sin\vartheta} = \frac{1}{4\pi}\int_0^\pi \sin\vartheta \, 2\pi \sin\vartheta \, d\vartheta = \frac{\pi}{4}. \tag{19}$$

Substituting $\overline{\sin\vartheta}$ and K_1, expressed through m_1/m using (16) and (17), into (18), we find

$$\frac{m_1}{m} = \frac{3-q}{2-q}\,\frac{160\sin^2\varepsilon}{3\pi^2(1+1/20)}\,\frac{K^2}{Gm^3r}. \tag{20}$$

For numerical estimates it is convenient to introduce the velocity of rotation at the equator v_r and the Keplerian angular velocity v_s at the surface of the planet:

$$K = \frac{2}{5}\mu m r v_r, \qquad v_c = \sqrt{Gm/r}. \tag{21}$$

Then from (20) we obtain

$$\frac{m_1}{m} = \frac{3-q}{2-q}\,\frac{10}{3(1+1/20)}\left(\frac{8\mu\sin\varepsilon}{5\pi}\,\frac{v_r}{v_c}\right)^2. \tag{22}$$

The masses of the largest bodies falling onto the planet, computed using this formula, on the assumption of a power function distribution for them with a value $q = 3/2$ (distribution by radii with the exponent $p = 3q - 2 = 2.5$) are given in the first column of Table 1.

For Uranus, in place of $\sin\varepsilon$ we used the ratio $\pi K_1/4K$, found on the assumption that the systematic component of the moment of Uranus K_0 corresponds to a period of rotation of 15 h (approximately the same as for Neptune).

When $\theta \geq 3$ the role of the parameter in (22) characterizing the relative velocities of bodies before encountering the planet is insignificant. We assumed $\theta = 3$. Change of q from $3/2$ to $5/3$ ($p = 3$) increases m_1/m by only $4/3$ times. Only when $q \to 2$ ($p \to 4$) does the result change appreciably. When $q = 2$ on the right-hand side of (22) in place of $(3-q)/(2-q)$ we have $\ln(m_1/m_0)$, where m_0 are the masses of the smallest particles in the used distribution. However, a distribution with $q = 2$, in which a large part of the mass is accounted for by large particles, apparently is unrealistic. Data on size distribution of asteroids [6, 7], meteors [8], and comets [9] indicate values p of about 2.6-3.4.

The second column of the table gives values m_1/m computed for $q = -\infty$, that is, for a case when all the falling bodies have identical masses. These values are smaller than the preceding values by a factor of three. Another limiting case can be considered, when the random component K_1 of the moment is imparted by only one body m_{11}, while all the remaining matter falling onto the planet imparts to it only regular rotation (K_0). Then

$$K\sin\varepsilon = \frac{\pi}{4}K_1 = \frac{\pi}{4}m_{11}lv = \frac{\pi}{4}m_{11}\frac{2}{3}l_0 v \tag{23}$$

and

$$\frac{m_{11}}{m} = \frac{6\sqrt{2}\,\mu}{5\pi\sqrt{1+1/2\theta}}\,\frac{v_r}{v_c}\sin\varepsilon. \tag{24}$$

The values m_{11}/m are given in the last column of the table. They are less variable from planet to planet than the values m_1/m and do not involve the assumption of a specific form of the size distribution function for the bodies. The values m_{11}/m can be considered the upper limit for masses of bodies falling onto the planets.

TABLE 1

Planet	m_1/m		m_{11}/m	Planet	m_1/m		m_{11}/m
	$q = 3/2$	$q = -\infty$			$q = 3/2$	$q = -\infty$	
Earth	$8\cdot10^{-4}$	$3\cdot10^{-4}$	$1\cdot10^{-2}$	Saturn	$3\cdot10^{-2}$	$1\cdot10^{-2}$	$6\cdot10^{-2}$
Mars	$2\cdot10^{-3}$	$6\cdot10^{-4}$	$1.3\cdot10^{-2}$	Uranus	$5\cdot10^{-2}$	$2\cdot10^{-2}$	$8\cdot10^{-2}$
Jupiter	$3\cdot10^{-4}$	$9\cdot10^{-5}$	$5\cdot10^{-3}$	Neptune	$5\cdot10^{-3}$	$2\cdot10^{-3}$	$2\cdot10^{-2}$

These results of computations reveal that despite the absence of final data on the size distribution function for the bodies the masses of the largest bodies falling onto the planets during the course of their formation are determined quite reliably: there is not more than a threefold deviation in either direction. The masses of the largest bodies falling onto the earth were about 10^{-3} of the mass of the earth. As a result of the lunar tidal effect the earth's rotation is slowed, and although the inclination ε of the axis increases, the value $v_r \sin \varepsilon$ decreases [10]. If the moon was formed considerably closer to the earth than its present position, $v_r \sin \varepsilon$ in the past was considerably greater. Therefore, it is not impossible that the value m_1/m for the earth determined above should be increased by a factor of 2-3.

The retrograde rotation of Uranus can be attributed naturally to the relatively greater sizes of the bodies forming the planet. The masses of the largest bodies falling onto Uranus attained 0.05 of the planetary mass. The bodies in the zone of formation of Saturn also were of considerable size. The largest of these bodies were 0.03 of the planetary mass. Therefore, the rotation of Saturn differs little in its anomalous character from that of Uranus. The apparent cause of the anomalies is the flight of larger bodies from the zone of Jupiter into the zones of these planets. Jupiter grew considerably more rapidly and attained a critical mass earlier. Upon attaining this critical mass the gravitational scattering of bodies from its neighborhood began. It should be noted that the estimates of m_1/m for Jupiter and Saturn made above are in need of appreciable refinement because they do not take into account the accretions of gaseous hydrogen in the final stage of growth of these planets. However, such a refinement is possible only on the basis of the theory of growth of the major planets, which requires consideration of a number of poorly studied factors, and such a theory has not yet been developed.

LITERATURE CITED

1. V. S. Safronov, Izv. AN SSSR, Ser. geofiz., No. 7, 1 (1965).
2. V. S. Safronov, Voprosy kosmogonii, 7, 59, 121 (1960).
3. B. V. Gnedenko, Course in the Theory of Probability [in Russian] (Fizmatgiz, Moscow, 1962), p. 181.
4. A. V. Artem'ev and V. V. Radzievskii, Astron. zh., 42, 124 (1965) [Soviet Astronomy – AJ, Vol. 9, p. 96].
5. E. L. Ruskol, Astron. zh., 37, 690 (1960) [Soviet Astronomy – AJ, Vol. 4, p. 657].
6. C. O. R. Jashek, Observatory, 80, 119 (1960).
7. S. Piotrowski, Acta Astron. ser. a, 5, 115 (1954).
8. H. Brown, J. Geophys. Res., 65, 1679 (1960).
9. E. J. Öpik, Monthly Notices Roy Astron. Soc., 120, 404 (1960).
10. H. Gerstenkorn, Z. Astrophys., 36, 245 (1955).

Planetary Rotation Induced by Elliptically Orbiting Particles (abstract)

A. V. Artem'ev
Translated from: Astron. Vestn. **3**, 18 (1969)
Sol. Syst. Res. **3**, 15 (1969)

A relationship was obtained between the moment of axial rotation of planets and the orbital eccentricity of protoplanetary particles in elliptic revolution. It is shown that a satisfactory explanation of the observed rotation of planets makes it necessary to take into account not only the orbital eccentricity but also the various inclinations and the orbital parameters of the particles in the environment of the Sun.

The Random Component of Planetary Rotation (abstract)

J. J. Lissauer and V. S. Safronov
Icarus **93**, 288 (1991)

We compute the root mean square angular momentum accumulated by a planet as a result of stochastic off-center impacts of large planetesimals during the accretionary epoch. Planets which accrete from large planetesimals have, on the average, shorter spin periods and greater obliquities than those which accrete exclusively from small bodies. We use the observed spin periods and obliquities of the planets in order to estimate the masses of the largest bodies to impact each planet during its history. Most, if not all, planets accreted one or more individual bodies having at least 1% of the final planetary mass. Several planets probably suffered at least one collision with a body having ⩾5% of the planet's final planetary mass. These results support giant impact theories for the origin of the Moon and Charon, and models which explain the excess iron in Mercury by the partial collisional stripping of that planet's mantle. The high angular momentum of the Pluto/Charon system suggests it formed as the result of a near-grazing collision of two bodies of comparable mass, although one or both of these constraints may be relaxed if the bodies approached each other at a speed significantly larger than their mutual escape velocity. If Saturn's obliquity was produced by stochastic impacts, then that planet must have accreted at least one "planetesimal" several times as massive as the Earth.

V. Formation of the Earth and Other Terrestrial Planets

On the Growth of Terrestrial Planets (abstract)

V. S. Safronov
Translated from: Vopr. Kosmog. **6**, 63 (1958)

The process of growth of the embryos of planets resulting from the fall of small particles and of larger bodies on them is examined. In the absence of crushing of bodies, small particles were swept out by the larger bodies at the early stage of planetary growth. The cloud transformed into a "swarm" and the further process of growth consisted principally in the fall of individual bodies on the embryos. The crushing of bodies at their collision prevented the swarm from a complete sweeping out of small particles. A considerable part of planetary matter was exposed to repeated crushings and cohesions. Gaseous matter did not take a great part in the process. The rate of the Earth's growth is considered, supposing the relative velocities of bodies to be determined by gravitational interaction with the largest bodies and the planet embryo. The heating of the Earth due to the shocks caused by the falling bodies, to the radioactive decay, and to the compression of the growing planet is estimated. The growth of the Earth had actually finished by the end of the first quarter of billion years. By that time the temperature in the Earth's center reached 1500 K. The surface of the Earth always remained cold. It is pointed out, in conclusion, that the process of growth of the major planets was essentially different.

Accumulation of Planets of the Earth Group

V. S. Safronov*

Problems of Cosmogony, Vol VII, pp. 63–70 (Joint Pubs. Res. Service, U.S. Dept. of Commerce, 1964)
Translated from: Vopr. Kosmog. **7**, 59 (1960)
Presented at General Meeting of International Astronomical Union, 1958

The accumulation process is calculated. The growth rate and the initial t^0 of the Earth are estimated. Some questions on gravitational instability in the dust component of the protoplanetary cloud, the disruption of colliding bodies and the rotation of the planets are discussed.

(1) Proceeding from the basic idea of the leading role of solid matter in the process of the planets' formation, in recent years the author has studied the mechanism of the accumulation (growth) of planets.

It is natural to consider that the original composition of the cloud was close to a solar one and that solid particles comprised but a small part of the cloud's mass. Moving along with the gas, they could not exert much effect on its movement and, in particular, could not appreciably diminish the critical density needed for the development of gravitational instability in the gas. However, in view of the absence in the cloud of conditions for the maintenance of turbulence[1–3] and for the rapid damping of the original irregular movements, there should have begun a separation of the dust and of the gas. The solid particles settled toward the equatorial plane of the cloud, forming there a layer of heightened density. When the density in the dust layer attained the Roche density, as a result of gravitational instability, it divided into a multitude of condensations. This process was described by Edgeworth[4] and in more detail by Gurevich and Lebedinskiy.[5]

From the very beginning, the condensations must have turned mostly in a forward direction, since on an average they had the positive momentum of condensed regions. The rotation of the condensations obstructed their unlimited compression, since at a reduction in the radius by several times, the rate of rotation increased to a Keplerian angular velocity. However they continued to compress in addition to combining during collisions. The most massive condensations compressed most quickly; an increase in their mass by 10^2 times was adequate for them to convert to solid bodies with a density of the order of unity.[6] There is no basis for assuming the existence of the "secondary condensations" (introduced by Gurevich and Lebedinskiy) with other regularities of combining and growth.

For the attainment in the dust layer of the Roche density $\rho_R \approx 10\rho^* = 10\ (3M_\odot/4\pi R^3)$, it is necessary that the particles' velocities did not exceed

$$v_{cr} \approx \sigma \sqrt{\frac{3\pi G}{2\rho_R}} \approx \sigma \sqrt{\frac{2GR^3}{M_\odot}},$$

where σ is the surface density of a layer. At Jupiter's distance from the Sun, this amounts to 2×10^2 cm/s, while at the distance of the Earth, it amounts to only several cm/s in all. The density needed for gravitational instability cannot be much less than the Roche density. Whatever the reasons were for the irregular movements of the particles in the cloud, it is natural to expect that they (the movements) were greater in the internal parts of the cloud than in the outer ones. Hence, in the zone of the Earth planets, the conditions for gravitational instability were much less favorable than in the zone of the major planets. One can scarcely consider that the cloud possessed an ideal symmetry and that in it there were not even very small irregular movements with velocities of several cm/s in all. Therefore the possibility is not excluded that in the region of the Earth planets, there was no gravitational instability in the dust layer, while the bodies grew directly from the particles as a result of their amalgamation during collisions. It may have been that as early as the initial stage of evolution, the protoplanetary cloud converted to a swarm of bodies, which originally were very flat, but gradually thickened in proportion to the growth of the mass of bodies and the intensification of their gravitational interaction. Basically, the gaseous material did not participate in the accumulation process. Only in the region of the major planets was there an accretion of gas by the most massive bodies during the last stage of their growth. The gas should have dissipated from the region of the Earth planets. The most effective is the mechanism of dissipation under the effect of corpuscular radiation, proposed by Kuiper.[7] Into the Earth's composition, there entered only those substances which had entered previously into the composition of solid particles and bodies.

(2) The growth of planetary embryos from bodies and particles falling on them was considered by Schmidt[8] and then by the author.[9,10] The rate of growth of an embryo with the mass m can be written as:

$$\frac{dm}{dt} = \gamma\pi r_e^2 \rho v = \frac{4\pi\gamma}{P} r_e^2 \sigma(t). \qquad (1)$$

*Affiliated with the O. Yu. Schmidt Institute of Earth Physics, USSR Academy of Sciences.

where m and r_e are the mass and the effective radius of the embryo, respectively, P is the period of revolution around the Sun, and γ is the probability of combining upon collision (it is usually assumed that $\gamma=1$). Expression (1) is fairly simple but contains a number of parameters that are not exactly known.

It can be considered that the amount of solid matter in the Earth's zone in the process of its growth remained constant, while for the surface density $\sigma(t)$, the simple relationship

$$\sigma(t)=\sigma_0\left(1-\frac{m}{Q}\right), \tag{2}$$

was typical, where Q is the present mass of the planet.

As an effective radius r_e, we can take the impact parameter, leading to collision:

$$r_e=r\sqrt{1+\frac{2Gm}{v^2 r}}.$$

Only for the very small particles with a radius less than 10^{-3}–10^{-2} cm could there occur a hydrodynamic accretion with a radius of capture larger by several times. Such particles comprised a relatively small part of the mass of the entire swarm and did not exert appreciable effect on the growth rate of the embryo. Since the number of bodies in the swarm was great, the fine particles are quickly depleted by these bodies. The value for r_e depends considerably on the relative velocities in the swarm. The value

$$v=\sqrt{\frac{Gm}{2r}},$$

adopted by Gurevich and Lebedinskiy has a simple physical meaning, namely the doubled kinetic energy v^2 equals the potential energy of interaction in case of the closest approach $Gm/2r$. This relationship correctly reflects the nature of the dependence v on the mass and radius of the embryo, but it disregards the factors leading to a reduction of the relative velocities; therefore the actual v value should be less. We have assumed

$$v=\sqrt{\frac{Gm}{\theta r}},$$

where θ equals several units. Then $r_e^2=(1+2\theta)r^2$ and from Eq. (1), we can find $m(t)$. A certain indeterminacy is introduced by the value for θ, which is not precisely known.

The results of the calculations were published in reports.[10,11] In the calculations, it was assumed that $\sigma_0=10$ g/cm^2. The Earth achieved 99% of its present mass in 10^8 yr, if $\theta=3$, and in 4×10^7 yr at $\theta=10$. (Previously, Schmidt obtained a much smaller growth rate, since he considered the velocities of the bodies as high and took the geometric radius of the embryo.) The heating of the growing Earth by impacts from falling bodies was slight; even during the maximum bombardment, the surface t^0 was not more than 350–400 K. (In a study made in 1954, the calculations were made for several constant values for v. The new results taking into account a gradually increasing v value correspond to lower t^0 values as compared to those obtained earlier.) The internal parts of the growing Earth were heated from the release of radioactive heat and from compression under the pressure of layers being added from above. In a composite examination of these effects,[11] we used the theoretical calculations of heating resulting from compression that were conducted by Lyubimova.[12] The original temperature of the Earth, i.e., the t^0 at the end of its growth was maximum in the center—in 10^8 yr, it increased to 1000 K. The temperature gradient was around -0.03 K per km. The age of the Earth could not be determined from the amount of meteoritic material falling on it at the present time, since this material is not a relic from the material in the Earth's "supply" zone.

The heating of Jupiter's surface, having occurred in a similar way in the process of its growth, was much greater and exceeded 1000 K. This permits us to explain the higher density of Jupiter's inner satellites.

The outer planets (Uranus, Neptune, and Pluto) grew much more slowly as a result of the large P and the smaller σ in formula (1). In order for them to grow to their present sizes in acceptable time periods, it is necessary to assume that the initial density of the matter in the region of these planets was much greater than the density computed from the contemporary masses of these planets. The modern theory of the origin of comets, according to which a cloud of comets originated as the result of the ejection of bodies by massive embryos of the major planets, arrives at the same conclusion.

(3) An important problem is the crushing of bodies during collisions. This grinding increased the amount of fine particles in the swarm and made it less transparent. For the most massive embryos, the collisions with other bodies were the least dangerous, since the kinetic force of an impact only slightly exceeded the potential energy at the surface of these bodies ($v^2/v_e^2=\frac{1}{2}\theta\ll1$).

During a collision, a considerable part of the energy converts to heat. Therefore after impact, a large part of the fragmented matter should fall once again on the embryo. This furnishes a basis for assuming that the grinding of the bodies did not change the basic tendencies in the growth of the planetary embryos. In spite of the unique complexity of the process of repeated grindings and combinings of bodies of various sizes, the growth of the most massive bodies, i.e., of the planetary embryos, can be described satisfactorily by the above-adduced relationships. The γ value should be close to unity and the indeterminacy of the product of $\gamma(1+2\theta)$ essentially reduces to the vagueness in θ. The slight increase in the growth rate, when we assume $\gamma=1$, is probably entirely compensated by the somewhat depressed value assumed for θ, and also by the failure to consider the more rapid accretion of the finer particles. For refining the growth rate of planets, at present of most importance is the refinement of the θ parameter, i.e., the refinement of the value for the relative velocity of bodies in the swarm.

For the less massive bodies, the collisions are a greater danger and their breaking up often occurs. One can arbitrarily differentiate two stages in the evolution of a swarm. The first was typified by small relative velocities of the bodies and by their combining, practically without crushing. At this stage, the number of fine particles quickly decreased. In proportion to the growth of the bodies, their velocities increased. In case of masses of the order of 10^{22} g, crushing began, while with masses of more than 10^{26} g, a partial or complete evaporation began. However, the evaporated material quickly condensed to fine particles. At this second stage, the greater part of the entire mass of the swarm passed through fragmentation. The number of fine particles in the swarm increased once again. The extent of the crushing process can be illustrated by the following rough estimation. The minimum amount of material that had undergone crushing during the entire process of growth can be found after we have examined the simplest case of a swarm consisting of homogeneous bodies. If β is the probability of fragmentation upon one collision, while $1-\beta$ is the probability of combining during the growth of the mass of bodies from m_0 to m, there will be fragmented the portion p from the total mass, equaling $p \approx 1-(m_0/m)^{\beta}$. Having assumed, e.g., $\beta=0.1$, $m_0 \approx 10^{22}$ g, and $m_0/m=10^{-5}$, we find $p \approx 0.7$. In reality, the proportion of the material having undergone fragmentation can be greater. The repeated fragmentation and combining of bodies of various sizes with different internal pressures and temperatures led to the formation of diverse and complex structures similar to those that we find in meteorites. The study of meteorites should yield valuable information on the properties of these intermediate bodies and the basic regularities in their formation. The physicochemical investigations of meteorites conducted in recent years by Urey[13] are of unique interest in this respect. The above-adduced concepts regarding the overall trend of the process agree fully with the results of these studies. The results of a detailed study of the chemical properties of meteorites testifies to their diverse qualities; this study was made by Yavnel'.[14]

For studying the process of the accumulation of protoplanetary bodies taking into account their fragmentations during collisions, it is necessary to develop the actual mechanism of fragmentation and also a more precise knowledge of the value for the relative velocity of the bodies. Specifically, this would permit us to find the distribution function of the bodies according to sizes and to follow its temporal variation.

(4) Certain data along these lines can be obtained from analyzing the rotation of the planets. The bodies falling onto a growing planet of mass m and radius r bring their own rotational moment, equaling

$$\Delta K = \alpha r v \Delta m,$$

where Δm, v, and αr are, respectively, the mass, the velocity prior to approach, and the impact parameter of the falling body. The randomly directed vectors of ΔK during the summation of the moments of many falling bodies for the most part dampen each other. Although the mean value for $\bar{\alpha}$ is small, it does not equal zero. It is combined from two components: a systematic and a random one. The random component is manifested in the inclination of the planet's equator to the plane of its orbit. Its value is closely associated with the masses of the largest falling bodies. Simple estimations demonstrate that at the final growth stage of a planet, the masses of falling bodies were less than 10^{-2} of the mass of the planet itself. In the opposite case (otherwise), the planets would not have had a forward rotation. Their axes would have been directed in the most haphazard way.

The systematic component determining the forward rotation of the planets testifies to the "asymmetry" of the impacts of the falling bodies, connected with the general rotation of the entire swarm. The simplest assumption as to the constancy of $\bar{\alpha}$ at $v \sim \sqrt{Gm/r}$ leads to $K \sim m^{5/3}$ and $\omega \approx$const, i.e., to the approximate permanence in the ω value during the planet's growth and to the approximate equality of the periods of the planets' rotation, which in general agrees with the factual data. One of the typical features of planets is the slight difference in their periods of rotation at a tremendous difference in masses. It would be desirable to find $\bar{\alpha}$ theoretically from a statistical examination of the restricted three-body problem and to compare it with the value obtained from the present rotation of planets. A quantitative agreement would be a good indication of the validity of our concepts on the mechanism of planets' formation.

(5) The process of the growth of major planets was much more complex as a result of a number of additional factors: (a) the dissipation of material from this zone; (b) the accretion of gas by the most massive embryos; and (c) the possible conversion of part of the hydrogen into a solid state. A solution of these questions is necessary to approach a quantitative consideration of the process of accumulation of large planets.

[1] H. Jeffreys, Proc. R. Soc. London Ser. A, **214**, 281 (1952).
[2] V. S. Safronov and Ye. L. Ruskol, (Dokl.) Akad. Nauk Sci. USSR **108**, 413 (1956); Vopr. Kosmog. **5**, 22 (1957).
[3] V. S. Safronov, in Proceedings of the Third Symposium on Cosmic Gas Dynamics, Rev. Mod. Phys. **30**, 1023 (1958).
[4] K. E. Edgeworth, Mon. Not. **109**, 600 (1949).
[5] L. E. Gurevich and A. I. Lebedinskiy, Izv. Akad. Sci. USSR Fiz. Ser. **14**, 765 (1950).
[6] V. S. Safronov, Vopr. Kosmog. 7 (1960), p. 121.
[7] G. P. Kuiper, Mem. Soc. R. Liege Collect. 4 **13**, 361 (1953).
[8] O. Yu. Schmidt, Dokl. Akad. Nauk Sci. USSR **46**, 392 (1945).
[9] V. S. Safronov, Astron. Zh. **31**, 409 (1954).
[10] V. S. Safronov, Vopr. Kosmog. **6**, 63 (1958).
[11] V. S. Safronov, Izv. Akad. Sci. USSR Ser. Geofiz. **1959**, (1), 139.
[12] Ye. A. Lyubimova, Izv. Akad. Sci. USSR Ser. Geofiz. **1955** (5), 416.
[13] H. C. Urey, Proc. Chem. Soc. March, 67 (1958).
[14] A. A. Yavnel', Astron. Zh. **34**, 445 (1957).

The Origin of the Earth (abstract)

B. Yu. Levin
Translated from: Izv. Akad. Nauk SSSR Fiz. Zemli **1972** (7), 5
Izv. Acad. Sci. USSR Phys. Solid Earth **1972** (7), 425

A few decades ago, evidence was obtained that the Earth was formed by accumulation of solid bodies and particles in a circumsolar swarm. Later it was shown that the accumulation must have lasted about 10^8 yr and that a large part of the Earth's mass was brought by bodies of asteroid size. During impacts of these large bodies several percent of the energy was liberated in the form of seismic waves, which heated the deep interior. In addition, heating of the neighborhoods of impact points created the initial temperature inhomogeneities of the interior which played a large role in its subsequent development.

Lack of knowledge concerning the origin of the circumsolar nebula, its early thermal history, and the condensation of gas into solid particles makes it difficult to study the Earth's chemical evolution and the nature of the Earth's core. It is almost impossible to reconcile the iron-core hypothesis with the contemporary solid state of the mantle. Such a state can be explained on the basis of the metallized-silicate hypothesis, but the latter is sharply criticized and is even rejected by many scientists.

Evidence exists that the upper mantle is not a residue from the separation of crustal material but is the product of a widespread differentiation process.

Time Scale for the Formation of the Earth and Planets and Its Role in Their Geochemical Evolution (abstract)

V. S. Safronov
Soviet–American Conference on Cosmochemistry of the Moon and Planets (June 1974), Report No. NASA SP-370, edited by J. H. Pomeroy and N. J. Hubbard (NASA, Washington, DC, 1977), pp. 797–803

The duration of the process of formation of the Earth and planets is discussed. A short time scale for formation of the Earth (10^4–10^5 yr) has been proposed, not from a consideration of the rate of its growth, but from geochemical and geophysical considerations. On the basis of the dynamics of a swarm of protoplanetary bodies and the process of accretion of the planets, the author has found an accumulation time for 98% of the mass of the Earth of 6×10^7–10^8 yr and a characteristic time for sweeping out of the protoplanetary cluster (reducing its mass by half) of 10^7 yr. It is shown that the shorter accretion time found by Opik, Cameron, Hallam, and Marcus is related to arbitrary assumptions about the parameters of the model for the swarm (e.g., the relative velocities of the bodies, the cluster density).

The initial mass of the solar nebula is discussed. Models of a massive nebula (two solar masses and more) encounter serious difficulties: an effective mechanism of transfer of the momentum from the central part of the nebula outward, capable of leading to formation of the Sun and removal of half the mass of the nebula from the solar system has not been found. As a consequence of the instability of these models, their evolution can end with the formation, not of a planetary system, but of a binary star. The possibility is demonstrated of obtaining acceptable growth rates for Uranus and Neptune by prolonging the thickening of preplanetary dust in the region of large masses.

The important role of large bodies in the process of formation of the planets is noted. The impacts of such bodies, moving in heliocentric orbits, could have imparted considerable additional energy to the forming Moon, which, together with the energy given off by the joining of a small number of large protomoons, could have led to a high initial temperature of the Moon.

Heating of the Earth by the Impact of Accreted Bodies (abstract)

V. S. Safronov and S. V. Kozlovskaya

Translated from: Izv. Akad. Nauk SSSR Fiz. Zemli **1977** No. 10
Izv. Acad. Sci. USSR Earth Phys. **13**, 677 (1977)

The thermal constitution of the Earth, accumulated from solid bodies, is traced, and the principal source of heat, the impact of fallen bodies, is studied. The quasistationary equation of thermal conductivity for a moving, spherically symmetrical surface layer of the Earth, heated by energy of impacts of bodies and cooled by the impact mixing of its material, is solved. The temperature at the base of the layer depends substantially on the dimensions of the fallen bodies and is almost independent of the rate of accumulation of the Earth. The heating from small bodies of radius r' less than some characteristic dimension, r_1, is insignificant. When $r' > r_1$, the dimensions of the craters are decreased substantially by the gravitational field, and the heating is proportional to the ratio r'/r_1. For the present Earth, $r_1 \sim 1$ km. The heating of the layer by impacts of bodies of different dimensions, distributed along the radii according to an inverse power law with an exponent $p=3.5$, is evaluated. In the final stage of growth of the Earth, the fusion point is reached in the layer if the radius of the largest body in the group exceeds 20 km. The presence of huge impact craters on all planets of the terrestrial group attests to the existence of such large-scale preplanetary bodies. It is thus highly probable that the outer layer of the Earth was fused for a thickness of some hundreds of kilometers in the final stage of its growth. Then, the beginning of the gravitational differentiation process and the formation of the core must also relate to this period. More confident conclusions will be possible when reliable data on the distribution of the dimensions of the large preplanetary bodies have been obtained. The possibility of the use of the method in the estimation of the initial temperature of the other planets is discussed.

Limiting masses, distances, and times for the accumulation of planets of the terrestrial group

A. V. Vityazev, G. V. Pechernikova, and V. S. Safronov

Institute of Earth Physics, Academy of Sciences of the USSR

(Submitted February 15, 1977)

Astron. Zh. **55**, 107–112 (January–February 1978)

The growth of a planetary embryo having a feeding zone which expands in proportion to the increase in the embryo's mass is discussed. The upper limit of the planet's mass and the width of its feeding zone are estimated when its growth is not limited by the presence of neighboring planets. For planets of the terrestrial group these quantities are close to the current masses of the planets and the distances between them.

PACS numbers: 96.30.–t, 95.10.Ce

1. According to modern concepts, in the course of the evolution of the protopolanetary cloud a swarm of solid bodies was formed rotating about the sun in orbits with relatively small eccentricities and inclinations. As the masses of the bodies increased owing to their amalgamation the gravitational interaction of the bodies was strengthened and their relative velocities grew. The properties of such a system of bodies have been discussed in Refs. 1-3. Mass distribution functions for the bodies and an expression for the parameter θ characterizing the average relative velocities in a system of bodies amalgamating and fragmenting during collisions were obtained, in particular. The conclusion that the largest individual bodies of the swarm grew considerably faster than the others was also drawn there. Their orbits approached circular orbits owing to the effect of averaging the eccentricities of the orbits of the small amalgamating bodies. According to Refs. 1 and 2, the rate of growth in the mass m of such a planetary "embryo" can be written in the following form:

$$\frac{dm}{dt}=\frac{4\pi(1+2\theta)}{P}\sigma r^2, \tag{1}$$

where P is the period of the orbital motion, r is the radius of the growing planet, and σ is the surface density of matter in its feeding zone. The masses of planets of the terrestrial group are too small to allow them to eject solid bodies from their zone through gravitational perturbations. Therefore, the initial surface density σ_0 of solid matter is determined from the current masses of the planets. The values of σ_0 are obtained in Ref. 1, where the boundaries of the planetary zones are taken in accordance with Shmidt (Ref. 4) and with Gurevich and Lebedinskii,[5] with σ_0 being taken as constant within the limits of each zone. In accordance with Shmidt, the decrease in σ with time owing to depletion of the matter by the growing planetary embryo was written in the form

$$\sigma(t)=\sigma_0\left(1-\frac{m(t)}{M}\right), \tag{2}$$

where M is the final mass of the planet, equal to the mass of all the matter in the zone. It is more natural to take σ not as a stepped but as a rather smooth function $\sigma(R)$ of the distance from the sun, in the form

$$\sigma(R,\ t=0)=\sigma_{0i}R^{-\nu}, \tag{3}$$

let us say, perhaps with each zone having its own value of ν. By definition we have

$$M=2\pi\int_{R'}^{R''}\sigma R dR, \tag{4}$$

where R' and R" are the inner and outer boundaries of a planetary zone, respectively:

$$R'=R-\Delta R,\quad R''=R+\Delta R. \tag{5}$$

Here R is the current distance of the planet from the sun and ΔR is the half-width of its feeding zone, estimated from the laws of planetary distances.[4,5]

On the other hand, the feeding zone of a growing embryo, according to Ref. 1, consisted of an annular region whose width 2ΔR was determined by the average eccentricity e of the orbits of the main mass of bodies:

$$\Delta R\approx eR\approx\sqrt{2}\frac{v}{V_c}R, \tag{6}$$

with the average relative velocity

$$v=\sqrt{\frac{Gm}{\theta r}}, \tag{7}$$

where $m(t) = (4/3)\pi\delta r^3(t)$ and δ is the density of the embryo, being taken as constant; V_c is the Keplerian circular velocity at the distance R from the sun. Substituting v and V_c into (6), we obtain

$$\Delta R(R,t)=2\sqrt{\frac{2\pi\delta}{3\theta M_\odot}}R^{3/2}r(t). \tag{8}$$

It is seen from (8) that ΔR increases with time in proportion to the radius of the growing embryo.

2. Let us consider the scheme of growth of an individual embryo when its feeding zone is determined by the mass of the embryo itself. We neglect the flux of matter through the boundaries of the zone, the "drift" of the embryo along R, the eccentricity of the embryo's orbit, and the variation in the surface density distribution

 0038-5301/78/2201-0060$02.70

(3) in the process of growth. According to Ref. 1, in the absence of gas θ depends on the mass distribution function of the bodies, i.e., it is a function of R and t, generally speaking. This dependence is weak, however, and for simplicity we will neglect it below, using some average values of θ found in Ref. 1 in the estimates.

Let Q be the mass of all the matter in the feeding zone (including the embryo m), which we write in the form

$$Q(R,t)=2\pi\int_{R-\Delta R(t)}^{R+\Delta R(t)}\sigma(R,0)\,RdR. \tag{9}$$

In the feeding zone σ(R, t), with allowance for (3), will be

$$\sigma(R,t)=\sigma_{0|}R^{-\nu}\left(1-\frac{m(t)}{Q(t)}\right). \tag{10}$$

Substituting (3) into (9) and integrating, we obtain

$$Q(R,t)=2\pi\sigma_{0|}\int_{R(1-e)}^{R(1+e)}R^{1-\nu}dR$$
$$=\begin{cases}\dfrac{2\pi\sigma_{0|}}{2-\nu}R^{2-\nu}[(1+e)^{2-\nu}-(1-e)^{2-\nu}] & \text{for } \nu\neq2,\\ 2\pi\sigma_{0|}\ln\dfrac{1+e}{1-e} & \text{for } \nu=2,\end{cases} \tag{11}$$

where

$$e=2\sqrt{\frac{2\pi\delta R}{3\theta M_\odot}}\,r(t).$$

From (1), with allowance for (10), we obtain

$$\frac{dr}{dt}=\begin{cases}\dfrac{(1+2\theta)\sigma_{0|}R^{-\nu}}{P\delta}\left\{1-\dfrac{2\delta(2-\nu)r^3(t)}{3\sigma_{0|}R^{2-\nu}[(1+e)^{2-\nu}-(1-e)^{2-\nu}]}\right\} & \text{for } \nu\neq2,\\ \dfrac{(1+2\theta)\sigma_{0|}R^{-\nu}}{P\delta}\left[1-2\delta r^3(t)/3\sigma_{0|}\ln\dfrac{1+e}{1-e}\right] & \text{for } \nu=2.\end{cases} \tag{12}$$

It follows from (12) that a planet ceases to grow when some maximum radius (mass) is reached, the value of which is determined in the present model mainly by the distance from the sun and the surface density of matter in the zone. Acutally, dr/dt = 0 when the expression in curly brackets is reduced to zero. From the latter condition we obtain, in a first approximation with respect to e,

$$r_{\max}=\left(\frac{6\pi}{\theta M_\odot\delta}\right)^{1/4}(2\sigma_{0|})^{1/2}R^{(5-2\nu)/4}, \tag{13}$$

$$m_{\max}=\frac{4}{3}\pi\delta\left(\frac{6\pi}{\theta M_\odot\delta}\right)^{3/4}(2\sigma_{0|})^{3/2}R^{(15-6\nu)/4}, \tag{14}$$

$$\Delta R_{\max}=2\left(\frac{\delta}{3}\right)^{1/4}\left(\frac{2\pi}{\theta M_\odot}\right)^{3/4}(2\sigma_{0|})^{1/2}R^{(11-2\nu)/4}. \tag{15}$$

The values of σ_0 obtained in Ref. 1 can be used for the case of ν = 0, i.e., σ(R, 0) = σ_0 = const. The values of m_{max} and ΔR_{max} calculated from (14) and (15) and their actual values of the planets of the terrestrial group are presented in Table I. One can see that the calculated values of m_{max} and ΔR_{max} for Venus with θ = 2 and for the earth with θ = 3 are close to their current values (ΔR_{cur} was calculated from Eq. (8) with the actual values of the parameters). The disagreements are considerable for Mercury and Mars, however. This is evidently due to two circumstances. First, the eccentricity of the planet's own orbit was neglected above. This will clearly not do for Mercury: ΔR_e due to its eccentricity is $\sim 1.2\cdot10^{12}$ cm, i.e., three to four times greater than the width of the zone due to its gravitational action. In determining the actual width of Mercury's feeding zone, therefore, one must sum these two effects:

$$\Delta R_{\text{eff}}=\Delta R_{\max}+\Delta R_e\approx1.3\cdot10^{12}\ \text{cm},$$

which corresponds to m_{max} = $1.42\cdot10^{26}$ g (with σ_0 = 1.5 g/cm^2 and θ = 2), i.e., still two times less than the current mass of Mercury. Second, σ_0 for the region of Mercury was estimated by smearing out its current mass over the entire region from the boundary of the zone of Venus to the sun. In reality, the inner boundary of the cloud evidently passed at some distance from the sun. Agreement between m_{max} and m_{cur} is achieved with σ_0 = 3 g/cm^2.

For Mars one must also allow for the broadening of its feeding zone due to the eccentricity of its orbit, since

$$\Delta R_{\max}=2.22\cdot10^{12}\ \text{cm (for } \theta=2),$$
$$\Delta R_e=2.11\cdot10^{12}\ \text{cm}.$$

For a total zone width ΔR_{eff} = $4.3\cdot10^{12}$ cm we have $m_{max}\approx3.7\cdot10^{26}$ g (with σ_0 = 0.3 g/cm^2). It should be noted that the value of σ_0 used for Mars is also somewhat understated. Only for σ_0 = 0.5 g/cm^2 do we have $m_{max}\approx m_{cur}$. According to Ref. 1, the initial surface density of matter in the region of the orbit of Mars could be even higher, since a considerable portion of the solid matter was extracted from the zones of Mars and the asteroids by the massive Jupiter. In explaining the effective broadening of the feeding zones of Mercury and Mars we assume that their eccentricities at the end of the accumulation were almost equal to the current values. The closeness of m_{cur} and m_{max} for Mercury and Mars, calculated with a correction for their eccentricities, indicates the reality of such an assumption.

3. The growth time of a planet is obtained by integrating (12) with allowance for (11):

$$\frac{dr}{dt}=(1-a^2r^2)/b. \tag{16}$$

For the simplest case of ν = 0 we have

$$a^2=\sqrt{\frac{\theta\delta M_\odot}{6\pi}}\frac{1}{2\sigma_0R^{5/2}},\quad b=\frac{\delta P}{(1+2\theta)\sigma_0}. \tag{17}$$

Making the change ar = z, we obtain

$$\frac{a}{b}dt=\frac{dz}{1-z^2} \tag{18}$$

and

TABLE I

Planets	R, cm	σ_0, g/cm²	δ, g/cm²	m_{cur}	m_{max}		$\theta=2$		$\theta=3$	
					$\theta=2$	$\theta=3$	ΔR_{cur}	ΔR_{max}	ΔR_{cur}	ΔR_{max}
Mercury	$0.58\cdot10^{13}$	1.5	5.45	$3.30\cdot10^{26}$	$1.38\cdot10^{25}$	$1.02\cdot10^{25}$	$3.62\cdot10^{11}$	$1.03\cdot10^{11}$	$2.97\cdot10^{11}$	$7.60\cdot10^{10}$
Venus	$1.08\cdot10^{13}$	16	5.25	$4.87\cdot10^{27}$	$4.89\cdot10^{27}$	$3.61\cdot10^{27}$	$2.25\cdot10^{12}$	$2.26\cdot10^{12}$	$1.84\cdot10^{12}$	$1.66\cdot10^{12}$
Earth	$1.5\cdot10^{13}$	10	5.52	$5.98\cdot10^{27}$	$8.32\cdot10^{27}$	$6.14\cdot10^{27}$	$3.96\cdot10^{12}$	$4.42\cdot10^{12}$	$3.25\cdot10^{12}$	$3.26\cdot10^{12}$
Mars	$2.27\cdot10^{13}$	0.3	3.94	$6.42\cdot10^{26}$	$1.90\cdot10^{26}$	$1.40\cdot10^{26}$	$3.33\cdot10^{12}$	$2.22\cdot10^{12}$	$2.78\cdot10^{12}$	$1.64\cdot10^{12}$

$$t=\frac{b}{2a}\ln\left|\frac{1+z}{1-z}\right|. \tag{19}$$

We estimate the time of growth of a planet up to the moment when its mass reaches 97% of m_{max}, i.e., $z = 0.99$ and $\ln|(1+z)/(1-z)| = 5.3$. Then

$$t=5.3\left(\frac{6\pi}{\theta M_{\odot}}\right)^{1/4}\frac{\delta^{3/4}R^{5/4}}{(1+2\theta)(2\sigma_0)^{1/2}}. \tag{20}$$

If we take $\bar{\delta} = 4.5$ g/cm³, $\theta = 3$, and $\sigma_0 = 10$ g/cm² for the earth, then $t = 1.15\cdot10^8$ yr.

For the same values of $\bar{\delta}$, θ, and σ_0 and for a constant $Q = M_{\oplus}$ Safronov[1] obtained $t = 0.88\cdot10^8$ yr. Consequently, the variation of Q with time hardly affects the estimate of the growth time of the earth. It is easy to verify that this conclusion remains in force for the other planets of the terrestrial group also.

We note that earlier Harris[6] proposed a correction to Eq. (2) in the form of an additional multiplier $r_{\oplus}/r$ to m/M, i.e.,

$$\sigma=\sigma_0\left(1-\frac{m(t)}{M_{\oplus}}\frac{r_{\oplus}}{r(t)}\right)=\sigma_0\left(1-\frac{r^2(t)}{r_{\oplus}^2}\right),$$

where $M_{\oplus}$ and r are the current mass and radius of the earth (only the case of $\nu = 0$ was considered). Harris's corresponding expression for the growth rate of the earth is analogous to (16), where $a = 1/r_{\oplus}$. In our analysis the quantity a is determined by the parameters of the protoplanetary cloud in the region of the forming planet [see (17)].

4. With a number of assumptions (see Sec. 2), the idealized model for the growth of embryos with nonintersecting feeding zones allows one to obtain simple expressions for some limiting characteristics (r_{max}, m_{max}, etc.). A fuller analysis allows one to bring out several other interesting features of the accumulation of planets, but no longer in such simple form. In this note we are confined to only pointing out the effects which must be taken into account in stricter estimates.

a) The eccentricities and the ΔR corresponding to them were taken as equal in the outer and inner zone above. For a wide enough zone, however, they may differ significantly. Taking θ as constant as before, we write the minimum of the possible eccentricities at the edge of the inner zone as

$$e'=\sqrt{\frac{2mR'}{\theta rM_{\odot}}}$$

and the corresponding maximum of the possible eccentricities at the edge of the outer zone as

$$e''=\sqrt{\frac{2mR''}{\theta rM_{\odot}}}.$$

It is easy to obtain the following relations:

$$R'=\frac{R}{1+e'},\quad R''=\frac{R}{1-e''},$$
$$\Delta'R=R\frac{e'}{1+e'},\quad \Delta''R=R\frac{e''}{1-e''}, \tag{21}$$
$$e'^2+e'^3=e^2=e''^2-e''^3.$$

Since $\Delta'R/\Delta''R = e'(1-e'')/e''(1+e') \le 1$, the width $\Delta'R$ of the inner zone is less than the width $\Delta''R$ of the outer zone. The limits R' and R'' in the interval (4) can be expressed through R as before and determined as

$$R=R'+\sqrt{\frac{2m}{\theta rM_{\odot}}}R'^{3/2},\quad R=R''-\sqrt{\frac{2m}{\theta rM_{\odot}}}R''^{3/2}. \tag{22}$$

For the parameters characterizing the zone of the earth, the estimates give

$$\frac{\Delta''R}{\Delta'R}\simeq1.7;\qquad\frac{2\Delta R_{max}}{\Delta'R+\Delta''R}\simeq0.97.$$

b) Earlier we neglected the possible movement of an embryo along R. The drift of bodies in the presence of a gas was discussed in Ref. 7. At the stage of large bodies it is insignificant. The drift of an embryo, an accumulating body with different angular momenta relative to the sun, can be estimated from the relation

$$R^{1/2}\approx\int_{R'}^{R''}R^{1/2-\nu}dR\Big/\int_{R'}^{R''}R^{1-\nu}dR, \tag{23}$$

where R is the radius of the orbit of the planet which is formed, having exhausted all the matter between R' and R''. Taking $\nu = 0$ and $\sigma = 10$ g/cm² with $\theta = 3$ for the zone of the earth, we find the movement of the embryo from the middle of the zone: $\delta R \approx 2\cdot10^{11}$ cm $\approx 0.06\cdot\Delta R_{max}$. For $\nu = 3$, corresponding to a very steep decline in surface density, $\delta R = -5\cdot10^{11}$ cm $\approx -0.14\Delta R_{max}$. It is clear that in the zone of planets of the terrestrial group such movements could not significantly affect the character of the accumulation.

5. Repeated attempts have been made to explain the law of planetary distances on the basis of one or another concepts about the formation of the planets. The statistical

significance of the "Titius–Bode law" has been subjected to doubt. The simplest form of this law is

$$R_n = R_0 b^n \quad \text{or} \quad \frac{R_{n+1}}{R_n} = b,$$

where R_n is the radius of the orbit of the n-th planet and R_0 and b are constants chosen empirically. According to Ref. 8, $\bar{b}$ = 1.73 for the planetary system.

If one assumes that the growth of the planets of the terrestrial group took place in accordance with the model of nonoverlapping growth zones and ended when these zones joined together, then for the distances between two adjacent planets one can write

$$R_{n+1} - R_n = \Delta_n R + \Delta_{n+1} R,$$

from which

$$R_{n+1}/R_n = (1+e_n)/(1-e_{n+1}). \tag{24}$$

In accordance with (6) and (15), and with allowance for (14), we have

$$e_n \propto m_{\max}^{(7-2\nu)/(15-6\nu)}. \tag{25}$$

These relations give the connection between the distances of the planets and their masses. In the case when the orbit of the n-th planet has a large enough eccentricity e_0, one must taken the sum $e_n + e_0$ instead of e_n in (24).

Even from Table I it is seen that with reasonable values of θ and σ the sizes of the feeding zones of the planets correspond to the current distances between them. Partial overlapping of the feeding zones occurs (for Venus and the earth, for example). However, these results naturally do not rule out the possibility of subsequent changes in the distances of the planets under the action of their mutual gravitational perturbations.

Summing up what has been said, one can conclude that planetary distances relatively close to the current ones (with an accuracy of the first dozen percent) arose in the very process of accumulation of the planets. Their subsequent alteration as a result of the prolonged evolution of the solar system probably was not so substantial as to disturb the basic regularities of the distribution.

[1]V. S. Safronov, Evolution of the Protoplanetary Cloud and Formation of the Earth and the Planets [in Russian], Nauka, Moscow (1969); English translation; Israel Programs for Scientific Translations, Jerusalem (1972).
[2]V. S. Safronov, On the Origin of the Solar System, CNRS, Paris (1972), p. 89.
[3]G. V. Pechernikova, V. S. Safronov, and E. V. Zvyagina, Astron. Zh. 53, 612 (1976) [Sov. Astron. 20, 346 (1976)].
[4]O. Yu. Shmidt, Four Lectures on the Theory of the Origin of the Earth [in Russian], 3rd ed., Izd. Akad. Nauk SSSR, Moscow (1957).
[5]L. E. Gurevich and A. I. Lebedinskii, Izv. Akad. Nauk SSSR, Ser. Fiz. 14, 765 (1950).
[6]A. Harris, Dissertation, UCLA, Los Angeles (1975).
[7]C. Hayashi, K. Nakazawa, and I. Adachi, Preprint KUNS 379, Kyoto (1976).
[8]M. A. Blagg, Mon. Not. R. Astron. Soc. 73, 414 (1913).

Translated by Edward U. Oldham

Influence of the Greenhouse Effect in Primordial Atmospheres on the Formation of the Earth and Venus (abstract)

D. D. Kvasov, A. P. Gal'tsev, and A. S. Safrai
Translated from: Astron. Vestn. **14**, No. 2, 72 (1980)
Sol. Syst. Res. **14**, 55 (1980)

The greenhouse effect has been calculated in the primordial atmospheres, which could have been formed in the case of the prompt degassing of part of the volatile materials that are contained in the interiors of Earth and Venus. These atmospheres would have consisted mainly of CO_2 and H_2O in the ratio of 1:4. In the case of $\gtrsim 1/500$ of the volatile materials which were there in the Earth and $\gtrsim 1/1000$ of those which were there in Venus going into the atmosphere, the high surface temperatures would have made the formation of liquid water impossible: development would have gone along the course of that of Venus. Assuming the existence of the hydrosphere, the Earth never had an atmosphere whose greenhouse effect would have caused the heating of its surface to higher than the temperature of boiling water. This creates insurmountable difficulties for the theories of the condensation of planets of the terrestrial group from a gaseous solar nebula, of the rapid accretion of meteoritic material, and also of the capture of the Moon by the Earth. In all these cases, at some stage in the Earth's development, a thick atmosphere would have existed for it, and the main atmospheric components would never have gone into the hydrosphere and the lithosphere. Only a theory of the slow accretion of the Earth and a theory of the formation of the Moon in a near-Earth orbit permit the existence of a primodial atmosphere of the Earth with a low pressure and which does not hinder the start of the hydrosphere. The occurrence of an Earth-type planet that is suitable for life is very unlikely.

Possibility of Formation of an Originally Inhomogeneous Earth (abstract)

A. B. Makalkin
Phys. Earth Plan. Int. **22**, 302 (1980)

A process of segregation of iron-rich from silicate-rich particulate matter in the protoplanetary nebula has been investigated in a quantitative model taking into account aerodynamics and intergrowth of dust grains. Segregation should have occurred if silicate-rich grains were smaller than iron-rich, the latter being about 10^{-2} cm in size or larger. In this paper the model is briefly described and geophysical implications are discussed.

The following conclusions are inferred. (1) The Earth and inner planets would form with an Fe/Si abundance ratio higher than the primitive ratio in the nebula (=solar ratio). (2) If the gas was in the Earth's zone of the solar system when the accumulation of the present-core-size embryo the Earth was finishing, the Earth could form as an inhomogeneous planet with the original iron-rich core and a silicate mantle. The same conclusion is applicable to Venus and Mercury.

Numerical Investigations of the Terrestrial Planets' Accumulation (abstract)

S. I. Ipatov
Institute of Applied Mathematics, Academy of Science, USSR Preprint No. 144 (1982)

The problem of the terrestrial planets' accumulation from a protoplanet disk is investigated. The investigations are carried out on the basis of computer simulations of flat rings consisting of gravitating bodies moving around the Sun and coagulating at collisions. The results show that the number of planets formed in the terrestrial feeding zone would be larger than the actual number of planets if we assume that all bodies move in the same plane and if we do not take into account the effect of the gas resistance force on the planetesimals' motion. The necessary number of planets in the terrestrial feeding zone could be obtained during the solid bodies' accumulation if we investigate a spatial model and admit the possibility of the solid bodies' coagulation under high velocity collisions.

State-of-the-Art for the Theory of the Origin of the Earth (abstract)

V. S. Safronov
Translated from: Izv. Akad. Nauk SSSR Fiz. Zemli (1982) No. 6, 5
Izv. Acad. Sci. USSR Phys. Solid Earth **18**, 399 (1983)

The fundamental stages of the process of formation of the solar system are discussed. Substantial progress is noted in the study of the initial stage of condensation (collapse) of a solar nebula, which is, in turn, today a problem of unifying the models of evolution for the protosolar nebula and the evolution of the preplanetary cloud into a single model of the origin of the solar system. The possibility is discussed of the formation, at the time of nebular collapse, of a turbulent, preplanetary disk around a nucleus and the expansion of this disk to the dimensions of the present-day solar system. The process of the formation of many planetary embryos, which grow more rapidly than all the remaining bodies in the zones of nourishment, is considered. Relative velocities of bodies in such a system are estimated. In the intermediate stage, in the zone of planets of the terrestrial group, the parameter θ, which characterized the velocity of the bodies, was increased to 10–15, and toward the end of accumulation was reduced to 1–2. The full duration of the growth of the Earth remains, as before, of the order of 10^8 yr. Heating of the Earth, which was being formed by means of large bodies, is discussed. The outer layer, which contains about half the mass of the Earth, was heated to the melting temperature (viscosity $\approx 10^{17}$ P). Convection prevented further heating. Differentiation may have begun at the end of accumulation.

The Fractionation of Matter in the Course of the Formation and Evolution of the Earth (abstract)

A. V. Vityazev
Translated from: Izv. Akad. Nauk SSSR Fiz. Zemli No. 6, 52 (1982)
Izv. Acad. Sci. USSR Phys. Solid Earth **18**, 434 (1982)

The possibility is considered of combining various trends in the chemistry of space and the theory of the accumulation of the planets which was developed at the Institute of the Physics of the Earth (IPE), USSR Academy of Sciences, on the basis of Schmidt's ideas. Fundamental processes in the evolution of the protoplanetary cloud are considered, and are accompanied by the evolution of the composition of the matter. The question is discussed of the loss of gas and volatiles, which make up more than 99% of the original mass of the matter in the zone of planets in the terrestrial group. An interpretation is given for the variety observed in the composition of the asteroids and planets, which is consistent with theoretical and experimental data on the impact reworking of matter in the course of the planetary system's formation. The possibility is mentioned of a lateral inhomogeneity in the mantle in chemical composition as a result of its incomplete homogenization in the course of the formation of the Earth's core and mantle.

Solar Wind and Noble Gases in Atmospheres of Mars and Earth (abstract)

A. V. Vitjazev
Lun. Plan. Sci. **16**, 883 (1985)

There is considerable evidence now in support of the hypothesis that the differentiation of the Earth and formation of the mantle and the core, as well as much degassing and formation of the significant fractions atmosphere and hydrosphere, have occurred during the later stages of accumulation.[1]

Here we shall analyze primordial noble gases data to estimate the formation interval of ionospheres and magnetospheres of Mars and Earth.

(1) The amount of the ith isotope ${}^iX$ in the atmosphere of the kth planet $M_k({}^iX)$ can be evaluated from the mass-balance equation:

$$\frac{d}{dt} M_k({}^iX) = \sum_j \frac{d}{dt} \epsilon_{k,j}({}^iX), \qquad (1)$$

where ϵ_j is the ith source (or sink) strength. Previously, the following ϵ_j were considered: (1) gas accretion from solar nebula, (2) outgassing during accumulation of the planets and their following evolution, (3) gas escape from atmospheres, and (4) gas loss to surface rocks. Capture of the solar wind particles into the atmospheres of planets during their evolution (T=4.6 billion years) has not been considered before. In view of the fact of "nonsolar" ratios of the isotopes it seems clear that the solar wind could not be an effective source of the noble gases for atmospheres. Nevertheless let us estimate the amounts of the noble gases which could be incorporated into atmospheres by the solar wind:

$$\epsilon_{k,5}({}^iX) = m_p \mu_i \delta_i S_k \int_0^T \zeta_k({}^iX) N_k V_k dt, \qquad (2)$$

where m_p is the mass of proton, μ_j the mass of the ith isotope, δ_i its relative (${}^iX/H$) abundances in the solar wind, S_k the geometrical cross section of the solar wind collision with the planet, ζ_k the capture efficiency of the particles in the atmosphere, N_k the quantity of protons, and V_k the solar wind velocity.

We suppose that $\zeta_k({}^iX)=1$ up to the moment ($t=T_k$) of the formation of the ionosphere or magnetosphere. After their formation ($T_k \leq t \leq T$) due to the shielding effect $0<\zeta_k({}^iX)\leq 1$. The averaged flux of solar protons in the interval $0<t\leq T$ can be denoted by $\alpha\langle N_k V_k\rangle$ where $\langle N_k V_k\rangle$ is its

TABLE 1.

^iX	M_3[a]	max $\epsilon_{3,5}$	M_4	max $\epsilon_{4,5}$
^{20}Ne	6.6×10^{16}	1.8×10^{17}	$(0.8–9.6)\times10^{13}$	2.2×10^{16}
^{36}Ar	2.1×10^{17}	8.9×10^{15}	$(1–1.7)\times10^{4}$	1.1×10^{15}
^{84}Kr	2.3×10^{16}	6.1×10^{12}	$(0.2–3.9)\times10^{13}$	7.6×10^{11}
^{132}Xe	2.0×10^{15}	4.3×10^{11}	$(0.36–6.3)\times10^{12}$	5.3×10^{10}

[a]Mass in grams.

present value (for the Earth $\langle N_3V_3\rangle=2.5\times10^8$ at/cm s). The data are summarized in Table 1. M_3 and M_4 were evaluated according to Ref. 2, maximum $\epsilon_{k,5}$ were estimated according to Ref. 2 for $\alpha=1$, $\zeta_k(^iX)=1$, and δ_i from Ref. 3.

(2) One can see that values of $\epsilon_{3,5}(^{20}\mathrm{Ne})$, $\epsilon_{4,5}(^{36}\mathrm{Ar})$, and $\epsilon_{4,5}(^{20}\mathrm{Ne})$ are rather high. Because

$$\epsilon_{4,3}(^{36}\mathrm{Ar}),\epsilon_{4,4}(^{36}\mathrm{Ar})\ll\epsilon_{4,1}(^{36}\mathrm{Ar}),\epsilon_{4,2}(^{36}\mathrm{Ar})$$

one of the possible explanations of the fact that solar wind did not incorporate a large amount of ^{36}Ar in the Martian atmosphere is the early formation of the ionosphere or the core with a magnetic field, which caused the origin of magnetosphere. They prevented the effective penetration of the solar wind into the atmosphere. From Eqs. (1) and (2) we can obtain a crude estimate for the moment of origin of shielding screen:

$$\frac{T_4}{T}=\left(\frac{M_4(^{36}\mathrm{Ar})-\Sigma\ \epsilon_{4,j}(^{36}\mathrm{Ar})}{\alpha\ \max\ \epsilon_{4,5}(^{36}\mathrm{Ar})}-\zeta_4(^{36}\mathrm{Ar})\right)\Bigg/[1-\zeta_4(^{36}\mathrm{Ar})]\leq\frac{M_4(^{36}\mathrm{Ar})}{\alpha\ \max\ \epsilon_{4,5}(^{36}\mathrm{Ar})}. \quad (3)$$

For $\alpha\geq1$, $T_4\leq(4\text{ to }7)\times10^8$ yr. For the Martian Ne such a simple estimate is not valid, because Ne dissipates rather fast (like He from the Earth's atmosphere). We can neglect the dissipation of Ne from the Earth's atmosphere and obtain a similar estimate for the time of origin of the Earth's shielding screen:

$$T_3\leq T\left(\frac{0.36}{\alpha}-\zeta_3(^{20}\mathrm{Ne})\right)\Bigg/[1-\zeta_3(^{20}\mathrm{Ne})]\leq1.6\times10^9\ \mathrm{yr}. \quad (4)$$

In the absence of the estimation of $\zeta_3(^{20}\mathrm{Ne})$ we cannot say anything about nature of shielding screen (as well as in the Martian case).

(3) The second possibility of the large values of max $\epsilon_{k,5}$ is our suggestion $\alpha\simeq1$. It may be somewhat less. More accurate estimates could be obtained after definite measurements of the noble gas abundances in the Martian atmosphere and solar wind. We can obtain the estimates of α and ζ_k based on the obvious condition $0\leq T_k/T\leq1$:

$$0.09+0.15\leq\alpha\leq(0.09+0.15)/\zeta_4(^{36}\mathrm{Ar}), \quad (5)$$

$$0.36\leq\alpha\leq0.36/\zeta_3(^{20}\mathrm{Ne}). \quad (6)$$

From Eqs. (5) and (6) it follows that $\zeta_4(^{36}\mathrm{Ar})\leq0.25–0.41$. Taking into account paleomagnetic data for the Earth's core origin 2.8 billion years ago and (less well confirmed data) 3.9 billion years ago, we obtain the following estimates:

$$\alpha\leq0.36/[0.37+0.63\zeta_3(^{20}\mathrm{Ne})],\quad \alpha\leq0.36/[0.14+0.86\zeta_3(^{20}\mathrm{Ne})].$$

[1]A. V. Vitjazev *et al.*, Abstracts 27th International Geological Congress, Moscow, 1984, Vol. 9, pp. 86–88.
[2]J. B. Pollack and D. C. Black, Icarus **51**, 169 (1982).
[3]E. Anders and M. Ebihara, Geochim. Cosmochim. Acta **46**, 2363 (1982).

The Origin and Early Evolution of the Terrestrial Planets (abstract)

V. S. Safronov and A. V. Vityazev
In: *Chemistry and Physics of the Terrestrial Planets*, edited by S. K. Saxena, (Springer, New York, 1986), pp. 1–29

The formation of the presolar nebula, of the Sun, and of the preplanetary disk, the evolution of the disk, and the formation of the planets are the topics considered in this chapter. The main attention is paid to the processes of formation of the terrestrial planets. The initial state and early evolution and differentiation of the planets are discussed. Present-day models lead to the conclusion that differentiation began during the formation of the planets.

Solid-Body Accumulation of Terrestrial Planets (abstract)

S. I. Ipatov
Translated from: Astron. Vestn. **21**, 207 (1987)
Sol. Syst. Res. **21**, 129 (1987)

Analysis of results of the evolution of three-dimensional disks of gravitating solid bodies accreting due to collision indicate the solid-body accumulation of the terrestrial planets. The number of large ($\gtrsim 0.05 m_{\oplus}$) embryos of terrestrial planets may have been greater than the actual number of these planets. The average orbital eccentricities of these bodies during accumulation may have exceeded 0.2. The relatively large eccentricities of the orbits of Mercury and Mars could have been due to the effect of bodies entering the orbital regions of the gas giants.

Erosion of Mercury Silicate Shell during its Accumulation (abstract)

G. V. Pechernikova and A. V. Vitjazev
Lun. Plan. Sci. **18**, 770 (1987)

The composition of the terrestrial planets is an intriguing problem of planetary cosmogony and comparative planetology.

We present the results of our calculations[1] within the framework of a model of terrestrial planet formation, taking into account early differentiation of the primitive planetary interiors. In this model we can explain the planetary variations of Fe/Si even with the assumption of a quasihomogeneous distribution of this quantity in the preplanetary disk up to the stage of the small body formation.

This model describes (1) the growth of the planets' masses during the accumulation of the preplanetary bodies, (2) the increase of the planetesimals' relative velocities, (3) the widening of feeding zones of planets, (4) the impact heating of the growing planets, (5) the creation of the melting zones, (6) the differentiation of iron and silicates, (7) the ejection of matter of the planet to heliocentric orbits by high velocity impacts of the large bodies, and (8) the exchange of the matter and the redistribution of angular momentum in overlapping feeding zones of the planets. This model can explain simultaneously Mercury's enrichment in iron, a small silicate excess (~1–2%) on Venus as compared to the Earth as well as the relative masses of the planets and irregularities in their relative distances.

The growth of the masses of Mercury (a) and Venus (b) are calculated according to Ref. 2; the average relative velocity of preplanetary bodies $V_{\rm rel}$ in the feeding zone of Venus (c) and the escape velocity V_e from Mercury's surface (d) during the accumulation process are given in Fig. 1. The calculations show that during the accumulation process of the planets the moment t^*, when Venus begins to outstrip Mercury in mass, is sure to come. The average eccentricity $\bar{e}$ of orbits of the bodies forming the planets reaches the value $\bar{e} \approx 0.1$–0.2 by then. Corresponding relative velocities of the bodies in the Mercury zone are $V_{\rm rel} \approx \bar{e} V_c \approx 4$–$9$ km/sec, i.e., they have the same values and even more than the escape velocity V_e from the surface of present Mercury ($V_e \simeq 4$ km/sec). The situation is the same as that in the asteroid zone—the accumulation process is changed by erosion. The impact energy proves enough for the partial ejection of the matter from the surface layers of Mercury to the

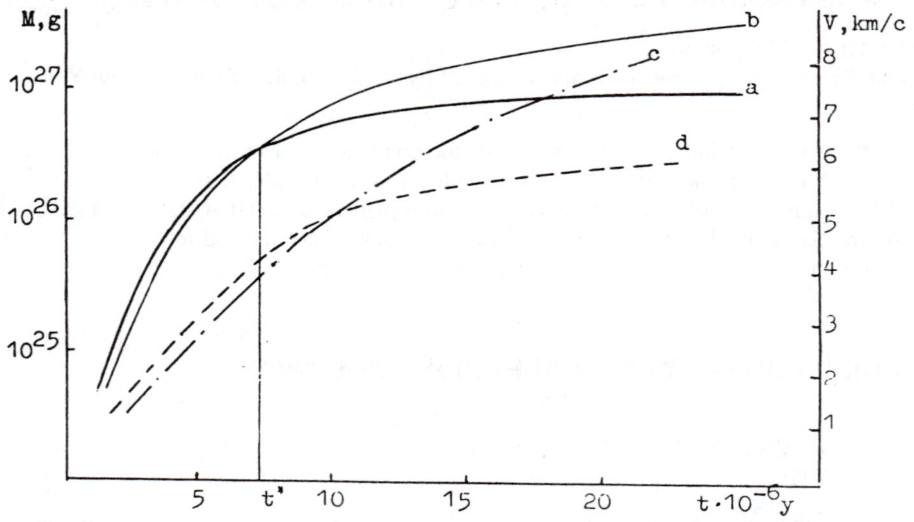

FIG. 1.

heliocentric orbits. The estimates show that the mass of the ejected matter m_∞ is approximately twice as much the mass m of the body projectile, having velocity ~10 km/sec before the encounter with the planet (we supposed that ~10% of the impact energy is converted into kinetic energy). The matter ejected from the Mercurian influence sphere is absorbed in part by Venus, and some part of it falls on the Sun. The calculations show that when the ejection of matter on the Sun proves large enough, due to the conservation of angular momentum Mercury moves slightly to the Sun and Venus outwards.

The largest part of the impact energy (~0.9) is converted into heat. In collisions with the above velocities part of the target material (up to several masses of the falling body) undergoes melting. These estimates as well as other known data about the early heating of the Earth and the Moon indicate the early beginning of the differentiation of the terrestrial planets already during their formation and the early emergence of the shells, enriched in silicate and depleted in iron. In the case of Mercury its silicate shell has been practically lost while Venus has received a small additional portion of the matter depleted in iron.

We have made similar estimates for the ejection process of matter from the Earth surface. Some part of the matter of the Earth's primitive silicate mantle is captured by a presatellite swarm during the ejection, and some part of this material is accumulated by the swarm from heliocentric orbits. As Pechernikova and Maeva[3] noted in such a model, we can explain many cosmochemical data including the depletion of the Moon in iron, taking into account the fractionation of the matter in the prelunar swarm.

The process of exchange of the matter between bodies in the solar system, which Shergottites, Nakhlites, and Chassignites, ALHA 81 005, and some others indicate, was apparently much more intensive in the past.

[1] A. V. Vitjazev and G. V. Pechernikova, Meteoritika **44**, 3 (1985).
[2] A. V. Vitjazev, G. V. Pechernikova, and V. S. Safronov, Sov. Astron. **22**, 60 (1978).
[3] G. V. Pechernikova and S. V. Maeva, in Abstracts of the Symposium on Thermodynamics in Geology, March, 1985, Suzdal, Moscow, pp. 203–204.

When [Was] the Gas Removed from the Zone of Terrestrial Planets? (abstract)

A. V. Vityazev and G. V. Pechernikova
Lun. Plan. Sci. **18**, 1044 (1987)

Some researchers believe that the gas was removed by a strong solar wind, or has escaped from the exosphere of the preplanetary disk heated by UV radiation. One can suppose also that substantial quantities of nebular gas may have been lost during the turbulent spreading of the preplanetary disk. But the parameters of the young Sun (the intensity of UV radiation, the strength of the solar wind) as well as the parameters of the disk (the coefficient of turbulent viscosity, etc.) are poorly defined.

Some constraints for the interval of gas escape from the zone of Jupiter and Saturn were introduced in the model of the disk with moderate mass ($\approx 0.05 M_{\odot}$). Such dissipation had to happen only after the preaccretion of solid cores of these planets ($3-5 m_{\oplus}$) had been accumulated, i.e., $\approx 10^7-10^8$ yr after the formation of the Sun and preplanetary disk.[1]

We have the unique possibility to estimate the moment of clearing of the disk in the zone of terrestrial planets from the data of precompaction irradiation of meteorite grains. Most researchers believe that such irradiation occurred on the surface of primitive bodies.[2] The low density of tracks is usually explained by the short time interval of exposure (10^5-10^6 yr) of the ancient planetesimal's regolith. Our estimates[3] have shown that the precompaction irradiation already occurred after considerable dissipation of gas. In fact, solar cosmic rays (SCR) could not penetrate into the central part of the gaseous disk with the initial surface density of gas $\sigma = 10^3-10^4$ g/cm^2. Only galactic cosmic rays could penetrate into the deep regions of the disk, but they produce a flux of secondary particles with a spectrum different from SCR. Goswami and Lal[2,4] supposed that precompaction irradiation of meteorite grains occurred in the regolith of primitive (C chondrite) spherical bodies with sizes 1–100 cm. But our estimates[3] have shown that even in the absence of the gas the preplanetary disk on the distance 1–5 A.U. is opaque for the SCR until a stage when the substantial quantities of solid matter accumulated in the bodies with sizes ~100–1000 km. In accordance with the theory of accumulation[1] the time $\approx 10^5-10^6$ yr is sufficient for the accumulation of such bodies. The same values give the estimates of interval of exposition of irradiated olivine grains.

In our model we investigated a swarm of bodies with mass spectrum $n(m,t) = c(t) m^{-q}$, where $c(t)$ and q ($1.3 \lesssim q \lesssim 1.8$) have been obtained from the coagulation theory. It was supposed that the grains in the regolith on a small depth (<0.1 cm) can be irradiated by SCR. We find that the major portion of irradiated material in accordance with Ref. 2 came from small bodies but only at the latest stages when the largest bodies in the swarm reached the sizes ~1000 km. On average, only several percent (~3–5%) of matter can obtain the irradiated grains. This estimate is in good agreement with laboratory investigations.[5] As distinct from the Goswami and Lal model our approach gives the possibility for the explanation of irradiation data of all classes of meteorites (both chondritic and achondritic).

We must conclude that although the great bulk of irradiated matter came from bodies with sizes indicated by Goswami and Lal,[2] such irradiation occurred at the stage of accumulation of large bodies. Just at this moment the characteristic time between collisions was great enough for a long exposure of surfaces, a substantial portion of gas was lost, and the necessary transparency of the swarm of preplanetary bodies was reached. It seems to us that the gas in the zone of terrestrial planets was lost before the planet completed their growth. This means that "Kyoto" model and hypothesis about preservation of gas in the inner zone up to the final stage of accumulation does not fit meteoritic data. Of course we could not exclude now the other possibility, that the irradiated grains of meteorites (including C chondrites) were exposed after accumulation of terrestrial planets. But in this case it is necessary to revise a generally accepted paradigm about the pristine origin of meteoritic grains and chondrules.

[1] V. S. Safronov and A. V. Vitjazev, Sov. Sci. Rev. E. Astrophys. Space Phys. **4**, 1 (1985).
[2] D. Lal, in Proceedings of the International School of Physics "Enrico Fermi," course 73, pp. 213-234 (1980).

[3] A. V. Vitjazev and G. V. Pechernikova, Meteoritika **44**, 3 (1985).
[4] J. N. Goswami and Doctoral Lal, Icarus **40**, 510 (1979).
[5] N. N. Korotkova, Doctoral Thesis (in Russian), Moscow (1983).

Formation of Mercury and Removal of its Silicate Shell (abstract)

A. V. Vityazev, G. V. Pechernikova, and V. S. Safronov
In: *Mercury*, edited by F. Vilas, C. R. Chapman, and M. S. Matthews, (University of Arizona Press, Tucson, 1988), pp. 667–9

The model for formation of Mercury by gas-free accumulation of planetesimals provides a natural way for understanding Mercury's low silicate-to-iron ratio. Impact heating would rapidly lead to differentiation of Mercury, a metallic core surrounded by a silicate mantle. Subsequent high-velocity impacts by late-arriving planetesimals would erode away much of the silicate crust and mantle; such silicates would be accumulated by Venus or fall into the Sun.

VI. Origin of the Moon and Other Planetary Satellites

THE ORIGIN OF THE MOON

I. Formation of a Swarm of Bodies Around the Earth

E. L. Ruskol

Earth Physics Institute, Academy of Sciences, USSR

Translated from Astronomicheskii Zhurnal, Vol. 37, No. 4, pp. 690-702
July-August, 1960
Original article submitted February 9, 1960

The basic hypotheses on the origin of the moon are discussed in detail, interest centering on the work of V. V. Radzievskii and E. P. Razbitnaya, which are based on modern concepts of the accumulation of planets and satellites from smaller solid bodies. These authors considered the restricted three-body problem (sun, moon, small particle) and assumed that the capture by the earth of particles leading to the formation of a satellite was due to an increase in the earth's mass and the resultant expansion of the zero-velocity surfaces. It is shown that this capture mechanism is ineffective for the formation of a moonlike body.

Another capture mechanism is considered: inelastic collisions between particles and bodies in the vicinity of the growing earth. It is shown that a swarm of bodies of total mass 0.01-0.1 $m_{\oplus}$ could have formed about the earth during its growth period. This would require the "effective" dimension of the colliding particles to be not more than 10-100 km, for the average probability of capture in a single collision to be of the order of 0.01. The size of the swarm would be about 100 earth radii, with the density of the matter strongly concentrated toward the earth. The rotation of such swarms would have to take place in the same direction as the rotation of the protoplanetary cloud.

The possibility of utilizing this collisional model in the study of the origin of other satellite systems and in the study of the Trojans is noted.

1. Three variants are discerned in the various hypotheses extant on the formation of the moon and other direct planetary satellites: 1) the simultaneous formation of a protoplanet from the planet itself, involving its extended envelope; 2) the cleavage of an already formed planet; and 3) the capture by the planet of satellites or presatellite matter from interplanetary space.

In the spirit of the first concept listed, the formation of the moon is taken up in the hypotheses advanced by Laplace, Kuiper [1], V. G. Fesenkov [2]. In V. G. Fesenkov's writings, the earth-moon system is viewed as a binary planet, in view of the fact that the moon/earth mass ratio exceeds all similar values for the satellites belonging to other planets, and the formation of the moon is thought of as the formation of two condensation centers into a primeval rotating protoplanet. The stipulation of tidal stability presents serious difficulties for any hypothesis of the formation of satellites from a continuous medium surrounding the protoplanet (including the "binary protoplanet" hypothesis put forth by V. G. Fesenkov). Here we may make the same critical remarks voiced on the subject of the protoplanet hypothesis [3]. The density of the protoplanetary condensation must be an order of magnitude larger than the "smeared-out" density of the sun at a given distance from the latter, otherwise it will prove unstable with respect to the tidal forces of the sun. The mass of the protoplanetary cloud required for protoplanets to form is inacceptably high (greater than the mass of the sun itself). The formation of protosatellites within the protoplanets would require a still greater mass, since the protoplanets would have to possess peripheral parts of extreme thickness. A density of the order of unity, i.e., the density of a solid, would hardly be sufficient for tidal stability in close satellites of giant planets. It is obvious that, with this obstacle in mind, Kuiper was constrained in his later writings [4] to abandon the hypothesis of protosatellites of cosmic composition (i.e., primarily gaseous protosatellites) formed in the envelopes of protoplanets. He assumes that the satellites formed by way of gradual accumulation of solid particles settling toward the equatorial planes of the protoplanets. Kuiper thus accepts, with respect to satellites and small bodies in the solar system, the mechanism of formation suggested by O. Yu. Shmidt for planets, i.e., he acknowledges the primary role of solid matter in the formation of celestial bodies.

A second viewpoint, ascribed toG. Darwin, holds that the moon was torn asunder from the earth, and has not been accepted, since it contradicts many known facts in astronomy, mechanics, and geophysics. On this point, cf. Jeffries [5].

A third point of view turns out to be the only acceptable one in the light of contemporary cosmogonic concepts. The satellites formed from the same material as the planets, i.e., from solid particles and solid bodies. The presatellite embryos grew in the vicinities of the growing planets, having collected in that region from interplanetary space. Different capture mechanisms have been considered in application to the problem of the origin of the satellites, by T. See [6]. O. Yu. Shmidt [7], K. M. Savchenko [8], V. V. Radziewskii [9], and E. P. Razbitnaya [10]. The capture of a satellite by a planet is usually investigated within the framework of the restricted three-body problem in celestial mechanics. The sun and the planet are approached as bodies of finite mass, while the satellite or presatellite is viewed as a third body of zero mass. See and Savchenko made reference to the possibility, in principle, of the capture of satellites by planets in the presence of a medium presenting drag, in the neighborhood of the planets involved. However, the quantitative problem was not considered by these authors. See suggested that the satellites were captured in ready-made form. This suggestion is unrealistic, as was pointed out by O. Yu. Shmidt. The regular character of the orbits of the basic satellites serves as evidence to the effect that the satellites passed through a protracted formative period in the vicinity of the planets, and were not captured in a finished state.

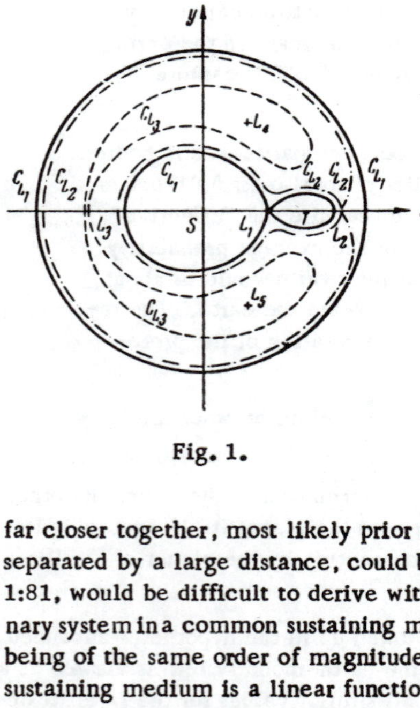

Fig. 1.

V. A. Krat advanced the suggestion that the earth-moon system may have come into being from two embryos formed practically on the same orbit, and constituting a binary system [11]. It must be noted that the possibility of such a binary embryo forming may present itself only at an early stage, such that many small bodies were still present in the zone of the earth, i.e., when both the earth and moon still faced the prospect of growing to several times the size they then possessed. But the subsequent increase in the mass of the two components would have to have brought them far closer together, most likely prior to their fusing. It is highly doubtful that the present earth-moon system, separated by a large distance, could have come about in this manner. The present mass ratio of the components, 1:81, would be difficult to derive within the framework of that hypothesis, because the growth of the embryo binary system in a common sustaining medium would, in all likelihood, result in the masses of the two components being of the same order of magnitude. Actually, the increase in the diameters of the bodies within a common sustaining medium is a linear function of time, and that is why the mass ratio of the bodies tends rapidly to unity, assuming, of course, that neither of the bodies experiences a catastrophic collision.

O. Yu. Shmidt described in qualitative form the process of formation of satellite swarms around growing planets, as a result of inelastic collisions of particles composing a protoplanetary cloud, in the vicinity of the growing planets.

The papers of V. V. Radzievskii and E. P. Razbitnaya deal with the problem of the capture of particles from interplanetary space, destined for the formation of satellites, by a growing planet. The problem was studied by these authors in quantitative form, and we propose to examine their work in greater detail at this point. As will be demonstrated later on, the mechanism suggested by these authors to account for the capture is not adequately effective for the formation of satellites.

2. In the restricted problem of three bodies, the motion of a zero-mass body is considered in a system of coordinates uniformly rotating about the sun at an angular velocity corresponding to the motion of the planet along a circular orbit. The region of the motion of the small body is uniquely defined by its position and the Jacobian constant C, which has the dimensionality of energy:

$$C = 2\Omega - v^2,$$
$$\Omega = \frac{\omega^2}{2}(x^2 + y^2) + \frac{GM}{R} + \frac{Gm}{r}, \quad (1)$$

where ω is the angular velocity of the orbital motion of the planet (m) about the sun (M); x and y are the coordinates of the particle relative to the center of gravity of M and m; the x axis joins M and m; v is the velocity of the particle in the rotating system of coordinates. Depending on C and on the coordinates, the body may be a satellite of the sun, a satellite of a planet, or may possess an orbit embracing both gravitating centers (Fig. 1). Under certain special conditions, other forms of motion may also arise, as for instance periodic orbits about centers of libration – singularities in the zero-velocity surface, restricting the possible regions of motion of a zero-mass body. If a small body, not a satellite belonging to a planet, should increase its Jacobian constant C upon passing rather close to the planet, it may become bound forever to the planet, i.e., it may become the planet's satellite. In the writings of V. V. Radzievskii and E. P. Razbitnaya, the primary role in the secular increase of C for the particles involved in the formation of the moon is ascribed to the increase in the earth's mass as the latter grew. The capture of the particles, according to this suggestion, took place by way of expansion of Hill's limiting closed cavity; in other words, the expansion of the zero-velocity surface corresponding to L_1. This expansion is proportional to the rate of growth of the earth's mass. The libration point L_1 then shifts sunward, while the particles possessing constants C, close to C_{L_1}, and located within the closed volume, are the captured particles.

The suggestion is voiced, in V. V. Radzievskii's paper, that the particles dwell for an indeterminately long period near L_1 prior to their capture, executing periodic motions about this one point. This model is certainly an idealized one, since, in the presence of encounters between particles, there would hardly be a large number of particles capable of remaining for any long period near the point L_1 (until the mass of the earth acquired the necessary increment Δm), i.e., they would travel in stable, almost circular orbits about the sun. Furthermore, the stability of periodic orbits near a point of libration L_1 has not been proved in celestial mechanics, even assuming the absence of collisional encounters.

In the paper by E. P. Razbitnaya, which is a further development of this trend, a less idealized model is invoked: the particles of the protoplanetary cloud are visualized in continuous interaction. It is assumed that, as a result of mutual collisions going on constantly, the distribution of the particles with respect to the constants C remains constant when the number of particles having C close to C_{L_1} remains consistently proportional to the total quantity of matter in the zone of the earth; Q − m, where Q is the contemporary mass of the earth, and m is its instantaneous mass at any time during its growth. The motion of the particles prior to their capture by the earth is thought to be as follows. If their constants C are less than C_{L_1}, but close to it, then the zero-velocity limiting surface bounding the region of possible motions of those particles acquires the form of an hourglass or pear, cutting out a figure-eight on the zero-velocity surface corresponding to C_L. The presence of the "neck" on this "pear" signifies that the particle may execute a certain number of revolutions about the earth, after which it passes through the "neck," to slip out of the control of the earth into the control of the sun, and vice versa. The increase in the C of the particle during the time the particle is a temporary satellite of the earth may result in the "neck" of the zero-velocity surface closing up with respect to that particle, so that the particle will be constrained to remain a stable satellite of the earth. This increment in the constant C must satisfy the restraints

$$C < C_{L_1},$$
$$C + \Delta C > C_{L_1} + \Delta C_{L_1}. \quad (2)$$

As demonstrated in the paper by V. V. Radzievskii, a necessary requirement for the fulfillment of these conditions is that the particle be closer to the earth than is the point of libration L_1, during the time that the required increment Δm is registered in the mass of the earth. This is a corollary of the equations

$$\Delta C = 2\,\frac{\Delta m}{r},$$

$$\Delta C_{L_1} = 2\,\frac{\Delta m}{r_{L_1}}. \qquad (3)$$

The rate of increase in the constant C with time was assumed by Razbitnaya to be proportional to the rate of growth of the mass of the earth with time, dm/dt. The quantity of the substance μ_1 flowing per unit time into Hill's closed cavity is, according to Razbitnaya, proportional to the quantity

$$(Q - m)\frac{dm}{dt}.$$

The concrete mechanism responsible for the combining of these captured particles to form a single satellite, the moon, was not considered in that paper. Since the particles having constants C close to C_{L_1} possess appreciable geocentric moments upon passing through the "neck," they cannot be drawn directly to the earth. In line with the assumption stated, their only recourse is to become incorporated into the moon, which is in process of formation at approximately the contemporary distance from the earth in this model. It is assumed that the earth is growing solely on account of the direct incidence of particles upon it, following O. Yu. Shmidt's formula. The same "direct" growth is being experienced by the moon. The instantaneous mass of the moon is designated as $\mu = \mu_1 + \mu_2$, where μ_1 is the mass captured via the "neck," and μ_2 is the mass accreted by the moon in consequence of direct collisions; $d\mu_2 = 1.4\left(\frac{\mu}{m}\right)^{2/3} dm$. Particles yielding μ_2 may have any constants C prior to colliding with the moon, including values considerably lower than C_{L_1}. It is assumed that the entire contemporary geocentric moment of the moon is introduced by the mass μ_1. From this, we get the proportionality factor between $d\mu/dt$ and $(Q - m)(dm/dt)$. From the constraint stating that both terms involved in μ must yield the contemporary mass of the moon ($m = Q$), Razbitnaya derives the initial mass of the earth, the mass at which the capture of the lunar substance was begun,

$$m_0 = 0.96Q \qquad (4)$$

and the age of the moon,

$$T_{☾} = 0.84T_{♁}.$$

Mention is made in her paper of the fact that this model of lunar formation is also suitable for all of the other direct satellites of planets comprising the solar system.

This approach to the solution of a problem is cosmogony may be viewed as a legitimate one only in the case where it allows no room for doubt that the mechanism invoked, and no other, is responsible for a real process. There are, however, difficulties standing in the way of acceptance of the capture model proposed by V. V. Radzievskii and E. P. Razbitnaya.

Consider now whether or not an increment in the earth's mass could bring about the required increase in constants C, so that the particles might be captured and serve to form the moon. According to Eqs. (3) and (4)

$$\Delta C = \frac{2\Delta m}{r} = \frac{2\cdot(1 - 0.96)\,Q}{r}\cdot\frac{t_{\text{earth}}}{t_{\text{total}}}, \qquad (5)$$

where $t_{\text{earth}}/t_{\text{total}}$ is the ratio of the time interval during which the particle revolves about the earth at a distance comparable to r_{L_1}, to the entire period of growth in the earth's mass by the amount Δm. When the particle escapes from under the earth's control and begins to revolve about the sun, the effect of the increment in the mass of the earth in the particle's motion will be vanishingly small. The ratio $t_{\text{earth}}/t_{\text{total}}$ is in general very small. The larger the number of revolutions of the temporarily captured particle about the earth, the smaller the probability of such an orbit. On the average, t_{earth} may be not more than several years, whereas t_{total} is equal, according to Razbitnaya's estimate, to 0.84 of the age of the earth, amounting to several billion years. Following the customary procedure in the restricted three-body problem, we take the masses of the earth and sun as the unit of mass, the distance between those bodies as the unit of length, and choose the unit of time such that the gravitational constant is 1, then we have, from Eq. (5)

$$\Delta C = 2.4\cdot 10^{-5}\cdot\frac{t_{\text{earth}}}{t_{\text{total}}} \ll 10^{-5}.$$

The contemporary value of C for the moon is 3.00117, with C_{L_1} equal to 3.00090. Constants C for the "semibound" particles considered by Razbitnaya are less than C_{L_1}, so that the required increment ΔC must satisfy the constraint

$$\Delta C > 3.00117 - 3.00090 = 2.7 \cdot 10^{-4}.$$

From this, we learn that Razbitnaya's mechanism cannot account for the capture of particles by the earth to form the moon.

Moreover, in view of the extremely low value found for the ratio t_{earth}/t_{total} for temporarily captured particles, we apparently will have $\Delta C \ll C_{L_1}$ on the average, despite the fact that ΔC will be slightly larger tha.ı ΔC_{L_1} during the time the particles are in the vicinity of the earth. And this, in turn, means that the necessary conditions for capture, Eq. (2), will, in general, not be fulfilled. The hypothesis under consideration cannot overcome this difficulty at any value of m_0, including even values where

$$m_0 \ll Q.$$

Consequently, tne earth was not capable of capturing matter for the creation of the moon from a protoplanetary cloud solely by reason of an increase in the earth's mass. This feat requires a still more effective mechanism, for example inelastic collisions between particles and bodies near the embryos of the planets, which O. Yu. Shmidt envisaged as the principal factor operating in the formation of satellite swarms. It is true that a process involving the participation of inelastic collisions cannot be described by the equations of celestial mechanics. That is perhaps the reason why the role of mutual collisions between particles is reduced, in E. P. Razbitnaya's paper, to one of maintaining a certain steady-state distribution of particles with respect to the constants C, "making available" particles with C smaller than C_{L_1}, but close to it. Razbitnaya also takes into account inelastic collisions between particles and the moon, which grows through direct incidence of particles upon it (mass μ_2). However, this direct growth can proceed only after the lunar embryo has been already captured or formed from captured particles. The effect of inelastic collisions on the capture process per se was not investigated in [9] or [10].

Inelastic collisions between particles and rigid bodies took place in any neighborhood of the growing planet, and capture could occur through the entire volume of Hill's cavity corresponding to the point of libration L_1, and not just in the vicinity of that point. We should, therefore, not approach the collisional encounters only as a secondary, auxiliary mechanism serving to facilitate capture via the point of libration.

The work of V. V. Radzievskii and E. P. Razbitnaya is of great interest in that they present, for the first time, and in quantitative form at that, a study of the mechanism responsible for satellite formation, from the vantage point of contemporary cosmogonic concepts of the accumulation of planets from smaller, cold, solid bodies. However, the concrete mechanism proposed by these authors to account for the capture of particles by the growing planet is not an effective one, so that the problem of the origin of the satellites is shown to require further study.

3. First, let us show that inelastic collisions may lead to significant increments in the values of C for the particles. At the instant of collision between two particles, the values $C = 2\Omega - v^2$ experience a stepwise change, completely determined by the change in v^2, i.e., by the loss of kinetic energy of motion in the frame of coordinates bound to the growing planet. The distance function Ω at the instant of collision does not change.

$$\Delta C = -2v\Delta v.$$

In an inelastic collision, the velocity loss Δv usually amounts to an appreciable fraction of the velocity v per se, so that the assumption that ΔC will be of the order of v^2 is warranted. According to L. E. Gurevich and A. I. Lebedinskii [12], the peculiar velocity of particles, superposed upon their velocity of their circular motion about the sun, is $v \approx \sqrt{\frac{Gm_p}{r_p}}$, where m_p and r_p are, respectively, the mass and radius of the largest embryo present in that zone. When the earth's mass is of the order of magnitude of its present-day value, the increment ΔC associated with a collision (in the systems of units selected above), is

$$v^2 \approx \frac{m_p}{r_p} \approx \frac{3 \cdot 10^{-6}}{5 \cdot 10^{-5}} = 6 \cdot 10^{-2}.$$

We see that a single inelastic collision may impart to the particle an increment of several orders of magnitude larger in C than any corresponding increase in the mass of the earth. Each body experiences one collision with some other body during a time interval of the order of $t = \lambda / v$, where λ is the mean-free-path length of the body, and $\underline{v}$ is the velocity. If we assume, for sake of simplicity, that all of the bodies in the zone of the earth have the same mass m_i, then, by performing fairly straightforward transformations, we get

$$t \approx \frac{10^{33}}{(Q - m)} \cdot (m_i)^{1/3} \text{ sec.}$$

If the zone of the earth has not yet been depleted, say $Q - m \geq 10^{27}$ g, then for $m_i = 10^{15}$ g, the time interval between collisions will average out at $3 \cdot 10^3$ years; for $m_i = 10^{18}$ g, $3 \cdot 10^4$ years, etc. In other words, collisions occur rather frequently. The increment ΔC will average yearly, for a body of mass 10^{15} g, at around the order of 10^{-5}, and about 10^{-6} for a body of mass 10^{18} g, with a correspondingly larger increment for smaller bodies. At the same time, Eq. (5) may be used to show with ease that the annual increase in C due to the increment in the earth's mass will only be of the order of 10^{-13}, even during the period of most impetuous growth. At the concluding stage of the earth's growth, say from 0.96Q to Q, the annual increase in C will still be many times smaller.

In the light of the foregoing, we may safely neglect capture of matter by the earth via the "neck" of Hill's limiting surface as a consequence of the increase in the earth's mass, and examine the formation of a presatellite swarm about the earth as the result of collisions between particles and larger bodies in the immediate vicinity of the planet.

4. Consider the totality of particles belonging to the protoplanetary cloud comprising the zone of sustenance for the embryonic earth. As in the restricted problem of three bodies, we shall consider motions of particles in a coordinate frame of reference rotating in unison with the earth about the sun at an angular velocity, but we shall use the CGS system. It is readily seen that a particle belonging to the cloud, circling past the earth, may become trapped as a satellite of the latter if it should experience an inelastic collision within the limiting closed Hill surface surrounding the earth. Beyond the confines of this surface, it would be impossible for the particle to become trapped as a satellite, no matter how small its velocity relative to the earth, although there always remains the possibility of the particle coming into direct contact with the earth or with the lunar embryo. The capture of particles colliding near the limiting Hill surface itself is highly improbable, since this would require practically zero residual velocity with respect to the earth. Capture of particles at the present-day distance separating the earth and moon would be possible at residual velocities of the order of 1 km/sec, and near the surface of the present earth, up to 11 km/sec. Thus, virtually any particle colliding with another particle of about the same size will be trapped in the presatellite swarm (or brought to the earth), should it venture near the earth's surface. However, the distribution of the captured matter over the distance from the earth may obviously have its maximum not at the very surface of the earth, but at some distance away, since the number of colliding bodies is proportional to the volume of the spherical zone.

It is difficult to get a clear picture of the distribution of matter trapped in the swarm with respect to the distance. What is required is to take into account the velocity distribution before and after collisions, energy losses in each collision in heat and in pulverization, and an estimate must also be made of the geocentric moment of the swarm. However, before tackling this problem, it would be advisable to estimate the mass of the matter trapped by the earth into the presatellite swarm, assuming in the process that some fraction of all the collisions occurring near the earth result in the capture of the colliding particles. Let us assume further, for the sake of simplicity, that random velocities of the particles in the earth's zone are identical, and equal to the mean value $v = \sqrt{\frac{Cm}{\theta r}}$, where $\underline{m}$ and $\underline{r}$ are, respectively, the mass and radius of the largest body, i.e., of the earth embryo, while θ is a numerical factor ranging from 1 to 3 [13].

Let the size distribution $\underline{a}$ of the particles and bodies be expressed in terms of the conventionally adopted formula $dN(a) = Ba^{-n}da$. On account of the presence of random velocities, collisions between particles with any size ratios must take place. But not every, even completely inelastic, collision actually results in sufficient deceleration of the particle. For example, a body of the dimensions of an asteroid, on colliding with a dust grain, does not lose its velocity. Collisions between coarse bodies and fine dust are not to be taken into account; consequently, in the sense of encounters capable of leading to the trapping of bodies into a presatellite swarm. As pertains to fine particles, of course, collisions between them and coarse bodies lead, in all probability, to their adhering to these bodies, i.e., in some cases resulting in capture, if the large body belongs to the swarm, in other cases resulting in the particle being swept out of the swarm. However, collisions between fine particles and large bodies will occur much less frequently than collisions between fine particles, so that we are fully justified in neglecting the former with preference for the latter. We shall take under consideration only those collisions which occur between particles of comparable linear dimensions, say in the case of particles of size a: only collisions with particles ranging from $a - \Delta a$ to $a + \Delta a$, where $\Delta a = \alpha a$, $\alpha \approx \frac{1}{2}$.

The total mass S_1 of all the particles and bodies experiencing similar collisions per cm^3 per sec is found by taking the integral

$$S_1 = \int_{a_{max}}^{a_{min}} da \int_{a-\Delta a}^{a+\Delta a} \frac{4}{3} \pi \delta r^3 \pi r^2 \cdot \bar{v} \cdot \frac{1}{2} (Br^{-n})^2 \, dr, \tag{9}$$

where δ is the density of the particles. Then, finding the constant B from the normalization conditions,

$$\rho_1 = \int_{a_{min}}^{a_{max}} Br^{-n} \cdot r^3 \cdot \frac{4}{3} \pi \delta dr = \frac{B}{4-n} \cdot \frac{4}{3} \pi \delta \left(a_{max}^{4-n} - a_{min}^{4-n}\right),$$

where ρ_1 is the density of matter in the zone of the earth,

$$B = \frac{3\rho_1 (4-n)}{4\pi\delta \left(a_{max}^{4-n} - a_{min}^{4-n}\right)} \tag{10}$$

and taking Eq. (10) into account, we find, by integrating Eq. (9), that

$$S_1 = \frac{3}{4} \frac{\bar{v}\rho_1^2}{\delta} \cdot \frac{\alpha (4-n)^2}{(7-2n)} \cdot \frac{a_{max}^{7-2n} - a_{min}^{7-2n}}{\left(a_{max}^{4-n} - a_{min}^{4-n}\right)^2} \,. \tag{11}$$

For brevity, we shall denote the multiplicative factor dependent on the size distribution function of the particles as K. It has the dimensionality of reciprocal length, and determines the effective dimension of the colliding particles in the following manner:

	n				
	3.0	3.5	3.75	4.0	4.25
$K = \frac{\alpha(4-n)^2}{7-2n} \frac{a_{max}^{7-2n} - a_{min}^{7-2n}}{\left(a_{max}^{4-n} - a_{min}^{4-n}\right)}$ $\alpha = 1/2$	$\frac{1}{2a_{max}}$	$\frac{\ln\frac{a_{max}}{a_{min}}}{8a_{max}}$	$\frac{1}{16\sqrt{a_{max}a_{min}}}$	$\frac{1}{2a_{min}\ln^2\frac{a_{max}}{a_{min}}}$	$\frac{1}{48a_{min}}$

The equations for K in the above table are approximate expressions, but yield an excellent picture of its magnitude, since, in view of the enormous dispersion of particle sizes within the protoplanetary cloud, $a_{max} \gg a_{min}$. The particle dimensions may range from a thousand kilometers to vanishingly small fractions of a centimeter. At $n \leq 3.5$, the value of K is determined by the largest dimension of the bodies; at $3.5 < n < 4.0$, by the combination of dimensions a_{max} and a_{min} in the form $a_{max}^{z} \cdot a_{min}^{1-z}$, $z < 1$; at $n \geq 4.0$, by the smallest dimension of the particles.

Let us assign to each collision a mean probability β_1 that the collision will result in the trapping of the particles in the swarm encircling the primary planet. As the presatellite swarm takes shape, collisions will also be occurring between particles flying into the swarm and particles already with membership in the swarm, which may, in turn, lead to the capture of the previously free particles. The mean probability of such a capture shall be designated β_2. By analogy with Eq. (11), we find that the colliding mass S_2 in this case is

$$S_2 = \frac{3}{2} \frac{\bar{v}\rho_1\rho_2}{\delta} \cdot K, \tag{12}$$

where ρ_2 is the density of the matter in the swarm encircling the earth. The captured mass per unit volume per second will equal the sum

$$\beta_1 S_1 + \beta_2 S_2.$$

Multiplying this quantity by the volume within which captures may take place (of the order of $r_{L_1}^3$), we obtain the rate of capture of matter, i.e., the rate at which matter from the protoplanetary cloud is trapped into the circumterrestrial swarm

$$\frac{d\mu}{dt} = (\beta_1 S_1 + \beta_2 S_2)\cdot r_{L_1}^3, \tag{13}$$

where

$$r_{L_1}^3 = R^3 \cdot \frac{m}{3M_\odot}.$$

To find the increase in μ relative to the earth's mass $\underline{m}$, we must divide the above expression by the rate of growth of the earth's mass dm/dt. According to O. Yu. Shmidt and V. S. Safronov,

$$\frac{dm}{dt} = \frac{4\pi\gamma}{P}\cdot r_e^2 \cdot \sigma, \tag{14}$$

where γ is a factor of the order of unity, P is the period of revolution of the planetary embryo about the sun, equal to $2\pi\sqrt{\frac{R^3}{GM}}$, r_e is the effective radius of the planet, $r_e^2 = (1 + 2\theta)r^2$, θ is the dispersion parameter of the velocities mentioned above, and σ is the surface density of matter in the zone of sustenance of the planet

$$\sigma = \frac{Q - m - \mu}{AR^2},$$

where A ≈ 3. Taking into account the fact that the volume densities of matter ρ_1 and ρ_2 are expressed in the form:

$$\rho_1 = \frac{\sigma}{H} = \frac{2\sigma v_c}{\pi R v}; \quad v_c = \sqrt{\frac{GM}{R}}, \tag{15}$$

and

$$\rho_2 = \frac{\mu}{r_{L_1}^3},$$

and dividing Eq. (13) by Eq. (14) term by term, we obtain, after some modification,

$$\frac{d\mu}{dm} = D_1(Q - m - \mu) + D_2\mu = D_1(Q - m) + (D_2 - D_1)\mu, \tag{16}$$

where

$$D_1 = \frac{\beta_1 \sqrt{\theta}\cdot K}{3\gamma\cdot A(1 + 2\theta)\cdot\sqrt{3\pi\delta MR}}$$

and

$$D_2 = \frac{\beta_2\cdot 3K}{\gamma(1 + 2\theta)\, r^2\delta}.$$

Here the term with coefficient D_1 represents the growth of the swarm due to mutual collisions between flying particles of the protoplanetary cloud, and the term with D_2 represents the trapping of these particles within the swarm. In order to avoid having the equation cluttered with fractional powers of $\underline{m}$, we assumed that, in the denominator of coefficient D_2,

$$r^2 = \text{const} = r^2_{\text{contemporary}}$$

over the entire interval of integration from m_0 to Q (although, explicitly, $r_0^2 < r^2_{\text{contemporary}}$). This will result in a value for D_2 which is slightly depressed below the true value, but only at those stages of the process where $\underline{m}$ is small and, consequently, μ is small, and the role of the term D_2 itself is small.

Equation (16) has the solution:

$$\mu = \frac{D_1 Q}{D_1 - D_2} - \frac{D_1 Q}{D_1 - D_2} \cdot e^{(D_2 - D_1)(m - m_0)} + \frac{D_1}{(D_1 - D_2)^2} - \frac{D_1}{(D_1 - D_2)^2} e^{(D_2 - D_1)(m - m_0)} - \tag{17}$$

$$- \frac{D_1 m}{D_1 - D_2} - \frac{D_1 m_0}{D_1 - D_2} e^{(D_2 - D_1)(m - m_0)}.$$

From this, we find that, at the end of the growth process of the planet, at $m = Q - \mu$

$$\mu = \frac{D_1}{D_2}(Q - m_0) \cdot e^{(D_2 - D_1)(Q - m_0 - \mu)} + \frac{D_1}{D_2 (D_2 - D_1)} \left(1 - e^{(D_2 - D_1)(Q - m_0 - \mu)}\right). \tag{18}$$

This algebraic equation for μ may, generally speaking, yield a solution for over the broad range extending from $\mu \ll \mu$ to $\mu \approx Q$, depending on coefficients D_2 and D_1. However, since there is no sense in obtaining μ of the order of Q (in all satellite systems, the mass is much smaller than the mass of the planet), we neglect μ compared to Q in the exponent. Nor do we know what the value of m_0 was when the formation of the lunar swarm commenced. It is clear from over-all considerations that m_0 must be rather large, that, in any case, the embryo of the planet must be larger than other competing bodies in its zone, otherwise the model of the statistical three-body problem employed here would have no validity. But, in all probability, m_0 is far smaller than Q, so that m_0 enters the equation only as the amount subtracted from Q, and may be safely neglected. Taking into account the fact that

$$\frac{D_1}{D_2} = \frac{\beta_1}{\beta_2} \frac{\sqrt{\theta} \cdot r^2 \delta}{9A \sqrt{3\pi\delta M_{\odot} R}} \approx 2 \cdot 10^{-7}$$

at

$$\frac{\beta_1}{\beta_2} = 2$$

we obtain an approximate solution for μ (observing the condition $\mu \ll Q$) in the form

$$\mu = \frac{D_1}{D_2}\left[\left(Q - \frac{1}{D_2}\right) e^{D_2 Q} + \frac{1}{D_2}\right]. \tag{19}$$

In the case where $D_2 Q < 1$, the role of the exponential multiplier is small, and we obtain values of μ of the order of $(D_1/D_2)Q \approx 2 \cdot 10^{-7} Q$, which is insufficient to contribute to the creation of the moon, assuming that the bulk of the lunar mass was derived from the swarm encircling the earth, rather than becoming attached to the lunar embryo from interplanetary space.

Consider the other case, namely $D_2 Q > 1$; at $D_2 Q$ equal to several times unity, we may neglect the quantity $1/D_2$ in both the parentheses and brackets appearing in Eq. (19). We then have

$$\mu \cong \frac{D_1}{D_2} Q e^{D_2 Q}. \tag{20}$$

We see now that the mass of the satellite swarm is very strongly dependent on the coefficient D_2, which includes the product $\beta_2 K$. What interests us here is the case where "collisional" capture may yield the mass of the swarm included, say, within the limits

$$\frac{Q}{100} < \mu < \frac{Q}{10},$$

which corresponds to the limits

$$11 < D_2 Q < 13,$$

hence, the product $\beta_2 K$ must be included within the interval

$$\frac{11\gamma(1 + 2\theta) r^2 \delta}{3Q} < \beta_2 K < \frac{13\gamma(1 + 2\theta) r^2 \delta}{3Q}.$$

This corresponds to $\beta_2 K = 5 \cdot 10^{-9}$. At a capture probability $10^{-2} < \beta_2 < 10^{-1}$, the value of K ranges from $5 \cdot 10^{-8}$ to $5 \cdot 10^{-7}$ cm^{-1}. The effective dimension of the bodies colliding in the vicinity of the earth must therefore constitute, for the most part, a transverse diameter of tens or of one- or two-hundred kilometers, i.e., a body of asteroidal dimensions. If the bulk of the mass of bodies found within the zone of the earth consisted of thousand-kilometer lumps, then "collisional" capture would be ineffective (this corresponds to the case considered above, $D_2Q < 1$). This means that such large bodies, having sizes on the order of the present moon, would not be admissible within the framework of this model unless as an exception, certainly not as a general rule. It may be inferred, then, that capture of a lunar embryo with a mass of the order of that of the present moon is highly improbable.

On the other hand, if the bulk of the mass of this matter were found in bodies of less than 10^6 cm dimension, and if this mass were not transferred to larger bodies during the process, we might possibly obtain the mass of the swarm encircling the earth relative to the greater mass of the earth, by finding their sum, equal to Q. However, this scheme is unrealistic. The size of the particles will increase rapidly in a dense medium, and the effectiveness of the capture mechanism will suffer. The process is a self-regulating one.

The dimension of bodies of the order of tens of kilometers required to obtain the required mass of the swarm encircling the earth is quite reasonable from several points of view. The size of the protoplanetary bodies in the zone of the earth should be of this order (according to V. S. Safronov [13] and H. Urey [14]). The size of the lunar craters may be due to the falling of similar bodies at the time when the moon, moving away from the earth, traversed a belt of earth satellites smaller than itself. This explanation for the origin of lunar craters formed after the epoch of maximum fusion of the lunar interior, was put forth by Kuiper [15]. The model of the moon constructed by Urey, Elsasser, and Rochester [16], with stresses at the center assumed absent in view of the irregular form, must represent an accumulation from such bodies of slightly differing densities.

We may conclude that capture by inelastic collisions provides a satisfactory explanation for the origin of the lunar swarm, under the highly plausible assumptions made with regard to the effective dimension of the bodies. As pertains to the distribution function of those bodies at that remote epoch, if the order of the maximum dimension of the bodies, not counting the earth, and possibly not counting the moon either, was not greater than 10^7 cm, then the exponent of the distribution function of the bodies with respect to size may have been close to 3. But if a_{max} is taken equal to 10^8 cm, then $\underline{n}$ must have been somewhere between 3.5 and 3.75, closer to the first value. As we know [17], the current value of $\underline{n}$ for asteroids is close to 3, so that the model is not an absurd one in this respect.

5. It is an exceedingly difficult job to make an estimate of the geocentric (or planetocentric) angular momentum of the satellite swarm, if we consider the motions of the individual particles. As in the case of the axial moment of the planets, the planetocentric moment of the swarm amounts to a negligibly low value, compared to its heliocentric moment. We should, however, make some attempt to arrive at an estimate of the planetocentric moment of the swarm from the standpoint of hydrodynamics. In a continuous medium in motion, the instantaneous angular velocity of any volume element of the medium relative to the center of that volume is $\frac{1}{2}\,\mathrm{rot}\,\vec{v}$. In the case of a Keplerian angular motion of the particles about the sun, $\frac{1}{2}\,\mathrm{rot}\,\vec{v} = \frac{1}{4}\,\vec{\omega}_c$, where $\vec{\omega}_c$ is the angular velocity of this motion. The circumsolar cloud of rather large discrete bodies having extended and inclined orbits may be viewed as a continuous medium only in a very crude approximation. However, a volume element of this cloud, of the order of the dimensions of a satellite swarm, is already quite large enough to justify the possible use of hydrodynamics. At the same time, it is small compared to the dimensions of the entire system (its linear dimensions being of the order of 0.01 of the distance from the sun). If we assume that the planetocentric moment of the swarm is of the same order as the instantaneous moment of any volume element in the cloud of like size, relative to the center of the cloud, then we obtain a rotation of the satellite swarm in the direct sense, with angular momentum of $\frac{1}{4}J\omega_c$, where J is the moment of inertia of the swarm. The same conclusion regarding the axial moment of extended objects of the protoplanetary cloud was obtained by Hoyle [18] by integrating the moment with respect to the area of the objects in the equatorial plane of the cloud. A comparison of the moment evaluated in this fashion with the observed moment of satellite systems yields agreement on the order of unity.

As stated by V. V. Radzievskii, it may be proved with greater rigor, within the framework of the statistical restricted three-body problem, that the matter captured by the planet must bring with it a positive planetocentric moment of the order of $J\omega_c$.*

---

*Address delivered at the seminar of the Department of Theoretical Physics, Yaroslavl' Pedagogical Institute, August 12, 1959.

6. Equation (20) states the exponential dependence of the mass of the satellite swarm on the finite mass of the planet. If we assume that the present mass of the satellites of a given planet is proportional to the mass which the swarm possessed at some time, then the relation derived, Eq. (20), goes far in explaining the difference in the masses of the terrestrial and Martian satellites. The absence of any satellite encircling Mercury is no surprise either when seen within this scheme, in view of its small mass and its proximity to the sun. The absence of satellites around Venus, which, after all, must have had a presatellite swarm intermediate between that of Mars and the earth, may be laid to the impacting of the previously extant satellites on the surface of Venus as a consequence of tidal evolution. This would require that the period of axial rotation of Venus be longer than the period of revolution of its satellites, i.e., the satellites must not have been remote from their primary. As will be shown in a subsequent article, the mechanism considered yields a strong concentration of the density of satellite swarms in the direction of the planets. The most probable pathway of formation of satellites at moderate distances from their planets (predominantly within the range of tens of planetary radii) is consequently found to be in agreement with the observed position of most of the satellites in the solar system.

Equation (20) is inapplicable for computing the mass of satellite swarms around giant planets, because the planets themselves grew not only at the expense of available solid matter, but also by accretions of gas. The peculiarities of the growth of giant planets have as yet received insufficient study in quantitative form. When this task will have been accomplished, the hypothesis under consideration for explaining the formation of presatellite swarms will apparently provide a satisfactory explanation, because the satellite systems of giant planets themselves are more massive than the earth by two orders of magnitude.

The hypothesis on the formation of satellite swarms around growing planets as a consequence of inelastic collisions between particles and bodies in the immediate vicinity of the planets may thus be laid at the basis of any investigations of the origin of the satellites belonging to those planets.

The capture of small bodies through collisions might also result in the formation of the Trojan group, which execute a motion in proximity to points of libration L_4 and L_5 in the Jupiter-Sun system. Their observed total mass and size are explained by the assumption that the capture probability in the neighborhood of those libration points is less by 1-2 orders of magnitude than the probability of being trapped in a presatellite swarm, and entirely natural variant.

I take this occasion to express my gratitude to Professor V. V. Radzievskii and to Doctor of Physical Mathematical Sciences, B. Yu. Levin, for their thorough discussion of the present paper.

LITERATURE CITED

[1] G. P. Kuiper, Astrophysics (ed. Hynek) (New York, 1951) Chap. 8, pp. 357-424.

[2] V. G. Fesenkov, Symposium: Origin of Life on Earth [in Russian] (Moscow, 1957) pp. 9-14.

[3] E. L. Ruskol, Symposium: Problems of Cosmogony [in Russian] (Moscow, 1960) No. 7.

[4] G. P. Kuiper, Vistas in Astronomy (1956) Vol. 2, pp. 1631-1666.

[5] H. Jeffreys, The Earth [Russian translation] (IL, Moscow, 1960).

[6] T. See, Astron. Nachr. 181, 182 (1909); Monthly Notices, Roy. Astron. Soc. 4341, 4367 (1910).

[7] O. Yu. Shmidt, Four Lectures on the Theory of the Origin of the Earth [in Russian] (Moscow, 1957) p. 57.

[8] K. N. Savchenko, Trudy Odesskogo Gos. Univ. 5, 59 (1953).

[9] V. V. Radzievskii, Byull. VAGO, No. 11 (18), 3 (1952).

[10] E. P. Razbitnaya, Doctoral dissertation [in Russian] (Leningrad, 1954).

[11] V. A. Krat, Symposium: Proceedings of the First Conference on Problems of Cosmogony [in Russian] (Moscow, 1951) p. 76.

[12] L. E. Gurevich and A. I. Lebedinskii, Izvest. Akad. Nauk SSSR, Ser. Fiz. 14, 765 (1950).

[13] V. S. Safronov, Symposium: Problems of Cosmogony (1958) [in Russian] (Moscow, 1958) No. 6.

[14] H. Urey, Symposium: Problems of Cosmogony (1960) [in Russian] (Moscow, 1960) No. 7.

[15] G. Kuiper, Symposium: Problems of Cosmogony [in Russian] (Moscow, 1960) No. 7.

[16] H. C. Urey, W. M. Elsasser, and M. G. Rochester, Astrophys. J. 129, 842 (1959).

[17] S. Piotrowski, Acta Astronomica A5, 115 (1954).

[18] F. Hoyle, Monthly Notices Roy. Astron. Soc. 106, 406 (1946).

SOVIET ASTRONOMY – AJ VOL. 7, NO. 2 SEPTEMBER-OCTOBER, 1963

ON THE ORIGIN OF THE MOON. II. THE GROWTH OF THE MOON IN THE CIRCUM-TERRESTRIAL SWARM OF SATELLITES

E. L. Ruskol

Institute of Earth Physics, Academy of Sciences, USSR
Translated from Astronomicheskii Zhurnal, Vol. 40, No. 2, pp. 288-296, March-April, 1963
Original article submitted July 21, 1962

The hypothesis recently advanced by Urey [2] on the origin of the moon through capture by the earth in ready-made form, flowing from assumptions of the different chemical compositions of the earth and moon, is discussed. Proofs are advanced in support of the like composition of the two bodies. Several variants in the capture of a ready-made moon by the earth are considered. It is shown that the probability of such a capture event was in any case several orders of magnitude less than 10^{-2}, thus arguing strongly against Urey's hypothesis.

On the basis of arguments developed in the first section of this paper [1], the growth of the moon by accretion in a circumterrestrial swarm of satellites which is postulated to have formed around the growing earth as a result of inelastic collisions between solid bodies of effective dimensions less than 10 to 100 km is discussed. It is demonstrated that the probability of capture of bodies into circumterrestrial orbits is greater the closer the collision occurred near the earth, so that the material density in the satellite swarm must have sharply increased in the direction toward the earth. This leads us to the conclusion that the moon formed preponderantly at a distance of 5 to 10 earth radii, and that its subsequent receding is due to tidal friction. The most massive satellites attending the other planets must also have formed close to their primaries in this scheme and this supposition turns out to be in excellent agreement with their observed positions.

1. In the first section of this article [1], it was demonstrated that the growth of the earth by accretion of small solid bodies must have been accompanied by the formation around the earth of a swarm of small satellites. This flows from the inevitability of mutual collisions between the bodies, on account of the spread of their velocities in the preplanetary cloud. As a result of irreversible losses of kinetic energy, the bodies or fragments of same transferred from circumsolar orbits to circumterrestrial orbits in the course of several collisions near the growing planet, and thus become earth satellites. The first section of this article [1] also provided an estimate of the mass of the circumterrestrial swarm of bodies in connection with the problem of the formation of the moon out of this swarm.

Accordingly, we laid a sound cosmogonical foundation for the view that the moon formed primarily from the same material as the earth. Until recently, this viewpoint was generally shared. However, recently Urey [2], in discussing the possible chemical composition of the moon compatible with the lunar density, arrived at the conclusion that the iron content in the moon was not greater than 14% in mass. The inference drawn therefrom holds that the moon is poorer in iron than chondrites (22-28% iron, according to Urey's estimate), while the earth and the other planets in the earth group may have a "chondritic" composition, according to Urey's present views [2a]. This inference as to a substantial difference between the composition of the moon and that of the earth led Urey to the conclusion that the moon must have formed under quite different conditions, i.e., in some region of the solar system quite removed from the earth, and stimulated him to advance the hypothesis of the capture by the earth of a moon in ready-made form, a moon which, in Urey's view, constituted one of many protoplanetary bodies.

First of all, we must take note of the fact that the composition of the earth, at least as pertains to iron, need not correspond exactly to the composition of chondrites, since the conditions of their formation are not the same. The solution of the problem of similarities between the chemical compositions of the earth and moon depends, as we know, on the solution of the problem of the nature of the earth's inner core. If this inner core is found to consist predominantly of iron, then the earth would be found to be far richer in iron than the moon, Mars, and even the chondrites. If, on the other hand, the Lodochnikov-Ramsey hypothesis of the phase transformation of silicate material in the core into the metal state is justified, then the composition of the moon and the earth may turn out to be identical [3]. Static pressures of approximately 1.5 million atm prevailing at the boundary of the earth's core being unattainable in our laboratories, it remains as yet impossible to get through to a definitive solution of this problem. The study of the properties of various materials under high pressures has only begun. It is interesting to note that the recently [4] discovered dense modification of quartz (a density jump from 2.65 to 4.35 g/cm^3 at a pressure of roughly 150 thousand atm) is a mineral which must form at relatively shallow depths inside the earth, but which cannot form even at the center of the moon, where the pressure amounts to only 50 thousand atm. We might surmise that the earth and moon models constructed of the same material will acquire experimental justification in future research. Urey's conclusion on the different chemical compositions of the earth and moon may not be viewed as definitive on that account.

Let us add that the difference in chemical composition of a planet and its satellite does not in and of itself constitute a reliable criterion for a "capture" origin of the satellite, an event which would be reflected most likely in an orbital irregularity. For example, now, a substantial difference in the composition of the planet Jupiter and its satellites is observed in Jupiter's system, and is ascribed to the fact that the planetary density in this case is several times less than the density of the nearest satellites. However, despite that fact, no one entertains any serious doubt that the Galilean satellites of Jupiter were formed in the vicinity of the planet, since the regular character of their motions around Jupiter could not possibly have resulted in the wake of any other avenues of satellite formation. Capture in a "ready-made" form is generally plausible only for peripheral and irregular satellites.

The moon is not as a regular a satellite of the earth as the Galilean satellites of Jupiter or the system of satellites belonging to Uranus, since the present orbit of the moon has a slight eccentricity and is inclined to the plane of the earth's equator. To date no definitive clarity has been obtained on the problem of whether or not this eccentricity and inclination would have to decline if we could go back in time, extrapolating the contemporary tidal friction effect into the past, i.e., to find out whether the moon was in earlier ages a more regular satellite than it is now [5, 6, 7, 8]. The inclination of the lunar orbit to the earth's equator was probably less than the contemporary inclination, which varies from 18°.3 to 28°.6. This result was obtained by Darwin [5] and confirmed by Gerstenkorn [6], at least for times as far back in the past as two billion years ago, when the moon must have been situated much closer to the earth than it is now. There are apparently no grounds for supposing the moon to have been a more irregular satellite in the past, a possibility which would argue in favor of the capture hypothesis.

Let us now discuss the problem of the possible capture of a ready-made moon. Urey himself does not indicate any concrete scheme for the capture, but we may suppose that a gravitational capture was involved, and this is realizable in principle when three bodies approach: the growing earth and two bodies similar in mass to the moon, i.e., having radii on the order of 1000 km. The close approach of two "moons" would have to have occurred within the earth's sphere of influence, and the mutual distance separating the "moons" would have to exceed their "gravitational" radius for one of them to be transferred to an orbit encircling the earth. The "gravitational" radius would be determined by the line-of-sight distance at which the velocity vector of one of the bodies would be rotated through an angle of the order of $\pi/2$. Since the spread of velocities of the protoplanetary bodies was determined by the perturbations of the largest body, i.e., the growing earth, and was of the order of $\bar{v} = \sqrt{Gm/\theta r}$, where m is the mass, r is the earth's radius, and θ is a factor ranging from 1 to 3, the dispersion would appreciably exceed the parabolic velocities at the surface of moonlike bodies. The gravitational radius of these bodies would be therefore less than their geometrical radius. Gravitational capture onto an almost circular orbit would be realized only under extremely special approach conditions. Hence, the probability of capture of a ready-made moon was less than the probability of a collision with some moonlike body. It may be demonstrated that the probability of such a collision was less than 10^{-2} during the entire growth period of the earth.

In fact, the number of collisions occurring between bodies of dimensions a_i in unit volume in one sec is

$$2n^2 \cdot \pi a_i^2 \bar{v},$$

where $n = \rho_1 / m_i$. The space density ρ_1 is determined by the amount of matter not swept up by the earth in its intake zone, i.e., $(Q - m)$, and by the volume of that zone. The radial boundaries of the zone are generally assumed to range from 0.8 to 1.3 AU. The thickness H of the zone is due to the dispersion of the peculiar velocities $\bar{v}$ of the bodies, i.e., $\bar{v} = \sqrt{\frac{Gm}{\theta r}}$; $H = \frac{\pi R \bar{v}}{2 v_c}$,

where R is the distance to the sun, and v_c is the angular velocity at the same distance.

Hence

$$\rho_1 \cong \frac{2(Q - m)v_c}{3\pi R^3 \bar{v}}.$$

The number of collisions per second through the entire sphere of influence of the earth is

$$\frac{dN}{dt} = 2\frac{\rho_1^2}{m_i^2}\cdot \pi a_i^2 \bar{v}\cdot R^3 \cdot \frac{m}{3M_\odot}, \qquad (1)$$

where $R^3m/3m_\odot$ is the volume of the sphere of influence.

Dividing dN/dt by $dm/dt = 2\gamma\sqrt{\frac{GM_\odot}{R^3}}\cdot r^2(1+2\theta)\rho_1 H$, where $\gamma \approx 1$ is the coefficient of accretion of bodies colliding with the earth [9], we obtain dN/dm, and then the total number of collisions between bodies of dimensions a_i during the growth of the earth mass from m_0 to Q:

$$N = \int_{m_0}^{Q} \frac{Q-m}{3\pi\delta\sqrt{3\pi\delta M_\odot R\, a_i^4}}\, dm, \qquad (2)$$

where δ is the average density of the bodies and of the earth. At a_i = 1000 km = 10^8cm and at mass Q = 6×10^{27}g, the number of collision events, independently of the value of m_0, is less than $8\times10^{-3} < 10^{-2}$.

As mentioned earlier, the gravitational capture hypothesis requires, in addition to close approach, an unusual and special assignment of velocities for the moon to be transferred to an almost circular orbit. Now according to Urey, the moon had to be not a body from the vicinity of the earth's orbit and exhibiting suitable dimensions, but rather a body arriving from remote regions of the solar system. The probability of gravitational capture of a ready-made moon even in the artificially contrived scheme of several moonlike bodies alone must be several orders of magnitude less than 10^{-2}.

In the calculations reported here, it is assumed that the total mass of solid matter Q in the zone of intake of the earth is equal to the present mass of the earth. It could not have greatly exceeded that value since the remnants of matter had no way of straying far away from this region in the ensuing time. And, in fact, there exists no sufficiently massive body in the environs of the earth's orbit which, by its perturbations, would be capable of diverting other bodies to great distances. Such a feat would require a body having a mass of the order of Jupiter's mass [10]. The mechanism of mutual fragmentations and subsequent precipitation of the fragmentation products into the sun, in the light of the Poynting-Robertson effect, would likewise fail to account for the desired sweeping-out of space, since even with the most optimistic estimates of loss of the solar mass due to radiation, this mechanism could not provide for the precipitation into the sun of more than a mass of the order of the earth's mass. This inference is drawn from V. S. Safronov's paper [11].

The method of satellite capture proposed by V. V. Radzievskii and E. P. Razbitnaya [12], and based on the expansion of zero-velocity Hill's surfaces surrounding the earth as a result of the growth in the earth's mass, is not effective for satellite bodies of any dimension [1]. The capture of a resisting medium, as described by T. Sea, would be possible only in the presence of a thick, extended satellite envelope around the earth. Capture of the moon as a result of vulcanic eruptions, as suggested by Boneff [13], is completely groundless.

We arrive at the conclusion that the only realistic view of the matter is to envision the growth of the moon in swarm of satellite bodies girdling the earth.

2. As was evaluated in our first article [1], the earth must have been surrounded during its growth process by a mass ranging from 1/10 to 1/100 of the earth's mass, if the effective dimensions of the colliding bodies were in the 10-100 km range, and at a mean capture probability of 1/100. The principal mass of the solid material was concentrated in bodies having that effective dimension. There is no point in attempting to ascertain whether there was some asteroid body serving as the lunar embryo or whether the lunar embryo grew from the material constituting the satellite swarm, since a "collisional" capture of bodies is accompanied by their fragmentation. We might attempt to find the density distribution of matter in the swarm as a function of distance from the earth. The formation of a large embryo is a more likely event in the region of maximum density. The resulting distance of the satellite is determined by the mean orbital momentum of the particles accreted to it and the tidal interaction between the satellite and its primary.

Assume, as earlier, that all the particles and bodies present in the earth's intake zone possessed a mean random velocity $\bar{v} = \sqrt{Gm/\theta r}$, determined by the perturbations on the part of the earth embryo, and superimposed on their ordered motion with the Keplerian velocity around the sun. Now assume that the mass of the sun and the distance separating the growing earth and the sun remained practically unchanged over the entire process of formation of the satellite swarm. To the degree that the earth embryo remained the largest body in the earth zone, its orbits would become more prolate as a consequence of the averaging out of the orbital characteristics of the particles becoming accreted to the earth embryo. For simplicity, we may assume that the earth embryo moved on a circular orbit, and we might consider the motions of the bodies in a system of coordinates rotating in unison with the earth, similar to the procedure followed in the restricted circular three-body problem.

Approaching the earth to within a distance l, the bodies would acquire an acceleration, and their peculiar velocities would be augmented to the value

$$\bar{v}_1 = \sqrt{\frac{Gm}{\theta r} + \frac{2Gm}{l}} = \sqrt{\frac{Gm}{\theta r}}\sqrt{1+\frac{2\theta r}{l}}. \qquad (3)$$

The directions of the velocities $\bar{v}_1$ are random, as are the directions $\bar{v}$, since the bodies are approaching the earth from all sides. Here we must take into account the spatial compression of the orbits of the particles near the earth itself, this phenomenon being responsible for the increase in particle density to the value

$$\rho_1\left(1+\frac{2GM}{\bar{v}^2 l}\right)=\rho_1\left(1+\frac{2\theta r}{l}\right).$$

If one body collides at a distance l from the earth with another similar body, then both will lose, on the average, a certain part of the velocity $\bar{v}_1$ per collision. The conditions for capture of bodies or fragments of bodies into circumterrestrial orbits are

$$w\cdot\bar{v}_1^2=w\left(\frac{Gm}{\theta r}+\frac{2Gm}{l}\right)<\frac{2Gm}{l}, \quad (4)$$

and hence

$$w<\frac{2\theta r}{l+2\theta r}, \quad (5)$$

i.e., the fraction w of the kinetic energy of the random motion of the bodies remaining after the collision must be less than the energy required to free the bodies from the earth's gravitational field at that distance l.

The distance l at which capture may take place might vary from the radius r of the growing planet to some distance on the order of r_{L_1}, i.e., out to the distance to the point of libration L_1:

$$r_{L_1}=R\sqrt[3]{\frac{m}{3M_\odot}}\approx r\frac{R}{R_\odot}\approx 200r.$$

Here, according to Eq. (5), the fraction of the kinetic energy which may still remain in the trapped particle after the collision is determined by the inequality

$$w<\frac{2\theta}{1+2\theta}$$

at the very surface of the planet, and by the inequality

$$w<\frac{\theta}{100+\theta}$$

near the limiting zero-velocity closed Hill's surface.

We see that the fraction of energy remaining may be close to unity near the planet, and nevertheless the particle will be trapped; as the distance increases, this fraction will diminish, and at great distances this fraction must be no greater than one-hundredth or several hundredths of the original energy for capture to be realized. In other words, at close distances the relative energy losses may be moderate and captures will be possible even in noncentral grazing collisions while, at some distance removed from the planet, capture will require almost a total loss of energy, which is possible only in the case of central head-on collisions which completely suppress the velocities of the bodies relative to the earth. Clearly, the probability of capture will also be greater in a single collision the closer the collision occurs to the planet. This probability must be close to unity at the planet's surface, and gradually declines to zero as we proceed out to the limiting zero-velocity closed surface, i.e., to the boundary of the planet's sphere of influence. A quantitative determination of this probability requires knowledge of the degree of inelasticity of the colliding bodies in space, and this we have no way of knowing. However, we may assume that in the case of central collisions, the energy losses comprise no less than several tens of percent of the energy of the bodies. We may therefore assume as an approximate estimate that the capture probability is equal to, say, half the maximum permissible value of w, i.e., we may assume

$$p_1(l)=\frac{w_{max}(l)}{2}=\frac{\theta r}{l+\theta r}. \quad (6)$$

Physically speaking, the value of p_1 determines the mean fraction of the mass of colliding bodies which is actually trapped into orbit. Aside from captures resulting from the collision of two "foreign" particles, collisions of foreign particles with already trapped particles (adhering of foreign or outside particles to the swarm) must also occur. Since the arrival of fresh matter here will average one-half that involved in collisions of previously free particles in the swarm, we may assume the probability of outside particles adhering to be

$$p_2(l)=\frac{1}{2}p_1(l)=\frac{\theta r}{2(l+\theta r)}. \quad (7)$$

We now apply the same method to the calculation of the number of collisions between bodies that we used in the first section of this article [1], where only collisions between bodies of comparable mass were considered, and the mass of all the bodies colliding with similar bodies in unit time was totalled. The size distribution of the colliding bodies was also formulated as $dN(a) \sim a^{-n}da$, and the assumption was made that the parameter n of this distribution did not vary as the earth increased in size, even in the vicinity of the earth's satellite swarm. This hypothesis is justified by the fact that the increase in the dimensions of the bodies due to their consolidation is to some extent compensated by the smaller dimensions brought about by fragmentation.

We designate the material density in the swarm at a specified distance as $\rho_2(l)$. It is natural to expect that the trapped particles will possess highly prolate orbits, which will become closer to circular and coplanar orbits as a result of mutual collisions of particles in the swarm. It might be surmised that particles trapped at a distance l will execute revolutions about the earth at, on the average, the same distance, until the increased mass of the earth brings about a corresponding contraction of the satellite orbits. This contraction of the swarm is a consequence of the constant geocentric momentum of the satellites: $ml = m_0 l_0$.

In the calculations below, we neglect the effect of consolidation of the swarm, which must take place in the course of the swarm's evolution, and we shall assume the swarm to be spherically symmetrical about the growing earth.

At a distance l from the earth's center, the variation of density in time may be expressed in the form

$$\frac{\partial \rho}{\partial t} = p_1 S_1 + p_2 S_2 - \frac{1}{l^2}\frac{d}{dl}(\rho v_l l^2), \qquad (8)$$

where the first term in the right-hand member designates the density increment of the swarm due to collisions, the second term indicates the density increment referable to adhering of outside particles, and the third term deals with the density variation at a given l due to constriction of the swarm to its center at a speed v_l. From the equation lm = const for the orbit of each particle, we may find

$$v_l = -\frac{l}{m}\frac{dm}{dt} = -\frac{3l}{r}\frac{dr}{dt}. \qquad (9)$$

Substitution of this equation into Eq. (8) yields

$$\frac{\partial \rho}{\partial t} = p_1 S_1 + p_2 S_2 + \frac{3}{r}\left[l\frac{\partial \rho}{\partial l} + 3\rho\right]\frac{dr}{dt}. \qquad (10)$$

Replacing the variable t in Eq. (10) by the variable r with the use of the formula for the rate of increase in the earth's radius r with time [9]:

$$\frac{dr}{dt} = \frac{(Q-m)\,\gamma\,(1+2\theta)\cdot\sqrt{GM_{\odot}}}{6\pi\delta R^3\sqrt{R}},$$

we obtain an equation for the value of ρ at a specified distance l from the earth as a function of r (the increase in r runs from r_0 to r_{cont}, where r_{cont} is the contemporary radius):

$$\frac{\partial \rho}{\partial r} = \frac{3l}{r}\cdot\frac{\partial \rho}{\partial l} + \left[\frac{3\theta r\,(l+2\theta r)^{1/2}}{l^{3/2}\gamma\,(1+2\theta)\,a_{eff}} + \frac{9}{r}\right]\rho$$

$$+\frac{3^{1/2}\theta^{3/2}(l+2\theta r)^{3/2}M_{\odot}^{1/2}(3Q-4\pi\delta r^3)}{\pi^{3/2}l^{5/2}R^{7/2}\delta^{1/2}\gamma\,(1+2\theta)\cdot a_{eff}}. \qquad (11)$$

On the basis of this equation, we may calculate the density increase in the circumterrestrial swarm during the growth period of the earth.

Figure 1 shows the density distribution through the swarm of bodies having effective diameter 10 km. In view of the complexity of the computational work, we assumed this diameter to be a constant dimension, even though it must have varied with time and consequently

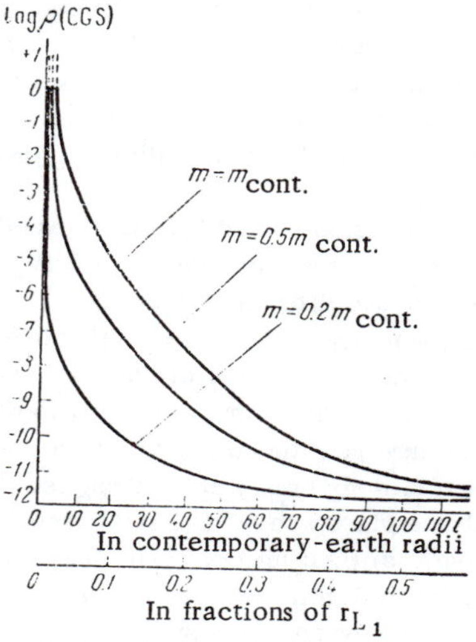

Fig. 1. Density distribution through the circumterrestrial swarm as a function of distance from the earth for various stages of the earth's growth. a_{eff} = 10 km; m_0 = 0.1 m_{cont}.

with the increase in r, i.e., with m, the variation necessarily occurring at different rates in the various portions of the swarm and in the preplanetary cloud. The probability of capture is assumed to decrease from 1/3 at the earth's surface to 1/200 at the periphery of the swarm in these calculations. The initial mass m_0 of the earth, the value at which the swarm commences its formation, is assumed to be 0.1 Q.

We perceive on the graph a sharp concentration of material density in the direction toward the planet. At the earth's surface proper, unreasonably high densities of the order of unity are actually obtained in these calculations. However, when we take into account the rapid consolidation of the particles in the dense portion of the swarm, then the effect of adhering of outside particles in the swarm will be reduced, and the increase in density near the earth will be drastically inhibited. The process must be a self-regulating one, as we see, and no such absurdly high densities may be attained in the actual swarm.

The inference as to an intense concentration of density in the satellite swarms in the direction toward their planet derives not so much from the numerical value assignments as from the properties associated with the capture of swarms as a result of inelastic collisions and first of all as a result of the high probability of becoming captured in a collision at a short distance from the planet. The reverse trend may yield nothing more than a significant consolidation of satellites near the primary, which would decrease the total surface of the satellite material and slow down the growth of the swarm. From the foregoing discussion, the point stands out that the formation of the largest and most regular satellites takes place most likely of all in the immediate vicinity of the planet, but beyond the Roche limit, of course, in some region where the tidal forces generated by the planet would not interfere with the growth of the satellites.

The last conclusion voiced here is in excellent agreement with the observed arrangement of the principal and most massive satellites near their primaries. Figure 2 depicts the mean distances of the satellites from their primaries, expressed in fractions of the distance r_{L_1} to the corresponding points of libration L_1. A most conspicuous fact is how near the satellites are to the primaries, and how relatively empty are the spheres of influence within which the planets may have stable satellites over the entire course of their history. Only the orbits of irregular satellites of Jupiter and Saturn, which are small in mass, are found beyond the range 0.2-0.4 r_{L_1}.

The moon-earth distance is about $^1/_4$ r_{L_1}. This is a large distance compared to similar distances observed in other systems. However, it was initially less, since we may assume as an incontrovertible fact that the effect of tides in the body of the earth brought about a considerable receding of the moon during the history of the earth-moon system [14].

From the discussion of the scheme of formation of the satellite swarm encircling the earth, we find that the growth of the moon must have begun at a short distance from the earth, since the density of matter in the swarm was at a peak near the earth. This distance is related to the mass of the earth at which the formation of the lunar embryo commenced. Of course, this mass could not possibly be computed to any accuracy. However, we may affirm that the earth-moon system formed did not form as a binary planet from some binary embryo. For, in that case, it would have been impossible to obtain the observed mass ratio of 1 : 81 of the two components since the growth of the bodies, started simultaneously for both components, would have resulted in masses of the same order of magnitude. On the other hand, formation of the moon at a time when the earth was already in completed form, practically speaking, is highly improbable, since the earth zone would have been swept free of matter in that case, and no material would have been left over from which to form satellites. The swarm would have to have grown most rapidly at an earth mass of 0.3 to 0.5 of its present mass; here again there was still a good deal of material in the zone and the earth's gravitational pull was still great. The commencement of the moon's formation apparently belongs to this period.

As a study of the earth's rate of accumulation shows [9], the growth of the earth proceeded at a far more rapid rate in the beginning than later. At most, 100 to 200 million years were required to form 99% of the earth's mass. During this same time interval, the formation of

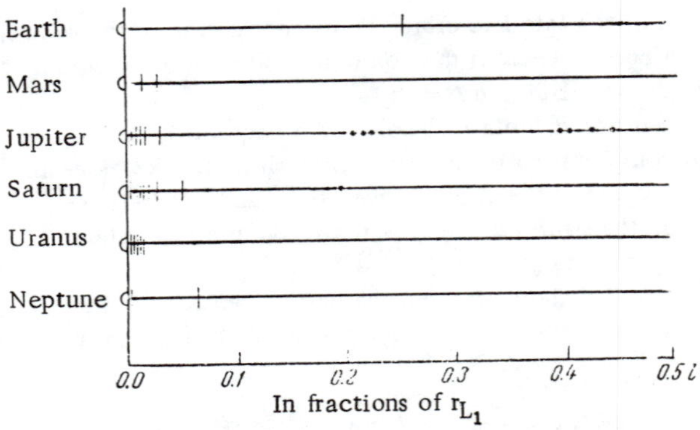

Fig. 2. Location of the satellites near their primaries, in fractions of the distance of the point of libration r_{L_1}.

the satellite swarm must have been completed for the most part. The accumulation of the major bulk of the moon's mass would have necessarily taken no longer a time than the accumulation of the earth. The difference in the ages of the earth and moon could not be greater than 200 million years in the light of this argument.

The earth-moon distance must have always exceeded the Roche limit. The increased mass of the earth caused a constriction of the satellite orbit. The initial distance of the moon must therefore have exceeded by at least 2 to 3 times the present Roche limit (which is about 3 earth-radii). Since the bulk of the mass of the swarm was included within a range of 10 earth-radii, we may arrive at the conclusion that the moon was formed basically at a distance of 5-10 earth-radii. Outside this distance in all likelihood there existed small earth satellites of the asteroidal type. As we know, Kuiper came to a similar conclusion on the presence of asteroidal earth satellites in the past on the basis of a study of the structure of the lunar surface [15]. He suggested that the formation of some of the lunar maria is due to the impacting of these satellites on the lunar surface, when the moon, pushed away from the earth in response to tidal forces, traversed a belt composed of such asteroidal bodies. This offers an explanation of why the moon is the earth's sole satellite, whereas there may have been several satellites in the past.

LITERATURE CITED

1. E. L. Ruskol, Astron. Zhur., 37, 690-702 (1960) [Soviet Astronomy — AJ, Vol. 4, p. 657].
2. H. C. Urey, Proc. of the First Internat. Space Symp., Amsterdam, 1114-1122 (1960); The moon, Z. Kopal and Z. Kadla, editors, Academic Press, N. Y., London (1962); The Origin and Evolution of the Solar System (1962), preprint.
3. S. V. Kozlovskaya, Voprosy Kosmogonii, 8, 145-149 (1962).
4. S. M. Stishov, Abstract of Thesis. IFVD, Academy of Sciences of the USSR, Moscow (1962).

5. G. H. Darwin, Scientific Papers, 2, 36-139, Cambridge (1908).
6. H. Gerstenkorn, Z. Astrophys., 36, 245-274 (1955).
7. G. W. Groves, Monthly Notices Roy. Astron. Soc., 121, 497-502 (1960).
8. H. Jeffreys, Monthly Notices Roy. Astron. Soc., 122, 339-344 (1961).
9. V. S. Safronov, Voprosy Kosmogonii, 7, 59-65 (1960).
10. B. Yu. Levin, Voprosy Kosmogonii, 7, 55-58 (1960).
11. V. S. Safronov, Doklady Akad. Nauk SSSR, 105, 6, 1184-1187 (1955).
12. V. V. Radzievskii, Byulletin VAGO. No. 11 (18), 3-8 (1952); E. P. Razbitanaya, Thesis, Leningrad (1954).
13. N. Boneff, The Moon, Z. Kopal and Z. Kadla, editors, Academic Press, New York-London (1962).
14. E. L. Ruskol, Izvestiya Akad. Nauk SSSR, Seriya Geofiz., 2, 216 (1963).
15. G. P. Kuiper, Vistas in Astronautics, Vol. 2, p. 273, Pergamon Press, New York (1959); Russian translation, in symposium: Éxperimental'noe Issledovanie Okolozemnogo Kosmicheskogo Prostranstva, Foreign Lit. Press (1961), pp. 118-147; Voprosy kosmogonii, 7. 89-90 (1960).

All abbreviations of periodicals in the above bibliography are letter-by-letter transliterations of the abbreviations as given in the original Russian journal. *Some or all of this periodical literature may well be available in English translation.* A complete list of the cover-to-cover English translations appears at the back of this issue.

The Tidal Evolution of the Earth–Moon System (abstract)

E. L. Ruskol

Translated from: Izv. Akad. Nauk SSSR Ser. Geofiz. **1963**, No. 2, 216
Bull. Acad. Sci. USSR Geophys. Ser. **1963**, No. 2, 129

It is shown that the present observed lag in the tides in the solid body of the Earth allows us to construct a consistent picture of the formation of the Moon from a swarm of small satellites moving about the Earth at distances of 5–20 Earth radii, with the subsequent recession of the Moon to its present distance being a result of tidal interaction with the Earth over a time of 4.5 billion years. An estimate is given of the generation of heat within the Earth as a consequence of tidal friction at various epochs, in comparison with the heat generated by the disintegration of radioactive elements.

History of the Lunar Atmosphere and the Possibility of Ice and Organic Compounds Existing on the Moon (abstract)

V. S. Safronov and E. L. Ruskol

Translated from: Vopr. Kosmog. **9**, 203 (1963)
Translation: NASA Report No. N64-29497
Also published in *Proceedings of the XIII International Astronautical Congress, Varna, 1962*, edited by N. Boneff and I. Hersey (Springer, New York, 1964), pp. 42–53

The maximum density and probable composition of the ancient lunar atmosphere are determined. It is assumed that the Moon had been formed as a cold, solid body with a relative abundance of volatile substances similar to those of the Earth. The total amount of volatiles released is assumed to be 100 kg H_2O, 5 kg CO_2, and 0.23 kg N_2 per 1 cm^2 of lunar surface. In accordance with the thermal history of the Moon heated by radioactive elements, it is supposed that the period of intense degasification of the Moon's interior coincided with its melting (some 2.5–3 billion years ago), and lasted about 10^9 y. As the atmosphere accumulated, its escape rate increased; when degasification ceased the density of the atmosphere diminished to its present value. The maximum density near the surface of the Moon is determined from the quality of the flow of escaping gases to the flow of gas released in the interior in the period of intense degasification, and is found to equal 10^{-8}–10^{-9} of the density of the present terrestrial atmosphere. This corresponds to the density at altitudes of about 150 km above the Earth's surface. Therefore, the most abundant components of the lunar atmosphere, H_2O and CO_2, were mainly dissociated into O and CO.

The Structure of the Moon (abstract)

B. Yu. Levin

Translated from: Astron Zh. **43**, 606 (1966)
Sov. Astron. AJ **10**: 479 (1966)

This paper presents a survey dealing with the origin of the Moon, the tidal evolution of the Earth–Moon system, the history of bombardment of the lunar surface, and the thermal history of its interior, the figure of the Moon, the density distribution along its radius, the composition of the Moon, and the history of its atmosphere.

On the Past History of the Earth–Moon System (abstract)

E. L. Ruskol
Icarus **5**, 221 (1966)

New calculations of secular variations of lunar orbit owing to tidal friction are discussed. It is concluded that in the past the lunar orbit should have been closer to the Earth, more circular, and less inclined with respect to the Earth's equator than it is now. This is in favor of the formation of the Moon in the vicinity of the Earth.

SOVIET ASTRONOMY — AJ VOL. 10, NO. 4 JANUARY-FEBRUARY, 1967

THE TIDAL HISTORY AND ORIGIN OF THE EARTH—MOON SYSTEM

E. L. Ruskol

Institute of Geophysics, Academy of Sciences of the USSR
Translated from Astronomicheskii Zhurnal, Vol. 43, No. 4, pp. 829-836, July-August, 1966
Original article submitted January 11, 1966

A survey is given of the results of recent calculations for the evolution of the lunar orbit through tidal friction in the lunar and terrestrial interiors. The orbit would originally have been considerably closer to the earth than now, with its plane less inclined to the earth's equatorial plane, although never coplanar with it. The eccentricity probably was orginally smaller than now, provided that tidal energy dissipation in the lunar interior was below a certain limit. Graphs illustrate the quantitative relationships. The past character of the lunar orbit attests to formation of the moon in the earth's vicinity.

Introduction

Our understanding of the formation process of the moon as a satellite of the earth would be greatly aided if we could explain the character of its orbit in the past. The moon's present orbit is intermediate in type between regular and irregular. Satellites are customarily called regular if they revolve in nearly circular orbits lying in the equatorial plane of their planet. Their distances from the planet usually are less than ten radii of the central body (the Galilean satellites of Jupiter; the satellites of Uranus). Evidently this type of orbit could arise if the satellites were formed in the neighborhood of their planet. The orbits of irregular satellites are characterized by large eccentricities; their inclinations to the equatorial plane of the planet are also large and sometimes even exceed 90°, so that the satellites are in retrograde motion relative to the rotating planet. They usually revolve at great distances from the planet and their masses are small compared to the masses of regular satellites (with the exception of Triton). The origin of irregular satellites is often interpreted as a capture by the central body after formation of the satellite itself.

The moon's present orbit is inclined to the earth's equator by a considerable angle, which varies over the range $\varepsilon = 23°27' \pm 5°9'$ because of the motion of its line of nodes in the ecliptic. The orbital eccentricity fluctuates periodically about the value $e = 0.055$. The mean value of the major semiaxis is $a = 60.3$ R, where R is the radius of the earth; this would be a large value for the distance of a regular satellite from its planet. At the present time it is still difficult to say how similar the current orbits for satellites of other planets are to their orbits during the formation period, but there can be no question that the primeval lunar orbit differed markedly from the current one, and that the main cause of the secular change has been tidal friction in the earth-moon system. If one were able to establish that the primeval lunar orbit was more regular than now, it would be an important argument for the hypothesis that the moon was formed in the vicinity of the earth from material similar in chemical composition to terrestrial material. But if the moon had been a more irregular satellite in the past, it would strengthen the hypothesis that the moon originated as a satellite through capture by the earth, a suggestion based on the still unproved assumption that the two bodies differ in chemical composition.

Dimensions and Orientation of the Initial Lunar Orbit

In the late 1870's, calculations by Darwin [1] based on tidal-friction theory yielded for past epochs a lunar orbit closer to the earth and less inclined to the earth's equator. The decrease in the moon's orbital angular momentum was compensated by a more rapid rotation of the earth. This circumstance was used by Darwin as an argument for the hypothesis that the moon was separated from the earth. The hypothesis itself was subsequently discarded as contrary to the mechanics of rotating masses, but the result that the lunar orbit was more regular in the past was long accepted without dispute. The duration of the tidal evolution of the orbit is measured by times of order 10^9 years, and depends on tidal friction in the oceans and on the dissipative properties of terrestrial material. Within the scope of low-friction theory, that is, for a small value of the lag angle of the tidal bulge relative to culmination of the tide-raising body, the relation among the orbital elements a, e, ε does not depend on the duration of the evolution. Hence, in tidal-evolution calculations one often adopts as independent variable the quantity a, with e and ε as functions of a.

In 1955 Gerstenkorn [2] performed calculations with practically the same physical simplifications in formulating the tidal-friction problem that Darwin had made, but obtained a different result: for some time into the past the lunar orbit turned out to be more regular than now, but it was found that during the epoch of "closest approach" between earth and moon the orbit was inclined to the earth's equator by 45°.7, while still earlier a different branch of tidal evolution could in principle have existed. The moon would then have possessed an extremely irregular orbit of nearly parabolic type, with a retrograde direction of motion relative to the earth. "Closest approach" here means the minimum possible distance a_{min}, at which point the moon would acquire a positive angular momentum from the earth. In Gerstenkorn's calculations this distance coincided with the limiting Roche distance (2.89 R); at closer distances the moon would have been disrupted by the tidal forces of the earth. The fact that a_{min} was not located inside the Roche limit gave further reason to construct a previous, "irregular" branch of tidal evolution.

Without verifying Gerstenkorn's calculations, Alfvén [3] in 1963 used them as a basis to advance the hypothesis that the moon originated through capture by the earth in a highly elongated retrograde orbit, which was then modified by tidal friction into a close circular orbit of large inclination; only afterward did it gradually begin to be transformed into the present orbit.

However, recent calculations of the tidal evolution of the earth-moon system, by Slichter [4], MacDonald [5], and Sorokin [6] (the last two calculations were made with far higher computational accuracy than those of [2]), failed to confirm Gerstenkorn's results regarding the orientation of the lunar orbit in the past. In the new calculations the distance a_{min} of "closest approach" did not coincide with the Roche limit, but was found to fall inside it (2.50 R for MacDonald, 2.40 R for Sorokin). The new results agree satisfactorily with one another regarding the orientation of the lunar orbit, and differ little from Darwin's old results: with increasing retreat into the past the lunar orbit would have become ever closer to the earth, and its inclination to the earth's equator would have diminished over almost the entire period, becoming 9-11° at minimum; only near a_{min} would the value of ε have increased slightly, reaching 10-13° (Fig. 1). On the basis of further analysis of [2, 5, 6], the author [7] has presented evidence that the moon has never had an orbit of irregular type. The probability that the earth would have captured a "ready-made" moon is extremely low [9], even under the most favorable assumptions, with all the protoplanetary material consisting of bodies of lunar dimensions. MacDonald offers still other considerations, not only against an irregular-satellite phase in the evolution of the lunar orbit, but also against the stage of a very close satellite ($a \cong a_{min}$), when the earth's rotation should have been very rapid: in the first case the earth would actually have been rotationally unstable, while in the second case, even though stable, it would have possessed a very large excess of rotational kinetic energy. The release

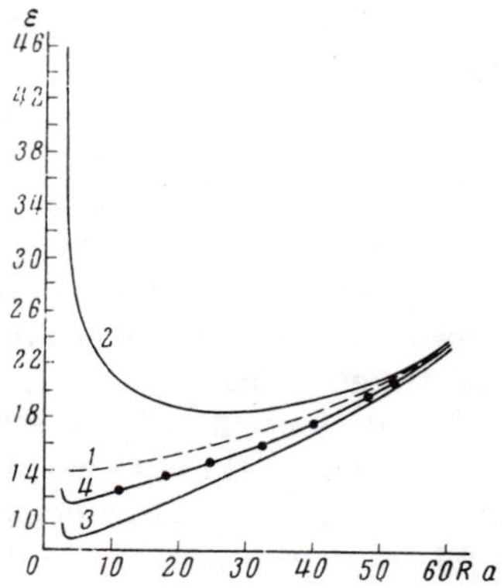

Fig. 1. Inclination angle of the lunar orbit plane to the earth's equatorial plane as a function of the earth-moon distance, according to calculations by various authors. 1) Darwin [1]; 2) Gerstenkorn [2]; 3) MacDonald [5]; 4) Sorokin [6]; dots, Slichter [4].

of this energy in the form of heat through tidal friction would have produced an unacceptably strong heating of the earth's interior, higher than the melting temperature. Moreover, an excessively rapid initial rotation of the earth would have distinguished it from various other planets of the solar system. MacDonald therefore believes that the primeval lunar orbit would have been located no closer than 30-40 R, and that the moon would have been formed there through merger of several smaller earth satellites. The original orbit, according to MacDonald, would have been more nearly circular ($e \approx 0.02$), with a mean inclination of 14-15° to the earth's equatorial plane.

Several years ago, assuming formation of the moon from a circumterrestrial swarm of small satellites, the author [8, 9] obtained the result that the moon would have had an even closer initial orbit (5-10 earth radii), since the density of the material in the swarm would have declined very rapidly with distance from the earth, and the primeval orbit of the massive satellite evidently would not have extended beyond a limit of 10 R. This initial distance for the moon would correspond to a rotation of the earth in a period of 6-7 h, while the integrated release of tidal heat in the earth's interior was found to be several times less than the heat that would have been released through the decay of radioactive elements [10]. The inclination of the initial lunar orbit to the earth's equator evidently would not have exceeded 12-14°. It was found to be more complicated to estimate the initial elongation of the orbit, since this parameter is appreciably influenced by tidal friction in the lunar interior, about which we know very little. This problem merits a special investigation.

Eccentricity of the Initial Lunar Orbit

The current eccentricity of the lunar orbit is small, and in many tidal-evolution calculations the orbit is considered circular, not only for the present epoch but for all times in the past [2, 4, 6]. In Gerstenkorn's calculations [2] a finite e was introduced solely for the epoch prior to close approach, but the forces operating to vary the eccentricity were not fully taken into account [5, 7]. Since in general there is no reason to recognize the existence of such a prior epoch in tidal evolution, we shall admit only variations in e that might have occurred from the present time to the time in the past corresponding to the distance a_{min}.

MacDonald [5] considers tidal variations of the lunar orbit, including changes in the eccentricity. Parallel calculations are also made for a circular orbit ($e = 0$). Depending on the time into the past, the results of the a and ε variations differ very little from each other in the circular and elliptical cases, while the value of e itself in the elliptical case declines into the past to very small values of order 10^{-3}; that is, the orbit became increasingly close to circular in the past. This implies that an approximate treatment of a circular lunar orbit would indeed be useful, at any rate under the assumptions on which MacDonald's calculations rested—in particular, his estimates for the dissipation of tidal energy in the lunar interior.

These assumptions are related to definite ideas about the internal structure of the moon, to which such authors as Jeffreys [11], Urey [12], Kaula [13, 14], and others have adhered. According to these views, the interior of the moon is solid almost to the center, and perhaps only its centermost parts are in a softened state. A fairly high rigidity is ascribed to lunar material – a value characteristic of the solid material in the earth's mantle, which is compressed to far higher pressures than the lunar material. These views are in turn based on the idea that the dynamical figure of the moon differs from the hydrostatic equilibrium figure for the case of a spherically symmetric temperature distribution within the moon. Recently, however, allowance for a latitude dependence of the temperature distribution in the lunar interior has made it possible to construct a quasi-equilibrium model for the moon having the observed flattening at the poles [15, 16]. Thus one of the arguments for a solid lunar model is excluded. It is also very difficult to reconcile calculations of the moon's thermal history with a primarily solid state in the lunar interior: the thickness of the solid envelope apparently should not exceed 500-800 km [17]. A final selection of a specific model for the internal structure of the moon can only be made on the basis of a direct study by geophysical methods. Until then, by estimating the influence of tides within the moon on its orbital motion one should examine a range of possible models for the internal structure, such as that given in Table 5[1] of Harrison's paper [18], including both completely solid models of varying rigidity, and models with a sizable liquid core.

Let us consider briefly the effect of tidal friction in the earth and in the moon upon the variation of the eccentricity e and the major semiaxis a of the lunar orbit; for simplicity we shall regard the

[1] This table is reproduced in the paper by V. N. Zharkov et al., Astron. zh., 43, 622 (1966) [Soviet Astronomy–AJ, Vol. 10, p. 492].

orbit plane as coinciding with the plane of the earth's equator. The presence of a comparatively small angle between these planes will not alter the character of the results.

The phase lag of the earth tide means, first, that the earth's rotation causes the tidal bulge to deviate from the earth-moon center line. This effect generates the familiar twisting moment or torque between the two bodies. The earth's axial rotational momentum will thereby be partially converted into the orbital momentum L of the moon, leading to a gradual increase in a. At the same time e should also increase. This will happen because the tidal interaction will increase inversely as the sixth power of the distance between the bodies, and even for a slight ellipticity of the orbit the moon should be pushed ahead appreciably more strongly at perigee than at apogee. In a general distension of the orbit its similarity will not be preserved: each successive apogee will be more distant from the preceding apogee than each successive perigee is from the preceding perigee. This effect will also lead to an increase in e with time.

On the other hand, the phase lag of the maximum tidal deformation, associated with the ellipticity of the lunar orbit, will cause e to decrease with time. The height of the tidal bulges on the earth and on the moon is inversely proportional to the cube of the distance between them, so that even a lunar tidal bulge oscillating slightly relative to the earth because of librations and the bulge traveling through the oceans and the solid earth will experience radial pulsations with a monthly period as the moon revolves in its orbit. Because of inelasticity the pulsation maxima should occur after the moon has already passed through perigee, and the minima after apogee. Since these pulsations are purely radial in character their lag will generate no torque; thus the orbital momentum L will not change, and the energy dissipated in the lunar interior will be derived from the orbital energy E. The decrease in the eccentricity during this process is graphically expressed by the formula [19]:

$$e = \sqrt{1 + \frac{2EL^2}{m^3_{☾} m^3_{♁} G}},$$

where L, the masses $m_{♁}$ and $m_{☾}$ of the earth and moon, and the gravitational constant G are constant quantities, while the value of E (negative in magnitude) declines. At the same time the major semiaxis a of the orbit will also decrease somewhat, in accordance with the formula

$$L = m_{☾} \sqrt{\frac{Gm^2_{♁}}{m_{♁} + m_{☾}} a(1 - e^2)} = \text{const}.$$

If we consider tidal friction in the earth only, the influence on a and e of the monthly-pulsation lag would be much less than the influence of the lag in the diurnal and semidiurnal tides; the energy dissipation in the earth associated with the monthly pulsations would also be small, three orders of magnitude smaller than that associated with the diurnal pulsations. However, in some cases the influence on the quantity e of the tides on the moon could be considerably stronger even than the basic effect of earth tides. This circumstance will depend primarily on the internal structure and the dissipative properties of the lunar material.

In geophysics wide use is now being made of the dissipative parameter Q, equal to the ratio of the maximum energy that is acquired by a system during a given cycle to the energy dissipated over the whole cycle. For large Q one has the relation $Q \approx \delta/2$, where δ is the angle of lag – in particular, for our case, the tidal phase lag.

In the papers mentioned above, Kaula [13, 14] has estimated the energy dissipated in the moon during a tidal cycle. Assuming that the moon consists entirely of solid matter with a rigidity of $7.38 \cdot 10^{11}$ dyn/cm^2, he evaluated the energy of elastic deformations in the solid moon (both constant and periodically varying deformations). Then, regarding the parameter Q for lunar material as equal to the value adopted for the earth's mantle [5], namely $Q \approx 13$, or equivalently $\delta = 2{.}^{\circ}16$, Kaula found the current dissipation of tidal energy in the moon to be $\Delta E_0 = 0.6 \cdot 10^{16}$ ergs/sec, and he also calculated how this energy is distributed within the moon. Furthermore, he derived formulas for ΔE as a function of the orbital parameters a, e, ε of the moon.

Using this dependence on a and e, Petrova [20] has computed several variants for the tidal history of the earth-moon system, taking into account all effects that influence the time variation of a and e, for selected values of the energy dissipation ΔE_0 in the moon ranging from $0.6 \cdot 10^{15}$ to $0.6 \cdot 10^{17}$ ergs/sec. Her results are presented in the form of graphs (see Figs. 2-4). Before discussing these results it is appropriate to examine the relation between the values of ΔE_0 and the assumed internal structure of the moon.

The values $\Delta E_0 < 0.6 \cdot 10^{16}$ ergs/sec correspond essentially to solid lunar models with angles of lag less than 2°. The values $0.6 \cdot 10^{16} < \Delta E_0 < 0.6 \cdot 10^{17}$ ergs/sec correspond to various lunar models containing a

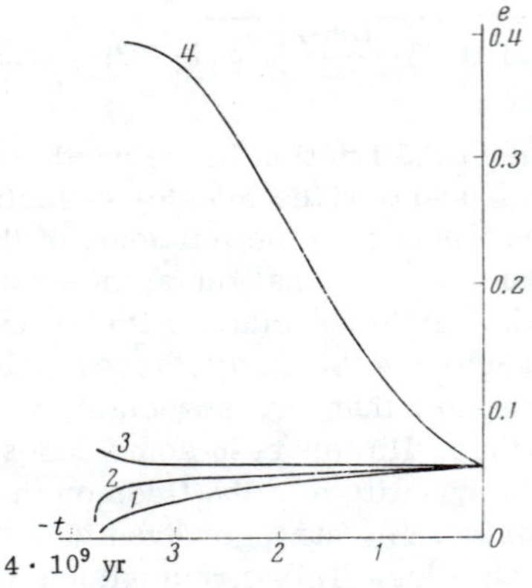

Fig. 2. Eccentricity of the lunar orbit as a function of time, according to G. G. Petrova. 1) $\Delta E_0 = 0.6 \cdot 10^{15}$ ergs/sec; 2) $\Delta E_0 = 0.6 \cdot 10^{16}$ ergs/sec; 3) $\Delta E_0 = 0.8 \cdot 10^{16}$ ergs/sec; 4) $\Delta E_0 = 0.6 \cdot 10^{17}$ ergs/sec; $\delta_{♁} = 1°$.

liquid core with a radius as great as 0.75 lunar radius and angles of lag of about 2-3°. If we accept Goldreich's claim [19] that the factor Q should in general be proportional to the square of the radius of a solid gravitating body, then even larger values for the angle of lag would be permissible for the moon, together with larger values for the quantity ΔE_0. However, too large a value for ΔE_0 would not be justified. In particular, we can hardly agree with the estimate $\Delta E_0 = 3.8 \cdot 10^{18}$ ergs/sec that Kopal [21] has obtained from a lunar model consisting of a viscous fluid with a coefficient of viscosity yielding a maximum effect. As Kaula correctly points out [14], Kopal strongly overestimated the radial displacements of the lunar surface under the action of tides produced by the earth, and since the quantity ΔE_0 is proportional to the square of the displacements, its value was also found to be too large. With so great a heat release the moon would have been heated to more than 1000°K by tides alone, with a marked effect on its thermal history. Moreover, Kopal has not considered the large influence of his assumed energy dissipation in the moon toward changing the lunar orbit. We believe that the most probable values of ΔE_0 presently fall in the range considered by Petrova.

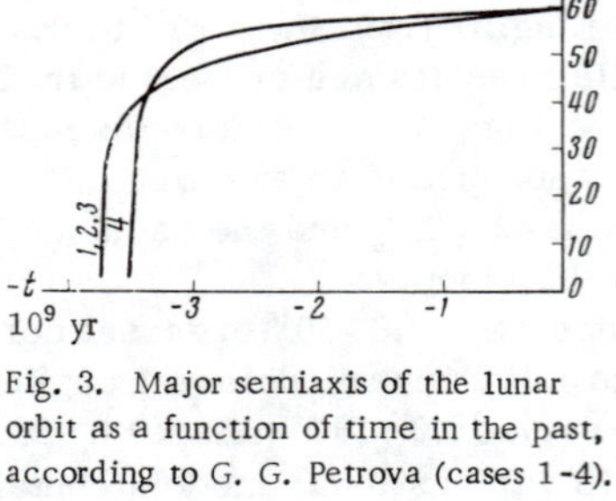

Fig. 3. Major semiaxis of the lunar orbit as a function of time in the past, according to G. G. Petrova (cases 1-4).

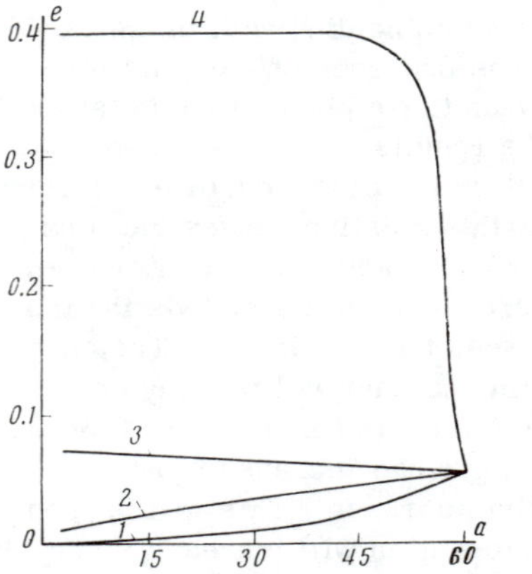

Fig. 4. Eccentricity of the lunar orbit as a function of major semiaxis, according to G. G. Petrova (cases 1-4).

The results of Petrova's calculations show that for comparatively small values of energy dissipation in the moon throughout its history (cases 1 and 2), the change in its orbital eccentricity would be determined primarily by earth tides, and the lunar orbit in the past would have been very nearly circular (see Figs. 2 and 4). For comparison, we reproduce MacDonald's result from [5] (Fig. 5); here the present value of the energy dissipation in the moon was assumed to be $0.3 \cdot 10^{16}$ ergs/sec. Petrova's case 3 is interesting: it corresponds to an approximately equal effect on e from earth and moon tides. Numerically, Petrova assumed $\delta_{♁} = 1°$, $\Delta E_0 = 0.8 \cdot 10^{16}$ ergs/sec. The same result would have been obtained for $\delta_{♁} = 2°$, $\Delta E_0 = 1.6 \cdot 10^{16}$ ergs/sec, except that the duration of tidal evolution of the orbit would have been half as long, since it is determined basically by the value of $\delta_{♁}$. In order not to complicate matters and without labori-

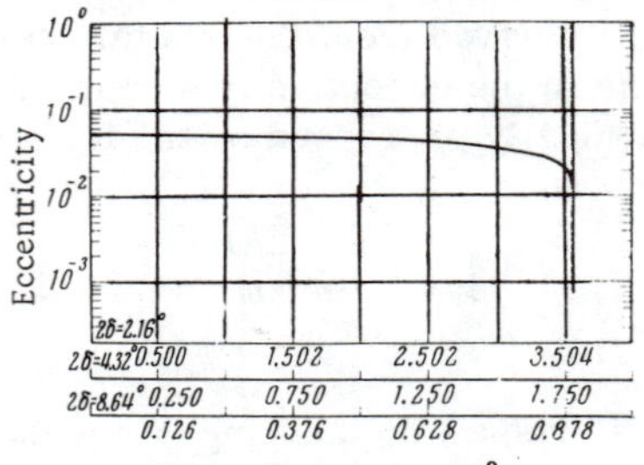

Fig. 5. Eccentricity of the lunar orbit as a function of time in the past for various δ, according to MacDonald [5] ($\Delta E_0 = 0.33 \cdot 10^{16}$ ergs/sec).

ous calculations Petrova assumed that both $\delta_{\text{♁}}$ and $\delta_{\text{☾}}$ are constant in time, although this is actually not true, and $\delta_{\text{♁}}$ should be considered as increasing slowly with time [10]. The behavior of $\delta_{\text{☾}}$ with time could be more complicated, depending on the thermal history of the moon. This quantity is perhaps less definite. For lunar models with large energy dissipation ($\Delta E_0 > 1.6 \cdot 10^{16}$ ergs /sec; $\delta_{\text{♁}} = \delta_{\text{☾}} = 2°$), the orbital eccentricity will vary primarily because of tidal friction in the moon, and one then finds that the orbit could initially have been more eccentric than now, with e reaching values of 0.2-0.3 (case 4, Figs. 2 and 4). On the other hand, the major semiaxis of the lunar orbit behaves in the same way in all the cases considered: in the past the orbit would always have been tighter than now (see Fig. 3).

The calculations of the tidal evolution of the earth—moon system demonstrate, then, that the primeval lunar orbit was located considerably closer to the earth than now, and that its plane was closer to the earth's equator, although never coplanar with it. As we have seen, the problem of the initial eccentricity of the lunar orbit is not solved uniquely within the scope of this type of calculation. Depending on the value of the energy dissipated in the lunar interior throughout its history, the lunar orbit could have been more nearly circular in the past than now, and also more eccentric, although in the latter case the eccentricity evidently would not have exceeded 0.2-0.3.

One interesting possibility for estimating the eccentricity of the lunar orbit in the past has been suggested in a recent note by Goldreich [22], in connection with a discussion of the newly observed fact that the axial rotation of Mercury (T = 59 days) is not synchronous with its orbital revolution (88 days) [23]. This phenomenon was quickly interpreted by Peale and Gold [24] as a confirmation of the large role of the sun in tidally decelerating Mercury. Because of Mercury's considerable orbital eccentricity (e = 0.2), the angular velocity of its motion at perihelion is approximately half again as great as the mean motion over its orbit: $2\pi / \omega_{\text{perih}} = 56$ days, while $2\pi/\omega_{\text{av}} = 88$ days. If the tidal bulge on Mercury is the only violation of the axial symmetry of the planet, then tidal friction tending to synchronize its rotation and revolution periods should in the limit lead to an axial rotation at an angular velocity intermediate between the "perihelion" and mean values, as is observed. The numerical value of the eventual angular velocity will be determined by the dependence of the dissipative parameter Q on the amplitude and frequency of the tidal pulsations (judging from geophysical data, this dependence is a very weak one). But if the figure of the planet differs from a body of revolution, and if its triaxiality is significantly greater than the triaxiality due to tidal deformation, then the final result would be an axial rotation with a period corresponding to ω_{av}.

If this theory is applied to the moon, for which the rotation period coincides with the revolution period and for which the triaxiality is known $[(B - A)/C = 2.05 \cdot 10^{-4}]$, one finds that for a fairly long period into the past the eccentricity of the lunar orbit would have been less than its present value of 0.055; more accurately, it would not have exceeded 0.043. This result of Goldreich argues in favor of Petrova's cases 1-3 for the eccentricity variation, although it probably cannot legitimately be extended to the entire past history of the moon, since the retardation of the moon's axial rotation would require time intervals 10^3-10^4 times as short as the time span for tidal evolution of its orbit.

We may summarize by concluding that extrapolation of tidal variations of the lunar orbit into the past furnishes no evidence to support the hypothesis that the moon was captured in ready-made form, but favors instead the hypothesis that it was formed in the vicinity of the earth.

LITERATURE CITED

1. G. H. Darwin, Phil. Trans. Roy. Soc., 11, 170, 447 (1879).
2. H. Gerstenkorn, Z. Astrophys., 36, 245 (1955).
3. H. Alfvén, Icarus, 1, 357 (1963).
4. L. B. Slichter, J. Geophys. Res., 68, 4281 (1963).
5. G. J. F. MacDonald, Rev. Geophys., 2, 467 (1964).
6. N. A. Sorokin, Astron. zh., 42, 1070 (1965) [Soviet Astronomy — AJ, Vol. 9, p. 826].
7. E. L. Ruskol, Icarus (in press).
8. E. L. Ruskol, Astron. zh., 37, 690 (1960) [Soviet Astronomy — AJ, Vol. 4, p. 657].
9. E. L. Ruskol, Astron. zh., 40, 288 (1965) [Soviet Astronomy — AJ, Vol. 7, p. 221].
10. E. L. Ruskol, Izv. Akad. Nauk SSSR, Ser. Geofiz., No. 2, 216 (1963).
11. H. Jeffreys, The Earth, 4th edn, Cambridge Univ. Press (1959).
12. H. C. Urey, W. M. Elsasser, and M. G. Rochester, Astrophys. J., 129, 842 (1959).
13. W. M. Kaula, J. Geophys. Res., 68, 4959 (1963).
14. W. M. Kaula, Rev. Geophys., 2, 661 (1964).
15. B. Yu. Levin, Astron. Tsirk., No. 285 (1964); Nature, 202, 1201 (1964).

16. V. S. Safronov, in The Figure and Motion of the Moon [in Russian] (Kiev, 1967, in press).
17. S. V. Maeva, DAN SSSR, 159, 294 (1964) [Soviet Physics – Doklady, Vol. 9, p. 945].
18. J. C. Harrison, J. Geophys. Res., 68, 4262 (1963).
19. P. Goldreich, Monthly Notices Roy. Astron. Soc., 126, 257 (1963).
20. G. G. Petrova, Dissertation, Physics Dept., Moscow Univ., June, 1965.
21. Z. Kopal, Icarus, 1, 412 (1963).
22. P. Goldreich, Nature, 208, 375 (1965).
23. G. H. Pettengill and R. B. Dyce, Nature, 206, 1240 (1965).
24. S. J. Peale and T. Gold, Nature, 206, 1240 (1965).

SOVIET ASTRONOMY – AJ VOL. 15, NO. 4 JANUARY-FEBRUARY, 1972

THE ORIGIN OF THE MOON
III. SOME ASPECTS OF THE DYNAMICS OF THE CIRCUMTERRESTRIAL SWARM

E. L. Ruskol

Institute of Earth Physics, Academy of Sciences of the USSR
Translated from Astronomicheskii Zhurnal, Vol. 48, No. 4, pp. 819-829, July-August, 1971
Original article submitted October 26, 1970

A discussion is given of the possibility that the moon may have been formed in the neighborhood of the earth during the accumulation process of the earth itself. More attention is devoted than in previous papers of this series [6, 7] to the following topics in the dynamics of the prelunar swarm of bodies: the role of incoming interplanetary particles in destroying satellites of the earth; the effect of collisions on the size distribution of particles captured into the circumterrestrial swarm; and various mechanisms for replenishing the swarm from the interplanetary medium. The general progress of evolution in the swarm involved the formation of a single or several large bodies (protomoons) comprising most of the swarm's mass, together with a cloud of fine-scale collisional debris which played a major role in capturing new interplanetary particles.

1. INTRODUCTION

In the past few years investigators have devoted much attention to the origin of the moon and the tidal evolution of the earth–moon system (see, for example, the surveys by Levin [1] and Kaula [2]). Hypotheses of three types have been discussed: a) detachment of the moon from the earth; b) capture of a preexisting moon by the earth; c) formation of the moon in the vicinity of the earth.

As for the detachment hypothesis, there appears to be no chance of putting it on a firm mechanical and physical foundation. Objections recently have also been raised against detachment as a result of chemical analysis of the lunar soil [3]. The capture hypothesis, as always, faces many mechanical difficulties. The Alfvén-Gerstenkorn argument, which has often been adduced to support capture, is no longer tenable – namely, the agreement between the "distance of closest approach" in the tidal evolution of the earth-moon system and the Roche limit (2.89 earth radii R), according to Gerstenkorn's 1955 calculations [4]; after a prolonged controversy Gerstenkorn admitted that the distance should actually be 2.50 R on the basis of the same assumptions that were adopted in his former calculations for a circular lunar orbit, and as little as (1.4-1.6)R in his new 1969 calculations for an elliptic orbit [5]. Shifting the "closest approach" inside the Roche limit makes it pointless to develop the preceding, "capture" branch of the tidal evolution. The hypothesis that the moon was formed near the earth out of a circumterrestrial swarm of bodies during the earth's active grown phase [6, 7] warrants more serious attention. It agrees better than the other hypotheses with the information on the composition and age of lunar rocks that has been gathered from the Apollo 11 and Apollo 12 voyages to the moon [3]. From an analysis of all models for the formation of the moon that have been proposed in the past decade, Kaula [2] concludes that this last model [6, 7] merits further development.

We shall discuss here several points concerning the dynamics of the circumterrestrial swarm.

2. COMPARATIVE ROLE OF INTERPLANETARY AND SWARM PARTICLES IN SUPPLYING A GROWING SATELLITE OF THE EARTH

That a circumterrestrial swarm should have been formed is indicated by the conditions for collisions between a satellite of the earth and particles from interplanetary space as well as particles moving in geocentric orbits.

The possibilities for growth of the satellite depend on the balance of the mass gained and lost in the collisions, that is, on the energy of the colliding particles and on the mass of the satellite. In an encounter with a protoearth having a mass m, the geocentric velocity of an interplanetary particle would be

$$\sqrt{\frac{2Gm}{l}+v^2},$$

where l is the distance from the earth and v is the particle's geocentric velocity "at infinity." A study of the velocity dispersion of the particles in the supply zone of the earth has shown [8] that the mean velocity v may be expressed as

$$\bar{v}=\sqrt{\frac{Gm}{\theta r}},$$

where r is the radius of the growing earth, and the dimensionless coefficient θ amounts to several units. For a random orientation of the velocities of the interplanetary particles, their velocity relative to a satellite revolving at a circular velocity $(Gm/l)^{1/2}$ about the earth would, on the average, be

$$\bar{v}_1=\sqrt{\frac{2Gm}{l}+\frac{Gm}{\theta r}+\frac{Gm}{l}}.$$

The dispersion in the values of v_1 is large, of the order of $\bar{v}_1$ itself. For $\theta = 3$, Table 1 gives the dependence of $\bar{v}_1$ on the distance to a growing earth having its present mass M as well as a mass of M/2. We point out for comparison that the escape velocity v_e at the surface of an embryonic satellite with density $\delta_s = 3.3$ g/cm^3 and radius 1000 km would be only 1.36 km/sec.

Evidently at all distances within the moon's present orbit the mean encounter velocities of interplanetary particles with satellites would have considerably exceeded the escape velocity from the surface of a satellite having a radius even as large as the moon's radius. The specific kinetic energy of the interplanetary particles, the energy with which they should have collided with the earth's satellite, would have been of order 10^{11} ergs/g. In such a collision a particle should have broken up a far larger mass than its own. From 10^8 to 10^9 ergs/g is required to fragment stony bodies [9], while 10^6–10^8 ergs/g suffices to overcome the gravitational field of a satellite with a radius of 10-100 km. Hence collisions with interplanetary particles would in most cases have been accompanied not only by fragmentation of the bodies and the particles themselves, but also by a dispersal of fragments into independent circumterrestrial orbits. For example, if we adopt a fragmentation energy of 10^8 ergs/g, corresponding to a disintegration into large pieces, then a satellite of 100-km radius should have been disrupted and dispersed in an impact of a body of 30-km radius colliding with it at 1.5-km/sec velocity, or a body of 15-km radius arriving at 5-km/sec velocity. In order for a 100-km satellite to be more forcibly disrupted (with an energy of 10^9 ergs/g), the radius of a body colliding with it at a velocity of 1.5 or 5 km/sec should have been 50 or 25 km, respectively, a size still considerably smaller than the radius of the satellite.

Let us estimate the time required for a satellite of radius a_s to be broken up by various interplanetary bodies colliding with it at the same velocity $\bar{v}_1$. All particles larger than a_d will be considered to demolish the satellite at once, while particles smaller than a_d will erode it gradually. As mentioned above, $a_d < a_s$.

We shall assume, as usual, that in the vicinity of the earth the size distribution of preplanetary

TABLE 1. Mean Particle Velocity $\bar{v}_1$ in the Gravitational Field of the Growing Earth as a Function of the Distance l (in units of the earth's present radius R)

l		1	3	5	10	20	30	40	50	60
$\bar{v}_1$, km/sec	for $m = M$	14.4	9.0	7.7	6.4	5.5	5.3	5.2	5.0	4.9
	for $m = \frac{M}{2}$	10.2	6.3	5.4	4.5	3.9	3.7	3.6	3.5	3.4

bodies and particles follows the power law

$$dN(a) = Ba^{-n}da.$$

Then the time within which the satellite will, on the average, suffer one disruptive collision will be

$$\tau_d = \frac{1}{\int_{a_d}^{a_{max}} Ba^{-n}\pi(a_s + a)^2\bar{v}_1\, da}, \qquad (1)$$

where a_{max} is the maximum radius of the bodies.

The time required for a satellite of mass m_s to collide with particles smaller than a_d whose combined mass is sufficient to destroy the satellite will be

$$\tau_c = \frac{m_s}{f\int_{a_{min}}^{a_d} \frac{4\pi}{3}\delta_s a^3 Ba^{-n}\pi(a_s + a)^2 v_1\, da}, \qquad (2)$$

where δ_s is the density of a particle, and the bombardment loss coefficient $f(v_1)$ indicates how many times the mass ejected from the satellite by the particle exceeds the mass of the particle itself. The average time τ_f required for disruption of the satellite is given by the expression

$$\frac{1}{\tau_f} = \frac{1}{\tau_d} + \frac{1}{\tau_c}. \qquad (3)$$

The coefficient f depends on the energy of the colliding particle and on the mass of the satellite. In a paper dealing with the conditions of asteroid growth, Hartmann [10] has computed f on the basis of experimental findings by D. E. Gault et al. regarding collisions at hypersonic velocities (5-7 km/sec). In the Gault experiments the ratio of the fragmented mass to the projectile mass was 10^2-10^3, but high ejection velocities, up to 19 km/sec, were observed for only a small portion of the material having a mass approximately equal to the mass of the projectile. A lower limit of 0.25 km/sec was recorded for the velocity of the "jet." This value corresponds to the parabolic velocity at the surface of a body of 176-km radius (for a density of 3.6 g/cm^3). Hartmann takes the 176 km as a lower bound on the radius of a nondisrupted asteroid subject to particle impacts at velocities of 5-7 km/sec. For smaller bodies, which do not grow but are only fragmented by collisions, Hartmann gives the following values for the coefficient f as a function of the radius a_s of the body (if the bodies have a density of 3.3 g/cm^3 the radius would be 5% larger):

a_s, km	176	100	10	1	$\leqslant 1$
f	$\geqslant 1$	4	26	10^3	10^3

Marcus [11] has used the same experimental results of Gault et al., but he includes the entire velocity distribution function for the ejected material and obtains higher values for the fragmentation coefficient. According to Marcus, for an impact velocity of 5-7 km/sec the value $f = 1$ would correspond to a body with a radius of about 400 km; for bodies with a radius less than 200 km the coefficient $f > 10$; for $a = 100$ km, $f \approx 30$; and for still smaller bodies f increases to several hundred.

The findings of Gault et al. are applicable to the conditions of circumterrestrial space, in which typical velocities of interplanetary particles would have been about 5 km/sec. It follows that for growth of satellites with a radius less than some value (400 km, according to Marcus), a circumterrestrial swarm of particles with far smaller peculiar velocities would have been needed. A calculation shows that in the absence of a swarm, during the earth's active growth period, when the volume density of interplanetary particles in the supply zone of the earth was 10^{-11}-10^{-12} g/cm^3 and their radius distribution had an exponent n in the range from three to four, a satellite with a radius of about 100 km ($f \simeq 30$) would have been destroyed in a time of order 10^5 yr. Bodies with a radius of 200 km, for which f is somewhat smaller, would have lasted 10^6-10^7 yr. Thus in the neighborhood of a sufficiently massive earth without the presence of a circumterrestrial swarm, no body of asteroidal size could have grown into the moon due to interplanetary bodies and particles; it could only have been destroyed by them.

Let us estimate what conditions should have been satisfied by the circumterrestrial swarm in order for the satellite to be able to grow. The average velocities $\bar{v}_s$ of the swarm particles relative to a satellite moving in a circular orbit would be determined by the deviations of the particle orbits from the satellite orbit:

$$\bar{v}_s \cong \bar{e}v_c \cong \bar{i}v_c, \qquad (4)$$

where the geocentric circular velocity $v_c = (Gm/l)^{1/2}$, e is the eccentricity of a particle orbit, and i is the inclination of that orbit to the satellite orbit (in radians). If the swarm particles have a sufficiently well-ordered motion, $e \ll 1$, $i \ll 1$, and the velocities $\bar{v}_s$ would be much smaller than the $\bar{v}_1$ given in

Table 1. The ordering of the motion would have been achieved in a comparatively short time, as shown in Sec. 4 below. The velocities v_s are uniquely related to the uniform thickness H_s of the swarm [8] by $\bar{v}_s = 4H_s/P$, where P is the geocentric revolution period of the swarm particles. The degree of organization or flattening of the circumterrestrial swarm would have been limited by the occurrence within it of large bodies, whose perturbations would have governed $\bar{e}$ and $\bar{i}$:

$$\bar{e} \approx \bar{i} \approx \frac{\bar{v}_s}{v_c} \frac{\sqrt{\frac{G(m_s)_{max}}{\theta a_{max}}}}{\sqrt{\frac{Gm}{l}}} = \frac{a_{max}}{r} \sqrt{\frac{l \cdot \delta_s}{r\theta\delta_\oplus}},$$

where r is the radius of the growing earth, $\delta_\oplus$ is its density, and δ_s is the density of the satellite. It is these bodies, the largest in size at a given epoch, that would have been capable of further growth due to the swarm particles and could have served as progenitors of satellites, while the smaller bodies would merely have been a source of supply for the medium.

The possibilities for growth of large bodies in the neighborhood of the growing earth would have depended on the density ratio of the circumterrestrial swarm and the interplanetary medium. Two components of the swarm should be distinguished when analyzing this process. One component would comprise the particles in organized orbits; the other would consist of particles newly captured into the swarm, whose orbits would initially have been strongly elongated and inclined to the central plane of the swarm, where the large satellites should have accumulated. We may also call such particles the disordered component of the swarm. The increment in the satellite mass would be

$$\frac{dm_s}{dt} = \pi a_s^2 [\rho_s v_s (1 - f_s) + \rho_i v_i (1 - f_i) + \rho_1 v_1 (1 - f_1)],$$

where f_s is the loss coefficient through collisions with the organized swarm component; ρ_s is the density of that component; and f_i and f_1, ρ_i and ρ_1, v_i and v_1 respectively denote the loss coefficient, density, and velocity of the particles in the disordered component and of interplanetary particles. The sum of ρ_s and ρ_i comprises the total density ρ_2 of the swarm. For a rough estimate let us assume that f_i and f_1 are the same and equal to f, and since $\bar{v}_s \ll \bar{v}_1$, we shall set $f_s = 0$ and let $\bar{v}_i \approx \bar{v}_1$ and $\rho_i \approx \rho_1$. Then the condition $dm_s/dt > 0$ for growth of the satellite may be written in the form

$$\rho_s \bar{v}_s > 2\rho_1 \bar{v}_1 (f - 1) \quad (5)$$

or with the substitution

$$\frac{\bar{v}_1}{\bar{v}_s} = \frac{r}{a_{max}} \sqrt{\frac{\delta_\oplus}{\delta_s}\left(1 + \frac{3\theta r}{l}\right)}$$

as

$$\frac{\rho_s}{\rho_1} > 2 \frac{r}{a_{max}} (f - 1) \sqrt{\frac{\delta_\oplus}{\delta_s}\left(1 + \frac{3\theta r}{l}\right)}, \quad (5a)$$

where the value of the square root ranges from approximately 2.5 for small l to 1.4 for large l. It is appropriate to consider this ratio near the plane of symmetry of the swarm, where the density ρ_s is greatest and the embryonic satellites would be formed. Let us apply this condition to protosatellites of various sizes a_s, taken to be the largest bodies in a given zone of the swarm, which thereby would govern the value of v_s. Satellites larger than some limit, having $f \leq 1$, would grow for all $\rho_s > 0$. According to Marcus this limit is 400 km, while according to Hartmann the limiting radius is twice as small. But for smaller satellites the condition (5a) imposes fairly stringent requirements on the ratio ρ_s. For example, for satellites with a radius of 100 km and for velocities v_s determined by their interaction, we have $2r/a_s \approx 100$; moreover, f is equal to several tens. Thus in order for a 100-km satellite to grow the density of the organized component of the swarm should be some three orders of magnitude higher than the density of the interplanetary medium. The growth of a smaller embryo would not demand not only a higher density in the plane of symmetry but also a larger mass for the swarm. In fact, the condition (5) represents a lower bound on the surface density of the swarm, if it is written as

$$\sigma_s > \frac{P}{2} \rho_1 \bar{v}_1 (f - 1). \quad (5b)$$

We may conclude that the swarm, which is proportional to σ_s, should increase as the size of the embryonic satellite decreases, because with decreasing size the coefficient f will increase. It therefore seems more likely that the satellite would grow from a rather large body captured into the swarm through collisions with similar bodies; it is doubtful whether a satellite could grow if it were to begin as an embryo of arbitrarily small radius.

A computation of the number of collisions among bodies of various sizes in the neighborhood of the earth throughout its period of growth – collisions that could have led to capture of a satellite – shows that the probability of a single collision between two bodies with radii of 1000 km is less than 10^{-2} [7]. The same calculation indicates that during the same period of time isolated collisions of bodies with radii of 200-300 km could have occurred, as well as dozens of collisions between bodies with 100-km radii (if the bulk of the material in the earth's supply zone is concentrated in such bodies). Accordingly, the maximum initial size of earth satellites captured into the swarm would probably have been no more than 100 km, and that would have been the case only if sufficiently large fragments had been left intact by collisions among the bodies. For such a satellite of the earth to grow a circumterrestrial swarm would be necessary with an initial density in its central plane some 100 times higher than the density of the interplanetary medium.

3. SIZE DISTRIBUTION OF THE BODIES AND PARTICLES IN THE CIRCUMTERRESTRIAL SWARM IMMEDIATELY AFTER THEIR CAPTURE

We shall now show that when particles are captured into the circumterrestrial swarm their size distribution law would change even if fragmentation were absent. Suppose that in the zone near the earth (outside the swarm) the size distribution of the particles follows the power law

$$dN_1(a) = B_1 a^{-n_1} da.$$

The number of particles with radii in the range from a to $a + da$ experiencing collisions in unit volume and unit time with particles having radii from a to $a + \Delta a$ will be

$$dN_2 = v\pi(2a)^2 \frac{B_1^2 a^{-2n_1}}{2} \Delta a\, da. \tag{6}$$

Since the number of captured particles is determined mainly by the number of particles colliding with other bodies of similar size, the interval Δa is proportional to a, that is, $\Delta a \approx \alpha a$, and the size distribution immediately after capture will, by Eq. (6), take the form

$$dN_2(a) = B_2 a^{-n_2} da, \tag{7}$$

where $n_2 = 2n_1 - 3$, so that for $n_1 > 3$ we will have $n_2 > n_1$. Different values of n_1 correspond to the following values of n_2:

n_1	2	2.5	3	3.5	3.75
n_2	1	2	3	4.0	4.5

The value $n_1 = 3.5$ may evidently be considered the most characteristic one during the stage of preplanetary bodies [8, 12]. In this event the change due to capture would correspond to a transition from $n_1 = 3.5$ to $n_2 = 4$; thus a fine-scale component of interplanetary particles would be discriminated and its role strengthened in the swarm. This discrimination would have practically no effect on the particle distribution over the zone of the earth as a whole (on the value of n_1), provided that the mass of the circumterrestrial swarm is small compared to the mass of all the particles.

The unavoidable fragments accompanying collisions would further increase the number of fine particles.

4. RELAXATION TIME OF THE SWARM

We shall next estimate the time τ_S during which a swarm particle (the satellite) will experience collisions with smaller particles, also belonging to the swarm, whose combined mass is equal to the mass m_S of the satellite. The fragments of the particles will be neglected. For small bodies incapable of growth, τ_S is the characteristic time for organization of motion in the swarm due to inelastic collisions. For large bodies that are capable of growth, τ_S would be the time required for doubling of the mass, or the characteristic growth time if the mass loss through collisions with particles is small ($f_S \approx 0$):

$$\Delta m_s \cong m_s = \pi a^2 \rho_s v_s \tau_s. \tag{8}$$

Here ρ_S is the density due to particles smaller than m_S. If we substitute here $m_S = {}^4/_3\pi\delta_S a^3$ and $\rho_S v_S = 4\sigma_S/P$, where σ_S is the surface density of the material in the swarm and includes only bodies with masses less than m_S, we obtain

$$\tau_s = \frac{a\delta_s P}{3\sigma_s},$$

$$\sigma_s = \frac{\int_{a_{min}}^{a} m_s B_3 a^{-n_2} da}{S} = \frac{\mu_s}{S} \frac{\int_{a_{min}}^{a} m_s B_3 a^{-n_2} da}{\int_{a_{min}}^{a_{max}} m_s B_3 a^{-n_2} da}, \tag{9}$$

in which it has been convenient to introduce explicitly the mass μ_S of the entire swarm, equal to the integral in the denominator; S is the area of the swarm. The normalization factor B_3 here refers to the full volume of the swarm.

If $n_2 \neq 4$ we then find

$$\sigma_s = \frac{\mu_s}{S} \frac{(a^{4-n_2} - a_{\min}^{4-n_2})}{(a_{\max}^{4-n_2} - a_{\min}^{4-n_2})};$$

$$\tau_s = \frac{\delta_s P a S}{3\mu_s} \cdot \frac{(a_{\max}^{4-n_2} - a_{\min}^{4-n_2})}{(a^{4-n_2} - a_{\min}^{4-n_2})}, \tag{10}$$

while if $n_2 = 4$,

$$\sigma_s = \frac{\mu_s}{S} \frac{\ln \dfrac{a}{a_{\min}}}{\ln \dfrac{a_{\max}}{a_{\min}}}; \quad \tau_s = \frac{\delta_s P a S}{3\mu_s} \cdot \frac{\ln \dfrac{a_{\max}}{a_{\min}}}{\ln \dfrac{a}{a_{\min}}}. \tag{10a}$$

It is apparent from Eqs. (10) that, for $n_2 = 3$ and $a \gg a_{\min}$, τ_s is independent of the particle size. If $n_2 < 3$, τ_s is smaller for the larger-sized particles; for $n_2 > 3$, on the other hand, the motion of fine-scale particles becomes organized more rapidly than the large bodies can grow and be organized. In the latter case, more favorable conditions prevail for the aggregation of large satellites. Table 2 gives the values of τ_s computed for $n_2 = 3.5$ and 4 as a function of μ_s for $P = 10^6$ sec ($l = 32$ R) and $S = \pi(60R)^2$ (the area of the circle bounded by the present lunar orbit). For other values of l, τ_s differs from the tabulated values by a factor of $l^{3/2}$, since $P \sim l^{3/2}$; for example, for $l = 60$R the time τ_s is 2.5 times as long.

According to Table 2, the relaxation times are much shorter than the characteristic time for accumulation of the earth, 10^8 yr [8]. This circumstance argues in favor of a rapid evolution for the swarm.

5. GROWTH OF LARGE SATELLITES FROM INTERPLANETARY PARTICLES

Let us now estimate the growth of the largest bodies, the embryonic satellites, due to direct infall upon them of particles from interplanetary space. The particles will tend to erode the surface of the satellites, but their fragments and ejection products would remain in circumterrestrial orbits and would ultimately fall onto the satellites and the earth. The simultaneous growth of the earth and of a satellite having mass μ and radius a are described by the relations

$$\frac{dm}{dt} = \frac{4\pi}{P} r_e^2 \sigma; \quad r_e^2 = r^2\left(1 + \frac{2Gm}{v^2 r}\right) \tag{11a}$$

$$\frac{d\mu}{dt} = \frac{4\pi(1 - f_1)}{P} a_e^2 \sigma; \quad a_e^2 = a^2\left(1 + \frac{2G\mu}{v^2 a}\right), \tag{11b}$$

where we shall set f_1 equal to zero in order to obtain the maximum value of $d\mu/dt$. As before, let $v^2 = Gm/\theta r$; then

$$r_e^2 = r^2(1 + 2\theta), \quad a_e^2 \approx a^2\left[1 + 2\theta\left(\frac{\mu}{m}\right)^{2/3}\right] \approx a^2.$$

Equating the expression for $d\mu/dm$ to that obtained by dividing Eq. (11b) by Eq. (11a), we find

$$\frac{d\mu}{dm} = \frac{4\pi\delta_s a^2 da}{4\pi\delta_\oplus r^2 dr} = \frac{a^2}{r^2(1 + 2\theta)}, \tag{12}$$

so that

$$\frac{da}{dr} = \frac{\delta_\oplus}{\delta_s} \frac{1}{1 + 2\theta}; \tag{13}$$

for $\delta_\oplus/\delta_s = 5/3$ and $\theta = 3$, this relation yields $da/dr = 5/21$, while for $\theta = 5$, $da/dr = 5/33$.

The only satellites that could have been preserved are those that had begun to grow when the mass of the protoearth reached about one half the

TABLE 2. Relaxation Time [yr]

$n_2 = 3.5$, $a_{\max} = 10^8$ cm, $a \gg a_{\min}$

in units of lunar mass	$a = 10^3$	10^5	10^6	10^7	10^8 cm
10^{-4}	7×10^3	7×10^4	2.2×10^5		
10^{-2}	70	700	2.2×10^3	7×10^3	
1.0	0.7	7	22	70	220 yr

$n_2 = 4$, $a_{\max} = 10^8$ cm, $a_{\min} = 10^{-4}$ cm

μ_s	$a = 10$	10^3	10^5	10^6	10^7	10^8 cm
10^{-6}	48	3.4×10^3	2.7×10^5	2.4×10^6		
10^{-4}	0.48	34	2.7×10^3	2.4×10^4		
10^{-2}	4.8×10^{-3}	0.34	27	240	2.2×10^3	
1.0	4.8×10^{-5}	3.4×10^{-3}	0.27	2.4	22	200

earth's present mass (for a smaller mass of the protoearth there would be no assurance that the system would not have disintegrated through impacts of other large masses in the earth's supply zone). Moreover, if the initial mass of the embryonic planet was small, then as it continued to show a substantial enlargement all satellites originally located at distances $l_0 < l_R M/m_0$ would have been drawn inside the Roche limit l_R and would have been destroyed. As the earth grew from M/2 to M its radius increased by 1300 km. According to Eq. (13), the increase in the radius of a large satellite would have been 200-300 km, at most. In order for the moon to have grown in this manner it would have been necessary, when the earth's mass was half as great as now, for a protomoon already to have existed with a radius of 1400-1500 km, that is, a satellite with a mass equal to 0.5-0.7 of the moon's present mass, or alternatively, three or four protomoons, each with a radius of 1000 km and having the same combined mass. Thus growth of large satellites from interplanetary particles would indeed have been possible, although it would not solve the problem of the original formation of those satellites, which themselves should have grown out of the swarm.

6. REPLENISHMENT OF THE SWARM BY PARTICLES FROM INTERPLANETARY SPACE

The swarm could be replenished in two different ways by free (interplanetary) particles: a) capture through free-free collisions (mass μ_1); b) capture through free-bound collisions, or adhesion (mass μ_2). An efficient mechanism for replenishing the swarm should provide for the capture there of a mass several times greater than the mass of the moon, since a considerable portion of the swarm material would evidently be expected to have fallen onto the earth.

a) The total mass of the particles participating in free-free collisions per unit volume and unit time is given [6] by

$$\Sigma_1 = \frac{3}{4}\frac{v_1\rho_1^2\alpha}{\delta_s a_{eff}}, \tag{14}$$

where

$$a_{eff} = \frac{7-2n_1}{(4-n_1)^2} \times \frac{(a_{max}^{4-n_1} - a_{min}^{4-n_1})^2}{a_{max}^{7-2n_1} - a_{min}^{7-2n_1}},$$

and the "comparability" interval α [see Eq. (6)] may be taken numerically to be $\alpha = \frac{1}{4} - \frac{1}{5}$. Multiplying Σ_1 by the total volume of the swarm ($\approx \Re^3 \cdot m/M_\odot$, where $\Re$ is the distance from the earth to the sun) and by the capture probability $\bar{p}_1$ averaged over the volume, we obtain $d\mu_1/dt$; and then if we divide this expression by the growth rate dm/dt of the earth's mass due to accretion of particles having a density ρ_1, we find the following expression for the function $\mu_1(m)$:

$$\frac{d\mu_1}{dm} = D_1(M - m - \mu_s) \approx D_1(M - m);$$

$$\mu_1 = D_1\left(\frac{M^2}{2} - Mm_0 + \frac{m_0^2}{2}\right), \tag{15}$$

where m_0 and M denote the earth's initial and present mass, and

$$D_1 = \frac{\bar{p}_1\sqrt{\theta}}{9a_{eff}(1+2\theta)\sqrt{3\pi\delta_s M_\odot \Re}}. \tag{16}$$

To capture ten times the mass of the moon, $D_1 \approx 1/3M$ would be needed, and by Eq. (16),

$$a_{eff} \approx 10^3 \cdot \bar{p}_1 \text{ cm}, \tag{17}$$

or no more than 10 cm if $\bar{p}_1 = 10^{-2}$ [7]. For particles with a power-law size distribution and an exponent $n \approx 3.5$ [8, 12], we would have $a_{max} = 5a_{eff} < 1$ m. However, an investigation of the earth's formation mechanism [8] indicates that $a_{max} \approx 10^3$ km. Thus capture through free-free collisions would yield only one millionth of the required mass of the swarm.

b) The frequency of free-bound collisions and the acceptance of free particles into the swarm would also have depended on a_{eff}. According to our previous estimate [6, 7], if the swarm had an initial density equal to the "background" density ρ_1 and if the free and captured particles had the same distribution, the requisite swarm mass μ_2, ranging from one to ten lunar masses, could have been captured within the time required for the earth's mass to grow if a_{eff} were equal to 10-100 km, independently of the exponent n_1 in the power law. For $d\rho_2/dt \sim \rho$, that is, $d\mu_2 \sim \mu_s$ (in the absence of protosatellites that could efficiently sweep up the swarm material), a rapid exponential growth of the swarm would be achieved if $a_{eff} \leq 10$ km.

As the fine-scale component becomes swept up by the protosatellites the swarm density will no longer increase exponentially.

Let us now estimate the mass $\mu_2(m)$ acquired by the swarm through its acquisition of free particles. The combined mass of all the particles participating in free-bound collisions within 1 cm^3

TABLE 3

n_2	3.5	3.75	4.0	4.5	5.0
$a'_{\text{eff}} =$	$\dfrac{4a_{\max}}{\ln \dfrac{a_{\max}}{a_{\min}}}$	$2a_{\max}^{3/4} a_{\min}^{1/4}$	$\ln\left(\dfrac{a_{\max}}{a_{\min}}\right) \times \sqrt{a_{\max}\, a_{\min}}$	$4\sqrt{a_{\max}\, a_{\min}}$	$3\sqrt{a_{\max} \cdot a_{\min}}$

each second will be

$$\sum_2 = \frac{3\bar{v}_1 \rho_1' \rho_2' \alpha}{2\delta_s a'_{\text{eff}}} = \frac{3}{2}\frac{\bar{v}_1 \alpha k_1 k_2 \rho_1^2}{\delta_s a'_{\text{eff}}}, \quad (18)$$

where ρ_1' represents the density of the fine particles in the interplanetary medium that participate in collisions with comparably sized particles in the fine component of the swarm; $k_1 = \rho_1'/\rho_1$, $k_2 = \rho_2'/\rho_1$, and a'_{eff} is the effective length for collisions between these two ensembles (having different n).

The function $\mu_2(m)$ here takes a form analogous to Eq. (15) but with D_1 replaced by D_1':

$$D_1' = \frac{2\bar{p}_2 \sqrt{\theta}\, k_1 k_2}{9a'_{\text{eff}}(1+2\theta)\sqrt{3\pi\delta_s M_{\odot}\mathfrak{R}}}, \quad (19)$$

where $\bar{p}_2 \approx \bar{p}_1/2$.

For different exponents n_1, n_2 in the distribution laws of the two ensembles, the effective length a'_{eff} becomes

$$a'_{\text{eff}} = \frac{7-n_1-n_2}{(4-n_1)(4-n_2)} \cdot \frac{(a_{\max}^{4-n_1} - a_{\min}^{4-n_1})(a_{\max}^{4-n_2} - a_{\min}^{4-n_2})}{a_{\max}^{7-n_1-n_2} - a_{\min}^{7-n_1-n_2}}, \quad (20)$$

and for $n_1 = 3.5$ and $n_2 \geq 3.5$ we obtain the values of a'_{eff} given in Table 3. We have shown in Sec. 3 above that for $n_1 = 3.5$ the captured particles will have $n_2 = 4.0$ if fragmentation is absent. For the preplanetary bodies, $a_{\max} \approx 10^8$ cm, while for the swarm, if the embryonic satellites are excluded, $a_{\max} \approx 10^7$ cm. The smaller of these values should be taken for interactions between the two ensembles, that is, $a'_{\max} \approx 10^7$ cm. For $a_{\min} = 10^{-4}$–10^{-5} cm and $n_2 = 4$ we obtain $a'_{\text{eff}} = 200$–600 cm. Taking $k_1 \approx 1/3$ in Eq. (19), corresponding to elimination from the swarm of the largest preplanetary bodies with radii from 10^8 to 10^7 cm, we find that capture into the swarm of five to ten times the mass of the moon would be possible for $k_2 \approx 10$, that is, when the density ρ_2' of the fine component of the swarm is an order of magnitude higher than the "background" density ρ_1. For $\rho_1 \approx 10^{-12}$ g/cm^3, which would have been the density of matter in the earth's supply zone during its active growth phase, we would accordingly have a swarm mass of about 10^{-4} times the lunar mass within a zone of 60 R. Equations (10) and (10a) imply that such a mass could have been swept up by two or three 1000-km bodies in 10^3 yr. This time interval represents the characteristic time for "renewal" of the fine component of the swarm through capture of new particles. After 10^8 yr of aggregation of the earth, the fine component of the swarm would have been renewed about 10^5 times, which would provide for about ten lunar masses.

The dynamical circumstances of swarm formation that we have discussed afford an explanation of certain differences between the chemical compositions of the earth and the moon [13].

CONCLUSIONS

The following conclusions may be drawn from this investigation: 1) The moon was most likely formed in the neighborhood of the earth through the amalgamation of several large protomoons that had grown out of the circumterrestrial swarm. 2) The principal mechanism for replenishment of the swarm was its acquisition of a fine-scale component of interplanetary particles and bodies.

I should like to thank B. Yu. Levin, V. V. Radzievskii, and V. S. Safronov for discussing the paper and offering valuable critical comments.

LITERATURE CITED

1. B. Yu. Levin, Astron. Zh., 43, 606 (1966) [Sov. Astron.–AJ, 10, 479 (1966)], see also Proc. CalTech–JPL Lunar Planet. Conf. (Sept. 1965), Jet Propul. Lab. Tech. Memo No. 33-266 (1966), p. 61.
2. W. M. Kaula, Inst. Geophys. Planet. Phys. UCLA Publ. No. 893 (1970); Rev. Geophys. Space Phys., 9, 217 (1971).
3. R. Ganapathy, R. R. Keays, J. C. Laul, and E. Anders, Proc. Apollo 11 Lunar Sci. Conf., Pergamon Press, 2, 1117 (1970).

4. H. Gerstenkorn, Z. Astrophys., 36, 245 (1955).
5. H. Gerstenkorn, Icarus, 9, 394 (1968); 11, 189 (1969).
6. E. L. Ruskol, Astron. Zh., 37, 690 (1960) [Sov. Astron.–AJ, 4, 657 (1961)](Paper I).
7. E. L. Ruskol, Astron. Zh., 40, 288 (1963) [Sov. Astron.–AJ, 7, 221 (1963)] (Paper II).
8. V. S. Safronov, Evolution of the Preplanetary Cloud and the Formation of the Earth and Planets [in Russian], Nauka, Moscow (1969).
9. E. J. Öpik, Irish Astron. J., 5, 14 (1958).
10. W. K. Hartmann, Astrophys. J., 152, 337 (1968).
11. A. H. Marcus, Icarus, 11, 76 (1969).
12. J. S. Dohnanyi, The Zodiacal Light and the Interplanetary Medium, Proc. Honolulu Sympos., NASA SP-150 (1967), p. 315.

SOVIET ASTRONOMY – AJ VOL. 15, NO. 6 MAY-JUNE, 1972

POSSIBLE DIFFERENCES IN THE CHEMICAL COMPOSITION OF THE EARTH AND MOON, FOR A MOON FORMED IN THE CIRCUMTERRESTRIAL SWARM

E. L. Ruskol

Shmidt Institute of Earth Physics, Academy of Sciences of the USSR
Translated from Astronomicheskii Zhurnal, Vol. 48, No. 6, pp. 1336-1338, November-December, 1971
Original article submitted April 19, 1971

A discussion is given of the possible depeletion of lunar material in volatile constituents and its relative enrichment with silicates as compared to the earth, assuming that the moon was formed in the circumterrestrial swarm.

In models in which the moon was formed from a circumterrestrial swarm of bodies and particles during the active growth period of the earth, the same protoplanetary medium would have served as the source of supply for both the earth and the moon. Nevertheless, there would have been two distinct processes for differentiation between the earth and the moon in chemical composition.

1. Partial loss of volatile elements in the circumterrestrial swarm. Direct analyses of lunar samples have shown that lunar basalts are relatively depleted compared to terrestrial basalts in many elements having the property of volatility in common; conversely, the lunar samples are enriched in refractory elements [4]. The loss from the lunar surface of volatile elements of low atomic weight, in particular, water molecules, would be possible. But the deficiency in lunar basalts of low melting elements with an atomic weight of about 200, as compared to terrestrial basalts, can be explained only through their loss from the protolunar material during the accumulation process of the moon. This situation applies particularly to the elements Pb, Bi, and Tl. The possibility that volatile elements may have been lost from the circumterrestrial swarm is evident from the circumstances of its origin. Active formation of the swarm should have proceeded until the mass of the growing earth reached approximately one half its present value. Typical collision velocities for particles captured into the swarm could then have been 3-7 km/sec throughout the entire sphere of influence of the earth [3]. During subsequent collisions within the swarm, the particle velocities would have diminished. The high energy of collision (originally 10^{11} ergs/g but subsequently lower) would have resulted in multiple fragmentations and in complete and selective vaporization of the particle material, followed by condensation of the vaporized atoms on solid particles. In this process different atoms and molecules could have been lost from the periphery of the swarm through the action of the solar wind, but this should have been particularly true of the most volatile substances (water and the low-melting elements Pb, Bi, Tl).

One can show that the protoplanetary cloud should have been quite transparent ($\tau \lessdot 1$) in a radial direction from the sun out to the earth's distance from the sun if the distribution of protoplanetary bodies with respect to radius followed a law $dN(r) \sim r^{-n}dr$, where $n \leq 3.5$, for a mass of material equal to the mass of the terrestrial planets, and for $r_{max} \approx 10^8$ cm. Thus the circumterrestrial swarm would also be expected to have been quite transparent, at least at its periphery. In this region every free lead atom with a cross section of $\approx 10^{-15}$ cm^2, even if the solar wind had its present intensity of $2 \cdot 10^8$ protons $\cdot$ cm^{-2} $\cdot$ sec^{-1}, would have had a chance of colliding with a high-energy proton and flying off from the swarm about once every two months. On the other hand, the condensation of an atom on a particle, for a swarm of optical thickness $\tau \approx 1$, would have required a time comparable to the revolution time about the earth, or once again several months at the periphery of the swarm. If in the past the solar wind was far stronger than now, then the ejection of volatile elements from the outer part of the swarm would have been more efficient than the condensation of atoms on particles. Volatile elements would not have been shifted out during the accumulation of the earth itself. One would also expect that in the inner, more opaque part of the swarm, most of the elements would have been preserved, since the density of the swarm, and thereby also the opacity, should have increased toward the center. In particular, considerably less water would presumably have been included in the primordial composition of the moon than in the composition of the earth. This circumstance merely strengthens the arguments given earlier [6] against the assumption of an identical relative abundance of water in both bodies.

The explanation that Anders and his coauthors [4] have given for the deficiency of volatiles in the moon seems less likely to us, as they presume that the very transitory epoch of condensation of low-melting elements during the earliest phase of the protoplanetary cloud (the first 10^5-10^6 yr) coincided with the epoch when the growth of the earth and moon was completed. The growth process itself, according to current ideas, lasted at least 10^8 yr [7]. Ringwood [5], like ourselves, has given an explanation for the deficiency of volatiles in terms of the solar wind. However, the model he adopts for the formation of the moon from a primordial heavy iron-silicate atmosphere of the earth appears to us unjustified (too short a time scale for formation of the earth, a vague mechanism for formation of the satellite from the atmosphere, and so on).

2. The sorting out of fine particles upon capture into the swarm, and the associated enrichment of the lunar material with silicates. Assuming that the core of the earth consists mainly of iron, the composition of the moon and the earth cannot be the same. According to recent estimates by Al'tshuler and Sharipdzhanov [9], the moon contains about 14% iron; the earth, about 35%. Other investigators have arrived at similar values [10].

One can seek to explain this disparity by the circumstances of formation of the circumterrestrial swarm, if the protoplanetary cloud separately contained iron and stony particles whose different physical and mechanical properties were accompanied by a systematic difference in average size.

As Orowan [11] has shown, stony particles would be friable and after collisions would break into a large number of fine fragments, whereas iron particles (provided, of course, that they did exist at the protoplanetary stage) would be malleable and ductile even at fairly low temperatures. They would not disintegrate upon collision but would stick together with stony particles. Al'tshuler and Sharipdzhanov [9] also mention several factors (durability, weldability) that would have stimulated a leading growth of iron particles in the protoplanetary cloud.

When particles were captured into the circumterrestrial swarm, inelastic two-body collisions would have altered their size distribution law, even in the event that fragmentation was absent [3]. If these size distributions are approximated by power functions of the form $dN(r) \sim r^{-n}dr$, then the value n_1 for interplanetary particles would, following their capture into the swarm, change over to the exponent n_2 given by

$$n_2 = 2n_1 - 3.$$

In particular, the value $n_1 = 3.5$ derived theoretically for protoplanetary bodies [7, 8] would change to $n_2 = 4$ so that the swarm would acquire a still finer component of interplanetary particles. Fragmentation upon capture could only have strengthened this selection effect. The selection process would have had practically no influence on the distribution of particles throughout the entire zone of the earth if the mass of the circumterrestrial swarm was small compared to the mass of all the particles in the zone. The sorting out of the finest particles in the swarm could have served to enrich it with the stony fraction of protoplanetary material. Since the moon was formed essentially from the circumterrestrial swarm, while the main contribution to the mass of the earth would have come from the direct influx of interplanetary particles, the chemical composition of the moon could have acquired a higher Si/Fe ratio than for the earth.

One can show that this conclusion does not at all depend on the assumption that pure iron particles were present in the cloud. A more general hypothesis would be that with increasing size of the cloud particles, there was a monotonic increase in the ratio of iron to silicon, so that the particle density δ would have increased with r. Our statement can then be illustrated by a numerical example. Suppose that for the finest particles, with radii of, say, 1 μ, the density of a particle was 2.5 g/cm^3 (about the same as for the lightest silicates), while the density of the very largest bodies, with radii of 1000 km, was 4.5 g/cm^3 (about the same as for the material of the earth with an iron core, at zero pressure). Suppose further that over the whole range in size for the protoplanetary particles and bodies (12 orders of magnitude) their density varied monotonically according to the law

$$\delta(r) = \delta(r_{\min})\left[1 + {}^1/_6 \log\left(\frac{r}{r_{\min}}\right)\right].$$

Then it can readily be shown (we omit the computations here) that the average density of material in an ensemble of particles with a distribution of exponent $n_1 = 3.5$ would have been

$$\bar{\delta}_1 = \delta(r_{\min})\left[1 + {}^1/_6 \log\left(\frac{r_{\max}}{r_{\min}}\right) - 0.145\right] \approx 4.36 \text{ g/cm}^3,$$

while for an ensemble with a distribution of exponent $n_2 = 4$,

$$\bar{\delta}_2 = \delta(r_{\min})\left[1 + {}^1/_{12} \log\left(\frac{r_{\max}}{r_{\min}}\right)\right] = 3.5 \text{ g/cm}^3.$$

Even though this numerical example is extremely primitive, the value obtained for $\bar{\delta}_1$ differs very little from the mean density of the earth (at zero pressure), and $\bar{\delta}_2$ is nearly the same as the mean density of the moon.

It has not been our intention in this note to enter into a discussion of the highly complex problem regarding the subsequent behavior of iron in planetary interiors, a problem that is still very far from being solved.

I should like to acknowledge conversations with L. V. Al'tshuler, B. Yu. Levin, and V. S. Safronov on this topic.

LITERATURE CITED

1. O. Yu. Shmidt, Four Lectures on the Theory of the Earth's Origin [in Russian], Izd. AN SSSR, Moscow (1950).
2. E. L. Ruskol, Astron. Zh., 37, 690 (1960); 40, 288 (1963) [Sov. Astron.–AJ, 4, 657 (1961); 7, 221 (1963)].
3. E. L. Ruskol, Astron. Zh., 48, 819 (1971) [Sov. Astron.–AJ, 15, 646 (1972)].
4. R. Ganapathy, R. R. Keays, J. C. Laul, and E. Anders, Proc. Apollo 11 Lunar Sci. Conf., Pergamon Press, 2, 1117 (1970).
5. A. E. Ringwood, Earth Planet. Sci. Letters, 8, 131 (1970).
6. V. S. Safronov and E. L. Ruskol, Voprosy Kosmogonii, 9, 203 (1963).
7. V. S. Safronov, Evolution of the Preplanetary Cloud and the Formation of the Earth and Planets [in Russian], Nauka, Moscow (1969).
8. J. S. Dohnanyi, in: The Zodiacal Light and the Interplanetary Medium, Proc. Honolulu Sympos., NASA SP-150 (1967), p. 315.
9. L. V. Al'tshuler and I. I. Sharipdzhanov, Izv. Akad. Nauk SSSR, Fiz. Zemli, No. 4, 3 (1971).
10. R. T. Reynolds and A. L. Summers, J. Geophys. Res., 74, 2494 (1969).
11. E. Orowan, Nature, 222, 867 (1969).

Formation of the Moon from a Cluster of Particles Encircling the Earth (abstract)

E. L. Ruskol
Translated from: Izv. Akad. Nauk SSSR Fiz. Zemli **1972** (7), 99 (1972)
Izv. Acad. Sci. USSR Phys. Solid Earth **1972**, 483 (1972)

A theoretical model of the formation of the Moon, based on an idea of O. Yu. Schmidt, is described. The origin is considered of a satellite-particle cloud during the period of formation of the Earth and the formation of the Moon from such a cloud; an attempt is made to explain the difference between the chemical constitution of the Moon and the Earth.

On the Initial Distance of the Moon Forming in the Circumterrestrial Swarm (abstract)

E. L. Ruskol
In: *The Moon*, edited by S. K. Runcorn and H. C. Urey (Reidel, Dordrecht, 1972), pp. 402–4

According to the Radzievskij–Artemjev hypothesis of the "locked" revolution of the circumplanetary swarms around the Sun, the initial Moon-to-Earth distance and the angular momentum acquired by the Earth through the accretion of the inner part of the swarm can be evaluated. Depending on the concentration of the density to the center of the swarm we obtain the initial distance for a single proto-Moon in the range 15–26 Earth radii R and for a system of 3–4 proto-Moons in the range (3–78) R, if the outer boundary of the swarm is equal to the radius of the Hill's sphere ($235R$). The total angular momentum acquired by the primitive Earth–Moon system through the accretion of the swarm particles is 1/2–2/3 of its present value. The rest of it should be acquired from the direct accretion of interplanetary particles by the Earth. The contribution of satellite swarms into the rotation of other planets is relatively less.

On the Possible Difference in the Bulk Chemical Composition of the Earth and the Moon Forming in the Circumterrestrial Swarm (abstract)

E. L. Ruskol
In: *The Moon*, edited by S. K. Runcorn and H. C. Urey (Reidel, Dordrecht, 1972), pp. 426–8

The model of the origin of the Moon in the circumterrestrial swarm during the active stage of the Earth's growth now receives support. In this model the feeding substance is the same for both bodies. However, two different processes may be mentioned of chemical fractionation between the Earth and the Moon.

On the Model of the Accumulation of the Moon Compatible with the Data on the Composition and the Age of Lunar Rocks (abstract)

E. L. Ruskol
The Moon **6**, 190 (1973)

It is suggested that overall early melting of the lunar surface is not necessary for the explanation of facts and that the structure of highlands is more complicated than a solidified anorthositic "plot." The early heating of the interior of the Moon up to 1000 K is really needed for the subsequent thermal history with the maximum melting 3.5×10^9 yr ago, to give the observed ages for mare basalts. This may be considered as an indication that the Moon during its accumulation retained a portion of its gravitational energy converted into heat, which may occur only in rapid processes. A rapid ($t<10^3$ yr) accretion of the Moon from the circumterrestrial swarm of small particles would give the necessary temperature, but it is not compatible with the characteristic time 10^8 yr of the replenishment of this swarm, which is the same as the time scale of the accumulation of the Earth. It is shown that there were conditions in the circumterrestrial swarm for the formation at a first stage of a few large proto-Moons. Their number and position is evaluated from the simple formal laws of the growth of satellites in the vicinity of a planet. Such "systems" of proto-Moons are compared with the observed multiple systems, and the conclusion is reached that there could have been not more than 2–3 large proto-Moons with the Earth. The tidal evolution of proto-Moon orbits was short not only for the present value of the tidal phase lag but also for a considerably smaller value. The coalescence of proto-Moons into a single Moon had to occur before the formation of the observed relief on the Moon. If we accept the age 3.9×10^9 yr for the excavation of the Imbrium basin and ascribe the latter to the impact of an Earth satellite, this collision had to be roughly at $30R$, where R is the radius of the Earth, because the Moon at that time had to be somewhere at this distance. Therefore, the proto-Moons had to be orbiting inside $(20–25)R$, and their coalescence had to occur more than 4.0×10^9 yr ago. The energy releases at coalescence is equivalent to several hundred Kelvins and even 1000 K. The process is very rapid (of the order of 1 h). Therefore, the model is valid for the initial conditions of the Moon.

Tidal Changes in the Orbital Inclinations of the Satellites of Uranus relative to its Equatorial Plane (abstract)

E. L. Ruskol
Translated from: Astron. Vestn. **7**, 150 (1973)
Sol. Syst. Res. **7**, 131 (1973)

The Uranus system of five satellites is remarkable for its regular configuration, in contrast to the exceptional 98° inclination of the planet's equator relative to the ecliptic. The maximum variations that tidal evolution of the system could produce in the satellite orbit inclinations relative to the equatorial plane are estimated. If the initial inclinations were at all appreciable, tidal friction could not have resulted in the present coplanar system. The regular structure presumably originated during the primordial aggregation process.

The Origin of the Moon (abstract)

E. L. Ruskol
Soviet–American Conference on Cosmochemistry of the Moon and Planets (Moscow, June 1974), edited by J. H. Pomeroy and N. J. Hubbard, NASA (Washington, DC: National Aeronautical Space Administration, 1977), pp. 815–22.
Published as NASA Report No. NASA SP 370 (1977)

This paper discusses fractionation of the chemical compositions of the Moon and the Earth and the thermal history of the Moon for formation of the Moon from an Earth-orbiting swarm of bodies, during the accumulation of the Earth.

Dynamical History of Coplanar Two-Satellite Systems (abstract)

E. L. Ruskol, E. V. Nikolajeva, and A. S. Syzdykov
Translated from: The Moon **12**, 3 (1975)
The Moon **12**, 11 (1975)

One of the possible early states of the Earth–Moon system was a system of several large satellites around the Earth. The dynamical evolution of coplanar three-body systems is studied: a planet (Earth) and two massive satellites (proto-Moons) with geocentric orbits of slightly different radii. Such configurations may arise in multiple satellite systems receding from a planet due to tidal friction. The numerical integration of the equations of motion shows that initially circular Keplerian orbits are soon transformed into disturbed elliptic orbits which are intersecting. The lifetime of such a coplanar system between two probable physical collisions of satellites is roughly from one day to one year for satellite systems with radii less than $20R_{\oplus}$, and may reach 100 yr for three-dimensional systems. This time scale is short in comparison with the duration of the removal of satellites due to tides raised on the planet, which is estimated as 10^6–10^8 yr for the same orbital dimensions. Therefore, the lifetime of a system of several proto-Moons is mainly determined by their tidal interactions with the Earth. For conditions which we have considered, the most probable result of the evolution was coalescence of satellites as the consequence of the collisions.

On the Formation of Satellites Near Giant Planets (abstract)

E. L. Ruskol and V. S. Safronov
Lun. Sci. **8**, 820 (1977)

The formation of satellites and the early evolution of their primary giant planets are closely connected. Many authors suggested a hot stage in Jupiter's history as an explanation of the composition change with distance among the Galilean satellites. But the hot stage may be obtained by several ways which are not indifferent for the satellite growth. One of us has shown that the growth of Jupiter and Saturn modeled as a spherically symmetrical accretion leads to the surface temperatures 17 000 and 3600 K, respectively.[1] High temperatures of the planets also result for the contraction models with constant mass found from the evolutionary tracks of stars with small masses.[2–4] However, the study of the evolution of the gas-dust protoplanetary cloud[5] has shown that the most reliable model of formation of giant planets had two stages: the accumulation of planetary embryos from condensed (solid) particles for 10^7–10^8 yr and the accretion of noncondensable gases (chiefly H and He) onto the embryos (10^5–10^6 yr). The mass of the embryo sufficient for the beginning of the accretion is about 1–3 Earth masses for Jupiter's distance from the Sun. One can distinguish also a third stage—the sweeping up of solid bodies remaining in the planetary zone with the simultaneous ejection of many others from the solar system and partly to its periphery ($>10^8$ yr).[6,7] Assuming that the massive satellites of all planets have originated from circumplanetary swarms,[8] we conclude that the main contribution of the material into the swarms is connected with the second stage, i.e., accretion of gas. Some part of the material could be also acquired at the third stage.

The accretion of gas onto the embryo signifies the flow of gas from the whole feeding zone, which in the case of Jupiter represents a circumsolar rotating torus with $R_{\min} \approx (5-6) \times 10^3 R_j$, $R_{\max} \approx 15 \times 10^3 R_j$, and the thickness $(1-2) \times 10^3 R_j$, where R_j is the radius of Jupiter. But only inside the sphere of action of Jupiter with the radius of $740 r_j$ (r_j is the current radius) is the gas motion governed by Jupiter's gravitational field, being directed toward the embryo. Outside the sphere the gas rotates differentially around the Sun with the heliocentric velocity slightly lower than the Keplerian velocity of the embryo. The models of gravitational contraction and of accretion are quite different and lead to a different time scale and temperature distribution along the radius. Another important difference is in the dynamics of satellite swarms because in the accretional model the increase of the mass of the planet is accompanied by the proportional decrease of the orbital radii of all particles from which the satellites accumulate. Therefore the conclusions about satellite formation made for the contraction model[9,10] cannot be extended to the accretional model favored by us. Only the last stage of gravitational contraction (after the end of accretion) is somewhat similar in the two models.

The hydrodynamics of the accretion of differentially rotating gas is very complicated. A simplified picture of gas motion may be found considering a steady spherically symmetric accretion perturbed by a slight rotation.[11] The boundary conditions "at infinity" must be replaced by the conditions at the radius of the sphere of action. Setting the averaged angular velocity around the planet equal to $\frac{1}{4}\,\boldsymbol{\omega}_c$, where $\boldsymbol{\omega}_c$ is the circumsolar angular velocity, we find that the gas is accreted mainly in polar regions and that in the equatorial plane the accretion is prevented, and a circumplanetary disk out to $\approx 20 R_j$ is formed. The conclusion about the accretion mainly in polar regions was also made earlier.[12] The rate of accretion is essentially less than the maximum one, usually assumed for the spherically symmetric accretion in infinite medium. The planet easily accretes the gas only from the region near its orbit. However, the gas from other heliocentric distances did not come to fill this region—due to small viscosity the gas conserved its angular momentum relative to the Sun. As a result the value of ρ_∞ (near the sphere of action of the planet) in the expression for the rate of accretion highly decreased with time. Nevertheless, the accretion at its maximum caused high temperatures of the surface of planets ($\approx$3000 K), securing the evaporation of volatiles in the region of the Galilean satellites.

The evaluation of the density and velocity of gas and of the maximum dimension of solid particles dragged by gas permits one to estimate the contribution of condensable matter into the satellite swarm due to drag by gas and due to "free–bound" collisions of particles. The contribution by "free–free" collisions is negligible. The evaporation–condensation process in the swarm resulted in the fractionation of chemical composition of satellites. The possibility of the origin of a gaseous torus on the orbit of Io as the remnant of gaseous disc during the accretion process is discussed.

[1] V. S. Safronov, Astron. Zh. **31**, 499 (1954).
[2] A. S. Grossman *et al.*, Phys. Earth Planet Int. **6**, 91 (1972).
[3] H. C. Graboske *et al.*, Ap. J. **199**, 265 (1975).
[4] J. B. Pollack *et al.*, Icarus (to be published).
[5] V. S. Safronov "Evolution of the protoplanetary cloud and formation of the earth and planets" (in Russian), translation, NASA (1972).
[6] E. J. Öpik, Moon **1**, 487 (1970).
[7] V. S. Safronov, in *The Motion, Evolution of Orbits and Origin of Comets* edited by G. A. Chebotarev and E. I. Kazimirchak-Polonskaya (Reidel, Dordrecht, 1972).
[8] V. S. Safronov and E. L. Ruskol, in *Planetary Satellites* edited by J. A. Burns (University of Arizona Press, Tucson, 1976).
[9] J. B. Pollack and R. T. Reynolds, Icarus **21**, 248 (1974).
[10] J. B. Pollack *et al.*, Icarus **29**, 35 (1976).
[11] P. Cassen and D. Pettibone, Astrophys. J. **208**, 500 (1976).
[12] V. S. Safronov, Astron. Vestn. **5**, 167 (1971).

The Accumulation of Satellites (abstract)

V. S. Safronov and E. L. Ruskol
In: *Planetary Satellites*, edited by J. A. Burns (University of Arizona Press, Tucson, 1977), pp. 501–12

An analysis of the formation processes of the regular satellites is given, following previous work (Ruskol, 1960, 1963, 1971a,b; Kaula, 1971; Kaula and Harris, 1973) on the accretion of the Moon. The natural satellites are considered to coalesce from a circumplanetary swarm which itself is fed by heliocentric particles which are captured following inelastic collisions. Characteristic times are calculated for the important physical processes: the replenishment time of the swarm, the growth time of the mass of the swarm, the accumulation time for the bodies in the swarm, and the orbital collapse time under the action of tides. During the capture of the swarm and its accumulation much material is lost on the planet's surface. Reasonable values for the mass of the satellite swarm can be found from this model.

Tidal Destruction of Bodies Near Planets (abstract)

I. N. Ziglina
Translated from: Izv. Akad. Nauk SSSR Fiz. Zemli **7**, 3 (1978)
Izv. Acad. Sci. USSR Phys. Solid Earth **14**, 467 (1979)

The Roche limit is investigated for brittle and plastic solid bodies approaching a planet. Depending on the size and rigidity of the body, its Roche limit can differ considerably from the classical Roche limit for fluid bodies. Sekiguchi's hypothesis is discussed, concerning the generation of double craters by the falling of bodies disintegrating near a planet, in connection with craters on Mars. Fragments of bodies, after passing close to a planet, move on independent proximate orbits about the Sun or, if the body is large enough, they remain gravitationally related. Approximate criteria are given for the case of a slowly rotating disintegrating body. For a very narrow range of values of the parameters of the problem some fragments can be captured by the planet.

Thermal Effect of the Collision of Two Massive Bodies and the Initial Temperature of the Moon (abstract)

E. L. Ruskol

Lun. Plan. Sci. **10**, 1042 (1979)

Recently Kaula[1] and Ransford and Kaula[2] calculated the initial temperature profiles of the Moon growing similarly to the Earth accretion, modeled after Safronov.[3,4] None of the profiles for the Moon gave a satisfactory initial temperature distribution with a maximum in the upper mantle reaching the melting point. Different variants in Ref. 1 had either the temperature lower than the melting point at all radii of the Moon or melting at a depth greater than desirable. In Ref. 2 the degree of melting occurred too low for the differentiation and the depth of the maximum melting was too large. At the same time the initial temperature for the Earth obtained in Ref. 1 exceeded the melting point in the upper mantle, favoring the early differentiation.

Here we consider a model of the Moon's growth which can give a satisfactory temperature distribution. This model was not included in Kaula's formalism. We assume the accumulation of the Moon from a few large Earth satellites. The dynamical aspects of such an origin were considered earlier.[5–7] Now we consider the thermal effects.

Suppose we have a collision of two massive bodies after their approach due to mutual gravitational attraction. The destruction of bodies and subsequent coalescence may be regarded as a limit case of crater formation when the crater dimension exceeds the dimensions of bodies. Let the bodies be spherical with masses μ_1 and μ_2 and radii r_1 and r_2. The total energy of the two-body system "at infinity" is the sum of their potential energies (negative) and of the energy of their relative motion:

$$E_1+E_2+E_\infty=-\frac{3}{5}\frac{G\mu_1^2}{r_1}-\frac{3}{5}\frac{G\mu_2^2}{r_2}+E_\infty .$$

At the coalescence of two bodies into one spherical body of mass $\mu=\mu_1+\mu_2$ and of radius r, the potential energy of the latter will be:

$$E_{12}=-\frac{3}{5}\frac{G\mu^2}{r}.$$

The energy will be released at the collision:

$$U=E_1+E_2+E_\infty-E_{12}, \tag{1}$$

with x denoting the part of this energy going on fragmentation of bodies, y the heating during the collision, and $z=1-x-y$, the expansion of the fragments to the maximum distance r_{max}, measured from the center of the newly forming body μ. For a spherically symmetric isotropic expansion we have:

$$zU=(1-x-y)\left(\frac{3}{5}\frac{G\mu^2}{r}-\frac{3}{5}\frac{G\mu_1^2}{r_1}-\frac{3}{5}\frac{G\mu_2^2}{r_2}+E_\infty\right)=\frac{3}{5}\left(\frac{G\mu^2}{r}-\frac{G\mu^2}{r_{max}}\right)(1-\beta), \tag{2}$$

where the value β is a part of the potential energy of the newly forming body μ which is concentrated in solid fragments, decreasing with the dimensions of the latter. In the limit case of the elastic rebound of two equal spherical bodies having $E_\infty=0$, i.e., at $x=0$, $y=0$, $z=1$, $r_{max}=\infty$, we have $\beta=0.63$. For real, inelastic collisions with fragmentation, β is lower depending on the mass distribution of debris. For example, the case of the power law distribution of $dN(m)=Cm^{-1.8}dm$ gives $\beta=0.14$.[8]

It may be shown that the value x is negligibly small in our case, because the specific energy in the impact is of the order of 10^{10} erg/g, while for fracturing it is sufficient to have 10^6 erg/g, if the mass distribution of debris is of the type cited above.

One can evaluate r_{max} for the collision of two equal masses $\mu_1=\mu_2=\frac{1}{2}\mu$. Putting $\frac{5}{3}E_\infty=\kappa G\mu^2/r$ and neglecting x in Eq. (2), we shall have

$$\frac{r_{max}}{r}=\frac{1-\beta}{0.63-\beta-\kappa(1-y)+0.37y}, \tag{3}$$

which gives for different values of y, κ, and β:

	$y=0$	$y=0.5$	$y=0.9$
$\kappa=0,\ \beta=0$	$\dfrac{r_{max}}{r}=1.6$	1.2	1.04
$\kappa=0,\ \beta=0.14$	1.74	1.3	1.05
$\kappa=0.25,\ \beta=0.14$	3.5	1.6	1.08

The maximum duration of both expansion and collapse is twice the free fall time from the distance r_{max}, i.e.,

$$2t_{ff}=2\ \frac{\pi}{2}\ \sqrt{\frac{r_{max}^3}{2G\mu}},$$

which for $r_{max}/r=1.5\text{–}2.0$ is $(4\text{–}6)\times 10^3$ s. It is evident that even in the maximum expansion phase the body μ is fully opaque, and it may be shown that the heat is lost only from an external layer 8–10 cm thick. Therefore almost all total energy U of the impact is transformed into heat at the end of the collapse.

Let us evaluate now the distribution of temperature within the body μ immediately after collapse. It can be assumed that the losses of energy during the impact are negligible and that the thermal energy of bodies μ_1 and μ_2 before the impact (equivalent to an average temperature T_0), as also the energy liberated during the impact, is uniformly distributed within the newly formed body, if we neglect thermal inhomogeneities. Thus at all radii of the body μ the temperature will have a constant term T_0+T_y, where $T_y=yU/c\mu$. On the other hand, the term T_z, connected with the heating due to the collapse, zU, is largest at the surface and vanishes in the interiors of the body. In fact,

$$T_z(r')=\frac{1}{c}\left(\frac{G\mu}{r'}-\frac{G\mu}{r'_{max}}\right)=\frac{4\pi G\rho r'^2}{3\quad c}\left(1-\frac{r'}{r'_{max}}\right), \tag{4}$$

where r' is the distance from the center of the body, c its heat capacity, and ρ the density. Supposing that the expansion after the impact is isotropic, we have $r'/r'_{max}=r/r_{max}=$ const, and therefore, from Eq. (4), $T_z\sim r^2$. In the case that the newly forming body has a lunar mass and expands up to $r_{max}=1.5r$, it is heated on the surface to 930 K by the term T_z only. The temperature at a given distance r' from the center will be $T(r')=T_0+T_y+T_z(r')$.

It is possible now to find a condition for the temperature distribution in which the melting point T_m is reached in a layer from the surface (r) down to the depth, say, 200 km $(r-200)$:

$$T(r-200)\leqslant T_m<T(r) \tag{5}$$

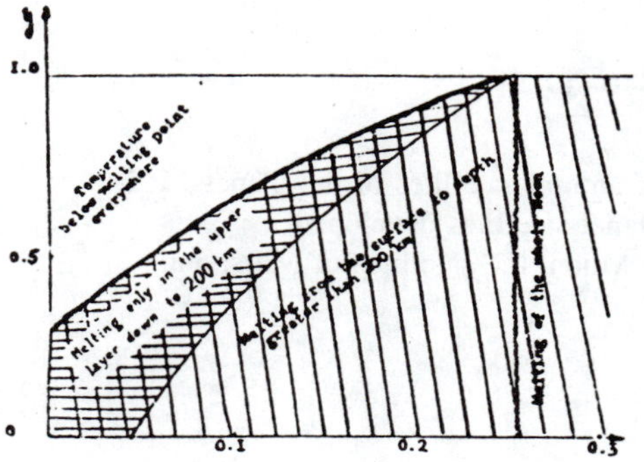

FIG. 1.

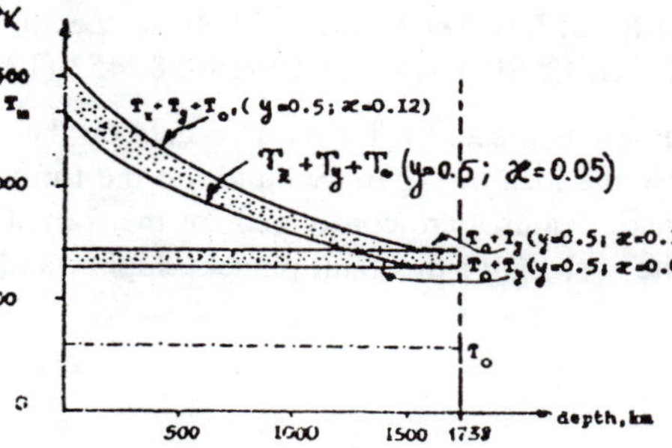

FIG. 2.

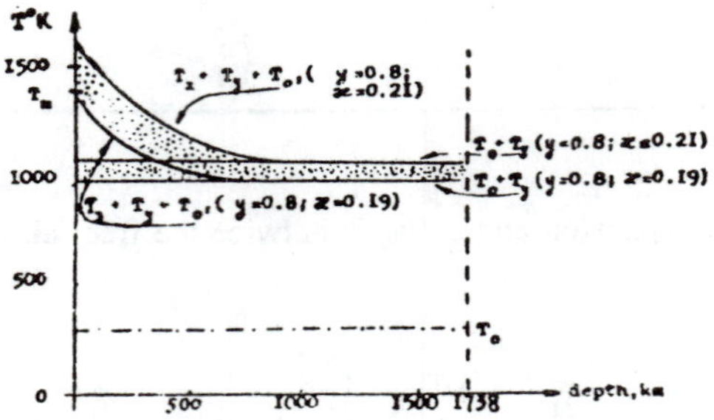

FIG. 3.

taking $T_m = 1350$ K, $c = 10^7$ erg/g, $T_0 = 300$ K, $\beta = 0.14$, and using Eqs. (3) and (4) for r_{max} and T_z, we transform Eq. (5) into a dimensionless condition:

$$\frac{0.78\kappa - 0.03}{0.27\kappa + 0.10} \leq y < \frac{0.05 + \kappa}{0.48 + 0.18}. \tag{5'}$$

Figure 1 presents a shaded area in y and κ coordinates where the condition (5′) is fulfilled. It is seen that the intervals of corresponding values of y and κ are rather broad. The data on heat losses in destructive cratering are not known. But it may be assumed that $y = 0.5 - 0.8$ is a reliable estimation. As to κ, we can roughly evaluate it from the perturbed motion of two-satellite systems[6]: $0 < \kappa \lesssim 0.5$.

Figures 2 and 3 present two temperature profiles for $y = 0.5$ and $y = 0.8$ and corresponding κ, taken from Eq. (5′). We see that the collision of two equal masses can give a satisfactory initial temperature of the Moon with a hot layer on the surface. It may be also shown that the collision of comparable masses can lead to similar effect.

[1] W. M. Kaula, J. Geophys. Res. 1978 (to be published).
[2] G. A. Ransford and W. M. Kaula, Lun. Plan. Sci. **9**, 931 (1978).
[3] V. S. Safronov, "Evolution of the protoplanetary cloud and the formation of the Earth and planets", translated in NASA Report No. NASA TT F-677, 1972.
[4] V. S. Safronov, Icarus **33**, 3 (1978).
[5] E. L. Ruskol, Moon **6**, 176 (1973).
[6] E. L. Ruskol *et al.*, Moon **12**, 3 (1975).
[7] E. L. Ruskol in *Soviet–American Conference on Cosmochemistry of the Moon and Planets*, NASA report SP370, Wash., 1977, pp. 815–822.
[8] E. V. Zvjagina, Astron. Zh. **50**, 1261 (1973).

Origin of Planetary Satellites (abstract)

E. L. Ruskol
Translated from: Izv. Akad. Nauk SSSR Fiz. Zemli, No. 6, p. 40 (1982)
Izv. Acad. Sci. USSR Phys. Solid Earth **18**, 452 (1983)

A brief review is presented for modern data on the systems of natural satellites of the planets. It is shown how the idea of O. Yu. Schmidt on the formation of planet satellites developed. Characteristic features are considered for the formation of the Moon, the satellites of Mars, and the satellite systems of the giant planets, Jupiter and Saturn.

Protosatellite swarm dynamics

G. V. Pechernikova, S. V. Maeva, and A. V. Vityazev

Shmidt Institute of Earth Physics, USSR Academy of Sciences, Moscow

(Submitted December 20, 1983; revised April 17, 1984)

Pis'ma Astron. Zh. **10**, 702–709 (September 1984)

The evolution of a model protosatellite swarm around a growing primeval planet is investigated. Equations are obtained to describe how the swarm surface density will be redistributed as material is accreted from the planet's feeding zone. Preliminary calculations indicate that the model can explain the origin of satellite systems comprising roughly 10^{-5}–10^{-4} the mass of the planet.

When the planets were in the making, they were surrounded by swarms of captured bodies and small particles. These swarms gave rise to the satellites accompanying the planets. One can treat the dynamics of the swarms in the same manner as the accretion and protoplanetary disks that may develop around stars.

In 1977 Safronov and Ruskol[1] discussed the accumulation of satellites in a circumplanetary swarm. Adopting a power-law dependence $\sigma(r) \propto r^{-b}$ for the surface density of the disk material upon the distance r from the planet, they showed that if the parameters have suitable values, swarms could form having a mass up to 1/10 that of the planet. The interrelation of the parameters was not investigated.

We shall here derive an equation describing how the surface density profile of the protosatellite swarm will change as the primeval circumsolar disk evolves, and we summarize the preliminary results of our calculations.

The mass and distribution of the swarm material may be altered by any of the following main processes: 1) an influx of material from the planet's feeding zone; 2) particles shooting through the swarm being swept up by large bodies in the zone; 3) infall of material onto the planet; 4) drift toward the planet as its growing mass causes the particle orbits to shrink (the "Jeans invariant"); 5) drift toward the planet because captured material has a smaller angular momentum than if in a circular orbit at the capture distance from the planet. Several other

 0360-0327/84/05 0293-05 $03.00

effects will come into play at certain stages in the formation of the swarm: 6) the loss of high-speed particles, moving faster than the escape velocity; 7) ejection into the swarm of material from the planet as it is impacted by planetesimals; 8) capture of fragments from large bodies that break up as they come within the Roche limit; 9) drift of small particles in any gas that may be present; 10) drift of large bodies due to tidal interaction with the planet. Still further effects may occur but should be less important.

Let us consider a model protolunar swarm around the growing earth. The change in the density $\rho(r, t)$ of matter in the swarm's central plane will be described by the equation

$$\frac{\partial\rho}{\partial t} + \frac{1}{r}\frac{\partial}{\partial r} ru\rho = I(\rho, r, t), \tag{1}$$

where u(r, t) represents the rate of flow of material at distance r from the planet at time t, and I designates the sources. In the thin-swarm approximation Eq. (1) will be equivalent to an equation for the surface density $\sigma(r, t)$:

$$\frac{\partial\sigma}{\partial t} + \frac{1}{r}\frac{\partial}{\partial r} ru\sigma = I(\sigma, r, t). \tag{1'}$$

Now let $u = u_1 + u_2$, where u_1 is the flow due to conservation of particle orbital momenta as the planet grows in mass (process 4) and u_2 is the flow due to the orbital-momentum difference between captured material and the swarm particles (process 5). From the condition rm_p = const that the momentum of a particle orbiting a growing planet of mass m_p at mean distance r be conserved, we have

$$m_p(t)\frac{dr}{dt} + r\frac{dm_p}{dt} = 0,$$

from which one finds[2] that

$$u_1 = \left(\frac{dr}{dt}\right)_1 = -3\,\frac{r}{r_p}\frac{dr_p}{dt}, \tag{2}$$

where r_p is the running value of the planet's radius.

If one takes an average over the mass and orbital elements of the infalling bodies, one may write the incremental angular momentum relative to the planet due to captures at distance r in the form[3]

$$\Delta K_1 = \bar{\beta}\bar{r}\bar{v}_t\Delta\rho, \tag{3}$$

$$\left.\begin{aligned} &\bar{\beta} = \frac{8}{\pi}\sqrt{\frac{3r\,(1 + 4v_e^2/e^2V_c^2)^{1/2}}{eR}}, \quad \bar{r} = \frac{3}{49}r\sqrt{1 + 4v_e^2/e^2V_c^2}, \\ &\bar{v}_t \simeq \frac{1}{2}eV_c \simeq \frac{v}{\sqrt{3}}; \quad v = \sqrt{\frac{Gm_p}{\theta r_p}}, \end{aligned}\right\} \tag{4}$$

where $\bar{\beta}$ denotes the proportion of near-tangential orbits, v_e is the escape velocity at distance r from the planet, e is the mean orbital eccentricity of the preponderance of bodies in the planet's feeding zone, v is their mean relative velocity, the dimensionless parameter $\theta \approx 2$, R is the planet's distance from the sun, and V_c is the Keplerian circular velocity at distance R. On substituting the expressions (4) into Eq. (3) we obtain

$$\Delta K_1 = c_1 r^{3/2}\Delta\rho, \tag{5}$$

where the coefficient

$$c_1 = \frac{64}{49\pi}\left(\frac{\pi\delta G}{\theta}\right)^{1/4}\left(\frac{V_c r_p}{R}\right)^{1/2}(1 + 6\theta r_p/r)^{3/4}.$$

At distance r, unit volume of the swarm will acquire an angular momentum

$$K_2 = \rho\sqrt{Gm_p r} = c_2 r^{1/2}\rho. \tag{6}$$

With Eqs. (5), (6), the momentum conservation condition implies that

$$c_1 r^{3/2}\,\Delta\rho + c_2 r^{1/2}\rho = c_2\,(r + \Delta r)^{1/2}\,(\rho + \Delta\rho),$$

and after passing to the limit $\Delta r/\Delta t \to (dr/dt)_2$, $\Delta\rho/\Delta t \to (d\rho/dt)_2$ changing from ρ to σ, and performing some manipulations we find that

$$u_2 = \left(\frac{dr}{dt}\right)_2 = -2\frac{r}{\sigma}\left[1 - A\left(1 + 6\theta\frac{r_p}{r}\right)^{3/4}\frac{r}{r_p}\right]I_2, \tag{7}$$

where $A = (32/49\pi)(\pi G\delta\theta)^{-1/4}\sqrt{V_c/R}$, and I_2 denotes the flux of particles from the planet's feeding zone that stick in the swarm.

Capture of zone material by the swarm (process 1) will occur through mutual collisions of zone bodies (mass m_z) within the planet's Hill sphere ($r < r_H$) and collisions of zone bodies with swarm members (mass m_s), that is, by free–free and free–bound collisions, respectively.[4]

Let us estimate the influx I_1 of matter into the swarm due to the free–free collisions. We shall adopt power-law mass distributions for the bodies in the planet's feeding zone and in the swarm: $n_1(m) = n_{01}m^{-q_1}$, $n_2(m) = n_{02}m^{-q_2}$. If $q_{1,2} < 2$, then we may write $n_{01} = (2 - q_1) \cdot M_1^{q_1-2}\rho_1$ and $n_{02} = (2 - q_2)M_2^{q_2-1}\rho_2$, where M_1, M_2 are up-

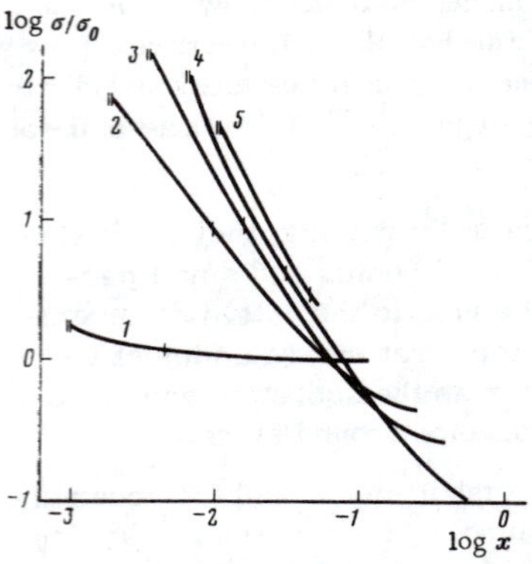

FIG. 1. Successive curves indicate the time evolution of the surface density profile $\sigma(x)$ of a protosatellite swarm. Table I gives the model parameters corresponding to each curve; $x = r/r_{H}{}^{\bullet}$, where $r_H{}^{\bullet}$ is the Hill-sphere radius when the planet has its present mass. A single tick mark corresponds to a radius 10 times the planet's running radius; a double tick mark, to the Roche limit.

TABLE I. Model Parameters at Successive Evolutionary Stages

Stage No.	$m_p/m_\oplus$	m_{swarm}, 10^{20}	m_{swarm}/m_p, 10^{-7}	t, 10^7 yr
1	0.001	0.5	83	0.62
2	0.010	1.2	20	1.3
3	0.043	1.9	7.2	2.3
4	0.30	2.5	1.4	5.0
5	0.94	1.3	0.22	14

per limits on the mass of the zone and swarm bodies, respectively. According to a previous calculation,[5] when colliding bodies fragment and amalgamate, the index $q \approx 11/6$.

Since collisions between bodies of comparable mass will release enough energy to put the bodies into heliocentric orbits, we shall consider collisions in the mass interval $1 - \gamma_1 \leqslant m/m_z \leqslant 1 + \gamma_2$, where $\gamma_1 \approx \gamma_2 \approx \gamma < 1$.

Now let μ_{z1} denote the mass of the zone bodies dispersed by a body m_z per unit time, that is,

$$\mu_{z1} = \xi\pi r_z^2 v \int\limits_{m_z(1-\gamma)}^{m_z(1+\gamma)} n_1(m)\, m\, dm = \xi\pi r_z^2 v \rho_1 \left(\frac{m_z}{M_1}\right)^{2-q_1} \times [(1+\gamma)^{2-q_1} - (1-\gamma)^{2-q_1}], \tag{8}$$

where $\xi \approx 2$ is a geometrical factor. If $q_1 = 11/6$, the total mass of the material dispersed per unit volume in unit time will be

$$\Delta\rho_1 = \int\limits_{m_0}^{M_1} \mu_{z1} n_1(m)\, dm = \xi\pi \left(\frac{3}{4\pi\delta}\right)^{2/3} \frac{\rho_1 v_1 \gamma}{18} \frac{\rho_1}{M_1^{1/3}} \ln(M_1/m_0), \tag{9}$$

where m_0 is a lower bound on the mass distribution and δ is the density of the bodies. The swarm will capture only a fraction $w(r)$ of the bodies passing through it, as expressed by the capture probability[4]

$$w(r) = \beta_1 \frac{2\theta r_p}{r + 2\theta r_p}\left(1 - \frac{r}{r_H}\right), \quad \beta_1 \simeq \frac{1}{2}. \tag{10}$$

For the planet's feeding zone[6] $\rho_1 v = 4\sigma_1/P_1$, where σ_1 is the surface density of zone material and P_1 is the orbit period about the sun. Let σ_f denote the surface density attributable to zone bodies within the planet's Hill sphere; then with Eqs. (9), (10) we will have

$$I_1 = \frac{2\xi\pi\gamma\beta_1}{9}\left(\frac{3}{4\pi\delta}\right)^{2/3} \frac{2\theta r_p}{r + 2\theta r_p}\left(1 - \frac{r}{r_H}\right) \frac{\sigma_1}{P_1} \frac{\ln(M_1/m_0)}{M_1^{1/3}} \sigma_f. \tag{11}$$

As for the collisions with bodies belonging to a swarm of radius r_s, all the smaller-sized objects will be captured, but their retention will be somewhat offset by the fact that small bodies passing through the swarm should be swept up by large bodies in the feeding zone (process 2). Following Safronov and Ruskol,[1] let us assume for simplicity that collisions between bodies of comparable mass will cause the objects to stick together. In the same manner as for μ_{z1}, we can determine the mass μ_{z2} of zone bodies dispersed by a body m_s per unit volume in unit time:

$$\mu_{z2} = \xi\pi r_s^2 v \int\limits_{m_s(1-\gamma)}^{m_s(1+\gamma)} n_1(m)\, m\, dm = \xi\pi\rho_1 v r_s^2 \left(\frac{m_s}{M_1}\right)^{2-q_1} \times [(1+\gamma)^{2-q_1} - (1-\gamma)^{2-q_1}]. \tag{12}$$

and if $q_1 = q_2 = 11/6$,

$$\Delta\rho_2 = \int\limits_{m_0}^{M_2} \mu_{z2} n_2(m)\, dm = \xi\pi \left(\frac{3}{4\pi\delta}\right)^{2/3} \frac{\rho_1 v \gamma}{18} \frac{\ln(M_2/m_0)}{(M_1 M_2)^{1/6}} \rho. \tag{13}$$

Introducing now the probability (10), we conclude that the flux of matter retained in the swarm because of free–bound collisions will be

$$I_2 = \frac{2\xi\pi\gamma\beta_1}{9}\left(\frac{3}{4\pi\delta}\right)^{2/3} \frac{2\theta r_p}{r + 2\theta r_p}\left(1 - \frac{r}{r_H}\right) \frac{\sigma_1}{P_1} \frac{\ln(M_2/m_0)}{(M_1 M_2)^{1/6}} \sigma. \tag{14}$$

We can allow for the infall of swarm material onto the planet (process 3) by imposing an appropriate inner boundary condition on the swarm. The loss of high-speed swarm particles ($I_3 < 0$; process 6) can be accommodated in terms of our analysis[7] of the dissipation of a rapidly rotating exosphere. We have also discussed the ejection of planetary surface material into the swarm[8] ($I_4 > 0$; process 7), showing that early in a planet's growth its axial angular momentum will be near-critical, so that when it is struck by planetesimals the planet will throw off some of its excess-momentum material into the surrounding swarm. Later on, ejecta can exit the surface of the growing planet only if it suffers noncentral impacts by isolated large bodies. In our present treatment, however, we shall limit attention to the first two sources described above (I_1, I_2), neglecting the effects involved in processes 6, 7, 8. Furthermore, several lines of evidence suggest that, at least for the terrestrial planets, we may neglect the influence of gas upon the phenomenon in the protosatellite swarm (process 9; see, for example, Safronov and one of us[9]).

Substituting now the expressions (2), (7), (11), (14) into Eq. (1'), we obtain

$$\frac{\partial\sigma}{\partial t} - \frac{1}{r}\frac{\partial}{\partial r} r\sigma \left\{3\frac{r}{r_p}\frac{dr_p}{dt} + 2\frac{r}{\sigma}\left[1 - A\left(1 + 6\theta\frac{r_p}{r}\right)^{3/4} \frac{r}{r_p}\right] I_2\right\} = I_1 + I_2. \tag{15}$$

In this equation the quantity r_p as well as M_1, M_2, q_1, which enter into I_1, I_2, vary with time.

Let $r_\oplus$ denote the present radius of the planet, and introduce the variable $z \equiv r_p/r_\oplus$. The equation for a planet's growth rate[10] gives

$$\frac{dz}{dt} = a\frac{1 - z^2}{2} \tag{16}$$

and

$$\sigma_1 = \sigma_0(1 - z^2), \tag{17}$$

where $a = 2(1 + 2\theta)\sigma_0/(P_1\delta r_\oplus)$, and σ_0 is the initial surface density of solid matter in the feeding zone.

For the mass of the largest body in the zone (other than the planet itself) we have obtained[11]

$$M_1/m_p \approx 3(1 - z^2)/4. \tag{18}$$

Changing to the dimensionless variables $x = r/r_H^*$, $y =$

σ/σ_{01} (σ_{01} = 1 g/cm^2, $r_H{}^*$ denotes the radius of the Hill sphere when the planet has its present mass $m_\oplus$) and using Eqs. (16)-(18), we find that

$$\frac{\partial y}{\partial z} - \frac{3}{x}\frac{\partial}{\partial x} y \frac{x^2}{z}\left\{1 + 2c_3 \frac{z-x}{(x+c_4 z)(1-z^2)^{1/4}} \times \left[1-(x+3c_4 z)^{3/4} x^{1/4}/z\right] f_1(M_2)\right\}$$
$$= \frac{3c_3}{z}\frac{z-x}{(x+c_4 z)(1-z^2)^{1/3}}\{[c_5 + \ln(z^3 - z^5)]\, y_f + (1-z^2)^{1/4} f_1(M_2)\, y\}, \quad (19)$$

where

$$c_3 = \frac{\xi\gamma\beta_1\theta r_\oplus}{18\,(^3/_4)^{1/4}(1+2\theta) r_H{}^*}; \quad c_4 = 2\theta r_\oplus / r_H{}^*$$
$$c_5 = \ln(0.75 m_\oplus / m_0); \; f_1(M_2) = \sqrt{z}\,\ln(M_2/m_0)/(M_2/m_\oplus)^{1/4}.$$

What values to assign the upper and lower limits M_2, m_0 in the mass distribution of the bodies is a rather perplexing question. However, in view of the weak dependence of Eq. (19) on M_2 and m_0 [the function $f_1(M_2)$], we may solve the equation using rough estimates for these limits. For m_0 we shall take a value of order 10^{-6} g, the mass of the smallest particles that will result from fragmentation.[12] Early in the growth process M_2 can be estimated from the probability of free–free collisions of large bodies within the planet's Hill sphere. During the intermediate stages one has to allow for the ability of the swarm to capture large bodies.

The likelihood that large objects may accumulate in the swarm itself will be determined by the relative velocities v_2 of the swarm particles. We may write for v_2 an energy-balance equation analogous to the standard equation for the relative velocities v of bodies in the feeding zone,[6] but containing two extra terms to describe the input of kinetic energy into the swarm by zone members and the removal of energy by fast particles leaving the swarm:

$$\frac{dv_2{}^2}{dt} = (a_1 - a_2 + a_3 - a_4)\, v_2{}^2. \quad (20)$$

Here the a_i are coefficients expressing the efficiency of energy transfer per unit time to unit mass: a_1, through gravitational interaction between the swarm members; a_2, through their mutual collisions; a_3, through their interaction with feeding-zone members; a_4, through the loss of fast particles from the swarm.

The energy-balance equation for collisions of swarm particles among themselves and with zone members (that is, neglecting a_1 and a_4) may be reduced to the form

$$\frac{4\sigma}{P_2}(2-q_2) M_2^{q_2-2} \int_0^{M_2/m_s} \frac{2(1+\sqrt{\zeta})(\sqrt{\zeta}x - 1)}{(1+x)^2}(1+x^{1/3})^2 x^{1-q_2} dx$$
$$+ \frac{4\sigma_1}{P_1}(2-q_1) M_1^{q_1-2} m_s^{q_2-q_1} \int_0^{M_1/m_s} \frac{f_2 - 2(1+\sqrt{\zeta})}{(1+x)^2}$$
$$\times (1+x^{1/3})^2 x^{1-q_1} dx = 0, \quad (21)$$

where $f_2 = (1+\sqrt{\zeta})^2 (v/v_2)^2 + \zeta - 1$, with ζ the proportion of the energy of impact that goes into kinetic energy. We are publishing elsewhere[13] an analysis of Eq. (21). Here we would merely point out that during the active planetary growth phase, especially in the outer part of the swarm, the external energy source will predominate ($a_3 \gg a_1$). Energy losses will result chiefly from incompletely elastic collisions (a_2) of bodies in the interior of the swarm, and from escapes a_4 near the periphery. Satellite accumulation in the swarm will be hindered during this stage by the bodies' large relative velocities, and if it does take place it will happen only in the zone closest to the planet, so the accumulation will not disturb the processes in the rest of the swarm.

Late in the planet's growth the influx of feeding-zone material will become much smaller ($a_3 \ll a_1$), and the relative velocities of the swarm members will be determined largely by gravitational interaction and mutual collisions. If the velocities are low, the biggest bodies in the swarm can grow through amalgamation of smaller ones. The drift of large bodies because of tidal interaction with the planet (our process 10) will become significant. One would expect the protosatellite growth dynamics to be analogous to the growth of embryonic planets that we have described previously.[10]

We have integrated Eq. (19) numerically on the BÉSM-6 computer. Figure 1 illustrates our preliminary results. Here we have plotted the evolution of the surface density profile $\sigma(x)$ for a model with M_2 = const and $\sigma(x, z_0) = \sigma_f(x, z_0) = \sigma_{01}\sqrt{1 + c_4 z_0/x}$. The model parameters at successive stages are given in Table I. It turns out that the distribution $\sigma(x)$ is close to a power law, $\sigma \propto x^{-b}$, with $b \gtrsim 2$ at distances $3 \lesssim r/r_p \lesssim 50$ and $b \lesssim 2$ for $r/r_p \gtrsim 50$. The swarm mass amounts to no more than 10^{-6}–10^{-5} the mass of the planet (Table I). Estimates suggest that the erosion of large bodies losing mass as they travel through the swarm may significantly raise the value of I_2, but even if the source I_2 were formally to increase by a factor 5-10, the swarm mass would not grow by the same amount; there would merely be an enhancement of about this factor of the flux of material through the swarm and onto the planet.

Our provisional results show that the development of a low-mass protosatellite swarm can satisfactorily be described by a thin circumplanetary model disk evolving as it accretes bodies from the planet's feeding zone. With such an approach one can trace the formation of satellite systems whose combined mass amounts to $\approx 10^{-5}$–10^{-4} that of the planet.

On the other hand our calculations tentatively suggest that the formation of massive satellites (10^{-2}–10^{-1} the mass of the planet) can hardly be a normal event. Such satellites, like the earth's moon, presumably would have originated from the breakup of one or more large bodies in the Roche zone, or even from the ejection of part of the planet into the swarm after a catastrophic collision. Safronov and Ruskol[1] maintain that the anomalously large angular momentum of the earth–moon system may have resulted from the flux of material from a massive protosatellite swarm onto the primordial earth, which would thereby have been put into rapid rotation. A plausible alternative might be to posit an interaction with some large body, for one could then simultaneously explain the high rotational velocity of the primeval earth, its axis tilt, and the large mass of the protolunar swarm in the (3-10)r_p zone. The rapid tidal recession of one or several protomoons at these distances could have been offset by the earthward drift due to the captured material's shortfall of angular momentum.

[1]V. S. Safronov and E. L. Ruskol, Astron. Zh. 54, 378 (1977) [Sov. Astron. 21, 211 (1977)].
[2]E. L. Ruskol, Astron. Zh. 40, 288 (1963) [Sov. Astron. 7, 221 (1963)].
[3]A. V. Vityazev and G. V. Pechernikova, Astron. Zh. 58, 869 (1981) [Sov. Astron. 25, 494 (1982)].
[4]E. L. Ruskol, The Origin of the Moon [in Russian], Nauka, Moscow (1975).
[5]G. V. Pechernikova, V. S. Safronov, and E. V. Zvyagina, Astron. Zh. 53, 612 (1976) [Sov. Astron. 20, 346 (1976)].
[6]V. S. Safronov, The Evolution of the Protoplanetary Cloud and the Formation of the Earth and Planets, Nauka, Moscow (1969) [NASA TT F-677 (1972)].
[7]G. V. Pechernikova and A. V. Vityazev, Adv. Space Res. 1, 55 (1981).
[8]A. V. Vityazev and G. V. Pechernikova, in: O. Yu. Shmidt and Soviet Geophysics in the 1980s [in Russian], ed. M. A. Sadovskii, Nauka, Moscow (1983), p. 244.
[9]V. S. Safronov and A. V. Vityazev, Itogi Nauki i Tekhniki, Ser. Astron. 24, 5 (1983) [Sov. Sci. Rev. E: Astrophys. Space Phys. Rev. 4, (in press)].
[10]A. V. Vityazev, G. V. Pechernikova, and V. S. Safronov, Astron. Zh. 55, 107 (1978) [Sov. Astron. 22, 60 (1978)].
[11]G. V. Pechernikova and A. V. Vityazev, Astron. Zh. 57, 799 (1980) [Sov. Astron. 24, 460 (1981)].
[12]A. Fujiwara, G. Kamimoto, and A. Tsukamoto, Icarus 31, 277 (1977).
[13]V. S. Safronov, G. V. Pechernikova, E. L. Ruskol, and A. V. Vityazev, Icarus (in press).

 0360-0327/84/05 0297-03 $03.00

A Model for Differentiation during Expansion of a Vapor Cloud in a Gravitational Field (abstract)

B. M. Manzon, I. L. Khrilev, and O. I. Yakovlev
Translated from: Geokhim. (2) **1990**, 163
Geochem. Int. **27** (9), 1 (1990)

A simple gas-dynamic model for dissipation has been examined in earlier papers, for application to the formation of the Moon. It was shown that an important differentiation mechanism at the accretion stage is selective shock evaporation accompanied by selective dissipation of the vapor during its adiabatic expansion in the gravitational field. Our model agrees well with qualitative features of the lunar composition, but we assumed that the components behave independently in the vapor. We now attempt sounder estimates of the selective loss of material during the accretion stage and generalize the model to a wider range of conditions.

VII. Formation of the Giant Planets

On the Hydrogen Content of the Major Planets

A. I. Lebedinsky*

Problems of Cosmog., Vol. VII, pp. 56–59 (Joint Publ. Res. Service, U.S. Dept. of Commerce, 1964)
Translated from: Vopr. Kosmog. **7**, xxx (1960)
Presented at the General Meeting of International Astronomical Union, 1958

The differences in chemical composition of the major planets are explained by the heating of the outer parts of the protoplanetary cloud due to motions of massive bodies.

In 1950,[1] the hypothesis was expressed explaining both the abundance of H on the major planets (Jupiter, Saturn, Uranus, and Neptune), as well as its low content on Earth, Venus, Mars, and Mercury and on the Moon. In brief, this hypothesis runs as follows.

The protoplanetary cloud with a composition close to that of stars and nebulas consisted mainly of H and perhaps He as well. The remaining elements were present as minor admixtures. Before the protoplanetary cloud attained the stage of intensive flattening, the H and all the sufficiently volatile compounds of it remained in gaseous form but the materials with high evaporation temperatures were of course condensed into dust grains.

During the stage of intense flattening of the preplanetary cloud, preceding the formation of the planets, the dust subsystem was concentrated very intensively toward the equatorial plane and surrounded the Sun in a ring, similar to the ring around Saturn. The outer regions of this ring were not heated by solar radiation and had a very low temperature, owing to which the volatile gases condensed there on the dust grains. As a result, the planets of the Earth's group originating from the material of the heated part of the dust subsystem consist chiefly of refractory substances (rocks and metals) and have relatively small masses, whereas the planets of the Jupiter group were formed mainly from the volatile substances that froze onto the dust grains and hence have large masses and a low density of the component material.

At first glance, this theory, explaining the essential nature of the planets' chemical composition, appears to contradict the fact that the density of the planets, starting with Saturn, increases with the distance from the Sun (Jupiter =1.31, Saturn=0.68, Uranus=1.09, and Neptune=1.61 g/cm^3). This effect becomes even more perceptible if we compare not the planets' densities, but the densities of the component material, reduced to equal pressure and temperature (t^0), based on an arbitrary theory of the planets' internal structure.

For an explanation of the difference in the chemical composition of the planets in the Jupiter group, it is necessary to examine the energy balance of the dust grains at the periphery of a greatly flattened cloud. The solar radiation scarcely reaches this far. The radiation flux from the stars constitutes 10^{-2} erg/cm^2 s and is sufficient to heat the grains to a t^0 of 3 K. Under such conditions, the most significant source of heat energy may be the work of friction during the passage of large protoplanetary bodies through the dust layer.

In Refs. 1 and 2, we demonstrated that the evolutionary tendencies of the subsystems of dustlike and massive bodies are in opposition.

At each collision with another grain, the dust grains lose a great part of the kinetic energy of their relative movement; therefore their subsystem quickly arrives at a state of minimum energy with almost stable angular momentum distribution between the particles, and a flat disk resembling Saturn's ring constitutes such a state. The massive bodies interact gravitationally and their "collision" can be regarded as elastic. As a result of multiple collisions, the subsystem of such bodies grows in size, given the absence of friction. More exactly, in it the following occurs: the main mass of the subsystem loses its angular momentum, and as a result it thickens, therein reducing the equatorial diameter and tending toward a spherical form. However, a small part of the bodies acquires angular momentum and because of this is ejected from the system or is removed to the periphery; in such a subsystem, the average eccentricities and inclinations of the bodies' orbits gradually increase.

In the case of interest to us, involving the existence of subsystems of dust and massive bodies, the following occurs: the solid (massive) bodies acquire unique velocities as a result of interaction with one another (i.e., the force of their relative motions originates from a decrease in the mean energy of the orbital movement), but are retarded (braked) by friction in the subsystem composed of dust grains. The energy being acquired by the grains as a result of this is converted to heat during the intercollisions and is radiated into space. The energy balance of this process essentially determines the grains' t^0, which we shall estimate here.

We will denote by m the mean value of the mass, V the velocities, r the radii of the bodies comprising the subsystem; we denote by H its effective thickness and by σ the surface density in it, i.e., the mass of a column of an individual section, perpendicular to the equatorial plane. By the indices 1 and 2, we connote the values referring, respectively, to the dust subsystem and to the subsystem of the large bodies.

The energy e, being obtained in a dust subsystem per

*Affiliated with the M. V. Lomonosov Moscow State University.

unit of time in the column of an individual section, is determined by the formula

$$e \approx \frac{\sigma_1 \sigma_2}{m_2 H_2} V_2^3 r_e^2, \tag{1}$$

where r_e is the effective radius of large bodies for collisions with bodies of the dust subsystem. In the case of gravitational interaction, at the distance

$$r_e = \alpha r_0 = \alpha \frac{\gamma m_2}{V_2^2}, \tag{2}$$

where α is the dimensionless coefficient equal to several tens, constituting the linear function of $\log H_1/r_0$, while γ is the gravitation constant.

In the case of direct collisions, which predominated in the process being considered, we can simply assume $r_e = r_2$. The chief difficulty involved in using these formulas is the lack of knowledge of the proportion of the material from protoplanetary bodies existing in fragmented form and comprising that which we are calling the "dust subsystem" in this discussion. In reality, the protoplanetary bodies not only interact elastically at a distance but also collided with one another, grinding into fragments that were added to the dust subsystem. Upon the impact of bodies of comparable mass, as a rule both bodies were crushed, since in most cases their relative velocities were not less than several kilometers per second (km/s). On the other hand, the falling of small bodies onto the surface of larger ones led to the growth of the protoplanetary bodies at the expense of depletion of the dust subsystem and in the final analysis led to the combining of the scattered matter to form the planets. Thus, the value σ_1 (the surface density of the dust subsystem) was determined by a complex dynamic balance, for which a theory has not yet been developed.

An estimation of the amount of released heat based on formula (1) yields, for various distances from the Sun, a value for e ranging from 0.1 to 100 erg/cm^2 s and correspondingly, a t^0 for the dust grains ranging from 5 up to 30 K.

In Table 1, we have adduced the evaporation energy (in electron volts per particle) and the t^0 of saturated vapors for the most volatile materials. The t^0 are computed for partial densities of $H = 10^{12}$ and for the remaining materials involving 10^{10} molecules per cm^3, i.e., for conditions resembling those that could have existed in a preplanetary cloud.

As is evident from the data in Table 1, at t^0 ranging from 5 to 30 K, all of the matter except H and He should condense. The hydrogen can be acquired by the planets only in the process of capture.

TABLE 1.

	Material					
	H_2	CH_4	OH	HCl	H_2S	NH_3
Evaporation force	0.01	0.10	0.17	0.20	0.22	0.32
Evaporation t^0	6	32	47	65	74	93

Thus we can say that the planet-forming process divides into two stages. Initially, there occurs the process of the accumulation of a solid component, and then, when the planets have become sufficiently massive, their chemical composition begins to change as a result of the capture of H molecules and He atoms.

The rate of the gas capture can be rapid only in that part of the preplanetary cloud where within the dust layer in the path of a planet's embryo, cold gas of high density occurred. This did not occur either in the region of the planets of the Earth group, where the concentration of gas in the plane of symmetry was obstructed by heating from solar radiation, or in the remote periphery in the zones of Uranus and Neptune where the gas density was low as a result of the low surface density of the material in the cloud. The maximal density of the gas near the symmetry plane should be somewhere near the planets of maximal mass or, more exactly, the maximum of the gas density should be located somewhat farther from the Sun than the maximum of the planetary masses. The coincidence of this maximum with the Jupiter–Saturn zone is quite natural from this standpoint.

In distinction from the particles of the dust ring, hardly altering its distance from the Sun, gas can readily flow over from one part of the solar system to another, filling the voids left as a result of capture.

Thus if in the composition of each planet there entered those solid particles that occurred in its zone from the very beginning of the cloud's flattening, then each planet scooped the gas out of the entire cloud, and the amount of gas acquired by a planet was basically determined by the rate of capture. Therefore Jupiter and Saturn, being under the most favorable conditions for capture, acquired the maximum hydrogen content and have the least density.

The monotonous dependence of density upon diameter of orbit also occurs for the Galilean satellites of Jupiter (Io, 2.71; Europa, 2.40; Ganymede, 1.74; and Callisto, 1.27 g/cm^3). The cause could have been the fact that the t^0 of the solid particles from which Jupiter's satellites formed was conditioned mainly by the bombardment by dust grains and molecules from which Jupiter itself was formed. The density of the energy flux of these bombarding particles increases with a decrease in the distance to Jupiter.

Assuming such an explanation of the densities of the satellites, we proceed from the concept that the planets' satellites originated in the general process of formation together with the corresponding planet from those particles, which, having undergone collision near a planet, were captured in the periodic orbits.

[1] L. E. Gurevich, and A. I. Lebedinskiy, Bull. Acad. Sci. USSR Phys. Ser. **14**, 765 (1950).
[2] A. I. Lebedinskiy, Voprosy. Kosmog. **2**, 5 (1954).

Numerical Investigations of a Planar Model of Accumulation of Nuclei of Giant Planets (abstract)

S. I. Ipatov
Institute of Applied Mathematics, USSR Academy of Sciences, Preprint No. 117 (1983)

The solid-body accumulation of giant planets' cores from a protoplanet cloud and planetesimals' ejection from these planets' feeding zones are investigated. The investigations are carried out by computer simulation of the evolution of bodies in a flat ring. The initial number of bodies is several hundred. These bodies move around the Sun and coagulate at collisions. The sphere method is used for estimating the mutual gravitational influence of bodies. The planar model shows a minimum of the orbits' gravitational changes as compared with the spatial model. The results show that in the case of very small inclinations the total mass of initial planetesimals should be almost the same as the mass of the solid substance in the present planets. At some stages of the ring evolution, heliocentric distances of the ring zones of matter concentration corresponded to a law of arithmetical progression.

Numerical Investigations of the Evolution of Spatial Rings of Gravitational Bodies Corresponding to the Feeding Zones of Giant Planets (abstract)

S. I. Ipatov
Institute of Applied Mathematics, USSR Academy of Sciences, Preprint No. 1 (1984)

The evolution of gravitating solid bodies' spatial rings corresponding to giant planets' feeding zones is investigated on the basis of computer simulation results. Bodies are assumed to coagulate at every collision. The initial number of bodies in the ring varies from 100 to 500. It is shown that more than 80% (or even 95% and more) of the bodies' mass is ejected in hyperbolic orbits from the feeding zones of Uranus and Neptune. During the evolution of the ring, consisting initially of nearly formed planets and identical bodies in the above zones, a considerable number of bodies migrated to Jupiter.

Planetesimal Migration During the Last Stages of Accumulation of Giant Planets (abstract)

S. I. Ipatov
Translated from: Astron. Vestn. **23**, 27 (1989)
Sol. Syst. Res. **23**, 16 (1989)

The migration and accumulation of bodies from giant planets' feeding zones, after the main part of the mass of these planets are formed, are investigated. The research is based on computer simulations for the evolving spatial disks which initially consisted of a few almost-formed planets and hundreds of identical bodies in the zones of Uranus and Neptune. It is shown that the total mass of bodies ejected to the asteroid zone from the giant-planet zones could be 10 times as large as the Earth's mass. The beyond-Neptune belt could have been formed during the accumulation of the giant planets. The evolution of planetary orbits resulting from encounters with planetesimals is studied.

Evolution of the Orbital Eccentricities of Planetesimals during Formation of the Great Planets (abstract)

S. I. Ipatov

Translated from: Astron. Vestn. **23**, 197 (1989)
Sol. Syst. Res. **23**, 119 (1989)

The changes of eccentricities during the evolution of disks containing a large number of bodies ($\approx 10^6$–10^{12}) are investigated. The calculations are carried out on the basis of the computer simulation results obtained for the evolution of disks initially consisting of hundreds of bodies which moved around the Sun. It is found that the Safronov number Θ can substantially change during the evolution. The total mass of bodies ejected into hyperbolic orbits from the giant-planet zone could exceed by an order of magnitude the mass of solids that entered into the planets.

Formation time of Jupiter

V. N. Zharkov and A. V. Kozenko

Institute of Earth Physics, Academy of Sciences of the USSR, Moscow

(Submitted April 20, 1989)

Pis'ma Astron. Zh. **15**, 745–749 (August 1989)

It is suggested that the accumulation time of Mars depends on the presence of proto-Jupiter. It is shown that the accumulation of Mars is limited by the time scale of growth of Jupiter. This limit has been roughly estimated to be $\sim 1.7 \cdot 10^7$ yr, which is approximately the time spent by the Sun in the T Tauri stage.

Introduction. The material from which a model of one of the giant planets, e.g., Jupiter or Saturn, is constructed may be divided into three components according to volatility: 1) glass (H_2, He, Ne...), the G-component, 2) ices (H_2O, CH_4, NH_3), the I-component, and 3) rock and iron—nickel + FeS, the S (solid) component. A model of the internal structure of Jupiter,[2] together with contemporary cosmological ideas[6] enables one tentatively to identify three stages in th eprocess of formation of a planet: 1) formation of an SI-core due to agglomeration of SI planetesimals; 2) when the core mass reaches a critical value the gas component in the Jupiter supply zone becomes unstable and there is accretion of gas onto the SI planetary embryo; 3) after formation of proto-Jupiter from the SI core and the G shell, the remaining SI planetesimals in the Jupiter supply zone are quickly swept up or expelled. Clearly, these three stages overlap in part.

Numerical information on the critical core mass M_{cr} and the mass of SI material in the planet's gas shell is obtained by calculating a model. The mass of the captured SI planetesimals has not been evaluated with comparable accuracy. On the basis of numerical modeling with highly simplified models,[4] it was estimated to be an order of magnitude larger than the SI mass captured by the planet. The present model of Jupiter[2] gives the value $M_{cr} \sim 5$ M and the following composition of Jupiter: $M_G \sim 251$ $M_\oplus$, $M_I \sim 50$ $M_\oplus$ and $M_S \sim 16.5$ $M_\oplus$, ($M_\oplus$ is the mass of the Earth). The model composition of Jupiter differs from the abundance of elements in the sun and indicates that about nine planetary masses of G-component were dispersed while the planet was forming. Allowing for the mass of the SI component expelled from the supply zone by Jupiter when forming, and estimating conservatively that this mass is of the order of the mass of Jupiter, we find the mass of the dispersed G-component to be ~ 55 M_J (M_J is the mass of Jupiter).

The formation of Jupiter had a great influence on the formation of the terrestrial planets. In fact, planetesimals from the Jupiter zone destroyed the supply zone of the asteroid belt, whose mass is $\sim 10^{-3}$ $M_\oplus$, and the Mars supply zone, which, at a minimal estimate, must have contained about 20 times more S material than the present mass of the planet. This brings up the question of the degree of destruction of the Earth supply zone, which thus far has been difficult to answer quantitatively. The formation time of Jupiter is determined by the time for the planet core to grow to the critical mass ~ 5 $M_\oplus$, since the time scale of the second stage (accretion of gas onto the core) is notably shorter, being $\sim 10^4$–10^5 yr,[9] according to estimates. The third stage also could not have been lengthy, since the giant planet rapidly sweeps its supply zone clean.

We can approach estimating the formation time of Jupiter frow two viewpoints. On the one hand this time cannot be greater than τ_T, the time that the sun spent in the T Tauri phase, when powerful corpuscular radiation led to dissipation of the G component of the protoplanetary nebula. One may take $\tau_T \sim 10^7$ yr. On the other hand, using the theory of growth of the terrestrial planets[1] we can estimate the formation time of Mars, which, according to what has been stated above, will also be the formation time of Jupiter. Agreement of the two estimates will indicate that the ideas on which these arguments are based mutually consistent.

Density of the TK component in the Mars supply zone. According to the theory of planetary growth, the balfwidth of the supply zone of a growing planet in the zone of the terrestrial planets is ~ 0.2 R, and is somewhat larger in the zone of the giant planets (R is the radius of the planet's orbit). Therefore, knowing the current mass of the planet, and in the case of the giant planets knowing also the mass of the heavy and light components on the basis of the model constructed, we can estimate the initial surface density of the solid component in the planet's supply zone from the following relation:

$$\sigma_0 = \frac{M_p}{0.8\xi\pi R^2}, \tag{1}$$

where M_p is the mass of the planet or its heavy component, and ξ is a coefficient of order unity.

However, Eq. (1) is not applicable for evaluating the initial surface density of the dust component in the supply zone of Mars. A substantial decrease of surface density occurs in this zone because of resonance perturbations of the giant planet when forming. We can evaluate the desired initial density by assuming that the surface density of the dust component is described by a power law and is a smooth continuous function of the distance from the center of the solar nebula. For example, $\sigma(r) = \sigma_0 r^{-3/2}$, where r is in A.U. (Ref. 7). This also

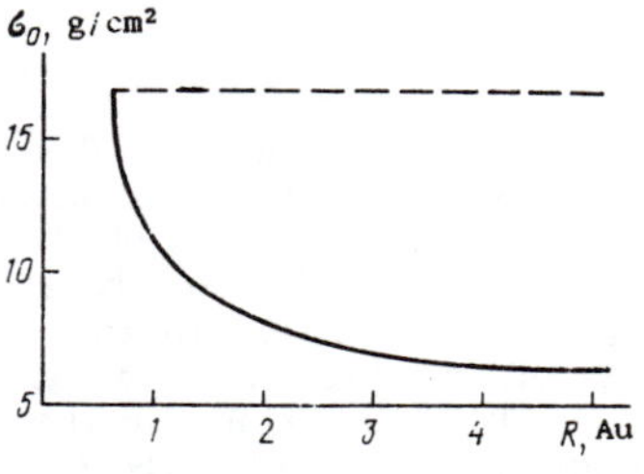

FIG. 1. Initial surface density of the S component in the protoplanetary cloud (see text).

can be done by simpler interpolation, drawing a smooth curve through the three points that determine the surface density of the S component in the zones of Venus, Earth and Jupiter using Eq. (1). The result obtained also agrees with the assertion [4] that the surface density of the dust layer in the asteroid zone may also have been of the same order as in the zones of Earth and Jupiter. We know the masses of Venus and Earth to high accuracy, and the mass of the silicate component of Jupiter, according to a recent model, is 16.5 $M_\oplus$. From Fig. 1 it can be seen that σ_0 ~ 9.5 g/cm in the zone of Mars. Then, assuming no influence of Jupiter, we find that Mars must have acquired a mass estimated from Eq. (1) to be ~$1.2 \cdot 10^{28}$ g. But the actual mass of Mars is ~$6.4 \cdot 10^{26}$ g. Thus, Mars was able to accumulate about 5% of the available mass in its zone. The influence of Jupiter then became appreciable. According ot the current hypothesis, this corresponds to the third stage of formation of Jupiter described abvoe.

Growth time of Mars. In the work of Vityazev et al. [1] an approximate analytical expression was derived to evaluaate the growth time of a planet with a given initial surface density of the dust layer. It is applicable to the terrestrial planets, in whose zone of formation we can neglect the influence of gas and the ejection of bodies.

Actually, during most of the time of formation of the terrestrial planets no gas remained in this zone. This evidence for this consists of the granules in the meteoritic material irradiated by the solar wind. These granules were fully exposed prior to the formation of the asteroids, the parent bodies of the meteorites. By that time there was no gas, since one thousandth of the initial amount would have sufficed to screen the solar irradiation completely.

The planet growth time to the point where its mass reaches 97% of the maximum possible value with the given surface density given by

$$t_{(97\%)} = 5.3D, \qquad (2)$$

where

$$D = \left(\frac{6\pi}{M_\odot}\right)^{1/4} \frac{\delta^{3/4} R^{3/4} P}{(2\sigma_0)^{1/4}\,\theta^{1/4}\,(1+\theta)}.$$

Here $M_\odot$ is the mass of the sun, R is the distance of the growing planet from the sun, P is the orbital period of the planet around the sun, δ is the density of the embryonic planet (assumed constant), σ_0 is the initial surface density of material in the planet's zone, and θ is a dimensionless parameter describing the mean relative speeds of bodies in the pre-planetary swarm during accumulation (θ ~ 3-5).

It is not difficult to obtain the relation required to determine the growth time of Mars

$$t_{(5\%)} = 0.77D. \qquad (3)$$

Equation (3) also limits the time of formation of Jupiter to ~$1.7 \cdot 10^7$ years. Because of condensation of volatiles (H_2O, CH_4, NH_3) in the region of the giant planets the surface density in the supply zone of Jupiter was rather high: ~26 g/cm². In that case the accumulation time of the Jupiter core was $t_{(7.5\%)}$ ~ $6 \cdot 10^8$ years. To obtain a value for the accumulation tim eof the Jupiter core not exceeding ~$6 \cdot 10^8$ years, we need to increase the parameter θ by a factor of 20-30 or introduce some kind of sticking efficiency factor in the gaseous medium ($\theta^*\theta$), since we cannot neglect the influence of gas in the region of the giant planets. Then θ^* ~ 4-10.

One should also bear in mind that in the last stage the large planetesimals might have had atmospheres, which would increase their effective collision cross sections.

In their paper, Hayashi et al. [9] proposed to take account of the effective collision cross section due to tidal destruction of planetesimals in the Hill sphere of the growing planet. Then the collision cross section is

$$S_{\rm col} = \pi\,(r + r')\,f^2\,(1 + 2\theta/f), \qquad (4)$$

where r and r' are the radii of the growing planet and the planetesimal, and f is some efficiency parameter, equa to 1.8 for the inner planets, since in this region the planetesimals are silicates, and 2.8 for the giant planets, since here icy planetesimals of lower density predominate. By allowing for this effect we can reduce the formation time of the Jupiter core.

Discussion and conclusions. A higher surface density of material in the supply zone can also substantially reduce the planet formation time. The probable scenario for the formation of Jupiter shows that if Jupiter expelled about the same amount of mass as was contained in its gas envelope, then the original density of condensate in the Jupiter zone could be a factor of 2 greater, while if the mass of condensate expelled was of the order of the mass of the planet, and it is 1.5 of the planet mass, then the surface density could be a factor of 5 larger. There have also been suggestions that the total mass of solid matter in the supply zones of the giant planets could be an orer of magnitude greater than the mass of solid matter that has gone into these planets, and could reach hundreds of times the mass of the Earth (Ref. 4). Therefore the solid curve in Fig. 1 is a lower bound on the surface density of silicate material in the solar system; the upper bound could be considerably higher, and the distribution represented by the dotted line is not at all unlikely. It can be seen that Jupiter could also partially disrupt the Earth's supply zone, and one indication of the creation of bodies of mass $\sim M_\male$ is the mega-impact hypothesis leading to the formation of the Moon. In that case the initial surface density in the Earth's supply zone could have been higher, and accordingly, its formation time could have been somewhat less than considered heretofore.

We note also that the mass of the gas component removed from the solar system could be a factor of 5-10 larger than the value obtained on the basis of the SI component entering into the composition of the planets, i.e., the mass of the protoplanetary cloud might have been ~0.1 $M_\odot$ or more.

In the work of Lissauer, [5] an increase of the dust dust component by a factor of 5-10 in the formation zone of the giant planets reduced the growth time of their cores to ~10^6 yr. However, in that case there is too large an increase of the amount of material in the supply zone of the terrestrial planets and of the total mass of the protoplanetary cloud.

In the work of Stevenson and Lunin, [8] the hypothesis was advanced that the increase of surface density required to reduce the accumulation time of the Jupiter nucleus results from diffusive redistribution of water vapor which is condensed within a narrow radial zone at ~5 A.U. from the center of the nebula, right in the region of formation of Jupiter. In their model, the accumulation time of the Jupiter nucleus is even less (or order 10^5-10^6 yr).

In conclusion, we stress once more that estimates of the formation time of Jupiter from two possible points of view coincide. It also follows from this estimate that the parameter $\theta \lesssim 9$-10. Previous estimates of θ in the zone of the giant planets (at least in the Jupiter region) should be regarded as overestimates by a factor of from tens to hundrds.

[1]A. V. Vityazev, G. V. Pechernikova, and V. S. Safronov, Astron. Zh. 55, 107 (1978) [Sov. Asron. 22, 60 (1978)].

[2]T. V. Gudkova, V. N. Zharkov, and V. V. Leont'ev, Astron. Vestnik 22, 252 (1988).

[3]T. V. Gudkova, V. N. Zharkov, and V. V. Leont'ev, Pis'ma Astron. Zh. 14, 371 (1988) [Sov. Astron. Lett. 14, 157 (1988)].

[4]S. I. Ipatov, Astron. Vestn. 23, 27 (1989).

[5]J. J. Lissauer, Icarus 69, 249 (1987).

[6]V. S. Safronov, and A. V. Vityazev, Itogi nauki i tekhniki. Ser. Astron., VINITI, 24, 5 (1983), Moscow.

[7]V. S. Safronov and E. L. Ruskol, Icarus 49, 284 (1982).

[8]D. J. Stevenson and J. I. Lunin, Icarus 75, 146 (1988).

[9]C. Hayashi, K. Nakazawa, and Y. Nakazawa, Protostars and Planets II, D. C. Black and M. S. Matthews (eds.), Univ. Arizona Press (1985), p. 1100.

Translated by A. D. McRonald

 0360-0327/89/04 0324-05 $02.00

On the role of Jupiter in the formation of the giant planets

V. N. Zharkov and A. V. Kozenko

Earth Physics Institute, Academy of Sciences of the USSR

(Submitted October 17, 1989)

Pis'ma Astron. Zh. **16**, 169–173 (February 1990)

A new scheme for the formation of Saturn, Uranus, and Neptune is proposed. After completion of the formation of Jupiter over a period of $\sim(1–2)\cdot 10^7$ yr that planet ejects a massive embryo with a mass of $\sim 5\ M_\oplus$ (where $M_\oplus$ is the mass of the Earth) into the Saturn feeding zone. This nucleus initiates the formation of Saturn. After the formation of Jupiter and Saturn, ejection of massive embryos into the Uranus and Neptune feeding zones occurs. These embryos lead to the formation of Uranus and Neptune over cosmologically realistic time intervals.

Introduction. As a result of the development of models of the giant planets (Zharkov et al., 1974a, b; Gudkova et al., 1988a, b) as well as cosmogonic concepts (Safronov and Vityazev, 1983; Hyashi et al., 1985) there has arisen the concept of formation of the giant planets (at least Jupiter and Saturn) in a four-stage process. In the first stage, after condensation toward the central plane and decay into dust clumps, the dust component of the protoplanetary disk completes the first stage of planetesimal formation. The second, lengthier stage, which essentially determines the time scale for planet formation, is associated with the amalgamation of planetesimals by collisions, with the resultant formation of a planetary embryo. When the mass of this embryo reaches a critical value of $\sim(5\text{-}10)\ M_\oplus$ (where $M_\oplus$ is the mass of the Earth), the third stage begins, with the gaseous component of the protoplanetary disk losing stability in the vicinity of the growing planet and precipitating onto that planet over a cosmogonically brief time of $\sim 10^4\text{-}10^5$ yr (Hayashi et al., 1984). After this, a quite brief fourth stage commences, in which the final formation of the planet occurs. In this stage the powerful gravitational center which has formed empties its feeding zone by sweeping up and scattering planetesimals ejected from the feeding zone of Jupiter indicate a mass of the same order as that of the planet itself (Ipatov, 1989).

Planetesimals expelled from the Jupiter zone exerted a significant effect on formation of the planetary system. It is generally accepted at present that they almost completely destroyed the feeding zone in the asteroid belt and severely (by a factor of approximately 20) reduced the amount of matter taking part in the formation of Mars (Zharkov and Kozenko, 1989). The present study will propose that bodies ejected from the feeding zone by Jupiter as the latter completed its formation played an even larger role in the process of formation of Saturn, Uranus, and Nepture.

Scenario for formation of the giant planets. The materials from which the giant planets are constructed can be divided into three classes by their volatility: 1) gases (H_2, He, Ne, ...), the G component; 2) ices (H_2O, CH_4, NH_3) the I component; 3) minerals and iron–nickel, the S component (solid condensate). It is assumed that the structure of Jupiter did not change over the course of its evolution. It consists of a SL core, in which the S and L components are found in their solar proportions. The mass of the core is $\sim 5\ M_\oplus$. The remainder of the planet consists of a G component with mass of $\sim 251\ M_\oplus$ having mixed within it S and L components with a total mass of $\sim 61.5\ M_\oplus$. This part of the planet has a definite structure (see Gudkova et al., 1988a, b). Such a planetary structure agrees well with the suggestion that a huge mass of condensate ($\sim 60\ M_\oplus$) was swept up during the final stage of formation and that the ejected mass of planetsimals is of the order of the mass of the planet.

The structure of Saturn has suffered changes over its evolution due to the fact that a significant fraction of the helium from the G-shell has precipitated onto an SL core. The mass of Saturn's SL core is $\sim 10\ M_\oplus$, while in the outer G-shell (with a mass of $\sim 58\ M_\oplus$) there is dissolved $\sim 18\ M_\oplus$ of the SL component. The quantity of helium that has settled to the center of the planet is $\sim 9\ M_\oplus$. Thus, on Saturn also a huge mass of condensate ($\sim 20\ M_\oplus$) is dissolved in the outer G shell of the planet.

Uranus and Neptune have very similar structures (Gudkova et al., 1988a, b; Gudkova and Zharkov, 1990), which however, differ markedly from those of Jupiter and Saturn. The structure of both planets is close to a two-layer one – an outer shell contains (1-1.5) $M_\oplus$ of the G-component and approximately the same amount of L and S components. The remainder of both planets is an SL core, in which ices and minerals are mixed in solar proportions.

According to cosmogonic concepts, formation of the embryos of the giant planets took place in a gaseous medium and over a characteristic time t_{for} (Hayashi et al., 1985)

$$t_{for} = 0.22\,\frac{10^4}{f^2(1+2\theta/f)}\left(\frac{m}{10^{18}\,\mathrm{r}}\right)^{1/3}\left(\frac{a}{1\ \mathrm{a.e.}}\right)^3, \qquad (1)$$

where f ~ 2.8 is a numerical coefficient appearing because of capture by the growing embryo of planetesimals entering the Roche sphere, θ is a numerical parameter (of the order of unity) which takes into account gravitational focusing upon collision of planetsimals with the growing planetary embryo; m is the critical mass of the planetary embryo, assumed equal to the mass of the planet's SL core: a is the distance from the sun in astronomical units. The planetary embryo formation (or growth) time is proportional to the Keplerian orbital period t_K of the planet about the sun ($t_K \sim a^{3/2}$) and inversely proportional to the surface density σ_S of the dust component in the protoplanetary disk. As a result in Eq. (1) $t_{for} \sim a^3$. Substitution of standard parameter values in Eq. (1) yields the following values for t_{for}: $\sim 10^9$ yr for Saturn, $\sim 10^{10}$ yr for Uranus, and $\sim 10^{11}$ yr for Neptune.

Horedt (1988) proposed a new model for the distribution of the condensate (S and L components) in the protoplanetary disk and carried out numerical modeling of the planet accumulation process. As a result, somewhat lower values of t_{for} were obtained: $8\cdot10^7$, $3\cdot10^8$, $1.3\cdot10^9$, and $1.8\cdot10^9$ yr, for Jupiter, Saturn, Uranus, and Nepture, respectively.

Construction of models of the giant planets (Gudkova et al., 1988a, b) has shown that the compositions of these planets differ markedly form the solar composition — the planets are enriched in the SL component. Hence it was concluded that during formation, Jupiter lost ~9 planetary masses of the G component. Analogous values for Saturn, Uranus, and Nepture are ~14, ~44, and ~46 planetary masses.

Current thought relates loss of the gaseous component during planet formation to the active state of the young sun in its T Tauri stage for a characteritic time of $\sim10^7$ yr. Zharkov and Kozenko (1989) estimated the time for formation of Jupiter from the formation time of Mars. The value obtained ($\sim1.7\cdot10^7$ yr) agrees well with the characteristic duration of the T Tauri stage. However, it is obvious tht within the framework of current cosmogonic concepts [Eq. (1)], reconciliation of the formation times of Saturn, Uranus, and Nepture with the characteristic duration of the young sun's active T Tauri stage is impossible.

To avoid this difficulty we turn to studies involving numerical modeling of the behavior of the ensemble of planetesimals (Ip and Fernandez, 1988; Ipatov, 1989). Those studies indicated that upon introductin of an embryo several hundred Earth masses in size into the ensemble of planetsimals, together with sweeping of planetesimals the new object perturbs their trajectories, so that many planetsimals commence to depart far from the orbit of the embryo, with this orbit deformation occurring over cosmogonically brief time intervals.

We therefore propose the following mechanism for the formation of Saturn, Uranus, and Nepture. After accretion of gas in the proto-Jupiter embryo, together with the sweeping up of planettesimals in its feeding zone, the embryo perturbs the orbits of a number of planetesimals in such a way that they begin to penetrate regions far from the Jupiter growth zone — for example, Saturn's formation zone. Some of the larger planetsimals, close in mass to the critical mass of the Saturn embryo, having impinged upon the Saturn supply zone over the cosmogonically brief time of $\sim10^4$-10^5 yr, meet with a gaseous component, which collects upon the embryo, forming a massive proto-Saturn. As was shown in the studies cited, the prsence of two powerful gravitational centers in Jupiter and Saturn very rapidly deforms the orbits the planetesimals so that they begin to penetrate first the feeding zone of Uranus, and then that of Nepture. Among these planetesimals are rather large ones, with a mass several times that of Earth, having appreciable gaseous shells acquired in the Jupiter and Saturn feeding zones before the dissipation of gas from the protoplanetary nebula. These large embryos of the future Uranus and Nepture play a dual role. On the one hand, the introduction of large embryos into the Uranus and Nepture feeding zones permits formation of both planets over a cosmogonically reasonable time period of $\sim10^9$ yr by the sweeping of planetesimals. On the other hand, the embryos have acquired hydrogen shells $\sim(1\text{-}1.5)\,M_\oplus$ in mass in the Jupiter and Saturn feeding zones even before the dissipationof the gas, thus eliminating the question of th eorigin of the gaseous component on Uranus and Nepture, despite the fact that formation of these planets occurred over a time interval markedly longer than the characteristic duration of the T Tauri stage.

The proposed hypothesis of giant planet formation is of course still schematic in nature and will take on more concrete features as it develops. Hubbard and MacFarlane (1980) called attention to the fact that at low temperatures, hydrogen compounds (the L component) of the protoplanetary disk should be enriched in deuterium, which should then be reflected in the isotopic composition of the outer G shells of the giant planets. At the present time, data have been obtained on D/H ratios in the atmospheres of Jupiter $\left(2.0\,{}^{+0.6}_{-0.6}\right)\cdot10^{-5}$, Saturn $\left(1.6\,{}^{+1.6}_{-1.0}\right)\cdot10^{-5}$, Uranus $\left(7.2\,{}^{+7.2}_{-3.6}\right)\cdot10^{-5}$ and $\left(12\,{}^{+12}_{-8}\right)\cdot10^{-5}$ (de Bergh et al., 1989). These data agree qualitatively with the giant planet models and formation scenario presented in the present study. Uranus and Nepture contain a markedly higher concentration of L componetn in the outer G shell than Jupiter and Saturn do, and the D/H ratio in their atmospheres is correspondingly elevated.

de Bergh, C., Jutz, B. L., Owen, T., and Maillard, J. P. (1989). Astrophys. J. (in press).

Gudkova, T. V., and Zharkov, V. N. (1990). Pis'ma Astron. Zh. 16, 174 [Sov. Astron. Lett. 16, 75 (1990)].

Gudkova, V. N., and Zharkov, V. N. (1988a), Astron. Vestn. 22, 252.

Gudkova, T. V., Zharkov, V. N., and Leont'ev, V. V. (1988b), Pis'ma Astron. Zh. 14, 371 [Sov. Astron. Lett. 14, 157 (1988)].

Hubbard, W. B., and MacFarlane, Z. Z. (1980), Icarus 44, 676.

Hayashi, C., Nakazawa, K., and Y. Nakagawa, Y. (1985), Protostars and Planets. II. D. C. Black and M. S. Matthews (eds.), University of Arizona Press, p. 1100.

Horedt, G. P. (1988), Astron. Astrophys. 202, 284.

Ip, W. H., and Fernandez, J. A. (1988), Icarus 74, 676.

Ipatov, S. I. (1989), Astron. Vestn. 23, 27.

Safronov, V. S., and Vityazev, A. V. (1983), Itogi Nauki i Tekhniki, 24, 5.

Zharkov, V. N., and Kozenko, A. V. (1989), Pis'ma Astron. Zh. 15, 745 [Sov. Astron. Lett. 15, 322 (1989)].

Zharkov, V. N., Makalkin, A. B., and Trubitsyn, V. P. (1974a), Astron. Zh. 51, 1288 [Sov. Astron. 18, 768 (1975)].

Zharkov, V. N., Trubitsyn, V. P., Tsarevskii, I. A., and Makalkin, A. B. (1974b), Fiz. Zemli, No. 12, 3.

Translated by J. F. Caffrey

Evolution of Initially Highly Eccentric Orbits of the Growing Nuclei of the Giant Planets (abstract)

S. I. Ipatov

Translated from: Pisma Astron. Zhu. **17**, 268 (1991)
Sov. Astron. Lett. **17**, 113 (1991)

Computer modeling supports the suggestion of Zharkov and Kozenko (1990) that if the initial orbits of the nuclei of Saturn, Uranus, and Neptune with masses of several Earth masses had been highly eccentric, their eccentricities could have decreased with evolution to the present values. The orbital eccentricity of the nucleus of Saturn could have decreased due to gas accretion, while that of the nuclei of Uranus and Neptune could have decreased due primarily to gravitational interactions with planetesimals. The investigation of models that take into account the migration toward Jupiter of some bodies initially located beyond the orbit of Saturn has shown that the nearly formed Saturn could have migrated from the zone of Jupiter, while the nuclei of Uranus and Neptune, with masses of several Earth masses, could have migrated from the zone of Saturn, moving continuously in low-eccentricity orbits.

VIII. Asteroids, Comets, and Meteorites

SOVIET ASTRONOMY – AJ VOLUME 5, NUMBER 2 SEPTEMBER-OCTOBER, 1961

ORIGIN OF RAPIDLY ROTATING ASTEROIDS

E. L. Ruskol and V. S. Safronov

Institute of Earth Physics, Academy of Sciences of the USSR
Translated from Astronomicheskii Zhurnal, Vol. 38, No. 2, pp. 273-277, March-April, 1961
Original article submitted August 25, 1960

The paper [1] by E. Rabe is discussed. It is shown that the process, proposed by Rabe, of the origin of rapidly rotating asteroids from double asteroid systems has an extremely small probability.

1. E. Rabe [1] has suggested that rapidly rotating asteroids may have been formed by the merger of pairs of asteroids in rotation about their center of gravity. This would also explain the elongated shape that many asteroids have. Rabe assumes that the asteroids have been built up by gradual accretion of finely dispersed matter from the protoplanetary cloud. The medium serving as a source of supply for the growing asteroids acted as a resisting medium for them at the same time.

Rabe has treated a problem in celestial mechanics analogous to the restricted three-body problem: the sun S acts as the main gravitator; the second body is represented by the center of gravity P of the two asteroids, which rotates about the sun in a circular Keplerian orbit; and the third body, of negligible mass, is one or the other of the two asteroids. For simplicity, it has been assumed that the masses of these two asteroids are the same ($m_1 = m_2 = m$). The distance PS is the unit of length, the sun serves as the unit of mass, and the unit of time is so chosen that the gravitational constant reduces to unity. Rabe gives the first approximation to the energy integral for such an asteroid pair as

$$V^2 = \frac{m}{\Delta} + \frac{1}{4}\Delta^2(3\cos^2\alpha - 1) - C. \quad (1)$$

where V is the velocity of each component relative to P, Δ is the distance between the components, and α is the angle between PS and Pm_2. The constant C plays the same role as the Jacobi constant in the restricted three-body problem. Different values of C are associated with different zero-velocity surfaces (V = 0) about the point P. For $\alpha = 0°$ or $180°$, the relation between C, m, and Δ for the zero-velocity surfaces becomes

$$C = \frac{m}{\Delta} + \frac{1}{2}\Delta^2. \quad (2)$$

The largest closed zero-velocity surface is given by the constant $C_0 = 3m/2\Delta_{0,\max}$. For asteroids with a density of 2.0 g/cm^3, this envelope has a dimension of $\Delta_{0,\max} = 725r$ in the direction toward the sun, and $\Delta_{0,\min} = 454r$ normal to this direction, where r is the radius of each asteroid. Rabe notes that the density in the asteroid belt was sufficient for bodies to approach each other at distances that were often smaller than Δ_0. He suggests that the dispersion in the velocities of asteroids concentrated along a circular heliocentric orbit is due to planetary perturbations, but does not give a quantitative estimate of the relative velocities.

In Rabe's view, the steady growth of asteroidal bodies in the medium from which they are formed, and their deceleration by this medium in close approaches, can lead to the formation of pairs. The increase in the mass of a body, and its deceleration, are reflected in a secular increase in the constant C, which not only results in the capture of one asteroid by another, but also in the transformation of wide, unstable pairs ($C \approx C_0$) into stable ones ($C \gg C_0$).* It is computed that an approach of the two asteroids of the pair from an initial orbit about P of semiaxis 360r will result in contact if each asteroid increases in radius by a factor of 3.7 (for a density of 2.0 g/cm^3). In this merger of two asteroids, the resulting body will be elongated in shape, rotating at the limit of rotational stability with a period of about 5 hours; this agrees with the observed rotational velocities of asteroids. Rabe considers that pair formation can hardly be occurring in the asteroid belt at the present time, for the bodies have practically ceased to grow there now; but in the past the process must have been very important. Nevertheless, Rabe does not deny the possibility that the rotation of the asteroids may have arisen through mutual collisions.

The authors of the present paper do not exclude the theoretical possibility that Rabe's process may have operated to produce asteroidal rotation as a relic of the orbital motion of steadily approaching pairs of bodies. However, the estimates we give below for pair formation in the asteroid belt at an arbitrary epoch in its evolution show that this process must have played an insignificantly small part.

*A similar capture problem has been treated by V. V. Radzievskii [2] for a three-body system consisting of the sun, a planet, and a small particle revolving about the planet as a satellite.

2. Let us consider two asteroids of mass m, not forming an associated pair. We shall make a numerical estimate of the secular increase of the constant C in the close approach of one asteroid to another.

Up until one asteroid captures the other, $C < C_0$. For capture, it is necessary for C to become greater than C_0 during the close approach. An increase in the masses of the bodies leads not only to an increase in C, but also to an increase in C_0. The increment in C which, according to Rabe is $dC = dm/\Delta$, must therefore be greater than $dC_0 = dm/\Delta_0$. This means that the asteroids must approach closer to each other than the distance Δ_0. Such approaches would have occurred frequently enough if there were many small bodies in the asteroid belt. The values of the constant C for chance approaches of asteroids would then have varied within limits depending on the relative velocity of the bodies. The ratio C/C_0 could have been negative (for large V), equal to zero, or greater than zero. The value $C/C_0 = 1$ separates the domain of nonassociated pairs from the domain of associated pairs; in the latter case, the ratio C/C_0 varies from unity for pairs of limiting (wide) separation to a few hundred for close pairs (since the greatest ratio Δ_0/Δ lies between 454 and 725, depending on α).

From the expression for the energy integral, it follows that the increase in C due to mass accretion, and the deceleration of the asteroids during close approaches is given by

$$dC = \frac{dm}{\bar{\Delta}} - d(V^2), \tag{3}$$

where $\bar{\Delta}$ denotes the mean value of Δ during the approach, and is of the order of $\frac{1}{2}\Delta_0$. Here both terms on the right-hand side give a positive contribution to dC, and are of the same order, since the resistance to the motion and the mass supply arise in the same medium. We may therefore set

$$dC \approx \frac{2dm}{\bar{\Delta}}. \tag{4}$$

The ratio of dC to C_0 will be equal to

$$\frac{dC}{C_0} \approx \frac{4\dot{d}m}{3m} \cdot \frac{\Delta_0}{\bar{\Delta}} = \frac{\pi r^2 \rho V dt}{\pi r^3 \delta} \frac{\Delta_0}{\bar{\Delta}}. \tag{5}$$

Taking dt to be the time for approach, we may set $Vdt \approx \Delta_0$. Then,

$$\frac{dC}{C_0} \approx 2 \frac{\Delta_0}{r} \frac{\rho}{\delta}. \tag{6}$$

With Rabe's values for the parameters of the asteroid belt, $a_{min} = 2$ a.u., $a_{max} = 3.5$ a.u., a thickness of 0.2 a.u., and a total mass of $5 \cdot 10^{24}$ g, the density of the matter is $\rho = 10^{-15}$ g/cm^3. Inserting this value of ρ into the preceding expression, and taking also $\delta = 2$ g/cm^3 and $\Delta_0 \approx 700r$, we obtain

$$\frac{dC}{C_0} \approx 10^{-12}. \tag{7}$$

This extremely small increment dC due to mass accretion and deceleration means that a pair could have passed from the domain of nonassociated to the domain of associated pairs only if, during the approach, its constants C agreed with C_0 to within a margin of the order of 10^{-12}, that is, if the asteroids in question possessed virtually zero relative velocity. The pair-formation process must evidently be of just as high an order of improbability, namely 10^{-12} or less, taking into account the wide range within which C/C_0 can vary. It is furthermore clear that, in the improbable events in which the mutual capture of two protoasteroids does occur, the bodies would form extremely wide and, therefore, unstable pairs ($C \approx C_0$). This conclusion has a simple physical interpretation. In the two-body problem, the relative velocities V_∞ of the bodies prior to the approach for which capture is possible under the influence of a resisting (or supplying) medium must satisfy the energy condition

$$\frac{mV_\infty^2}{2} \lesssim \int FV dt, \tag{8}$$

where F is the resisting force of the medium, determined by the momentum which the body imparts to the medium each second. It is equal to the mass $\pi r^2 \rho V$ of the matter encountered by the body each second, multiplied by the velocity V of the body.

Consequently,

$$\frac{mV_\infty^2}{2} \lesssim \int \pi r^2 \rho V^3 dt \approx \pi r^2 \rho \bar{V}^2 \Delta_0, \tag{9}$$

or

$$\frac{V_\infty^2}{\bar{V}^2} \lesssim \frac{2\pi r^2 \rho \Delta_0}{m} \approx 10^3 \rho \simeq 10^{-12}. \tag{10}$$

The presence of a third body, the sun, does not essentially relax this condition for capture. We must therefore have, in order of magnitude,

$$V_\infty^2 \lesssim 10^{-12} \bar{V}^2 < 10^{-12} \cdot 2 \frac{Gm}{r}. \tag{11}$$

It is not possible to accept so small a relative velocity for the bodies. According to L. É. Gurevich and A. I. Lebedinskii [3], mutual perturbations between the bodies upon close approach will set up relative velocities $\approx \sqrt{Gm/2r}$, a value six orders greater than that needed for capture.

3. The result obtained above implies that the probability of pair formation in binary approaches is very small. But pairs can be formed in ternary and other approaches.

Statistical mechanics leads to the following expression for the relative fraction of pairs in the case of dissociative equilibrium [4] for the interval da in the major semiaxis:

$$\frac{dn_2}{n_1} = 4(\pi a_0)^{3/2} e^{a_0/a} \sqrt{a} \cdot n_1 da, \tag{12}$$

where n_1 and n_2 are the numbers of single and double systems per cm^3, and $a_0 = 3Gm/2V^2$. It is here assumed that all the bodies have the same mass m, and that their relative velocities are equal to V. These velocities are determined by gravitational perturbations between the approaching bodies, and may be written in the form

$$V^2 = \frac{Gm}{\theta r}. \qquad (13)$$

Then $a_0 = 3\theta r/2$, where θ is of the order of unity.

We may obtain the fraction of double systems with semiaxes $a_0 < a < a_2$ by integrating V, noting that $e^{a_0/a} \approx 1$:

$$\frac{n_2}{n_1} \approx 4(\pi a_0)^{3/2} a_2^{3/2} n_1 \qquad (14)$$

For the widest pairs, Rabe has taken $a_2 \approx 360r$. Then,

$$\frac{n_2}{n_1} \approx 2 \cdot 10^5 n_1 r^3 \approx 10^5 \rho. \qquad (15)$$

Since the mean density ρ of the medium in the asteroid zone is very small, the fraction of asteroid pairs is also seen to be very small. For $\rho \approx 10^{-15}$ g/cm^3, we obtain

$$\frac{n_2}{n_1} \approx 10^{-10}. \qquad (16)$$

4. Let us now consider a pair of asteroids, however it may have been formed (in a triple approach, or in a double approach with the effect of a resisting medium). We shall show that the probability of the method Rabe has proposed for the evolution of a pair – a gradual approach and eventual union into a single body – is negligibly small. Before this happens, the pair will be demolished in chance close encounters with other bodies (or in collisions).

Indeed, the mean time for the destruction of a loosely bound pair ($a > a_0$) is given by [4]:

$$t_1 = \frac{3V}{16\pi G m n_1 a \ln\left(1 + \frac{a_2 V^4}{4G^2 m^2}\right)}. \qquad (17)$$

For $V^2 = Gm/\theta r$, the second term in the argument of the logarithm is equal to $a^2/4\theta^2 r^2 \gg 1$. The density $mn_1 = \rho$ may be expressed in the form [5]:

$$\rho = \frac{4\sigma}{PV}, \qquad (18)$$

where σ is the surface density and P the period of revolution about the sun. Then the mean lifetime t_1 of the pair is equal to

$$t_1 = \frac{3mP}{400\theta\sigma r a \ln\frac{a}{2\theta r}} \qquad (19)$$

Since

$$dm = 4\pi r^2 \delta\, dr = \frac{4\pi r^2 \sigma}{P}\, dt, \qquad (20)$$

we have

$$r = r_0 + \frac{\sigma}{\delta P} t. \qquad (21)$$

Over the time t_1, the body will grow to the radius

$$r_1 = r_0 + \frac{3m}{400\theta\delta r a \ln(a/2\theta r)} \approx r_0\left(1 + 10^{-2}\frac{r_0}{a}\right), \qquad (22)$$

a value which, for a wide pair, is only 10^{-4} over the original radius. However, if the pair of bodies is to merge into a single body by Rabe's process, the radii of the bodies must increase by a factor of 3.7. It is evident that this condition will hardly ever be fulfilled. As a result, the fraction of asteroid pairs must be determined by the dissociative equilibrium condition and, hence, as shown above, must be extremely small.

In our opinion, the rotation of the asteroids, and their irregular shapes as well, can be explained in a natural way by direct collisions and fragmentation of the bodies, to which they will have been subject repeatedly during the course of their evolution.

LITERATURE CITED

1. E. Rabe, Astrophys. J. 131, 231 (1960).
2. V. V. Radzievskii, Byull. BAGO, No. 11 (18), 3 (1952).
3. L. É. Gurevich and A. I. Lebedinskii, Izv. AN SSSR, Ser. fiz. 14, 765 (1950).
4. L. É. Gurevich and B. Yu. Levin, Astron. zh. 27, 5 (1950).
5. V. S. Safronov, Astron. zh. 31, 499 (1954); Voprosy kosmogonii 6, 63 (1958).

All abbreviations of periodicals in the above bibliography are letter-by-letter transliterations of the abbreviations as given in the original Russian journal. *Some or all of this periodical literature may well be available in English translation.* A complete list of the cover-to-cover English translations appears at the back of this issue.

Origin of Meteorites and Planetary Cosmogony (abstract)

B. Yu. Levin
In: *Meteorite Research* edited by P. M. Millman (Reidel, Dordrecht, 1969), pp. 16–30

In the solar system there can be no solid bodies which have been preserved since the time of a previous separation of the material of this system in the processes of galactic nucleosynthesis. Therefore, ages of iron meteorites which exceed the so-called "uranium age" (6.5×10^9 yr) are clearly unrealistic.

Formation intervals determined from ^{129}I–^{129}Xe and ^{244}Pu–^{136}Xe, and the isochronism of the beginning of the retention of xenon in chondrites, show that this retention had its origin about the time of the formation of the parent bodies. Later, a partial loss of gases, particularly ^{136}Xe, occurred.

The parent bodies were moderate in size and could not be heated by long-lived radioactive elements. This could only be done by ^{26}Al, which was formed during an additional nucleosynthesis in the protoplanetary cloud. The existence of an additional nucleosynthesis indicates that the Sun and the protoplanetary cloud were formed together, since only in this case was it possible for the cloud to be irradiated by the intensive corpuscular radiation of the new Sun.

Considerable progress has been made in the study of condensation in the protoplanetary cloud. However, existing works have not taken into account the opacity of the cloud due to the presence of dust in it, a factor which determined the extremely low temperature of the dust component of the cloud.

The possibility of chondrules forming in the protoplanetary cloud seems remote to the present writer. During recent years new evidence has accumulated which confirms the important role of diffusion in solid matter during the evolution of meteoritic matter and thermal metamorphism. In particular, new arguments have appeared in support of the hypothesis of the formation of chondrules *in situ.*

The difference in the cooling rates of iron meteorites definitely disproves the hypothesis of their formation in the nucleus of a large parent planet. For a long time this hypothesis served as one of the fundamentals of the theory that the Earth's core is composed of iron. Actually, iron meteorites were formed at different points in different parental bodies.

According to the author's hypothesis, iron meteorites measuring up to several meters were formed by diffusion in the solid state, i.e., in the same way as metallic inclusions in chondrites. The meteorites which formed the Arizona and other craters were probably not iron bodies tens or hundreds of meters in diameters, but stony bodies with metallic inclusions.

Asteroids, Comets, Meteor Matter—Their Place and Role in the Cosmogony of the Solar System (abstract)

B. Yu. Levin
Translated from: Izv. Akad. Nauk SSSR Fiz. Zemli (6), 25 (1982)
Izv. Acad. Sci. USSR Phys. Solid Earth **18**, 414 (1983)

The main differences between asteroid and comet nuclei and the place of the formation process are briefly discussed for these and other bodies in the overall process of the formation of the planetary system. The hypothesis of the cometary origin of the asteroids in the Apollo and Amor groups is critically considered, and it is emphasized that the possibility of their cometary enrichment is not definitely established. It is noted that there is an absence of observational confirmation for the possibility of the "deactivation" of comet cores and their transformation into asteroidlike objects. At the same time, judging from all the data, the asteroids in the Apollo and Amor groups are the latest parental bodies of meteorites, which constitutes weighty evidence against the cometary origin of these asteroids. The association of meteor matter with asteroids and comets is briefly discussed and the role of meteorite research in the cosmogony of the solar system is considered. Studies of the isotopic composition of chemical elements from meteorites indicate that O. Yu. Schmidt's ideas concerning the interstellar source of the solid component in the matter of a protoplanetary cloud were correct.

Formation of Solid Materials in the Preplanetary Nebula and the Composition of Chondrites (abstract)

M. N. Izakov
Translated from: Astron. Vestn. **20**, 35 (1986)
Sol. Syst. Res. **20**, 22 (1986)

On the basis of the model of the formation of the preplanetary nebula as an accretion disk during the formation of the Sun, the hypothesis is proposed that a significant fraction of the solid materials of the preplanetary nebula was formed by the successive condensation of the components of the gas of solar composition during its motion from the hot, dense region near the proto-Sun to regions of ever decreasing values of temperature and pressure in the periphery of the nebula. The hypothesis removes the contradiction between the presence of traces of high-temperature phenomena in chondrite materials, and the conclusion that there were never high temperatures in the preplanetary nebula at distances of 2–4 A.U. from the Sun, where meteorites encountering the Earth originate, and also explains a number of properties of chrondrites. It follows from this hypothesis that the mass and angular momentum of the nebula were close to their minimum possible values and that the loss of the nebular gas had already begun in the final stage of its formation.

On the Origin of Comets (abstract)

V. S. Safronov
In: *The Evolution of the Small Bodies of the Solar System*, edited by M. Fulchignoni and L. Kresak (North-Holland, Amsterdam, 1987), pp. 217–26

The origin of minor bodies of the solar system—comets, asteroids, meteorites—should be considered as a common problem, because in spite of differences between these bodies there are many similarities, a kind of "genetic relationship." But, besides the significance of this problem itself, the investigation of minor bodies becomes increasingly important for the cosmogony of the whole solar system. Asteroids and meteorites have recorded a large amount of valuable data on primitive matter in the zone of the terrestrial planets and on the physical conditions at the early stages of their formation. On the other hand, the comets help us to understand the character of the process of formation of the giant planets because there is much evidence that they have been formed in the same process.

On the Origin of SNC Meteorites (abstract)

L. K. Levskiy and Ye. R. Drubetskoy
Translated from: Meteoritika **47**, 134 (1988)
Abstract reprinted from: Meteoritics **26**, 253 (1991)

Analysis of new experimental data and of those already existing for the content and isotopic compositions of noble gases and nitrogen in the shergottite–nakhlite–chassignite meteorites has permitted the hypothesis of their Martian origin to be investigated. It has been established that their source was not the surface of Mars but parent bodies analogous to those from which the terrestrial-group planets formed.

Formation of the Geminid Meteor Stream with the Disintegration of a Comet Nucleus (abstract)

O. I. Bel'kovich and G. O. Ryabova
Translated from: Astron. Vestn. **23**, 157 (1989)
Sol. Syst. Res. **23**, 98 (1989)

The formation of the Geminid meteor stream is examined; the stream is modeled by 5000 particles. Several models of particle ejection from a comet nucleus are investigated. The purpose of the work is to study the effect of the ejection velocity distribution on the structure of the stream.

Transplutonian Comet Families (abstract)

A. S. Guliev and A. S. Dadashov
Translated from: Astron. Vestn. **23**, 88 (1989)
Sol. Syst. Res. **23**, 57 (1989)

Five transplutonian comet families are distinguished by the closeness of their aphelion distance. The possibility of the formation of these groups by hypothetical planets is examined. It is shown that the existence of parent planets at distances of 55 and 110 A.U. with the same inclination of 30° is quite possible for two groups. Such an assumption is unlikely for the other three groups.

Origin of Chondrules (abstract)

A. K. Lavrukhina
Translated from: Geokhim. 1989, No. 10, 1407
Geochem. Int. **27** (5), 26 (1990)

The parameters of the processes of chondrule formation have been estimated from a survey of all the major characteristics of chondrules in the various chondrite groups: droplet chondrule crystallization rates, density of chondrules in the chondrule formation regions, oxygen partial pressures and O/H ratios there, and physical conditions such as magnetic field strength and irradiation by low-energy VH nuclei. The following origins have been identified: melting of chondrule-precursor material condensation from hot gas in the protosolar nebula and from silicate vapor at various temperatures, chondrule collision and crushing, cold brecciation of aggregates of mineral grains (chondrule fragments), thermal metamorphism, abrasion, and low-temperature metamorphism and metasomatism in parent bodies. The main chondrule precursors were interstellar dust and condensates at various temperatures from the hot nebular gas and from silicate vapor formed by dust melting during chondrule formation. A sequence of chondrite-formation processes has been identified for the various stages in the evolution of the nebula, in the planetesimal agglomeration, and in the parent bodies. The differences in chondrule characteristics between the chemical groups are caused by the different astrophysical conditions of chondrite formation by the chemical and isotope compositions of the precursor materials, by the initial heating temperatures, and by the cooling rates. Chondrules from CV chondrites in the main were formed during collapse of the nebula, while those in O and E chondrites were formed in the dense gas-dust disk near the equator of the nebula, and those in CO ones in collisions on friable planetesimals.

Search for a Mechanism for the Origin of Clearly Defined Hyperbolic Meteor Orbits (abstract)

A. M. Kazantsev and L. M. Sherbaum
Translated from: Astron. Vestn. **24**, 72 (1990)
Sol. Syst. Res. **24**, 47 (1990)

The following properties of hyperbolic orbits of photographic meteors with heliocentric velocities V_h>50 km/s have been established: (1) the relative numbers of hyperbolas with i>90° decrease with an increase in the lower limit of V_h of the meteor sample (all meteors with V_h>57 km/sare prograde); (2) most hyperbolic meteors are observed near perihelion. It is shown that these properties contradict not only the hypothesis of an interstellar origin of hyperbolic meteors, but also any other hypothesis of their natural origin. It is concluded that all clearly defined hyperbolic orbits are a result of errors in the radiant determination.

Formation of Trojan Asteroids (abstract)

E. L. Ruskol
Translated from: Astron. Vestn. **24**, 244 (1990)
Sol. Syst. Res. **24**, 157 (1991)

The possibility is considered of capture into "Trojan" orbits of bodies which populated the asteroid zone in the initial epoch when many of them could have intersected the orbit of Jupiter. The total mass of all bodies experiencing collisions with similar bodies in the vicinities of the Lagrangian points L_4 and L_5, where the modern Trojan asteroids orbit, is 3–4 orders of magnitude greater than the contemporary mass of Trojan asteroids. If only one-thousandth of such collisions was not accompanied by disintegration and departure at velocities greater than 1 km/s relative to the Keplerian orbital velocity, both clusters of Trojans could have been formed by such collisions. The far distant irregular satellites of Jupiter could also have been formed by inelastic collisions of bodies in the asteroid belt in the vicinity of Jupiter with less severe limitations on residual velocity.

Problems of Origin of Asteroids and Comets (abstract)

V. S. Safronov
Meteoritics **25**, 243 (1990)

Various hypotheses of the origin of asteroids and comets are briefly discussed. Interaction of planetesimals in the asteroid zone (AZ) with the gas, their perturbations by proto-Jupiter, and sweeping them out by more massive Jupiter zone bodies when they penetrated the AZ are considered. If the gas was turbulent, it could prevent a settling of dust particles to the equatorial plane of the disk and formation of dust condensations due to gravitational instability. Then particles grew by sticking upon collision. Gas moved radially due to turbulent viscosity and its dissipation. Small particles moved more or less together with the gas. As a result of gas drag, larger particles and bodies moved relative to the gas in the direction of increasing gas pressure. Gas would remove much of the solid material from the AZ if most bodies larger than a few kilometers disintegrated by collisions into fragments smaller than a few tens of meters. Most of these fragments would then move into the Martian zone, and the small mass of Mars would have no explanation.

Resonant perturbations of asteroids by Jupiter are discussed. In the model of a small mass disk they could scan through the asteroid belt due to changes in Jupiter's distance from the Sun that occurred when this planet accreted the gas and ejected the bodies from the solar system. Such a scanning considerably accelerated the removal of asteroids from the AZ. Massive Jupiter zone bodies with large orbital eccentricities that crossed the AZ were probably efficient at sweeping out bodies. Larger bodies increased the random velocities of the remaining asteroids at close encounters to the present values ~5 km/s.

Restrictions on the runaway growth of giant planets and on the relative velocities of bodies and the disk surface density that follow from the consideration of the origin of the asteroid belt and the cometary cloud are considered.

Possible Formation of Comets *in Situ* (abstract)

V. S. Safronov and K. M. Guseinov
Translated from: Astron. Vestn. **24**, 248 (1990)
Sol. Syst. Res. **24**, 159 (1991)

The hypothesis of comet formation *in situ* in the transplutonic region is examined. The growth rate of initially small particles in the nonturbulent disk during their descent toward its central plane and radial drift toward the Sun is evaluated. Since the collisions and growth of particles are possible only with a difference in their sizes, the particle size distribution is taken into account as an inverse power law. The motion and growth of the larger particles relative to the other smaller particles are calculated. For most of the particles, $|v_z|>|v_R|$ at first, and their growth is determined primarily by the velocity difference $\overline{\Delta v_z}$. The ratio v_z/v_R decreases with decreasing z and the particles grow predominantly because of the differential drift $\overline{\Delta v_R}$. The descending particles from a dust layer with a density close to the critical density required for the onset of its gravitational instability. The particles grow to $r\sim1$ cm at $R=50$ A.U. in $\sim10^5$ yr and to 10^{-3} cm at 100 A.U. in $\sim10^6$ yr. Further growth of the largest particles is evaluated for subcritical density of the sublayer. This situation could have arisen with weak turbulence initiated by the difference in rotational velocities of the sublayer and overlying gas. The sizes of the particles and bodies with $r<10^4$ cm are less than the mean free path of molecules, and their drag in the gas is described by the Epstein law. As a result of radial drift, the larger particles are able to move from initial distances $R<1000$ A.U. to the region of the planets ($R<40$ A.U.) in $\sim10^6$ yr, while growing to $\lesssim10$ m. One cannot rule out the possibility of the formation of comet-size bodies as a result of gravitational instability with gas present, when only a small fraction of all the dust available in the transplutonic region could have reached the dust layer.

Transformation of the Orbits of Small Bodies by Close Encounters with the Terrestrial Planets (abstract)

G. V. Andreev, A. K. Terent'eva, and O. A. Bayuk
Translated from: Astron. Vestn. **25**, 177 (1991)
Sol. Syst. Res. **25**, 129 (1991)

The Laplace method of the unperturbed two-body problem in the spheres of influence of the planets and Sun is used to show that close encounters of small bodies with the terrestrial planets could be one of the sources of short-period orbits (the Aten asteroids and meteoroids of the Eccentrid system). At least some of the Amur and Apollo asteroids and meteor swarms like the Geminids and Arietids could be of cometary origin.

Statistical Test of the Hypothesis of Comet Ejection (abstract)

V. P. Tomanov
Translated from Astron. Vestn. **25**, 312 (1991)
Sol. Syst. Res. **25**, 230 (1991)

Some parameters of the system of nearly parabolic comets are analyzed statistically. No evidence is found to confirm the hypothesis (Vsekhsvyatskii, 1967) of comet ejection from the surface of giant planets or their satellites.

IX. Other Planetary Systems

The angular momentum of a protostar engendering a protoplanetary disk

T. V. Ruzmaĭkina

Shmidt Institute of Earth Physics, USSR Academy of Sciences, Moscow

(Submitted October 8, 1980)

Pis'ma Astron. Zh. 7, 188–192 (March 1981)

The contraction of a protostar of mass M and angular momentum J in the range $J(M/M_{\odot})^{-5/3} \approx (0.5\text{–}10)\times 10^{51}$ g cm^2/sec could result in a star with a protoplanetary disk. As the protostar collapses a single starlike core would form and grow by accreting envelope material. The angular momentum in the core would be redistributed by turbulence and the magnetic field, enlarging the outer parts of the core in its equatorial plane until an extended circumstellar disk develops.

PACS numbers: 97.10.Bt

1. INTRODUCTION

It is currently popular to maintain that one possible way for a newborn star to throw off its excess angular momentum is to form a protoplanetary disk—an alternative to breaking up into a double or multiple system of stars.

The purpose of this letter is to point out that conditions favoring the development of a star with a protoplanetary disk will arise if the contracting protostar has an angular momentum confined to a certain range of values (Sec. 2). The upper limit of this interval is set by the requirement that an intact stellar core be able to form, with the protoplanetary disk developing as the core grows; the lower limit is determined by the condition that even before the core has completely formed, gravitational and

 0360-0327/81/02 0104-04 $02.00

centrifugal forces will have become the main factors governing the equilibrium of its equatorial surface layer (Sec. 3). If this is the case, redistribution of angular momentum will cause a steady increase in the equatorial radius of the core and lead to the formation of a disk.

2. DISK FORMATION DURING GROWTH OF CORE

A preliminary analysis has shown that angular momentum will be redistributed with greatly differing efficiency at different stages in the contraction of the protostar.[1-3] The conditions best suited to redistributing angular momentum will arise in a starlike core, such as the quasistatic core of a nonrotating protostar. Angular momentum could be transferred within it either by turbulence[3] or by the magnetic field[4] on a time scale shorter than the core evolution time. A single stellar core will develop at the center of a sufficiently slowly rotating protostar once the dissociation of molecular hydrogen has been completed, and initially it will comprise only a small fraction of the mass. Subsequently the mass and central density of the core will continue to grow through accretion of material from the protostellar envelope.

Let us consider how the core will evolve if its angular momentum is efficiently redistributed.

The angular momentum per unit mass of material the specific angular momentum) will diminish in the central zone of the core and increase in the layers near the surface. This process will help to overcome the tendency for the core to fragment, and will cause its equatorial radius to grow. The manner in which the outer part of the core evolves will depend on the ratio between the pressure gradient and the centrifugal force.[5] If the pressure predominates, the conditions of hydrostatic equilibrium will allow the state most favorable from an energy standpoint to be established: solid-body rotation. Once this state is achieved, the outward flow of angular momentum and hence the growth of the equatorial core radius will cease. The final configuration will then presumably be a single rotating star.

If on the other hand centrifugal forces surpass the pressure gradient in the outer part of the core, the equilibrium conditions will compel differential rotation to be maintained there.[5,6] In this event provided a magnetic field or turbulent viscosity is present, the outward flow of angular momentum and the growth of the equatorial radius may continue indefinitely, resulting in the formation of an extended disk. For example, a disk whose radius is comparable with that of the solar system ($R \sim 10^{15}$ cm) could have developed within the time required for the sun itself to form [$\tau \sim 10^6$ yr (Larson[7])] or on the time scale of turbulent viscosity[4] with $\nu_t \approx 3 \cdot 10^{15}$ cm^2/sec. In a protoplanetary disk such viscosity would be produced by very weak subsonic turbulence, which probably could have been sustained by the energy of the accreting envelope. The mechanism of supporting the turbulence in the disk involves the mismatch of the specific angular momenta of the disk and the accreting envelope.[8]

According to the plan outlined above, the outer parts of the disk would receive not only material arriving from its inner regions, where high temperatures, magnetic fields, and so on would have prevailed, but also matter accreted directly from the protostellar envelope and retaining the interstellar composition. Perhaps this circumstance might account for the heterogeneous chemical, mineralogical, and isotopic composition observed in certain meteorites.

3. ANGULAR MOMENTUM OF PROTOSTAR GIVING RISE TO DISK

We proceed now to estimate the range of angular-momentum values for a protostar in which a disk can develop by the method described above.

For this purpose we shall utilize Bodenheimer and Ostriker's model calculations[6] for a rotating, gravitating body in hydrostatic equilibrium with a polytropic density distribution. In such a body the relation

$$J = \frac{G^{1/2} M^{5/3}}{\rho_0^{1/6}} f(\beta, n, n') \tag{1}$$

will hold, where J is the angular momentum, M is the mass of the body, $f(\beta, n, n')$ is a dimensionless angular momentum (β denotes the ratio of the rotational and gravitational energies of the body, n is the polytropic index, and the parameter n' describes the distribution of angular momentum in the body; if $n = n'$, most of a body with polytropic index n will be in rigid rotation), ρ_0 is the central density, and G is the gravitational constant.

The minimum protostar angular momentum J_{min} such that a disk can develop will be estimated from the condition that by the time the formation of the core has ended the pressure gradient at its equator will have become smaller than the centrifugal force (this is the case in which the conditions will favor the production of a disk by the mechanism discussed in Sec. 2). The density distribution in the core may approximately be described by a polytrope whose index n rises with the growth of mass[9] to a value in the range $1.5 < n < 3$. If angular momentum is to be efficiently redistributed in the core, the bulk of the core will be kept approximately in solid-body rotation, corresponding to $n' \approx n$. In this event n' will become greater than unity as the core mass grows. According to Bodenheimer and Ostriker's results the centrifugal force will become equal to the pressure gradient at the equator for $f \lessdot 0.13$. The corresponding values of the ratio β are ≈ 0.06, or less than the critical value $\beta_{cr} \approx 0.14$ required for development of the lowest mode of instability of equilibrium gravitating bodies [that is, secular instability of gravitating bodies against nonaxisymmetric perturbations (Ref. 10)]. Hence the disk will start to grow before instability leading to fragmentation has had a chance to build up, and fragmentation of the core will be averted. Furthermore, according to the plan we have discussed the disk will begin to develop before the equatorial rotational velocity of the core has reached its limiting value.

Substituting $f = 0.13$ into Eq. (1), we obtain

$$J_{min} \approx 5 \cdot 10^{50} (M/M_{\odot})^{5/3} (\rho_0/10^2)^{-1/6}. \tag{2}$$

Let us estimate the numerical value of J_{min} for a star with $M = 1\ M_{\odot}$. We shall take the central density to be $\rho_0 \approx 10^2$ g/cm^3 at the time the formation of the star has been completed[9]; then Eq. (1) gives $J_{min} \approx 5 \cdot 10^{50}$ g $\cdot$ cm^2/sec.

It is noteworthy that this estimate of J_{min} is at least six times lower than the limiting angular momentum of the solar nebula,[11,12] $J_{SN} \gtrsim 3 \cdot 10^{51}$ g · cm^2/sec. Thus while the embryonic sun was growing, enough differential rotation could have become established in its outer parts for a disk to develop by the method described above.

We would point out that the estimate (2) rests on the assumption that the redistribution of angular momentum in the core will ultimately lead to an effective parameter n' in the range $1 \lesssim n' \lesssim 1.5$. If $n' > 1.5$, then J_{min} should be somewhat smaller than given by Eq. (2), although it is hard to give an estimate for this case since no calculations have been published for equilibrium configurations with $n' > 1.5$. But if the redistribution of angular momentum in the core is inefficient, that is, if $n' \approx 0$, then the quantity J_{min} will be about twice as great as indicated by Eq. (2). This last case, however, is of no interest to us, because the formation of a protoplanetary disk without angular-momentum redistribution is evidently impossible. Of course our estimate of J_{min} is only intended as a rough guide: an accurate value for J_{min} cannot be obtained until the disk formation process has been studied in detail.

Let us now estimate an upper limit on the angular-momentum interval. We shall find the maximum value J_{max} of the protostar angular momentum such that as the protostar contracts a single starlike core can form whose parameters resemble those of the quasistatic core of a nonrotating protostar: mass $M = 10^{-2}\ M_\odot$, density $\rho \approx 10^{-2}$ g/cm^3. The angular-momentum distribution in the protostar prior to collapse will be regarded as corresponding to solid-body rotation for a uniform density distribution (for such a rotation law, $n' = 0$), with the angular momentum of each separate volume element conserved until the starlike core develops. Then as the collapse proceeds the ratio β of the rotational and gravitational energies in the central part of the protostar will increase and instability can set in, leading to fragmentation. At present it is unclear at what critical value of β the instability will arise at the center of the protostar during the dynamical collapse phase.[13] We will start with the premise[1] that as the protostar contracts it becomes unstable when $\beta \geq 0.14$. Then the criterion $\beta = 0.14$ in the newly formed core will represent a sufficient condition for a single starlike core to form at the center of the protostar. For $\beta = 0.14$, $n' = 0$, and $1.5 \lesssim n \lesssim 3$, the value of f will be confined to the interval $0.24 \gtrsim f \geq 0.18$.

To estimate the angular momentum J of the protostar from the core angular momentum J_c one must consider the collapse geometry at the stage prior to formation of the core. We shall take two limiting cases: 1) the core has acquired matter from a sphere containing mass M_c; 2) the core has acquired matter from a cylinder of mass M_c whose axis is parallel to the rotation axis. In the former case one can easily show that the angular momenta J and J_c will be related by

$$J = J_c\,(M/M_c)^{5/3}. \qquad (3)$$

In the second case,

$$J = \alpha\, J_c\,(M/M_c)^2 \text{ for } M_c \ll M, \qquad (4)$$

where $1.2 \lesssim \alpha \lesssim 1.7$ if $1.5 < n < 3$. Substituting the expression (1) for J_c into Eqs. (3) and (4), setting $J \equiv J_{max}$, and adopting the numerical values $M = 1\ M_\odot$, $M_c/M = 10^{-2}$, $\rho_0 = 10^{-2}$ g/cm^3, $f = 0.2$, and $\alpha = 1.5$, we obtain

$$3.5 \cdot 10^{51} \lesssim J_{max} \lesssim 2.5 \cdot 10^{52} \text{ g} \cdot \text{cm}^2/\text{sec}. \qquad (5)$$

These values of J_{max} are somewhat greater than given by the preceding estimate, because during the stage prior to core formation angular momentum could have been removed from the center of the protostar.[3] In view of this circumstance we shall adopt as an order-of-magnitude estimate

$$J_{max} \sim 10^{52} \text{ g} \cdot \text{cm}^2/\text{sec}.$$

4. CONCLUSIONS

We have proposed in this letter a mechanism for the formation of a single star with a protoplanetary disk in the course of collapse of the rotating protostar. The disk will develop as the equatorial radius of the stellar core grows and material is accreted from the protostellar envelope. Stars formed in this manner with a protoplanetary disk would occupy a place intermediate between single and double stars. Perhaps they may fill the gap that Kraicheva et al.[14] have found in the distribution of stars with respect to angular momentum.

The probable angular-momentum interval for a protostar of solar mass which can give rise to a disk as it contracts is comparatively narrow. For protostars with $M = 1\ M_\odot$ the permissible range would be $5 \cdot 10^{50} \lesssim J \lesssim 10^{52}$ g · cm^2/sec. The lower limit on the possible values for the angular momentum of the solar nebula falls within this interval. The mechanism described above for the formation of stars with disks may therefore be relevant to the cosmogony of the solar system.

The author is indebted to V. S. Safronov, A. V. Tutukov, and A. V. Vityazev for useful discussions, and to S. I. Blinnikov and B. Yu. Levin for valuable critical comments.

[1] W. M. Tscharnuter, "Collapse of the presolar nebula," Moon and Planets 19, 229-236 (1978).
[2] V. S. Safronov and T. V. Ruzmaikina, "Angular-momentum transfer and accumulation of solid bodies in the solar nebula," in: Protostars and Planets (IAU Colloq. No. 52), ed. T. Gehrels, Univ. Arizona Press (1978), pp. 545-564.
[3] P. J. Wiita, D. N. Schramm, and E. M. D. Symbalisty, "Star and planetary-system formation in collapsing, viscous, rotating clouds," Proc. 10th Lunar Planet. Sci. Conf. (Houston, March 1979), Geochim. Cosmochim. Acta Suppl. No. 11, Pergamon (1979), pp. 1849-65.
[4] T. V. Ruzmaikina, Report to 23rd Plenary Meeting of COSPAR, Budapest (June 1980) [Adv. Space Res. (in press)].
[5] D. Lynden-Bell and J. E. Pringle, "The evolution of viscous disks and the origin of the nebular variables," Mon. Not. R. Astron. Soc. 168, 603-637 (1974).
[6] P. Bodenheimer and J. P. Ostriker, "Rapidly rotating stars: zero-viscosity polytropic sequences," Astrophys. J. 180, 159-169 (1973).
[7] R. B. Larson, "Numerical calculations of the dynamics of a collapsing protostar," Mon. Not. R. Astron. Soc. 145, 271-295 (1969).
[8] A. G. W. Cameron, "Physics of the primitive solar accretion disk," Moon and Planets 18, 5-40 (1978).
[9] K.-H. A. Winkler, "Late stages of solar-type protostars," Moon and Planets 19, 237-244 (1978).
[10] J. P. Ostriker and P. Bodenheimer, "Oscillations and stability of rapidly rotating stellar models," Astrophys. J. 180, 171-180 (1973).

[11]F. Hoyle, "On the origin of the solar nebula" (report to IAU Sympos. on Origin of Earth and Planets, Aug. 1958), Quart. J. Roy. Astron. Soc. 1, 28-55 (1960) [Vopr. Kosmog. 7, 15-49 (1960)].
[12]S. J. Weidenschilling, "Iron/silicate fractionation and the origin of Mercury," Icarus 35, 99-111 (1977).
[13]P. Bodenheimer and D. C. Black, "Numerical calculations of protostellar hydrodynamic collapse," op. cit.,[2] pp. 288-322.
[14]Z. T. Kraicheva, E. I. Popova, A. V. Tutukov, and L. R. Yungel'son, "Some properties of spectroscopic binary stars," Astron. Zh. 55, 1176-89 (1978) [Sov. Astron. 22, 670-677 (1979)].

Model protoplanetary disks around F–G stars

A. V. Vityazev and G. V. Pechernikova

Shmidt Institute of Earth Physics, USSR Academy of Sciences, Moscow

(Submitted December 30, 1981)

Pis'ma Astron. Zh. 8, 371–377 (June 1982)

Some topics in the formation and early evolution of protoplanetary gas–dust disks around single solar-type stars are discussed. A new class of disk models is proposed, whose parameters are the disk mass and angular momentum as well as the mass and luminosity of the central star. Since disks may persist for fully 10^8 yr, a search for them among the members of decaying clusters would be worthwhile.

PACS numbers: 97.10.Fy, 97.20.Ge, 97.20.Jg, 96.10. + i

1. INTRODUCTION

Today almost everyone accepts the idea that the planets of the solar system were formed from a cloud of gas and dust rotating about the young sun. Astrophysical evidence indicates that many stars in process of formation are surrounded by gas–dust envelopes. Strel'nitskii[1] has given arguments to show that disk structures ought to be present. However, the question of how the circumsolar protoplanetary cloud originated, and the more general question of what mechanism is responsible for producing circumstellar disks and candidate planetary systems, are still a long way from being answered.

Many believe that as a rotating protostar collapses a disk may develop if the initial angular momentum K_0 is neither too small nor too large. According to Ruzmaikina,[2] for the protosolar nebula with its mass of $\approx 1\ M_\odot$ one would expect K_0 to have been in the range $K_0 \approx (0.5\text{–}10) \cdot 10^{51}$ g · cm^2/sec. If K_0 had been larger, a multiple system (not necessarily stable) would more likely have formed. Unfortunately no calculation has yet been made of a collapse (even for an isolated fragment) which would have given rise to a $1\ M_\odot$ star and a $0.01\text{–}0.1\ M_\odot$ disk. Kobrick and Kaula's results[3] show that in studying the disk formation process one has to allow for tidal interactions from close encounters between protostars.

The fragmentation of massive gas–dust nebulae has not been adequately investigated, and the mass spectrum of protostellar objects in the range $M_* < 0.5\ M_\odot$ is practically unknown. Hence one cannot yet rule out that small fragments might have experienced close encounters or even coalescence.

There can hardly be any doubt now that star formation in complexes of gas and dust bears some relationship to massive stars which go rapidly through their evolution and explode. The expanding supernova envelopes will naturally affect the formation of low-mass stars and their environment in different ways, depending on the frequency and the proximity of the exploding supernovae. Various processes could presumably have resulted in the creation of circumstellar disks, and the problem ought to be treated in the context of the whole star-formation process in stellar associations.

The study of many astrophysical objects – stars, star clusters, galaxies – concurrently involves discussion of their origins. In order to compare theory against observation one requires models for the anticipated circumstellar disks. We describe in Sec. 2 a model of the circumsolar disk which has already come to be regarded as standard. Then in Sec. 3 we develop and examine a new class of models for low-mass disks around solar-type stars, in terms of a more general approach to the problem.

2. STANDARD MODEL OF CIRCUMSOLAR PROTOPLANETARY CLOUD

When the first data on the cosmic abundance of the elements become available in the 1940s, attempts were begun to reconstruct the primordial circumsolar disk. To this end the material in the existing planets was smeared out into contiguous rings and supplemented to make up the cosmic abundance. Figure 1 illustrates the latest reconstruction of this kind, carried out by Weidenschilling.[4] Vertical bars represent the uncertainty in the composition of each planet; horizontal bars, the width of the ring from which the bulk of its material is assumed to have come. Allowing for the ejection of a certain amount of solid matter, the mass of the cloud obtained in this way is $0.01\text{–}0.07\ M_\odot$, and its surface density σ [g/cm^2] depends on the distance from the sun as $r^{-3/2}$.

It is widely believed that after a formative stage lasting $\approx 10^6$ yr the temperature in the cloud would have dropped

 0360-0327/82/03 0201-05 $03.00

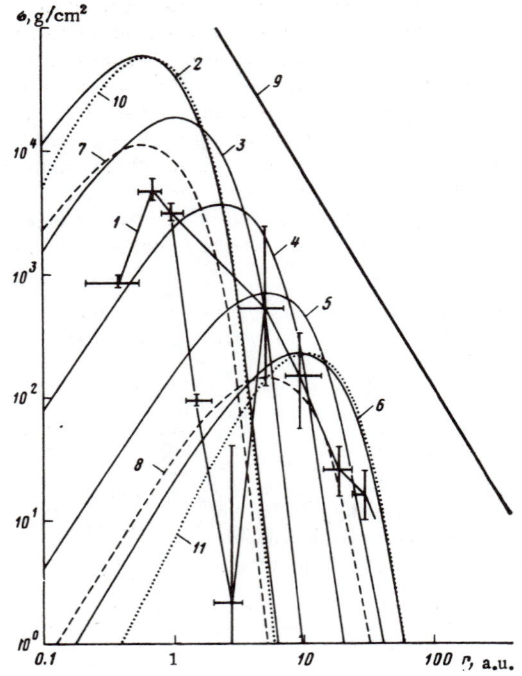

FIG. 1. Surface-density distribution $\sigma(r)$ in the protoplanetary disk around the sun. 1) The standard model of Weidenschilling[4]; 2) a model disk of mass $m = M/(10^{-2} M_\odot) = 5$ and angular momentum $k = K/(10^{51}\ g \cdot cm^2 \cdot sec^{-1})$ such that $m/k = 1$; 3) model with $m = 5$, $m/k = 3/4$; 4) $m = 5$, $m/k = 1/2$; 5) $m = 5$, $m/k = 1/3$; 6) $m = 5$, $m/k = 1/4$; 7) $m = 1$, $m/k = 1$; 8) $m = 1$, $1/3$; 9) the critical density σ_{cr} for gravitational instability; 10) a turbulent disk with $m = 5$, $m/k = 1$; 11) turbulent disk with $m = 5$, $m/k = 1/4$.

to values close to the blackbody temperature at given distances from the sun: $T = 270°K/\sqrt{r}$, with r expressed in a astronomical units.

The gas in the central plane is thought to have had low conductivity and little ionization ($< 10^{-11}$), and the influence of the magnetic field is customarily neglected. Since the cloud would not have been very cold and since its mass was low, its own gravitation may be neglected compared with the attraction of the sun.

In the z direction, normal to the central plane, equilibrium would have been maintained by the pressure gradient:

$$\frac{dP}{dz} = -\rho \frac{GM_\odot z}{R^3}, \quad R = \sqrt{r^2 + z^2}. \tag{1}$$

Here G is the gravitational constant; P and ρ are the pressure and density of the gas. The dust would have had so low a mass compared to the gas ($\approx 10^{-2}$) that we shall not take its influence into account. Using the equation of state for a perfect gas of molecular weight μ and regarding the temperature as independent of z, we obtain for the density distribution with respect to z:

$$\rho(z) \approx \rho_e(r) \exp(-z^2/h^2); \quad h^2 = 2kTr^3/GM_\odot\mu, \tag{2}$$

with k the Boltzmann constant. The subscript e will refer to quantities in the central (ecliptic) plane. By adopting $\sigma(r) = \sqrt{\pi}\rho_e(r)h(r)$ we arrive at the "standard" model:

$$\rho_e(r) \propto r^{\alpha-\beta}, \quad P(r) \propto r^{-\beta}, \quad T(r) \propto r^{-\alpha}, \quad \alpha \lesssim 1, \quad \beta \simeq 3. \tag{3}$$

The cloud will have been very strongly flattened:

$$\gamma = h/r \propto (rT)^{1/2}, \quad \gamma \lesssim 10^{-1}. \tag{4}$$

Its rotation would have been differential, deviating little from Keplerian motion:

$$\omega = \omega_K (1 + \varepsilon)^{1/2}; \quad \omega_K = v_K/r = \sqrt{GM_\odot/r^3};$$
$$\varepsilon \simeq (c^2/v_K^2)(d \ln P/d \ln r) \lesssim 10^{-1}; \quad c^2 = kT/\mu; \quad c^2 \ll v_K^2. \tag{5}$$

A reconstruction of the radial mass distribution basec on the present distribution fails to allow for the variation in $\sigma(r)$ as the cloud evolves. In Sec. 3 we shall develop models which do <u>not</u> make use of the present $\sigma(r)$ distribution. These models will be determined by the mass M and angular momentum K of the disk, so we shall call them MK models. Our approach represents a modification of early ideas due to Berlage, Chandrasekhar, ter Haar, and von Weizsäcker.

3. MK MODELS OF PROTOPLANETARY DISKS

Suppose that some violent process accompanying the formation of the sun gave rise to a cloud of gas and dust with a mass $M \ll M_\odot$ and an angular momentum $K \lesssim K_0$. One can show[5,6] that as the action of disturbing factors weakened – when the protostar stopped collapsing and when strong interactions with any adjacent stars or clouds ceased – the turbulent motions induced by them should have been damped within a few revolutions. However, the resultant cloud would not necessarily have had an equilibrium distribution of the parameters ω, ρ, P. Two questions arise: 1) What would have been the equilibrium ω, ρ, P distributions in a model cloud warmed by solar radiation and subject to centrifugal forces, gravitation, and gas pressure? 2) How soon would the cloud have arrived at some equilibrium state?

We shall here consider some very simple models for a gravitationally stable gaseous disk ($\rho_e \ll M_\odot/r^3$). We neglect meridional circulation and flows induced by Goldreich–Schubert instability, as well as weak turbulence during the stages of dust settling (when $\rho_{dust} \approx \rho_{gas}$) and planetesimal motion. The first two effects will be unimportant because of the strong flattening of the model disk, and furthermore they will partially offset each other. The questions of the turbulence generated in differentially rotating disks and the effects due to the motion of dust and planetesimals will be explored in a separate paper.

The system of equations for a cloud in steady rotation will be written in the form

$$\frac{GM_\odot z}{R^3} + \frac{1}{\rho}\frac{\partial P}{\partial z} = 0, \tag{6}$$

$$\frac{GM_\odot r}{R^3} + \frac{1}{\rho}\frac{\partial P}{\partial r} = \omega^2 r, \tag{7}$$

$$P = k\rho T/\mu, \tag{8}$$

$$T = T_0 R^{-\alpha} f(z). \tag{9}$$

For an opaque cloud whose dust has not yet settled to the central plane, we may, to sufficient accuracy for our purposes (see Horedt's discussion[7]), take $T = T_0 R^{-3/7}$, where $T_0 = (L_\odot/7\pi a_s c)^{2/7}(2k/G\mu M_\odot)^{1/7}$; $L_\odot$ denotes the solar luminosity, a_s is the Stefan–Boltzmann constant, c is the speed of light. For a transparent cloud, $T = T_0 R^{-1/2}$, with $T_0 = (L_\odot/16\pi a_s c)^{1/4}$. Generally speaking, the transparency of the cloud will differ at different r, z. In the exosphere allowance should be made for heating by short-wavelength solar radiation,[8] and at the periphery of the cloud by cosmic rays as well. Neglecting these effects here, we shall simply use a relation $T = T_0 R^{-\alpha}$ $(0 < \alpha < 1)$, setting the function $f \equiv 1$.

It is well recognized[9,10] that Eqs. (6)–(9) serve to determine the required parameters ω, ρ, P to within an unknown function of r, which can be established only by invoking supplementary arguments. First let us find a relation between ρ and ω.

Substituting the expression (9) into Eq. (7) and integrating over z, we obtain

$$\ln P = \ln P_e - \frac{GM_\odot\mu}{kT_0}\int_0^z \frac{\xi\, d\xi}{(r^2+\xi)^{\frac{3-\alpha}{2}}}. \tag{10}$$

Differentiating Eq. (10) with respect to r and substituting the result into Eq. (7), we find

$$\omega^2 = \frac{GM_\odot}{R^3} + \frac{kT_0 R^{-\alpha}}{\mu r}\left[\frac{d\ln\rho_e}{dr} - \frac{\alpha}{r}\right] + GM_\odot R^{-\alpha}(3-\alpha)\int_0^z \frac{\xi\, d\xi}{(r^2+\xi)^{\frac{5-\alpha}{2}}}. \tag{11}$$

In particular, for ρ_e and ω_e we have

$$\omega_e = \omega_K(1+\varepsilon)^{1/2};\quad \varepsilon = \frac{c^2}{v_K^2}\left(\frac{d\ln\rho_e}{d\ln r} - \alpha\right). \tag{12}$$

Among all the functions $\omega_e(r)$, $\rho_e(r)$ specified by Eqs. (12), we are primarily interested in the ones that correspond to an absence of angular-momentum transfer by viscosity. Furthermore, to determine a steady state we shall require[11] that the entropy production or, what is equivalent in our case, the dissipation function be minimized.

We shall supplement Eqs. (12) by conditions first introduced by Berlage.[12] Berlage made the assumption that in its steady state the cloud had parameters (definite values $\tilde\rho$, $\tilde\omega$, $\tilde P$, say) corresponding to an absence of radial angular-momentum transfer through friction and a mimumum energy loss by viscous dissipation. In cylindrical coordinates these conditions take the form

$$\frac{d}{dt}K_\eta = \frac{d}{dr}\left(\eta r^3 \frac{d\tilde\omega}{dr}\right) dr\, d\theta\, dz = 0, \tag{13}$$

$$\delta\left(\frac{dE}{dt}\right) = \delta\iiint \eta\left(\frac{d\tilde\omega}{dr}\right)^2 r^3\, dr\, d\theta\, dz = 0. \tag{14}$$

Here η is the viscosity coefficient. Berlage set it equal to the molecular value: $\eta_m = \rho v l$, where v is the rms thermal velocity and l is the molecular mean free path.

Since $\eta_m \approx 3\cdot 10^{-5}\, T^{1/2}\, g\cdot cm^{-1}\cdot sec^{-1}$, it is not hard to see that the corresponding viscosity term should be many orders of magnitude smaller than any of the terms in Eq. (7), and the time scale for viscous dissipation of the energy of the main motion would be much longer than 10^{10} yr. It was incomprehensible how so weak a mechanism could be responsible for singling out preferred solutions, and the conditions (13), (14) have not been used since Berlage's time. We will show presently that the changeover to the equilibrium values $\tilde\omega$, $\tilde\rho$, $\tilde P$ would have been determined by turbulent viscosity.

From Eqs. (13), (14) we have $\delta(d\tilde\omega/dr) = 0$. Varying $\tilde\omega$ from ω_K to ω_e, we obtain from Eq. (12) in the first approximation with respect to ε:

$$r^2\frac{d^2\ln\rho_e}{dr^2} + \left(\frac{1}{2} - \alpha\right) r\frac{d\ln\rho_e}{dr} + \alpha^2 + \frac{1}{2}\alpha = 0. \tag{15}$$

For $\alpha = 0$, $\tfrac{1}{2}$ we have the solutions obtained by Berlage: $\rho_e \propto e^{-a\sqrt{r}}$, $\rho_e \propto r^{1/2}e^{-ar}$. The general solution for $\alpha = \tfrac{1}{2}$ will clearly be:

$$\rho = \rho_0\left(\frac{R}{r_0}\right)^{1/2}\exp\left[-ar - \frac{2GM_\odot\mu}{kT_0}\left(\frac{1}{\sqrt{r}} - \frac{1}{\sqrt{R}}\right)\right]. \tag{16}$$

Following von Weizsäcker,[9] ter Haar[5,13] regarded a disk as turbulent, taking the turbulent velocities to be $v_t = v_K/3$ and the size of the largest eddies to be of order h. The pressure in the disk was considered to be made up of molecular and turbulent pressure, with $P_m \ll P_t$. In place of Eq. (8), ter Haar introduced $P_t = (1/9)\, GM_\odot/r$, and under certain assumptions having nothing to do with conditions of the form (13), (14) he obtained $\rho_e \propto re^{-ar}$ for the density in the ecliptic. Ter Haar himself estimated the relaxation time of the primordial turbulent motions and found that the turbulence would decay in a few revolutions unless it were sustained by some generation mechanism. It was felt that the lifetime of the turbulence stage in the disk certainly could not be identified with the planet formation time, which should have lasted some 10^7–10^8 yr.

One will recall that von Weizsäcker, ter Haar, and Chandrasekhar sought to explain the Titius–Bode law by attributing it to a configuration of turbulence cells. Current models no longer make any use of the artificial premise that the planets developed at the edges or in the centers of vortical cells.

Both von Weizsäcker and Berlage were right, however, in the sense that they correctly described the separate stages in the evolution of the disk. In fact, if the initial ω, ρ, P in the developing disk had deviated from the equilibrium values $\tilde\omega$, $\tilde\rho$, $\tilde P$, then hydrodynamic flows would inevitably have arisen; the greater the deviations, the more intensive the flows would have been. If the deviations from the equilibrium density were not too large, then to order of magnitude $v_t \sim (\delta\rho/\tilde\rho)v_K$. If $\delta\rho \sim \tilde\rho$, sonic velocities could have been reached $(v_K > v_s)$, but the energy of such motions would rapidly have been converted into heat and radiated away. For deviations of the density from equilibrium on scales $\lambda \sim h$, the quantity $\eta_t = \rho v_t \lambda$ would have been orders of magnitude larger than the molecular value: $\eta_t/\eta_m \sim (\delta\rho/\tilde\rho)(v_K/v)h/l$, where $h/l \sim 10^{11}$. At this stage the solar heating would have been accompanied by a heat source of a comparable strength resulting from the dissipation of turbulent motions: $E \sim \rho v_t^3/\lambda$. For $v_t \lessdot v_s$, $\lambda \sim h$ we have $E \lessdot 10^{-6}$ erg · cm^{-3} · sec^{-1}.

It is difficult to make a dependable estimate of the time required to regain equilibrium and damp the turbulence. In the first place, the Kolmogorov–Loitsyanskii–Millionshchikov results would have to be extended to the case of a compressible gas in a differentially rotating disk. Second, in the absence of other factors the gross changeover to an equilibrium state described by Eqs. (12)-(14) should have taken place in an interval comparable to a few times the quantity $\tau \approx r^2/v_t\lambda$, which is much shorter than 10^6 yr. In other words, the disk would have begun to reorganize itself as soon as its formation had concluded, when external perturbations were important and there might still have been considerable influence from the magnetic field.

The following points suggest that to a good approximation the solution (16) would have been achieved within $t \sim 10^6$ yr: a) the solutions for a turbulent disk are similar to those deduced from Eqs. (6)-(14); b) beginning with $v_t < v_s$ the pressure $P_m > P_t$, although, as before, the viscosity due to turbulent motion would have dominated the molecular viscosity. It is plain from Eqs. (13), (14) that the character of the conclusion will not be changed if the turbulent viscosity is used.

We still have to evaluate the constants in Eq. (16). Berlage, von Weizsäcker, and ter Haar selected the constants in the $\rho(r)$ distribution so that $\sigma(r)$ would fit the profile reconstructed from the present mass distribution in the planetary system. It is more natural, however, to introduce the normalization

$$M = 4\pi \iint \rho(r, z)\, r\, dr\, dz,$$
$$K = 4\pi \iint \rho(r, z)\, \omega(r, z)\, r^3\, dr\, dz. \qquad (17)$$

Let us set $m \equiv M/(10^{-2} M_\odot)$, $k \equiv K/(10^{51}$ g · cm^2 · sec$^{-1})$, express r in astronomical units, and denote the mass and luminosity of the star in question by M_*, L_*. The gas density in the ecliptic plane will then be given by

$$\rho_e = 1.12 \cdot 10^{-7} m\, (m/k)^{7.5} (M_*/M_\odot)^{12.75} (L_\odot/L_*)^{0.125}$$
$$\times\, r^{0.5} \exp[-2.8\, (m/k)^2 (M_*/M_\odot)^3 r]\ \text{g/cm}^3, \qquad (18)$$

and the surface density

$$\sigma = 1.55 \cdot 10^5\ m\, (m/k)^{7.5} (M_*/M_\odot)^{12.25} r^{1.75}$$
$$\times\, \exp[-2.8\, (m/k)^2 (M_*/M_\odot)^3 r]\ \text{g/cm}^2. \qquad (19)$$

By varying m, k, M_*, L_*, we can obtain a range of models for equilibrium circumstellar disks. The same expression may be used for disk parameters undergoing gradual change (through dissipation of the gas), as well as to allow for minor changes in the mass of the star and its luminosity when near the main sequence. For turbulent disks (Refs. 5, 13) with $\rho \propto re^{-ar}$, one can obtain an expression for $\sigma(r)$ analogous to Eq. (19). In the case of the solar disk,

$$\sigma_t = 2.45 \cdot 10^5 m\, (m/k)^{8.5} r^{2.25} \exp[-3.2\, (m/k)^2 r]. \qquad (20)$$

Figure 1 shows the $\sigma(r)$ curves calculated for several models of the solar protoplanetary disk. Notable features are: a) the qualitative similarity of the surface-density distributions in the standard and MK model disks; b) the appreciable deficiency of material in the far and/or near zones in the MK models compared with the standard model. This last circumstance has a natural interpretation in light of current theory for the evolution of the disk: it appears[14–16] that when the planets were accumulating, gravitational interaction among the planetesimals would have caused a diffusion of the semiaxes in space, spreading the disk out. Hence the present planetary system is larger than the gas–dust disk had been.

The heavy line in Fig. 1 represents the critical surface density $\sigma_{cr} = \sqrt{\pi} h \rho_{cr}$ for which the gaseous disk would have become gravitationally unstable. According to Genkin and Safronov,[17] $\rho_{cr} = 2.1\rho^*$, where $\rho^* = 3M_\odot/4\pi r^3$. In the low-mass model disks that we have considered, gravitational instability would develop only in the dust subdisks formed after the dust has settled to the central plane.

Independent estimates for the growth time of the preaccretion cores of Jupiter and Saturn[6] and the time required to dissipate gas from the disks[8] suggest that the gaseous disks would have a lifetime of order 10^8 yr. That is longer than the lifetime of open star clusters; thus disks around stars might best be sought at the periphery of clusters and even in galactic field stars. We have shown elsewhere[8] that in the exospheres of such disks the temperature would be high enough for dissociation to occur, and close to the star it would even be adequate for partial hydrogen ionization. The total radiant power of a disk in the infrared range should be of order $0.1L_*$.

In future it would be desirable to renounce Eq. (9) and make a direct calculation of the absorbed and scattered stellar radiation. The likelihood that turbulence will be generated in gaseous disks is often ascribed to their differential rotation or to large-scale circulatory flows. It is interesting that in the short-wavelength range $\nu/v_s \ll \lambda \ll v_s/\omega_K$ (ν denotes the kinematic gas viscosity), practically solid-body rotation will prevail. In the short-wave approximation for plane axisymmetric disturbances, the curvature of the flow could serve to induce an oscillatory instability beginning with $\lambda > \lambda_{cr} \approx 2\pi(\nu r/v_s)^{1/2}$. Introducing the critical Reynolds number $Re_{cr} = v_K\lambda_{cr}/\nu$, we obtain $Re_{cr} = 2\pi(Re \cdot M)^{1/2}$ (here $M = v_K/v_s$ is the Mach number); for $r \approx 1$ AU we obtain $Re_{cr} \approx 2\pi \cdot 10^7$ and an effective viscosity $\nu_{eff} \sim Re/Re_{cr} \lessdot 10^{12}$ in the disk. For a viscosity of this magnitude the distributions (18), (19) should be considered quasistationary.

The authors are grateful to Yu. A. Kukharenko, B. Yu. Levin, V. P. Myasnikov, T. V. Ruzmaikina, and V. S. Safronov for discussing particular points dealt with in this letter.

[1]V. S. Strel'nitskii, "Detection of disklike gas–dust stellar envelopes," in: Early Stages of Stellar Evolution [in Russian], ed. I. G. Kolesnik (Kiev conf., Oct. 1975), Central Astron. Obs. Ukrain. Acad. Sci., Naukova Dumka, Kiev (1977), pp. 118-127.
[2]T. V. Ruzmaikina, "The angular momentum of a protostar engendering a protoplanetary disk," Pis'ma Astron. Zh. 7, 188-192 (1981) [Sov. Astron. Lett. 7, 104-107 (1981)].

[3]M. Kobrick and W. M. Kaula, "A tidal theory for the origin of the solar nebula," Moon and Planets 20, 61-101 (1979).
[4]S. J. Weidenschilling, "The distribution of mass in the planetary system and solar nebula," Astrophys. Space Sci. 51, 153-158 (1977).
[5]D. ter Haar, "Further studies on the origin of the solar system," Astrophys. J. 111, 179-190 (1950).
[6]V. S. Safronov, Evolution of the Protoplanetary Cloud and the Formation of the Earth and Planets, Nauka, Moscow (1969) [NASA TT F-677 (1972)].
[7]G. P. Horedt, "Cosmogony of the solar system," Moon and Planets 21, 63-121 (1979).
[8]G. V. Pechernikova and A. V. Vityazev, "Thermal dissipation of gas from the protoplanetary cloud," in: Planetary Interiors (COSPAR Workshop, Budapest, June 1980), Adv. Space Res. 1, No. 7, 55-60 (1981).
[9]C. F. von Weizsäcker, "Über die Entstehung des Planetensystems," Z. Astrophys. 22, 319-355 (1943).
[10]E. A. Spiegel, "Some fluid dynamical problems in cosmogony," in: Symposium on the Origin of the Solar System (Nice, April 1972), Cen. Natl. Rech. Sci., Paris (1972), pp. 165-178 [Mir, Moscow (1976), p. 234].
[11]P. Glensdorf and I. Prigozhin, Thermodynamic Theory of Structure, Stability, and Fluctuations [in Russian], Mir, Moscow (1973).
[12]H. P. Berlage, "The disk theory of the origin of the solar system," Proc. K. Ned. Akad. Wet. 51, 796-806 (1948).
[13]D. ter Haar, "Some remarks on solar-nebula type theories of the origin of the solar system," op. cit.,[10] pp. 71-79.
[14]J. G. Hills, "Dynamic relaxation of planetary systems and Bode's law," Nature 225, 840-842 (1970).
[15]S. I. Ipatov, "Evolution of a flat ring of gravitating bodies coalescing through collisions" [in Russian], Preprint Inst. Prikl. Mat. Keldysha Akad. Nauk SSSR No. 101 (1978), 66 pp.
[16]G. W. Wetherill, "Formation of the terrestrial planets," Ann. Rev. Astron. Astrophys. 18, 77-113 (1980).
[17]I. L. Genkin and V. S. Safronov, "Gravitational instability in rotating systems with radial perturbations," Astron. Zh. 52, 306-315 (1975) [Sov. Astron. 19, 189-194 (1975)].

 0360-0327/82/03 0205-02 $03.00

Clearing of Planetesimal Disks around T Tauri Stars (abstract)

K. M. Guseinov
Translated from: Astron. Vestn. **22**, 319 (1988)
Sol. Syst. Res. **22**, 200 (1988)

We discuss the growth and collapse of dust condensations which form as a result of gravitational instability in a dusty subdisk. This work is set in the context of the widely accepted scenario for the evolution of a protoplanetary disk around the Sun. We estimate the time scale required for the aggregates to grow: this is necessary for clearing the hypothetical disks around young solar-type stars.

Formation of Planetary Systems during the Evolution of Close Binary Stars (abstract)

A. V. Tutukov
Translated from: Astron. Zh. **68**, 837 (1991)
Sov. Astron. **35**, 415 (1991)

Present-day scenarios of the formation of planetary systems around single stars and the products of component mergers in close binaries are described briefly. The frequency of occurrence of the various scenarios in our galaxy is estimated.

Soviet Publications on the Origin of the Solar System

(We thank Dr. Vladimir Kirsanov, and the staff of the Schmidt Institute of Earth Physics, for assistance in compiling this bibliography.)

ABBREVIATIONS

(For abbreviations not in this list, consult the reference in AJb or AAA.)
AAA=*Astronomy and Astrophysics Abstracts* 1969–. Supersedes AJb
Adv. Sp. Res.=*Advances in Space Research*
AJb=*Astronomisches' Jachresbericht* 1899–1968. Superseded by AAA
Ann. Astrophys.=*Annales d'Astrophysique*
Astrophys. Lett.=*Astrophysical Letters*
Astron. Nachr.=*Astronomische Nachrichten*
Astron. Tsirk.=*Astronomicheskii Tsirkulyar*. Byuro Astronomicheskiikh Soobshchenii Akademii Nauk SSSR
Astron. Vestn.=*Astronomicheskii Vestnik*
Astron. Zh.=*Astronomicheskii Zhurnal*
Astrophys. J.=*Astrophysical Journal*
Bull. Acad. Sci. Geophys. Ser.=*Bulletin of the Academy of Sciences, USSR, Geophysics Series*. Translation of Izv. Akad. Nauk SSSR, Ser. Geofiz. 1957–64. Pergamon Press. Superseded by Izv. Acad. Sci. Phys. Solid Earth
Byull. Abast. Astrofiz. Obs.=*Byulletin, Abastomanskaya Astrofizicheskaya Observatoriya*, Gora Kanobili. Akademiya Nauk Gruzinskoi SSR (Tiblisi)
Chem. Ab.=*Chemical Abstracts*
Cosmic Res. (USSR)=*Cosmic Research*. Translation of *Kosm. Issled.* Consultants Bureau, New York.
Dokl. Akad. Nauk Azerb. SSR=*Doklady Akademii Nauk Azerbaidzhanskoi SSR* (Baku)
Dokl. Akad. Nauk Kazakh. SSR=*Doklady Akademii Nauk Kazakhskoi SSR* (Alma-Ata)
Dokl. Akad. Nauk SSSR=*Doklady Akademii Nauk SSSR* (Ser. Mat., Fiz.)
Geochem. Cosmochem.=*Geochemistry and Cosmochemistry*
Geochem. Int.=*Geochemistry International*. Translation of *Geokhimiya*. Scripta Technica (subsidiary of Wiley), in cooperation with American Geological Institute and American Geophysical Union.
Geol. Zh.=*Geologicheskii Zhurnal*
Istoriko–Astron. Issled.=*Istoriko–Astronomicheskie Issledovaniya*
Itogi Naukii Tehk. Ser. Astron.=*Itogi Nauki i Tekhniki–Seriya Astronomiya*
Izv. Akad. Nauk Kazach. SSR=*Izvestiya, Akademii Nauk Kazakhskoi SSR*
Izv. Acad. Sci. Phys. Solid Earth=*Izvestiya Academy of Science, USSR—Physics of the Solid Earth*. Translation of Izv. Akad. Nauk Fiz. Zemli. American Geophysical Union.
Izv. Akad. Nauk SSSR Fiz. Zemli=*Izvestiya, Akademii Nauk SSSR, Fizika Zemli*
Izv. Akad. Nauk SSSR Ser. Geofiz.=*Izvestiya, Akademii Nauk SSSR, Seria Geofizicheskaya.*
Izv. Vsesoyuznogo Geograf. Obshch.=*Izvestiya Vsesoyuznogo Geograficheskogo Obshchestra*
J. Br. Astron. Assoc.=*Journal of the British Astronomical Association*
Kin. Fiz. Neb. Tel=*Kinematika i Fizika Nebesnykh Tel*, Akademiya Nauk Ukrainskoi SSR.
Kin. Phys. Cel. Bodies=*Kinematics and Physics of Celestial Bodies*, translation of Kin. Fiz. Neb. Tel.
Kosm. Issled.=*Kosmocheskie Issledovaniya*
Lun. Plan. Sci.=*Lunar and Planetary Science*, abstracts of papers presented at annual conferences on Lunar and Planetary Science, Houston, Texas.
Lun. Sci.=*Lunar Science*, abstracts of papers presented at annual conferences on Lunar Science, Houston, Texas. Superseded by Lun. Plan. Sci.
Math. Rev.=*Mathematical Reviews*
Mem. Soc. R. Sci. Liege=*Memoires de la Societe Royale des Sciences de Liege, Collection in 8°*
Mon. Not. R. Astron. Soc.=*Monthly Notices of the Royal Astronomical Society*, London.
Moon=*The Moon.* superseded by Moon Plan.
Moon Plan.=*Moon and Planets*, superseded by *Earth, Moon, and Planets*
Phys. Earth Plan. Int.=*Physics of the Earth and Planetary Interiors*
Pisma Astron. Zh.=*Pisma v Astronomicheskii Zhurnal*
Probl. Cosmog.=*Problems of Cosmogony* (Vols. 1, 5, and 7 only). Translation of Vopr. Kosmog. Published by U.S. Department of Commerce, Office of Technical Services, Joint Publications Research Service, May 1964 (Reports JPRS 24735; 24358; 24734).
Probl. Kosm. Fiz.=*Problemy Kosmicheskoi Fiziki*
Pokroky Mat. Fys. Astron.=*Pokroky Matematiky Fysiky a Astronomie*
Sol. Syst. Res.=translation of *Astronomicheskii Vestnik*. Consultants Bureau, New York.
Soobshch. Kharkov. Matemat. Obshch.=*Soobshcheniiya Kharkovshogo Matematicheskogo Obshchestra*
Sov. Astron. AJ=translation of Astron. Zh. American Institute of Physics. ["AJ" was dropped in 1973.]

Sov. Astron. Lett.=translation of Pisma Astron. Zh. American Institute of Physics.

Sov. Sci. Rev.=*Soviet Science Review: Scientific Developments in the USSR*, Vols. 1–3 (1970–2). Published by Iliffe Science and Technology Publications, Guildford, Surrey, England.

Theory Probab. Its Appl.=*Theory of Probability and its Applications*. Translated from *Teoriya Veroyatnostei i ee primeneniya*. Society for Industrial and Applied Mathematics, Philadelphia.

Tr. Geofiz. Inst. Akad. Nauk SSSR=*Trudy Geofizicheskii Institut, Akademii Nauk SSSR*

Tr. Gos. Astron. Inst. P. K. Shternberga Mosk.=*Trudy Gosudarstvennogo Astronicheskogo Instituta im P. K. Shternberga (Moscow)*

Tr. Inst. Fiz. Zemli Akad. Nauk SSSR=*Trudy Instituta Fiziki Zemli Adakemii Nauk SSSR*

Tr. Pervogo Sovesh. Vopr. Kosmog.=*Trudy Pervogo Soveshchanie po Voprosam Kosmogonii*

Ukr. Mat. Zh.=*Ukrainskii Matematischeskii Zhurnal*

Vestn. Akad. Nauk SSR=*Vestnik, Akademii Nauk SSR*

Vopr. Filos.=*Voprosy Filosofi*

Vopr. Ist. Estestvozn. Tekh.=*Voprosy Istorii Estestvoznaniya i Tekhniki*

Vopr. Kosmog.=*Voprosy Kosmogoni*

BIBLIOGRAPHY

Aliev, V. A. 1969. "The Titius-Bode rule and inclinations of the axes of rotation of the earth and planets." (in Russian) Dokl. Akad. Nauk Azerb. SSR **25** (7), 3–5.

Andreev, G. V., Terent'eva, A. K. and Bayuk, O. A. 1991. "Transformation of the orbits of small bodies by close encounters with the terrestrial planets." Sol. Syst. Res. **25**, 129–31. Translated from Astron. Vestn. **25**, 177–80. Abstract in Part IV.

Arsen'ev, A. S. 1954a. "The theoretical meaning of the cosmogonic ideas of Kant and Laplace" (in Russian). Dissertation, Moscow.

———. 1954b "On subjectivism in contemporary cosmogony" (in Russian). Priroda (Moscow) (6), 47–56.

———. 1955. "Some methodological problems of cosmogony" (in Russian). Vopr. Filos. (3), 32–44.

Artemev, A. V. 1969. "Planetary rotation induced by elliptically orbiting particles." Sol. Syst. Res. **3**, 15–21. Translated from Astron. Vestn. **3**, 18–25. Abstract in Part IV.

——— and Radzievsky, V. V. 1965. "The origin of the axial rotation of the planets." Sov. Astron. AJ **9**, 96–99. Translated from Astron. Zh., **42**, 124–28. Reprint in Part IV.

Ballakh, I. Ya. 1964/1968. "The role of explosive phenomena in cosmogonic processes." In *The Earth in the Universe*, edited by V. V. Fedynskii (Israel Program for Scientific Translations, Jerusalem, 1968), pp. 88–93. Translated from 1964 book.

Barabashov, N. P. 1953. *The Development of Views on the Cosmogony of the Solar System* (in Russian) (Izd-vo Khark. Univ. Gor'kogo, Kharkov).

———. 1955. "On the role of the study of the physical conditions of the moon and planets in cosmogony" (in Russian). Abh. Univ. Charkow **3** (55), 5–11. [German trans. of Russian journal title—reference from AJb.]

———. 1955. *On the origin of the earth and the other celestial bodies* (in Russian) Moscow, 107 pp. [Ref. from AJb.]

Baranov, A. S. 1990. "On a passage of particles through a ring resonant zone of a protoplanet cloud," in: *Astrophysical Processes and Structures in the Universe*, edited by D. Kiselman and C.-I. Lagerkvist (Uppsala Astronomiska Observatoriet, Uppsala), Report No. UAO-52, p. 7. [Reference from AAA.]

Baranov, V. I. 1969. "The age of bodies of the solar system" (in Russian). *Zemlya i Vselennaya* (1), 22–27. [Reference from AAA.]

——— and Knorre, K. G. 1969. "Consolidation and differentiation in the development of the solar system" (in Russian). In: *Proceedings of a Symposium on Meteorite Research, Vienna, 1968*, edited by P. M. Millman (Reidel, Dordrecht), pp. 31–40.

Barsukov, V. L. 1981. "Comparative planetology and the earth's early history." Geochem. Int. **18** (6), 1–12. Translated from Geokhimiya (11), 1603–14.

———. 1982. "Comparative planetology and the early history of the Earth" (in Russian) Vestn. Akad. Nauk SSSR (4), 52–4.

———. 1985. "Comparative planetology and early history of the earth" (in Russian). Geokhimiya (1), 3–19.

Bayuk, O. A. 1991. See Andreev *et al.*

Beletskii, V. V. 1983, "The Ehneev-Kozlov cosmogonic theory and evolution of rotations and accumulation of celestial bodies" (in Russian). *O. Yu. Schmidt i sov. geofiz goefiz. 80-kh Godov*, Moscow, pp. 234–38. [Reference from AAA.]

——— and Grushevskii, A. V. 1990. "Model of the Formation of the Rotational Motions of Celestial Bodies with Restrictions on the Order of Resonances." Sol. Syst. Res. **24**, 89–94. Translated from Astron. Vestn. **24**, 140–7.

Bel'kovich, O. I. and Ryabova, G. A. 1989. "Formation of the Geminid meteor stream with the disintegration of a comet nucleus." Sol. Syst. Res. **23**, 98–102. Translated from Astron. Vestn. **23**, 157–63. Abstract in Part VIII.

Bischoff, A. 1989. See Metzler and Bischoff

Borunov, S. P. 1985. See Dorofeeva *et al.*

Budtov, V. P. and Gladyshev, G. P. 1979. "On the processes of mass transport in rotating gas-dust clouds." Moon Plan. **20**, 213–8.

Burns, J. A. and Safronov, V. S. 1973. "Asteroid nutation angles." Mon. Not. R. Astron. Soc. **165**, 403–11.

Charadse, E. See Kharadse, E
Dadashov, A. S. 1989. See Guliev and Dadashov.
Dang, V. M. 1986. See Shukolyukov, Ya. A.
Denisik, S. A. 1978. See Ferronsky *et al.*
Divari, N. B. 1989. "Vassily Grigorievich Fesenkov—An outstanding astrophysicist (1889–1972). On the occasion of the 100th anniversary of his birthday" (in Russian). Vestn. Akad. Nauk SSSR (3), 101. Enlarged version in Istoriko-Astron. Issled. **21**, 302–26.
Djakov, B. B. and Reznikov, B. I. 1980. "Computer simulation of planet formation in a binary star system: Terrestrial planets." Moon Plan. **23**, 429–43.
——— and ———. 1981. "Computer simulation of the formation of asteroid belt structure." Moon Plan. **25**, 113–28.
Dorofeeva, V. A. 1984. "Physico-chemical model of the evolution of the protoplanetary cloud." *Reports of The 27th International Geological Congress, Moscow, 1984* (Nauka, Moscow), Vol. 5, Sec. 10-11, pp. 251–53. [Reference from AAA.]
Dorofeeva, V. A., Makalkin, A. B., and Borunov, S. P. 1985. "Distribution of Thermodynamical Parameters in the Protoplanetary Cloud" (in Russian). *Termodinam. i geol. I. Vses. Simp., Suzdal', Marta, 1985*. Chernogolovka, Tom 2, pp. 144–5. [Reference from AAA.]
Dorofeeva, V. A. 1985, 1991. See Makalkin and Dorofeeva.
———. 1985. See Mendybaev *et al.*
———. 1989. See Makalkin *et al.*
Drobyshevskii, E. M. 1978. "The origin of the solar system. Implications for trans-Neptunian planets and the nature of long-period comets." Moon Plan. **18**, 145–94.
———. 1988. "Planetary systems as the end- or by product of double-star formation." Sol. Syst. Res. **21**, 200–3. Translated from Astron. Vestn. **21**, 313–7.
Drubetskoy, Ye. R. 1991. See Levskii and Drubetskoy.
Dzhaparidze, D. R. 1985. See Kiladze and Dzhaparidze
Eneev [Ehneev], T. M. 1980. "On the equation of the accumulation process in the formation of planetary systems" (in Russian). Dokl. Akad. Nauk **253**, 69–73.
——— and Kozlov, N. N. 1980. "The problem of simulation of planetary systems accumulation processes." In: *Progress in Planetary Exploration* (COSPAR meeting, Budapest, 1980), edited by R. W. Shorthill *et al.*, pp. 201–15. Adv. Space Res. **1**, (8).
——— and ———. 1981. "Model of the accumulation process in the formation of planetary systems. I. Numerical experiments." Sol. Syst. Res. **15** (2), 59–70. Translated from Astron. Vestn. **15** (2), 80–94.
——— and ———. 1981/1982. "A model of the accumulation process in the formation of planetary systems. II. Rotation of the planets and the relation of the model to the theory of gravitational instability." Sol. Syst. Res. **15**, 97–104. Translated from Astron. Vestn. **15**, (3), 131–40.
——— and ———. 1983. "On the formation of the rotational motion of planets in a boundary model of the accumulation process" (in Russian). In: *O. Yu. Schmidt i sov. geofiz. 80-kh godov, Moskva*, pp. 228–33. [Reference from AAA.]
Fedotov, B. V. and Zelenev, V. M. 1984. "On regularities of formation of planetary systems." Voronezh. Gos. Ped. Inst. Voronezh, 28 pp. [Reference from AAA.]
Ferronsky, S. V. 1978. See Ferronsky, V. I. *et al.*
Ferronsky, V. I., Denisik, S. A., and Ferronsky, S. V. 1978. "The solution of Jacobi's virial equation for celestial bodies." Celest. Mech. **18**, 113–40.
Fesenkov, V. G. 1917. "On the angular momentum of the solar system from the point of view of the Laplace cosmogonic hypothesis." Soobshch. Kharkov. Matemat. Obshch. Ser. 2, **15**, (5-6), 278–87.
———. 1922. "Evolution du système solaire (partie 1ere) (Resume)." Astron. Nachr. **216**, 361–8.
———. 1930. "On the origin of the solar system" (in Russian). Astron. Zh. **7**, 130–51.
———. 1941. "The cosmogonic characteristics of the solar system" (in Russian). Usp. Astron. **2**, 67–100.
———. 1944. *The Cosmogony of the Solar System* (in Russian). (Akademiya Nauk SSSR, Moscow). Reviewed by O. Struve in Astrophys. J. **102**, 264–66 (1945).
———. 1945. "Sur l'origine du système solaire." Astron. Zh. **22**, 231–40 [Reference from AJb.]
———. 1948. "On the origin of the earth." (in Russian) *Priroda* (Moscow) **37**, (9), 5–16.
———. 1948/1954. "On the origin of meteorites." *Meteoritics* **1**, 208–27 (1954). Translated from Meteoritika **4**, 38 (1948).
———. 1949a. *Sovremenie Predstavleniya o Vselennoi* (Modern Views on the Universe) II (Izd-vo Akad. Nauk SSSR, Moscow). Includes a chapter on evolution of earth and planets. [Chapter titles are listed in AJb **49**, 17.]
———. 1949b. "Cosmogonic problems in contemporary astronomy" (in Russian). Astron. Zh. **26**, 67–83. [Abstract in AJb **49**, 112.]
———. 1950. *The Problem of the Development of the Earth and Planets* (in Russian). (Akademii Nauk Kazakhskoi SSR, Alma-Ata). [Chapter titles listed in AJb **50**, 123.]
———. 1951a. "On the origin of comets" (in Russian). Astron. Zh. **28**, 98–111. [Abstract in AJb **51**, 130.]
———. 1951b. "The criterion of tidal stability and its application in cosmogony" (in Russian). Astron. Zh. **28**, 492–517. [Abstracts in AJb **51**, 107 and Math. Rev. **13**, 498 (1952).]
———. 1951c. "The problem of the development of the earth and planets" (in Russian). Dokl. Akad. Nauk. Kazakh. SSR **104**, 19–30. [Reference from AJb.]

Fesenkov, V. G. 1952a. *Dnesni predstavy o vesmira*. (Osveta, Prague). Czechoslovakian translation of (1949a). [Reference from AJb.]

———. 1952b. "The origin of the planetary system" (in Czechoslovakian). Rise Hvezd **33**, 111–3, 123–6, 147–50. [Reference from AJb.]

———. 1952c. "The nature and possible origin of the meteorites, zodiacal light, and the asteroids." Prob. Cosmog. **1**, 89 (1964). Translated from Vopr. Kosmog. **1**, 92–130.

———. 1953. *Proiskhoshdenie i Razvitie nebesnykh tel po sovremennym Dannym* (Origin and development of the celestial bodies according to modern knowledge). (Akademii Nauk SSSR, Moscow). [Reference from AJb.]

———. 1956a. "Meteorites and their role in the cosmogony of the solar system" (in Russian). Astron. Zh. **33**, 767–77. [Abstract in AJb **56**, 321.]

———. 1956b. "The origin of the solar system and the problem of life in the universe" (in Russian). Izv. [?] Akad. Nauk Kazach. SSR (2), 3–13.

———. 1957a. "Meteorites and their significance for the cosmogony of the solar system" (in Czechoslovakian). Pokroky Mat. Fys. Astron. **2**, 588–97. [Reference from AJb.]

———. 1957b. "The early thermal history of the earth." Sov. Astron. AJ **1**, 112–23. Translated from Astron. Zh. **34**, 105–19.

———. 1958. "On the development of stars and the origin of planetary systems" (in Russian). Izv. Astrofiz. Inst. Alma-Ata **7**, 3–10. [Reference from AJb.]

———. 1960. "The origin of the solar system" (in Russian). Izdatel'stvo Znanie, Ser. 9, 1.

———. 1964. "Meteorites and the origin of the solar system" (in Russian). Priroda (Moscow) **53**, (10), 2–7.

———. 1965. "The significance of meteorites for the solution of the problem of the origin of the solar system" (in Russian). *Meteoritika* **26**, 69–76.

———. 1967. "Development and physical properties of the earth and planets as a consequence of the circumstances of their origin" (in Russian). In *Physics of the Planets* (in Russian), edited by W. G. Teifel ("Nauka" Kasakhskoi SSR, Alma-Ata), pp. 109–20. [Reference from AJb.]

———. 1970/1972. "On the origin of comets and their importance for the cosmogony of the solar system." In: *The Motion, Evolution of Orbits, and Origin of Comets* (IAU Symposium No. 45, Leningrad, 1970), edited by G. A. Chebotarev *et al.* (Reidel, Dordrecht, 1972), pp. 409–12.

——— and Massevitch, A. G. 1951/1953. "Zur Frage des Aufbaus und der chemischen Zusammensetzung der grosse Planeten." In: *Abhandlungen aus der Sowjetischen Astronomie und Astrophysik* (Verlag Kultur und Fortschritt, Berlin, 1953), Folge III, pp. 169–92. Translated from Astron. Zh. **28**, 317–37 (1951).

Fesenkov, V. G. See also Divari; Karyagina *et al.*; and Sitnik

Fesenkova, L. V. 1989. See Karyagina *et al.*

Frank-Kamenetskii, D. A. 1963. "Origin of solar-system elements." Priroda (Moscow) **52** (11), 17–26. [Reference from Chem. Abstrs.]

Fridman, A. M. 1972. See Polyachenko and Fridman.

———. 1990. See Gor'kavyi *et al.*

Galibina, I. V., Simonenko, A. N., and Levin, B. Yu. 1980. "A search for the last parent body of the Farmington meteorite" (in Russian). Meteoritika **39**, 114–20. English abstract in Meteoritics **19**, 64 (1984).

Genkin, I. L., and Safronov, V. S. 1975. "Gravitational instability in rotating systems with radial perturbations." Sov. Astron. **19**, 189–94. Translated from Astron. Zh. **52**, 306–15. Reprinted in Part III.

Gerasimov, M. V. 1979. "On mechanisms for shock degassing of planetesimals." Sov. Astron. Lett. **5**, 133–6. Translated from Pisma Astron. Zh. **5**, 251–6.

Gevorkian, O. S. 1956. "Bor'ba osnovnykh filosofskikh napravlenii v sovremennoi kosmogonii." In: *Voprosy dialekticheskogo i istoricheskogo materializma*, edited by P. N. Gapochka *et al.*, Moscow, pp. 244–82.

———. 1974. *Kosmogonicheskaya Gipoteza: Opyt istoriko-metodologicheskogo Issledovaniya* (Cosmogonical Hypotheses) (Nauka, Moscow).

Gladyshev, G. 1978. "The physicochemical mechanism of the formation of planetary systems." Moon Plan. **18**, 217–21.

———. 1978. "On the mechanism of reactions in rarified gas: Processes in the solar system." Moon Plan. **19**, 89–98.

——— and Budtov, V. P. 1981. "The solar system evolution." Moon Plan. **25**, 413–25.

Gladyshev, G. 1990. See Budtov and Gladyshev.

Gor'kavy, N. N., Polyachenko, V. L., and Fridman, A. M. 1990. "Dissipative instability of the protoplanetary disk and the law of planetary distances." Sov. Astron. Lett. **16**, 79–82. Translated from Pisma Astron. Zh. **16**, 183–90 (1990).

Grechinskii, A. D. 1989. "Analytical model of the evolution of the protoplanetary accretion disk." Sol. Syst. Res. **23**, 77–83. Translated from Astron. Vestn. **23**, 125–33. Abstract in Part III.

———. 1990. "Dust particle transport in protoplanetary accretion disks." Sol. Syst. Res. **24**, 85–89. Translated from Astron. Vestn. **24**, 134–9. Abstract in Part III.

Gruzinskaya, B. 1989. See Makalkin *et al.*

Guliev, A. S. and Dadashov, A. S. 1989. "Transplutonian comet families." Sol. Syst. Res. **23**, 57–62. Translated from Astron. Vestn. **23**, 88–96. Abstract in Part VIII.

Gurevich, L. E. and Lebedinskii, A. I. 1950a. "The formation of the planets. I. Gravitational condensation." Indian National Scientific Documentation Center, New Delhi, Report No. TR-1574/14378 (INSD). Translated from Izv. Akad. Nauk Ser. Fiz. **14**, 765–75. Reprinted with revisions in Part III.

Gurevich, L. E. and Lebedinskii, A. I. 1950b. "The formation of the planets. II. The law of planetary distances and the rotation of the planets." Indian National Scientific Documentation Center, New Delhi, Report No. TR-1575/14379 (INSD). Translated from Izv. Akad. Nauk Ser. Fiz. **14**, 776–89. Reprinted with revisions in Part III.

——— and ———. 1950c. "The formation of the planets. III (in Russian). Izv. Akad. Nauk Ser. Fiz. **14**, 790–9. Translation in Part III.

Guseinov, K. M. 1988. "Clearing of planetesimal disks around T Tauri stars." Sol. Syst. Res. **22**, 220–4. Translated from Astron. Vestn. **22**, 319–25. Abstract in Part III.

———. 1991. See Safronov and Guseinov.

Idlis, G. M. 1952. "Application of the tidal stability criterion to the problem of the distribution of planetary satellites" (in Russian). Astron. Zh. **29**, 556–62.

———. 1989. See Karyagina *et al.*

Ipatov, S. I. 1981a. "On the gravitational interaction of two planetesimals." Sov. Astron. **25**, 352–7. Translated from Astron. Zh. **58**, 620–9.

———. 1981b. "Computer modeling of the evolution of plane rings of gravitating particles moving around the sun." Sov. Astron. **25**, 617–23. Translated from Astron. Zh. **58**, 1085–94.

———. 1981c. "Some aspects of the spin formation of planets" (in Russian). Inst. Prikl. Mat. Akad. Nauk SSSR Report No. 102. [Reference from AAA.]

———. 1982. "Numerical investigations of the terrestrial planets' accumulation" (in Russian). Inst. Prikl. Mat. Akad. Nauk SSSR Report No. 144. Abstract in Part V.

———. 1983a. "Numerical investigations of a planar model of accumulation of nuclei of giant planets" (in Russian). Inst. Prikl. Mat. Akad. Nauk SSSR Report No. 117. Abstract in Part VII.

———. 1983b. "Axial rotations of accumulating planets" (in Russian). *O. Yu. Schmidt i sov. Geofiz. 80-kh Godov*, Moscow, pp. 239–43. [Reference from AAA.]

———. 1984. "Numerical investigations of the evolution of gravitating bodies' spatial rings corresponding to giant planets' feeding zones" (in Russian). Inst. Prikl. Mat. Akad. Nauk SSSR Report No. 1. Abstract in Part VII.

———. 1985. "Evolution of the eccentricities of orbits at the initial stage of solid-body accumulation of planets" (in Russian). Inst. Prikl. Mat. Akad. Nauk SSSR Report No. 4. [Reference from AAA.]

———. 1987a. "Accumulation and migration of the bodies from the zones of giant planets." Earth Moon Plan. **39**, 101–28.

———. 1987b. "Solid-body accumulation of terrestrial planets." Sol. Syst. Res. **21**, 129–35. Translated from Astron. Vestn. **21**, 207–15. Abstract in Part V.

———. 1988. "Evolution times for disks of planetesimals." Sov. Astron. **32**, 560–6. Translated from Astron. Zh. **65**, 1075–85. Reprinted in Part III.

———. 1989a. "Planetesimal migration during the last stages of accumulation of the giant planets." Sol. Syst. Res. **23**, 16–23. Translated from Astron. Vestn. **23**, 27–38. Abstract in Part VII.

———. 1989b. "Evolution of the orbital eccentricities of planetesimals during formation of the giant planets." Sol. Syst. Res. **23**, 119–25. Translated from Astron. Vestn. **23**, 197–206. Abstract in Part VII.

———. 1991a. "Evolution of initially highly eccentric orbits of the growing nuclei of the giant planets." Sov. Astron. Lett. **17**, 113–8. Translated from Pisma Astron. Zh. **17**, 268–80. Abstract in Part VII.

———. 1991b. "Possible migration of the giant planets' embryos." Lun. Plan. Sci. **22**, 607–8.

———. 1991c. "Computer simulation of the bodies migration in the forming solar system." In: *Abstracts of International Conference, "Origin and Evolution of the Solar System" (Moscow, August, 1991)*, p. 31.

Ivanov, A. V. 1986. See Shukolyukov *et al.*

Ivanovskiy, M. P. 1951. *Rozhdenie Mirov: Ocherk sovremennykh predstavlenii o vozniknovenii i razvitii solechnoi system* (The origin of the world: Contemporary ideas on the evolution of the solar system) (Molodoyae Gvardiya, Leningrad).

Izakov, M. N. 1979/1980. "Inert gases in the atmosphere of Venus, Earth and Mars and the origin of planetary atmospheres." *Cosmic Res.* **17**, 493–501. Translated from Kosm. Issled.

———. 1980. "Volatile matter in meteorites, in the protoplanetary nebula, and formation of planetary atmospheres." Cosmic Res. **18**, 651–62. Translated from Kosm. Issled. **18**, 918–32.

———. 1986. "Formation of solid materials in the preplanetary nebula and the composition of chondrites." Sol. Syst. Res. **20**, 22–32. Translated from Astron. Vestn. **20**, 35–49. Abstract in Part VIII.

———. 1987. "Refinement of the model for the formation of solid materials in the preplanetary disk." Sol. Syst. Res. **21**, 212–4. Translated from Astron. Vestn. **21**, 327–31.

Kagan, B. A. and Maslova, N. B. 1984. "Stochastic model of the tidal evolution of the earth-moon system" (in Russian). Dokl. Akad. Nauk SSSR Ser. Mat. Fiz. **276**, 88–91. [Reference from AAA.]

——— and ———. 1988. "Tidal evolution of the earth-moon system in the presence of random perturbations of the resonant frequency of the ocean." Sol. Sys. Res. **22**, 150–8. Translated from Astron. Vestn. **22**, 240–51.

Kalinina, G. V. 1990. See Kashkarov and Kalinina.

Kaplan, S. and Knyazhitsky, B. Ya. 1973. "Numerical experiment in the problem of accumulation of planets and their satellites from protoplanetary bodies." Sol. Syst. Res. **7**, 202–5. Translated from Astron. Vestn. **7**, 237–41.

Karyagina, Z. V., Idlis, G. M., Grigor'eva, N. B., and Fesenkova, L. V. 1989. "V. G. Fesenkov, 1989, January 1 (13)–1972, March 12" (in Russian). Zemlya i Vselennaya (1), 27–39. [Reference from AAA.]

Kashkarov, L. L. and Kalinina, G. V. 1990. "Pre-accretion irradiation of crystals in the Tieschitz H3.6 chondrules." Lun. Plan. Sci. **21**, 601–2. [See also the following abstracts by Kashkarov *et al.*]

Kasyutinski, V. 1961. See Vsekhsvyatskii and Kasyutinski.

Katterfel'd, G. N. 1962/1969. *The Face of the Earth and its Origin*. NASA Technical Translation F-533 (Clearinghouse for Federal Scientific and Technical Information, Springfield, VA), 1969. Translated from *Lik Zemli i ego Proiskhoshdenie* (Moscow, 1962).

Kazantsev, A. M. and Sherbaum, L. M. 1989. "The role of mutual collisions in the formation of resonance structures in the asteroid belt." Sol. Syst. Res. **23**, 135–8. Translated from Astron. Vestn. **23**, 220–5.

——— and ———. 1990. "Search for a mechanism for the origin of clearly defined hyperbolic meteor orbits." Sol. Syst. Res. **24**, 47–51. Translated from Astron. Vestn. **24**, 72–8. Abstract in Part VIII.

Kesarev, V. V. 1967. *Motive Forces in the Development of the Earth and Planets* (National Aeronautics and Space Admin., Washington, DC) (distributed by Clearinghouse for Federal Science and Technical Information, Springfield, VA). Translation of *Dvishushchie Sily Razvitiya Zemli i Planet* (Nedras, Leningrad).

Kessel'man, V. S. 1981. "The settling of grains in a contracting protoplanetary cloud." Sov. Astron. **25**, 33–7. Translated from Astron. Zh. **58**, 58–66. Abstract in Part III.

——— 1982. "Solid particles at an early stage of evolution of a protoplanetary cloud." Sov. Astron. **26**, 489–93. Translated from Astron. Zh. **59**, 810–6. Abstract in Part III.

Kharadse, E. 1955. *The Achievements of Soviet Cosmogony* (in Russian). Tiflis. [Reference from AJb **56**, 155 (1956).]

Khilmi, G. F. 1948. "On the possibility of capture in the three body problem" (in Russian). Dokl. Akad. Nauk SSSR **62**, 39–42.

———. 1951a. *Problema n tel v nebesnoi mekhanike i kosmogonii* (N body problem in celestial mechanics and cosmogony). (Izdvo Akademii Nauk, Moscow).

———. 1951b. "Evolution of a system of gravitating bodies by nonelastic collisions" (in Russian). Dokl. Akad. Nauk SSSR **77**, 589–92.

———. 1955. *Dvesti Let Nauchnoi Kosmogonii* (200 Years of Scientific Cosmogony). (Nauka, Moscow).

———. 1957. "Akademician Otto Yulevich Schmidt (obituary)" (in Russian). Fizika v shkole (1), 28–31.

———. 1958/1961. *Qualitative Methods in the Many-Body Problem*. (Gordon & Breach, New York, 1961). Translated from *Kachestvennye Metody v Probleme n tel* (Izd-vo Akademii Nauk, Moscow, 1958) by B. D. Seckler.

Khodakovskii, I. L., Mendybayev, R. A., and Lavrukhina, A. K. 1981. "On kinetic limits in physico-chemical models of the condensation process of the protoplanetary cloud" (in Russian). *Vses. Soveshch. po geokhimii ugleroda, Moskva, Dek., 1981* Tez. dokl. Moskva, pp. 300–4.

Khodakovskii, I. L. 1985, 1989. See Mendybaev *et al.*

Khramchikin, A. A. 1989. See Novikov and Khramchikin.

Khrilev, I. I. 1990. See Manzon *et al.*

Kiladze, R. I. 1977. "On the role of near-planetary particle swarms in the origin of spin" (in Russian). Byull. Abast. Astrofiz. Obs. (48) 191–212.

———. 1983. "Connection between the axial rotation of planets and the existence of protoplanetary swarms of small particles." (in Russian) *O. Yu. Schmidt io sov. geofiz. 80-kh godov, Moskva*, pp. 205–14. [Reference from AAA.]

——— and Dzhaparidze, D. R. 1985. "Influence of the oblateness of the early sun on the inclination of planetary orbits. I. Linear theory." Sol. Syst. Res. **19**, 22–8. Translated from Astron. Vestn. **19**, (4), 44–53.

Knorre, K. G. 1968/1969. See Baranov and Knorre.

Kochemasov, G. G. 1984. "Hypothesis of the Accretion of stratified planets of various compositions in the light of the discoveries of Venera 13 and 14, Voyager 1 and 2" (in Russian). *Report of the 27th International Geological Congress*, Vol. 5, Sec. 10-11, pp. 315–6. [Reference from AAA.]

Kostitzin, V. A., editor 1923. *Klassicheskie Kosmogonicheskie Gipotezy* (Classical Cosmogonic Hypotheses) (Gosizdat, Moscow).

Kozlov, N. N. 1980, 1981, 1981/1982. See Eneev and Kozlov.

Kozenko, A. V. 1989, 1990. See Zharkov and Kozenko.

Kramer, E. N. 1954. "On the role of the formation process of the planets in the formation of small bodies of the solar system" (in Russian). Publ. Astron. Obs. Stalinabad **4**, 114–24.

Krat, V. A. 1952/1964. "On the origin of the solar system." Probl. Cosmog. **1**, 31–88 (1964). Translated from Vopr. Kosmog. **1**, 34–91 (1952).

———. 1954. "On the mass of the sun at the time of the formation of the planets" (in Russian). Dokl. Akad. Nauk SSSR **95**, 481–4.

———. 1956. "The origin of the solar system" (in Russian). Mitt. Astron. Hauptobs. Pulkovo **20**, (3) (156), 1–15. Abstract in Astron. Jahresber. 1956.

———. 1957/1964. "The origin of the earth group of planets." Prob. Cosmog. **5**, 39–46 (1964). Translated from Vopr. Kosmog. **5**.

———. 1958/1964. "Process of formation of planets in the earth group." Prob. Cosmog. **7**, 69–70 (1964). Translated from Vopr. Kosmog. **7**.

———. 1960/1964. "Origin of the earth." Prob. Cosmog. **7**, 100–19 (1964). Translated from Vopr. Kosmog. **7**.

Krinov, E. L. 1982. "Problems and methods of investigation of cosmic matter on earth" (in Russian). Meteoritika **40**, 3–5. Abstract in Meteoritics **21**, 151 (1986).

Kropotkin, P. N. 1950. "O. Yu. Schmidt's cosmogonic theory and the structure of the earth" (in Russian). Izv. Akad. Nauk. SSSR **14**, (1), 37–63.

Kuskov, O. L. and Khitarov, N. I. 1982. "Initial stage of evolution of the earth, problems of geochemistry." Izv. Acad. Sci. USSR Phys. Solid Earth **18**, 447–56. Translated from Izv. Akad. Nauk SSSR Fiz. Zemli.

Kuskov, O. L. and Miklishanskiy, V. A. 1990. "Profiles of elastic properties and density for a chondritic model of a nondifferentiated earth." Izv. Phys. Solid Earth **25**, 739–41. Translated from Izv. Akad. Nauk SSSR Fiz. Zemli.

Kuyunko, N. S. 1985, 1989, 1990. See Mendybaev *et al.*

Kuznetsov, V. V. 1984. "Physics of the earth and of the solar system (models of formation and evolution)" (in Russian). Tr. Inst. Geol. Geofiz. SO Akad. Nauk SSSR (639) [Reference from AAA.]

Kvasov, D. D., Gal'tsev, A. P., and Safrai, A. S. 1980. "Influence of the greenhouse effect in primordial atmospheres on the formation of the earth and Venus" (in Russian). Astron. Vestn. **14**, 72–9. Abstract in Part V.

Lavrukhina, A. K. 1973. "On the differentiation of elements in the protoplanetary cloud" (in Russian). Meteoritika (32), 7–24.

———. 1981. "Cosmogenic isotopes in the early solar system." Bull. Acad. Sci. USSR Phys. Ser. **45** (4), 64–79. Translated from Izv. Akad. Nauk SSSR Ser. Fiz. **45** (4), 522–38.

———. 1983. "Characteristics of physi[c]ochemical processes of the evolution of cosmic matter" (in Russian). Meteoritika **42**, 3–22. Abstract in Meteoritics **22**, 165 (1987).

———. 1989. "Origin of chondrules." Geochem. Int. **27** (5), 26–35. Translated from Geokhimiya (10), 1407–16. Abstract in Part VIII.

Lavrukhina, A. K. 1985, 1989, 1990. See Mendybaev *et al.*

———. 1991. See Ustinova and Lavrukhina.

Lebedinsky, A. 1951a. "Sur la stabilite gravitationele d'une nebuleuse isotherme." Ann. Astrophys. **14**, 438–47.

———. 1951b. "Vystuplenie A. I. Lebedinskogo" (in Russian). *Trudy Pervogo Soveshchaniya po Voprosam Kosmogonii* (Izd-vo Akademii Nauk SSSR, Moscow), pp. 151–67

———. 1952. "Novel results in the teaching on the origin of planets" (in Russian). Izvestia, 17 February, p. 2.

——— 1953. "Recent Soviet theories on the origin of the solar system." J. Br. Astron. Assoc. **63**, 274–7.

———. 1958. "On the hydrogen content of the major planets." Probl. Cosmog. **7**, 56–9 (1964). Translated from Vopr. Kosmog. **7**. Reprinted in Part VII.

———. 1950. See Gurevich and Lebedinsky

Leliwa-Kopystynski, J. 1990. "Compaction of the icy/rocky granular mixtures and the physics of the icy satellites." In: *24th ESLAB Symposium: Formation of Stars and Planets and Evolution of the Solar System*, edited by B. Battrick (European Space Agency, Paris), pp. 97–102. [Abstract in AAA **52**(107), 030.]

Levin, B. Yu. 1946. "The Cosmogony of Jeans and contemporary astronomy" (in Russian). Priroda (Moscow) **35**(9), 3–10.

———. 1947. "The cosmogonic hypothesis of Academician Otto Julius Schmidt" (in Russian) Wiss. Leben [German title of Russian journal] (12), 2–6. [Reference from AJb.]

———. 1948. *Proiskhozhdenie Zemli i planet* (Origin of Earth and Planets) (Pravda, Moscow).

———. "The structure of the earth and planets and the meteorite hypothesis of their origin" (in Russian). Priroda (Moscow) **38**(10), 3–6.

———. 1952. "Novel results in the science of the origin of celestial bodies" (in Russian). Komsomolskaia Pravda, 1 March, p. 3.

———. 1953a "Cosmogony of the planetary system and the evolution of the sun." Dokl. Akad. Nauk SSSR **91**, 471–4. Also in Vopr. Kosmog. **3**, 20–32 (1954). Translation in Part III.

———. 1953b. "Some questions on the development, structure and consolidation of the earth" (in Russian). Dokl. Akad. Nauk SSSR Geofiz. (4), 289–306.

———. 1954. *Proiskhozhdenie Zemli i Planet* (Origin of Earth and Planets) (Gos. Izd-vo Tekhn.-Teoret. Lit., Moscow).

———. 1955. "Constitution of the earth" (in Russian). Tr. Geofiz. Inst. Akad. Nauk SSSR (26) (153), 11–38.

———. 1956. *Origin of the Earth and Planets* (Foreign Languages Publishing House, Moscow). Translated from 2nd Russian ed.

———. 1956/1957. "On the character and causes of the separation of molecules during planet formation." In: *Les Molecules dans les Astres, Communications presentees au septieme Colloque Internationale d'Astrophysique tenu à Liege, Juillet 1956*. Mem. Soc. Sci. Liege Collect. 4 **18**, 186–97 (1957).

———. 1957. "Origin and composition of the earth." Izv. Acad. Sci. USSR Geophys. ser. pp. 11–20. Translated from Izv. Akad. Nauk SSSR Ser. Geofiz. (11), 1323–31.

———. 1959a. "Creator of the theory of the origin of the earth" (in Russian). In: *Otto Iulevich Schmidt, Zhizn' i Deiatel'nost'* (Akademii Nauk SSSR, Moscow), pp. 64–94.

———. 1959b. "The development of planetary cosmogony." Space Technology Laboratories Report No. STL-TR-61-5110-37; ASTIA Document No. 264161. Translated from Priroda (Moscow) **48**(10), 19–26. Reprinted in Part III.

———. 1960/1962. "Thermal history of the moon." In: *The Moon* (IAU Symposium No. 14, Pulkovo Observatory, Dec. 1960), edited by Z. Kopal and Z. K. Mikhailov, pp. 157–67. (Academic Press, New York, 1962).

———. 1962a. "The origin of the solar system." *New Scientist* **13**, 323–5.

———. 1962b. "The origin of the earth, its structure and consolidation" (in Russian). Izv. [?] Akad. Nauk, SSSR (2), 9–16.

———. 1962c. "The origin of the earth: On the 70th birthday of O. Yu. Schmidt" (in Russian). Bull. Naturforsch. Ges. Moskau Abt. Geol. **37**: 159–60. [Reference from AJb.]

———. 1962d. "Comparative analysis of the internal constitution and development of planets." Mem. Soc. R. Sci. Liege [Ser. 5] **7**, spec. no., 39–46 ("Physics of Planets" symposium paper). Abstract in Part III.

Levin, B. Yu. 1963. "On the origin of comets" (in Russian). Vopr. Kosmog. **9**, 215f.

———. 1964a. "The problem of densities and composition of terrestrial planets in the light of modern ideas on the origin of meteorites." Icarus **3**, 498–9.

———. 1964b. *Proiskhozhdenie Zemli i Planet* (The Origin of Earth and Planets), 4th ed. (Nauka, Moscow).

———. 1966. "The structure of the moon." Sov. Astron. AJ **10**, 479–91. Translated from Astron. Zh. **43**, 606–21. Abstract in Part VI.

———. 1967a. "Accumulation in the solar nebula." International Dictionary of Geophysics, edited by S. K. Runcorn (Pergamon, Oxford), Vol. 1, pp. 5–8.

———. 1967. "Contemporary questions of planetary cosmogony" (in Russian) Zemlya i Vselennaya **3**(6), 49. [Reference from AJb.]

———. 1968/1969. "Origin of meteorites and planetary cosmogony." In: *Proceedings of a Symposium on Meteorite Research, Vienna, 1968*, edited by P. M. Millman (Reidel, Dordrecht, 1969), pp. 16–30.

———. 1968/1972. "Cosmogony of the planets" (in Russian). *Physics of the Moon and Planets* (in Russian), International Symposium in Kiev, 1968, edited by D. Ya. Martynov and V. A. Bronshtein (Nauka, Moscow, 1972), pp. 209–19.

———. 1969. "Four unsolved problems in planetary cosmogony" (in Russian). Priroda (Moscow) (6), 22–34.

———. 1970/1971. "Internal constitution and thermal histories of the terrestrial planets." In *Highlights of Astronomy* (IAU XIVth General Assembly, 1970), edited by C. de Jager (Reidel, Dordrecht, 1971), pp. 204–19.

———. 1972a. "Origin of the Earth." Tectonophys. **13**, 7–29. Also in *Developments in Geotectonics*, Vol. 4: *The Upper Mantle*, edited by A. R. Ritsema (Elsevier, Amsterdam), pp. 7–30.

———. 1972b. "The origin of the earth." Izv. Acad. Sci. Phys. Solid Earth 425–34. Translated from Izv. Akad. Nauk SSSR Fiz. Zemli (7), 5–21. Abstract in Part V.

———. 1972/1974. "Revision of initial size, mass and angular momentum of the solar nebula and the problem of its origin." In *L'Origin du System Solaire* (Symposium at Nice, 1972), edited by H. Reeves (Editions du CNRS, Paris, 1974), pp. 341–60.

———. 1976. "The cosmogonic hypothesis of Laplace: The history of its creation and publication" (in Russian). Vopr. Ist. Est. Tekh. **54**, 18–30.

———. 1978a. "Relative velocities of planetesimals and the early accumulation of planets." Moon Plan. **19**, 289–96.

———. 1978b. "Some problems concerning the accumulation of planets." Sov. Astron. Lett. **4**, 54–7. Translated from Pisma Astron. Zh. **4**, 102–7. Reprinted in Part III.

———. 1979. "Problems of planetary cosmogony" (in Russian). Zemlya i Vselennaya (3), 2–6.

———. 1982/1983. "Asteroids, comets, meteor matter—their place and role in the cosmogony of the solar system." Izv. Acad. Sci. Phys. Solid Earth **18**(6), 414–24 (1983). Translated from Izv. Akad. Nauk SSSR Fiz. Zemli (6), 25f. (1982). Abstract in Part VII.

——— and Lyubimova, E. A. 1955. "Thermal history of the moon" (in Russian). Priroda (Moscow) **44**(10), 81–4.

——— and Mayeva, S. V. 1960. "Thermal history of the earth." Bull. Acad. Sci. USSR Geophys. Ser. (2), 163–69. Translated from Izv. Akad. Nauk SSSR Ser. Geofiz. (2), 243–52.

——— and Mayeva, S. V. 1974/1977. "Riddles about the origin and thermal history of the moon." In: *The Soviet–American Conference on Cosmochemistry and the Moon and Planets*, Moscow, June, 1974, edited by J. H. Pomeroy and N. J. Hubbard (NASA, Washington, DC) Report No. NASA SP 370 (1977), pp. 367–88.

———. and Safronov, V. S. 1959. "Some statistical problems concerning the accumulation of planets." Theor. Probl. Appl. **4**, 220–1. Translated from Teor. Ver. Prim.

——— and Simonenko, A. N. 1984. "On the implausibility of cometary origin of most Apollo–Amor asteroids" (in Russian). Byull. Inst. Teor. Astron. **15**(6), 320–3. [Reference from AAA.]

Levin, B. Yu. 1980. See Galibana *et al.*

[Levin, B. Yu.] 1989a. "Obituary. Boris Yul'evich Levin (1912–1989)." Sol. Syst. Res. **23**, 2007–8. Translated from Astron. Vestn. **23**, 345–6.

[Levin, B. Yu.] 1989b. "Boris Yul'evich Levin (1912–1989)." Sov. Astron. Lett. **15**, 332. Translated from Pisma Astron. Zh. **15**, 767.

Levin, E. M. 1988. "Properties of the drop model of a protoplanetary Disk." Sov. Astron. **32**, 38–44. Translated from Astron. Zh. **65**, 73–85.

Levskii, L. K. 1978. "Meteorites, formation of the solar system, synthesis of elements" (in Russian). Vestn. LGU (24), 54–62. [Reference from AAA.]

———. 1983a. "Variable meaning of the [concept of] formation interval' [of meteorites]" (in Russian). Meteoritika **42**, 28–33. Abstract in Meteoritics **22**, 1965 (1987).

———. 1983b. "Isotope-cosmochemical models of formation of matter of the solar system" (in Russian). Geokhim. radiog. izot. na ran. stadiyakh. ehvol. Zemli, Moskva, pp. 5–24.

———. 1984. "Isotopes of rare gases and the origin of the earth and other planets" (in Russian?). *Reports, 27th International Geological Congress*, Vol. 5, Sec. 10-11, pp. 336–7.

———. 1984/1985. "The earth's interval of formation." Geochem. Int. **22**(4) 19–25 (1985). Translated from Geokhimiya (11), 1667–73 (1984).

Levskii, L. K. and Drubetskoy, Ye. R. 1988. "On the origin of SNC Meteorites" (in Russian). Meteoritika **47**, 134–37. Abstract in Meteoritics **26**, 253 (1991). Abstract in Part VIII.

Lissauer, J. J. and Safronov, V. S. 1991. "The random component of planetary rotation." Icarus **93**, 288–297. Abstract in Part IV.

Lupishko, D. F. and Velichko, F. P. 1991. "What is the cosmogonic meaning of the asteroid diameter 100–125 km?" Lun. Plan. Sci. **22**, 837–38.

Lupishko, D. F. 1991. See Velichko and Lupishko.

Lyustikh, E. N. 1977 see Vityazev *et al.*

Maeva, S. V. See Mayeva, S. V.

Makalkin, A. B. 1974. "Abundances of common substances in the protoplanetary cloud." Sov. Astron. **18**, 243–47. Translated from Astron. Zh. **51**, 417–24 (1974).

———. 1980. "Possibility of formation of an originally inhomogeneous earth." Phys. Earth Plan. Int. **22**, 302–12. Abstract in Part V.

———. 1987. "Thermal conditions of the protoplanetary disk." Sol. Syst. Res. **21**, 209–11. Translated from Astron. Vestn. **21**, 324–7. Abstract in Part III.

——— and Dorofeeva, V. A. 1985. "The Influence of *P-T* conditions and transport of dust particles in the protoplanetary cloud on the C/O ratio in the gaseous phase" (in Russian). *Termodinam. i geol. I. Vses. Simp., Suzdal', Marta, 1985*, Vol. 2, pp. 146–47. Chernoglovka. [Reference from AAA.]

——— and Dorofeeva, V. A. 1991. "Influence of solar radiation on temperature conditions in the solar nebula." Lun. Plan. Sci. **22**, 843–4.

———, Gruzinskaya, B., and Dorofeeva, V. A. 1989. "Temperature conditions in the preplanetary disk, implication for meteorites and planets," Meteoritics **24**, 297–8. (Abstracts of the 52nd Meeting of the Meteoritical Society, Vienna, July 1989.)

Makalkin, A. B. 1985. See Dorofeeva *et al.*; Mendybaev *et al.*

Manzon, B. M., Khrilev, I. L., and Yakovlev, O. I. 1990. "A model for differentiation during expansion of a vapor cloud in a gravitational field." Geochem. Int. **27** (9), 1–9. Translated from Geokhimiya. Abstract in Part VI.

Manzon, B. M. 1987. See Yakovlev *et al.*

Markova, O. M. 1987. See Yakovlev *et al.*

Marochnik, L. A. 1981. "On the origin of the solar system and the exceptional position of the sun in the galaxy." Astrophys. Space Sci. **89**, 61–75.

Maslova, N. B. 1984, 1988. See Kagan and Maslova.

Mayeva, S. V. 1969 "The thermal history of the terrestrial planets." Astrophys. Lett. **4**, 11–6.

———. 1971. "Thermal history of an iron-core earth." Bull. Acad. Sci. USSR Phys. Earth (1), 1–7; translated from Izv. Akad. Nauk SSSR Fiz. Zemli **1**, 3–12.

——— and Ruskol, E. L. 1977. "On the thermal history of Venus." Izv. Acad. Sci. USSR Phys. Solid Earth **13** (4), 239–42. Translated from Izv. Akad. Nauk. SSSR Fiz. Zemli **4**, 3–7.

Mayeva, S. V. 1984. See Pechernikova *et al.*

———. 1986. See Ruzmaikina and Mayeva.

Mendybayev, R. A., Kuyunko, N. S. and Lavrukhina, A. K. 1985. "Physico-chemical analysis of the stability of interstellar carbon in the protoplanetary cloud" (in Russian). *Termodinam. i geol. I. Vses. simpl., Suzdal', 12–14 Marta, 1985*, Vol. 2, pp. 142–3. Chernogolovka. [Reference from AAA.]

———, ———, and ———. 1989. "Interaction of Fe, Ni-metal with preplanetary nebula gases (H_2O, H_2S, CO, CO_2); physicochemical aspect." Meteoritics **24**, 303. (Abstracts of the 52nd Meeting of the Meteoritical Society, Vienna, July 1989.)

———, ———, ———, and Khodakovskiy, I. L. 1989. "Stability of interstellar carbon in the protoplanetary nebula." Geochem. Int. **26** (11), 1–10. Translated from Geokhimiya (4), 467–77.

Mendybaev, R. A., Kukuyenko, N. S., Mironenko, M. V., and Lavrukhina, A. K. 1990. "On the origin of the carbon-rich aggregates in the ordinary chondrites." Lun. Plan. Sci. **21**, 785–6.

Mendybayev, R. A., Makalkin, A. B., Dorofeyeva, V. A., Khodakovskiy, I. L., and Lavrukhina, A. K. 1985. "The role of CO and N_2 reduction kinetics in the chemical evolution of the protoplanetary cloud." Geochem. Int. **23**, 105–16. Translated from Geokhimiya (8), 1206–17.

Mendybaev, R. A. 1981. See Khodakovskii *et al.*

Metzler, K. and Bischoff, A. 1989. "Formation of accretionary dust mantles in the solar nebula as confirmed by noble gas data of CM chondrites." Meteoritics **24**, 303–4.

Miesserov, K. G. 1982a. "Physico-chemical approach to the problem of origin of the solar system" (in Russian). *Inst. Neftekhim. sinteza AN SSSR*, Moskva.

———. 1982b. "Formation of solar system as a result of evolution of protostar of second or subsequent generation." Moon Plan. **27**, 13–25.

Miklishanskiy, V. A. 1990. See Kuskov and Miklishanskiy.

Mironenko, M. V. 1990. See Mendybaev *et al.*

Myasnikov, V. P. and Titarenko, V. I. 1989a. "Evolution of self-gravitating clumps of a gas-dust nebula participating in the accumulation of planetary bodies." Sol. Syst. Res. **23**, 7–15. Translated from Astron. Vest. **23**, 14–26. Abstract in Part III.

Myasnikov, V. P. and Titarenko, V. I. 1989b. "Evolution of a self-gravitating gas-dust clump with allowance for radiative transfer in a diffusional approximation." Sol. Syst. Res. **23**, 126–35. Translated from Astron. Vestn. **23**, 207–19. Abstract in Part III.

Nikolaichik, V. V. 1957. See Vityazev and Nikolaichik.

Nikolajeva, E. V. 1975. See Ruskol *et al.*

Novikov, V. V. and Khramchikhin, A. A. 1989. "On the primary centers of accretion in the solar system" (in Russian). Tr. Gos. Astron. Inst. P. K. Shternberga Mosk. **61**, 316–29. [Abstract in AAA **52**, #107.035 (1990).]

Ogorodnikov, K. F. 1949. "On the probability of stellar encounters in the cosmogony of Jeans" (in Russian). Dokl. Akad. Nauk SSSR **66**, 357–8.

Pariyskiy, N. N. 1943. "On the origin of the solar system. Solution of the Russell problem" (in Russian). Russ. Astron. Zh. **20** (2), 9–29. Also in Astron. Circ. USSR (32) (1944), 1 [Reference from AJb.]

———. 1944a. "On the origin of the solar system. II. Classification of orbits in restricted hyperbolic three-body problem (Russell problem)" (in Russian). Russ. Astron. Zh. **21**, 71–9. Also in Astron. Tsirk. (33), 4 (1944). [Reference from AJb.]

———. 1944b. "On the origin of the solar system and tidal action on the size of the planet's orbits" (in Russian). Astron. Tsirk. (34), 1. [Reference from AJb.]

———. 1946. "Problems of cosmogony of the solar system" (in Russian). Publ. Kiev Astron. Obs. (1), 197–200. [Reference from AJb.]

———. 1955. "On the sun's angular momentum" (in Russian). Vopr. Kosmog. **4**, 5–33.

Pariyskiy, N. N. 1941. See Rein and Pariyskiy.

Pechernikova, G. V. 1974. "Mass distribution of protoplanetary bodies. I. Initial Data for Numerical Solution." Sov. Astron. **18**, 778–83. Translated from Astron. Zh. **52**, 1305–15 (1974). Reprinted in Part III.

———. 1986. "On the asymptotic mass spectrum of interacting particles" (in Russian). Astron. Tsirk. (1468), 5–8. Abstract in AAA 44.061.156.

———. 1987. "On the intermediate mass spectrum asymptotics in the system of coagulating bodies." Kin. Phys. Cel. Bodies **3** (5), 98–100. Translated from Kin. Fiz. Neb. Tel **3** (5), 85–7. Abstract in AAA **44**, #131.083.

——— and Maeva, S. V. 1985. "Fractionation of matter during the formation of the earth-moon system" (in Russian). *Termodinam. i geol. I Vses. Simp., Suzdal', Marta, 1985*, Vol. 1, pp. 203–4. Chernogolovka [Reference from AAA.]

———, ———, and Vityazev, A. V. 1984. "Protosatellite swarm dynamics." Sov. Astron. Lett. **10**, 293–7. Translated from Pisma Astron. Zh. 702–9. Reprinted in Part VI.

Pechernikova, G. V., Safronov, V. S., and Zvyagina, E. V. 1976. "Mass distribution of protoplanetary bodies. II. Numerical solution of generalized coagulation equation." Sov. Astron. **20**, 346–50. Translated from Astron. Zh. **53**, 612–9 (1976). Reprinted in Part III.

Pechernikova, G. V. and Vityazev, A. V. 1979. "Mass of the largest bodies and the velocity dispersion during the accumulation of planets." Sov. Astron. Lett. **5**, 31–4. Translated from Pisma Astron. Zh. **5**, 54–9. Reprinted in Part III.

——— and ———. 1980. "Evolution of orbital eccentricities of the planets in the process of their formation." Sov. Astron. **24**, 460–7. Translated from Astron. Zh. **57**, 799–811. Reprinted in Part III.

——— and ———. 1981. "Thermal dissipation of gas from the protoplanetary cloud." In: *Planetary Interiors* (COSPAR meeting in Budapest, June, 1980), edited by H. Stiller and R. Z. Sagdeev, (Pergamon, New York), pp. 55–60 (Advances in Space Research, Vol. 1, No. 7).

——— and ———. 1987. "Erosion of Mercury silicate shell during its accumulation." Lun. Plan. Sci. **18**, 770–1. Abstract in Part V.

——— and ———. 1988. "Evolution of dust clusters in the preplanetary disk." Sov. Astron. **32**, 31–8. Translated from Astron. Zh. **65**, 58–72. Reprinted in Part III.

Pechernikova, G. V. 1981, 1982, 1983, 1985, 1988, 1989. See Vityazev and Pechernikova.

———. 1986. See Safronov *et al.*

Polyachenko, V. L. 1989. "On the resonant nature of the planetary distances' law" (in Russian). Astron. Tsirk. (1538), 27–8. Abstract in AAA **50**, #107.027.

——— and Fridman, A. M. 1972. "The law of planetary distances." Sov. Astron. **16**, 123–8. Translated from Astron. Zh. **49**, 157–64.

Polyachenko, V. L. 1990. See Gor'kavyi *et al.*

Potapov, I. I. 1962 "On the origin of the earth" (in Russian). Hochschulnachr. Geol. [German trans. of Russian journal title] (1), 3–18. [Reference from AJb.]

Radzievskii, V. V. 1950. "Planetocentric radiation braking effect." Report No. 62-10800 (Clearinghouse for Federal Sci. and Tech. Info, Springfield, VA) Translation from Dokl. Akad. Nauk SSSR **74**, 197–200.

———. 1952a. "Origin of the moon in the light of the cosmogonic theory of O. Schmidt." Bull. Astron. Geol. Ges. USSR **11** (18), 3–8. [Reference from AJb.]

———. 1952b. "Novel results in the teaching on the origin and development of the solar system" (in Russian). *Bloknot Agitatora* (Yaroslavl'), No. 9 (63).

———. 1953. "Origin of the protoplanetary cloud according to the cosmogonical theory of O. Yu. Schmidt" (in Russian). Dokl. Akad. Nauk SSSR **90**, 517–20.

Razbitnaya, E. P. 1954. "On the origin of the moon" (in Russian). Dissertation, Leningrad State Pedagogical Institute.

Rein, N. F. and Pariyskiy, N. N. 1941a. "The present state of the methodological apparatus of dynamical cosmogony" (in Russian). Usp. Astron. Nauk **2**, 5–66.

Rein, N. F. and Pariyskiy, N. N. 1941b. "The catastrophe hypothesis of the origin of the solar system" (in Russian). Usp. Astron. Nauk **2**, 137–56.

Rudnitskii, G. M. 1987. "Observational display of protoplanetary disks around young stars." Sol. Syst. Res. **21**, 198–9. Translated from Astron. Vestn. **21**, 311–13.

Ruskol, E. L. 1958/1964. "The formation of a protoplanet." Probl. Cosmog. **7**, 5–13. Translated from Vopr. Kosmog. **7**. Reprinted in Part III.

———. 1960. "The Origin of the moon. I. Formation of a swarm of bodies around the earth." Sov. Astron. AJ **4**, 657–68. Translated from Astron. Zh. **37**, 690–702. Reprinted in Part VI.

———. 1960/1962. "The origin of the moon." In: *The Moon*, IAU Symposium No. 14 (Pulkovo Observatory, December, 1960), edited by Z. Kopal and A. K. Mikhailov (Academic, New York, 1962), pp. 149–55.

———. 1963a. "On the origin of the moon. II. The growth of the moon in the circumterrestrial swarm of satellites." Sov. Astron. AJ **7**, 221–7. Translated from Astron. Zh. **40**, 288–96. Reprinted in Part VI.

———. 1963b. "The tidal evolution of the earth-moon system." Bull. Acad. Sci. Geophys. Ser. (2), 129–33. Translated from Izv. Akad. Nauk SSSR Ser. Geofiz. (2), 216–22. Abstract in Part VI.

———. 1966a. On the past history of the earth-moon system. Icarus **5**, 221–7. Abstract in Part VI.

———. 1966b. "The tidal history and origin of the earth-moon system." Sov. Astron. AJ **10**, 659–65. Translated from Astron. Zh. **43**, 829–36. Reprinted in Part VI.

———. 1971a. "The origin of the moon. III. Some aspects of the dynamics of the circumterrestrial swarm." Sov. Astron. AJ **15**, 646–54 (1972). Translated from Astron. Zh. **48**, 819–29. Reprinted in Part VI.

———. 1971b. "Possible differences in the chemical composition of the earth and moon, for a moon formed in the circumterrestrial swarm." Sov. Astron. AJ **15**, 1061–3. Translated from Astron. Zh. **48**, 1336–8. Reprinted in Part VI.

———. 1972a. "Formation of the moon from a cluster of particles encircling the earth." Izv. Acad. Sci. Phys. Solid Earth (7), 483–8. Translated from Izv. Akad. Nauk SSSR Fiz. Zemli (7), 99–108. Abstract in Part VI.

———. 1972b. "On the initial distance of the moon forming in the circumterrestrial swarm." In: *The Moon*, edited by S. K. Runcorn and H. C. Urey (IAU Symposium No. 47) (Reidel, Dordrecht), pp. 402–4. Abstract in Part VI.

———. 1972c. "On the possible differences in the bulk chemical composition of the earth and the moon forming in the circumterrestrial swarm." *Ibid.*, pp. 426–28. Abstract in Part VI.

———. 1972d. "The role of the satellite swarm in the origin of the earth's rotation." Sol. Syst. Res. **6**, 80–3. Translated from Astron. Vestn. **6**, 91–5.

———. 1972e. "Cosmogony of the moon" (in Russian). In: *Fizika Luny i Planet*, edited by D. Ya. Martynov and V. A. Bronshten (Nauka, Moscow), pp. 160–7.

———. 1973a. "On the model of the accumulation of the moon compatible with the data on the composition and the age of lunar rocks." Moon **6**, 190–201. Abstract in Part VI.

———. 1973b. "Tidal changes in the orbital inclinations of the satellites of Uranus relative to its equatorial plane." Sol. Syst. Res. **7**, 131–3. Translated from Astron. Vestn. **7**, 150–3. Abstract in Part VI.

———. 1974. "The origin of the moon." In: *Soviet–American Conference on Cosmochemistry of the Moon and Planets* Moscow, 1974, edited by J. H. Pomeroy and N. J. Hubbard (NASA, Washington, DC, 1977), pp. 815–22 Report No. NASA SP 370.

———. 1975. *Origin of the Moon* (NASA, Washington, DC) Report No. NASA TT 16,623. Translated from *Proiskhozhdeniye Luny* (Nauka, Moscow).

———. 1979. "Thermal effect of the collision of two massive bodies and the initial temperature of the moon." Lun. Plan. Sci. **10**, 1042–4. Abstract in Part VI.

———. 1981. "Formation of Planets." In: *The Solar System and Its Exploration*, edited by W. R. Burke, pp. 107–13. Abstract in Part III.

———. 1982. "Origin of planetary satellites." Izv. Acad. Sci. Phys. Solid Earth **18**, 425–33. Translated from Izv. Akad. Nauk SSSR Fiz. Zemli (6), 40–51. Abstract in Part VI

———. 1986. *Natural Satellites of Planets* (in Russian). (Vsesoyuznyi Institut Nauchn. i Techn. Informatsi, Moscow) [Reference from AAA.]

———. 1990. "The origin of Trojan asteroids." Sol. Syst. Res. **24**, 157–9. Translated from Astron. Vestn. **24**, 244–7. Abstract in Part VIII.

———, Nikolajeva, E. V., and Syzdykov, A. S. 1975. "Dynamical history of coplanar two-satellite systems." Moon **12**, 11–8. Translated from Moon **12**, 3–10. Abstract in VI

Ruskol, E. L. and Safronov, V. S. 1961. "Origin of rapidly rotating asteroids." Sov. Astron. AJ **5**, 203–5. Translated from Astron. Zh. **38**, 273–7. Reprinted in Part VIII.

——— and ———. 1977. "Formation of satellites near giant planets." Lun. Sci. **8**, 820–2. Abstract in Part VI.

Ruskol, E. L. 1963. See Safronov and Ruskol.

———. 1977. See Safronov and Ruskol; Mayeva and Ruskol.

———. 1986. See Safronov *et al.*

Ruzmaikina, T. V. 1981a. "The role of the magnetic field and turbulence in the evolution of the presolar nebula." In: *Planetary Interiors*, edited by H. Stiller and R. Z. Sagdeev (Pergamon, New York), pp. 49–53 (Advances in Space Research, Vol. 1, No. 7).

———. 1981b. "The angular momentum of a protostar engendering a protoplanetary disk." Sov. Astron. Lett. **7**, 104–7. Translated from Pisma Astron. Zh. **7**, 188–92. Reprinted in Part IX.

Ruzmaikina, T. V. 1985. "The magnetic field in the collapsing protosolar nebula." Sol. Syst. Res. **19**, 65–73. Translated from Astron. Vestn. **19**, 101–12. Abstract in Part III.

———. 1988. "Formation of a protoplanetary disk." Sol. Syst. Res. **21**, 192–5. Translated from Astron. Vestn. **21**, 303–8.

——— and Maeva, S. V. 1986. "Process of formation of the protoplanetary disk." Sol. Syst. Res. **20**, 132–43. Translated from Astron. Vestn. **20**, 212–26. Abstract in Part III.

Ruzmaikina, T. V. and Safronov, V. S. 1976. "Transfer of angular momentum in the protoplanetary cloud by meridional currents." Sov. Astron. **20**, 486–9. Translated from Astron. Zh. **53**, 860–86.

——— and ———. 1986. "Premature particles in the solar nebula." Lun. Plan. Sci. **17**, 720–1. Abstract in Part III.

Ruzmaikina, T. V., Safronov, V. S. and Weidenschilling, S. J. 1989. "Radial mixing of material in the asteroidal zone." In: *Asteroids II*, edited by R. P. Binzel, T. Gehrels, and M. S. Matthews (University of Arizona Press, Tucson), pp. 681–700.

Ruzmaikina, T. V. 1991. See Ziglina and Ruzmaikina.

Ryabova, G. O. 1989. "Effect of secular perturbations and the Poynting-Robertson effect on structure of the Geminid meteor stream." Sol. Syst. Res. **23**, 158–65. Translated from Astron. Vestn. **23**, 254–64.

——— 1989. See Bel'kovich and Ryabova.

Safronov, V. 1951a. "Decrease in the sun's rotational momentum in connection with the shrinking of its mass in the process of evolution" (in Russian). Astron. Zh. **28**, 244–52. Long abstract in Probl. Cosmog. **1**, 268–70.

———. 1951b. "Problems of the origin of the earth and planets (in Russian)." Vestn. Akad. Nauk SSSR (10), 94–102.

———. 1951c. "Meeting on questions of cosmogony of the solar system" (in Russian). Astron. Zh. **28**, 535–43. Abstract in Astron. News Lett. **62**, 8–9. [Reference from AJb.]

———. 1954. "On the growth of planets in a protoplanetary cloud." (in Russian) Astron. Zh. **31**, 499–510.

———. 1955a. "The change in the sun's rotation resulting from material falling on it due to the Poynting-Robertson effect." English Translation No. 61-23076, John Crerar Library. Translated from Dokl. Akad. Nauk SSSR **105**, 1184–7.

———. 1955b. "Density of matter in the vicinity of the sun and the problem of star formation" (in Russian). Priroda (Moscow), **44** (12), 74–76. Abstract in AJb **55**, 465.

———. 1958a. "On the growth of the planets of the earth group" (in Russian). Vopr. Kosmog. **6**, 63–77. Abstract in Part V.

———. 1958b. "On the turbulence in the protoplanetary cloud." Rev. Mod. Phys. **30**, 1023–4. [Proceedings of the 3rd Symposium on Cosmical Gas Dynamics, Smithsonian Astrophysical Observatory, Cambridge, June, 1957; IAU Symposium No. 8) Reprinted in Part III.

———. 1959a. "On the primeval temperature of the earth." Bull. Acad. Sci. Geophys. Ser. (1), 85–9. Translated from Izv. Akad. Nauk SSSR Ser. Geofiz. (1), 139–43.

———. 1959b. "The results of planetary cosmogony" (in Russian). Priroda (Moscow) **48** (7), 58–60. [Reference from AJb.]

———. 1958/1964. "Accumulation of planets of the earth group." Probl. Cosmog. **7**, 63–70. Translated from Vopr. Kosmog. **7**, 59f. Reprinted in Part V.

———. 1960/1964. "The formation of protoplanetary dust clouds." Probl. Cosmog. **7**, 120–45 (1964). Translated from Vopr. Kosmog. **7**, 121f. (1960) Reprinted in Part III.

———. 1960. "On the gravitational instability in flattened systems with axial symmetry and nonuniform rotation." Ann. Astrophys. **23**, 979–82. Translated from Dokl. Akad. Nauk SSSR **130**, 53–6.

———. 1962a. "The temperature of the dust component of the protoplanetary cloud." Sov. Astron. AJ **6**, 217–25. Translated from Astron. Zh. **39**, 278–89. Reprinted in Part III.

———. 1962b. "A particular case of the solution of the coagulation equations" (in Russian). Dokl. Akad. Nauk SSSR **147**, 64–7.

———. 1962c. "On the problem of the rotation of planets" (in Russian). Vopr. Kosmog. **8**, 150–67. Abstract in Part IV.

———. 1962d. "On the velocity dispersion in rotating systems of gravitating bodies with inelastic collisions" (in Russian). Vopr. Kosmog. **8**, 168–79. Translation in Part III.

———. 1964a. "On gravitational instability and further development of perturbations" (in Russian). Vopr. Kosmog. **10**, 181. Abstract in Part III.

———. 1964b. "The primary inhomogeneities of the earth's mantle." Tectonophys. **1**, 217–21.

———. 1965a. "Original inhomogeneities in the Earth's mantle." Izv. Acad. Sci. Phys. Solid Earth (7), 425–9. Translated from Izv. Akad. Nauk SSSR Ser. Geofiz. (7), 1–8.

———. 1965b. "Sizes of the largest bodies falling on the planets during their formation." Sov. Astron. AJ **9**, 987–91. Translated from Astron. Zh. **42**, 1270–6. Reprinted in Part IV.

———. 1966. "The protoplanetary cloud and its evolution." Sov. Astron. AJ **10**, 650–8. Translated from Astron. Zh. **43**, 817–28. Reprinted in Part III.

———. 1969. "Relative sizes of the largest bodies during the accumulation of planets." Icarus **10**, 109–15.

———. 1969/1972. *Evolution of the Protoplanetary Cloud and Formation of the Earth and Planets*. (Israel Program for Scientific Translations, Jerusalem), Document No. NASA TT-F-677 (1972). Translated from *Evolyutsiya doplanetnogo oblaka i obrazovanie zemli i planet* (Izdatel'stvo "Nauka", Moscow, 1969).

———. 1971a. "Rotation of giant planets while accreting gas." Sol. Syst. Res. **5**, 139–44. Translated from Astron. Vestn. **5**, 167–73.

———. 1971b. "O. Iu. Schmidt and Cosmogony" (in Russian). Ukr. Mat. Zh. **23**, 707–16.

———. 1972a. "The initial state of the earth and certain features of its evolution." Izv. Acad. Sci. Earth Phys. 444–7. Translated from Izv. Akad. Nauk SSSR Fiz. Zemli (7) 35–41.

———. 1972b. "Development of Schmidt's theory" (in Russian). Zemlya i Vselennaya **4**, 18–23.

Safronov, V. 1972c. "Ejection of bodies from the solar system in the course of the accumulation of the giant planets and the formation of the cometary cloud." In: *The Motion, Evolution of Orbits, and Origin of Comets*, IAU Symposium No. 45, edited by G. A. Chebotarev *et al.* (Reidel, Dordrecht), pp. 329–34.

———. 1972d. "Accumulation of the planets." In: *On the Origin of the Solar System* (Symposium, Nice, 1972), edited by H. Reeves (CNRS, Paris), pp. 89–113 (reprinted 1974).

———. 1972e. "On the mass transport in the model of solar nebula by A. G. W. Cameron." In: *On the Origin of the Solar System* (Symposium, Nice, 1972), edited by H. Reeves (CNRS, Paris), pp. 361–6 (reprinted 1974).

———. 1974. "Time scale for the formation of the earth and planets and its role in their geochemical evolution." In: *The Soviet–American Conference on Cosmochemistry of the Moon and Planets*, Moscow, 1974, edited by J. H. Pomeroy and N. J. Hubbard, (NASA, Washington, DC), pp. 797–803, Report No. NASA SP 370 (1977). See also "Dlitel'nost' protsessa formirovaniya zemli i planet i ee rol' v ikh geokhimicheskoi evolutsii," in *Kosmokhimiya Luny i Planet* (Nauka, Moscow, 1976), pp. 624f. Abstract in Part V.

——— 1976/1977. "Oort's cometary cloud in the light of modern cosmogony." In: *Comets, Asteroids, Meteorites* (IAU Colloquium No. 39, Lyon, August 1976), edited by A. H. Delsemme (University of Toledo, Toledo, Ohio, 1977), pp. 483–4.

———. 1978. "The heating of the earth during its formation." Icarus **33**, 3–12.

———. 1979. "On the origin of asteroids." In: *Asteroids*, edited by T. Gehrels and M. S. Matthews Tucson: (University of Arizona Press, Tucson), pp. 975–91.

———. 1980a. "Accumulation of the protoplanetary bodies." In: *Early Solar System Processes and the Present Solar System*, edited by D. Lal (North-Holland, Amsterdam), pp. 58–72.

———. 1980b. "Some problems in evolution of the solar nebula and of the protoplanetary cloud." In: *Early Solar System Processes and the Present Solar System*, edited by D. Lal (North-Holland, Amsterdam), pp. 73–81.

———. 1981. "Initial state of the earth and its early evolution." In: *Evolution of the Earth* (Geodynamics Series, Vol. 5), edited by R. J. O'Connell and W. S. Fyfe (Geological Society of America, Boulder, Colorado), pp. 249–55.

———. 1982. "State-of-the-art for the theory of the origin of the earth." Izv. Acad. Sci. Phys. Solid Earth **18**, 399–413. Translated from Izv. Akad. Nauk SSSR Fiz. Zemli (6), 5–24. Abstract in Part V.

[Safronov, V. S.] Völk, H. J., editor 1982. "Diskussionsforum: Ursprung des Sonnensystems." Mitt. Astron. Ges. (57), 45–63. Includes contribution by Safronov.

Safronov, V. S. 1983. "The Development of Soviet Planetary Cosmogony" (in Russian). *O. Yu. Schmidt i Sovetskaya Geofizika 80-kh Godov* (Izdatalstvo "Nauka," Moscow), pp. 41–57. Translation in Part III.

———. 1984a. "Current problems of the cosmogony of the solar system." Sol. Syst. Res. **18**, 208–20. Translated from Astron. Vestn. **18**, 322–41.

———. 1984b. "Formation of the solar system." In *Planetary Rings*, edited by A. Brahic (IAU Colloquium No. 75), (Cepadues-Editions, Toulouse), pp. 647–59.

———. 1987a. *Proiskoshdenie Zemli* Nauki o Zemli, Podlisnaya nauchnopopulyarnaya Seria, No. 12. 48 pp.

———. 1987b. "Evolution of the dusty component of the circumsolar protoplanetary disk." Sol. Syst. Res. **21**, 135–8. Translated from Astron. Vestn. **21**, 216–20.

———. 1987c "On the origin of comets." In: *The evolution of the Small Bodies of the Solar System*, edited by M. Fulchignoni and L. Kresak (Proceedings of the International School of Physics "Enrico Fermi," Course XCVIII, Varenna, 1985), (North-Holland, Amsterdam), pp. 217–26. Abstract in Part VIII.

———. 1990. "Problems of origin of asteroids and comets." Meteoritics **25**, 243–8. Abstract in Part VIII.

———. 1991. "Kuiper Prize Lecture: Some problems in the formation of the planets." Icarus **94**, 260–71.

——— and Guseinov, K. M. 1990. "Possible formation of comets in situ." Sol. Syst. Res. **24**, 159–64. Translated from Astron. Vestn. **24**, 248–56. Abstract in Part VIII.

———, Pechernikova, G. V., Ruskol, E. L., and Vityazev, A. V. 1986. "Protosatellite swarms." In: *Satellites*, edited by J. A. Burns and M. S. Matthews (University of Arizona Press, Tucson), pp. 89–116.

——— and Kozlovskaya, S. V. 1977. "Heating of the earth by the impact of accreted bodies." Izv. Acad. Sci. Phys. Solid Earth **13**, 677–84. Translated from Izv. Akad. Nauk SSSR Fiz. Zemli (10). Abstract in Part V.

——— and Ruskol, E. L. 1956. "On the possibility of turbulence in the protoplanetary cloud" (in Russian). Dokl. Akad. Nauk SSSR **108**, 413–6.

——— and ———. 1957/1964. "The hypothesis of turbulence in the protoplanetary cloud." Probl. Cosmog. **5**, 16–38. Translated from Vopr. Kosmog. **5**, 22f. Reprinted in Part III.

——— and ———. 1963. "History of the lunar atmosphere and the possibility of ice and organic compounds existing on the moon." In: *Proceedings of the XIII International Astronautical Congress, Varna, 1962*, edited by N. Booneff and I. Hersey (Springer, New York, 1964) pp. 42–53. NASA Translation No. N64-29497. Also published in Russian in Vopr. Kosmog. **9**, 203–13 (1963). Abstract in Part VI.

——— and ———. 1977. "The accumulation of satellites." Sov. Astron. **21**, 211–17. Also in: *Planetary Satellites*, edited by J. A. Burns, (University of Arizona Press, Tucson), pp. 501–12. Reprinted in Part VI.

——— and ———. 1982a. "On the origin and initial temperature of Jupiter and Saturn." Icarus **49**, 284–96.

——— and ———. 1982b. "Origin of the earth and planets" (in Russian). Zemlya i Vselennaya (3), 6–11.

——— and Ruzmaikina, T. V. 1975/1977. "General problems of planetary and stellar cosmogony" (in Russian). In: *Rannie stadii evolutsii zvezd* (Early Stages of Stellar Evolution, conference in Kiev, 1975), edited by I. G. Kolesnik (Akad. Nauk Ukrainsk. SSSR, Kiev, 1977), pp. 147–50.

Safronov, V. S. and Ruzmaikina, T. V. 1978. "On angular momentum transfer and accumulation of solid bodies in the solar nebula." In: *Protostars and Planets*, edited by T. Gehrels (Univ. of Arizona Press, Tucson), pp. 545–64.

———. and ———. 1985. "Formation of the solar nebula and the planets." In: *Protostars and Planets II*, edited by D. C. Black and M. S. Matthews (University of Arizona Press, Tucson), pp. 959–80.

——— and Vityazev, A. V. 1983/1985. "Origin of the solar system." Sov. Sci. Rev. Sec. E (Harwood Academic Publishers, 1985), **4**, 1–98. Translated from Itogi Nauki i Tekh. Ser. Astron. **24**, 5–93 (1983).

——— and ———. 1986. "The origin and early evolution of the terrestrial planets." In: *Chemistry and Physics of the Terrestrial Planets*, edited by S. K. Saxena, (Springer, New York), pp. 1–29. Abstract in Part V.

———, ———, and Mayeva, S. V. 1978. "The original state and early evolution of the earth." Geochem. Int. **15** (6), 102–7. Translated from Geokhimiya (12), 1763–9 (1978).

——— and Zvyagina, E. V. 1968/1972. "Peculiarities of growth of planetary nuclei" (in Russian). In: *Physics of the Moon and Planets*, Symposium at Kiev, 1968, edited by D. Ya. Martynov and V. A. Bronshtehn (Izdatel'stvo "Nauka," Moscow, 1972), pp. 219–23.

Safronov, V. S. 1959. See Levin and Safronov.

———. 1961, 1977. See Ruskol and Safronov.

———. 1971. See Zvyagina and Safronov.

———. 1973. See Burns and Safronov; Zvyagina *et al.*

———. 1975. See Genkin and Safronov.

———. 1976. See Pechernikova *et al.*; Ruzmaikina and Safronov; Ziglina and Safronov.

———. 1978, 1988. See Vityazev *et al.*

———. 1989. See Ruzmaikina *et al.*

———. 1991. See Lissauer and Safronov.

———. See also Wetherill (1989).

Savchenko, K. N. 1953. "Some questions of nonclassical celestial mechanics and cosmogony" (in Russian). Trudy Odesskogo Gos. Univ. **5**, 59–147. [=Sammelwerk Math. Abt. Phys.-Math. Fakultät Odessa.]

Schmidt, O. Yu. 1925. "On the stability of planetary motions" (in Russian). Unpublished notes, in Schmidt (1960), p. 181. Translation in Part II.

———. 1942. "Origin of the Planetary System" (in Russian). Unpublished notes, in Schmidt (1960), pp. 183–4. Translation in Part II.

———. 1944. "A meteoric theory of the origin of the earth and planets." Comptes Rendus Acad. Sci. URSS **45**, 229–33. Translated from Dokl. Akad. Nauk **45**, 245–9. Reprinted in Part II.

———. 1945a. "The age of the Earth as derived from an astronomical theory." Comptes Rendus Acad. Sci. URSS **46**, 355–8. Translated from Dokl. Akad. Nauk SSSR **46**, 392–5.

———. 1945b. "On the origin of comets." Comptes Rendus Acad. Sci. URSS **49**, 404–7. Translated from Dokl. Akad. Nauk **49**, 413–6. Reprinted in Part II.

———. 1946a. "The cosmogonic significance of the position of the ecliptic plane in the galaxy." Comptes Rendus Acad. Sci. URSS **52**, 577–80. Translated from Dokl. Akad. Nauk SSSR **52**, 581–4. Reprinted in Part II.

———. 1946b. "The law of planetary distances." Comptes Rendus Acad. Sci. URSS **52**, 667–72. Translated from Dokl. Akad. Nauk. SSSR **52**, 673–8. Reprinted in Part II.

———. 1946c. "On the origin of the sun's rotation" (in Russian). Dokl. Akad. Nauk SSSR **54**, 15–8. Translation in Part II.

———. 1946d. "New theory of the origin of earth and planets" (in Russian). Priroda (Moscow) **35** (7), 6–18.

———. 1947a. "On the possibility of a capture in celestial mechanics" (in Russian). Dokl. Akad. Nauk **58**, 213–6. Translation in Part II.

———. 1947b. "New theory of the origin of earth and planets." (in Russian). Izv. Vsesoyuznogo Geograf. Obshch. **79** (5), 265–74. Also in *Trudy 2-go Vsesoyuznogo Geograficheskogo S'ezda* (Moscow, 1948), Vol. 1, pp. 210–22.

———. 1948. "Problem of capture in the three-body problem" (in Russian). Usp. Mat. Nauk **3** (4) (26), 157–9.

———. 1949. *Chetyre Lektsie o Teorie Proiskhozhdenie Zemli* (Four Lectures on the Theory of the Origin of the Earth) Moscow (Izdatelstvo Akad. Nauk SSSR, Moscow).

———. 1950a. *Chetyre Lektsie o Teorie Proiskhozhdenie Zemli* (Four Lectures on the Theory of the Origin of the Earth), 2nd ed. (Izdatelstvo Akad. Nauk SSSR, Moscow).

———. 1950b. "The Formation of planets and their satellites" (in Russian). Izv. Akad. Nauk. SSSR Ser. Fiz. **14**, (1), 29–45; Tr. Inst. Fiz. Zemli Akad. Nauk SSSR (11), 3–20.

———. 1951a. "The problem of the origin of the earth and planets" (in Russian). Tr. Pervogo Sovesh. Vopr. Kosmog. (Moscow), pp. 9–30.

———. 1951b. "The problem of the origin of the earth and planets" (in Russian). [Abridged version of 1951a.] *Vopr. Filos.* (4), 120–33. Translation in Part II.

———. 1952. "New achievements in the science of the origin of the earth" (in Russian). Bolshevik (5), 23–33. Translation in Part II.

———. 1953a. "What the new theory of the origin of the earth contributes to the theory of the origin of life" (in Russian). Unpublished notes, in Schmidt (1960).

———. 1953b. "On the origin of the earth and planets. Replies to readers' letters" (in Russian). Vopr. Filos. (5), 267–70. Translation in Part II.

———. 1954a. "On the origin of the asteroids" (in Russian). Dokl. Akad. Nauk **96**, 449–51. Translation in Part II.

Schmidt, O. Yu. 1954b. "Role des particules solides dans la cosmogonie planetaire." In *Les Particules solides dans les astres*, Communications pres. 6^e Colloque International d'Astrophysique, Liege, July, 1954. Mem. Soc. R. Sci. Liege Collect. 4 (15), 638–49.

———. 1954c. "On the Origin of Comets" (in Russian). Unpublished notes, in Schmidt (1960). Translation in Part II.

———. 1955. "The origin and early evolution of the earth" (in Russian). Tr. Inst. Geofiz. Akad. Nauk SSSR **26**, 5–10.

———. 1956. "The role of solid particles in planetary cosmogony." (in Russian). Priroda (Moscow) (11), 3–6.

———. 1957/1958. *Four Lectures on the Theory of the Origin of the Earth*, translated by G. H. Hanna from the third Russian edition (1957). (Foreign Languages Publishing House, Moscow). Also published by Lawrence and Wishart, London, 1959. 139 pp.

———. 1960. *Izbrannie Trudy—Geofizika i Kosmogoniya* (Izdatelstvo Akademiya Nauk SSSR, Moscow).

———. Undated. "Some new consequences from the theory, pertaining to geosciences" (in Russian). Unpublished notes, in Schmidt (1960).

Schmidt, O. Yu. See also Khilmi (1957); Levin (1959a, 1962c); Safronov (1971); Shcherbakov and Levin (1957).

Semenenko, N. P. 1982. "Origin of oxidized matter of the protoplanets (silicates in space)" (in Russian). Geokhimiya i Rudoobraz. (10), 3–14.

Shcherbakov, D. I., and Levin, B. Yu. 1957. "Fighter for materialistic science. On the anniversary of O. Yu. Schmidt's death)" (in Russian). Priroda (Moscow) (9), 40–6.

Sherbaum, L. M. 1989, 1990. See Kazantsev and Sherbaum.

Shilo, N. A. 1982. "On the mechanism of formation of the solar system" (in Russian). Tikhookean. Geol. (6), 20–27.

Shklovsky, I. S. 1951. "O vozmozhnostii ob'yasneniya razlichii v khimicheskom sostave zemli i solntsa termicheskoi dissipatsei legkikh gazov." Astron. Zh. **28**, 234–43.

Shukolyukov, Ya. A. 1984. "Xe isotopes and the origin of the volatile elements in the earth." Geochem. Cosmochem. 233–52. See a similar paper in *Report of the 27th International Geological Congress*, Vol. 11, Sec. S 11, pp. 94–103 (in Russian).

———, Dang, V. M., and Ivanov, A. V. 1986. "Isotopic heterogeneity of noble gases in the Kaidun carbonaceous chondrite" (in Russian). Meteoritika **45**, 31–7. Abstract in Meteoritics **22**, 292 (1987).

Simakov, G. V., Podurets, M. A., and Trunin, R. F. 1973. "Novye dannye o szhimaemosti okislov i ftopidov i gipoteza ob odnorodnom sostave zemli" (New data on the compressibility of oxides and fluorides and the hypothesis of a homogeneous composition of the earth) Dokl. Akad. Nauk SSSR **211**, 1330–2.

Simonenko, A. N. 1978. "Some features of the orbits of meteorites and their parent bodies" (in Russian). Meteoritika **37**, 266–70. Abstract in Meteoritics **15**, 241 (1980).

——— and Levin, B. Yu. 1983. "On a possible mechanism of transport of matter in the protoplanetary cloud" (in Russian). Meteoritika **42**, 23–27.

——— and ———. 1984. "Radial drift of dust particles moving in quasicircular orbits in the vicinity of the protosun at the Hayashi stage" (in Russian). Komety Meteory **35**, 3–7.

Simonenko, A. N. 1980. See Galibina *et al.*

Sitnik, G. F. 1976. "The life and activities of Vassily Grigorievich Fesenkov" (in Russian). In: *V. G. Fesenkov, Izbrannye Trudy. Solntse i Solnechnaya System* (Nauka, Moscow), pp. 5–16.

Slodkevich, V. S. 1956. *The Origin of the Earth and of Life* (in Russian). (State Pub. House of the Karel. Autonom. Rep., Petrosavodsk). [Reference from AJb.]

Sobotovich, E. V. 1981. "Isotopic data on the heterogeneity of the protoplanetary material." Geochem. Int. **18** (6), 126–37. Translated from Geokhimiya.

———. 1984. "Cosmochemical model of the origin of the earth" (in Russian). Geol. Zh. **44**, 112–23. Chem. Abstr. **101**, #10448v.

Stanyukovich, K. P. 1952. "On the problem of the origin of the solar system" (in Russian). Astron. Zh. **29**, 288–305.

Strel'nitskii, V. S. 1987. "Magnetic fields in regions of possible planetary system formation (from H_2O maser observations)." Sol. Syst. Res. **21**, 190–2. Translated from Astron. Vestn. **21**, 301–3.

Sytinskaya, N. N. 1956. *The Modern Science of the Origin of the Solar System* (in Russian). (Publishing House of the Academy of Pedagogical Science, Russian Soviet Republic, Moscow). [Reference from AJb.]

Syzdykov, A. S. 1975. See Nikolajeva *et al.*

Terent'eva, A. K. 1989. "Minor bodies of the solar system: Meteorite orbits, interrelations, and a mirror symmetry in the Tisserand-constant distribution." Sov. Astron. Lett. **15**, 112–6. Translated from *Pisma Astron. Zh.* **15**, 258–9.

———. 1991. See Andreev *et al.*

Titarenko, V. I. 1990. "Self-gravitating clot evolution of gas-dust nebulae with masses equivalent to those of small planets." In: *Asteroids, Comets, Meteors III* (Meeting at Uppsala, June 1989), edited by C. I. Lagerkvist *et al.* (Uppsala, Astronomiska Observatoriet, Uppsala), pp. 195–8. Abstract in AAA **51**, #107.005 (1990).

———. 1989. See Myasnikov and Titarenko.

Tomanov, V. P. 1991. "Statistical test of the hypothesis of comet ejection." Sol. Syst. Res. **25**, 230–3. Translated from Astron. Vestn. **25**, 312–6. Abstract in Part VIII.

Trubnikov, B. A. 1971. "Solution of the Coagulation Equations in the Case of a Bilinear Coefficient of Adhesion of Particles" (in Russian). Dokl. Akad. Nauk **196**, 1316–9.

Tutukov, A. V. 1987. "Planetary systems and their central stars." Sol. Syst. Res. **21**, 189–90. Translated from Astron. Vestn. **21**, 299–301.

Tutukov, A. V. 1991. "Formation of planetary systems during the evolution of close binary stars." Sov. Astron. **35**, 415–7. Translated from Astron. Zh. **68**, 837–42. Abstract in Part IX.

Ul'yanov, A. A. 1983. "Experimental study of the vaporization and condensation of silicate systems" (in Russian). Meteoritika **42**, 72–85. Abstract in Meteoritics **22**, 167 (1987).

Ustinova, G. K. and Lavrukhina, A. K. 1991. "On origin of cosmogenic Ne-21 excess in SiC of chondrites." Lun. Plan. Sci. **22**, 1425–6.

Velichko, F. P. and Lupishko, D. F. 1991. "Rotation of asteroids." Sol. Syst. Res. **25**, 189–202. Translated from Astron. Vestn. **25**, 259–76.

Velichko, F. P. 1991. See Lupishko and Velikov.

Vinogradov, A. P. 1971. "High-temperature protoplanetary processes (on the problem of formation of metallic cores of planets)." Geochem. Int. **8**, 799–812. Translated from Geokhimiya (11), 1283–96.

Vinogradova, V. P. 1962 "On the distribution of angular momentum in the solar system" (in Russian). Abh. Technol. Inst. Jaroslaw **7**, 143–54. [Reference from AJb.]

Vityazev, A. V. 1972. "On the dynamic-statistical theory of the earth's evolution." Izv. Acad. Sci. Phys. Solid Earth 448–51. Translated from Izv. Akad. Nauk SSSR Fiz. Zemli (7), 42–47.

———. 1980. "Heat generation and heat-mass transfer in the early evolution of the earth." Phys. Earth Plan. Int. **22**, 289–95.

———. 1982. "The fractionation of matter in the course of the formation and evolution of the earth." Izv. Acad. Sci. USSR Phys. Solid Earth **18**, 434–6. Translated from Izv. Akad. Nauk SSSR Fiz. Zemli (6), 52–68. Abstract in Part V.

———. 1983a. "Models of formation and early evolution of the terrestrial planets" (in Russian). In: *Geokhimiya Radiogennykh Izotopor na Rannikh Stadiyakh Evolyutsioi Zemli*, Moskva, pp. 42–61.

———. 1983b. *Geokhimiya Radiogennykh Izotopov na Rannikh Etapakh Evolyutsioi Zemil* (Geochemistry of radiogenic isotopes in the early stages of the evolution of the Earth) (Nauka, Moscow).

———. 1985. "Solar wind and noble gases in atmospheres of Mars and Earth." Lun. Plan. Sci. **16**, 883–4. Abstract in Part V.

———. 1989. "New results of planetary cosmogony and the reconsideration of the ideas on the early earth" (in Russian). In: *Planetnaia Cosmogoniia i Nauki o Zemle* (Nauka, Moscow), pp. 6–12.

——— and Kozenko, A. V. 1988. "Origin of the solar system" (in Russian). Zemlya i Vselennaya (2), 25–32.

———, Lyustikh, E. N., and Nikolaichik, V. V. 1977. "The problem of the formation of the core and mantle of the earth." Izv. Phys. Solid Earth **13**, 527–35 (1978). Translated from Izv. Akad. Nauk SSSR Fiz. Zemli (8), 3–14 (1977).

——— and Mayeva, S. V. 1976. "Model of the early evolution of the Earth." Izv. Acad. Sci. USSR Phys. Solid Earth **12**, 79–85. Translated from Izv. Akad. Nauk SSSR Fiz. Zemli (2). Also in Tectonophys. **41**, 217–25 (1977).

——— and ———.1980. "Simulation of the earth's core and mantle formation." Phys. Earth Plan. Int. **22**, 296–301.

——— and Pechernikova, G. V. 1981. "Solution of the problem of the rotation of the planets within the framework of the statistical theory of accumulation." Sov. Astron. **25**, 494–99. Translated from Astron. Zh. **58**, 869–78.

——— and ———. 1982. "Model of protoplanetary disks around F-G stars." Sov. Astron. Lett. **8**, 201–5. Translated from Pisma v. Astron. Zh. **8**, 371–7. Reprinted in Part IX.

——— and ———. 1983a. "On the regular and arbitrary components of angular momentum of planets in the process of formation" (in Russian). *O. Yu. Schmidt i sov. geofiz. 80-kh godov, Moskva*, pp. 244–9. [Reference from AAA.]

——— and ———. 1983b. "On the dynamics of a protosatellite swarm." *O. Yu. Schmidt i sov. geofiz. 80-kh godov, Moskva*, pp. 250–7.

——— and ———. 1985a. "Thermal conditions during differentiation of matter in the interior of planets" (in Russian). *Termodinam. i geol. I. Vses. Simp., Suzdal', Marta, 1985*, Chernogolovka, Tom 1, p. 201–2. [Reference from AAA.]

——— and ———. 1985b. "Toward a synthesis of cosmochemical and dynamical approaches in planetary cosmogony" (in Russian). Meteoritika (44), 3–20. Abstract in Meteoritics **23**, 383 (1988).

——— and ———. 1985c. "On the evaporation [of] interstellar dust during the preplanetary disk formation." Lun. Plan. Sci. **16**, 885–6. Abstract in Part III.

——— and ———. 1987. "Formation of planets and satellites." Sol. Syst. Res. **21**, 208–9. Translated from Astron. Vestn. **21**, 322–4.

——— and ———. 1987. "When the gas was removed from the zone of terrestrial planets?" Lun. Plan. Sci. **18**, 1044–5. Abstract in Part V.

Vityazev, A. V., Pechernikova, G. V., and Safronov, V. S. 1978. "Limiting masses, distances, and times for the accumulation of planets of the terrestrial group." Sov. Astron. **22**, 60–3. Translated from Astron. Zh. **55**, 107–12. Reprinted in Part V.

———, ——— and ———. 1988. "Formation of Mercury and removal of its silicate shell." In: *Mercury*, edited by F. Vilas, C. R. Chapman, and M. S. Matthews, (University of Arizona Press, Tucson), pp. 667–9. Abstract in Part V.

Vityazev, A. V. 1983. See Safronov and Vityazev.

———. 1980, 1981, 1987, 1988. See Pechernikova and Vityazev.

———. 1984. See Pechernikova *et al.*

———. 1986. See Safronov *et al.*

Voitkevich, G. V. 1951. "Concerning the age of the earth." In: *Report of the Committee on the Measurement of Geologic Time, 1951–1952*, pp. 122-8 (Washington, DC: National Academy of Sciences/National Research Council, Washington, DC), pp. 122–8 Publication No. 245 (1953). Translated from Dokl. Akad. Nauk SSSR **77**, 461–4.

Voitkevich, G. V. 1979/1986. *Die stoffliche Geschichte der Erde.* (VEB Deutscher Verlag fuer Grundstoffindustrie, Leipzig, 1986). Translated from *Osnovyi Teorii Proiskhoshdeniya Zemli* (Nedra, Moscow, 1979).

Voprosy Kosmogoni Editorial Board. 1952. "Advances and Problems of Soviet Cosmogony." Probl. Cosmog. **1**, 1–10. Translated from Vopr. Kosmog. **1**.

Vsekhsviatskii, S. K. 1966. "Comet cosmogony of Lagrange and the problem of the solar system." Mem. Soc. R. Sci. Liege Collect. 8^0, vol. **12**, 495–515.

———. 1969. "Cosmogony of the solar system" (in Russian). In: *Problemy Sovremennoi Kosmogonii*, edited by V. A. Ambartsumian (Nauka, Moscow), pp. 316–413. German translation: "Der gegenwärtige Stand der Kosmogonie," in *Probleme der Modern Kosmogonie*, 2nd ed., edited by V. A. Ambarzumjan (Birkhäuser, Basel, 1976), pp. 270–309. See also the French edition, *Problèmes de cosmogonie contemporaine* (Editions Mir, Moscow, 1971).

———. 1970/1972. "The origin and evolution of the comets and other small bodies in the solar system." In: *The Motion, Evolution of Orbits, and Origin of Comets* (IAU Symposium No. 45, Leningrad, 1970), edited by G. A. Chebotarev, E. I. Kazimirchak-Polonskaya, and B. G. Marsden (Reidel, Dordrecht), pp. 413–8.

———. 1971. "Indications of eruptive evolution of planets" (in Russian). In: *Problems of Cosmic Physics* (in Russian), edited by S. K. Vsekhsvyatskii, (Izdatel'stvo Kievskogo Universiteta, Kiev), No. 6, pp. 73–94. [Reference from AAA.]

———. 1972. "Volcanism and the history of the planets." Sov. Sci. Rev. **3**, 239–45.

———. 1974. "Cosmogonical problems in the W. Kaula compendium" (in Russian). Probl. Kosm. Fiz. **9**, 165–73. (Review of Russian translation of Kaula's *Introduction to Planetary Physics*)

———. 1977. "Comets and the cosmogony of the solar system." In: *Comets, Asteroids, Meteorites*, edited by A. H. Delsemme (University of Toledo, Toledo, Ohio), pp. 469–73.

———. 1979. "The rings of planets and the cosmogony of the solar system." In: *Dynamics of the Solar System* (IAU Symposium No. 81), edited by R. L. Duncombe (Reidel, Boston), p. 203 [abstract only].

———. 1980/1981. "The importance of latest discoveries in the solar system." In: *Planetary Interiors* (COSPAR meeting in Budapest, June 1980), edited by H. Stiller and R. H. Sagdeev, (Pergamon, New York), pp. 75–82.

——— and Kasyutinski, V. 1961. *Rozhdenie Mirov: Filosofskie Problemy Sovremennoii Kosmogenii* (The Formation of the World: Philosophical Problems of Modern Cosmogony), Moscow.

Wetherill, G. W. 1989. "Leonard Medal citation for Victor Sergeivitch Safronov." *Meteoritics* **24**, 347.

Yakovlev, O. I., Markova, O. M. and Manzon, B. M. 1987. "The roles of evaporation and dissipation in the formation of the moon." Geochem. Int. **24**, 1–15. Translated from Geokhimiya (4), 467–82 (1987).

Yakovlev, O. I. 1990. See Manzon *et al.*

Yarov-Yarovoy, M. S. 1963. "The development of the planetary system" (in Russian). Priroda (Moscow) **52** (12), 104–6.

Zelenev, V. M. 1984. See Fedetov and Zelenev.

Zharkov, V. N. and Kozenko, A. V. 1989. "Formation time of Jupiter." Sov. Astron. Lett. **15**, 322–4. Translated from Pisma Astron. Zh. **15**, 745–9. Reprinted in Part VII.

——— and ———. 1990. "On the role of Jupiter in the formation of the giant planets." Sov. Astron. Lett. **16**, 73–4. Translated from Pisma Astron. Zh. **16**, 169–73. Reprinted in Part VII.

Ziglina, I. N. 1976. "Effect on eccentricity of a planet's orbit of its encounters with bodies of the swarm." Sov. Astron. **20**, 730–33 (1976). Translated from Astron. Zh. **53**, 1288–94. Reprinted in Part III.

———. 1978. "Tidal destruction of bodies near a planet." Izv. Acad. Sci. Phys. Solid Earth **14**, 467–71 (1979). Translated from Izv. Akad. Nauk. SSSR Fiz. Zemli (7), 3–10. Abstract in Part VI.

———. 1985. "Eccentricities and inclinations of the orbits of growing planets." Sov. Astron. **29**, 81–7. Translated from Astron. Zh. **62**, 141–52. Reprinted in Part III.

———. 1986. "Changes in the semimajor axis of a planet's orbit in the accumulation process." Sol. Syst. Res. **20**, 194–8. Translated from Astron. Vestn. **20**, 334–42. Abstract in Part III.

———. 1989. "Kinetic analysis of an evolving protoplanetary disk: gravitational interactions." Sov. Astron. Lett. **15**, 285–7. Translated from Pisma Astron. Zh. **15**, 661–6. Reprinted in Part III.

——— and Ruzmaikina, T. V. 1991. "Influx of interstellar material onto the protoplanetary disk." Sol. Syst. Res. **25**, 40–5. Translated from Astron. Vestn. **25**, 53–9. Abstract in Part III.

——— and Safronov, S. 1976. "Averaging of the orbital eccentricities of bodies which are accumulating into a planet." Sov. Astron. **20**, 244–8. Translated from Astron. Zh. **53**, 429–35.

Zvyagina, E. V., Pechernikova, G. V., and Safronov, V. S. 1973. "Qualitative solution of the fragmentation equation with allowance for fragmentation." Sov. Astron. **17**, 793–800. Translated from Astron. Zh. **50**, 1261–73.

Zvyagina, E. V. and Safronov, V. S. 1971. "Mass distribution of protoplanetary bodies." Sov. Astron. AJ **15**, 810–7. Translated from Astron. Zh. **48**, 1023–32. Reprinted in Part III.

Zvyagina, E. V. 1976. See Pechernikova *et al.*

Index*

*Boldface numbers indicate pages of articles or abstracts for which the person is an author.